One of the greatest problems hydrology research faces today is how to quantify uncertainty, which is inherent in every hydrological process. This modern overview of uncertainty emphasises non-orthodox concepts, such as random fields, fractals and fuzziness. This book comprehensively reviews alternative and conventional methods of risk and uncertainty representation in hydrology and water resources. The water-related applications discussed in the book pertain to areas of strong recent interest, such as multifractals and climate change impacts.

The authors represent a variety of research backgrounds, achieving a broad subject coverage. The material covered provides an important insight into new theories of uncertainty related to the field of hydrology. The book is international in scope and will be welcomed by researchers and graduate students of hydrology and water resources.

New Uncertainty Concepts in Hydrology and Water Resources

INTERNATIONAL HYDROLOGY SERIES

New Uncertainty Concepts in Hydrology and Water Resources

Edited by
Zbigniew W. Kundzewicz *(Polish Academy of Sciences, Poland)*

CAMBRIDGE UNIVERSITY PRESS
Cambridge, New York, Melbourne, Madrid, Cape Town, Singapore, São Paulo

Cambridge University Press
The Edinburgh Building, Cambridge CB2 2RU, UK

Published in the United States of America by Cambridge University Press, New York

www.cambridge.org
Information on this title: www.cambridge.org/9780521461184

First published 1995
This digitally printed first paperback version 2006

A catalogue record for this publication is available from the British Library

Library of Congress Cataloguing in Publication data

New uncertainty concepts in hydrology and water resources / edited by
Z.W. Kundzewicz.
p. cm. – (International hydrology series)
Proceedings of the International Workshop on New Uncertainty
Concepts in Hydrology and Water Concepts, held Sept. 24–28, 1990 in
Madralin, near Warsaw, Poland.
ISBN 0 521 46118 9
1. Hydrology – Statistical methods – Congresses. 2. Uncertainty
(Information theory) – Congresses. I. Kundzewicz, Zbigniew.
II. International Workshop on New Uncertainty Concepts in Hydrology
and Water Concepts (1990: Madralin, Poland) III. Series.
GB656.2.S7N49 1995
551.48′015195–dc20 94-36393 CIP

ISBN-13 978-0-521-46118-4 hardback
ISBN-10 0-521-46118-9 hardback

ISBN-13 978-0-521-03673-3 paperback
ISBN-10 0-521-03673-9 paperback

Contents

List of Authors

A. Bárdossy
Institute for Hydrology and Water Management, University of Karlsruhe, Germany

J. J. Bogardi
Wageningen Agricultural University, Department of Water Resources, Wageningen, The Netherlands

J. P. Carbonnel
CNRS, Université P. & M. Curie, Paris, France

J. C. I. Dooge
Centre for Water Resources Research, University College Dublin, Ireland

J. W. Eheart
Department of Civil Engineering, University of Illinois at Urbana-Champaign, USA

W. Feluch
Technical University of Warsaw, Institute of Environmental Engineering, Warsaw, Poland

F. Friggit
EIIER, Ouagadougou, Burkina Faso

J. H. Garrett Jr
Department of Civil Engineering, University of Illinois at Urbana-Champaign, USA

K. P. Georgakakos
Hydrologic Research Center, San Diego, California, and Scripps Institution of Oceanography, UCSD, La Jolla, California, USA

L. Gottschalk
Department of Geophysics, University of Oslo, Norway

Guo Sheng Lian
Wuhan University of Hydraulic and Electric Engineering, Wuhan, People's Republic of China

P. Hubert
CIG, Ecole des Mines de Paris, Fontainebleau, France

W. Jakubowski
Chair of Mathematics, Agriculture University of Wrocław, Poland

A. Karbowski
Institute of Automatic Control, Warsaw University of Technology, Warsaw, Poland

M. W. Kemblowski
Department of Civil and Environmental Engineering, Utah State University, Logan, Utah, USA

J. Kindler
Institute of Environmental Engineering, Warsaw University of Technology, Warsaw, Poland
At present: World Bank, Washington D.C., USA

F. Konecny
Institute of Mathematics and Applied Statistics, Universität für Bodenkultur, Vienna, Austria

A. Kozłowski
Institute of Geophysics, Polish Academy of Sciences, Warsaw, Poland

V. Kotwicki
Department of Environment and Natural Resources, Adelaide, Australia

K. Kowalski
Research Centre of Agricultural and Forest Environment Studies, Polish Academy of Sciences, Poznań, Poland

W. F. Krajewski
Department of Civil and Environmental Engineering and Iowa Institute of Hydraulic Research, The University of Iowa, Iowa City, Iowa, USA

I. Krasovskaia
HYDROCONSULT AB, Uppsala, Sweden

Z. W. Kundzewicz
Research Centre of Agricultural and Forest Environmental Studies, Polish Academy of Sciences, Poznań, Poland
At present: World Meteorological Organization, Geneva, Switzerland

A. Łaski
HYDROPROJEKT Consulting Engineers, Warsaw, Poland

A. Łodziński
Technical University of Warsaw, Warsaw, Poland

S. Lovejoy
Physics Department, McGill University, Montreal, Canada

O. Manoliadis
Laboratory of Rural Technology, National Technical University of Athens, Greece

P. D. Meyer
Department of Civil Engineering, University of Illinois at Urbana-Champaign, USA

H. T. Mitosek
Institute of Geophysics, Polish Academy of Sciences, Warsaw, Poland

K. Mizumura
Civil Engineering Department, Kanazawa Institute of Technology, Ishikawa, Japan

M. E. Moss
US Geological Survey, Tucson, Arizona, USA

H.-P. Nachtnebel
Institute of Water Resources Management, Hydrology and Hydraulic Construction, Universität für Bodenkultur, Vienna, Austria

J. J. Napiórkowski
Institute of Geophysics, Polish Academy of Sciences, Warsaw, Poland

J. P. O'Kane
University College Cork, Ireland

E. J. Plate
Institute for Hydrology and Water Resources Planning, University of Karlsruhe, Germany

S. Ranjithan
Department of Civil Engineering, University of Illinois at Urbana-Champaign, USA

P. F. Rasmussen
Institute of Hydrodynamics and Hydraulic Engineering, Technical University of Denmark, Lyngby, Denmark

R. Romanowicz
Institute of Environmental and Biological Sciences, Lancaster University, Lancaster LA1 4YQ, UK

D. Rosbjerg
Institute of Hydrodynamics and Hydraulic Engineering, Technical University of Denmark, Lyngby, Denmark

D. Schertzer
Laboratoire de Météorologie Dynamique, C.N.R.S., Paris, France

M. B. Sharifi
Department of Civil Engineering, Mashhad University, Mashhad, Iran

J. A. Smith
Department of Civil Engineering and Operations Research, Princeton University, Princeton, New Jersey, USA

M. Sowiński
Department of Water Resources and Environment Engineering, Ahmadu Bello University, Zaria, Nigeria

W. G. Strupczewski
Institute of Geophysics, Polish Academy of Sciences, Warsaw, Poland

P. L. Sturdevant
Department of Civil Engineering, Princeton University, Princeton, New Jersey, USA

G. Tsakiris
Laboratory of Rural Technology, National Technical University of Athens, Greece

S. Tyszewski
Institute of Environmental Engineering, Warsaw University of Technology, Warsaw, Poland

A. J. Valocchi
Department of Civil Engineering, University of Illinois at Urbana-Champaign, USA

A. Verhoef
Wageningen Agricultural University, Department of Water Resources, Wageningen, The Netherlands

S. Węglarczyk
Institute of Water Engineering and Water Management, Cracow Technical University, Cracow, Poland

J.-C. Wen
Department of Civil and Environmental Engineering, Utah State University, Logan, Utah, USA

M. I. Yusuf
Civil Engineering Programme, Abukabar Tafawa Balewa University, Bauchi, Nigeria

I. Zawadzki
Department of Physics, University of Québec in Montreal, Canada

Preface

The present volume contains the edited proceedings of the International Workshop on New Uncertainty Concepts in Hydrology and Water Resources, held in Madralin near Warsaw, Poland from 24 to 26 September 1990. It was organized under the auspices of the Institute of Geophysics, Polish Academy of Sciences, Warsaw, Poland, and the International Commission on Water Resources Systems (ICWRS) – a body within the International Association of Hydrological Sciences (IAHS). The Organization and Programme Committee for the Workshop consisted of the following individuals: Professor Lars Gottschalk (Norway/ICWRS/IAHS), Professor Zdzisław Kaczmarek (Poland/IIASA), Professor Janusz Kindler (Poland), Professor Zbigniew W. Kundzewicz (Poland), who acted as the Secretary, Professor Uri Shamir (Israel/ICWRS/IAHS) and Professor Witold Strupczewski (Poland).

The Workshop was a continuation of series of meetings organized under the aegis of the International Commission of Water Resources Systems (ICWRS) within the IAHS. This series of meetings was initiated by the former ICWRS President, Professor Mike Hamlin in Birmingham, 1984. Last Workshop of similar character was organized by the ICWRS Secretary, Professor Lars Gottschalk in Oslo (1989).

The Workshop was primarily devoted to recent methods of representation of uncertainty in hydrology and water resources. This embraces newly introduced methods and approaches that, albeit not new, have raised considerable recent interest. In the menu of topics tackled at the Workshop were, among others, such diverse items, as fractals, risk and reliability-related criteria, fuzzy sets, pattern recognition, random fields, time series, outliers detection, non-parametric methods, etc. The apparent side effect of the Workshop was also putting different methods into perspective. It possibly helped assessing methodologies and answering the question, whether the apparent attractivity of particular methods is based on permanent values or it is just a band-wagon effect and the methods are likely to pass as a short-lasting fashion. The Workshop attracted 44 registered participants from 16 countries, who presented 44 oral contributions during nine technical sessions. The set of participants was highly heterogeneous, as regards their backgrounds, institutions represented, theoretical and practical experiences and research philosophies. The participants were, by background, hydrologists, civil, environmental and agricultural engineers, foresters, geographers, geologists, geophysicists, system scientists, mathematicians, computer scientists and physicists. The institutions, where participants worked ranged from universities, through non-university research institutes (e.g. academies of sciences), administration (government agencies) to consulting engineers. The variety of backgrounds, research orientations and preferences is clearly visible in this volume, where more descriptive contributions are neighbours to papers stuffed with heavy mathematical developments. The heterogeneity and multidisciplinarity is believed to have contributed to a broad subject coverage and to have caused a welcome cross-fertilization effect.

The idea of the Workshop was to report on recent research, to present and discuss work at different stages of progress. Some entries in the discussion were indeed thought-provoking and surely helped the presenters and the audience to shape their further research.

It is a pleasure of the editor of this volume (and also secretary of the Organization and Programme Committee) to thank the participants in the Workshop and the contributors to this volume for their fine work that made the Workshop an undoubtful success. Thanks are extended to the organizing institutions mentioned. The financial support provided by the Institute of Geophysics, Polish Academy of Sciences and by the International Institute for Applied Systems Analysis (IIASA) in Laxenburg, Austria, is gratefully acknowledged. Last, but not least, thanks are due to the UNESCO and its Director of Division of Water Sciences, Dr Andras Szollosi-Nagy, for the invitation to publish this volume within the International Hydrology Series and for support of the editorial work.

It is believed that the present contribution contains a wealth of illuminating and stimulating material. It may be useful for researchers, lecturers and graduate students in hydrology and water resources.

Z. W. Kundzewicz

I

Introduction

1 Hydrological uncertainty in perspective

Z. W. KUNDZEWICZ

Research Centre of Agricultural and Forest Environment Studies, Polish Academy of Sciences, Poznań, Poland

ABSTRACT Different aspects and meanings of uncertainty are reviewed. This introductory review forms a basis for putting recent developments in hydrological and water resources applications of uncertainty concepts into perspective. The understanding of the term uncertainty followed herein is a logical sum of all the notions discussed. An attempt is made to justify the structure of the present volume and to sketch the areas of particular contributions in the volume and to point out their connections to different facets of uncertainty.

INTRODUCTION

It seems that there is no consensus within the profession about the very term of uncertainty, which is conceived with differing degrees of generality. Moreover, the word has several meanings and connotations in different areas, that are not always consistent with the colloquial understanding.

In the following section the notions and concepts of uncertainty both beyond and within the water resources research are discussed. Further, particular contributions in this volume are reviewed in the context of their connections to different facets of uncertainty. This is done in the systematic way, following the structure of the book.

NOTIONS OF UNCERTAINTY

Let us take recourse to established dictionaries and see how the words 'uncertain' and 'uncertainty' are explained. Among the meanings of the word 'uncertain', given by Hornby (1974) and Webster's (1987) dictionaries, are the following: not certain to occur, problematical, not certainly knowing or known, doubtful or dubious, not reliable, untrustworthy, not clearly identified or defined, indefinite, indeterminate, changeable, variable (not constant).

The noun uncertainty results from the above concepts and can be summarized as the state (quality) of being uncertain, with the word uncertain attaining one of meanings listed above.

There is a plethora of single words that are synonymous of the word uncertainty. The meaning of the term uncertainty partly overlaps with the contents of such words as doubt, dubiety, skepticism, suspicion, mistrust, inconstancy.

Uncertainty is obviously opposed to certainty, where the complete information is available. One sometimes makes a distinction between risk and uncertainty. In the former case, i.e. when talking of risk, one tacitly assumes that a probability distribution of outcomes exists, made on a meaningful basis (i.e. agreed upon by a set of relevant experts). In the latter case, if there is an absence of information on prior probabilities, i.e. nothing (or little) is known as to the likelihood of particular outcomes (or a consensus among experts cannot be achieved), one can talk about uncertainty. In other convention of risk and uncertainty, risk embraces both uncertainty and the results of uncertainty, and means lack of predictability about outcomes, structure etc.

Sometimes authors distinguish between uncertainty and randomness. In this context, uncertainty results from insufficient information and may be reduced if more information is available. This is to be distinguished from the concept of randomness related to quantities fluctuating in a non-controllable way.

There are several practical approaches to dealing with some forms of uncertainty. One possibility is the Laplacean postulate, called also principle of insufficient reason. It states that if the probabilities of an event are not known, they should be assumed equal. A simple, though typically non-satisfactory method is – to replace the uncertain quantities with the worst case values (most pessimistic scenario) or with some measures of central tendency (expected value, median). By a sensitivity analysis the importance of uncertainties can be traced. The other simple way of coping with some types of uncertainty is by interval analysis, i.e. assuming ranges of parameters rather than a number. In a more advanced approach that is commonly used, uncertainty can be encoded with probability methods or with fuzzy sets methods. The former approach is most useful if the estimating functions have a statistical form, i.e. if standard deviations, standard errors of forecasts etc. are available. The measures of uncertainty are typically – probabilities, approximated by frequencies or via the geometric definition. The fuzzy sets approach is a powerful tool where insufficient data exist, or where it is difficult or even impossible to attribute probabilities. That is, the areas of dominance of concepts of randomness and fuzziness can be defined as follows. The former apparatus is used if the event is clearly defined but its occurrence is uncertain. In the latter approach the very event may not be strictly defined and no additivity property is present.

There are two basis attitudes to uncertainty in hydrological and water resources research. Either the world is considered as being basically indeterministic (i.e. must be modelled in terms of stochastic systems) or the stochasticity is a necessary evil (i.e. cannot be avoided at present, when the understanding is not sufficient, but would give floor to increasingly deterministic descriptions when our understanding improves).

It is not only uncertainty about numbers (e.g. inaccuracy of measurements). If one does not know whether some variable attains the value of 1.03 or 1.05, it is a very trivial lack of certainty, though it may be quite critical in some cases. The uncertainties in hydrology are much stronger and pertain to the directions of change, dominating mechanisms, and understanding of processes. Moreover, it follows from the theory of chaotic systems, that the time series of hydrological variables are inpredictable over a longer time horizon, hitherto inherently uncertain, unknown.

Uncertainty in hydrology results from natural complexity and variability of hydrological systems and processes and from deficiency in our knowledge. The uncertainty may pertain to magnitudes and space-time (i.e areal location and temporal frequency) attributes of signals and states of hydrological systems (storages).

Yet more uncertain (unexpected and unforeseeable) are the variables of relevance to water resources management. One may identify, among others, the following uncertainties:

- uncertainty in knowledge of the external environment (structure of the world, future changes of the environment);
- uncertainty as to future intentions in the related fields of choices;
- uncertainty as to appropriate value judgments of consequences.

Consider the water demands as an example. They depend on several demographical, economical, technological, social, political, and regional development factors, each of which is itself uncertain and non-predictable (e.g. forecast of future population, water use rates, priorities, irrigation patterns). On top of this there is also an uncertainty on the side of available water supply, whose natural variations have been typically considered known. The main uncertainty there falls in the category known under the collective name of the climate change.

There were numerous attempts in the water related literature to distinguish different types of uncertainty. Plate & Duckstein (1987) identified the groups of uncertainties in a hydraulic design. They distinguished hydrological uncertainties, embracing data uncertainties (e.g. measurements), sample uncertainties (e.g. number of data) and model uncertainty (density function). Further they identified hydraulic uncertainties in the process of hydraulic transformation of hydrological data, embracing parameter uncertainties (Manning's n), model uncertainty (empirical equations) and scaling laws (physical models). Finally, structural uncertainties were associated with the material, design models, external effects, etc. However, in another convention, structural uncertainty is understood as considerable lack of sureness about the formulation of the problem in terms of structure and assumptions.

Bernier (1987) distinguished natural uncertainty related to a random nature of physical processes and technological uncertainties embracing sampling errors and the model uncertainties. Uncertainty of the former category is linked with the total duration of the period of records, whereas the latter category may result from the model choice or imprecise identification of parameters.

Beck (1983) distinguished uncertainty and error in the field data, inadequate amount of data, uncertainty in relationships between variables, and uncertainty in model parameters estimation. After a successful calibration exercise it would be expected that the degree of uncertainty in a parameter estimate would be less than the uncertainty associated with the prior estimate before calibration. Reduction of uncertainty is the measure of relevance of a parameter. However, the uncertainty of parameter estimates is inversely related to the amount of field observations and to errors associated with these observations. Posterior estimates are also uncertain (fingerprint of the calibration method that can propagate forward).

Klir (1989) considered uncertainty versus complexity. Both categories are in conflict: if complexity decreases, the uncertainty grows. Uncertainty is related to information, being reduced by an act of acquiring information (observation, performing an experiment, receiving a message). The amount of information received quantifies the reduction of uncertainty. Klir (1989) considered the principle of maximum uncertainty – use all, but no more, information than available. Our ignorance should be fully recognized when we try to enlarge our claims beyond the given premises. At the same time – all information contained in premises should be fully utilized. Another aspect is the principle of minimum uncertainty, actual when simplifying a system. The loss of relevant information should be minimized at any desirable level of complexity.

The understanding of the word uncertainty in the present volume will follow the logical sum of all the uncertainty aspects discussed above.

FACETS OF UNCERTAINTY

Although the Workshop was primarily method-oriented, some contributions were rather problem-oriented. It is worth mentioning a couple of examples, starting from contributions pertaining to the very timely area of hydrological consequences of climatic change.

Moss (1995) studied the concept of Bayesian relative information measure, applied to evaluate the outputs of general circulation models (GCM). The relative information was understood as the ratio of information contents of the model and of the data, e.g. of the model-based histogram and the data-based histogram). The approach allows the complexity connected with disparate temporal and spatial scales of outputs of GCM simulations and the actual observations to be resolved. The methodology devised by Moss (1995) is capable of giving preliminary answers to several practical questions connected with comparisons of models, effects of the grid size and evaluation of the soundness of disaggregation.

Bardossy (1995) extended the classical hydrological perspective of transformation of rainfall into runoff by treating the atmospheric circulation pattern as the primary input signal. He confirmed that daily precipitation depths and mean temperatures are strongly linked to the atmospheric circulation patterns and developed relevant mathematical models. As GCMs produce air pressure maps with relatively good credibility, the model lends itself well to applications in stochastic simulation studies of climate change.

Another example of problem-oriented research was the contribution of Kotwicki & Kundzewicz (1995), studying the process of floods of Lake Eyre. This is quite a convincing manifestation of hydrological uncertainty. The available observational records are not long. Therefore one is forced to use a model and proxy data in order to reconstruct river flows.

Strupczewski & Mitosek (1995) showed that the hydrological uncertainty influencing a design in the stationary case, will significantly increase in the non-stationary case. One can only hypothetize the mechanism and structure of non-stationarity. Strupczewski & Mitosek (1995) developed a method of estimation of time dependent parameters of a distribution from the available non-stationary data set. The uncertainty in the design process is magnified due to the necessity of identification of additional parameters. This is likely to lead to the increase of the quantile estimation error, growing with the length of the time horizon of extrapolation.

Guo (1995) devised and examined a new plotting position rule. The new formula is applicable for the presented case, where flood records obtained in two different ways (i.e. of different accuracies) should be blended. In addition to the systematically recorded data (observation period) there may exist also historical data and palaeologic information related to flows over some threshold of perception.

NOVEL APPROACHES TO UNCERTAINTY

There is a number of novel methodological approaches to uncertainty, originating in areas outside hydrology (typically – applied mathematics, systems theory, physics) but relevant to water sciences. Although the origin and applicability of these approaches largely differ, they are treated collectively in one part of this volume.

The concept of fractals, as developed by Mandelbrot (1977), has found a strong resonance in hydrological sciences. This methodology made it possible to analytically describe complicated natural objects, without the need to approximate them via the constructs of the classical geometry.

Kemblowski & Wen (1995) tested the assumption of fractal permeability distribution in their study of infiltration soils. This represents one of the challenging avenues of application of fractal framework, gaining increasing recent interest in the theory of flow in porous media, groundwater and petroleum engineering. Kemblowski & Wen (1995) postulated that the fractal nature of the permeability distribution strongly influences the mixing processes in groundwater, with major impact on the value of asymptotic macrodispersivity. A higher fractal dimension results in higher vertical mixing and less longitudinal spreading of the plume, while for the fractal dimension approaching two the horizontal spreading disappears. It was also indicated that the travel distance necessary to reach the asymptotic conditions was scale-dependent and related to the thickness of the plume and to the pore-level transverse dispersivity. These findings of Kemblowski & Wen (1995) are believed to narrow the gap in the understanding of subsurface transport processes, and in particular of the process of mixing of soluble plumes with surrounding groundwater.

The contribution of Lovejoy & Schertzer (1995) differs considerably from the rest of the volume. It is a solicited extensive review paper reaching much further than the oral presentation at the Workshop. It summarizes the research work that has been done in the area of multifractal analysis of rain. The bulk of the material stems from the original research of the authors. Lovejoy & Schertzer (1995) established and tested the scaling ideas of description of the process of rain. The concept of continuous turbulent cascades is thoroughly discussed. Different applications of the concepts of multifractals at the interface between meteorology and hydrology are reviewed. The contribution by Lovejoy & Schertzer (1995) is possibly the most extensive lumped coverage of the problem available in the literature.

Zawadzki (1995) studied scaling properties of the spatial distribution of rainfall rate, for a broad spectrum of scales ranging from the radar coverage area down to individual rain drops. The results show that no scaling or multiscaling properties could be detected for scales exceeding the size of a mature precipitating cumulus. Preferential scales were found in the range of a few tens of kilometers. Within the scales of the order of a cumulus size some multiscaling properties were found.

Hubert *et al.* (1995) reported the evidence of a multifractal structure of temporal occurrence of rainfalls at particular locations of the Soudano-Sahelian region for a range of temporal scales of two orders of magnitude, ranging from days to months. Fractal dimensions of the data analyzed, varying between zero and one, were estimated with the help of the functional box counting method. As can be expected, fractal dimensions of rainfall occurrence depended strongly on the chosen rainfall intensity threshold, decreasing with the rise of the threshold. Attempts to find regional patterns and trends were also undertaken.

It seems that the fractal concepts are of permanent value in hydrological sciences, as they provide a new insight into the nature of processes, describing apparently irregular natural forms in a straightforward, novel way.

The contribution by Georgakakos *et al.* (1995) dealt with the area of chaotic dynamics that has risen considerable general interest. They reported on their analysis of very fine increment point rainfall data. High-resolution rainfall data recorded by a fast-responding optical raingauge was analyzed via classical statistical and recent fractal and chaotic dynamics methods. The analysis showed the evidence of scaling and chaotic dynamics.

An assemblage of four papers presented in Madralin, two of which are published in the present volume, dealt with the problem of applications of the fuzzy theory in hydrology.

Mizumura (1995) presented a model of snowmelt runoff based on the theory of fuzzy sets. The contribution shows that the combined approach using the conceptual tank model and a fuzzy logic model yields satisfactory results. The effect of different membership functions on the prediction is tested.

Kindler & Tyszewski (1995) objectively studied the applicability of the fuzzy sets theory to hydrological and water resources problems. They elucidated why the practical applications of the methodology in the water field are so rare. The fundamental problems are – how to identify membership functions, and how to interpret fuzzy results. According to Kindler & Tyszewski (1995), the theory seems more applicable for diagnostic problems rather than in the context of decision making.

Another novelty that has attracted much interest of hydrologists was the concept of pattern recognition, belonging to the area called artificial intelligence (AI).

Mizumura (1995) used a technique originating from pattern recognition methodology to forecast the ranges of runoff values likely to occur. The forecast made use of the values of rainfall and runoff recorded in former time steps. The Bayesian methodology used does not require the detailed physical information on the watershed. Use of the tanks model, and prediction of runoff by the pattern recognition method, from the past observed data and the past errors of the tank models improves the accuracy of forecasts.

Ranjithan *et al.* (1995) devised a method potentially useful for hydraulic gradient control design of plume migration. The neural network technique was used, i.e. a branch of the artificial intelligence (AI) framework. This method can capture information that is imprecise, complex, vague and inexpressable by mathematical or symbolic means. Pessimistic realizations of the uncertain field of hydrogeologic parameters that would influence the final design were identified. Although the process of training the neural network (pattern association task, i.e. learning the links between each spatial distributions of hydraulic conductivity values and the impact upon the final groundwater remediation design) was found difficult, a trained network screens a set of realizations with little computational effort.

One of the promising techniques raising recent interest in hydrology has been the non-parametric approach. This could help in objectivization of the design procedures. Non-parametric methods allow the bulk of information present in the data to be extracted, without forcing it to fit a tight uniform. They let the data speak for themselves, so to say, thus decreasing the degree of subjectivity. The non-parametric estimation methods enable one to estimate an unknown p.d.f. without a prior assumption of the underlying parent distribution (that is, in fact, never known in hydrology). This is particularly important for extreme value statistics used in the hydrological design that typically dwelt on a number of standard distributions. As the values used in design were obtained via extrapolation (e.g. assessment of the 100-years flood), they do heavily depend on the type of distribution used. The non-parametric methods are capable of inferring complicated densities or relationships. They allow, for instance, a bimodal form to be approximated, what can occur in hydrological data forming a superposition of two distinct mechanisms of generation.

Feluch (1995) applied the non-parametric estimation methods to two classes of practical hydrological problems. He presented the multivariate estimation of annual low flow and high flow characteristics (understood as: maximum flow and volume of the flood wave; or minimum flow and low flow period, respectively). He also used the non-parametric regression to establish the linkage between the river discharge and the water stage, and to find the relationships between concurrent time series of runoff and groundwater level. Variable kernel estimator was found better when the sample skewness was greater than one. Simulation showed that quantiles estimated by a non-parametric method compare favourably to the parametric estimator. In extrapolating exercises the non-parametric method places a stronger weight on a few large observations. Thus the problem of the tails of parameters distribution fitted to the whole sample is weakened.

Guo (1995) studied non-parametric methods of flood frequency analysis (FFA). He compared application of fixed and variable kernel estimators (FKE and VKE, respectively) for analysis of a design flood with pre-gauging data and information. He gave also the guidelines on the choice of the kernel type, depending on the sample skewness.

RANDOM FIELDS

The era of univariates in hydrology has been passing. It gives room to the era of random fields, where hydrological variables are treated as functions of location (in one, two, or three dimensions) or both location and time (spatial-temporal fields). Random fields approach better represents the nature of processes. Examples of typical applications of random fields in hydrology range from rainfall, through distributed runoff, to groundwater and water quality.

Krasovskaia & Gottschalk (1995) analyzed regional drought characteristics (deficit and extent). They used empirical orthogonal functions (EOF) in their quantification of regional meteorological droughts. It is again a type of technique that draws much information from the data, without the need to dwell on assumptions.

Georgakakos & Krajewski (1995) analyzed the worth of radar data in a real time prediction of areal rainfall. The methodology used was the covariance analysis of a linear, physically based model. The improvement of estimators due to the presence of radar data was quantitively evaluated.

Krajewski & Smith (1995) addressed the problem of rainfall estimation via radar, and in particular adjusting radar rainfall estimates to raingauge estimates.

Meyer *et al.* (1995) considered the risk of exposure to contaminated groundwater caused by leakage from a municipal solid waste landfill. In order to reduce this risk by early detection of the contaminant and appropriate prevention action, a groundwater quality monitoring network is to be designed. The methodology used contains numerical modelling of groundwater flow and contaminant transport and optimization. The main uncertainties pertain to the contaminant source location and variations in hydraulic conductivity. The issue of strong practical flavour reads – find the well locations in the neighbourhood of the landfill, that maximize the probability, that an unknown (here randomly generated) plume is detected.

Romanowicz *et al.* (1995) studied the effect of the spatial variation of the initial soil moisture contents on the distribution of the soil moisture and on the evaporation rate from the land surface. They used a lumped nonlinear model based on thermodynamics and evaluated distributions, means and variances of the soil moisture contents, time to desaturation and actual evapotranspiration. The case of lognormal distribution is considered in more details.

Gottschalk *et al.* (1995) addressed the problem of determining outliers in the data on floods in Norwegian rivers. Point kriging, i.e. the methodology belonging to the geostatistical framework, was exploited.

TIME SERIES AND STOCHASTIC PROCESSES

Several contributions presented at the Workshop can be clustered around the heading of time series and stochastic processes, although typically in a non-classical context.

Rasmussen & Rosbjerg (1995) analyzed the applicability of seasonal models for representation of extreme hydrological events. In the case of a prediction task it may well be that a simplified non-seasonal model performs superior to a seasonal model due to parsimony in parameters. Rasmussen & Rosbjerg (1995) recommended that in case of weak seasonalities the available data should be pooled. If the seasonality is pronounced, the most dominant season should be selected for the non-seasonal approach, and the data on the other seasons discarded. In the cases analyzed by Rasmussen & Rosbjerg (1995) optimal estimates of T-year events were always obtained with the non-seasonal approach.

Jakubowski (1995) presented analytical derivations of the general form of distribution of the l-day total precipitation. This has been achieved within the framework of the alternating renewal methodology. The assumptions made in the development are – dependence of total precipitation of the wet days sequence on the length of this sequence and independence of successive dry and wet spells.

Konecny & Nachtnebel (1995) presented a stochastic differential equation model, based on the mass balance of a linear reservoir, to describe daily streamflow series. The jump process for the input was based on the concept of intensity function randomly alternating between two levels. The model applied to an Austrian river yielded a good reproduction of the observed characteristics.

Weglarczyk (1995) performed extensive analysis of the impact of temporal discretization on the analysis of different stochastic characteristics of the point rainfall. The material used is a continuous record of precipitation depth at a gauge in Kraków, covering the time span of 25 years. The range of time intervals from five minutes to one day, i.e. over two orders of magnitude were considered. The modelling of distribution of particular characteristics and study of interrelations between them was made.

Gottschalk & Kundzewicz (1995) analyzed the series of maximum annual flows with respect to outlying values.

Tsakiris & Manoliadis (1995) modelled the hydrants operation in irrigation networks, using alternating renewal process in continuous time. It seems that the renewal theory is a good basis to assess the probability of hydrant operation and to aid in design of pressurized irrigation systems.

Kowalski (1995) compared the correlation time of hydroclimatical processes (flow of few European rivers, temperature and precipitation in Poznań). The correlation time was considered as a measure of the order or disorder of geophysical processes.

Napiórkowski & Strupczewski (1995) incorporated physical structure into the study of stochastic processes of river flow. They used a rapid flow model originating from a physically sound hydrodynamic description of the process of open channel flow.

RISK, RELIABILITY AND RELATED CRITERIA

This part of the book groups six contributions covering a broad spectrum of topics.

Plate (1995) sketched the outline of a gigantic research project into non-point pollution of surface waters in agricultural landscape (Weiherbach project). The models consist of four parts, each of which contains uncertainty elements. The input model provides the pollutant outflow along the river for a given rainfall and pollutant input. The process describes the transport of pollutants in a river. Finally, the decision model quantifies the consequences of excess pollution of the river. Plate (1995) discussed also risk as a figure of merit and uncertainty in the decision model.

Karbowski (1995) dwelt upon the idea of application of statistically safe sets and optimal operation of water storage reservoirs. The methodology seems to be quite a general tool for solving storage management problems under risk. Inflows are assumed to be independent random variables of known distribution or the Markovian chain. A set of constraints (chance-constraint or expected value-constraints) need to be fulfilled. The approach separated two basic elements of a control problem, i.e.

(a) optimization of a performance index (PI); and
(b) fulfillment of reliability constraints.

Applicability of different variants of dynamic programming is discussed.

Kozłowski & Łodziński (1995) proposed a method of risk assessment in the problem of management of a system of storage reservoirs in Poland. The risk was estimated on the basis of probability distributions of inflows, conceptualized as a non-stationary Markovian chain. Kozłowski & Łodziński (1995) advocated that two indices related to risk, i.e. the probability of failure and the magnitude of loss should be considered in the decision making process.

Kundzewicz & Laski (1995) reported on a subset of criteria of evaluation of performance of water supply systems, embracing different notions of reliability and related concepts. The analysis was presented for two case studies in Poland, for which system simulation for historical series of flows and hypothetical assessments of future water demands were performed. The criteria studied were related to frequency, duration, and severity of non-satisfactory system performance (reliability, resilience, vulnerability). Straightforward results and links between criteria are due to the application of the renewal theory, where exponential distribution of periods of nonsatisfactory and satisfactory system performance were assumed.

Bogardi & Verhoef (1995) analyzed a number of performance indices describing the operational behaviour of the multi-unit multipurpose reservoir system situated on the Mahaweli River, Sri Lanka. They formulated several conclusions on the relative importance of particular performance indices. It was shown, for example, that maximum vulnerability and frequency of failures are more important characteristics of municipal water supply systems than mean vulnerability or duration.

Sowinski & Yusuf (1995) addressed another problem of risk analysis in hydraulic engineering. They studied a composite risk model of the Ogee type spillway.

CONCLUSIONS

Uncertainty means lack of sureness about something (or somebody). It may range from the complete lack of definite knowledge (about facts, mechanisms, processes, results) to small doubts (falling short of certainty, imprecision). The very term of uncertainty has quite fuzzy borders.

Uncertainty is inherent in water resources research as observations, measurements, theories, assumptions, models, equations, predictions, estimators, parameters do not closely reproduce the reality. And sometimes no observations, measurements, theories and models exist at all.

It is shown in the present volume that the problems of uncertainty are by no means simple. Therefore much effort must be directed to this problem area. The importance of the topic can be illustrated by an excerpt of rare beauty, formulated by an ancient Chinese philosopher – Tsu, and quoted by Klir (1989): *Knowing ignorance is strength. Ignoring knowledge is sickness.*

The coverage of uncertainty problems in the book is by no means complete, nor uniform. Some areas are presented in detail, the others, though undoubtedly important, remain untouched.

The contributions gathered in this volume range from rigorous analytical developments, where under some simplifying assumptions a closed-form formula can be obtained, through mixed analytical-numerical approaches to numerical studies, where a large number of variants are calculated.

In all contributions quantitative methods, i.e. various classes of mathematical models are explored.

The apparent side effect of the Workshop was also putting different methods into perspective. It possibly helped assessing methodologies and answering the question whether the apparent attractivity of particular methods is based on permanent values or it is just a band-wagon effect and the methods are likely to pass as a short-lasting fashion.

As an example of promising methodologies the renewal theory can be mentioned. It is being increasingly applied in hydrology and water resources. This has been proved in the present volume in three distinct areas; i.e. precipitation totals; reliability properties of irrigation systems; and water supply systems by Jakubowski (1995), Tsakiris & Manoliadis (1995) and Kundzewicz & Łaski (1995), respectively.

The Workshop was primarily devoted to recent methods of representation of uncertainty in hydrology and water resources. This embraces newly introduced methods and approaches that, albeit not new, have raised considerable recent interest. In the menu of topics tackled at the Workshop were, among others, such diverse items as fractals, risk and reliability related criteria, fuzzy sets, pattern recognition, random fields, time series, outliers detection, non-parametric methods, etc.

Hydrologists claim to have mastered uncertainty. This statement may be relatively valid in the sense that hydrologists have always had to deal with uncertainty and they have developed some tools (e.g. flood frequency studies). The numbers for design (e.g. 100-years flood) are produced. The critical question may occur – how uncertain are these numbers? This rhymes with Klemes's rhetoric – unreliability of reliability estimates.

It is believed that the material presented in the Workshop and contained in the present volume is an important contribution to our knowledge of uncertainty in hydrology and water resources, showing us how to deal with uncertainty.

REFERENCES

Bardossy, A. (1995) Stochastic weather generator using atmospheric circulation patterns and its use to evaluate climate change effects, present volume.

Beck, M. B. (1983) Uncertainty, system identification and prediction of water quality. In: Beck, M. B. & van Straten, G. (Eds.), *Uncertainty and Forecasting in Water Quality*, Springer, Berlin.

Bernier, J. M. (1987) Elements of Bayesian analysis of uncertainty in hydrological reliability and risk models. In: Duckstein, L. & Plate, E. J. (Eds.), *Engineering Reliability and Risk in Water Resources, NATO ASI Series, Series E: Applied Sci.*, No. 124, Nijhoff, Dordrecht.

Bogardi, J. J. & Verhoef, A. (1995) Reliability analysis of reservoir operation, present volume.

Feluch, W. (1995) Nonparametric estimation of multivariate density and nonparametric regression, present volume.

Georgakakos, K. P. & Krajewski, W. F. (1995) Worth of radar data in real time prediction of areal rainfall, present volume.

Georgakakos, K. P., Sharifi, M. B. & Sturdevant, P. (1995) Analysis of fine increment point rainfall, present volume.

Gottschalk, L., Krasovskaia, I. & Kundzewicz, Z. W. (1995) Detection of outliers in flood data with geostatistics, present volume.

Gottschalk, L. & Kundzewicz, Z. W. (1995) Outliers in Norwegian flood data, present volume.

Guo Sheng Lian (1995) Non-parametric approach for design flood estimation with pre-gauging data and information, present volume.

Guo Sheng Lian (1995) New plotting position rule for flood records considering historical data and paleologic information, present volume.

Hornby, A. S. (1974) *Oxford Advanced Learner's Dictionary of Current English*, Oxford Univ. Press, Third Ed.

Hubert, P., Friggit, F. & Carbonnel, J. P. (1995) Multifractal structure of rainfall occurence in West Africa, present volume.

Jakubowski, W. (1995) The distribution of the l-day total precipitation amount – a renewal model, present volume.

Karbowski, A. (1995) Statistically safe sets methodology for optimal management of reservoirs in risk situations, present volume.

Kemblowski, M. W. & Jet-Chau Wen (1995) Dispersion in stratified soils with fractal permeability distribution, present volume.

Kindler, J. & Tyszewski, S. (1995) On the value of fuzzy concepts in hydrology and water resources management, present volume.

Klir, G. J. (1989) Methodological principles of uncertainty in inductive modelling: a new perspective. In Erickson, G. J. & Smith, C. R. (Eds.) *Maximum-Entropy and Bayesian Methods in Science and Engineering*, Vol I.

Konecny, F. & Nachtnebel, H. P. (1995) A daily streamflow model based on a jump-diffusion process, present volume.

Kotwicki, W. & Kundzewicz, Z. W. (1995) Hydrological uncertainty – floods of Lake Eyre, present volume.

Kowalski, K. (1995) Order and disorder in hydroclimatological processes, present volume.

Kozłowski, A. & Łodziński, A. (1995) The risk evaluation in water system control, present volume.

Krajewski, W. F. & Smith, J. A. (1995) Estimation of the mean field bias of radar rainfall estimates, present volume.

Krasovskaia, I. & Gottschalk, L. (1995) Analysis of regional drought characteristic with empirical orthogonal functions, present volume.

Kundzewicz, Z. W. & Łaski, A. (1995) Reliability-related criteria in water supply system studies, present volume.

Lovejoy, S. & Schertzer, D. (1995) Multifractals and rain, present volume.
Mandelbrot, B. B. (1977) *Fractal Geometry of Nature*, Freeman, New York.
Meyer, P. D., Eheart, J. W., Ranjithan, R. S. & Valocchi, A. J. (1995) Design of groundwater monitoring networks for landfills, present volume.
Mizumura, K. (1995) Application of fuzzy theory to snowmelt-runoff, present volume.
Mizumura, K. (1995) Application of pattern recognition to rainfall-runoff analysis, present volume.
Moss, M. E. (1995) Bayesian relative information measure – a tool for analyzing the outputs of general circulation models, present volume.
Napiórkowski, J. J. & Strupczewski, W. G. (1995) Towards the physical structure of river flow stochastic process, present volume.
Plate, E. J. (1995) Stochastic approach to non-point pollution of surface waters, present volume.
Plate, E. J. & Duckstein, L. (1987) Reliability in hydraulic design. In: Duckstein, L. & Plate E. J. (Eds.), *Engineering Reliability and Risk in Water Resources, NATO ASI Series, Series E: Applied Sci.*, No. 124, Nijhoff, Dordrecht.
Ranjithan, R. S., Eheart, J. W. & Garrett, J. H. Jr. (1995) Application of neural network in groundwater remediation under conditions of uncertainty, present volume.
Rasmussen, P. F. & Rosbjerg, D. (1995) Prediction uncertainty in seasonal partial duration series, present volume.
Romanowicz, R. J., Dooge, J. C. I. & O'Kane, P. (1995) The spatial variability of the evaporation from the land surface, present volume.
Sowiński, M. & Yusuf, M. I. (1995) Composite risk model of Ogee type spillway, present volume.
Strupczewski, W. G. & Mitosek, T. M. (1995) Some aspects of hydrological design under non-stationarity, present volume.
Tsakiris, G. & Manoliadis, O. (1995) Stochastic modelling of the operation of hydrants in an irrigation network, present volume.
Webster's Ninth New Collegiate Dictionary (1987) Merriam-Webster Inc., Springfield, Massachusetts, USA.
Węglarczyk, S. (1995) The influence of time discretization on inferred stochastic properties of point rainfall, present volume.
Zawadzki, I. (1995) Is rain fractal?, present volume.

II

Facets of uncertainty

1 Bayesian relative information measure – a tool for analyzing the outputs of general circulation models

M. E. MOSS

U.S. Geological Survey, Tucson, Arizona, USA

ABSTRACT Mathematical models that generate scenarios containing no temporal correspondence to time series of actual occurrences are difficult to evaluate. One such class of models consists of atmospheric General Circulation Models (GCM), which have an additional drawback that the temporal and spatial scales of their outputs do not match those of the actual observations of the simulated phenomena. The problem of disparate scales can be ameliorated by aggregating both the model output and the observed data to commensurate scales. However, this approach does not permit quantitative testing at scales less than the least common level of aggregation.

The lack of paired observations in the aggregated time series makes standard statistical methods either invalid or ineffective in testing the validity of a GCM. One approach to resolving this quandary is the use of a relative information measure, which is based on the uncertainties contained in the histograms of the aggregations of both the model output and the data base. Each interval of each histogram is analyzed, from a Bayesian perspective, as a binomial probability. For the data-based histogram, the reciprocal of the sum of the variances of the posterior distributions of probability in each interval is denoted as its information content. For the model-based histogram, the reciprocal of the sum of the expected mean squared errors of the posterior distributions in each interval likewise is its information content. The expected mean squared error, which accounts for potential biases in the GCM, is computed as the expectation of the squared differences between the data-based and the model-based posterior distributions for each interval. The ratio of the information content of the model-based histogram to that of the data-based histogram is the relative information of the model.

A five-year monthly precipitation time series for January at a single node of the current version of the Community Climate Model including the Biosphere Atmosphere Transfer Scheme contains 7.5 percent of the information in the most recent 30 years of data in the Climatological Data Set assembled by the National Climate Data Center (NCDC). If the five-year simulation were extended, its relative information content could be expected to approach 11.5 percent. For July monthly precipitation, the five-year simulation contained 13.7 percent of the information of NCDC data base and had a limit of 39.3 percent for extremely long simulations. Aggregation of two adjacent cells at the same latitude showed some improvements in relative information, but longitudinal aggregation with two additional adjacent cells showed improvement for January precipitation, but degraded information for July precipitation.

An extended version of this contribution (with Appendix) was published in *Journal of Geophysical Research* (*Atmospheric Sciences*), American Geophysical Union.

NOTATION

D_G Gaussian discrepancy used to define an optimum histogram
E[a] Expected value of the random variable contained within the brackets
I_D Information content of data-based time series
I_M Information content of GCM output
I_R Relative information content of the GCM output to the data-based time series
M_i Mean squared error of the model-based estimate of the probability of an observation in the ith interval of a histogram
N Number of observations in a time series
P[a] Probability of the event described within the brackets
U Uncertainty
X Random variable
Y Logarithmic transform of the random variable, X
d Derivative symbol
e Base of natural logarithm
f Probability density function of a random variable
$\bar{f}_i$ Average probability density function within an interval, i, of a histogram
i Index of the interval of a histogram
i^o Left-most non-empty interval of a sample histogram
i^* Right-most non-empty interval of a sample histogram
m_i Expected value of the model-based Bayesian estimator of the probability of an event in the ith interval of a histogram
n_i Number of observations in the ith interval of a histogram
p_i Probability of an event in the ith interval of a histogram
s_i^2 Variance of the model-based Bayesian estimator of the probability of an event in the ith interval of a histogram
s_X Estimate of the standard deviation of the random variable X
u_i Limit of integration
x Realization of the random variable X
y Logarithm of a realization of the random variable X
z_i Probability of an event in an expanded interval (to account for empty intervals adjacent to interval i)
$\Gamma[a]$ Complete gamma function of the argument contained in the brackets
Σ Summation
α Ratio of the half length of an empty string adjacent to and left of the ith interval to the interval width
β Ratio of the half length of an empty string adjacent to and right of the ith interval to the interval width
ε Interval width
ζ_i The Bayesian estimator of z_i
λ The Bayesian estimator of the parameter of an exponential probability distribution
λ^o The parameter of an exponential probability distribution of the logarithms of observations on the low end of sample histogram
λ^* The parameter of an exponential probability distribution of observations on the high end of a sample histogram
v Condition in which the random variable X is known to be within the range of an expanded interval of a histogram
π_i Bayesian estimator of the probability of an event in the ith interval of a histogram
$\dot{\pi}_i$ Bayesian estimator of the probability of a data-based observation in the ith interval of a histogram
$\ddot{\pi}_i$ Bayesian estimator of the probability of a model-based observation in the ith interval of a histogram
τ_i Mathematical expression used to compute the expected square of the Bayesian estimator of the probability in the ith interval of an exponential probability distribution
ϕ_i^o Bayesian estimator of the probability of an event of magnitudes less than ε_i^o
ψ_i Mathematical expression used to compute the expected value of the Bayesian estimator of the probability in the ith interval of an exponential probability distribution
ω_i Mathematical expression used to compute the expected square of the Bayesian estimator of the probability in the ith interval of an exponential probability distribution
Var[a] Variance of the random variable contained in the brackets
$\tilde{}$ Indicator of conditioning on the random variable X being less than i^o
$\hat{}$ Indicator of conditioning on the random variable X being in an expanded interval of a histogram
| Indicator of conditioning on the event defined to its right.

INTRODUCTION

In many scientific investigations, mathematical models are developed to capture understanding about physical, chemical, or biological processes in a form that subsequently can be used in actual decision making. However, prudent decision makers frequently question the goodness of such models before accepting their output. For certain classes of models in which observations of the modeled phenomena are putatively concurrent with the output of the model, goodness-of-fit measures are available (e.g., Troutman, 1985). For other classes that lack concurrence of measurement and model output, the evaluation of the validity of a model is not a straight-forward exercise.

One class of models in the latter category is the general circulation model (GCM) as employed in climate analysis (Washington & Parkinson, 1986). GCMs are being used to develop scenarios of atmospheric variables for hypothetical conditions of doubled atmospheric carbon dioxide (e.g. Schlesinger & Zhao, 1989), nuclear war (e.g., Malone *et al.*, 1986), and denudation of tropical rain forests (e.g., Dickinson, 1989; Lean & Warrilow, 1989) that can be compared with the climate of today. However, because of inherent inaccuracies in GCMs, such comparisons usually are made between the simulation of the altered environment and a simulation of today's climate. By comparing simulated scenarios, it is hoped that errors will be compensating and that the changes experienced between the simulations will approximate those that would occur if the hypothetical condition came to pass.

The utility of the comparative analysis and of the simulated changes could be enhanced if an adequate measure of the validity of the underlying simulation of today's climate were available. One approach that has been used (Chervin, 1981; Katz, 1982, 1988) to address this need is that of statistical hypothesis testing in which the statistics of the simulation and those of actual climate data are compared to see if there is a significant difference between the two. Even the most sophisticated model is not a perfect representation of the real world. Therefore, the results of hypothesis testing simply indicate whether sufficient data are incorporated to detect the differences that exist and do not serve as a validation or invalidation of the model. Thus, such attempts to instill objectivity into the investigation of the validity of GCMs have not been fully successful. Therefore, an alternative approach is presented here to address this problem.

Traditionally, climatology has been a statistical component of the atmospheric sciences (Guttman, 1989) in which suites of variables are described probabilistically in time and space. However, from this suite, only a single variable, monthly precipitation over a single grid cell, will be used to illustrate the proposed approach for measuring the goodness of a GCM's output. Extension of the approach to multi-cell, spatial dimensions will be provided in an example; it is hoped that extensions in time and to other atmospheric variables will be evident.

APPROACH

The seasonal-precipitation climatology for a particular grid cell can be described partially by the probability density function (pdf) of the spatially averaged depth of precipitation during the season over the grid cell. Neither the mathematical form nor the parameter values of such a pdf are known, but they can be estimated from actual data from within the geographical region containing the cell. Actual measurements of precipitation exist only at a finite number of points within or near the grid cell; actual measurements of spatially averaged depth of precipitation do not exist, but can be constructed mathematically from the point data by weighted averages, by Thiessen polygons (Thiessen & Alter, 1911), or by more sophisticated approaches such as Kriging (Bras & Rodriguez-Iturbe, 1985). In the absence of persistence, the unknown pdf would contain perfect information about the seasonal-precipitation climatology. However, even in the absence of persistence, the estimate of the pdf will contain only limited information because (1) the time series of measurements upon which it is based is of finite length and (2) each computed value of the series contains a combination of measurement and interpolation errors. In this paper, the measurement and interpolation errors are assumed to be negligible; in actuality, they usually are not. A subsequent paper will incorporate the effects of errors in the data into the approach.

The data from the time series of grid-cell precipitation can be manipulated to form a histogram that approximates the unknown pdf. Within any interval, i, of the histogram, the count, n_i, of precipitation events divided by the total number of events, N, is one estimate of the probability, p_i, of a random precipitation event of a magnitude within the range of the interval. The true probability, p_i, is a function of the unknown pdf:

$$p_i = \int_{\varepsilon(i-1)}^{\varepsilon i} f_X(x)\,\mathrm{d}x \tag{1}$$

where $f_X(x)$ is the pdf of a precipation event, X, and ε is the interval width of the histogram. Therefore,

$$\bar{f}_i = p_i/\varepsilon \tag{2}$$

where $\bar{f}_i$ is the average value of the pdf within the interval i.

With a finite sample of N events, it is possible to estimate p_i only with limited accuracy. The uncertainty in p_i can be quantified by the variance of the conditional (posterior) probability density function of its Bayesian estimator, $\dot{\pi}_i$, given the n_i observations in the ith interval. As is discussed subsequently in more detail, the conditional pdf of $\dot{\pi}_i$ can be developed by considering the numbers of observations in consecutive intervals of the data-based histogram as realizations of binomial random variables. The total uncertainty about spatially averaged precipitation over the grid cell can be approximated by the sum over all intervals of the variances of the individual conditional pdfs of $\dot{\pi}_i$:

$$U = \sum_{i=1}^{\infty} \text{Var}[\dot{\pi}_i] \tag{3}$$

where U is the total data-based uncertainty about spatially averaged precipitation and Var[*] is the variance of the random variable contained within the brackets. A measure of the information content, I_D, of the data base pertaining to spatially averaged precipitation is the reciprocal of the total uncertainty (Fisher, 1960):

$$I_D = 1/U \tag{4}$$

A comparable measure for the information contained in the GCM output also can be derived. With the GCM output, the binomial analogy is used to estimate posterior means and variances for the model-based Bayesian estimators. However, the posterior variances are inappropriate measures of the uncertainties in the GCM output. The variance is a measure of the dispersion of likelihood of a value of a random variable about its mean, which, in the absence of any other information, is its best estimate, in a least-squares sense, of the unknown value. However, in comparing the GCM output with the measured data, it is assumed that the data provide the best estimate of today's climate. Therefore, the dispersion of the GCM output should be measured relative to the information contained in the data. If p_i were known with certainty, the GCM dispersion could be evaluated by using the transfer of moments equation from basic mechanics:

$$M_i = s_i^2 + (m_i - p_i)^2 \tag{5}$$

where M_i is the mean squared error of $\ddot{\pi}_i$, the model-based Bayesian estimator of p_i, s_i^2 and m_i are the variance and mean, respectively, of the posterior distributions of $\ddot{\pi}_i$. To account for the imperfect knowledge concerning p_i in the data-based time series, an expected value of M_i is developed by integrating the product of m_i and the posterior pdf of $\dot{\pi}_i$ derived from the measured data:

$$\text{E}[M_i] = \int_0^1 [s_i^2 + (m_i - \dot{\pi}_i)^2] f(\dot{\pi}_i|n_i)\, d\pi_i \tag{6}$$

where $f(\dot{\pi}_i|n_i)$ is the data-based posterior pdf. The summation of $\text{E}[M_i]$ over all intervals is an estimate of the total uncertainty about the spatially averaged precipitation in the GCM output; the reciprocal of the total uncertainty is a measure of the information content of the model output, I_M:

$$I_M = 1/\sum_{i=1}^{\infty} \text{E}[M_i] \tag{7}$$

The information measures defined by equations 4 and 7 have the units of the reciprocal of the variance of probability. To facilitate interpretation of the model's information content, its ratio to that of the data-based time series is presented as relative information, I_R, a concept first introduced to hydrology by Matalas & Langbein (1962):

$$I_R = I_M/I_D = \sum_{i=1}^{\infty} \text{Var}[\dot{\pi}_i]/\sum_{i=1}^{\infty} \text{E}[M_i] \tag{8}$$

Because the data are assumed to provide the best estimates of the p_i, I_R always will be contained within the range from zero to one.

FUNDAMENTAL CONSIDERATIONS

The following sections expand the descriptions of the key elements of the relative-information approach.

Data-based time series

The average precipitation on the land surface is assumed to be analogous to the precipitation leaving the lowest elements of the stacks that represent the atmosphere above each cell of interest. To construct commensurate data-based time series, point precipitation data from a collection of actual weather stations must be converted to areal averages over identical cells to those used in the GCM. As stated above, several procedures exist for performing this task.

For purposes of illustration, this study uses the precipitation data from the monthly climatic averages compiled by the National Climatic Data Center of the U.S. National Oceanic and Atmospheric Administration for each climatic division in the contiguous 48 States. The climatic divisions are shown in Fig. 1 as are the borders of a GCM grid cell for which a time series of monthly mean precipitation is to be constructed. The precipitation series, one for each climatic division, were constructed by simple averaging of the monthly precipitation for each station within the climatic division that reports both monthly temperature and precipitation (Karl & Knight, 1985). To construct a monthly time series of precipitation over the grid cell, the reported precipitation magnitude for a given month and year for each climatic division is weighted by the ratio of that portion of its area contained within the cell to the total land area of the cell, and these weighted magnitudes are summed to obtain a single entry in the time series. Fig. 2 shows such a time series of January precipitation for the grid cell delineated in Fig. 1 for the period from 1960 to 1989. The units of the time series have been transformed into average daily precipitation

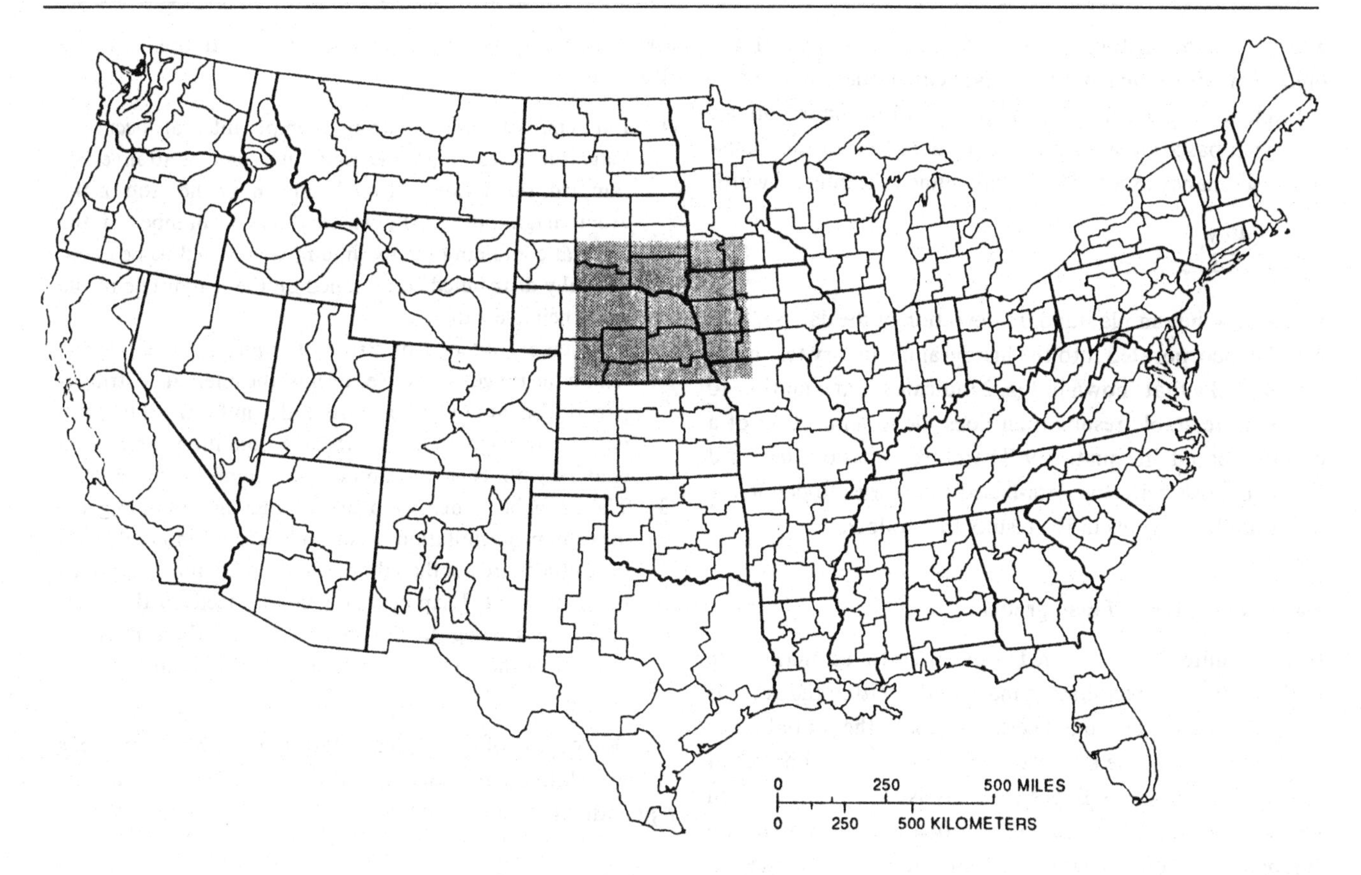

Fig. 1 A GCM cell superimposed on the map of the State Climatic Divisions.

rates to be consistent with the GCM output described subsequently.

Although precipitation data for the State climatic divisions are available for all years since 1895, climatic norms traditionally have been based on the most recent three decades of data. This is a tacit recognition by climatologists that climate is nonstationary. Therefore, for the sake of consistency with tradition, only data from the most recent 30 years will be used to define the optimum histogram and the data-based uncertainties. Use of the most recent data also avoids potential problems of reconstruction errors in the time series prior to 1931 (Karl & Knight, 1985).

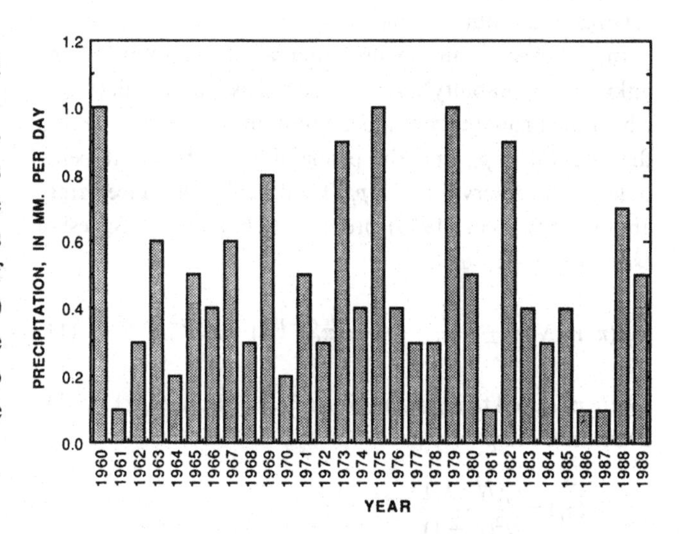

Fig. 2 Time series of January precipitation for a grid cell.

Optimum histogram

Information content, determined by either equation 4 or equation 7, will vary with the locations of the histogram intervals that underpin its definition, as can be readily seen in equation 1. Therefore, for the sake of consistency, the histogram definitions used to compute the information contents both of the data-based time series and of the model output should be identical. Because of the assumption that the data base contains significantly more information than does the model, the preferable definition for the computation of relative information is that which minimizes the expected discrepancy between the data-based histogram and the unknown pdf of spatially averaged depth of precipitation. Linhart & Zucchini (1986) advocate the use of a Gaussian discrepancy, D_G, in defining the optimum histogram for a given set of observations:

$$D_G = \sum_{i=1}^{\infty} \int_{\varepsilon(i-1)}^{\varepsilon i} [f_X(x) - \bar{f}_i]^2 \, dx \tag{9}$$

where $\bar{f}_i$ is the histogram estimate of the average of the unknown pdf in the ith interval (see equations 1 and 2).

Linhart & Zucchini (1986, p. 31) also report on an approach based on asymptotic considerations of the Gaussian discrepancy (Scott, 1979) that defined the interval width

$$\varepsilon = \frac{3.49\ s_X}{N^{1/3}} \tag{10}$$

where s_X is the sample standard deviation of the data set that is to be incorporated into the histogram. Cursory testing of data sets like that shown in Fig. 2 indicated that equation 10 yields more stable results than does the minimization of a criterion based on equation 9. Therefore, equation 10 is used in conjunction with the data-based time series to define the optimum histograms in the results that follow.

Bayesian analysis of histograms

Because finite time series are the sole basis for estimating the probabilities of precipitation magnitudes within the various intervals of a histogram, uncertainty about the probabilities will be inherent. One means of dealing with uncertainties of this sort is known as Bayesian analysis (Epstein, 1985), in which a constant, but unknown, probability is treated as a surrogate random variable. The uncertainty about the unknown probability can be quantified by the variance of the surrogate random variable.

In the case of an isolated interval of a histogram, the unknown probability can be described as a Bernoulli trial in which the probability of a precipitation event falling within the interval is p_i, and the probability of the event being outside the interval is $1-p_i$. To describe the uncertainty about p_i, Epstein (1985) provides a pdf of the Bayesian estimator of p_i:

$$f(\pi_i|n_i, N) = \frac{\Gamma(N)}{\Gamma(n_i)\,\Gamma(N-n_i)}\,\pi_i^{(n_i-1)}\,(1-\pi_i)^{N-n_i-1} \tag{11}$$

where π_i is used to represent either $\dot{\pi}_i$ or $\ddot{\pi}_i$. Equation 11 yields a variance of π_i:

$$\mathrm{Var}(\pi_i) = \frac{n_i(N-n_i)}{N^2(N+1)} \tag{12}$$

Equation 12 is an appropriate measure of the uncertainty of p_i only when the ith interval and the adjacent intervals on its right and left are all non-empty. A non-empty interval is one for which n_i is not equal to zero. When equation 11 is used in conjunction with an empty interval, the mean and variance degenerate to zero, which indicates with certainty that no precipitation events with magnitudes within the interval could occur. It is very likely that the occurrence of an empty interval is a relic of a finite sample size and does not represent a physical impossibility of such an event. Therefore, three assumptions are made to avoid this unlikely situation.

(1) If an empty interval or string of empty intervals occurs at magnitudes less than the minimum precipitation event sampled, the negative of the logarithm of the ratio of the event magnitude to εi^0, where i^0 is the number of the interval containing the minimum, is assumed to be exponentially distributed conditioned on the magnitude of the event being less than εi^0.
(2) If an empty interval or string of empty intervals occurs within the range of observed events, the interval or string is partitioned into two equal parts and a uniform pdf for each part is assumed to derive from a conditional binomial analysis with its nearest non-empty neighboring interval.
(3) The probability density over the interval containing the maximum precipitation event and over all intervals with magnitudes greater than the observed maximum is assumed to be distributed exponentially conditioned on the event magnitudes being greater than $\varepsilon(i^*-1)$, where i^* is the number of the interval containing the maximum sampled event.

A description of the implementation of these assumptions in the relative-information approach is provided in the Appendix in Moss (1991).

Expected mean squared error and relative information

Equation 6 defines the expected mean squared error, $\mathrm{E}[M_i]$, of the Bayesian estimator of the probability in the ith interval based on the output of a GCM.

As shown in equation 7, the information content of the GCM output is the reciprocal of an infinite sum of the $\mathrm{E}[M_i]$s. If the information content is to be a non-trivial measure, this infinite series must converge. Attempts to prove this convergence, using the relations provided earlier and in the Appendix in Moss (1991), failed; therefore, only empirical evidence for convergence can be offered herein.

A set of computations, using the examples described in the next section, were executed in which the sum of $\mathrm{E}[M_i]$ was truncated at 10, 50, 100, and 200 intervals beyond the last non-empty interval of the model-based histograms. In no case did the sum increase by more than 0.05 percent between the truncation levels of 10 and 200 intervals. To test this measure further, one summation was extended to 1000 intervals, and, to four significant figures, no further increment to the sum was obtained. Thus, no evidence was found to indicate lack of convergence. Similar experiments pertaining to the information content of the data-based time series resulted in an identical conclusion. The results presented subsequently are based on truncating the sums of $\mathrm{E}[M_i]$ and Var $[\pi_i]$ after 10 intervals beyond the highest non-empty interval of either the data-based or the model-based histograms.

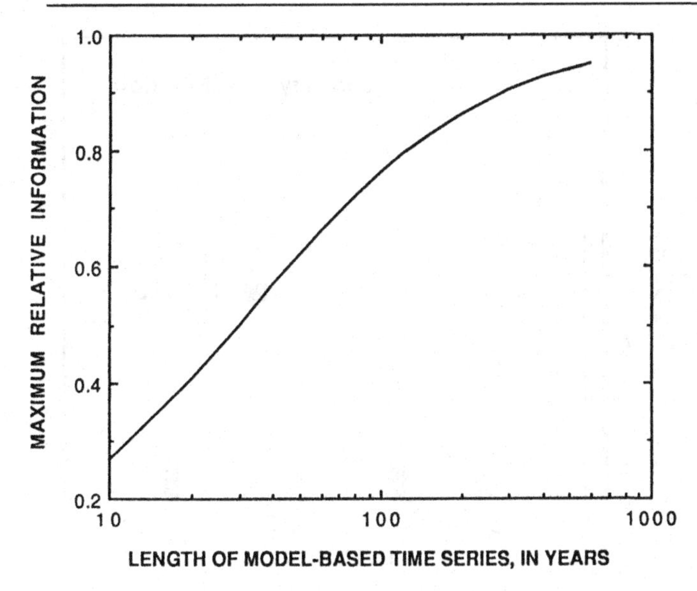

Fig. 3 Maximum relative information of a model's output based on a 30-year record.

By carrying out the integration on the right-hand side of equation 6 and rearranging terms, it can be seen that

$$E[M_i] = s_i^2 + \mathrm{Var}[\pi_i] + (m_i - E[\pi_i])^2 \qquad (13)$$

Thus, if a model output of equivalent length to the data-based time series results in an n_i equal to that of the data-based time series, m_i will equal $E[\pi_i]$, s_i^2 will equal Var $[\pi_i]$, and $E[M_i]$ will equal 2 $\mathrm{Var}[\pi_i]$. If the time series of the model output is identical with that derived from the data base, $E[M_i]$ will equal 2 Var $[\pi_i]$ for each interval, and the relative information of the model output will be half of that of the data base. If the two series were putatively concurrent, their cross-correlation coefficient would be unity; the relative information of the model would equal one. Thus, the lack of concurrence reduces the information content of the model by a factor of two. This reduction is directly attributable to the unknown biases, that is the inability of the model to exactly mimic the real world. Such model errors, which cannot be expected to be reduced by extending the length of the simulation, can only be addressed by improved conceptualization, parameterization, or resolution of the underlying GCM.

It can be shown that the maximum value of I_R attainable for two time series of equal length is 0.5, which once again is a manifestation of the unknown model errors. For I_R to exceed 0.5, the length of the time series of the model output must exceed that derived from the data base, and the resulting pairs of values of $E[\pi_i]$ and m_i for each interval must be approximately equal. To approach the upper limit of one for I_R, the model output must be very long and the differences in $E[\pi_i]$ and m_i must be very small within each interval. For example, Fig. 3 shows the relation of the maximum attainable value of I_R to the length of the model-based time series when the data-based time series is thirty years long. The maximum I_R for given lengths of the two time series is obtained when the differences in $E[\pi_i]$ and m_i are minimum for each interval.

The expected mean squared error is comprised of three separate components: (1) time-sampling error of the data-based histogram represented by Var $[\pi_i]$ in equation 13, (2) time-sampling error in the model-based histogram represented by s_i^2 in equation 13, and (3) squared bias represented by $(m_i - E[\pi_i])^2$. Extension of the output time series of the model can be expected to reduce only the second component, model time-sampling error. Therefore, by subtracting the sum of the s_i^2s from the total uncertainty of the model-based histogram, an estimate of the uncertainty and concomitantly the I_R contained in the output as the run length goes to infinity can be obtained.

EXAMPLES

The Biosphere-Atmosphere Transfer Scheme (BATS), developed by Dickinson *et al.* (1986), contains one of the more sophisticated representations in a GCM of the interactions between the atmosphere and the land surface. BATS is designed to function either interactively with a GCM or as a post processor of GCM outputs.

It has been most commonly used in conjunction with the Community Climate Models of the National Center for Atmospheric Research, where BATS was developed. Monthly precipitation time series for grid cells in the Central Plains of the U.S. from an interactive run of BATS and CCM1, the most current version of the Community Climate Model, were provided to illustrate the use of the relative information measure (P.J. Kennedy; written communication). The model output was five years in length.

A single grid cell

The model-based time series of monthly precipitation for January and July for the grid cell depicted in Fig. 1, which will henceforth be designated as cell A, are shown in Fig. 4 along with the most recent 30 values of the data-based series for the same cell. Sample statistics for these data are provided in Table 1. By examining the data-based time series and its statistics, it can be seen that the climatology of cell A has a relatively dry winter with little variability of precipitation and a dramatically heavier and more variable summer precipitation regime. On a qualitative basis, the model does reasonably well in mimicking the July precipitation character, but generates a much wetter and more variable January regime than the observations indicate.

Table 1. *Sample statistics for January and July precipitation for cell A*

	Data	Model
January precipitation [mm/day]		
Expected value	0.47	1.22
Standard deviation	0.28	1.00
First-order serial correlation coefficient	−0.31	−0.23
July precipitation [mm/day]		
Expected value	2.73	2.52
Standard deviation	0.94	1.06
First-order serial correlation coefficient	0.11	−0.23

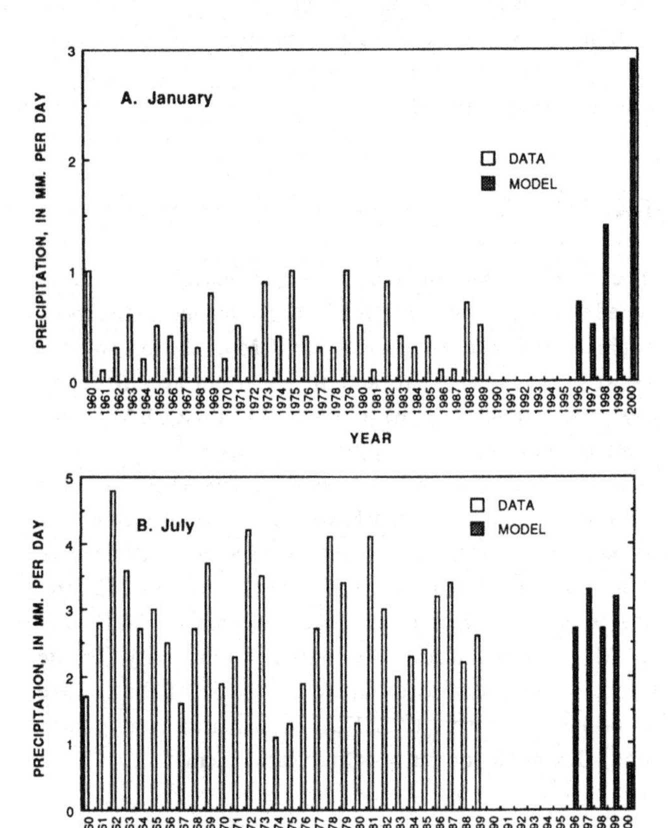

Fig. 4 Data-based and model-based monthly precipitation time series for January and July for cell A.

With the sample sizes available for the time series, none of the serial correlation coefficients in Table 1 are statistically significant. Furthermore, if there were persistence at the level indicated by these coefficients, the estimates of standard deviations would tend to be biased only slightly. Therefore, the analysis will proceed under the assumption of negligible persistence in the time series.

The individual pairs of time series, for January and for July, were analyzed on the basis of the concepts and assumptions presented above. Fig. 5 shows the histograms of the

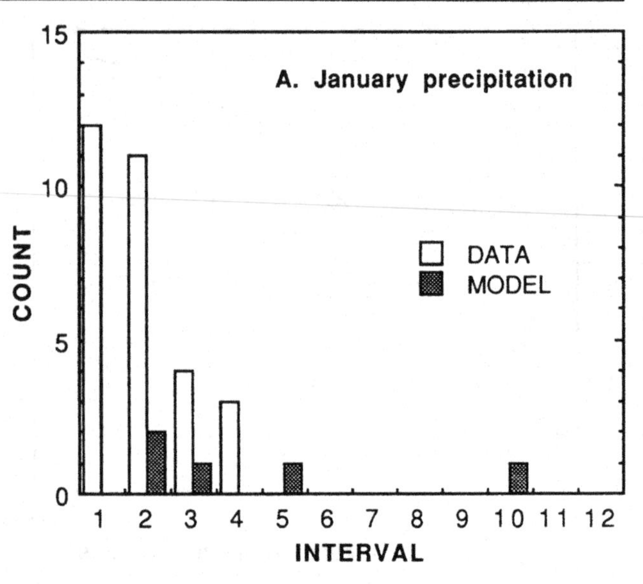

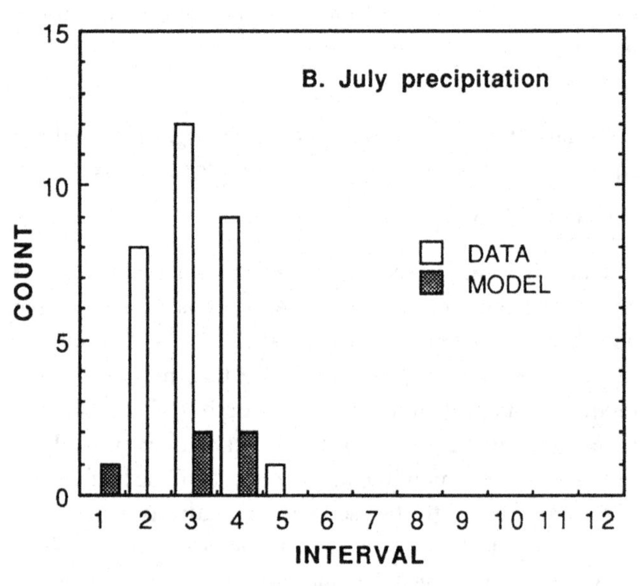

Fig. 5 Histograms of January and July precipitation for cell A.

four input time series, and Tables 2 and 3 present the salient results. Consistent with the statistics in Table 1, the model's performance in January is not as good as it is in July – January I_R equals 7.5 percent, while July's is 13.7 percent.

The total uncertainty in the data-based histograms, about 0.02, is very similar for January and July in spite of the large difference in their interval widths. Therefore, the difference in I_R between the two months can be related almost solely to the characteristics of the model-based histograms. It can be seen that the sum of the posterior variances of the January histogram is about equal to that of July. Thus, the potential biases of the model seem to be the major cause of the differences in the two months. The bias component of histogram uncertainty is the sum over all intervals of the squared differences in the expected probabilities of the data-based and model-based posterior distributions. The expected probabilities for each month are presented in Fig. 6. The sum

Table 2. *Posterior analysis of January precipitation in cell A (interval width = 0.32 mm/day)*

	Data-based probabilities		Model-based probabilities		
Interval	Mean	Variance	Mean	Variance	Expected mean squared error
1	0.400	0.0077	0.035	0.0048	0.1459
2	0.367	0.0075	0.365	0.0378	0.0453
3	0.133	0.0037	0.133	0.0156	0.0193
4	0.097	0.0028	0.080	0.0069	0.0100
5	0.002	0.0000	0.133	0.0156	0.0327
6	0.000	0.0000	0.027	0.0011	0.0018
7	0.000	0.0000	0.027	0.0011	0.0018
8	0.000	0.0000	0.033	0.0017	0.0028
9	0.000	0.0000	0.033	0.0017	0.0028
10	0.000	0.0000	0.115	0.0134	0.0266
11	0.000	0.0000	0.001	0.0003	0.0004
12	0.000	0.0000	0.003	0.0001	0.0001
13	0.000	0.0000	0.002	0.0000	0.0000
...	...	...	...	...	...
20	0.000	0.0000	0.000	0.0000	0.0000
SUM		0.0218		0.0999	0.2893

Data-based information = 45.9
Model-based information = 3.4
Relative information = 0.075

Table 3. *Posterior analysis of July precipitation in cell A (interval width = 1.06 mm/day)*

	Data-based probabilities		Model-based probabilities		
Interval	Mean	Variance	Mean	Variance	Expected mean squared error
1	0.034	0.0008	0.133	0.0156	0.0261
2	0.232	0.0054	0.167	0.0167	0.0264
3	0.400	0.0077	0.300	0.0300	0.0477
4	0.300	0.0068	0.392	0.0395	0.0548
5	0.022	0.0006	0.005	0.0003	0.0012
6	0.005	0.0000	0.001	0.0000	0.0001
7	0.002	0.0000	0.000	0.0000	0.0000
8	0.001	0.0000	0.000	0.0000	0.0000
9	0.001	0.0000	0.000	0.0000	0.0000
...	...	...	...	...	...
20	0.000	0.0000	0.000	0.0000	0.0000
SUM		0.0214		0.1021	0.1886

Data-based information = 6.7
Model-based information = 6.4
Relative information = 0.137

of the squared differences for January is 0.168 of which 0.133 is from the first interval. The sum of the squared differences for July is 0.096.

By deducting the time-sampling error from the total uncertainty of the model-based histogram, an estimate of the relative information contained in the output of a model run approaching infinite length can be obtained. For January, the infinite run could be expected to have an I_R of 11.5 percent, an increase from 7.5 percent for the 5-year run. For July, the infinite run should attain an I_R of 39.3 percent, an increase from 13.7 percent for the 5-year output time series.

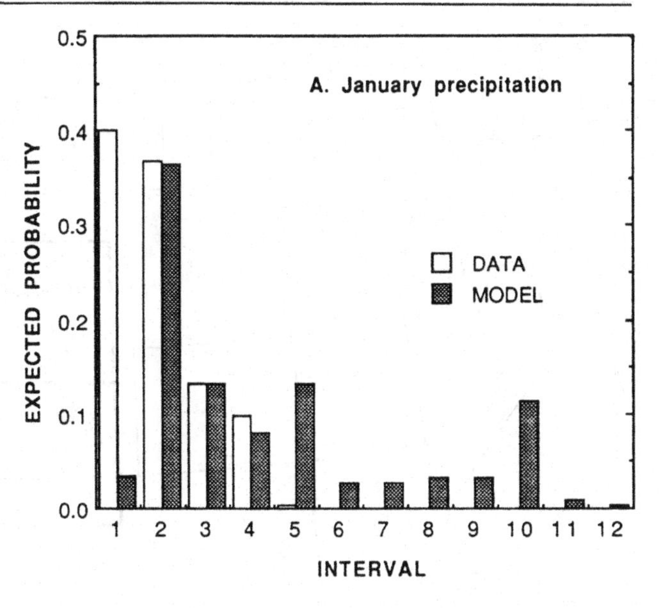

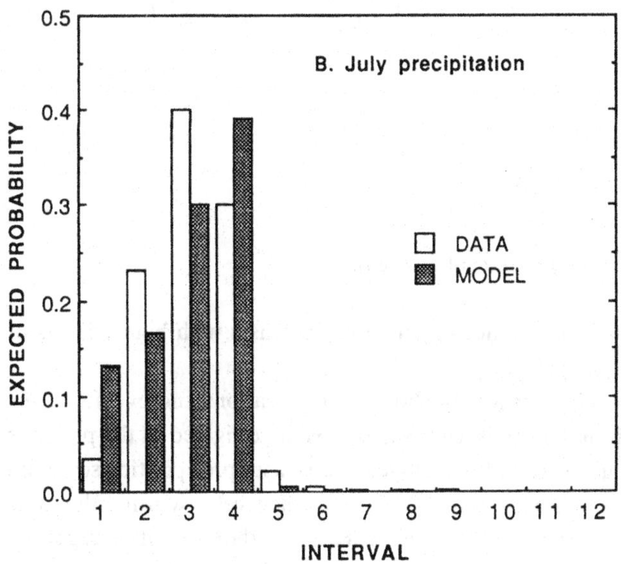

Fig. 6 Expected probabilities for cell A.

Aggregation of cells

A common approach for presenting and analyzing the outputs of GCMs is to average or aggregate time series across latitudinal bands, usually encompassing the globe. Two examples, one aggregating two adjacent cells at the same latitude and another aggregating two additional cells from the next tier to the south, are presented to illustrate the use of the relative-information approach in exploring

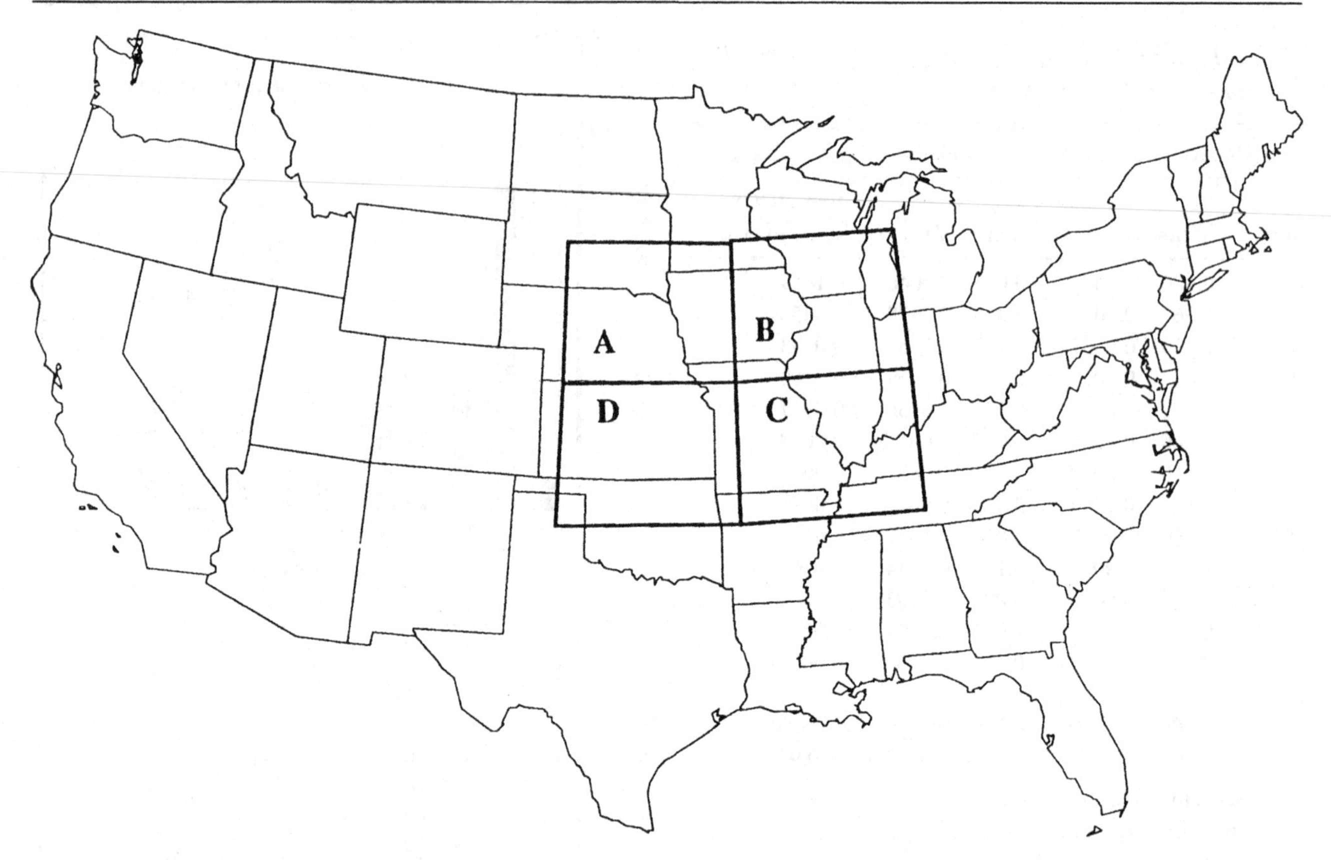

Fig. 7 Location of grid cells.

whether or not information increases with higher levels of aggregation.

Fig. 7 is a map showing the locations and extents of the four grid cells; cell A is the cell investigated in the previous sub-section. First, 30-year data-based precipitation series are generated for each of the three new cells by using the same procedure used for cell A. Then, for the two-cell aggregation, the data-based time series for cells A and B are averaged for each month to compute an aggregated time series, and likewise, the model outputs for the two cells are averaged to obtain a model-based time series. The results of this two-cell aggregation are shown in Table 4, where it can be seen that latitudinal aggregation provided a relative-information increase for January from 7.5 percent to 8.1 percent and caused an increase for July from 13.7 percent to 16.1 percent. There also are improvements in the limits of I_R as the run length goes to infinity. For January, the limit increased from 11.5 percent to 12.4 percent, while, for July, it went from 39.3 percent to 57.5 percent. These increases are derived primarily from decreases in the bias components of the model-based histograms for each month. The changes in biases can be seen by comparing Figs. 6 and 8.

The second aggregation example was devised by averaging for each month the data-based precipitation series for cells C and D of Fig. 7 and then averaging the resulting time series with the aggregated series for cells A and B. An aggregated model-based time series was generated in the same manner. The results of the relative-information analysis of the four-cell aggregation are given in Table 5, where it can be seen that the results for January improve, but those for July degrade. The degradation is particularly obvious for the sum of the squared biases for July, which increases from 0.017 for the two-cell aggregation to 0.046. This increase reduces the limit of the relative information from 57.5 percent to 33.2 percent. Thus, the very tentative conclusion from the four-cell aggregation is that, for the Central Plains of the U.S., longitudinal aggregation of GCM precipitation output may not increase its information content.

CONCLUSIONS

A new approach for estimating the information content of the outputs of GCMs has been presented. The underlying concepts account for the uncertainties in both the model outputs and in the data base that is used as a metric of comparison. However, the current version of the approach incorporates only uncertainties in the time-sampling domain of the data base and does not include components of measurement and computation errors that degrade its information content. These are elements for future study.

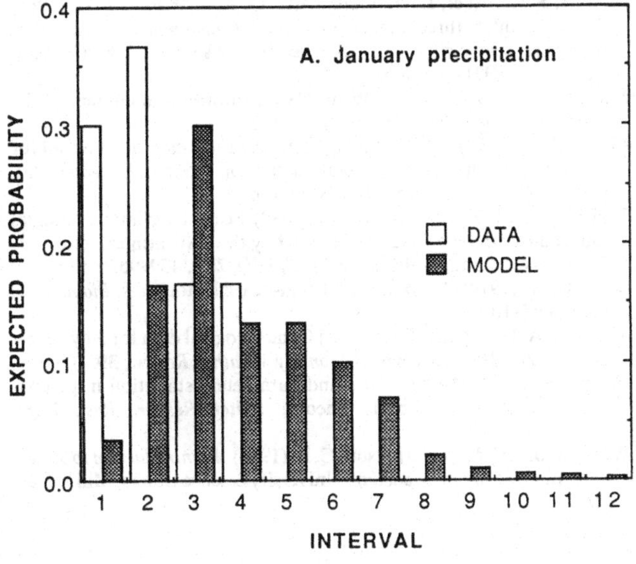

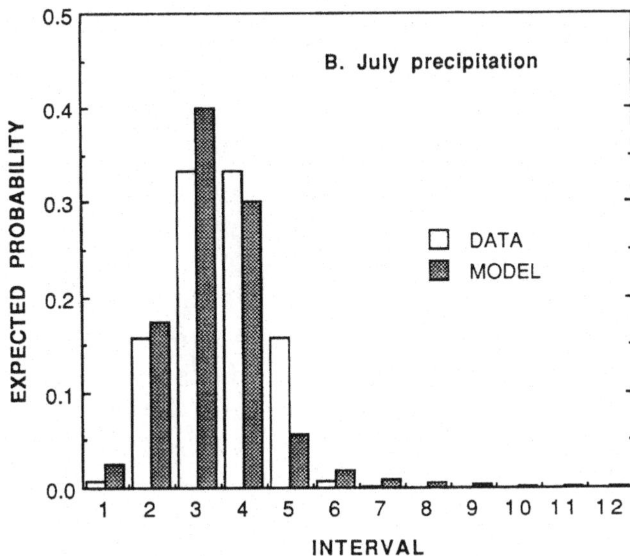

Fig. 8 Expected probabilities for cells A and B aggregated.

Table 4. *Analysis of aggregated monthly precipitation for cells A & B*

	Data	Model
January:		
Mean*	0.74	1.64
Standard Deviation*	0.41	0.98
Serial Correlation	−0.33	−0.34
Σ Variance	0.023	0.101
Σ E[M_i]		0.288
Σ Squared bias		0.164
Information	43.1	3.5
Relative information (5–year simulation) =		0.081
Relative information limit =		0.124
Interval width* =		0.46
July:		
Mean*	2.98	2.80
Standard Deviation*	0.89	0.76
Serial Correlation	0.20	−0.44
Σ Variance	0.023	0.103
Σ E[M_i]		0.143
Σ Squared bias		0.017
Information	43.4	7.0
Relative information (5-year simulation) =		0.161
Relative information limit =		0.575
Interval width* =		1.00

Note:
* Units are mm/day.

The relative-information measure is a statistic with unknown sampling properties; confidence bands cannot be placed on any particular realization at this time. Nor can inferences about the underlying biases of a GCM be tested for significance. These also are topics for further study.

Nevertheless, even in its infancy, the approach should be capable of giving preliminary answers to questions like:

(a) Does a finer grid of the same model generate more information than its coarser predecessor?
(b) Does model X provide more information about a given variable or suite of variables than model Y?
(c) At what level of aggregation of a GCMs output is there sufficient information to undertake its statistical disaggregation to smaller spatial or temporal scales?

Such studies are planned to be concurrent with the continued development of the approach.

REFERENCES

Bras, R. L. & Rodriguez-Iturbe, I. (1985) *Random Functions and Hydrology*, Addison-Wesley Publishing Co., Reading, MA.

Chervin, R. M. (1981) On the comparison of observed and GCM simulated climate ensembles, *Jour. Atmos. Sci.*, **38**(5), 885–901.

Dickinson, R. E. (1989) Predicting climate effects, *Nature*, **342**, 343–4.

Dickinson, R. E., Henderson-Sellers, A., Kennedy, P. J. & Wilson, M. F. (1986) *Biosphere-atmosphere transfer scheme (BATS) for the NCAR Community Climate Model*, NCAR Technical Note 275+STR, National Center for Atmospheric Research, Boulder, CO.

Epstein, E. S. (1985) Statistical inference and prediction in climatology: a Bayesian approach, *Meteorological Monographs*, **20**(42).

Fisher, R. A. (1960) *The Design of Experiments*; 8th Edition, Hafner Publishing Co., New York, NY.

Guttman, N. B. (1989) Statistical descriptors of climate, *Bull. Amer. Met. Soc.*, **70**(6), 602–7.

Karl, T. R. & Knight, R. W. (1985) Atlas of monthly and seasonal precipitation departures from normal (1895–1985) for the contiguous United States: *Historical Climatology Series 3–13*, National Climatic Data Center, Ashville, NC.

Katz, R. W. (1982) Statistical evaluation of climate experiments with general circulation models: a parametric time series modeling approach, *Jour. Atmos. Sci.*, **39**, 1446–55.

Katz, R. W. (1988) Statistical procedures for making inferences about climate variability, *Jour. of Climate*, **1**(11), 1057–64.

Lean, J. & Warrilow, D. A. (1989) Simulation of regional climatic impact on Amazon deforestation, *Nature*, **342**, 411–13.

Linhart, H. & Zucchini, W. (1986) *Model Selection*, Wiley, New York, NY.

Table 5. *Analysis of aggregated monthly precipitation for cells A, B, C & D*

	Data	Model
January:		
Mean*	1.08	1.72
Standard Deviation*	0.56	0.87
Serial Correlation	−0.24	−0.58
Σ Variance	0.021	0.116
Σ E[M_i]		0.215
Σ Squared bias		0.078
Information	47.6	4.6
Relative information (5–year simulation) =		0.098
Relative information limit =		0.212
Interval width* =		0.63
July:		
Mean*	2.93	2.48
Standard Deviation*	0.83	0.48
Serial Correlation	0.48	−0.30
Σ Variance	0.023	0.086
Σ E[M_i]		0.155
Σ Squared bias		0.046
Information	43.9	6.4
Relative information (5-year simulation) =		0.147
Relative information limit =		0.332
Interval width* =		0.93

Note:
* Units are mm/day.

Malone, R. C., Auer, L. H., Glatzmaier, G. A. & Wood, M. C. (1986) Nuclear winter: three-dimensional simulations including interactive transport, scavenging, and solar heating of smoke, *Jour. Geophys. Research*, **91**(D1), 1039–53.

Matalas, N. C. & Langbein, W. B. (1962) Information content of the mean, *Jour. Geophys. Res.*, **67**(9), 3441–8.

Moss, M. E. (1991) Bayesian relative information measure – a tool for analyzing the outputs of general circulation models, *J. Geoph. Res. (Atmosph. Sci.)*, Amer. Geophys. Union.

Schlesinger, M. E. & Zhao, Z. C. (1989) Seasonal climatic changes induced by doubled CO2 as simulated by the OSU atmospheric CM/mixed-layer ocean model, *Jour. of Climate*, **2**(5), 459–95.

Scott, D. W. (1979) On optimal and data-based histograms, *Biometrika*, **66**(3), 605–10.

Thiessen, A. H. & Alter, J. C. (1911) Climatological data for July 1911 – District No. 10, Great Basin, *Monthly Weather Review*, **39**, 1082–4.

Troutman, B. M. (1985) Errors and parameter estimation in precipitation-runoff modeling, 1. Theory, *Water Resour. Res.*, **21**(8), 1195–213.

Washington, W. M. & Parkinson, C. L. (1986) *An Introduction to Three-dimensional Climate Modeling*, University Science Books, Mill Valley, CA.

2 A stochastic weather generator using atmospheric circulation patterns and its use to evaluate climate change effects

A. BÁRDOSSY

Institute for Hydrology and Water Management, University of Karlsruhe, Karlsruhe, Germany

ABSTRACT The daily rainfall and the daily mean temperature are modelled as processes coupled to atmospheric circulation. Atmospheric circulations are classified into a finite number of circulation patterns. Rainfall occurrence is linked to the circulation patterns using conditional probabilities. Rainfall Z is modelled using a conditional distribution (exponential or gamma) for the rainfall amount, and a separate process for rainfall occurrence using a normal process which is then transformed, delivering both rainfall occurrences and rainfall amounts with parameters depending on the actual circulation pattern. Temperature is modelled using a simple autoregressive approach, conditioned on atmospheric circulation. The simulation of other climatic variables like daily maximum and minimum temperature, and radiation, is briefly discussed. The model is applied using the classification scheme of the German Weather Service for the time period 1881–1989. Precipitation and temperature data measured at different locations for a period of 30 years are linked to the circulation patterns. Circulation pattern occurrence frequencies are analyzed, and anomalies due to a possible climate change are presented. A stationary model uses a semi-Markov chain representation of circulation pattern occurrence. The possibility of developing a non-stationary process representation using General Circulation Models is also presented.

INTRODUCTION

Daily weather data are required for many different hydrological applications, such as hydraulic engineering design, water quality and erosion modelling, evaluation of different watershed management options etc. Observed weather events are often insufficient to get useful model responses. Particularly in the case of climate change investigations there are no observed data at all. Therefore it is useful to generate weather series which either reproduce the statistics of the observed data or reflect possible changes in climate.

Most known precipitation and weather generation procedures are purely stochastic point models (Jones *et al.* 1972, Richardson 1981). The assumption of stationarity is absolutely necessary for these models.

There appears to be a close relationship between atmospheric circulation and climatic variables. Burger (1958) studied the relationship between the atmospheric circulation patterns and mean, maximum and minimum daily temperatures, precipitation amounts and cloudiness using the time series from 1890 to 1950 measured at four German cities (Berlin, Bremen, Karlsruhe and Munich). He found a good correspondence between climatic variables and atmospheric circulation. Lamb (1977) stated that even the highly varying precipitation is strongly linked to the atmospheric circulation.

Recently McCabe *et al.* (1989) classified nine weather types for Philadelphia by using seven climatological parameters for a period of 35 years (1954–88) as a basis for stochastic precipitation modeling. Wilson *et al.* (1990) developed a daily precipitation model using a weather classification scheme for the Pacific Northwest US. Wilks (1989) developed a regression based method for generation of local weather elements (temperature, precipitation), using the large scale information.

In Bárdossy & Plate (1991) a model was developed for the precipitation occurrence at a selected site conditioned on the actual atmospheric circulation pattern. A multivariate model for the spatial distribution of rainfall depending on atmospheric circulation patterns was developed in Bárdossy & Plate (1992).

The purpose of this paper is to develop a mathematical model for the daily precipitation and daily mean temperature based on atmospheric circulation patterns. Daily precipitations are described using two different approaches:

(a) The rainfall occurrences are considered as primary processes, and precipitation depths on rainy days are taken from an appropriate distribution.
(b) Rainfall occurrences and amounts are modelled within a single process.

Mean daily temperature is modelled with the help of an autoregressive type approach. In all cases the atmospheric circulation pattern is a conditioning factor for the processes.

The model can be applied in stationary and non-stationary cases. In Bárdossy & Caspary (1990b) time series of circulation patterns were analysed, and it was found that frequencies of several circulation patterns changed considerably during the last years, showing extremes never reached before. The link to atmospheric circulation patterns makes it suitable for local weather simulation under climate change (Bárdossy & Caspary, 1990a). The model is applied using the classification scheme of the German Weather Service which is available for the time period 1881 to 1989.

MATHEMATICAL MODEL

Let $Z(t)$ be the daily precipitation amount on day t and $T(t)$ be the mean daily temperature on day t. Both $Z(t)$ and $T(t)$ are considered to be random variables, depending on the actual atmospheric circulation pattern A_t. A_t is also a random variable, with possible values $\{\alpha_1,\ldots,\alpha_k\}$. The random process A_t will not be analysed in this paper: for example semi-Markov process model for A_t developed in Bárdossy and Plate (1990b) may be used.

Precipitation

A major problem in the mathematical description of precipitation is that it has a mixed distribution, dry days occurring with high probability, and a continuous distribution for rainfall amounts on rainy days. To describe the precipitation occurrence process one can define the process $Y(t)$ as the indicator of $\{Z(t)>0\}$:

$$Y(t)=\begin{cases}1; & \text{if } Z(t)>0\\ 0; & \text{if } Z(t)=0\end{cases} \tag{1}$$

The probability of rainfall at time t depends on the atmospheric circulation pattern A_t

$$P[Y(t)=1|A_t=\alpha_i]=p_i \tag{2}$$

The distribution of rainfall amount also depends on A_t:

$$P[Z(t)<z|A_t=\alpha_i,\ Y(t)=1]=F_i(z) \tag{3}$$

Let $f_i(z)$ be the density function corresponding to $F_i(z)$.

Rainfall amounts using rainfall occurrences

The link between circulation patterns and wet days is obtained through the probabilities p_i. For simplicity let $p_{1i}=p_i$ and $p_{0i}=1-p_i$.

There is a weak persistence property of $Y(t)$ which may be expressed using conditional probabilities as follows:

$$P[Y(t)=k|\,Y(t-1)=l \text{ and } A_{t-1}=A_t=\alpha_i] \tag{4}$$

$$=\begin{cases}\rho+(1-\rho)p_{ki}; & \text{if } k=1\\ (1-\rho)p_{ki}; & \text{if } k\neq 1\end{cases} \tag{5}$$

However if the circulation pattern changes the precipitation occurrence of the following day is independent of that on the previous day.

The inclusion of this Markovian property within a period of constant circulation patterns does not influence the probabilities p_i.

Rainfall amounts $Z(t)$ are then related to $Y(t)$ through the conditional distributions $F_i(z)$ given as (3).

Daily rainfall amount distributions $F_i(z)$ are usually assumed to be exponential (Katz, 1977), or in some special cases geometrical (Duckstein *et al.*, 1972) or gamma. Distribution parameters were estimated using both the moment method and the maximum likelihood method (Bowman & Shenton, 1989).

Rainfall amounts of adjacent days were usually found to be independent (Katz, 1977). Therefore daily rainfall amounts are modelled independently, only in connection with the atmospheric circulation pattern through the conditional distribution $F_i(z)$.

Transformed normal process

Another possibility is to describe both precipitation occurrence and amount with the help of a single process $W(t)$. Let $W(t)$ be a random variable which is related to $Z(t)$ for a given circulation pattern $A_t=\alpha_i$ through the power transformation relationship:

$$Y(t)=\begin{cases}0; & \text{if } W(t)\leq 0\\ W^{\beta}(t); & \text{if } W(t)>0\end{cases} \tag{6}$$

Here β is an appropriate positive exponent. There are infinitely many $W(t)$ random variables which satisfy (6). Suppose that $W(t)$ is normally distributed. This way the

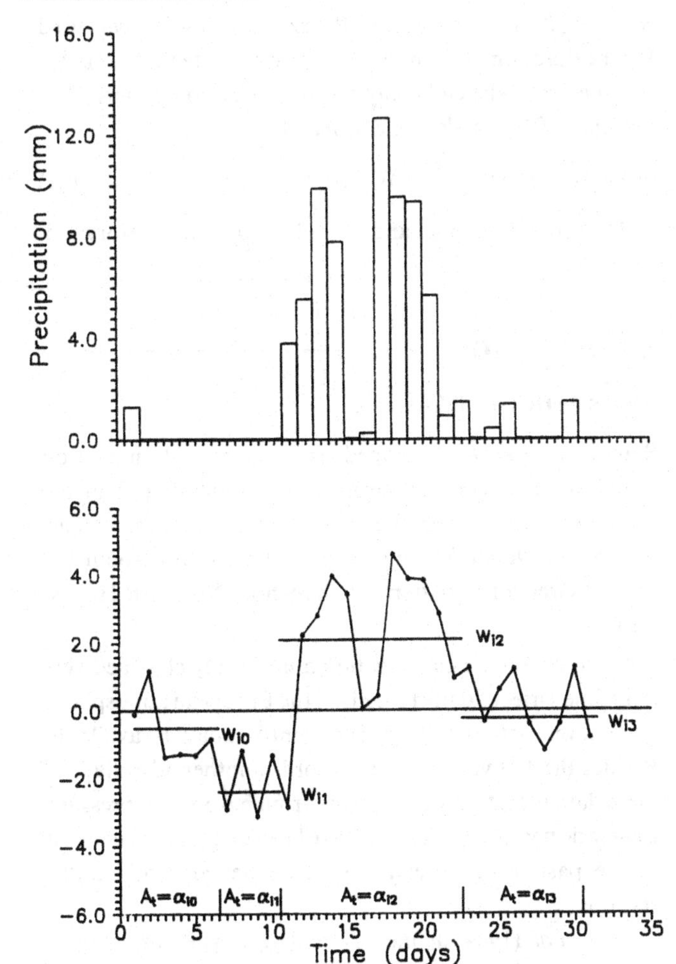

Fig. 1 Relationship between intermittent rainfall process, multivariate normal and transformed temporal rainfall process.

mixed (discrete – continuous) distribution of $Z(t)$ is related to a normal distribution. Fig. 1 displays the relationship between $Z(t)$, $Y(t)$ and $W(t)$. The relation between the normal density function of $W(t)$, p_i and $f_i(z)$ is explained in Fig. 2. As a first step, parameters μ_i and σ_i of $W(t)$ have to be estimated for each given α_i. Here the following equation should hold for the expected rainfall amount:

$$E[Z(t)|A_t=\alpha_i, Y(t)=1] = \frac{1}{\sqrt{2\pi}\,\sigma_i}\int_0^{\infty} t^{\beta}\exp\left[-\frac{(t-\mu_i)^2}{2\sigma_i^2}\right]dt \quad (7)$$

Let

$$m_i(t)=E[Z(t)|A_t=\alpha_i, Y(t)=1]$$

The second moment of the rainfall is given by:

$$E[Z^2(t)|A_t=\alpha_i, Y(t)=1] = \frac{1}{\sqrt{2\pi}\,\sigma_i}\int_0^{\infty} t^{2\beta}\exp\left[-\frac{(t-\mu_i)^2}{2\sigma_i^2}\right]dt \quad (8)$$

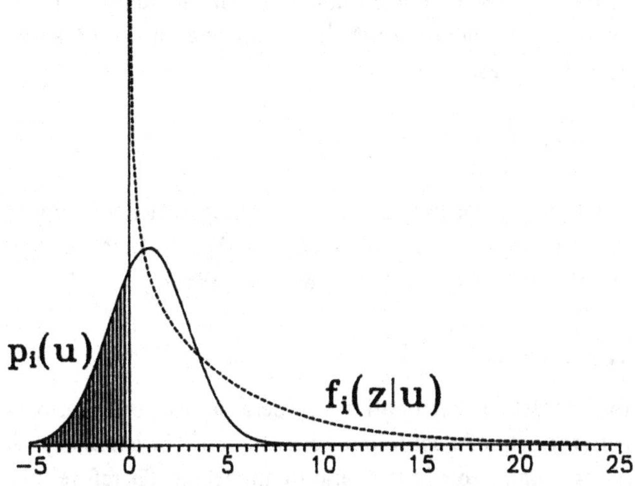

Fig. 2 Relationship between the normal density function of $W(t)$, p_i and $f_i(z)$.

The rainfall occurrence probability p_i satisfies equation (2). Therefore:

$$\mu_i+\Phi^{-1}(1-p_i)\sigma_i=0 \quad (9)$$

where $\Phi^{-1}(.)$ is the inverse of the standard normal distribution function.

If the circulation pattern changes then $W(t)$ of the following day is independent of that on the previous day $W(t-1)$:

$$P[W(t)<w\,|\,W(t-1) \text{ and } A_t=\alpha_i\neq A_{t-1}]=\Phi_i(w, \mu_i, \sigma_i) \quad (10)$$

If there is no change in the circulation pattern then $W(t)$ is described with the help of a simple AR(1) process:

$$W(t)-\mu_i=\Phi_1[W(t-1)-\mu_i]+V_i(t); \text{ if } A_t=A_{t-1}=\alpha_i \quad (11)$$

where V_i is a standard normal random variable.

The parameter ϕ_1 which is equal to the step 1 autocorrelation ρ_1 of $W(t)$ for α_i (Bras & Rodriguez-Iturbe, 1985) has to be estimated. The precipitation series $Z(t)$ only provide the positive values of $W(t)$, therefore a straightforward estimation is not possible. For this purpose the indicator series $I^q(t)$ is defined for any $0<q<1$ as:

$$I^q(t)=\begin{cases}1; & \text{if } F_i[Z(t)]+p_i\geq q\\ 0; & \text{if } F_i[Z(t)]+p_i<q\end{cases} \quad (12)$$

Note that the indicator series for $Z(t)$ and $W(t)$ are the same. Let ρ be the indicator correlation for the partial indicator series $A_t=\alpha_i$.

The required correlation ρ is related to the indicator correlation through the following relationship (after Abramowitz & Stegun, 1962):

$$\rho_q^{(i)}=\frac{1}{2\pi\,q\,(1-q)}\int_0^{\arcsin\rho_i}\exp\left[\frac{y_q^2}{1+\sin t}\right]dt \quad (13)$$

y_q is the value of the q quantile of the standard normal distribution function $y=\Phi^{-1}(q)$. Applying this formula for $q=0.5$ one has:

$$\rho_i=\sin\frac{\pi\rho_{0.5}^{(i)}}{2} \tag{14}$$

This makes the estimation of ρ possible. Other techniques for estimating ρ using formulas for the bivariate normal distribution are given in Patel & Read (1982).

Temperature

In contrast to precipitation, temperature can be described with the help of a continuous distribution. The annual cycle plays a major role in temperature modeling. Therefore this annual cycle has to be separated and the residual has to be modeled. Atmospheric circulation patterns also play a different role in different seasons. For example air flow from the Atlantic ocean causes mild weather, up to 5 °C warmer than normal in winter, and cold weather, up to 5 °C colder than normal in summer. Let

$$T(t)=M(t)+R(t) \tag{15}$$

where $M(t)$ is the expected value of $T(t)$ and $R(t)$ is the residual.

$$E[T(t)]=M(t) \text{ and } E[R(t)]=0 \tag{16}$$

This residual is dependent on the actual atmospheric circulation pattern.

$$E[R(t)|A_t=\alpha_i]=\tau_i(t) \tag{17}$$

In order to fulfill (15) for the above defined $\tau_i(t)$s one has:

$$\sum_{i=1}^{I} \mathrm{P}[A_t=\alpha_i]\,\tau_i(t)=0 \tag{18}$$

Here continuity in temperature is assumed, whereas independence of temperature residuals corresponding to adjacent days before and after a change in the atmospheric circulation pattern is not assumed.

$$R(t)=\psi_i R(t-1)+U(t|A_t) \tag{19}$$

In order to satisfy equation (16) the expectation of the random variable $U(t|\mathrm{A}_t)$ has to be:

$$\mathrm{E}[U(t|A_t=\alpha_i)]=\tau_i(t)\,\frac{1+\mathrm{P}[A_t=\alpha_i]\sum_{k=1}^{\infty}h_k^{(i)}\psi_1^k}{\sum_{k=1}^{\infty}h_k^{(i)}\sum_{l=0}^{k-1}\psi_1^l} \tag{20}$$

Here

$$h_k^{(i)}=\mathrm{P}(A_t=A_{t-1}=\ldots=A_{t-k+1}=\alpha_i\neq A_{t-k})$$

The actual values h depend on the stochastic model used for A_t. In the semi-Markov approach presented in Bárdossy & Plate (1991) a generalized Poisson distribution was used for the duration of a circulation pattern. Note that $h_k^{(i)}$ is not site specific, it depends only on the circulation pattern. The variance of U can also be calculated:

$$\mathrm{Var}[U(.)]=(1-\psi_1^2)\,\mathrm{Var}[R(t)] \tag{21}$$

The estimation of ψ_1 and τ_i; $i=1,\ldots,I$ is straightforward.

APPLICATION

Atmospheric circulation

Baur *et al.* (1944) developed an atmospheric circulation classification scheme for European conditions describing the circulation types within the area of Europe and the eastern part of the North Atlantic Ocean taking into account the general circulation pattern of the whole Northern Hemisphere.

Based on Baur's classification a uniformly classified very long time series of daily records of the European atmospheric circulation patterns from 1881 until today is available. Besides the 109 years of data records another advantage of these data is that they are derived from barometric pressure, a climatic parameter likely to have been measured accurately in the past using a relatively dense network of weather stations.

Baur *et al.* (1944) defined a circulation type as a mean air pressure distribution over an area at least as large as Europe. Any given circulation type persists for several days (normally at least 3 days) and during this time the main features of weather remain mostly constant across Europe. After this there is a rapid transition to another circulation type. Only large-scale features of the general circulation are included in Baur's classification scheme, namely:

(a) the location of sea level semi permanent pressure centers, (i.e. Azores high/Iceland low);
(b) the position and paths of frontal zones; and
(c) the existence of cyclonic and anticyclonic circulation types.

Using the classification of Baur *et al.* (1944), Hess & Brezowsky (1969) presented a catalogue of classified daily European circulation patterns from 1881 till 1966. Since 1948 the European circulation patterns and types are classified and published monthly by the German Weather Service (Deutscher Wetterdienst).

Precipitation

Thirty-four years of daily precipitation data measured at Essen were used for the model. Rainfall probabilities and amounts were supposed to depend both on the circulation

Table 1. *Observed and fitted statistics for different circulation patterns for daily precipitation in Essen (1952–87)*

		Observed			Fitted		
Circulation pattern	Season	Mean	Standard deviation	Probability of rain	Mean	Standard deviation	Probability of rain
West anticyclonic	Winter	2.4	2.8	44.8	2.4	2.9	44.8
	Spring	3.1	2.9	46.4	2.9	3.5	46.6
	Summer	3.4	4.2	34.9	3.4	4.3	34.9
	Fall	3.3	3.8	38.2	3.2	4.1	38.4
West cyclonic	Winter	5.9	5.6	85.6	5.9	5.5	85.4
	Spring	5.7	4.7	78.2	5.5	5.5	79.0
	Summer	5.7	6.0	73.4	5.7	6.0	73.3
	Fall	5.8	6.4	76.0	5.9	6.0	75.5
Central European high	Winter	1.1	1.8	8.2	1.2	1.7	8.2
	Spring	2.2	3.1	13.1	2.2	3.1	13.1
	Summer	4.5	5.7	19.2	4.4	6.0	19.4
	Fall	1.5	1.9	9.5	1.4	2.0	9.5
Central European low	Winter	2.6	3.3	47.7	2.6	3.1	47.6
	Spring	4.2	5.8	60.6	4.5	5.0	60.0
	Summer	6.7	6.3	72.3	6.5	6.8	72.9
	Fall	4.4	4.9	68.3	4.4	4.8	68.2
British Islands low	Winter	2.0	2.0	65.3	2.0	2.2	65.3
	Spring	4.2	4.6	71.7	4.2	4.5	71.6
	Summer	4.6	5.3	76.8	4.7	4.8	76.3
	Fall	3.9	5.4	69.0	4.2	4.5	68.3

pattern and on the season. Therefore these quantities were calculated for each season separately.

The parameters of the first model of rainfall occurrence were estimated according to the method described in Bárdossy & Plate (1991). Then for each season and for each circulation pattern the parameters of the conditional rainfall distributions were estimated. Three distributions were considered: exponential, lognormal and gamma. The first two gave very poor fits in more than 25% of the cases. The gamma distribution fitted the experimental distributions quite well, with only a very few exceptions.

Two different time series of circulation patterns were used:

(a) a semi-Markov chain based simulated time series;
(b) observed values of 34 years daily series (the same that occured in the observed precipitation series).

Only the gamma distribution was used for rainfall amount generation.

For the second model parameters for the distribution of $W(t)$ were estimated using (7), (8) and (9) for each circulation pattern. As there are only two unknowns (μ_i, σ_i), the weighted sum of the squared differences between the right and left hand side of equations (7)–(9) was minimized. Table 1 shows the estimated and observed statistics for different circulation patterns. The transformed normal distribution fitted the observed data in 75% of the cases better than any other distribution (gamma, exponential, lognormal) used. The comparisons were made using a Kolmogorov–Smirnov distance.

The indicator correlations were calculated for two different sets of circulation patterns: (a) with $p_i > 0.5$, (b) with $p_i \leq 0.5$. For the first set $q = 0.5$ was selected for the indicator correlation, for the second $q = 0.75$ or $q = 0.9$. The final correlation coefficient ρ was then calculated as a weighted sum of the individual ρ_i values. The values of ρ_i varied between 0.1 and 0.3, the overall ρ was found to be 0.215.

After the model parameters were estimated, the same two series of circulation patterns were used for rainfall simulation. Different statistics were calculated to evaluate the model results. Fig. 3 shows the autocorrelation functions for winter. Fig. 4 shows the observed and the simulated distributions of dry periods.

Temperature

Table 2 gives, as an example, the deviations of winter (December, January and February) mean daily temperatures

Table 2. *Deviations of winter (December, January and February) mean daily temperatures from the mean monthly temperatures for selected circulation patterns at four German cities (Berlin, Bremen, Karlsruhe and Munich, after Bürger, 1958, modified)*

	Warmer				Colder			
Location	≥5°	3.0° to 4.9°	1.0° to 2.9°	0.0° to 0.9°	−0.0° to −0.9°	−1.0° to −2.9°	−3.0° to −4.9°	≤−5°
Berlin	—	Wa, Wz, SWz	NWa, NWz, Ws SWa, TB, TrW Ww	TrM, Sz, HM	HB, TM	HNa, Nz, Sa NEz, NEa, BM	SEz	SEa, HFa
Bremen	SWz	Wa, Wz, SWa	NWa, NWz, Ws Sz, TB, TrW Ww	—	HB, TrM, HM	Nz, Sa BM, TM	HNa SEz, NEz, NEa	SEa, HFa
Karlsruhe	SWz	Wz, Ws	NWz, Wa, SWa Sz, TB, TrW Ww	—	TrM, NWa Sa, SEz, TM	HB, Nz, NEa NEz, HM, BM	HNa, SEa, HFa	—
Munich	SWz	Wz, Ws, TB	NWz, Wa SWa, Sz, TrW Ww	TrM, NWa	SEz, TM	Nz, Sa, NEa NEz, HM, BM	HNa, HB, SEa	Hfa

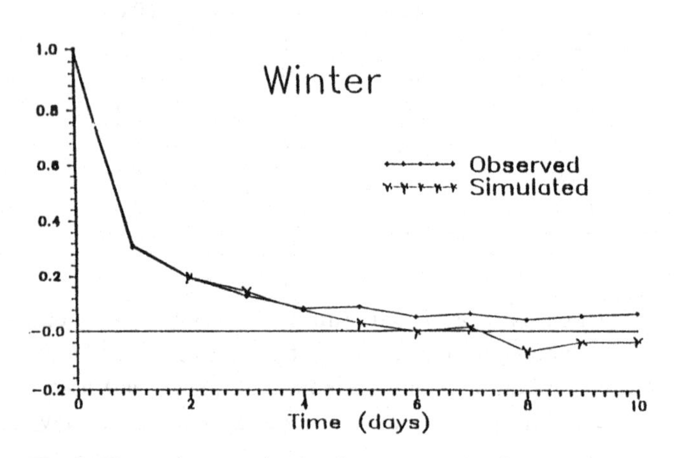

Fig. 3 Observed versus simulated autocorrelation function for winter precipitation.

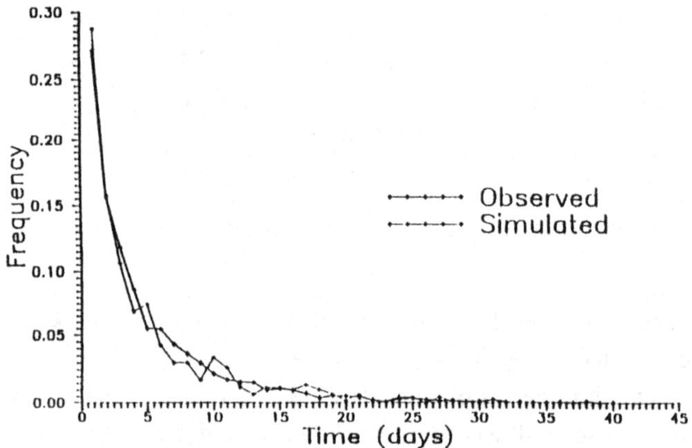

Fig. 4 Observed versus simulated distributions of dry periods.

from the mean monthly temperatures for selected circulation patterns. A comparison using data collected at Karlsruhe over the period 1951 to 1989 indicated no significant changes from the results of Burger (1958). These later data were used for calibrating the model.

DISCUSSION AND CONCLUSIONS

The results of this paper may be summarized as follows.

(a) Daily precipitation amounts and mean daily temperature are strongly linked to the atmospheric circulation and can be modeled as coupled processes.
(b) Two different models were presented for the daily rainfall amount
  (i) based on rainfall occurence and corresponding conditional distributions;
  (ii) based on a transformed normal process.
(c) Daily mean temperature can be modeled as a non-stationary autoregressive process with parameters depending on the actual atmospheric circulation pattern.
(d) The model is well suited for daily rainfall and mean daily temperature simulation, coupled to a deterministic or stochastic atmospheric circulation model.
(e) GCM air pressure maps can be used for circulation pattern classification, so the model could be applied to simulate stochastic climate change effects over small scale regions.

ACKNOWLEDGEMENTS

Research leading to this paper was partly supported German Science Foundation (DFG). Precipitation time series were received from the Ruhrverband.

REFERENCES

Abramowitz, M. & Stegun, I. (1964) *Handbook of Mathematical Functions*, Dover Publ.

Bárdossy, A. & Caspary, H. (1990a) *Modellkonzept zur Folgeabschätzung von Klimaänderungen für den regionalen Wasserhaushalt*, presented at 5. Wissenschaftliche Tagung 'Hydrologie und Wasserwirtschaft', München, April 2–4, 1990.

Bárdossy, A. & Caspary, H. (1990b) Detection of climate change in Europe by analyzing European circulation patterns from 1881 to 1989, *Theoretical and Applied Climatology*, **42**(4), 155–67.

Bárdossy A. & Plate, E. J. (1991) Modelling daily rainfall using a semi-Markov representation of circulation pattern occurrence, *J. of Hydrol.*, **122**, 33–47.

Bárdossy A. & Plate, E. J. (1992) Space-time model for daily rainfall using atmospheric circulation patterns, *Water Resour. Res.*, **28**(5), 1247–59.

Baur, F., Hess, P. & Nagel, H. (1944) *Kalender der Großwetterlagen Europas 1881–1939*, Bad Homburg.

Bowman, K. O. & Shenton, L. R. (1989) *Properties of Estimators for the Gamma Distribution*, Marcel Dekker, New York.

Bras, R. S. & Rodriguez-Iturbe, I. (1985) *Random Functions in Hydrology*, Addison-Wesley.

Bürger, K. (1958) *Zur Klimatologie der Großwetterlagen, Berichte des Deutschen Wetterdienstes Nr. 45, Bd. 6*, Selbstverlag des Deutschen Wetterdienstes, Offenbach a. Main.

Duckstein, L., Fogel, M. & Kisiel, C. C. (1972) A stochastic model of runoff producing rainfall for summer type storms, *Water Resour. Res.*, **8**, 410–21.

Hess, P. & Brezowsky, H. (1969) *Katalog der Großwetterlagen Europas, Berichte des Deutschen Wetterdienstes Nr. 113, Bd. 15*, 2nd ed., Selbstverlag des Deutschen Wetterdienstes, Offenbach a. Main.

Jones, J. W., Colwick, R. F. & Threadgill, E. D. (1972) A simulated environment model for temperature, evaporation, rainfall, and soil moisture, *Transactions ASAE*, **15**, 366–72.

Katz, R. W. (1977) Precipitation as a chain dependent process, *Journal of Climate and Applied Meteorology*, **16**, 671–6.

Lamb, H. H. (1977) *Climate, present, past and future, Vol. 2: Climatic history and the future*, Methuen, London.

McCabe, G. J., Hay, L. E., Kalkstein, L. S., Ayers, M. A. & Wolock, D. M. (1989) *Simulation of precipitation by weather-type analysis*. Int. Symp. on Sediment Transport modeling. Aug. 14–18, 1989, New Orleans.

Patel, J. K. & Read, C. B. (1982) *Handbook of the Normal Distribution*, Marcel Dekker, New York.

Richardson, C. W. (1981) Stochastic simulation of daily precipitation, temperature, and solar radiation, *Water Resour. Res.*, 17, 182–190.

Wilks, D. S. (1989) Statistical specification of local surface weather elements from large-scale information, *Theoretical and Applied Climatology*, **40**, 119–34.

Wilson, L. L., Lettenmaier, D. P. & Wood, E. F. (1990) Simulation of precipitation in the Pacific Northwest using a weather classification scheme, *Surveys in Geophysics*, in press.

3 Hydrological uncertainty – floods of Lake Eyre

V. KOTWICKI

Water Resources Branch, Engineering and Water Supply Department, Adelaide, Australia

Z. W. KUNDZEWICZ

Research Centre of Agricultural and Forest Environment Studies, Pol. Acad. Sci., Poznań and Institute of Geophysics, Pol. Acad. Sci., Warsaw, Poland

ABSTRACT The uncertainty aspects of the process of floods of Lake Eyre are examined. The available records of floods cover the time span of 40 years only. As longer time series of precipitation records are available, one has extended the observed series of inflows to Lake Eyre with the help of a rainfall–runoff model. Further reconstruction of the inflow series has been achieved with the help of proxy data of coral fluorescence intensity. However the limitations of these extensions and reconstructions of inflows are severe. The process of inflows to Lake Eyre could be considered one of the most convincing manifestations of hydrological uncertainty.

LAKE EYRE AND ITS BASIN

Lake Eyre, a large depression in arid Australia, rarely filled with water, attracts the interest of limnologists, hydrologists, geomorphologists and ecologists all over the world. The process of inflows to Lake Eyre has been recently studied by Kotwicki (1986). The following general information draws from the data assembled there.

The Lake Eyre basin (Fig. 1) spreads over 1.14 million km^2 of arid central Australia. Almost half of the basin area receives as little rainfall as 150 mm per year or less. The higher rainfalls of the order of 400 mm per year occur in the northern and eastern margins of the basin, influenced by the southern edges of the summer monsoon.

The annual potential evaporation as measured by Class A evaporometer ranges from 2400 to 3600 mm, with the value of pan coefficient for the Lake Eyre basin of the order of 0.6. The annual evaporation rate for the filled Lake Eyre ranges from 1800 to 2000 mm.

Since discovery of Lake Eyre in 1840 until its first recorded filling in 1949 the lake was considered permanently dry and eventual reports on the existence of water in the lake were dismissed as observation errors. After 1949 a sequence of wet and dry spells have been observed. Amidst minor isolated floodings a major flood event was recorded, which began in 1973, reached its peak in 1974 and persisted until 1977. It is estimated that the peak water storage in the lake during this event read 32.5 km^3. Until recently one used to consider fillings of Lake Eyre as rare and independent events. Now they are increasingly being looked at as a predictable manifestation of the global circulation patterns (e.g. El Niño–Southern Oscillations phenomena).

The Lake Eyre drainage basin is quite well developed and of persistent nature, due to favourable structural conditions. Much of this drainage pattern are disconnected relics from linked river systems which developed under the wetter past climate and which became disorganized under the arid conditions.

The lake is mainly fed by its eastern tributaries, the Cooper Creek and the rivers Diamantina and Georgina (Fig. 1), featuring extreme variability in discharge and flow duration. Mean annual runoff of the Lake Eyre basin, of the order of 3.5 mm depth (i.e. 4 km^3 volume) is the lowest of any major drainage basin in the world. This is some six per cent only of the value for the whole waterless Australian continent. A good demonstration of the aridity of the basin is its specific yield of 10 m^3 km^{-2} day^{-1} in comparison to the value of 115 m^3 km^{-2} day^{-1} for the Nile. In the conditions of arid central Australia rainfall of the volume of 50 mm is required to sustain a full channel flow and the frequency of such an event is less than once a year. Major events of filling Lake Eyre are associated with rare cases of annual rainfall in exceedence of 500 mm, or, as happened in 1984 and 1989, by heavy

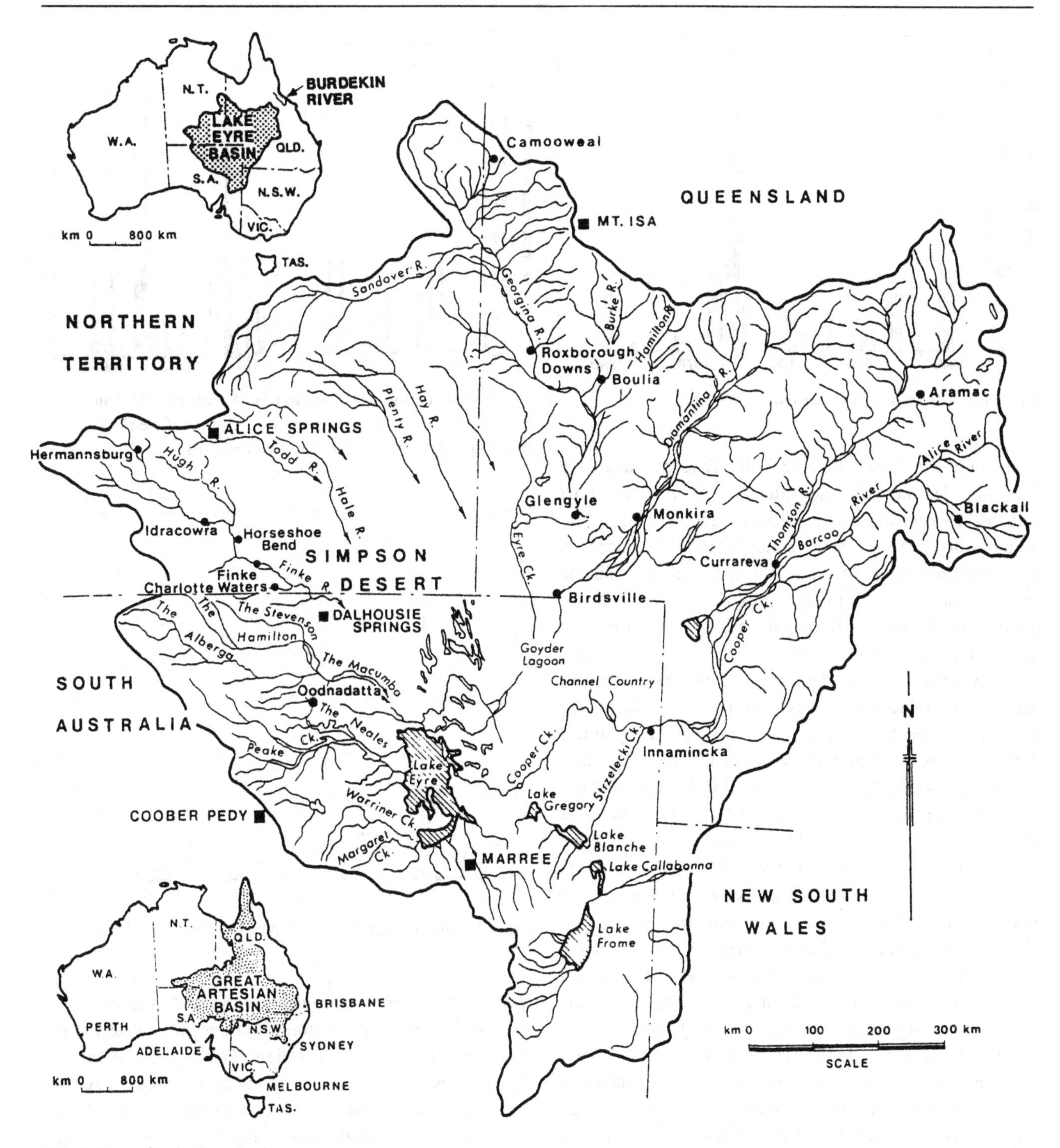

Fig. 1 Map of Lake Eyre Basin.

localized storms with precipitation of some 200–300 mm in the vicinity of Lake Eyre.

FLOODS OF LAKE EYRE – RECORDS AND RECONSTRUCTIONS

Time series of recorded fillings of Lake Eyre, shown in Fig. 2 embrace a short span of four decades. This short record is also subject to significant uncertainties. The existing instrumentation and observation network is not adequate. Even now the inflows to Lake Eyre are not measured directly in the lower course of either of its tributaries. The existing gauges measure runoff from as little as 40 per cent of the catchment area. The information on fillings before 1949 is practically non-existent. Therefore some means of extension of the observational records have been urgently required.

Kotwicki (1986) used a rainfall–runoff model for determi-

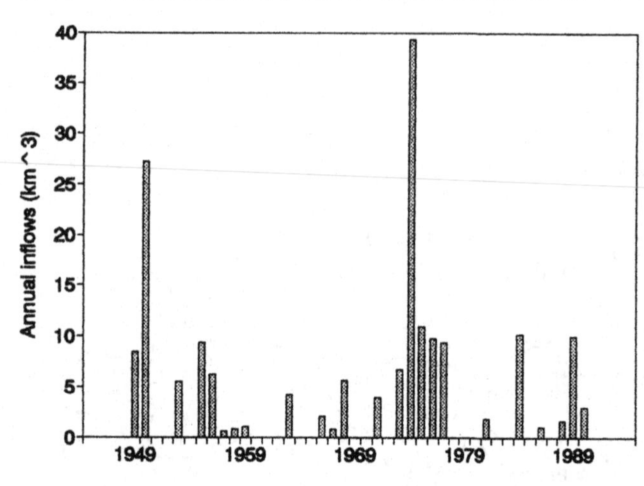

Fig. 2 Recorded inflows to Lake Eyre (1949–90).

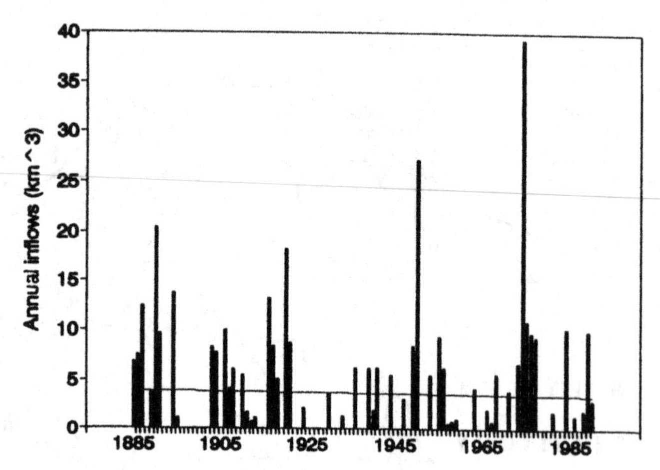

Fig. 3 Observed annual inflows to Lake Eyre (since 1949) augmented with inflows reconstructed with the help of the rainfall–runoff model (1885–1990). Linear trend marked.

nation of past inflows to the lake. Using 40 years of available data for identification and validation of the model he managed to reconstruct a time series of inflows to Lake Eyre for a century (1885–1984). Reliability of the results depends largely on the adequacy of the mathematical model available for transformation of rainfall into runoff. The idea of rainfall–runoff models has been developed for areas of humid or moderate climates. Therefore most of these models function satisfactorily under such climatic conditions and may not account for the processes of water losses, essential in the Lake Eyre basin, with sufficiently good accuracy. Although it is believed that the particular rainfall–runoff model used (RORB3, cf. Laurenson & Mein, 1983), that has been developed and tested for the arid Australian conditions may be the best available method, it is still a source of some uncertainty. The results of Kotwicki (1986) are shown in Fig. 3. Fig. 4 shows the same data, that look rather erratic in the raw plot, in the moving average (11 terms) framework.

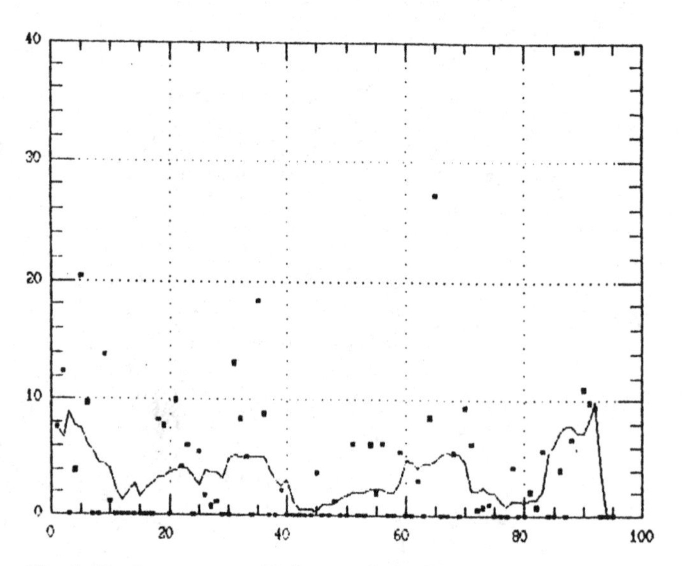

Fig. 4 Moving average of inflows to Lake Eyre (1885–1980).

One of the possibilities of further extension of the available time series of inflows is to use the El Niño–Southern Oscillation link that manifests itself via some proxy data, thus allowing reconstruction of longer series of records.

Isdale & Kotwicki (1987) and Kotwicki & Isdale (1991) used coral proxy data to further reconstruct the inflows to Lake Eyre. This is possible as the coral data reflect in some way the flows of the Burdekin River, draining a catchment of around 130 thousand km^2, directly adjoining the much larger Lake Eyre system. The process of the flow of the River Burdekin is strongly nonstationary, both in the yearly and over-yearly scale. The annual flows range from 3 to 300 per cent of the long term mean. During high flow periods the discharge of the Burdekin River moves northwards from the river outlet due to a longshore drift, and eventually reaches the shelf-edge reefs 250 km north of the mouth. The land-derived organic compounds, like humic and fulvic acids, are transported by the river and introduced to the marine system. These compounds taken up by corals and accomodated in their growing skeleton structures can be detected as they fluoresce under ultraviolet light. This fluorescence intensity provides a proxy measure of adjacent river discharge in the region of high flows. Dendrometric measurements along the depth of the core allowed dating of skeletal carbonate growth bands since 1724. The technique of dating the core resembles the method of dating yearly tree rings.

Although the proxy data of the River Burdekin are shown to be statistically linked to inflows to Lake Eyre, there are again significant uncertainty elements involved. The basin of the River Burdekin, though adjacent to the Lake Eyre basin, may have been behaving quite differently for particular events, as pointed out by Isdale & Kotwicki (1987). The anomalies have been caused by the non-uniform storm coverage (heavy local rains). Moreover, the place where the

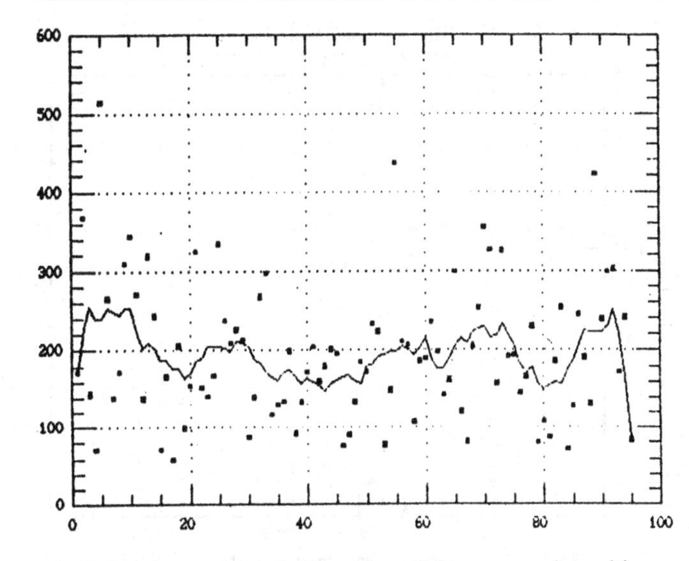

Fig. 5 Moving average of observed coral fluorescence intensities (1885–1980).

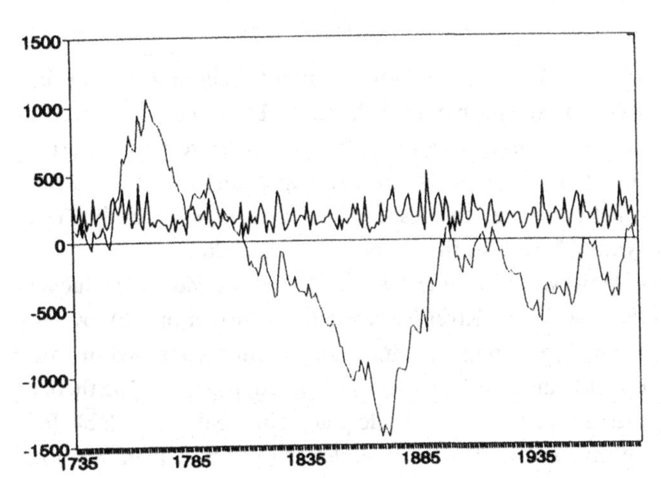

Fig. 6 Time series of coral fluorescence data (1735–1980) and its cumulative departure from mean.

coral reefs are analyzed is quite distant from the river outlet, bringing an additional contribution to uncertainty.

Kotwicki & Isdale (1991) observed a significant correlation between the time series of inflows to the Lake, the coral fluorescence intensities and the El Niño–Southern Oscillations (ENSO) index. The time series of coral fluorescence intensity for the period 1885–1980 in the form of moving average (11 terms) is shown in Fig. 5. It is clearly visible from Figs. 4 and 5, that there is some similarity of the two processes. The link between the two time series can be used to establish a relationship between the coral proxy data and the inflows to Lake Eyre. This could help drawing from the entire coral proxy record available (Fig. 6) since 1735. There is, however, also a significant difference in behaviour of the 100-years series of inflows to Lake Eyre and of coral fluorescence intensities. Coral intensity may be characterized by a conti-

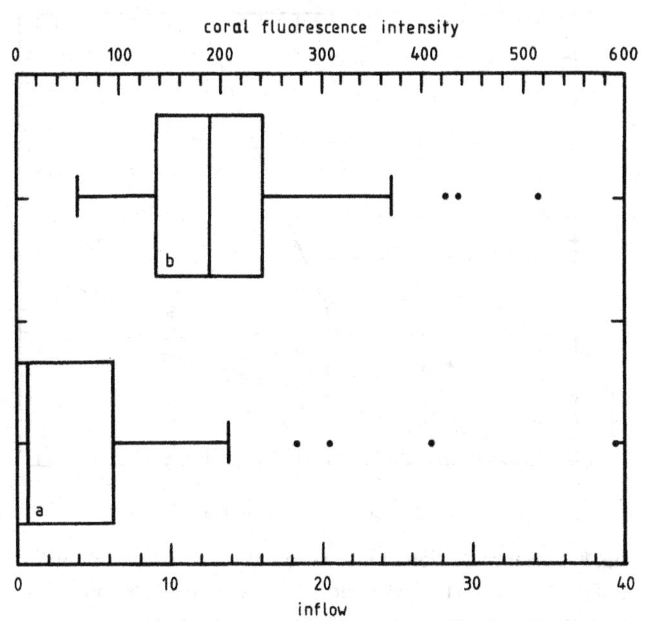

Fig. 7 Box-plots of (a) inflows to Lake Eyre; (b) coral fluorescence data.

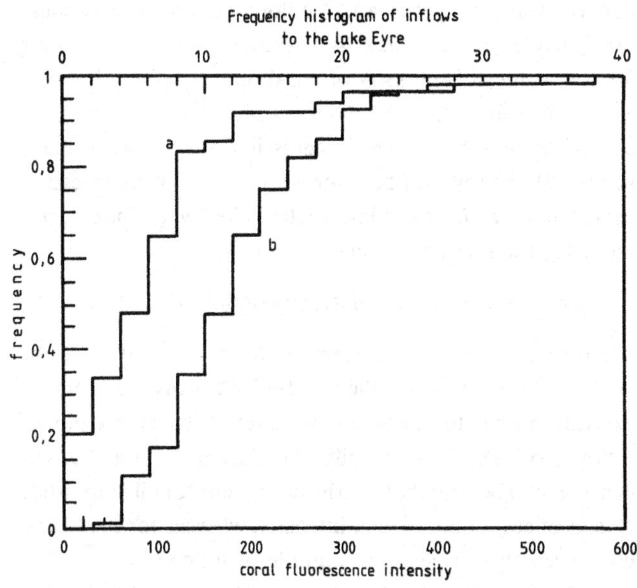

Fig. 8 Frequency histograms of (a) inflows to Lake Eyre; (b) coral fluorescence data.

nuous distribution, whereas the process of inflows is described by a mixed (discrete-continuous) distribution with a large part of the population (including the lower quartile) attaining zero value. This difference in behaviour of both series is shown in the form of box-plots (Fig. 7) and in the form of frequency histograms (Fig. 8). Both series contain a few (three or four) points lying significantly outside the upper hinge of the box-plots.

It seems that the standard two-parametric linear regression (line a in Fig. 9) is, in general, not the proper link

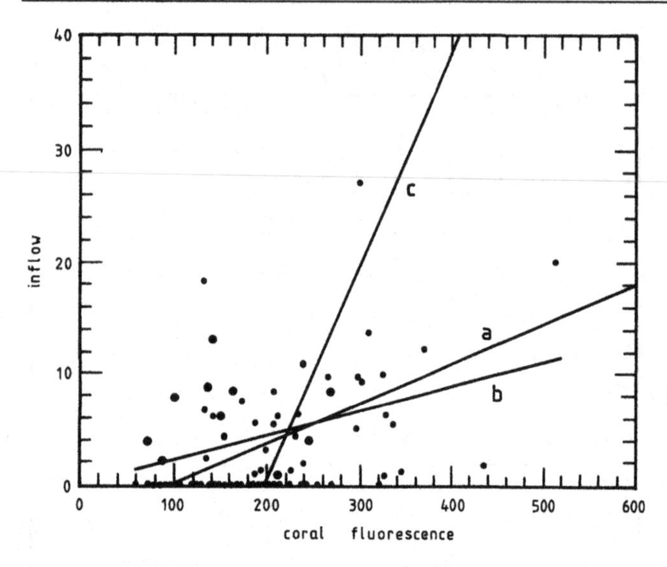

Fig. 9 Linear regression of inflows to Lake Eyre vs coral fluorescence intensity; (a) two-parametric linear regression; (b) one-parametric linear regression with supressed intercept; and (c) two-parametric linear regression for non-zero inflows.

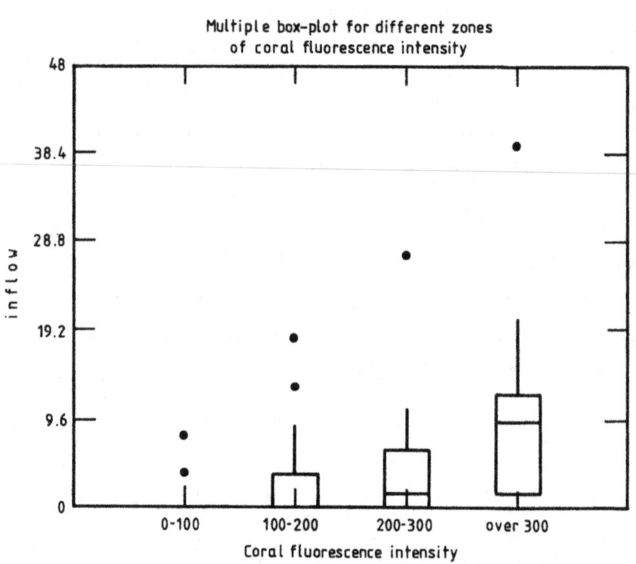

Fig. 10 Multiple box-plot for different zones of coral fluorescence intensities.

between the coral proxy and the flood data. This is because one is likely to obtain negative values of inflows to the lake for low values of intensity of coral fluorescence. This deficiency is eliminated if the intercept is set to zero (i.e. in the case of one-parametric linear regression, also shown as line b in Fig. 9). The fit can be improved if the linear regression excludes the zero yearly inflows to Lake Eyre. That is, one looks for the linear relation (line c in Fig. 9):

inflow = function (coral intensity | inflow > zero).

However, it does not seem possible to identify if the condition of non-zero inflow is fulfilled, drawing from an exterior information. Fig. 10 shows the multiple box-plot of inflows to Lake Eyre for different classes of coral fluorescence. It can be seen that zero inflows occur for all magnitude classes of coral fluorescence intensity, whereas for the lowest class even the 75 per cent quantile of inflows is zero. The probabilities of zero inflow in particular coral fluorescence classes read: 0.79 for the class from 0 to 100, 0.58 for the class from 100 to 200, 0.41 for the class from 200 to 300, and 0.08 for the class over 300.

The relation between the coral proxy data and the inflows to Lake Eyre is the result of some dynamical process. Causal relationships call for an input–output dynamical model linking the flows of the River Burdekin (model input) and the coral fluorescence intensity (model output). If linear formulation is used, one gets the following convolution integral valid for an initially relaxed case:

$$y(t) = \int_0^t h(t-\tau)\, x(\tau)\, d\tau = h(t) * x(t) \qquad (1)$$

where $x(t)$, $h(t)$, $y(t)$ denote the input function (flows of the River Burdekin that are believed to be closely linked to the process of inflows to Lake Eyre), impulse response (kernel function of the linear integral operator) and the output function (coral fluorescence intensity), respectively. The symbol * denotes the operation of convolution.

However, what one needs is the inverse model producing the flows of the River Burdekin (and, further on – inflows to Lake Eyre) from the coral data. Input reconstruction is a difficult, and mathematically ill-posed, problem. The theory warns that even small inadequacies in the data available for an inverse problem may render the result of identification unstable. Moreover, the available yearly data are not sufficient for identification of such dynamics. This results from the analysis of cross-correlation between the two series (Fig. 11), where a significant value is attained only for the lag zero.

CONCLUSIONS

The process of inflows to Lake Eyre involves a very high degree of hydrological uncertainties. The most essential uncertainty aspects in the process of floods of Lake Eyre are as follows.

(a) The observations gathered until present (gauge records) pertain to runoff from a portion (some 40 per cent only) of the area of Lake Eyre basin.
(b) The available records (biased as noted under (a)) cover the time span of 40 years only, that is the period of data for identification and validation of the rainfall–runoff model is very short. Therefore the recommended split-sample technique cannot be used.

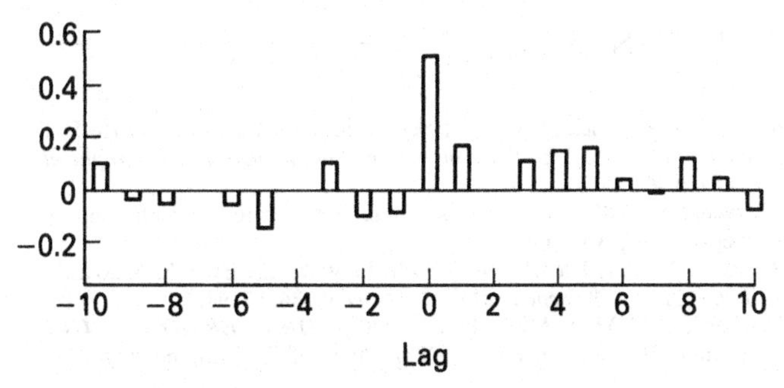

Fig. 11 Estimated cross-correlations between coral fluorescence and inflows to Lake Eyre.

(c) The accuracy of rainfall–runoff model (used for augmenting the available time series) for arid conditions may be lower than for humid or moderate conditions, where the idea of the unit hydrograph and alike concepts has been developed and widely applied.

(d) In order to validate the coral fluorescence–inflow relationship one disposes with 40 years of observations (cf. (a)) that can be augmented by results of rainfall–runoff modelling (see remarks (b)–(c)) to 100 years. This may be still too little for the rigorous split-sample approach.

(e) The coral proxy analysis is based on the assumption of the same climatological forcing of both the River Burdekin and Lake Eyre basins. However, the River Burdekin basin, although adjacent to the Lake Eyre basin, has not always been subject to a similar precipitation regime (anomalies identified by Isdale & Kotwicki, 1987). There were numerous periods of different behaviour of the process of flows of the River Burdekin and of the inflows to Lake Eyre. It was not uncommon that the spatial coverage of rainfalls did not embrace both basins. As the Lake Eyre basin itself is huge, different climatic conditions can occur simultaneously in various parts of the basin.

(f) It is only in cases of high flow that the land-derived organic compounds are transported to Pandora Reef.

(g) There is a large distance between the gauge at the River Burdekin and the site of coral colonies studies. There may be additional uncertainty factors influencing the long range transport process.

(h) The process of inflows to Lake Eyre is extremely complex. It is driven by several mechanisms and therefore the sample is heterogenous. This is clearly seen in the examples of 1984 and 1989, when contrary to most of historical records, the bulk of inflows was provided by the ephemeral rivers west of Lake Eyre. These rivers were typically dry during other events. It is believed that after having removed the heterogenity of available records, a significally better correlation with coral proxy data would be achieved.

Fig. 12 shows the cause–effect structures used for extension of available records. There are uncertainties contained in the rainfall–runoff analysis shown as the line 1 (e.g. difficulties in obtaining the average rainfall, questionable

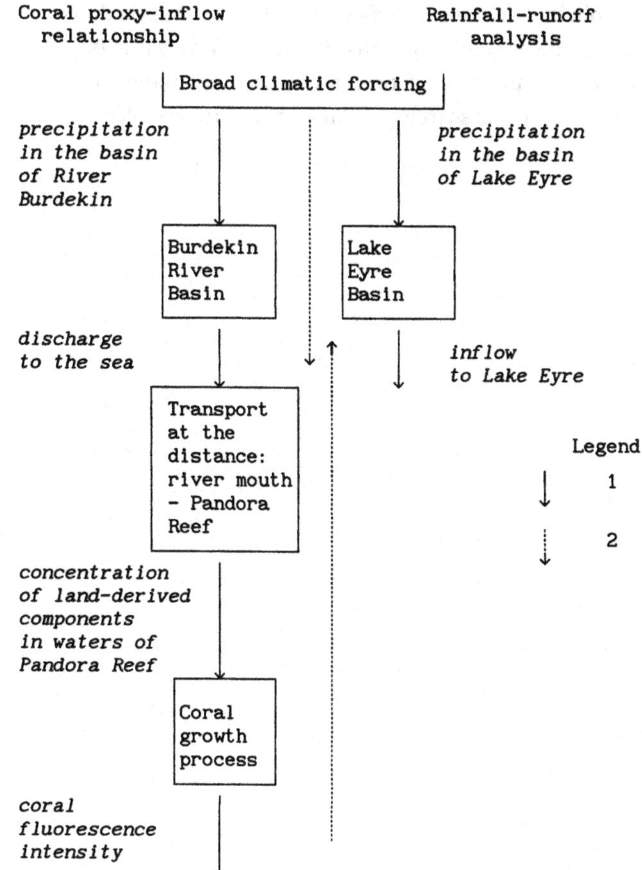

Fig. 12 Cause–effect structures for extension of available records.

applicability of a lumped model with average precipitation as the lumped input). The cause–effect structure of the relationships between the coral proxy data and the inflows to Lake Eyre is far more complicated (line 2 in Fig. 12). It is an inverse problem combined with inferring on an adjacent system. That is, coral proxy reconstruction of past inflows to Lake Eyre is almost a hopeless task. However, despite all the uncertainties in the cause–effect links, Isdale & Kotwicki (1987) had the right to draw the following corollaries from their studies.

(a) The vast area of the Australian landmass comprising both Lake Eyre and the Burdekin River basins endures a common broad climatic forcing, which is the Southern Oscillations or some function of it; and

(b) the coral proxy record may be used to hindcast annual scale paleohydrological sequences (and ENSO periodicities) for several centuries before the modern instrument period.

Considering the process of inflows to Lake Eyre one can possibly notice some analogy to the Galilean statement on the infinitely complex movement of a single droplet of water. This complexity does not hamper the specialists to forecast routinely the movement of water masses (flood waves) in open channels. The complexity at the microscale turns into simplicity at the macroscale. It is not to say that on this scale the problem suddenly turns simple. However, it is possibly closer to our scale of perception and its solution is more adequate to the scarcity of data which can be collected.

REFERENCES

Isdale, P. & Kotwicki, V. (1987) Lake Eyre and the Great Barrier Reef: A paleohydrological ENSO connection, *South Australia Geographical Journal*, **87**, 44–55.

Kotwicki, V. (1986) *Floods of Lake Eyre*, Engineering and Water Supply Department, Adelaide.

Kotwicki, V. & Isdale, P. (1991) Hydrology of Lake Eyre: El Nino link, *Palaeogeogr., Palaeoclimatol., Palaeoecol.*, **84**, 87–98.

Laurenson, E. M. & Mein, R. G. (1983) *RORB VERSION 3 – User Manual*, Monash University, Department of Civil Engineering.

4 Some aspects of hydrological design under non-stationarity

W. G. STRUPCZEWSKI

Civil Engineering Department, The University of Calgary, Alberta, Canada

H. T. MITOSEK

Institute of Geophysics, Polish Academy of Sciences, Warsaw, Poland

ABSTRACT If hydrological systems and properties of their input are changing year to year the stationarity assumption of hydrological time series could not hold any longer. The non-stationarity may occur due to the global climate changes and continued man-induced transformations of river basins. The assumption is made that the time series of annual values of hydrological variables are non-stationary, which in fact can be tested statistically. Then the question arises how to use such series in hydrological design if the structure is to be dimensioned on the statistical base. It seems reasonable to try to extend the existing statistical procedure to cover such a case. Following this line it is proposed to keep the type of probabilistic distribution constant and to allow its parameters to vary in time. Estimation of time dependent parameters is described. Various hypotheses regarding the form of time dependence can be compared and significance of the dependence tested. In order for a hydraulic structure to be dimensioned, its period of life shall be defined, while probability of exceedance or economic risk shall refer to this period. If the same trend in the parameter change is to be preserved in future, it will enable extrapolation of parameter values to every year of the life period and then to evaluate the probability distribution for the whole period of life.

INTRODUCTION

Many works devoted to problems of the global climatic change and of the anthropopression within the scale of the drainage area, have appeared in recent years. This may advocate postulating the existence of a trend of the hydrological régime, which directly influences the procedure to be taken in hydrological design. The fact has to be accounted that recorded time series of annual values of hydrological variables cannot be further treated as stationary. Therefore a question arises how to deal with the non-stationary realization within the design process, if dimensioning of a structure results of a statistical analysis.

The presented work is an attempt to expand the existing design procedures, by the presumed non-stationarity, which so far have been based on the stationary assumption.

PROBABILITY DISTRIBUTIONS

In the hitherto existing practice a probability distribution of occurrence of the phenomenon was assumed unchangeable from year to year. Thus parameters g_j of the distribution of the variable X

$$f(x, g_1, \ldots, g_m) \tag{1}$$

can be estimated from its random sample of $\boldsymbol{x} = (x_1, x_2, \ldots, x_i, \ldots, x_n)$, by using one of the standard estimation techniques, e.g., maximum likelihood method with the condition of maximization of likelihood function (or its logarithm):

$$L = \prod_{i=1}^{n} f(x_i, g_1, \ldots, g_m) \tag{2}$$

A very convenient consequence of the assumption of stationarity is the non-significance of random gaps in observational series and the validity of the determined distribution in an arbitrarily long time horizon.

Rejection of the stationarity assumption means that the probability distribution:

$$f(x, g_{t,1}, g_{t,2}, \ldots, g_{t,i}, \ldots, g_{t,m}) \quad (3)$$

is time dependent. Assume here, that the type of distribution remains unchanged. The immediate problems are the one of estimation of parameters in every observation year, as it only gives one realization of the random variable X, as well as the one of prediction of distribution for the period corresponding to our practical or research interests. If we eliminate the possibility of physical estimation of parameters, arbitrary assumption of time dependence structure of distribution parameters is essential to solve the problem. It can have a variant character and concern only one or more parameters, whereas the remaining parameters are left as invariable with time, and the decision about the choice of a variant should be based on statistical tests.

Lower order statistical moments have a clear interpretation, which makes it easier to build hypotheses on the temporal trend of the distribution. Therefore, we shall replace the original distribution parameters $g_{t,i}$ with moments using relations for each type of distribution, available in the literature. Here the moments will be estimated by the maximum likelihood method, and not by the method of moments. In order to simplify the notation the cumulants k_i will be used, which are functions of moments (Fisz, 1963). It can be easily shown that the first cumulant is equal to the first moment about the origin, whereas the second and third cumulants are equal to the second and third central moments, respectively. Thus

$$f_{x,t} = f(x, k_{t,1}, k_{t,2}, \ldots, k_{t,i}, \ldots, k_{t,m}) \quad (4)$$

Assume, for simplicity, that only one cumulant, e.g. the first one, is time dependent

$$f_{x,t} = f(x, k_{t,1}, k_2, \ldots, k_m) \quad (5)$$

and let the dependence be linear

$$k_{t,1} = k_{0,1} + a_1 t, \quad (6)$$

where the notion $k_{0,1}$ refers to the first cumulant in the year preceding the first year of observation, i.e., for $t=0$. Then the probability density function may be written as

$$f_{x,t} = f(x, t, k_{0,1}, k_2, \ldots, k_m, a_1) \quad (7)$$

It has only one parameter (a_1) more than the appropriate density function for a stationary case (equation (1)).

PARAMETER ESTIMATION

The maximum likelihood method for parameter estimation will be applied here. In case of gaps in observations, t is the time index and not the successive number in the observational series. Therefore, if the observations (also with gaps) cover a period T years, then estimated parameters will be obtained from the condition:

$$\max\left[\sum_{=1}^{T} \ln f(x, t, k_{0,1}, k_2, \ldots, k_m, a_1)\right] \quad (8)$$

whereas, according to the idea of creating the likelihood function, no observation in the year t_i is represented in the likelihood function as the probability of the certain event, that is

$$\int_{-\infty}^{+\infty} f(x, t_i)\mathrm{d}x = 1 \quad (9)$$

ACCURACY OF QUANTILE ESTIMATION FOR AN INDIVIDUAL YEAR

The variance–covariance matrix of distribution parameters estimated by the maximum likelihood method may be obtained (e.g. Wilks, 1950; Schmetterer, 1956; Kaczmarek, 1960) by determining the following Jacobian

$$J = \left[-\frac{\partial^2 \ln L}{\partial g_i \partial g_j}\right] \quad (10)$$

where $(g_i, g_j) = \{k_{0,1}, k_2, \ldots, k_m, a_1\}$. Substituting the estimates of parameters obtained from equation (8), and then inverting it one gets the variance–covariance matrix:

$$\mathbf{M} = [a_{ij}] = [r\sigma(g_i)\sigma(g_j)] = \left[-\mathrm{E}\left(\frac{\partial^2 \ln L}{\partial g_i \partial g_j}\right)\right]^{-1} = [\mathrm{E}(J)]^{-1} \quad (11)$$

where r is the correlation coefficient between the estimation errors of appropriate parameters and $\sigma(.)$ is the standard error of the parameter in question. In practice the accuracy of the estimated quantile of defined probability of exceedance is used. In the non-stationary case this may concern the individual year t, or a period being its multiplicity. Let the probability distribution (7) in the individual year t be expressed in the form:

$$\hat{x}_{p,t} = \phi(p, t, \hat{k}_{0,1}, \hat{k}_2, \ldots, \hat{k}_m, \hat{a}_1) \quad (12)$$

and then let us determine the functional matrix

$$\mathbf{N}=[b_{ij}]=\left[\left(\frac{\partial\phi}{\partial g_i}\right)\left(\frac{\partial\phi}{\partial g_j}\right)\right] \tag{13}$$

where g_i and g_j are the same as in equation (10), and the partial derivatives are calculated in points corresponding to estimated values of parameters. The variance of quantile $x_{p,t}$ may approximately be defined (e.g. Hald, 1952) as the sum of products of appropriate matrix terms defined by equations (11) and (13).

$$\sigma^2(x_{p,t})=\sum_{i,j}a_{ij}b_{ij} \tag{14}$$

Note that accuracy of estimation is derived under assumption of knowledge of the 'true' form of the probability density function and of the exact observation data. We shall soon return to the problem of defining the quantile and its accuracy for a period of several years.

COMPARISON OF HYPOTHESES ON NON-STATIONARITY

Let us assume two hypotheses concerning the type of non-stationary functions (4):

$$H_0(f_{x,t}=f^{(0)}_{x,t})$$
$$H_1(f_{x,t}=f^{(1)}_{x,t}) \tag{15}$$

whereas e.g., $f^{(0)}$ may be expressed by (7), and $f^{(1)}$ may contain a linear trend in standard deviation

$$f^{(1)}=f(x,k_1,k_{t,2},\ldots,k_m,a_2) \tag{16}$$

where

$$k_{t,2}=(k_{0,2}^{1/2}+a_2t)^2 \tag{17}$$

Another simple possibility, producing one extra parameter only, is to accept a linear trend in the average value k_1 and in standard deviation $k_{2,2}^{1/2}$ maintaining a constant variability coefficient C_v:

$$m_t=m_0+a_1t$$
$$\sigma_t=C_v(m_0+a_1t) \tag{18}$$

A very satisfactory method of comparison of hypotheses is the quotient test (Fisz, 1963; Kaczmarek, 1977) defined as the ratio of likelihood functions calculated for both discussed hypotheses:

$$U=L^{(1)}/L^{(0)}=\frac{\Pi f^{(1)}(x,t)}{\Pi f^{(0)}(x,t)} \tag{19}$$

Verification of alternative hypotheses involves defining to which of three areas (H_0, no basis to make a decision, or H_1) the value of the ratio (19) belongs.

Competing with the above quotient test is the information criterion proposed by Akaike (1971). It states that when several competing hypotheses are being estimated by the method of maximum likelihood the one with the smallest value of the Akaike's Information Criterion (AIC):

$$\text{AIC}=-2(\text{minimum log likelihood})+2(\text{number of independently adjusted parameters}) \tag{20}$$

should be chosen as the best one. If some hypotheses give the same minimal value of the AIC then the one with the smallest number of free parameters is the best.

HYDROLOGICAL DESIGN

Hydrological design involves consideration of risks. A water-control structure might fail if the magnitude of a hydrological variable exceeds the design value within the expected span of life of the structure.

Assume that the design life period of a hydraulic structure is known, i.e., the year t_s when it is planned that it will come into operation and the design life is T_e years. This construction must be dimensioned with regard to the magnitudes of the hydrological variable X expected within the period T_e years (e.g. the maximum annual discharge). Assume also that the probability distribution $f(x,\tau)$ of the random variable X in each exploitation year $\tau=1,2,\ldots,T_e$ is known in the meaning of the structure of the density function and its parameters. The relation, essential for extrapolation, between the time index of the observational series t, and the time index of the exploitation period τ is in the form:

$$\tau=t-T-\Delta \tag{21}$$

where Δ is the length of the period in years between the last included observation in the time series and the beginning of the exploitation period of the structure t_s. Then, assuming, similarly as in the stationary case, the independence in the time series, the probability of exceedance of the level x by maximum values X in the period of T_e years attains the form:

$$P_{T_e}(X>x)=1-\prod_{\tau=1}^{T_e}\int_{-\infty}^{x}f(x,\tau)\,\mathrm{d}x \tag{22}$$

while probability of nonexceedance

$$P_{T_e}(X<x)=1-\prod_{\tau=1}^{T_e}\int_{x}^{\infty}f(x,t)\,\mathrm{d}x \tag{23}$$

Probability distributions prepared for the life period of the structure would serve to determine the design value for the given risk of failure during the expected life of structure.

Using the principles of the combinatorial analysis probability of m-fold exceedance ($m=1,2,\ldots,\tau_e$) of the assumed level x in the period of T_e years may be additionally determined.

Proceeding in a similar way, the optimal economic risk-based design procedure (cf. Prichett, 1964) can be adjusted to the non-stationary case.

QUANTILE ESTIMATION ERROR

Determination of accuracy of quantile estimation with a given probability of exceedance in the life period of the water-control structure is of essential practical significance. Let us assume, for example, that only the first cumulant is non-stationary, as in equation (6). Then, using an inverse transformation we may write equation (22) in a form analogous to equation (12):

$$\hat{x}_{p,T_e,t_s}=\psi(p,T_e,t_s,\hat{k}_{0,1},\hat{k}_2,\ldots,\hat{k}_m,\hat{a}_1) \tag{24}$$

Transformation of equation (22) to the form of equation (24) and determination of the functional matrix (equation (13)) will, in general, be possible in the numerical way only. Knowing the matrix **M** (equation (11)) and the matrix **N** (equation (13)) the quantile estimation error may be determined from equation (14).

EXAMPLES

If a normal, log-normal, exponential, or Fisher–Tippett type I (max) distribution is assumed, then depending on the hypothesis concerning non-stationarity, one can obtain analytical results. Although, because of asymmetry of distributions the normal distribution has limited application in hydrology, it will be used for exemplification. There are several reasons for that. This simple distribution allows one to get analytical results. The estimations obtained by the method of moments and by the maximum likelihood method are identical, and the linear trend in the mean value is expressible by the known regression equation. For convenience only we shall assume no breaks in the time series.

Take the hypothesis of linear trend in a mean value, i.e., $k_{t,1}=k_{0,1}+a_1t$ (equation 6), whereas $k_2=\text{const.}$:

$$f_{x,t}=\frac{1}{\sigma(2\pi)^{1/2}}\exp\left\{-\frac{[x-m_0+a_1t)]^2}{2\sigma^2}\right\} \tag{25}$$

where $m_0=k_{0,1}$, $\sigma^2=k_2$, $t=1,2,\ldots,T$.

System of equations of the maximum likelihood method:

$$\frac{\partial \ln L}{\partial m_0}=\frac{1}{\sigma^2}\sum_{t=1}^{T}(x_t-m_0-a_1t)=0 \tag{26a}$$

$$\frac{\partial \ln L}{\partial a_1}=\frac{1}{\sigma^2}\sum_{t=1}^{T}(x_t-m_0-a_1t)=0 \tag{26b}$$

$$\frac{\partial \ln L}{\partial \sigma}=-\frac{T}{\sigma}+\frac{1}{\sigma^3}\sum_{t=1}^{T}(x_t-m_0-a_1t)^2=0 \tag{26c}$$

and its solution gives following estimators of the parameters:

$$\hat{a}_1=\frac{\overline{xt}-\bar{x}\bar{t}}{\overline{t^2}-(\bar{t})^2} \tag{27}$$

$$\hat{m}_0=\bar{x}-a_1\bar{t} \tag{28}$$

and

$$\hat{\sigma}^2=\frac{1}{T}\sum_{t=1}^{T}[x_t-(\hat{m}_0+\hat{a}_1t)]^2=\overline{x^2}-\hat{m}_0^2+\hat{a}_1(\hat{a}_1\overline{t^2}-2\overline{xt}) \tag{29}$$

Considering in equations (27)–(29) that

$$\bar{t}=\frac{T+1}{2} \tag{30}$$

and

$$\overline{t^2}=\frac{(T+1)(2T+1)}{6} \tag{31}$$

one obtains

$$\hat{a}_1=\frac{12}{T^2-1}\left(\overline{xt}-\frac{T+1}{2}\bar{x}\right) \tag{32}$$

$$\hat{m}_0=\bar{x}-\frac{T+1}{2}\hat{a}_1 \tag{33}$$

$$\hat{\sigma}^2=\overline{x^2}-\hat{m}_0^2+\hat{a}_1\left[\hat{a}_1\frac{(T+1)(2T+1)}{6}-2\overline{xt}\right] \tag{34}$$

The variance–covariance matrix reads:

$$\mathbf{M}=\begin{pmatrix} D^2(m_0) & \mu_{11}(m_0,a_1) & \mu_{11}(m_0,\sigma) \\ & D^2(a_1) & \mu_{11}(a_1,\sigma) \\ & & D^2(\sigma)\end{pmatrix}=$$

$$=\begin{pmatrix} \frac{2(2T+1)\sigma^2}{T(T-1)} & -\frac{6\sigma^2}{T(T-1)} & 0 \\ & \frac{12\sigma^2}{T(T^2-1)} & 0 \\ & & \frac{\sigma^2}{2T}\end{pmatrix} \tag{35}$$

Equation (12) takes the form

$$\hat{x}_{p,t}=\hat{m}_0+\hat{a}_1t+z_p\hat{\sigma} \tag{36}$$

where z_p is the lower integral limit:

$$P=\frac{1}{(2\pi)^{1/2}}\int_{z_p}^{+\infty}\exp(-z^2/2)\,dz. \tag{37}$$

The functional matrix reads:

$$\mathbf{N}=[b_{ij}]=\begin{pmatrix}1 & t & z_p \\ & t^2 & tz_p \\ & & z_p^2\end{pmatrix} \quad (38)$$

Thus, the accuracy of the quantile estimation (36) can be expressed, as:

$$\sigma^2(\hat{x}_{p,t})=\left\{\frac{2}{T-1}(2T-6t+1)+\frac{12t^2}{T^2-1}+\frac{z_p^2}{2}\right\}\frac{\hat{\sigma}^2}{T} \quad (39)$$

The parameter $\hat{\sigma}$, on the right hand side of equation (39), can be expressed in terms of the parameter

$$\hat{\sigma}_s\equiv\overline{x^2}-(\bar{x})^2$$

under the stationarity assumption, the drift parameter ($\hat{a}_1$), and the length of records (T):

$$\hat{\sigma}^2=\hat{\sigma}_s^2-\frac{\hat{a}_1^2}{12}(T^2-1) \quad (40)$$

In a limiting case all variability of the process x_t is caused by the time drift and then $\hat{\sigma}$ is equal to zero while $\hat{\sigma}_s$ takes its lowest possible value

$$\hat{\sigma}_s^2=\frac{\hat{a}_1^2}{12}(T^2-1) \quad (41)$$

In the stationary case the mean-square error of the quantile estimation is given by

$$\sigma^2(\hat{x}_p)=(1+0.5z_p^2)\frac{\hat{\sigma}_s^2}{T} \quad (42)$$

The ratio, R, of the mean-squared errors can serve to assess the consequence of the non-stationary hypothesis in regard to the mean value

$$R=\frac{\sigma^2(\hat{x}_{p,t})}{\sigma^2(\hat{x}_p)}=\frac{1}{1+0.5z_p^2}\left[\frac{2}{T-1}(2T-6t+1)+\frac{12t^2}{T^2-1}+0.5z_p^2\right]\left[1-\frac{1}{12}\left(\frac{\hat{a}_1}{\hat{\sigma}_s}\right)^2(T^2-1)\right] \quad (43)$$

with minimum for

$$t=(T+1)/2 \quad (44)$$

Fig. 1 presents, for $p=50\%$ and $T=30$ years, and for various ratios $(\hat{a}_1/\hat{\sigma}_s)^2$, the value of R versus t. Note that a loss in accuracy in reference to the stationary case does not depend on the sign of the parameter $\hat{a}_1$ and is greatest for $\hat{a}_1=0$.

Equation (22) for the normal distribution takes the form:

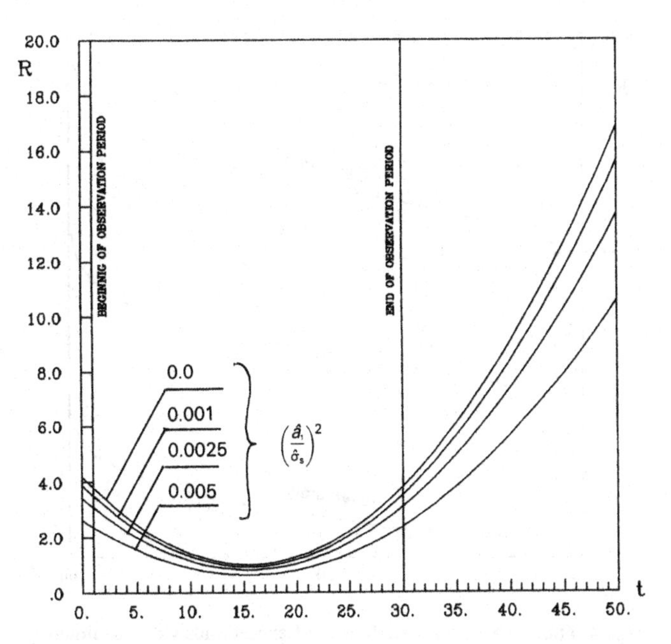

Fig. 1 The ratio of mean squared errors $R=\hat{\sigma}^2(\hat{x}_{p,t})/\sigma^2(\hat{x}_p)$ for $p=50\%$, $T=30$, and various ratios $(\hat{a}_1/\hat{\sigma}_s)^2$.

$$P_{T_e}(X>x)=1-\prod_{\tau=1}^{T_e}\int_{-\infty}^{x}\frac{1}{\sigma(2\pi)^{1/2}}\exp\left[-\frac{(x-m_\tau)^2}{2\sigma^2}\right]dx \quad (45)$$

Introducing a standardized variable

$$z_\tau=\frac{x-m_\tau}{\sigma} \quad (46)$$

one obtains the probability distribution for the design life of a structure in the form:

$$P_{T_e}(X>x)=1-\prod_{\tau=1}^{T_e}\int_{-\infty}^{(x-m_\tau)/\sigma}\frac{1}{(2\pi)^{1/2}}\exp\left[-\frac{z^2}{2}\right]dz \quad (47)$$

Assuming $T=30$ years, $\Delta=5$ years (i.e. $t_s=35$ years), $T_e=50$ years, $m_0=1000$, $a_1=2$, and $\sigma=300$, we obtain the following estimations of quantiles $\hat{x}_{p,T_e,t_s}$: 1785, 1983, and 2187 for $P_{T_e}=50\%$, 10%, and 1%, respectively. The functional matrix (equation (13)) was estimated numerically and for $P_{T_e}=50\%$ it reads:

$$\mathbf{N}(P_{T_e}=50\%)=\begin{pmatrix}1.00 & 63.53 & 2.20 \\ & 4036.65 & 139.67 \\ & & 4.83\end{pmatrix} \quad (48)$$

Multiplying appropriate terms of the functional matrix (equation (48)) and of the variance–covariance matrix (equation (35)) we obtain the mean square error of the quantile estimate. Fig. 2 presents the probability distribution

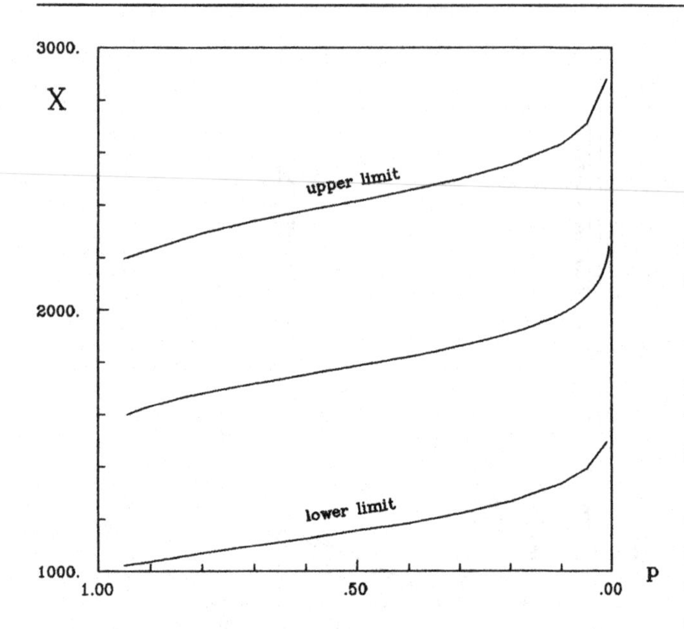

Fig. 2 The probability distribution of exceedance with confidence intervals for the confidence level $\beta = 95\%$.

exceedance with the confidence intervals for a confidence level $\beta = 95\%$.

CONCLUSIONS

In the stationary case, the hydrological uncertainty of water-control structure design, is the result of the non-representativeness of a short random sample available, randomness of the phenomenon concerning the life period of the structure, error of measurement, and arbitrariness of the choice of the type of probability distribution. In the non-stationary case this uncertainty is increased because:

(a) one does not know in what way it is manifested (this can be only estimated by statistical tests); and

(b) even if make a proper estimation and choice, the modelling of non-stationarity is connected with the introduction of additional parameters that are estimated for the time series available.

The latter increases the error of the quantile estimation, and this error grows with the increasing time horizon of extrapolation.

Existing design procedures based on the risk expressed in the terms of the probability of failure or in the terms of the expected economic losses during the life time can be adjusted to the non-stationary case and applied if only the form of non-stationarity has been identified.

However, the problem remains open, whether the description of the phenomenon in the form of a stochastic process is to be included in the project, or (as done here) the random variable framework is to be used. The available realizations of annual values can be treated as a Markov type process in a procedure of estimation by the maximum likelihood method, developed here.

REFERENCES

Akaike, H. (1971) A new look at the statistical model identification, *IEEE Trans. Automat. Contr.*, **AC-19**, 716–23.
Fisz, M. (1963) *Probability Theory and Mathematical Statistics*, New York.
Hald, A. (1952) *Statistical Theory with Engineering Applications*, New York–London.
Kaczmarek, Z. (1960) Przedzial ufnosci jako miara dokladnosci oszacowania prawdopodobnych przeplywow powodziowych (Confidence interval as a measure of estimating accuracy of probabilities of flood discharges), *Wiad. Sluz. Hydrol. Meteo.*, **VII**, 4.
Kaczmarek, Z. (1977) *Statistical Methods in Hydrology and Meteorology*, Warsaw.
Prichett, H. D. (1964) *Application of the principles of engineering economy to the selection of highway culverts*, Stanford Univ., Report EEP-13.
Schmetterer, L. (1956) *Einführung in die mathematische Statistik*, Wien.
Wilks, S. S. (1950) *Mathematical Statistics*, Princ. Univ. Press.

5 New plotting position rule for flood records considering historical data and palaeologic information

GUO SHENG LIAN

Department of Engineering Hydrology, University College Galway, Ireland, on leave from Wuhan University of Hydraulic and Electric Engineering, Wuhan, People's Republic of China

ABSTRACT The graphical curve fitting procedure has been favoured by many hydrologists and engineers, and the plotting positions are required both for the display of flood records and for the quantile estimation. The existing plotting position formulae which consider historical floods and palaeologic information are reviewed and discussed. The plotting positions for systematically recorded floods below the threshold of perception must be adjusted to reflect the additional information provided by the pre-gauging period if the historical flood data and the systematic records are to be analyzed jointly in a consistent and statistically efficient manner. However, all available formulae are unlikely to adjust these plotting positions properly. It is felt that the traditional rule and exceedance rule assumptions are inconsistent with the floods over and below the threshold of perception of historical floods. A new type of formula is proposed and examined. Simulation studies and numerical examples show that the new formula type performs better than the traditional rule and competitive to the exceedance rule. The Weibull based formulae result in large bias in quantile estimation. If an unbiased plotting position formula were required, then the proposed modified exceedance Cunnane formula would be the best selection.

INTRODUCTION

Probability plots are much used in hydrology as a diagnostic tool to indicate the degree to which data conform to a specific probability distribution, as a means of identifying outlier, in order to infer quantile values. Probability plotting positions are often used for the graphical display of annual maximum flood series and serve as estimates of the probability of exceedance of those values. They also provide a non-parametric means of forming an estimate of the data's probability distribution by drawing a line or a curve by hand or automated means through the plotted points. Because of these attractive characteristics, the graphical approach has been favoured by many hydrologists and engineers. It has been widely used both in hydraulic engineering and water resources planning. For example, a graphical curve fitting procedure which depends on plotting position formula was recommended as a standard method of design flood estimation in China (MWR, 1980). Probability plots were recently recommended by the National Research Council of U.S.A. (1985) as a basis for extrapolation of flood frequency curves in dam safety evaluation. Although the U.S. Interagency Advisory Committee on Water Data (IACWD, 1982) advocated the use of the method of moments to fit the log Pearson type 3 distribution to observed flood data, their recommendation also include the use of probability plots. Clearly, probability plots play an important role in statistical hydrology.

The plotting position formulae express the relationship between order number of order statistic and corresponding average frequency value of that statistics over a large number of samples. Most of them originated from the theory of order statistics and they measure the central tendency of the distributions of either $F(x_i)$ or x_i. Probability plotting of hydrologic data requires that individual observations or data points be independent of each other and that the sample data be representative of the population.

To date, more than ten plotting position formulae have

appeared in the literature. Most of them, in situations where no pre-gauging floods are considered, may be expressed as special cases of the general form

$$P_i = \frac{i-\alpha}{s+1-2\alpha} \qquad i=1,\ldots,s \tag{1}$$

where P_i is the plotting probability of the ith largest value, s is the sample size (systematic record) and α is a constant.

For the Weibull (1939) formula $\alpha=0$, for the Cunnane (1978) formula $\alpha=0.4$, while $\alpha=0.44$, and $\alpha=0.5$ are Gringorten (1963), and Hazen (1914) formulae, respectively. Cunnane (1978) and Harter (1984) provided detailed reviews and discussion on this subject. Much of the disagreement and confusion as to the choice of plotting positions is due to the fact that the cumulative distribution function (cdf) at the expected value of the ith order statistic is not equal to the expected value of the cdf at the ith order statistic, except in the case of the uniform distribution. i.e. $F[\mathrm{E}(x_i)] \neq \mathrm{E}[F(x_i)] = i/(s+1)$, in which i is the rank of the sample, and the expression $i/(s+1)$ is the familiar Weibull (1939) formula. The plotting position $F[\mathrm{E}(x_i)]$ leads to unbiased estimates of the values of a measured variable corresponding to particular values of the cumulative probability. This criterion should be considered as the best one for the purpose of flood quantile estimation (Cunnane, 1978). Only the Weibull and Cunnane based formulae will be discussed in this paper due to their widespread use in practice. The Gringorten (1963) and Hazen (1914) formulae are not considered in this study since they are similar to Cunnane formula.

The methods and the value of historical floods or palaeologic information, for improving estimates of flood quantiles, have been discussed by many authors (Stedinger & Cohn, 1986; Guo, 1990). Historical flood peaks reflect the frequency of large floods and thus should be incorporated into flood frequency analysis (NERC, 1975; MWR, 1980; IACWD, 1982 and Hirsch, 1987). They can also help to judge the adequacy of an estimated flood frequency relationship. Appropriate plotting position formulae are required for the graphical display of real data and serve as the estimates of design floods.

The existing plotting position formulae for historical floods and palaeologic information are reviewed and discussed. This paper emphasizes the development of a probabilistic model of flood records, which include pre-gauging information, the use of that model to develop reasonable plotting position formulae and the evaluation of these estimators in comparison with others. Finally, the unbiased plotting position formulae for historical data and palaeofloods are identified and recommended.

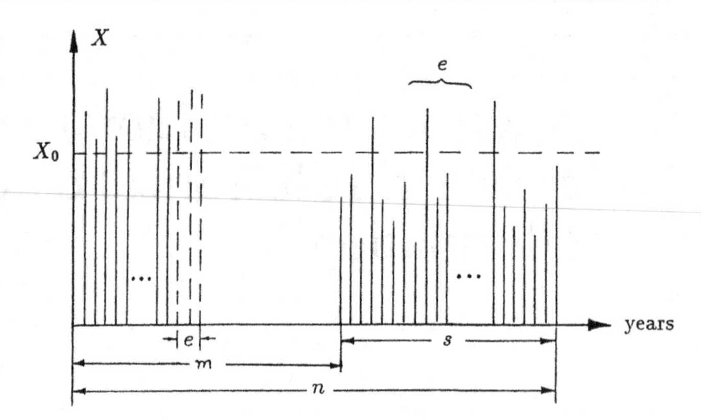

Fig. 1 Sketch of the annual maximum flood series when the historical information is available, in which the total number of known floods $g=k+s-e$.

REVIEW OF EXISTING PLOTTING POSITION FORMULAE FOR PRE-GAUGING FLOOD DATA

Various formulae have been proposed for estimating the exceedance probability of flood discharges in annual maximum flood series which include systematic records as well as historical information. The major citations on this subject include Benson (1950), Qian (1964), NERC (1975), IACWD (1982), Hirsch (1987) and Hirsch & Stedinger (1987). Some useful notation is introduced in Fig. 1, largely following the one presented by Hirsch (1987). There are a total of g known flood magnitudes in n years. Of these floods, k are known to be the largest k values in the period of n years. The n year period contains within it some systematic (gauged) record period of s years ($s \leq n$). Of the k largest floods, e of them occurred during the systematic record ($e \leq k, e < s$). Note also that $g=k+s-e$. We assume that there is a threshold of perception X_0 such that the k largest floods are larger than or equal to it and the remainder are smaller that it and that all floods greater or equal to X_0 over the n years should be known.

Notation used reads:

n – historical period;
s – length of recently recorded series;
m – pre-gauging period ($m=n-s$);
X_0 – threshold of perception;
r – number of floods exceeding X_0 within m years;
e – number of floods exceeding X_0 within s years;
k – total number of floods exceeding X_0 within n years ($k=r+e$);
g – total known flood magnitudes in n years ($g=s+k-e$).

Benson (1950) first discussed the use of historical information to improve flood estimates on the Susquehanna river at Harrisburg. He proposed the following plotting position formula to display historical data and to estimate flood quantiles.

$$P_i=\begin{cases}\dfrac{i}{n+1} & i=1,\dots,k\\[2ex] \dfrac{k}{n+1}+\dfrac{n-k}{n+1}\dfrac{i-k}{s-e} & i=k+1,\dots,g\end{cases} \quad (2)$$

Qian (1964) derived a plotting position formula which is more closely based on the Weibull rule

$$P_i=\begin{cases}\dfrac{i}{n+1} & i=1,\dots,k\\[2ex] \dfrac{k}{n+1}+\dfrac{n-k+1}{n+1}\dfrac{i-k}{s-e+1} & i=k+1,\dots,g\end{cases} \quad (3)$$

This latter formula is recommended by Chinese authorities (MWR, 1980) as a standard formula to calculate probability plotting positions in the graphical curve fitting procedure. Equation (3) implies that the probabilities are uniformly distributed over the range $k/(n+1)$ (the plotting position of the smallest extraordinary flood) and 1 for the floods below the threshold. NERC (1975) suggested using Gringorten formula (1963)

$$P_i=\frac{i-0.44}{n+0.12}; \quad i=1,,\dots,k \quad (4)$$

to calculate plotting positions for extraordinary floods (above the threshold of perception X_0 and formula

$$P_i=\frac{i-k+e-0.44}{s+0.12}; \quad i=k+1,,\dots,g \quad (5)$$

for the floods below the threshold X_0. This rule at least has two problems. One is non-monotonicity when these floods are displayed on the probability paper. Another problem is that large gaps can occur between the probabilities assigned to the floods above or below the threshold of perception.

IACWD (1982) recommended the following formula, based on Hazen's formula, for determining flood flow frequency

$$P_i=\begin{cases}\dfrac{i-0.5}{n} & i=1,\dots,k\\[2ex] \dfrac{k+0.5}{n+1}+\dfrac{n-k}{n+1}\dfrac{i-k-0.5}{s-e} & i=k+1,\dots,g\end{cases} \quad (6)$$

Some formulae simply extend equation (1) to form a new formula when pre-gauging information is available and can be characterized by the form

$$P_i=\begin{cases}\dfrac{i-\alpha}{n+1-2\alpha} & i=1,\dots,k\\[2ex] P_k+(1-P_k)\dfrac{i-k-\alpha}{s-e+1-2\alpha} & i=k+1,\dots,g\end{cases} \quad (7)$$

where

$$P_k=\frac{k-\alpha}{n+1-2\alpha}$$

Equation (7) is a general form of traditional rules defined by Hirsch (1987). These formulae are referred to herein as the Weibull (W) formula when $\alpha=0$, Cunnane (C) formula when $\alpha=0.4$, etc.

Recently, Hirsch (1987) and Hirsch & Stedinger (1987) introduced the concept of exceedance probability P_e and proposed a group of formulae which can be expressed in a general form as

$$P_i=\begin{cases}\dfrac{i-\alpha}{k+1-2\alpha}P_e & i=1,\dots,k\\[2ex] P_e+(1-P_e)\dfrac{i-k-\alpha}{s-e+1-2\alpha} & i=k+1,\dots,g\end{cases} \quad (8)$$

which is the same as equation (7) but with P_e estimated by a particular P_k. Equation (8) is referred to as exceedance rules. This result is based on the fact that P_i is an order statistic of a uniformly distributed random variable. These order statistics follow a beta distribution regardless of the distribution of discharge. The exceedance probability (P_e) for X_0 can be estimated by maximum likelihood (and method of moments) both of which estimate is as k/n. From equation (8), one can form an exceedance Weibull (E-W) formula by setting $\alpha=0$, an exceedance Cunnane (E-C) formula with $\alpha=0.4$, etc.

A PROPOSED NEW FORMULA GROUPING

The plotting positions for systematic record floods below the threshold must be adjusted to reflect the additional information provided by the pre-gauging period if the historical flood data and the systematic record are to be analyzed jointly in a consistent and statistically efficient manner. However, all available formulae are unlikely to adjust these plotting positions properly. Consider the plotting positions for the floods less than X_0. Both the traditional rules (equation 7) and the exceedance rules (equation 8) simply assume that the probabilities of $s-e$ floods are distributed either in the range of P_k to 1.0 or P_e to 1.0 (see Fig. 2). This assumption is inconsistent with the plotting positions over and below the threshold, because the plotting positions of the k largest floods are calculated on the historical period n, and the values range either from 0.0 to P_k or from 0.0 to P_e. The inconsistency of the formulae in treating the two parts of the sample may cause the separation between pre-gauging floods and systematic records when these flood data are plotted on the same probability paper. Therefore, it would be more reasonable to distribute $n-k$ floods (less than X_0) over the range of P_k to 1.0 for the traditional rule or P_e to 1.0 for the exceedance rule. Since $n-s-k+e$ pre-gauging floods which are less than threshold X_0 are unknown, the concept of historically weighted moments, recommended by USWRC (1981), can be used to form a new formula grouping. It has been shown by Hirsch (1987) as well as by Hirsch and Stedinger (1987) that the exceedance rule formulae (equation 8) are better than the traditional rule (equation 7). Hence, a group of modified exceedance rule formulae is proposed here with the following form

$$P_i=\begin{cases}\dfrac{i-\alpha}{k+1-2\alpha}P_e & i=1,,\ldots,k\\ P_e+(1-P_e)\dfrac{i-k-\alpha}{n-k+1-2\alpha} & i=k+1,\ldots,g\end{cases} \tag{9}$$

in which the weighting factor (W) is defined as

$$W=\frac{n-k}{s-e} \tag{10}$$

As before, one can form a modified exceedance Weibull (M-E-W) formula by setting $\alpha=0$ and substituting equation (10) into equation (9), which results in

$$P_i=\begin{cases}\dfrac{i}{k+1}\dfrac{k}{n} & i=1,\ldots,k\\ \dfrac{k}{n}+\dfrac{n-k}{n}+\dfrac{i-k}{n-k+1}+\dfrac{n-k}{s-e} & i=k+1,\ldots,g\end{cases} \tag{11}$$

A modified exceedance Cunnane (M-E-C) formula can be obtained by setting $\alpha=0.4$.

$$P_i=\begin{cases}\dfrac{i-0.4}{k+0.2}\dfrac{k}{n} & i=1,\ldots,k\\ \dfrac{k}{n}+\dfrac{n-k}{n}+\dfrac{i-k-0.4}{n-k+0.2}+\dfrac{n-k}{s-e} & i=k+1,\ldots,g\end{cases} \tag{12}$$

COMPARISON BETWEEN PROPOSED FORMULAE WITH TRADITIONAL RULE AND EXCEEDANCE RULE

Hirsch (1987) and Hirsch & Stedinger (1987) considered the bias in probability, and bias in discharge of both traditional rule and exceedance rule based plotting position formulae for the case of $i\leq k$ (floods greater than or equal to X_0, which are the cases of greatest interest in terms of design flood estimation). In terms of bias with respect to probability, a desirable plotting position is that $E[P_i]-E[\hat{P}_i]$ should be small, where $E[\hat{P}_i]$ is the expected value of the probability plotting position of the ith largest flood and $E[P_i]$ is the expected value of true exceedance probability of the ith largest flood.

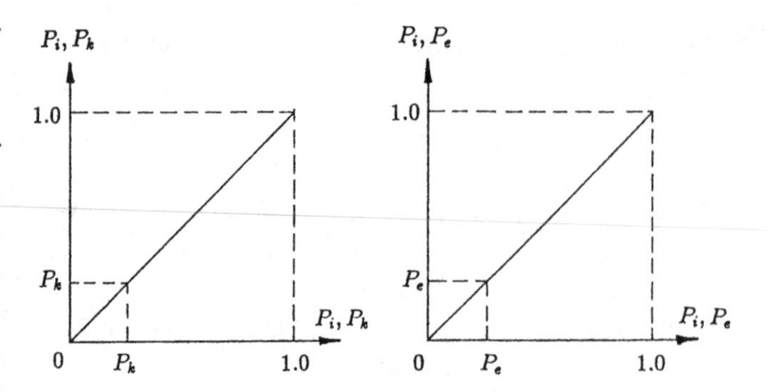

Fig. 2 Sketch of the probability plotting position values for traditional rule (P_k) and exceedance rule (P_e).

While bias in discharge is defined as the difference between $E[Q_i]$ (the expected value of the ith largest flood) and $E[F^{-1}(1-\hat{P}_i)]$, in which F^{-1} is inverse distribution form and $\hat{P}_i$ is estimated by plotting position formula. With limited Monte Carlo experiments and numerical analysis, they concluded that with combined historical and systematic records, no rule appears unbiased in terms of probability, while the Gringorten and Hazen based formulae all appear relatively unbiased in terms of discharge. The appropriate choice of historical period n and the consequences of values often employed, the impact of uncertainty as to the value of X_0 and the effect of multiple historical record lengths have also been discussed by Hirsch and Stedinger (1987) in their appendix A, B and C.

Since the proposed formulae are exactly the same as the exceedance rule for the case of floods over the threshold of perception X_0, the conclusions made by Hirsch (1987), Hirsch & Stedinger (1987) are also valid for the new proposed formulae. However, the criteria used by them to compare different plotting position formulae have some limitations. As we know in practice, the floods over threshold of perception are quite few, and systematic records also affect design flood estimates. Therefore, it is desirable to consider all the sample values (both systematic records and pre-gauging data, i.e. $i=1,k,\ldots,g$) rather than using the floods over the threshold ($i\leq k$) only in comparison.

In order to determine which group of plotting position formulae are the most suitable for pre-gauging floods, another two experiments based on the criteria of the descriptive ability and predictive ability of the formulae are proposed. The former criterion relates to the ability of a chosen formula to describe the flood data, while the latter criterion relates to a procedure's statistical ability to achieve its assigned task, such as minimum bias and maximum efficiency of quantile estimation (Cunnane, 1987). In this study, five plotting position formulae are considered: Weibull (W), exceedance Weibull (E-W), modified exceedance Weibull (M-E-W), exceedance Cunnane (E-C) and modified exceedance Cunnane (M-E-C).

Experiment 1 (test of descriptive ability)

The graphical data display method is used to compare the fitting ability of different plotting position formulae. The general extreme value (GEV) distribution is commonly used for flood frequency analysis (NERC, 1975), and the Gumbel (EV1) distribution is a special case. Therefore the GEV distribution is chosen for this experiment. Under the assumption of GEV parent parameters, 10000 traces of data are generated by Monte Carlo experiments. It was decided to test the fitting for two cases of data which often meet in practice:

(a) $n=50, s=30, k=2$, and $e=0$
(b) $n=100, s=20, k=10$, and $e=0$.

Two types of censored sample are defined by statisticians and the distinction between them depends on the process that created the sample. With type II censoring, a fixed number of the smallest or largest observations are removed regardless of their magnitudes. The type II censoring method is used to simulate a sample series from case (1). Fifty random variates are generated and the largest and second largest values are considered as historical floods. The first 20 values generated are then censored, excluding the two largest values if they fell in that group. Similarly, we can simulate case (2) data series. Each censored sample series is ranked in descending order, i.e. $X_1 > X_2 > \ldots > X_{s-e} > \ldots > X_g$. The ranked values are then averaged over the 10000 samples to obtain $\bar{X}_{(i)}$ values which represent the parent population $E[X_i]$

$$\bar{X}_{(i)} = \frac{1}{10000} \sum_{i=1}^{10000} X_{(i),j} \tag{13}$$

One can obtain g values of exceedance plotting probability $P_{(i),j}$ from each of the formulae, $j=1,2,3,4,5$. Then for the GEV parent population, the corresponding variate value of $X_{(i),j}$ can be calculated by

$$X_{(i),j} = \begin{cases} u + \frac{\alpha}{\beta}\left(1 - [\ln(1-P_{(i),j})]^{\beta}\right) & \beta \neq 0 \\ u + \alpha\left(1 - \ln(1-P_{(i),j})\right) & \beta = 0 \end{cases} \tag{14}$$

in which u, α and β are location, scale and shape parameters of the GEV distribution, i,j represent sample order and formula used respectively. $X_{(i),j}$ is the ith order statistic value from the GEV population corresponding to the jth formula. The values ($X_{(i),j}$; $i=1,2,\ldots,g$) are considered to be a representative sample (not necessarily random) from the population. If the jth formula is unbiased then $X_{(i),j}$ should equal $\bar{X}_i$ of equation (13).

Since the probability paper for the GEV distribution is commercially unavailable, instead the relationships between $\bar{X}_{(i)}$ and $X_{(i),j}$ were plotted in Figs. 3 and 4, in which the 45° theoretical line is based on the plots of $\bar{X}_{(i)}$ vs $X_{(i),j}$. Because the difference between E-W and M-E-W, E-C and M-E-C formulae are relatively small in these experiments, only the results of W, M-E-W and M-E-C formulae are presented for comparison. Figs. 3 and 4 show the points of the M-E-C formula are correctly located on the 45° theoretical line and the points of the M-E-W formula are more close than that of the Weibull formula at high flood values. The difference between W, E-W and M-E-W formulae are reduced as the historical period n is increased.

Experiment 2 (test of predictive ability)

Four populations are considered in this experiment. These are EV1, two parameter log-normal (LN2), GEV and Pearson type 3 (P3) distributions. The theoretical quantile values, Q_T, can be easily calculated at given return periods ($T=50$, 100, 200, 500, 1000) for each parent. The historically weighted moments method (USWRC, 1981) was used to calculate sample statistics, such as mean, C_v and C_s. Then the parameters and quantiles ($\hat{Q}_{T,j}$) of a particular distribution can be obtained from the corresponding $X_{(i),j}$ representative samples. It should be noted here that the choice of the method of moments is only for the purpose of comparison. The variate values of $X_{(i),j}$ are calculated for each parent population and formula

$$X_{(i),j} = F^{-1}(1 - P_{(i),j}); \qquad i=1,\ldots,k,\ldots,g \tag{15}$$

in which F^{-1} is the inverse distribution form; i, j represent sample order and formula used respectively. $X_{(i),j}$ is the ith order statistic value from the same parent population corresponding to the jth formula. The values ($X_{(i),j}$; $i=1,2,\ldots,g$) are considered to be a representative sample (not necessarily random) from the population. The relative bias (RB) of quantile estimates for jth formula can be calculated by

$$RB_{Tj} = \frac{Q - \hat{Q}_{Tj}}{Q_T} \tag{16}$$

where T is return period. Two different sample groups are considered:

(1) $n=50, s=20, k=2$ and $e=1$; and
(2) $n=200, s=20, k=10$ and $e=0$.

Monte Carlo method was used to generate these two sample groups from the EV1, LN2, GEV and P3 distributions respectively. The relative biases of quantile estimates are plotted on Figs. 5, 6, 7 and 8 for the convenience of visual comparison. It is clear that the Weibull formula has the largest bias in quantile estimation for all parents and samples. In the case of two parameter distributions (EV1 and LN2), the proposed modified exceedance formulae (equation 9) perform better both than the traditional rule formulae (equation 7) and the exceedance rule formulae (equation 8).

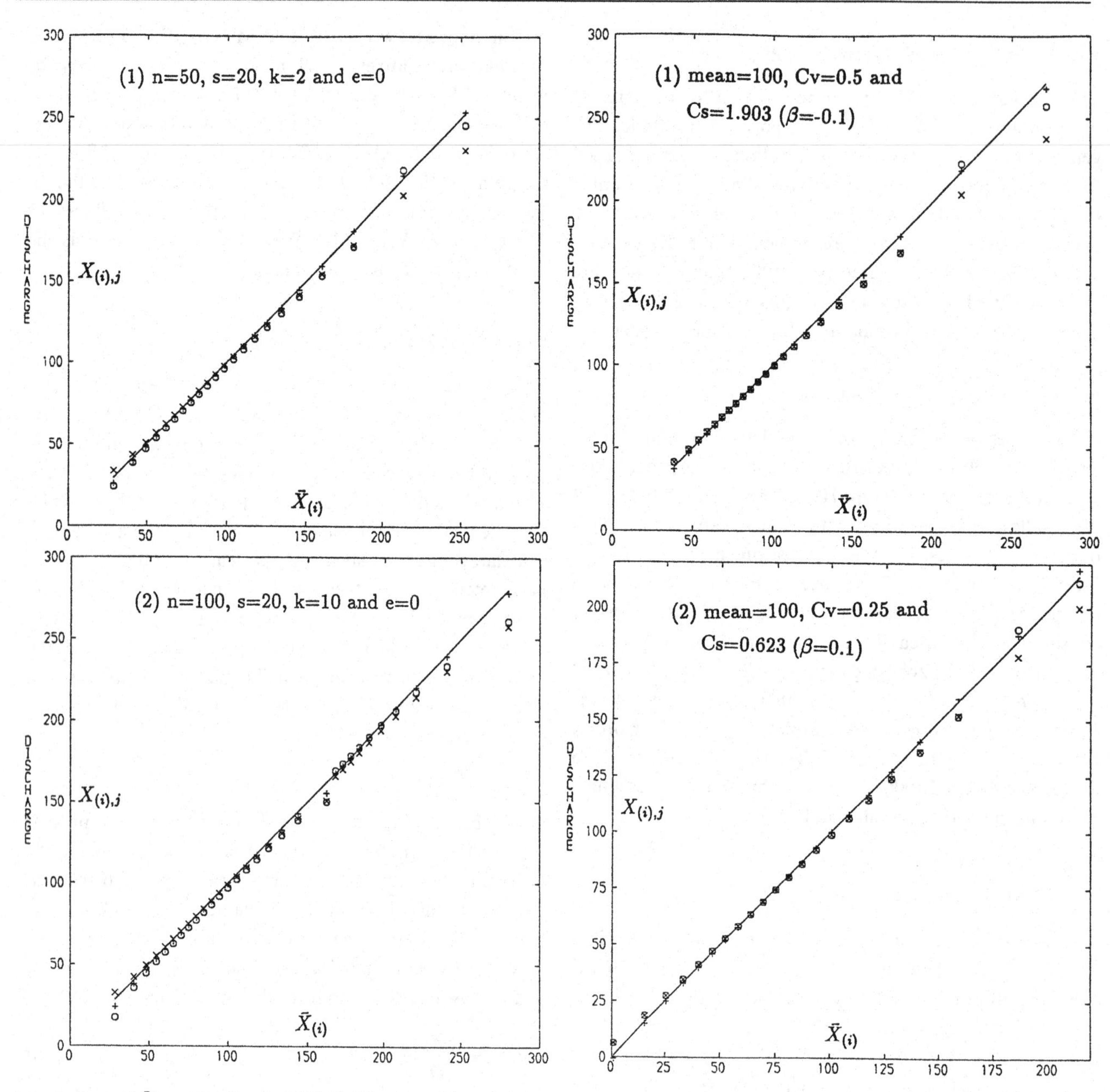

Fig. 3 Plots of $\bar{X}_{(i)}$ vs $X_{(i),j}$ for the EV1 distribution with parent mean = 100 and $C_v = 0.5$.

Fig. 4 Plots of $\bar{X}_{(i)}$ vs $X_{(i),j}$ for the EV1 distribution with different parent populations and sample size $n = 50$, $s = 20$, $k = 2$, and $e = 0$.

Legend: xxx, Weibull formula; + + +, modified exceedance Cunnane formula; and ooo, modified exceedance Weibull formula.

The modified exceedance Cunnane formula (equation 12) is nearly unbiased in quantile estimation (Figs. 5 and 6). When dealing with three parameter distributions (P3 and GEV), both the proposed rule and exceedance rule formulae perform better than the traditional rule formulae. The differences between modified exceedance rule and exceedance rule are very small and could be neglected (Figs. 7 and 8).

SUMMARY AND CONCLUSIONS

Historical data and palaeofloods provide a useful source of information in addition to the recently recorded series. The graphical curve fitting procedure has been favoured by many hydrologists and engineers, and the plotting positions are required both for the display of flood data and for the quantile estimation. The existing traditional rule and exceedance rule formulae were reviewed and discussed. The main objective of this study was to develop new plotting position formulae when historical data and palaeologic information are available. The comparison of proposed formulae with other plotting position rules were conducted by Monte Carlo experiments. The main conclusions are summarized as follows:

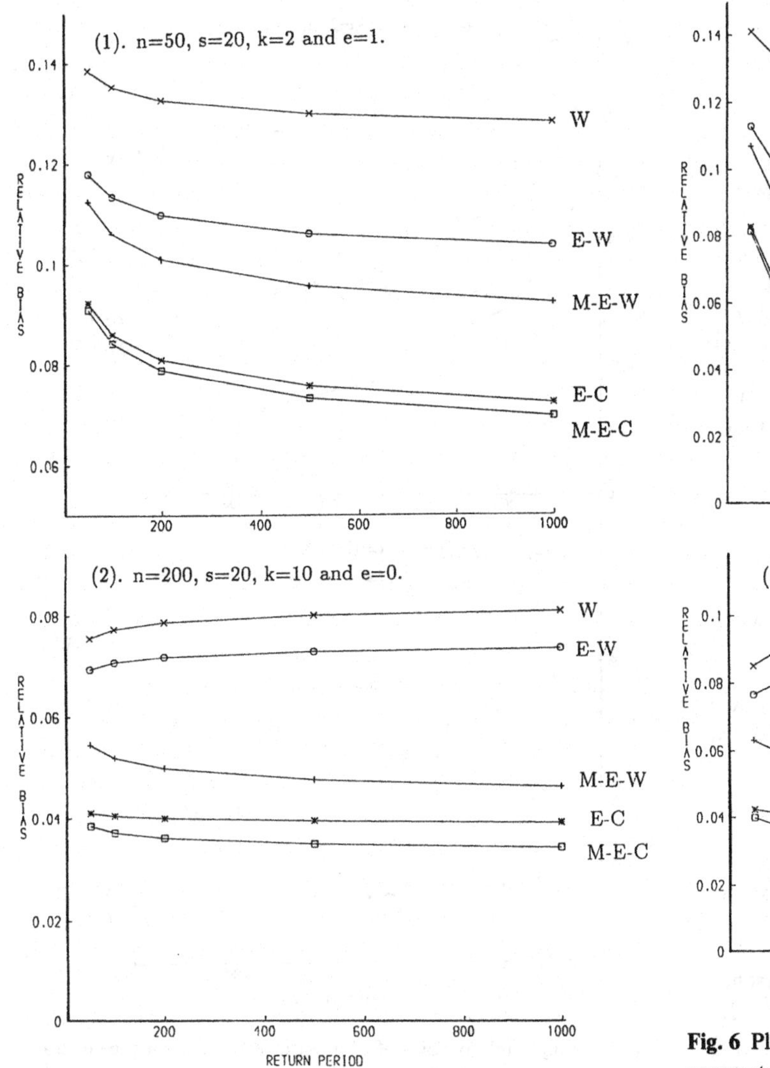

Fig. 5 Plots of relative bias quantile estimates for the EV1 distribution with parent $C_v = 0.5$.

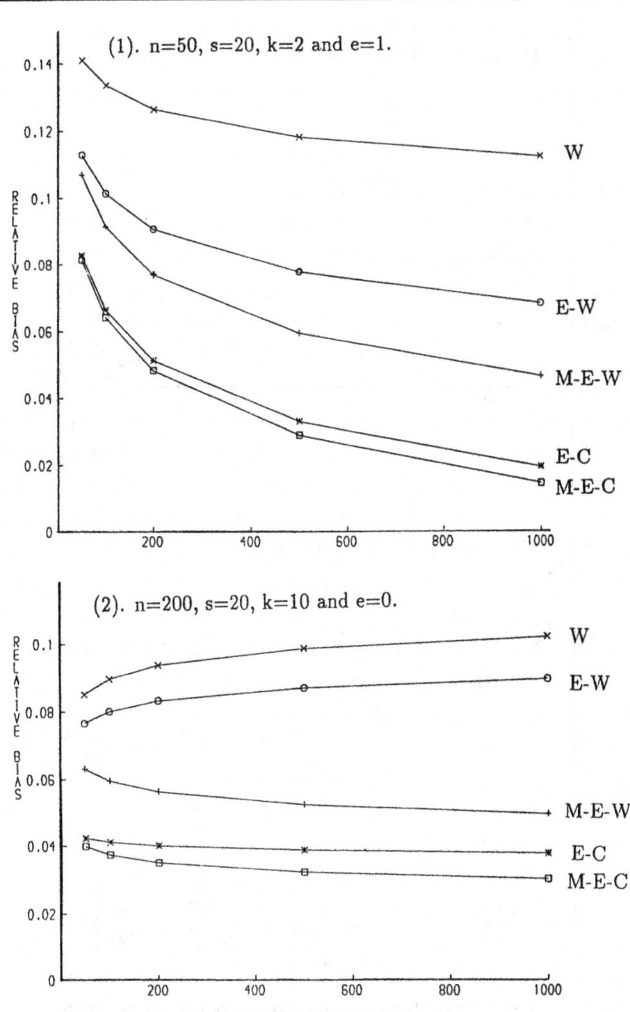

Fig. 6 Plots of relative bias of of quantile estimates for the two parameter log normal distribution with parent mean = 100 and $C_v = 0.5$.

(a) The assumption of traditional rule (equation 7) and exceedance rule (equation 8) formulae is inconsistent with the flood above and below the threshold X_0, which may result in the separation between extraordinary floods and systematic records when displayed on probability paper. The recognition of these limitations leads to the consideration of a new group of plotting position formulae, which make use of the same concept as historically weighted moments. The proposed modified exceedance rule (equation 9) is more reasonable since its assumptions about floods above or below the threshold are consistent.

(b) Hirsch (1987) and Hirsch & Stedinger (1987) have shown that the exceedance rule formulae are better than corresponding traditional rule formulae in flood quantile estimation. This conclusion has been confirmed in these studies.

(c) Based on the results of experiment 1 (test of descriptive ability), the modified exceedance Cunnane formula always performs best. The plots of traditional Weibull formula are separated from theoretical line at high discharge values. The results of two simulation experiments show that the performance of Cunnane based formulae are much better than that of Weibull based formulae both in the descriptive ability and the predictive ability. Therefore, there is not any good reason to continue using Weibull rule in flood frequency analysis, particularly when historical data and palaeofloods are included.

(d) The modified exceedance Cunnane formula (equation 12) is the least biased plotting position formula for EV1 and LN2 distributions. The performance of M-E-C and E-C formulae are very similar for three parameter distributions (GEV and P3). It should be noted here that the performances of modified exceedance Gringorten formula (set $\alpha = 0.44$ in equation 9) and modified exceedance Hazen formula (set $\alpha = 0.5$ in equation 9) are competitive with modified exceedance Cunnane formula. If an unbiased plotting position formula were required for the historical records and palaeoflood data, then the proposed modified exceedance Cunnane formula (equation 12) would be the best selection.

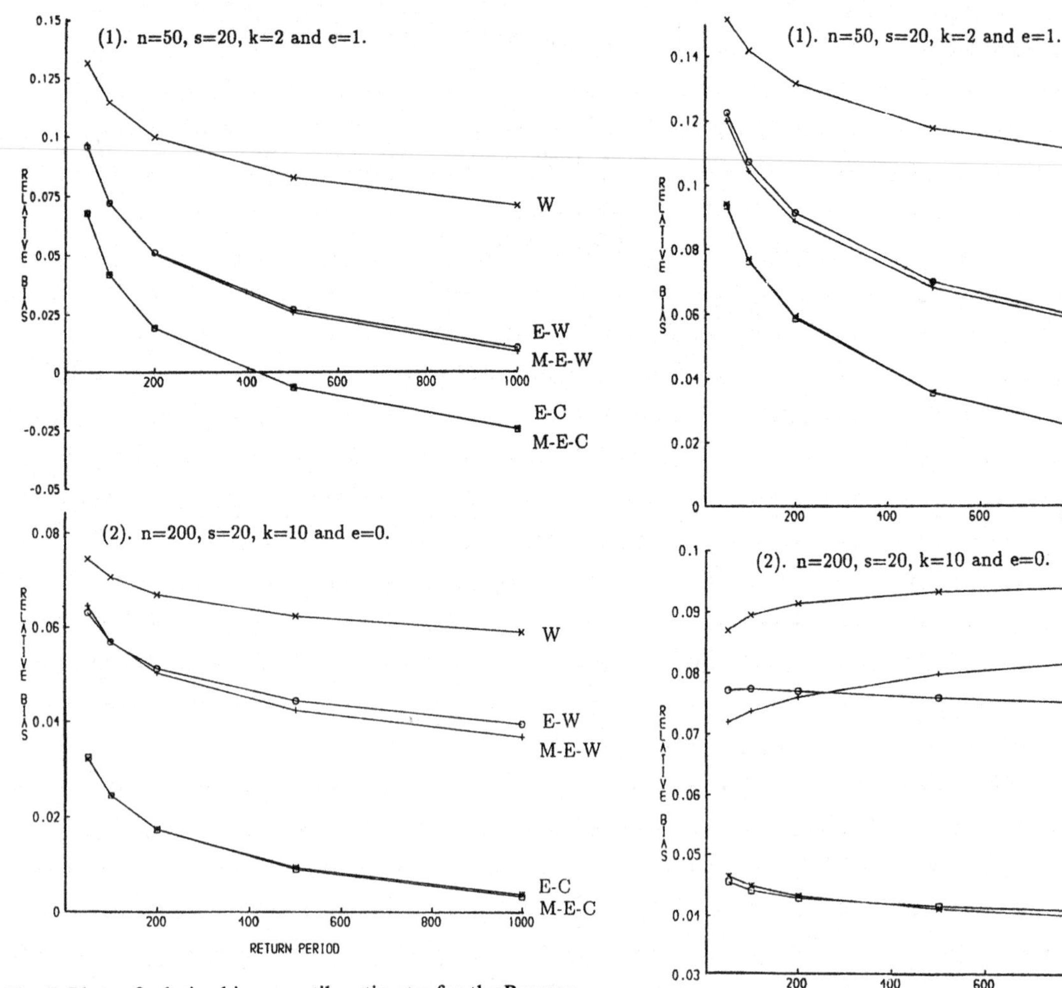

Fig. 7 Plots of relative bias quantile estimates for the Pearson type 3 distribution with parent mean = 100, $C_v = 0.5$ and $C_s = 2.0$.

Fig. 8 Plots of relative bias of of quantile estimates for the general extreme value distribution with parent mean = 100, $C_v = 0.5$ and $C_s = 1.903$.

ACKNOWLEDGEMENTS

The author wishes to record his thanks to Professor C. Cunnane for his supervision. Grateful acknowledgements for general encouragement and help are due to Professors J. E. Nash and Ye Shouze.

REFERENCES

Benson, M. A. (1950) Use of historical data in flood frequency analysis. *Trans. Am. Geophys. Union*, **31**(3), 419–24.

Cunnane, C. (1978) Unbiased plotting positions – A review, *J. Hydrol.*

Cunnane, C. (1987) Review of statistical models for flood frequency estimation. In: Singh, V. P. (Ed.), *Hydrologic Frequency Modelling*, Reidel, Dordrecht.

Gringorten, I. I. (1963) A plotting rule for extreme probability paper, *J. Geophys. Res.*, **68**(3), 813–14.

Guo, S. L. (1990) *Flood frequency analysis based on parametric and nonparametric statistics*, Ph. D. Thesis, National University of Ireland, Galway.

Harter, H. L. (1984) Another look at plotting position, *Commun. Stat., Theory and Methods*, **13**(13), 1613–33.

Hazen, A. (1914) Storage to be provided in impounding reservoirs for municipal water supply, *Trans. Amer. Soc. of Civil Eng.*, **77**, 626–32.

Hirsch, R. M. (1987) Probability plotting position formulas for flood records with historical information, *J. Hydrol.*, **96**, 185–99.

Hirsch, R. M. & Stedinger, J. R. (1987) Plotting positions for historical floods and their precision, *Water Resour. Res.*, **23**(4) 715–27.

IACWD (Interagency Advisory Committee on Water Data) (1982) *Guidelines for determining flood flow frequency*, Bull. 17B, U.S. Geol. Surv.

MWR (Ministry of Water Resources) (1980) *Standard method for flood quantile estimation procedure*, Ministry of Water Resources and Ministry of Power Industry, SDJ 22–79, Beijing, China (in Chinese).

National Research Council (1985) *Safety of dams – flood and earthquake criteria*, National Academy Press, Washington, D.C., U.S.

NERC (Natural Environmental Research Council) (1975) *Flood Studies Report*, London, England, U.K.

Qian, T. (1964) The determination of empirical frequency of the flood in the presence of historical flood data, *Shui Li Xue Bao*, **2**, 50–4 (in Chinese).

Stedinger, J. R. & Cohn, T. A. (1986) The value of historical and paleoflood information in flood frequency analysis, *Water Resour. Res.* **22**(5), 785–93.

U.S.W.R.C. (1976, 1977 & 1981) *Guidelines for determining flood flow frequency*, Bull. 17, Hydrology Committee, Water Resources Council, Washington, D.C.

Weibull, W. (1939) *A statistical theory of strength of materials*, Ing. Vet. Ak. Handl., 151, generalstabens litografiska Anstals Forlag, Stockholm, Sweden.

III

Novel approaches to uncertainty: fractals, fuzzy sets and pattern recognition, non-parametric methods

1 Dispersion in stratified soil with fractal permeability distribution

M. W. KEMBLOWSKI and JET-CHAU WEN

Department of Civil and Environmental Engineering, Utah State University, Logan, Utah, USA

ABSTRACT Stochastic analysis of flow and transport in subsurface usually assumes that the soil permeability is a stationary, homogeneous stochastic process with a finite variance. Some field data suggest, however, that the permeability distributions may have a fractal character with long range correlations. It is of interest to investigate how the fractal character of permeability distribution influences the spreading process in porous media. Dispersion in perfectly stratified media with fractal distribution of permeability along the vertical was analyzed. Results were obtained for the transient and asymptotic longitudinal dispersivities. The results show that the macroscopic asymptotic dispersivity depends strongly on the fractal dimension of vertical permeability distribution. Macroscopic dispersivity was found to be problem-scale dependent in development and asymptotic phases.

INTRODUCTION

The impact of heterogeneities on flow and mass transport in groundwater has been investigated for some two decades. Usually this type of investigation is performed using a stochastic, as opposed to deterministic, framework. This choice is not based on the assumption that the flow process itself is stochastic, but rather on the recognition of the fact that the deterministic description of the parameter distributions would be impractical, if not impossible.

Initial research in this area did not consider the spatial structure of flow properties, assuming that either they behaved like the white noise process (lack of spatial correlation), or had a layered structure in the direction parallel or perpendicular to the flow (perfect correlation in one direction). The next step was to consider spatial correlation of flow properties. Various autocovariance functions, including anisotropic ones, were used to describe the spatial correlation (Dagan, 1984; Gelhar & Axness, 1983). Excellent review of this research is given by Gelhar (1986). However, in most of these efforts it was assumed that the correlation structure of parameter fluctuations is such that the fluctuation variance is bounded. The validity of this assumption for geologic formations has yet to be demonstrated. The field data from the rather homogeneous Borden site indicate that, at least for this site, the assumption is acceptable. Other sites, however, show scale-dependent variance and long-range correlations of subsurface properties (Burrough, 1981, 1983; Hewett, 1986; Kemblowski and Chang, 1993). This evidence prompted us to investigate the statistical behavior of solute transport in heterogeneous systems whose properties exhibit long-range correlations. The statistics of such properties is described using the concept of fractal, self-similar objects. Following that, we examine the behavior of a relatively simple transport problem: two-dimensional (vertical cross section) solute transport in a perfectly stratified medium (a medium in which hydraulic conductivity varies only along the vertical, and is uniform in the horizontal plane).

FRACTAL PERMEABILITY DISTRIBUTION

Fractal objects are characterized by their self-similar structure at theoretically all scales, and thus have partial correlations over long ranges. Such self-similarity may be exact, as in the case of Koch Snow Flake, or statistical, as in the case of natural objects (Mandelbrot, 1983). The basic assumption of the classical stochastic analysis of subsurface properties is that the variance of increments, or variogram, is bounded by the property's variance (Journel & Huijbregts, 1978).

$$\lim_{l \Rightarrow \infty} \gamma(l) = 0.5 \, \mathrm{E}\{[X(z+l) - X(z)]^2\} = \sigma_x^2 \quad (1)$$

One such variogram, used frequently in mining exploration, is the spherical variogram

$$\gamma(l)=\sigma^2\left[\frac{3}{2}\left(\frac{l}{a}\right)-\frac{1}{2}\left(\frac{l}{a}\right)^3\right]; \qquad l\leq a$$

$$\gamma(l)=\sigma^2; \qquad l>a \tag{2}$$

The interval a, where the variogram reaches its maximum value ($\gamma(a)=\sigma^2$), is called the range. For fractal distributions, the variogram is not bounded because of the correlations over all scales, and is described by a power law (Hewett, 1988)

$$\gamma(l)=\gamma_0 l^{2H} \tag{3}$$

where γ_0 is the variogram value at $l=1$, and H is the fractal co-dimension, which is equal to the difference between the Euclidean dimension in which the fractal distribution is described and the fractal dimension of this distribution, D. Thus, for the vertical distribution of hydraulic conductivity K, the co-dimension is given by

$$H=2-D \tag{4}$$

since such distribution is described in the two-dimensional space $K-z$. The statistical self-similarity of distributions whose variogram is given by equation 3 is indicated by the fact that variations over any scale $r\,l$ may be expressed in terms of the variations over a scale l by

$$\gamma(rl)=\gamma(l)r^{2H} \tag{5}$$

where: r = scaling factor.

An important conclusion from this relationship is that for fractal distributions the variance at any scale can be defined by the variance estimated at any other scale. This also implies that for the fractal distribution the variance is scale dependent. The variogram may be related to the two-point autocovariance function. The definition of the autocovariance function of process X is given by

$$C_X(t)=\mathrm{E}\{X(z+l)X(z)\}-\mathrm{E}^2\{X(z)\} \tag{6}$$

The relationship between the variogram and the autocovariance function is defined as follows:

$$\gamma_X(l)=\mathrm{E}\{X^2(z)\}-\mathrm{E}^2\{X(z)\}-C_X(l) \tag{7}$$

Using this relationship and the Wiener–Khintchine theorem (Hewett, 1986)

$$C_X(l)=\int_0^\infty S_X(f)\cos(2\pi fl)\,\mathrm{d}f \tag{8}$$

where: $S_X(f)$ = spectral density of $X(z)$, it can be demonstrated that the spectral density of the fractal objects will also have the power-law form,

$$S_X(f)=S_0 f^{-\beta} \tag{9}$$

where: S_0 = spectral density at $f=1$, and $\beta=2H+1$.

Thus, the fractal dimension of a given process may be estimated by approximating the variogram and the spectral density of the process with linear functions in logarithmic coordinates. The slopes of these functions are related to the process' fractal dimension ($2H=4-2D$ and $\beta=5-2D$).

Several sets of hydraulic conductivity data have been analyzed by Ababou & Gelhar (1989). Their analysis indicates that the vertical distribution of hydraulic conductivity is characterized by a highly irregular character, with the fractal dimension used varying to 2.0. In the next section we will explore the relationship between this fractal dimension and solute spreading.

ANALYSIS OF SPREADING IN PERFECTLY LAYERED MEDIA

In this section, we will deal with the impact of the fractal nature of hydraulic conductivity distribution on the spreading of soluble plumes. The specific scenario considered in this paper assumes that groundwater flows horizontally in a perfectly stratified aquifer. Thus, in the horizontal direction the correlation scale of permeability is infinite, whereas in the vertical direction permeability is described as a stochastic process. Mass transport of a conservative tracer in such a situation consists of horizontal advection, and horizontal and vertical local (pore-level) dispersion. This transport problem is described in detail by Gelhar *et al.* (1979), who also derived for this scenario a general relationship between the stochastic structure of hydraulic conductivity and the longitudinal macrodispersivity. Using their general results, we will investigate the behavior of macrodispersivity for the case when the vertical distribution of permeability is fractal.

For a detailed description of the derivation of longitudinal macrodispersivity the reader is referred to Gelhar *et al.* (1979). However, for the reader's convenience, we will review the major assumption involved in the derivation. Mass transport of a conservative tracer is described by:

$$\frac{\partial \boldsymbol{C}}{\partial t}+\frac{\partial}{\partial x}(\boldsymbol{UC})=\frac{\partial}{\partial x}\left(\boldsymbol{D}_\mathrm{L}\frac{\partial \boldsymbol{C}}{\partial x}\right)+\frac{\partial}{\partial z}\left(\boldsymbol{D}_\mathrm{T}\frac{\partial \boldsymbol{C}}{\partial z}\right) \tag{10}$$

where $\boldsymbol{D}_\mathrm{L}$ and $\boldsymbol{D}_\mathrm{T}$ are local (pore-level) longitudinal and transverse dispersion coefficients, $\boldsymbol{C}$ is the concentration, and $\boldsymbol{U}$ is pore-water velocity. These quantities are considered random processes with:

$$\boldsymbol{C}(x,z,t)=C(x,t)+c(x,z,t), \quad C=\mathrm{E}[\boldsymbol{C}], \quad \mathrm{E}[c]=0$$
$$\boldsymbol{D}_\mathrm{L}=D_\mathrm{L}+d_\mathrm{L}, \quad D_\mathrm{L}=\mathrm{E}[\boldsymbol{D}_\mathrm{L}], \quad \mathrm{E}[d_\mathrm{L}]=0$$
$$\boldsymbol{D}_\mathrm{T}=D_\mathrm{T}+d_\mathrm{T}, \quad D_\mathrm{T}=\mathrm{E}[\boldsymbol{D}_\mathrm{T}], \quad \mathrm{E}[d_\mathrm{T}]=0$$
$$\boldsymbol{U}=U+u=\boldsymbol{K}\,J/n=(K+k)J/n \tag{11}$$

where: $C(x,t)$ = average vertical concentration, U = seepage velocity, $\mathbf{K}$ = hydraulic conductivity, K = mean hydraulic conductivity, k = hydraulic conductivity variations, J = constant hydraulic gradient, and n = porosity.

These equations and the spectral representations of the variations of hydraulic conductivity and concentration are used to derive a mass transport equation for the average concentration $C(x,t)$. The derivation, based on the approach first presented by Taylor (1953), neglects the second order terms in the mass transport equation, and assumes that the fluctuations in the local longitudinal coefficient of dispersion are proportional to the fluctuations of hydraulic conductivity, namely:

$$d_L/D_L = 3/2(k/K) \tag{12}$$

Based on these assumptions, Gelhar *et al.* (1979) derived the following mass balance equation in terms of the vertically averaged concentration C:

$$\frac{\partial C}{\partial t} = (A+\alpha_L)U\frac{\partial^2 C}{\partial \xi^2} - B\frac{\partial^3 C}{\partial t \partial \xi^2} - 3AU\alpha_L\frac{\partial^3 C}{\partial \xi^3} + \ldots + \text{higher order terms} \tag{13}$$

where: $x = x - Ut$, A = longitudinal macrodispersivity,

$$A = \int_{-\infty}^{+\infty} \frac{S(f)}{K^2}\frac{(1-e^{-bt})}{\alpha_T f^2}\,df \tag{14}$$

$S(f)$ = spectral density of k, $b = \alpha_T U f^2$, α_T = local transverse dispersivity, and f = wave number. Parameter B is given by:

$$B = \int_{-\infty}^{+\infty} \frac{S(f)}{K^2}\frac{[1-(1+b)e^{-bt}]}{\alpha_T^2 f^4}\,df \tag{15}$$

Gelhar *et al.* (1979) concluded that, for the large-time behavior, the second derivative term in equation (13): $((A+\alpha_T)\,U\delta^2 C/d\xi^2)$ is the most important term on the right hand side. The large-time limit of macrodispersivity A is given by:

$$A_\infty = \int_{-\infty}^{+\infty} \frac{S(f)}{K^2}\frac{df}{\alpha_T f^2} \tag{16}$$

At this point we can start considering the behavior of solute transport spreading in media with fractal hydraulic conductivity distribution. Such distributions are characterized by the spectrum:

$$S(f) = S_0 f^{-\beta} \tag{17}$$

It is apparent that the spectrum is not bounded when the frequency approaches zero ($S(0) \rightarrow$ Infinity). This can be easily understood when one remembers that the low frequencies are associated with large distances, and that as the lag approaches infinity, so does the variogram of a fractal process. However, in practice size of a transport domain is finite. Denoting the characteristic vertical dimension of a soluble plume as L_0, we can redefine the spectral density as follows:

$$S(f) = \begin{cases} S_0|f|^{-\beta} & \text{for } f > f_0 \\ 0 & \text{otherwise} \end{cases} \tag{18}$$

where $f_0 = 1/L_0$. Substituting the fractal spectrum into equation (16) leads to the following relationship for the asymptotic macrodispersivity:

$$A_{\text{inf}} = \frac{2S_0 L_0^{1+\beta}}{\alpha_T K^2(1+\beta)} = \frac{2S_0 L_0^{6-2D}}{\alpha_T K^2(6-2D)} \tag{19}$$

It is of interest to investigate the impact of fractal dimension D on macroscopic dispersivity. The derivative of A with respect to D is given by:

$$\frac{\partial A_{\text{inf}}}{\partial D} = -2A_{\text{inf}}\left[\log(L_0) - \frac{1}{6-2D}\right] \tag{20}$$

It can be seen that, except for very small values of L_0, macroscopic dispersion decreases when the fractal dimension of K increases. This agrees with our physical intuition. As the fractal dimension increases, the front of the plume becomes 'rougher', which leads to more mixing between the 'layers'. This enhanced vertical mixing reduces the horizontal spreading. Note that for the same reason macrodispersivity is inversely correlated with local transverse dispersivity. The positive correlation between macrodispersivity and the fractal dimension for small values of L_0 is an artifact which reflects the fact that by keeping S_0 constant and increasing D (which is equivalent to decreasing β), we increase the power (variance) of the process associated with higher frequencies (Fig. 1).

A more appropriate way to analyze the correlation between the fractal dimension and macrodispersivity is to assume that the variance of K over L_0 remains constant. This variance can be estimated as follows for $D < 2$:

$$\sigma^2(L_0) = \sigma_0^2 = \int_{-\infty}^{\infty} S(f)df = \int_{-\infty}^{\infty} S_0 f^{-\beta}df = \frac{S_0 L_0^{\beta-1}}{\beta-1} \tag{21}$$

Substituting this equation into equation (19) leads to:

$$A_{\text{inf}} = \frac{2\sigma^2(L_0)L_0^2(\beta-1)}{\alpha_T K^2(1+\beta)} = \frac{2\sigma_0^2 L_0^2(4-2D)}{\alpha_T K^2(6-2D)} \tag{22}$$

Thus, the asymptotic macrodispersivity appears to depend on the problem scale. In particular, it depends on the characteristic plume thickness and the variance of K over this thickness. The derivative of A_{inf} with respect to D can be estimated as follows

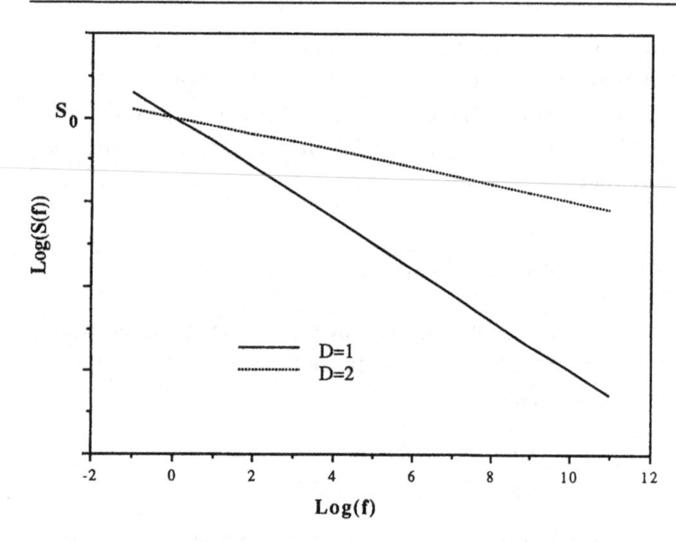

Fig. 1 Spectral density for $D=1$ and $D=2$, $S_0=\text{constant}$.

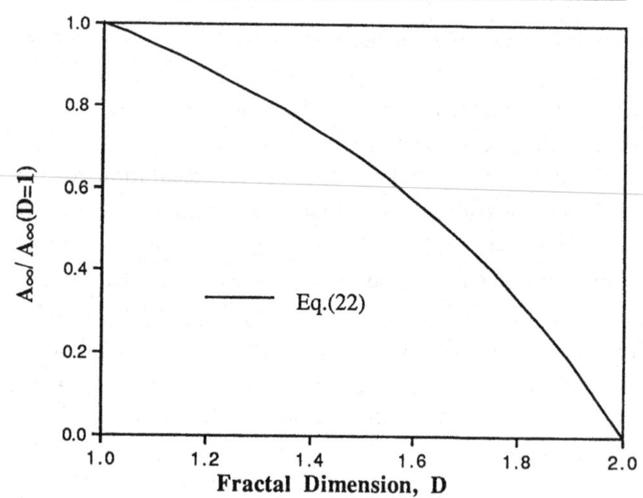

Fig. 3 Normalized asymptotic macrodispersivity as a function of fractal dimension of K.

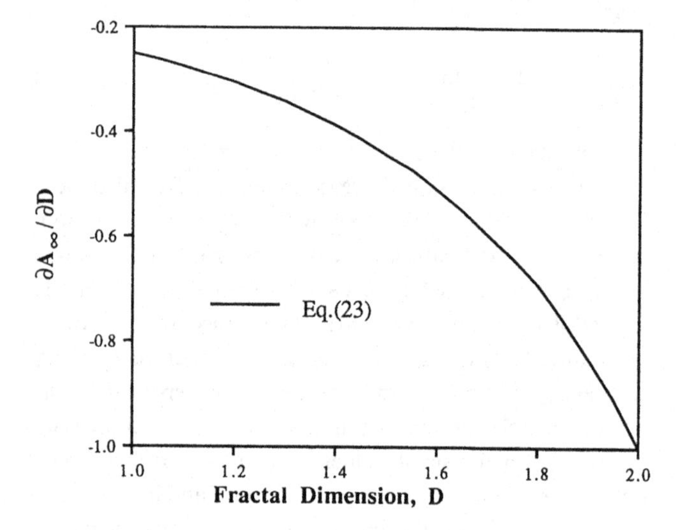

Fig. 2 dA_{inf}/dD for constant variance of K over L_0.

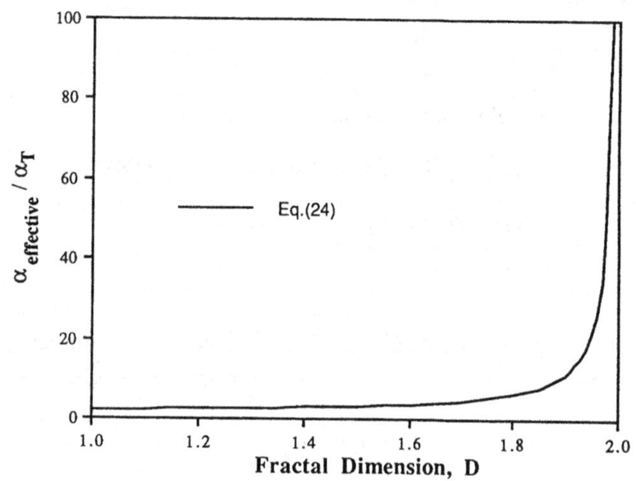

Fig. 4 Effective mixing coefficient as a function of D.

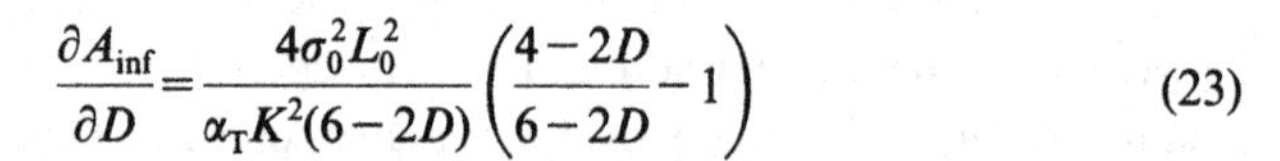

$$\frac{\partial A_{inf}}{\partial D}=\frac{4\sigma_0^2L_0^2}{\alpha_T K^2(6-2D)}\left(\frac{4-2D}{6-2D}-1\right) \tag{23}$$

The behavior of this function for $\sigma_0^2L_0^2/\alpha_T K^2=1$ is shown in Fig. 2. It can be seen that $\delta A/\delta D$ is in this case (for constant variance σ_0^2) always negative, and its absolute magnitude increases with D. This again demonstrates that the higher fractal dimension enhances the vertical mixing process and therefore decreases the horizontal spreading.

Fig. 3 depicts the behavior of the normalized asymptotic dispersivity (the dispersivity is normalized with regard to the asymptotic dispersivity at $D=1$) as a function fractal dimension D. It can be clearly seen that the asymptotic dispersivity decreases as the fractal dimension decreases. In fact, for the fractal dimension D approaching 2, we have the case of perfect mixing (which is similar to the case of α_T approaching infinity, although in this case the mixing is caused by the fact that the vertical distribution of K completely fills the $x-z$ plane). Due to the perfect mixing effect there is no longitudinal spreading related to the heterogeneous velocity field. This problem can be illustrated by introducing the concept of effective mixing coefficient, $\alpha_{effective}$, defined as follows

$$\alpha_{effective}=\frac{\alpha_T(6-2D)}{(4-2D)} \tag{24}$$

Using this definition, the asymptotic macrodispersivity may be defined as

$$A_{inf}=\frac{2\sigma_0^2L_0^2}{\alpha_{effective}K^2} \tag{25}$$

The behavior of the normalized effective mixing coefficient, $\alpha_{effective}/\alpha_T$, as a function of the fractal dimension is shown in Fig. 4. The effective mixing coefficient seems to be most sensitive to the changes in the fractal dimension in the region $1.8<D<2$. Field data suggest that the fractal dimension of the vertical distribution of hydraulic conductivity lies

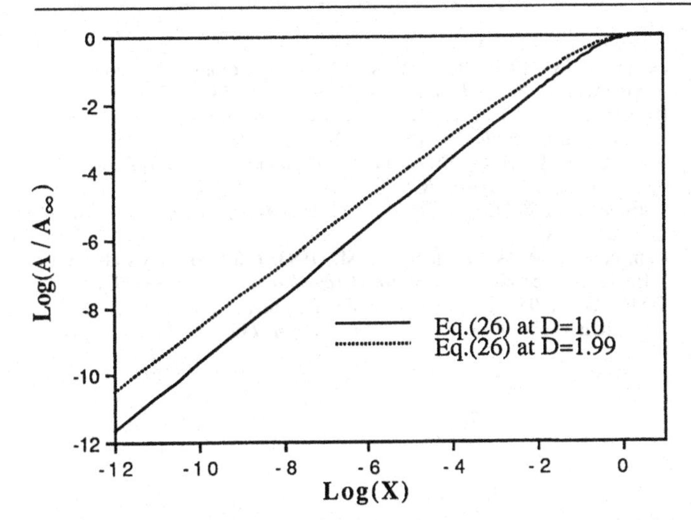

Fig. 5 Development of the spreading process.

precisely in this region (Ababou & Gelhar, 1989; Kemblowski & Chang, 1993). It may be therefore diffficult to estimate the effective mixing coefficient with adequate accuracy, since this would require robust estimation procedures for D. This problem will be exacerbated when one considers the actual three-dimensional field situation.

It is worthwhile noting that equation (22) is similar to the one obtained by Gelhar *et al.* (1979). There are, however, two significant differences. First of all, in the fractal case, the variance of K is scale-dependent, i.e. it depends on the characteristic plume thickness. Secondly, the correlation scale in this case is related to the plume thickness itself, since the vertical hydraulic conductivity distribution has the correlation scale theoretically equal to infinity. To better understand the behavior of the dispersion process, parameter A has to be evaluated as a function of travel time, or more specifically, travel distance. Substituting equation (18) into equation (14) leads to the following result:

$$\frac{A}{A_{\inf}} = 1 - \exp\left(\frac{\alpha_T x}{L_0^2}\right) + \left(\frac{\alpha_T x}{L_0^2}\right)^{\frac{1+\beta}{2}} \Gamma\left(\frac{1-\beta}{2}, \frac{\alpha_T x}{L_0^2}\right) \tag{26}$$

where x = average travel distance, $x = Ut$, and $\Gamma(\mu, \zeta)$ = incomplete Gamma function. The behavior of this ratio as a function of normalized travel distance $X = \alpha_T x / L_0^2$ and fractal dimension D is shown in Fig. 5.

It appears that the asymptotic value of macrodispersivity is reached at X equal approximately to 1, regardless of the magnitude of fractal dimension. The fractal dimension influences the development of the spreading process to a limited degree only. Specifically, the asymptotic value is reached slightly faster for the higher fractal dimension of K. It is interesting to note that the travel distance required to reach the asymptotic behavior depends not only on the pore-level transverse dispersivity, but also on the scale of the problem, namely the characteristic vertical dimension of the plume L_0. Thus, the spreading process in fractal porous media appears to be scale dependent in the development and asymptotic phases. The approach to the asymptotic regime may be rather slow. For typical values of $L_0 = 300$ cm and $\alpha_T = 1$ cm, the distance required to reach the asymptotic conditions is $x = 900$ m.

CONCLUDING REMARKS

The impact of the fractal dimension of the vertical distribution of hydraulic conductivity on the behavior of soluble plume spreading was investigated. It was found that the fractal dimension has a major impact on the value of asymptotic macrodispersivity. A higher fractal dimension enhances vertical mixing and results in less longitudinal advective spreading of the plume. For the fractal dimension approaching 2, the longitudinal spreading of the plume disappears altogether. Our results indicate also that the asymptotic dispersivity is a scale-dependent parameter. In particular, its value depends on the characteristic thickness of the soluble plume and the variance of the K process over this thickness. The transient development of the spreading process does not appear to depend on the fractal dimension. Our analysis indicates that the travel distance necessary to reach the asymptotic conditions is also scale-dependent, and is directly proportional to the squared characteristic thickness of the plume (L_0^2), and inversely proportional to the pore-level transverse dispersivity (α_T). The approach to the asymptotic conditions may be for practical cases quite slow. It is recognized that these results were obtained for a rather idealized transport scenario. However, it is felt that this approach may lead to narrowing the gap in our understanding of subsurface transport processes, particularly regarding the actual mixing of soluble plumes with surrounding groundwater. It is a very important phenomenon which has not been given enough attention in the past. It is generally agreed that the mixing of soluble contaminant plumes with the oxygen-containing groundwater is, along with the actual aerobic biodegradation process, the major mechanism contributing to and limiting the biodegradation of soluble hydrocarbon plumes (see for example Frind *et al.*, 1989). Our results clearly indicate that there is a strong connection between the fractal dimension of K and the mixing process. We are now in the process of developing this theory to deal with more realistic hydrogeologic situations.

REFERENCES

Ababou, R. & Gelhar, L. W. (1989) Self-similar randomness and spectral conditioning: analysis of scale effects in subsurface hydrology. In: Cushman, J. H. (Ed.) *Dynamics of Fluids in Hierarchical Porous Formations*, Academic Press Ltd., London.

Burrough, P. A. (1981) Fractal dimension of landscapes and other environmental data, *Nature*, **294**, 240–2.

Burrough, P. A. (1983a) Multiscale sources of spatial variation in soil, I. The application of fractal concepts to nested levels of soil variation, *Journal of Soil Science*, **34**, 577–97.
Burrough, P. A. (1983b) Multiscale sources of spatial variation in soil, II. A non-Brownian fractal model and its application in soil survey, *Journal of Soil Science*, **34**, 599–620.
Dagan, G. (1984) Solute transport in heterogeneous porous formations, *Journal of Fluid Mechanics*, **145**, 151–77.
Gelhar, L. W., Gutjahr, A. L. & Naff, R. L. (1979) Stochastic analysis of macrodispersion in a stratified aquifer, *Water Resour. Res.*, **15**(6), 1387–97.
Gelhar, L. W. & Axness, C. L. (1983) Three-dimensional stochastic analysis of macrodispersion in aquifers, *Water Resour. Res.*, **19**(1), 161–80.
Gelhar, L. W. (1986) Stochastic subsurface hydrology from theory to applications, *Water Resour. Res.*, **22**(9), 135S–45S.
Hewett, T. A. (1986) *Fractal distributions of reservoir heterogeneity and their influence on fluid transport*, SPE 15386, 1986.
Hewett, T. A. & Behrens, R. A. (1988) *Conditional simulation of reservoir heterogeneity with fractals*, SPE 18326.
Journel, A. G. & Huijbregts, Ch. J. (1989) *Mining Geostatistics*, Academic Press.
Kemblowski, M. W. & Chang, C.-M. (1993) Infiltration in soils with fractal permeability distribution, *Ground Water*, **31**(2), 187–92.
Taylor, G. I. (1953) The dispersion of matter in a solvent flowing slowly through a tube, *Proc. R. Soc. Lond.*, Ser. A, **219**, 189–203.

2 Multifractals and rain

S. LOVEJOY

Physics Department, McGill University, Montreal, Canada

D. SCHERTZER

Laboratoire de Météorologie Dynamique, C.N.R.S., Paris, France

ABSTRACT Scaling models and analyses of rain have now been around for over ten years, a period in which the corresponding scale invariant notions have seen rapid development. We review these developments concentrating on multifractals that are believed to provide the appropriate theoretical framework for scaling nonlinear dynamical systems. Although early scaling notions were geometric rather than dynamic, they contributed towards establishing and testing scaling ideas in rain and in determining the limits of scaling in both time and space. The problematic of passive scalar clouds and (continuous) turbulent cascades, provided them with a sound physical basis. Building on these advances, later analysis methods (particularly Double Trace Moment technique) made it possible to obtain robust estimates of the basic multifractal parameters. Continuous (and universal) cascades allow us to exploit these parameters to make dynamical models. We also discuss various applications of multifractals to rain including multifractal objective analysis, statistics of extreme values, multifractal modelling, space-time transformations, the multifractal radar observer's problem, stratification, and texture of rain.

INTRODUCTION

Stochastic models of rain, atmospheric scaling and multifractals

The atmosphere is probably the most familiar highly nonlinear dynamical system; the nonlinear terms are roughly $\approx 10^{12}$ (the Reynolds number) times larger than the linear (dissipation) terms, and structures vary over 9–10 orders of magnitude in space (≈ 1 mm to 10^4 km) and at least as much in time ($\approx 10^{-3}$ s on up). The nonlinearity involves many fields: rain is dynamically coupled with the velocity, temperature, radiation, humidity, liquid (and solid) water fields. Because it so palpably impinges on the human senses, it is undoubtedly subjectively experienced as the most extremely variable atmospheric field. For similar reasons, in terms of accuracy of measurements over the widest range of space and time scales, the associated radar ('effective') reflectivity field is likely to be the best measured turbulent field in geophysics or elsewhere.

While this extreme variability is undeniable, traditional modelling approaches have been limited by lack of knowledge of the nonlinear partial differential equations governing rain. Since the 1960s, these two circumstances have combined to lead to the development of stochastic[1] rain models[2]. In the 1980s, with the growing recognition of the fundamental importance of 'scaling' (especially associated with the fractal geometry of sets, Mandelbrot, 1983); scale invariant symmetries and fractals, it was natural to construct stochastic models that respected such symmetries (Lovejoy & Mandelbrot, 1985, Lovejoy & Schertzer, 1985). Unfortunately, the first scaling models were totally *ad hoc*, designed only to respect a purely statistical scaling symmetry (i.e., with no direct connection either with physics or phenomenology), and worse still, were restricted to a very simple kind of scaling

[1] Influenced by the rapid pace of developments in deterministic chaos, the idea was recently suggested (e.g. Tsonis & Elsner, 1989, Rodriguez-Iturbe *et al.*, 1989) that only a very small number of degrees of freedom were dynamically important, and that in rain deterministic rather than stochastic models would be appropriate. As argued by Osborne & Provenzale (1989), Ghilardi (1990) and Schertzer & Lovejoy (1991a) (section on stochastic chaos vs. deterministic chaos), such conclusions are based on overinterpretations of the data; in our view, there is no compelling reason for abandoning stochastic (large number of degrees of freedom) models. See also Visvanathan *et al.* (1991) for a discussion of stochastic behaviour of deterministic models.

[2] Early models include Cole (1964), Arajimo (1966) and Bras & Rodriguez-Iturbe (1976).

now known as 'simple' scaling[3]. This was all the more true since evidence had been accumulating since the 1960s suggesting that rather than being qualitatively distinct, the large and small scale regimes of the atmosphere were actually both part of a very wide single scaling regime. Rather than consisting of an isotropic two dimensional turbulent regime at large scales, and an isotropic three dimensional regime at small scales, the atmosphere is apparently scaling but anisotropic throughout[4].

In parallel with the development of these geometric 'monofractal' models, work in turbulent cascade processes and strange attractors showed that real dynamical systems were much more likely to be 'multifractal[5]' (Hentschel & Proccacia, 1983; Grassberger, 1983; Schertzer & Lovejoy, 1983, 1984, 1985b; Parisi & Frisch, 1985). They therefore require an infinite number of scaling exponents for their specification, a fact soon empirically confirmed in rain with radar data[6].

The multiscaling/multifractal problematic provided much more than just an improved empirical fit with the data. The bold proposal (Schertzer & Lovejoy, 1987a) that rain variability could be directly modelled as a turbulent cascade process for the first time provided the physical basis for stochastic rain modelling. This proposal was all the more attractive since such cascade processes were found to generically yield multifractals. In the same way that Gaussian noises frequently occur in linear (sums) of random variables, cascade processes generically produce special (universal) multifractals by nonlinear mixing of scaling noises. The existence of stable and attractive universality classes implies that the infinite number of multifractal dimensions can be described by just three basic exponents. This finding greatly simplifies analysis and simulation of multifractal fields.

Monofractal analyses, scaling and intermittency

In the following sections, we will argue that scaling systems will generally involve universal multifractals. This current understanding was the result of many years of research during which simpler (geometric monofractal) scaling analyses and models were developed and criticized. In order to understand these developments, we briefly review some early results.

The simplest scaling of relevance to rain is the following 'simple scaling[7]' or 'scaling of the increments': before the discovery of multifractals, it was thought to be quite generally associated with fractal fields. For the rainrate R, it can be defined as follows:

$$\Delta R(\lambda^{-1}\Delta x) \overset{d.}{=} \lambda^{-H}\Delta R(\Delta x) \tag{1}$$

where the small scale difference is $\Delta R(\lambda^{-1}\Delta x)=R(x_1+\lambda^{-1}\Delta x)-R(x_1)$ and the large scale difference is $\Delta R(\Delta x)=R(x_2+\Delta x)-R(x_2)$ where x_1, x_2 are arbitrary, λ is a reduction ratio, and H is the (unique) scaling parameter. The equality '$\overset{d.}{=}$' means equality in probability distributions viz. $a \overset{d.}{=} b$ if and only if $\Pr(a>q)=\Pr(b>q)$ for all q, where 'Pr' indicates 'probability'. The special case of equation 1 where the probability distributions are Gaussian is Brownian motion ($H=\frac{1}{2}$, increments are independent), and fractional Brownian motion ($H\neq\frac{1}{2}$, Kolmogorov, 1940; Mandelbrot & Van Ness 1968). Fractional Brownian motion was proposed as a streamflow model by Mandelbrot & Wallis (1969); the nontrivial exponent H was to account for the 'Hurst phenomenon' of long range dependence in streamflow (Hurst, 1951 empirically found $H\approx0.7$ in many streamflow records over scales up to millenia).

In rain, Lovejoy (1981) hypothesized that simple scaling holds – although due to the extreme variability of rain – the probability distributions were expected to have algebraic ('fat') tails instead of ('thin') Gaussian tails[8]. Below, we show that such hyperbolic tails (associated with the divergence of the corresponding statistical moments) can be considered as multifractal phase transitions. Since then, Bak *et al.* (1987) have considered the combination of scaling with hyperbolic tails as the basic features of 'self-organized criticality' (S.O.C.) and Schertzer & Lovejoy (1994) have shown how multifractals generically lead to S.O.C. Probability distributions were used to test empirically both the simple scaling and the 'fatness' of the tails in space using Montreal radar rain data (with $\Delta x=0.25$, 0.5, 1 km). Equation 1 was reasonably well followed (see Fig. 1), especially for the extreme tails, and

[3] Keddem & Chiu (1987) discuss an even simpler scaling which we called 'very simple' scaling, Lovejoy & Schertzer (1989), but it does not seem to be relevant to rain.

[4] For early discussion and reviews of scaling and its limits in the atmosphere, see Schertzer & Lovejoy (1985a), Lovejoy & Schertzer (1986a); for more recent discussion see many of the papers in the book *Nonlinear Variability in Geophysics; Scaling and Fractals*, (Schertzer & Lovejoy, 1991).

[5] This expression was coined somewhat later by Parisi & Frisch (1985). In a paper devoted in considerable part to defending the 'unicity' of fractal dimensions, Mandelbrot (1984), for the first time admitted the possibility of multiple fractal dimensions.

[6] Rain data provided the first determination of multifractal dimensions in any empirically measured field. Furthermore, the original analysis was done in one, two, three, four (x, y, z, t) and 1.5 dimensions (a simulated measuring network, see Fig. 5), showing the utility of radar rain reflectivities (Schertzer & Lovejoy, 1985b). Later, when similar analysis techniques were applied to other turbulent fields (the turbulent velocity field, Meneveau & Sreenivasan, 1987, Schmitt *et al.*, 1991), the data were only one dimensional (time series at a single point).

[7] This type of scaling was first introduced by Lamperti (1962) under the name 'semistable'. It was called 'self-similarity' by Mandelbrot & Van Ness (1968). However, this name turned out to be a misnomer since the actual functions were not self-similar but self-affine, and self similarity is a much wider concept anyway. We use the expression 'simple scaling', which contrasts it with the more general and interesting multiple scaling discussed later. For more on rain applications, see Waymire (1985).

[8] This terminology was introduced by Waymire (1985). Schertzer & Lovejoy (1985a) use the expression 'hyperbolic intermittency' for the 'fat' algebraic tails.

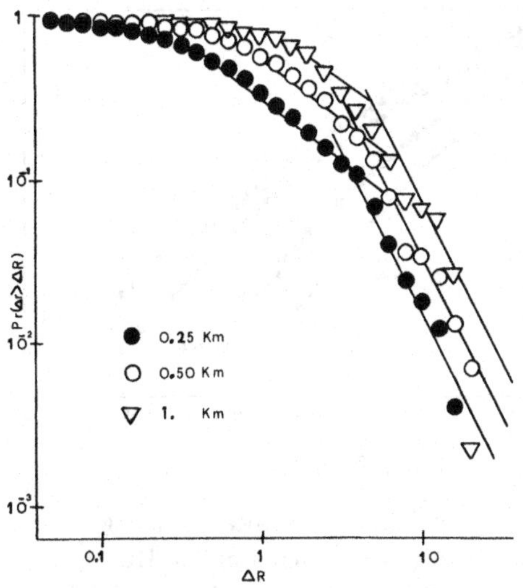

Fig. 1 The first direct empirical test of simple scaling. The probability ($\Pr(\Delta r > \Delta R)$ of a random (absolute) rain rate difference Δr, exceeding a fixed ΔR for spatial increments in 0.25 × 0.25 × 1 km averaged rain rates. The curves shown are for $\Delta x = 0.25$, 0.5, 1 km respectively. Data are from the tropical Atlantic (GATE experiment, phase III) radar reflectivities from a single radar scan converted to rainrates using standard reflectivity/rain rate relations. The straight reference lines correspond to $q_D = 0.75$, 2 respectively, and indicates that the variance (the moment $q = 2$) barely converges. The roughly linear left–right shift between the curves indicates – at least for the extreme gradients – that simple scaling roughly holds with $H \approx 0.5$. From Lovejoy (1981).

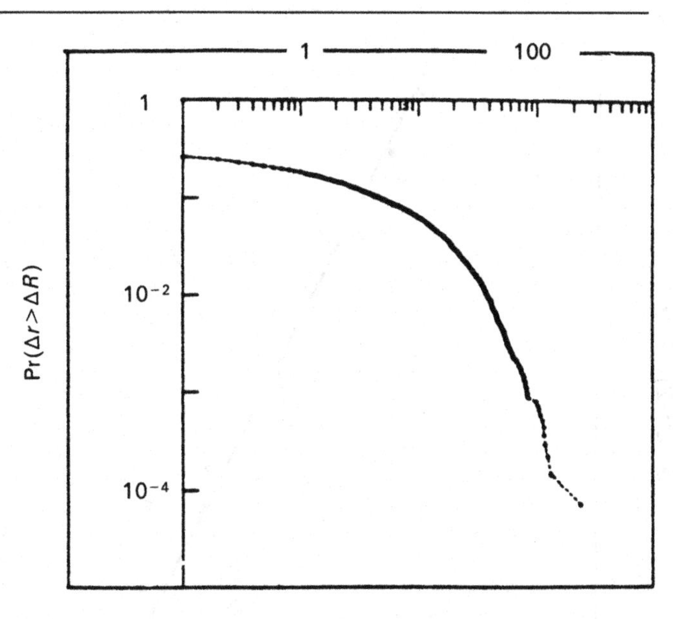

Fig. 2a The probability ($\Pr(\Delta r > \Delta R)$ of a random (absolute) rain rate difference Δr, exceeding a fixed ΔR for daily differences in daily rain accumulations at Nimes-Courbessac (France) from 1949–88 (14245 days). The tail is nearly straight with exponent $q_D \approx 3.5$. From Ladoy *et al.* (1991).

the value[9] of H was estimated as ≈ 0.5. In time, instead of using Eulerian differences, isolated storms were tracked, their total rain flux was determined every five minutes for 100 minutes. Here, simple scaling was again found to hold reasonably well (Montreal, Spain and tropical Atlantic data yielded similar results with value $H \approx 0.7$); in addition, the extreme tail of the distribution was roughly hyperbolic: $\Pr(\Delta r > \Delta R) \approx \Delta R^{-q_D}$ ($\Delta R >> 1$) for the probability of a random rainfall fluctuation Δr exceeding a fixed value ΔR. The subscript D is necessary since the value of the exponent is expected to depend on the dimension of space over which averages are performed[10]. It was found[11] that $q_D \approx 1.7$, (with $D = 2$; the integration is over areas). For comparison, Fig. 2a (from Ladoy *et al.*, 1991) shows the probability distributions for daily rainfall accumulations from a station at Nimes from 1949–88, with $q_D \approx 3.5$ (see Ladoy *et al.* (1993) for an interpretation in terms of multifractal phase transitions). Similarly, Fig. 2b shows $q_D \approx 1.1$ and Fig. 2c shows $q_D \approx 3.0$ for radar rain reflectivities of rain, 2.4 for snow, and 3.9 for melting snow ('bright band')[12]. A related result is Fig. 2d, the probability distribution of raindrops with volumes in various Hawaiian rains (replotted from Blanchard, 1953). The tails are nearly hyperbolic with[13] $q_D \approx 1.9 \pm 0.5$. The long tails on these distributions point to the extremely variable, highly intermittent non-Gaussian nature of rain.

Perhaps the most systematic and voluminous study of high resolution (tipping bucket) raingage data to date is described in Segal (1979). He digitized 90000 tipping bucket records from 47 recording stations across Canada, each 5–15 years in length, seeking to obtain statistics on rare, high rain rate events that effect microwave transmission. After comparing regressions of a variety of functional forms (including the log-normal, see Fig. 2e and discussion in the next section) for one minute averaged rain rates greater than ≈ 3 mm/h he concluded that 'a power law relationship ... provided the best fit except in the low-intensity (drizzle) region', with (for seven of the stations for which the parameters were given), $q_D \approx 2.5 \pm 0.5$. Table 1 compares all the values cited above. It is still not clear whether the dispersion in values of q_D is due to the difficulty in obtaining accurate estimates (very large samples are needed), differences in the effective D of averag-

[9] In the same paper, a similar value of H was obtained via another method (R/S analysis) over the range 0.25–13 km.

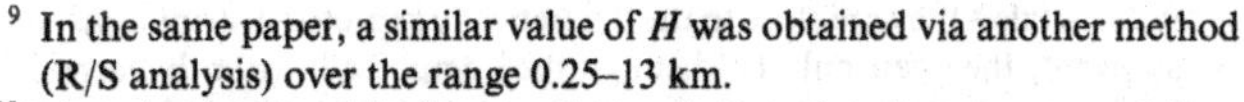

[10] Note that in the original paper, the symbol α rather than q_D was used since α is the corresponding divergence exponent for Levy variables (the rain process was thought to be an additive, simple scaling Levy process).

[11] In Fig. 1, $q_D \approx 2$, although the evidence for this asymptotic behaviour is not conclusive. To our knowledge, other strictly comparable analyses do not exist. A related result was obtained by Zawadzki (1987) who found some some evidence for hyperbolic behaviour (with $q_D \approx 2$) in tipping bucket rain rain measurements rain with roughly the same exponent, although (as expected) the sample sized required to empirically see it was quite large. Table 1 summarizes related results.

[12] Further, we show that these fat tails cannot arise due to fluctuations in the reflectivities due to drop 'rearrangement'; the latter is a thin tailed (exponential) effect.

[13] This is significant for radar measurements of rain, since (among other things) standard theory requires $q_D > 2$ so that the variance converges.

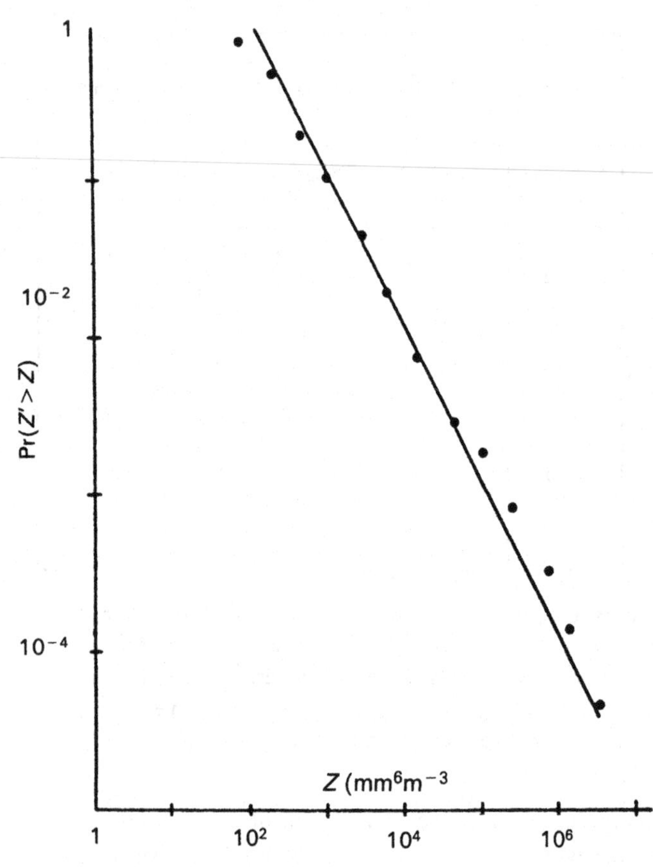

Fig. 2b The probability $Pr(Z' > Z)$ of a random radar rain reflectivity Z' exceeding a fixed threshold Z for 10 CAZLORs (Constant Altitude Z LOg Range maps), taken at the McGill weather radar, Montreal (data from 1984). The resolution varies over the map from ≈ 0.25 to 2.5 km at ranges 20 to 200 km. Each value is determined from the maximum of several consecutive pulses (the 'peak detection' method – necessary at the time (1984) due to limitations on the speed of digitizers). The reference line corresponds to $q_D = 1.06$. From Schertzer & Lovejoy (1987a).

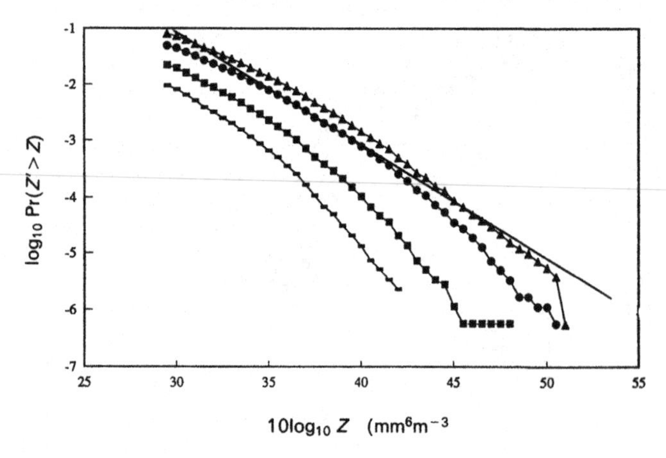

Fig. 2c The same as Fig. 2b except for a vertically pointing (nonscanning) radar at a pulse repetition rate of 1300 Hz, with each pulse return digitized, for a single Montreal storm (October 15, 1991) for 1380 seconds (1.8×10^6 points per histogram). The curves from top to bottom are bright band (melting snow and ice, 2.3 km altitude), rain (2.0, 2.15 km) and snow (2.45 km). The differences in reflectivities are largely explained by the low dielectric constant of ice compared to water, and by the large size of the water coated ice/snow particles. The asymptotic slopes yield estimates of $q_D \approx 2.4$ for snow, ≈ 3.9 for the bright band, and ≈ 3.0 for rain. The reference line corresponds to $q_D = 2$. From Duncan *et al.* (1992), Duncan, (1993).

ing, or due to true variations for different locations, climatological regimes[14] etc.

Other evidence for scaling was the area-perimeter[15] relation for radar rain and satellite cloud areas over the range 1 km^2 to $\approx 1.2 \times 10^6$ km^2, and the distribution of radar determined rain areas that was argued to be hyperbolic not log-normal[16]: $\Pr(A > a) \approx a^{-B}$ (with $B \approx 0.8$ over the range ≈ 1 to 10 km), for the probability of large areas A exceeding a fixed threshold a. A related result is the finding by Cahalan (1991) that over the range ≈ 80 m to ≈ 1 km, $B \approx 0.75$ for stratocumulus and intertropical convergence zone clouds.

The evidence suggesting that radar rain data could be approximated by simple scaling – although with highly non-Gaussian (hyperbolic) probability distributions[17] – was reviewed by Lovejoy & Mandelbrot (1985), where the 'Fractal Sums of Pulses' (FSP) process (an additive compound Poisson process involving pulses) was developed as a model. Although it had features common with other existing stochastic rain models such as those proposed by Waymire &

[14] These issues are discussed in the section on basic properties of multifractal fields; according to multifractal theory, a finite q_D is not necessary, indeed, there is evidence based on estimates of universality parameters that it may be very large (or even infinite) in time. It may well be that the dispersion of estimates for q_D are simply the result of undersampling a distribution with a much larger q_D; if we have an insufficient number of independent samples we really estimate the 'sampling moment' q_s; see equation 15a.

[15] See also Lovejoy (1982, 1983), Lovejoy *et al.* (1983), Lovejoy & Schertzer (1985b), Rhys & Waldvogel (1986), Come (1988) for more rain analyses of this type. For more recent empirical area-perimeter results, (for clouds) see Welch *et al.* (1988), Seze & Smith (1990), Cahalan (1991) and Yano & Takeuchi (1991). Other highly geometric (and, compared to statistical methods, indirect) type analyses are possible including analyses of fractal sets associated with graphs of rain series (Boucquillon & Moussa, 1991). Originally, area-perimeter exponents were interpreted as fractal dimensions of the perimeters. Since rain and clouds are in fact multifractals, a correction is necessary: for this as well as a detailed criticism of these geometric approaches to multifractals, see Lovejoy & Schertzer (1990a, Appendix A).

[16] The log-normal phenomenology of rain and cloud areas goes back to at least Lopez (1976, 1977a, b). Since lognormal distributions are long tailed – and except for the problem of 'dressing', correspond to universal multifractals – they are close to the theoretically expected distributions. In any event, they can only be distinguished empirically from hyperbolic distributions by carefully examining their tails corresponding to extremely rare large areas. The lognormal fits to rain area histograms could be profitably re-examined in this light.

[17] Such models exhibit what Mandelbrot & Wallis (1968) called the 'Noah' effect i.e. stochastic realizations of the corresponding processes involve extreme fluctuations the largest of which dominate the others. In multifractals, the effect is generalized to moments of fluctuations higher than the first. These authors also introduced the term 'Joseph' effect to denote the phenomenon of long range correlations; all multifractals exhibit this effect.

Table 1. *A comparison of various empirical estimates of the divergence of moments exponent q_D*

Data Type	Radar rain differences (space)	Radar rain differences (time)	Radar reflectivity	Vertical pointing radar reflectivity	Daily rain gauge accumulations	Tipping bucket gauges	Rain drop volumes	
Location	Tropical Atlantic	Tropical Atlantic	Montreal	Montreal	Nimes	Montreal	Western Canada	Hawaii
q_D	2	1.7	1.1	3.0 (rain) 2.4 (snow) 3.9 (bright band)	3.5	2	2.5 ± 0.5	1.9 ± 0.5
References	Lovejoy, 1981	Lovejoy, 1981; Lovejoy & Mandelbrot, 1985	Schertzer & Lovejoy, 1987a	Duncan *et al.*, 1992; Duncan, 1993	Ladoy *et al.*, 1991, 1993	Zawadzki, 1987	Segal, 1979	Blanchard 1953

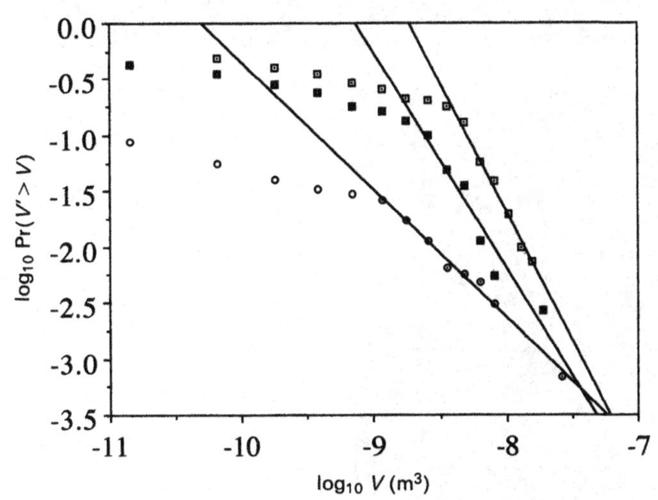

Fig. 2d Three rain drop distributions replotted from original data published in Blanchard (1953) from three different Hawaiian orographic rain events showing that the extreme tails have from top to bottom $q_D \approx 2.3$, 1.9, 1.1 in rains with rain rates 127 mm/h, 21 mm/h, 9 mm/h respectively.

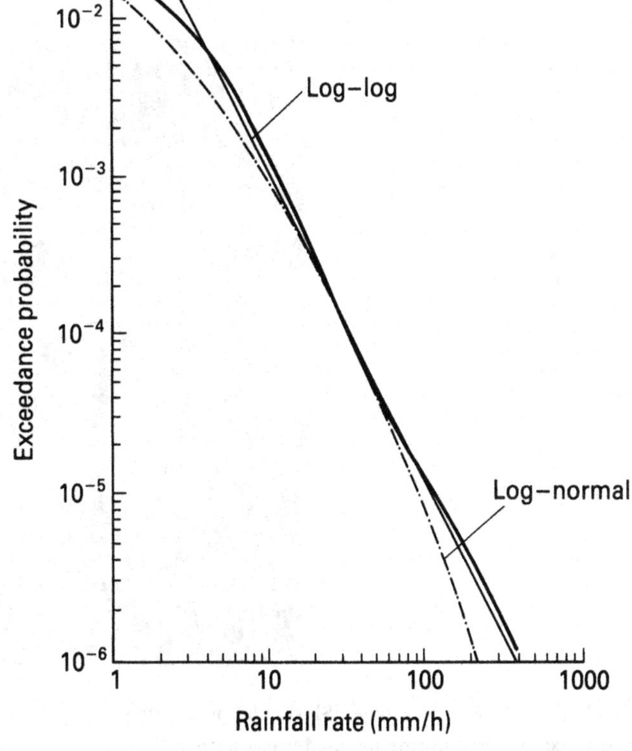

Fig. 2e An example (from 10 years of tipping bucket raingauge data at St. John, New Brunswick) of the extreme rainrate end of one minute resolution rainrate probability distributions from Segal (1979). The straight reference line corresponds to $q_D = 1.9$, the curved reference line is the best fit lognormal for comparison.

Gupta (1984) and Rodriguez-Iturbe *et al.* (1984), both its philosophy and properties are different. Instead of basing itself on an *ad hoc* division of the atmosphere into a hierarchy of qualitatively different regimes, each occuring at different scales, and each requiring a different set of modelling parameters[18], the FSP involved the linear superposition of structures whose relative size and frequency of occurence were related so that the resulting process lacked characteris-

[18] The better known of these scale dependent models the 'Waymire–Gupta–Rodriguez-Iturbe' (WGR) model involved over 10 empirically adjustable parameters; and even then only provided plausible statistical properties over a relatively narrow range of scales (see Rodriguez-Iturbe *et al.*, 1987, Eagleson *et al.*, 1987). Another 'nearly' scaling model (Bell, 1987) had similar problems.

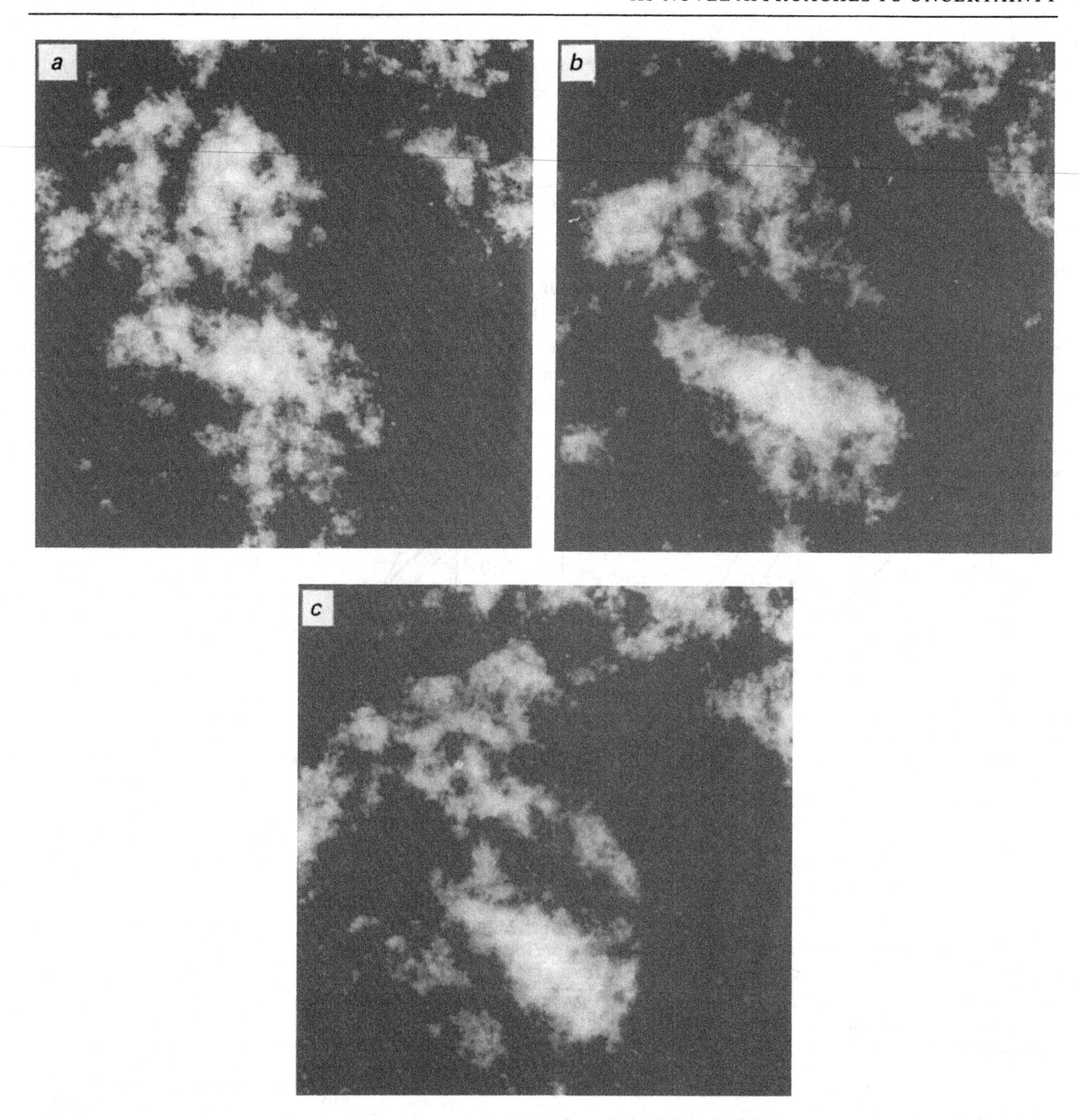

Fig. 3a–3c A single space-time FSP simulation of rain on an 800 × 800 grid showing three simulated fields seperated by 80 time units. The space/time transformation used was a statistical (and isotropic) version of Taylor's hypothesis of frozen turbulence. As expected, small structures live the shortest time, large ones longer, here (on average) linearly increasing with duration. The grey scale is proportional to the logarithm of the rain. From Lovejoy & Mandelbrot (1985).

tic scale[19]. It yielded simple scaling[20] with $q_D = H^{-1}$ (with $1 < q_D < 2, \frac{1}{2} < H < 1$). Two dimensional models on large grids were produced, and time series were modelled by making simulations in three dimensional (x, y, t) space[21]. By varying the shape of the 'pulses' from circles to annuli, more or less 'fragmented' or 'lacunar' rain fields could be produced. Using the same scaling parameters to model the concentration of liquid water, surprizingly realistic cloud fields were produced (see e.g. Figs. 3a, b, c for a temporal sequence).

[19] Similar models were discussed in Rosso & Burlando (1990).

[20] Lovejoy & Schertzer (1985a) proposed a variant on this model called the Scaling Sums of Pulses process (SCP) in which q_D, H could be varied separately. Another related model is (Wilson *et al.*, 1986), the Wave Intermittent Packet (WIP) model; in current parlance, the packets are essentially 'wavelets'.

[21] This relies on a generalization of Taylor's hypothesis of 'frozen turbulence'.

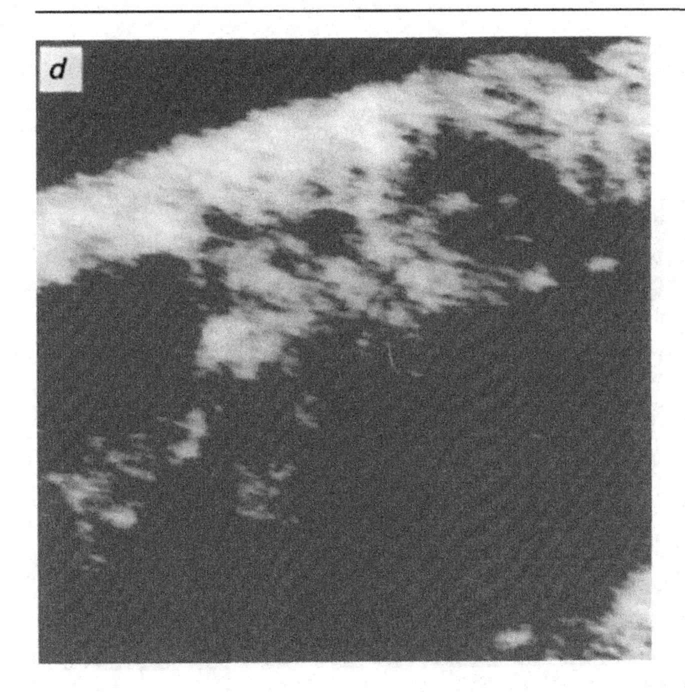

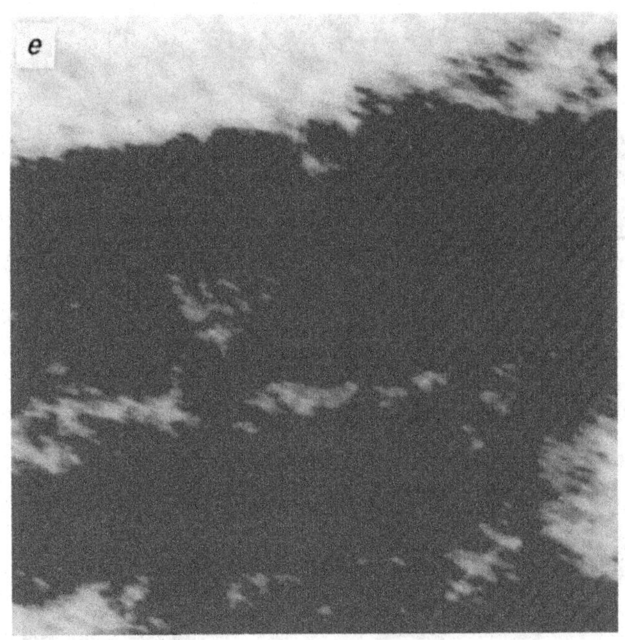

Fig. 3d, e Non self-similar (anisotropic scaling) FSP rain models on a 400 × 400 point grid, using linear Generalized Scale Invariance with generators with off-diagonal elements to yield differential rotation. This results in differences in orientation of structures as functions of scale which is clearly perceived as cloud 'texture'. From Lovejoy & Schertzer (1985a).

Realistic textures/cloud types also can be obtained by using 'Generalized Scale Invariance' – see Figs. 3d, e, as can vertical cross-sections with appropriate 'elliptical dimensions' (Figs. 3 f, g, h, i).

Another analysis method that can be used to investigate scaling is the energy spectrum $E(k)$. For statistically isotropic scaling fields[22] $E(k)$ will be of the power law form $k^{-\beta}$ where k is a wave vector modulus, and β is the spectral exponent. In time, the spectrum as a function of frequency ω will be of the same form but not necessarily with the same exponent. The most impressive single analysis of this sort to date is found in Duncan *et al.* (1992). These authors used a high resolution vertically pointing radar to perform a time series of 7×10^6 pulses at 1.3 kHz from a single pulse volume 30 × 37 × 37m in size. For computational reasons, the total range $\approx 10^{-3}$–10^4 s was split up into two regions, with average spectra calculated in each (Fig. 4a, 4b). One notices two scaling regimes with $\beta \approx 1.66$ (roughly the same in each) corresponding to time periods $2 \times 10^{-3}\,\mathrm{s} < t < 10^{-2}$, and $t > 3$ s. Duncan *et al.* (1992) and Duncan (1993) argue (with the help of multifractal models) that the breaks at 10^{-2} s, 3 s separating the flat 'spectral plateau' are both due to instrumental effects; they are simply the time scales associated with the spatial scales of the radar wavelength (3 cm), and the pulse volume[23] size (≈ 30 m). The rain itself is apparently scaling over almost the entire regime: only the high frequency ($t < 2 \times 10^{-3}$) regime is believed to be a real break associated with dissipation[24].

Other relevant temporal spectra are found in Ladoy et al. (1991) who examined daily raingauge accumulations, finding $\beta \approx 0.3$ (Fig. 4c) for periods of 1 day to 4 years at a station in Nimes (France). Fraedrich & Larnder (1993) find (Fig. 4d) the corresponding spectrum for a 45 year period for an average of 13 stations in Germany, showing roughly similar behaviour although for frequencies lower than $\approx$ (3 years)$^{-1}$ the spectrum rises more quickly[25]. The only relevant spatial spectra of which we are aware are shown in Figs. 4e and 4f from Tessier *et al.* (1993) using radar reflectivities, show $\beta \approx 0.3$ over the range 2–256 km in the tropical Atlantic but $\beta \approx 1.45$ in Montreal (over the range 150 m to 19.2 km; indicating the possibility of significant climatological differences[26]). Other relevant power law spectral analyses

[22] Self-similar fields – see section on generalized scale invariance.

[23] The corresponding velocities are $3\,\mathrm{cm}/10^{-2}\,\mathrm{s} = 3$ m/s, and $30\,\mathrm{m}/3\mathrm{s} \sim 10$ m/s respectively which are quite plausible fall speeds for rain. Further below we see that the velocity is expected to be a function of scale, so that the small difference in the two velocity values is not surprising.

[24] As expected, the exact breakpoint depends on the meteorological situation and precipitation type, although contrary to the standard theory of radar measurements, for all the cases studied, it is apparently smaller than the radar wavelength (3 cm here).

[25] Fig. 4d actually seems to have low and high frequency scaling regimes separated by a 'spectral plateau' of the sort found in temperature series by Lovejoy & Schertzer (1986b), Ladoy *et al.* (1986) and Ladoy *et al.* (1991). In this case, the difference with Fig. 4c could be due to differences in climatological regimes, and breaks delimiting the plateau might be time scales corresponding to structures of global spatial extent (see Lovejoy & Schertzer, 1986b for more discussion on this possibility).

[26] Another source of variation is the possibility of significant scatter of the estimated β from one scan to the next – this is expected since the asymptotic logarithmic probability distribution slopes (q_D or q_S) are frequently ~ 2, and the spectrum is a second order statistic.

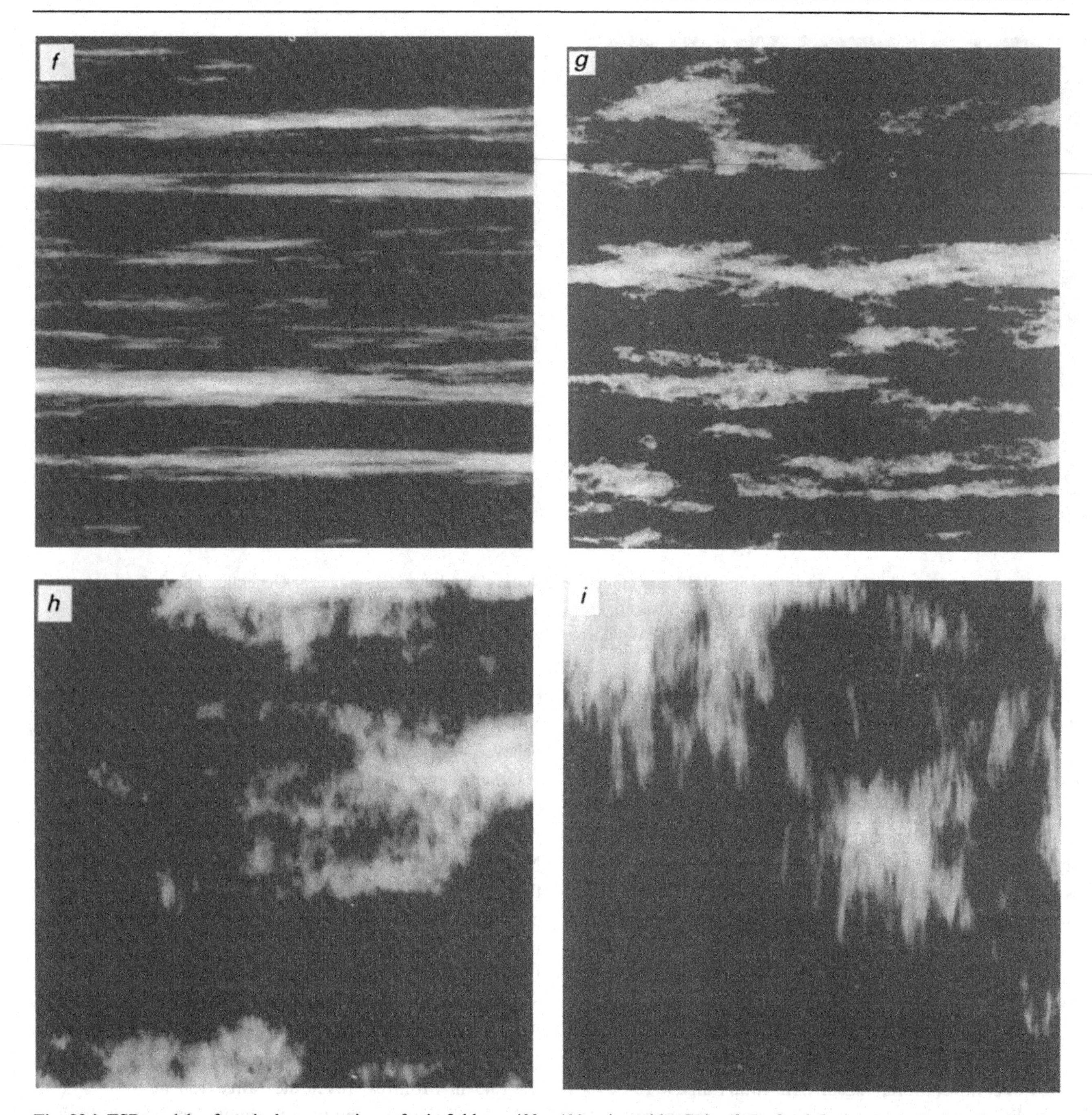

Fig. 3f–i FSP models of vertical cross sections of rain fields on 400 × 400 point grids. Going from f to i the isotropic ('sphero') scale increases from one pixel to 10, 100, 1000 (equivalent to 'zooming' in at random positions). In Fig. 3f, stratification dominates completely, as we zoom in, more and more vertical structure is visible, finally, at highest resolution (Fig. 3i), structures are vertically aligned, simulating convective rain 'shafts'. The elliptical dimension used for these cross-sections was 1.5. From Lovejoy & Schertzer (1985a).

are, Crane, 1990 (log radar reflectivity in space[27]), and Rodriguez-Iturbe *et al.*, 1989 (15 second averaged rain gage rain rates[28]).

[27] He obtains $\beta \sim 5/3$ over the range 1 minute to 1 hour. The scaling of the log reflectivities is not related in a simple way to the scaling of the reflectivities.

[28] From their Fig. 4, we estimate $\beta \sim 1.3$ over periods of ~ 1 minute to 2 hours. It is worth noting that gauge rain rates are frequently estimated from tipping buckets which mark equal accumulation times; this leads to a nontrivial bias in rainrate statistics, especially for the (very frequent) lower rain rates.

With the development of multifractals, it was realized that the apparent visual success of the FSP process masked a basic shortcoming: Lovejoy & Schertzer (1985a) criticized its monodimensional character, calling for the development of 'multidimensional'[29] alternatives. Nearly simultaneously, the first empirical multifractal analyses were performed using radar rain data (Schertzer & Lovejoy, 1985b, Fig. 5). An

[29] This cumbersome expression was a forerunner of the term 'multifractal'.

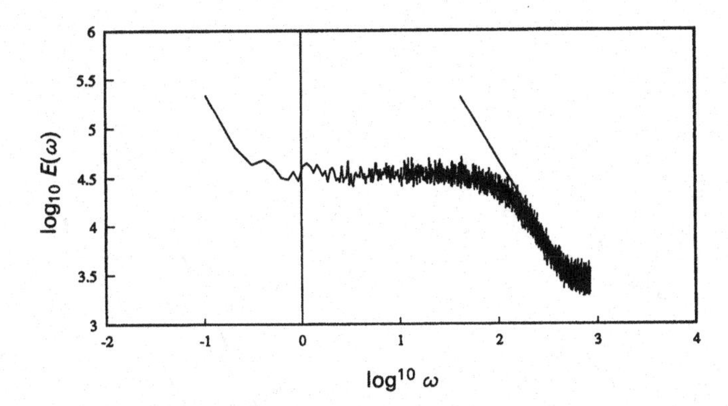

Fig. 4a Average power spectra $E(\omega)$ from 896 consecutive 8192 point sections of a time series from a vertically pointing, 3 cm wavelength radar at McGill taken from a single (30 × 37 × 37 m) pulse volume at 1 km altitude on Sept. 19, 1990 as a function of frequency w (in units of rad/s, from Duncan *et al.* (1992). The data was sampled at 1.3 kHz, so the entire $\approx 7 \times 10^6$ point data set spanned the range $\approx 10^{-3}$ to 10^4 s. The straight reference line shows an exponent $\beta = 1.66$.

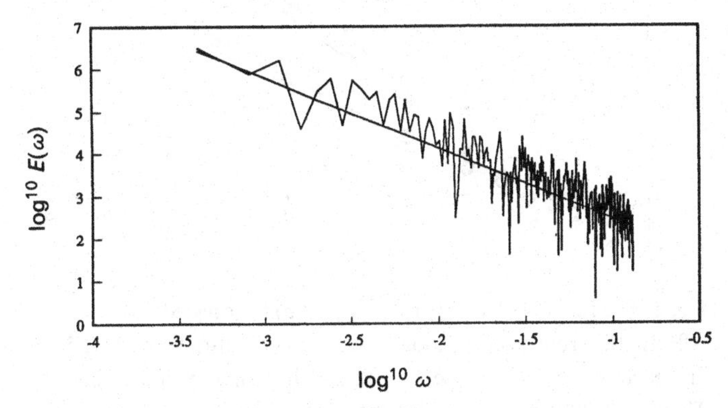

Fig. 4b Same as Fig. 4a except that the series was averaged over 512 consecutive points before the spectrum was taken. Here we obtain scaling over the entire range shown here (with the $\beta = 1.66$ line, same as in Fig. 4a, the beginning of this regime is ≈ 3 s, and is seen on Fig. 4a) shown for reference.

entire codimension function was necessary to specify the scaling of the reflectivities, not just the small number of exponents[30] (q_D, H) involved in simple scaling. The *ad hoc*, geometric FSP construction had to be replaced by a physically based multiscaling/multifractal model. There were two main obstacles to doing this. The first was the establishment of a sound connection between passive scalar concentrations and multifractal energy fluxes (via fractional integration, see below), and the second, was that then, multifractal cascades were discrete, i.e. they involved horrible artificial straight line structures; continuous cascades were needed. While the situation was apparently better as far as data analysis was concerned, it was soon to become evident that it too,

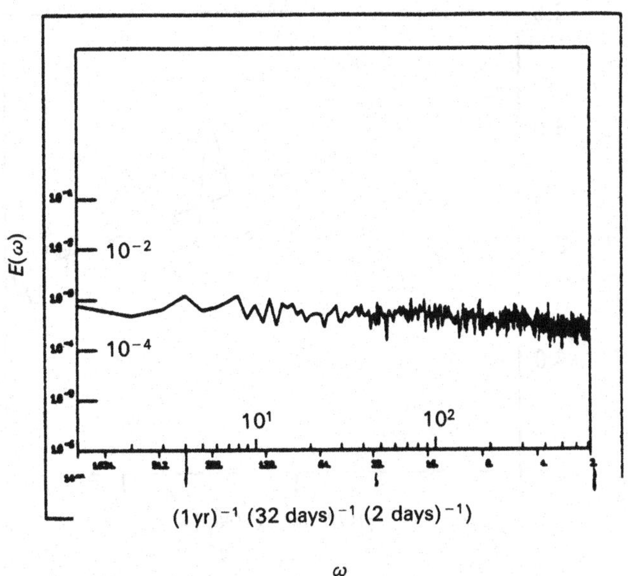

Fig. 4c The average of six consecutive 4 year spectra of the daily rainfall accumulations at Nimes-Courbessac. The annual peak is fairly weak, the scaling holds over most of the regime with slope $(= -\beta) \approx -0.3$. There is no clear evidence for the 'synoptic maximum' (i.e. a break at periods of a few weeks). From Ladoy *et al.* (1991).

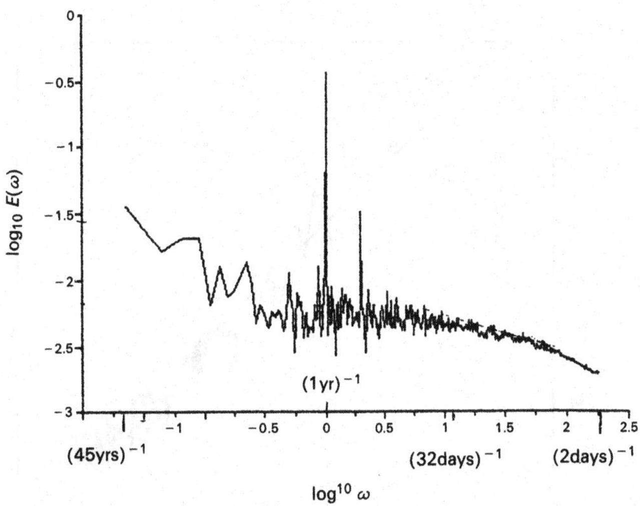

Fig. 4d Energy spectrum of daily rainfall accumulations over a 45 year period in Germany. The spectrum is an average of that obtained from 13 stations. The annual peak is much more pronounced than Fig. 4c, with evidence for a 'spectral plateau' from ≈ 20 days to ≈ 3 years. The overall spectral shape, including the low frequency rise ($\beta \approx 0.5$) is very similar to the temperature spectra analyzed in Lovejoy & Schertzer (1986b). The high frequency fall-off (also with $\beta \approx 0.5$) may be due to smoothing introduced by the spatial averaging (the 13 stations had correlated temperatures). At high frequencies, the power was averaged over logarithmically spaced frequency bins to decrease statistical scatter (Fraedrich & Larnder, 1993).

[30] In the multifractal models, we shall see that q_D, H are independent.

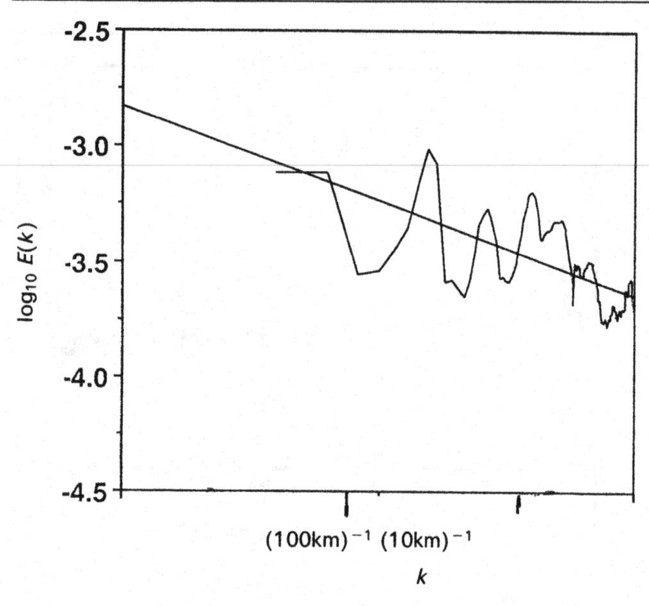

Fig. 4e Horizontal spectrum of tropical Atlantic (GATE experiment) radar reflectivities for 14 radar scans at 15 minute intervals, each scan with 360 radials (1°), 1 km downrange resolution. The (one dimensional) spectra were taken downrange (over 256 pulse volumes) and averaged over all the radials and scans. The reference line has $\beta = 0.3$. From Tessier *et al.* (1993).

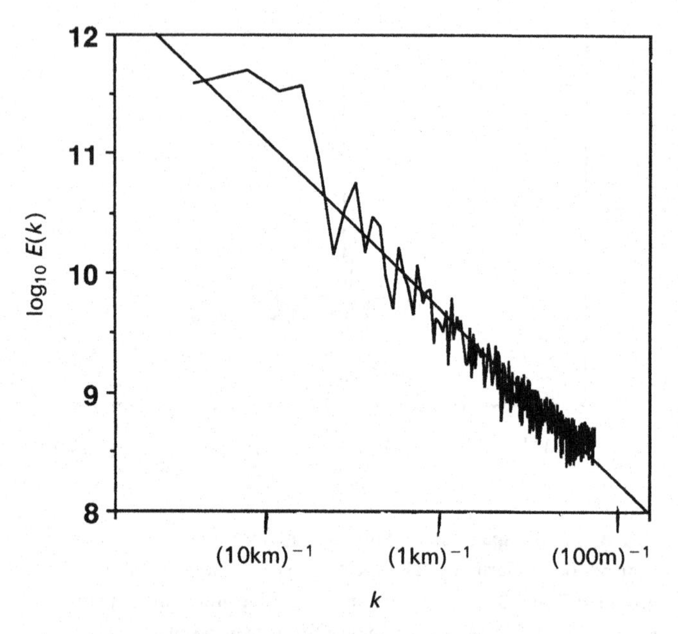

Fig. 4f Horizontal spectrum of McGill radar reflectivities for a 2.2° elevation radar scan with reflectivities averaged over 75 × 75 m grids (PPI). The isotropic two dimensional spectrum was taken over 256 × 256 grid points. The reference line has $\beta = 1.45$, (quite different from GATE). From Tessier *et al.* (1993).

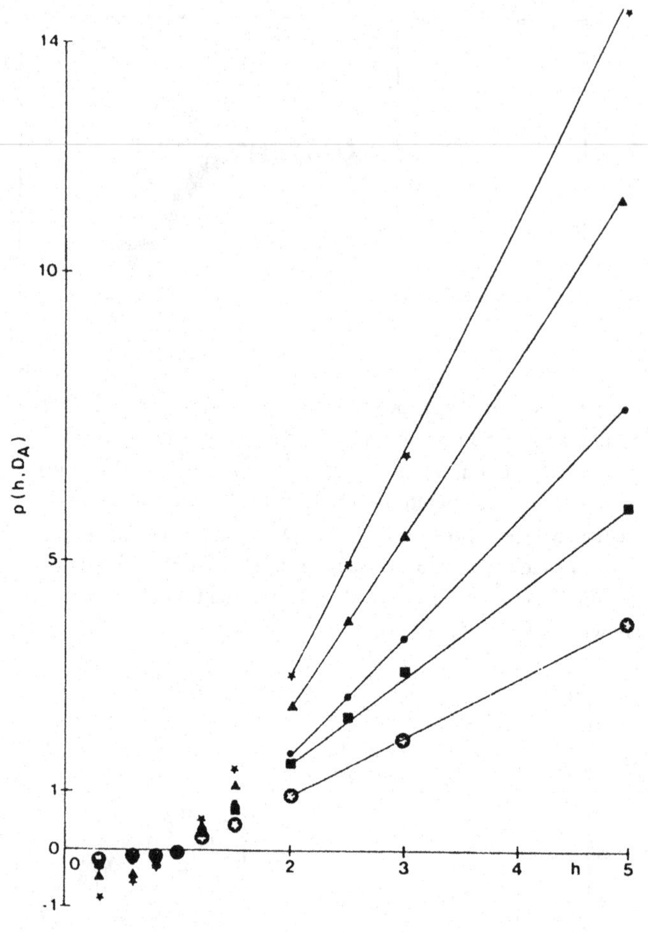

Fig. 5 The function $p(h, D_A)$ $(= K(q), h = q)$ for dressed radar reflectivities from 5 sets of radar CAZLORs at altitudes of 3, 4, 5 km, each set involving 14 scans taken at thirty minute intervals. There were 200 downrange elements, 375 azimuthal elements; the total data set involved $5 \times 3 \times 14 \times 200 \times 375 \approx 1.5 \times 107$ points. The data was dressed (averaged) over sets with various dimensions D_A: over the downrange only (bottom curve), downrange and cross range (third from bottom), downrange, cross-range and in altitude (second from top), space/time (top), as well as over a simulated measuring network, dimension 1.5 (second from the bottom). This curvature clearly shows the multiscaling, multifractal nature of rain. From Schertzer & Lovejoy (1985b).

involved nontrivial difficulties.... The remainder of this paper will concentrate on these developments.

PROPERTIES AND CLASSIFICATION OF MULTIFRACTALS

An explicit multifractal process, the α model

Multifractals arise when cascade processes concentrate energy, water, or other fluxes into smaller and smaller regions. To understand cascades intuitively and to see their relevance to rain, consider the daily rainfall accumulations

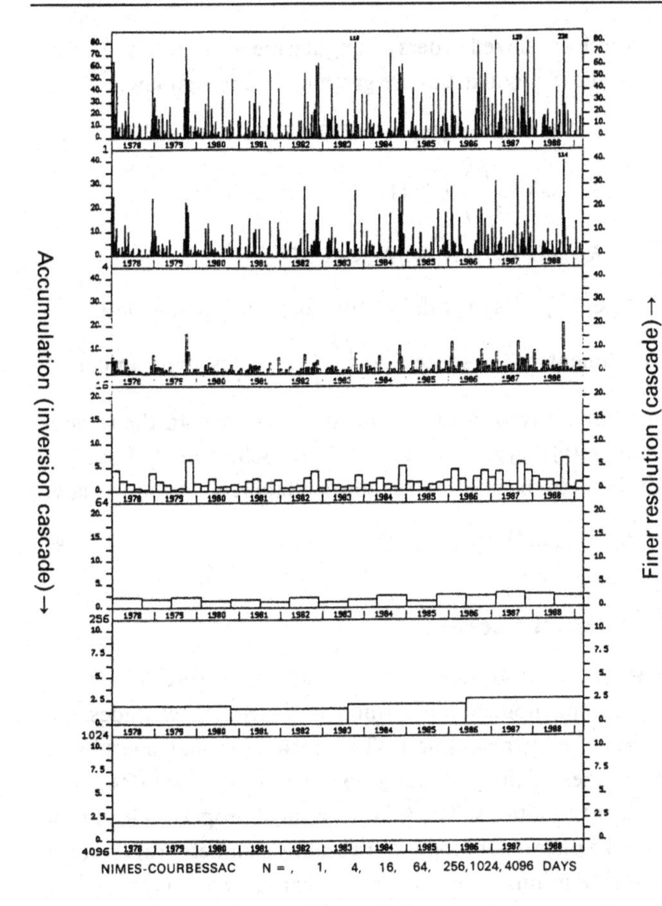

Fig. 6 The 'inverse cascade' produced by averaging daily rainfall from Nimes over longer and longer periods. Accumulation/averaging periods from top to bottom: 1, 4, 16, 64, 256, 1024, 4096 days, 11 years. From Ladoy *et al.* (1993).

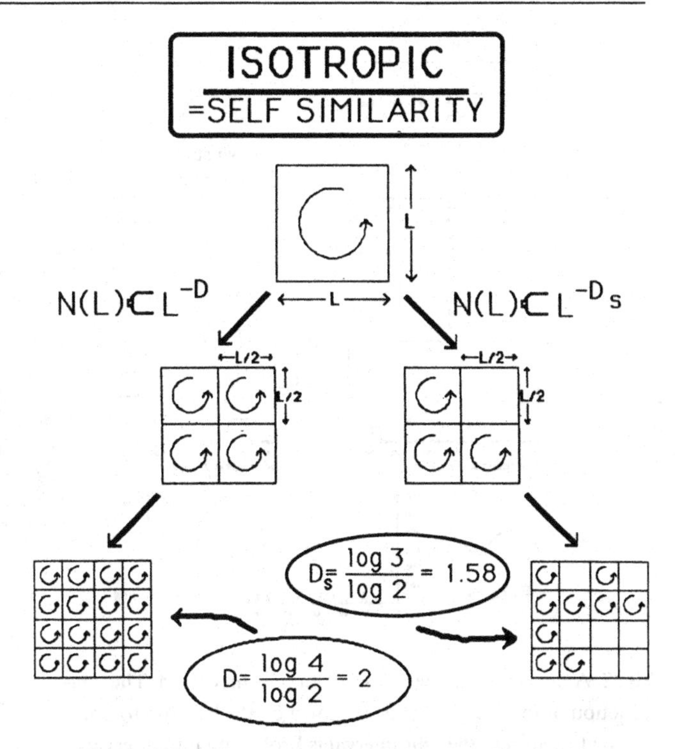

Fig. 7 A schematic diagram showing a two dimensional cascade process at different levels of its construction to smaller scales. Each eddy is broken up into four subeddies, transferring a part or all its energy flux to the sub-eddies. The left hand side shows a homogeneous cascade as originally proposed by Kolmogorov (1941); the nonlinearities simply redistribute energy flux density to smaller scales of motion, the overall density stays uniform. On the right hand side, the β model involving occasional dead eddies (here one in four) is shown, simulating intermittency; it already leads to a monofractal support. From Lovejoy & Schertzer (1986a).

for Nice shown in Fig. 6. Over the 11 year period (1978–88), we can see that several extreme events stand out; in particular, notice the record holder in October 1988 that had a 24 hour accumulation of 228 mm (compared to a mean of ≈ 1.5 mm)[31]. This extreme rainfall event is sufficiently violent that it stands out as the temporal resolution is degraded by averaging the series over four days (second row), 16 days (third row), 64 days (fourth row); even at 256 days (fifth row) its effect is still noticeable (see even ≈ 35 months (sixth row) or the entire 11 year period (bottom row)). We can see that the same type of behaviour is true (although to a lesser degree) of the less extreme 'spikes'. If the analysis sequence high resolution $\Rightarrow$ low resolution is inverted, we have a cascade that can be thought of as a dynamical production process by which rain water is concentrated from a low resolution 'climatological' average value into wet/dry years, wet/dry seasons, months, weeks, days etc. Since the lifetime of atmospheric structures including storms depends on their spatial scale[32], the actual cascade is a space/time process with analogous mechanisms concentrating water fluxes into smaller and smaller regions of space, yielding the observed high spatial variability.

As an example of the inverse low $\Rightarrow$ high resolution process that corresponds to the actual dynamics, consider a cascade produced by dividing the 11 year period with initial rainrate $R_1 = 1$ into sub-periods each of scale λ^{-1} where $\lambda\,(=2$ here) is the scale ratio (see the schematic diagram Fig. 7, and simulation Fig. 8). The fraction of the rain flux concentrated from a long period (large interval) into one of its sub-intervals is given by independent random factors[33] (μR) given by the Bernoulli law shown in equation 2.

[31] This extreme behavior is quite typical; using tipping bucket gages, Hubert & Carbonnel (1989) have even determined in the Sahel that half of the yearly rainfall occurs in less than three hours!

[32] The time scale of structures of global extent seems to be of the order of two to three weeks; in temperature series, it is associated with a spectral break called the 'synoptic maximum'; see Koleshnikova & Monin (1965), Lovejoy & Schertzer (1986b). There is some evidence of this in Figs. 4d, see also Fig. 23c, d.

[33] These multiplicative 'increments' are denoted 'μ' in analogy with the 'Δ' used for the usual increments in additive processes.

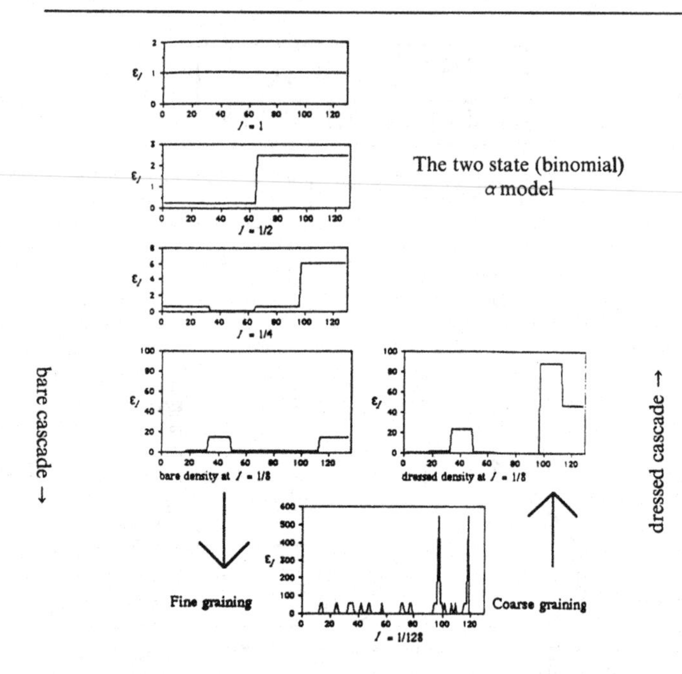

Fig. 8 A discrete (α model) cascade in one dimension. The construction of the 'bare' cascade is shown on the left (top to bottom), at each step, the unit interval is broken up into intervals half the size of the previous step and the energy flux density (vertical axis) is multiplied by a random factor. In the α model, there are only two possibilities – a boost or a decrease with probabilities chosen to respect ensemble conservation $\langle \varepsilon_l \rangle = 1$. As the scale ratio increases, the flux density is increasingly dominated by a few large spikes, the singularities. The right hand side shows the corresponding 'dressed' cascade obtained by averaging over the corresponding scale. The dressed cascade is more variable due to the high resolution modulations. From Schertzer & Lovejoy (1989a).

$$\Pr(\mu R = \lambda^{\gamma+}) = \lambda^{-c}, \qquad \Pr(\mu R = \lambda^{\gamma-}) = 1 - \lambda^{-c} \tag{2}$$

The parameters $\gamma+, \gamma-, c$ are usually constrained so that the ensemble average $\langle \mu R \rangle = 1$, $\lambda^{\gamma+} > 1$ $(\gamma^+ > 0)$ corresponds to strong (wet) intervals, $\lambda^{\gamma} < 1$ $(\gamma^- < 0)$ to weak (dry) subintervals. This pedagogical model (Schertzer & Lovejoy, 1983, 1984; Levich & Tzvetkov, 1985; Bialas & Peschansky, 1986; Meneveau & Sreenivasan[34], 1991) was introduced and called the 'α model' because of the divergence of moment exponent[35] α it introduced (in the notation used here, the corresponding divergence parameter is q_D). The dead/alive β model[36] is recovered with $\gamma^- = -\infty$, $\gamma^+ = c$, c being the codimension of the support $(= D - D_S$, D is the dimension of space in which the cascade occurs). As the cascade proceeds, the pure orders of singularities γ^-, γ^+ yield an infinite hierarchy of mixed orders of singularities $(\gamma^- < \gamma < \gamma^+)$, after steps these singularities are given by a binomial law:

$$\gamma = (n^+\gamma^+ + n^-\gamma^-)/n; \qquad n^+ + n^- = n$$

$$\Pr(n^+ = k) = \binom{k}{n} \lambda^{-ck}(1 - \lambda^{-c})^{n-k}$$

$$\Pr(R_{\lambda^n} \geq (\lambda^n)^\gamma) \approx N_n(\gamma)/N_n \approx (\lambda^n)^{-c_n(\gamma)} \tag{3}$$

$N_n \approx (\lambda^n)^{-D}$ is the total number of intervals at scale λ^{-n}, D the dimension of space and $\binom{k}{n}$ indicates the number of combinations of n objects taken k at a time. In the large n limit, $c_n(\gamma) \approx c(\gamma)$, and we are lead (Schertzer & Lovejoy, 1987a, b) to the multiple scaling probability distribution law:

$$\Pr(R_{\lambda^n} \geq (\lambda^n)^\gamma) \approx (\lambda^n)^{-c_n(\gamma)} \tag{4}$$

Multifractal processes

The multifractal processes discussed here were first developed as phenomenological models of turbulent cascades, the α model being the simplest. They are designed to respect basic properties of the governing nonlinear dynamical ('Navier–Stokes') equations. The following three properties lead to a cascade phenomenology[37]: a) a scaling symmetry (invariance under dilations, 'zooms'), b) a quantity conserved by the cascade (energy fluxes from large to small scale), c) localness in Fourier space (the dynamics are most effective between neighbouring scales: direct transfer of energy from large to small scale structures is inefficient). Cascade models are relevant in the atmosphere in general and in rain and hydrology in particular since (as argued in Schertzer & Lovejoy, 1987a), although the full nonlinear partial differential equations governing the atmosphere will be more complex than those of hydrodynamic turbulence, they are nonetheless still likely to respect properties a, b, c. In other words we expect the complete dynamics to involve coupled cascades. There are now a whole series of phenomenological models: the 'pulse in pulse' model (Novikov & Stewart, 1964), the 'lognormal' model (Kolmogorov, 1962; Obukhov, 1962; Yaglom, 1966), 'weighted curdling' (Mandelbrot, 1974), the 'β model' (Frisch *et al.*, 1978), 'the α model' (Schertzer & Lovejoy, 1983b, 1985a), the 'random β model' (Benzi *et al.*, 1984), the '*p* model'[38] (Meneveau & Sreenivasan, 1987) and the 'continuous' and 'universal' cascade models (Schertzer & Lovejoy, 1987a, b). It is now clear that scale invariant multiplicative processes generically yield multifractals and – due to the existence of stable and attractive multifractal generators – to universal multifractals in which many details of the dynamics are unimportant. These results are important in hydrology and geophysics since they show

[34] Although it was never intended to be more than pedagogical, these authors attempt a detailed comparison with turbulence data.

[35] The choice α for this exponent seemed natural at the time since it generalized the Lévy exponent α.

[36] This monofractal model was studied in various slightly different forms at different times (Novikov & Stewart, 1964; Mandelbrot, 1974; Frisch *et al.*, 1978), the parameter $\beta = \lambda^{-c}$ in the notation used here.

[37] First proposed by Richardson (1922) in his now celebrated poem.

[38] This is a microcanonical version of the α model.

that while geometrical fractals are sufficient to study many aspects of scaling sets, that multifractals (with their statistical exponents) provide the general framework for scaling fields (measures). In models of hydrodynamic turbulence, the energy flux ε from large to small scales is conserved (i.e. its ensemble average $\langle\varepsilon\rangle$ is independent of scale), therefore it is the basic cascade quantity. Directly observable fields such as the velocity shear (Δv_l) for two points separated by distance l are related to the energy flux via dimensional arguments:[39]

$$\Delta v_l \approx \varepsilon_l^{1/3} l^{1/3} \tag{5}$$

This equation should be understood statistically. A straightforward interpretation useful in modelling is to view the scaling $l^{1/3}$ as a power law filter ($k^{-1/3}$, fractional integral) of $\varepsilon_l^{1/3}$ (Schertzer & Lovejoy, 1987a; Wilson, 1991; Wilson *et al.*, 1991).

In contrast to the well studied case of hydrodynamic turbulence, the dynamical equations responsible for the distribution of rain and cloud radiances are not known;[40] the best we can do now is to speculate on the appropriate fundamental dynamical quantities analogous[41] to ε. Since *a priori*, there is no obvious reason the rainrate or cloud radiance fields themselves should be conservative, in analogy with turbulence, we introduce a fundamental field φ_l that has the conservation property $\langle\varphi_l\rangle$ = constant (independent of scale). The observable (nonconserved) rainfall (or cloud radiance) fluctuations (ΔR_l) is then given by:

$$\Delta R_l \approx \varphi_l^a l^H \tag{6}$$

Since we have yet no proper dynamical theory for rain, we do not know the appropriate fields φ_l nor the corresponding values of a. We shall see that changing a essentially corresponds to changing C_1 defined below. Therefore, the scaling parameter H has a straightforward interpretation: it specifies how far the measured field R is from the conserved field φ: $\langle|\Delta R_l|\rangle \approx l^H$. H therefore specifies the exponent of the power law filter (the order of fractional integration) required to obtain R from φ.

Basic properties of multifractal fields

We now focus our attention on the conserved quantity φ_l. Early scaling ideas were associated with additive (linear) processes, and unique scaling exponents H (which – only in these special cases) were related to unique fractal dimensions by simple formulae. The properties of φ_l were more straightforward, and were usually understood implicitly. We have already discussed 'simple scaling'. This is a special case of equation 6 in which φ_l is simply a scale invariant noise ($\langle\varphi_l^q\rangle$ are all constants, independent of scale).

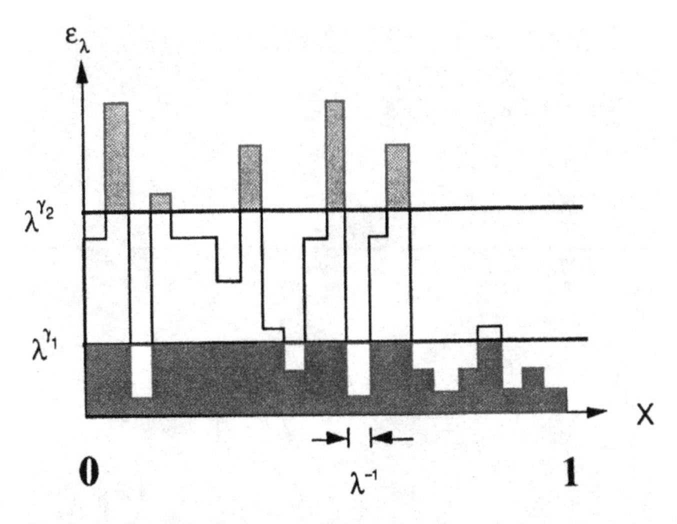

Fig. 9 A schematic diagram showing a multifractal energy flux density with smallest resolution λ^{-1}, and indicating the exceedance sets corresponding to two orders of singularities, γ_1, γ_2. From Tessier *et al.* (1993).

Turning our attention to (nonlinear) multiplicative processes we can consider some properties of φ that will generically result from cascades. We have already discussed the example of the α model, including the form of the probability distribution after n cascade steps. In fact, denoting the entire range of scales from the largest to smallest by λ, and considering the cascade of φ (rather than of R directly), we obtain the following general multifractal relation:

$$\Pr(\varphi_\lambda \geq \lambda^\gamma) \approx \lambda^{-c(\gamma)} \tag{7}$$

(equality is to within slowly varying functions of λ such as logs). $c(\gamma)$ is therefore the (statistical) scaling exponent of the probability distribution (see Fig. 9 for an illustration). However, when the process is observed on a low dimensional cut of dimension D (such as the $D=2$ dimensional simulation shown in Fig. 10) it can often be given a simple geometric interpretation. When $D > c(\gamma)$, we may introduce the (positive) dimension function $D(\gamma) = D - c(\gamma)$ which is the set with singularities γ.

This geometric interpretation can be useful in data analysis. For example, consider a data set consisting of N_s radar scans (assumed to be statistically independent realizations from the same statistical ensemble). A single D dimensional scan ($D=2$ in this example) will enable us to explore structures with dimension $D \geq D(\gamma) \geq 0$; structures with $c(\gamma) > D$ (which would correspond to impossible negative[42] values of

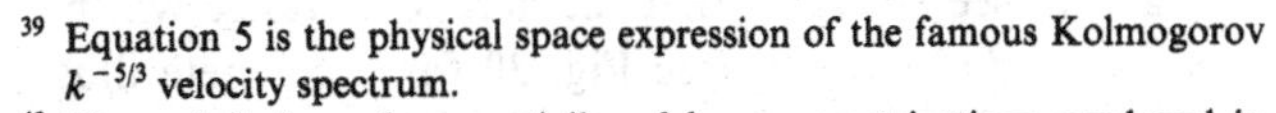

[39] Equation 5 is the physical space expression of the famous Kolmogorov $k^{-5/3}$ velocity spectrum.

[40] We exclude here the essentially *ad hoc* parametrizations employed in numerical cloud and weather models.

[41] These will be various conserved fluxes such as the humidity variance flux and bouyancy force variance flux.

[42] Mandelbrot (1984) introduced the expression 'latent' for these nonstandard dimensions. If the (intrinsic) codimensions are used, this artificial problem is entirely avoided.

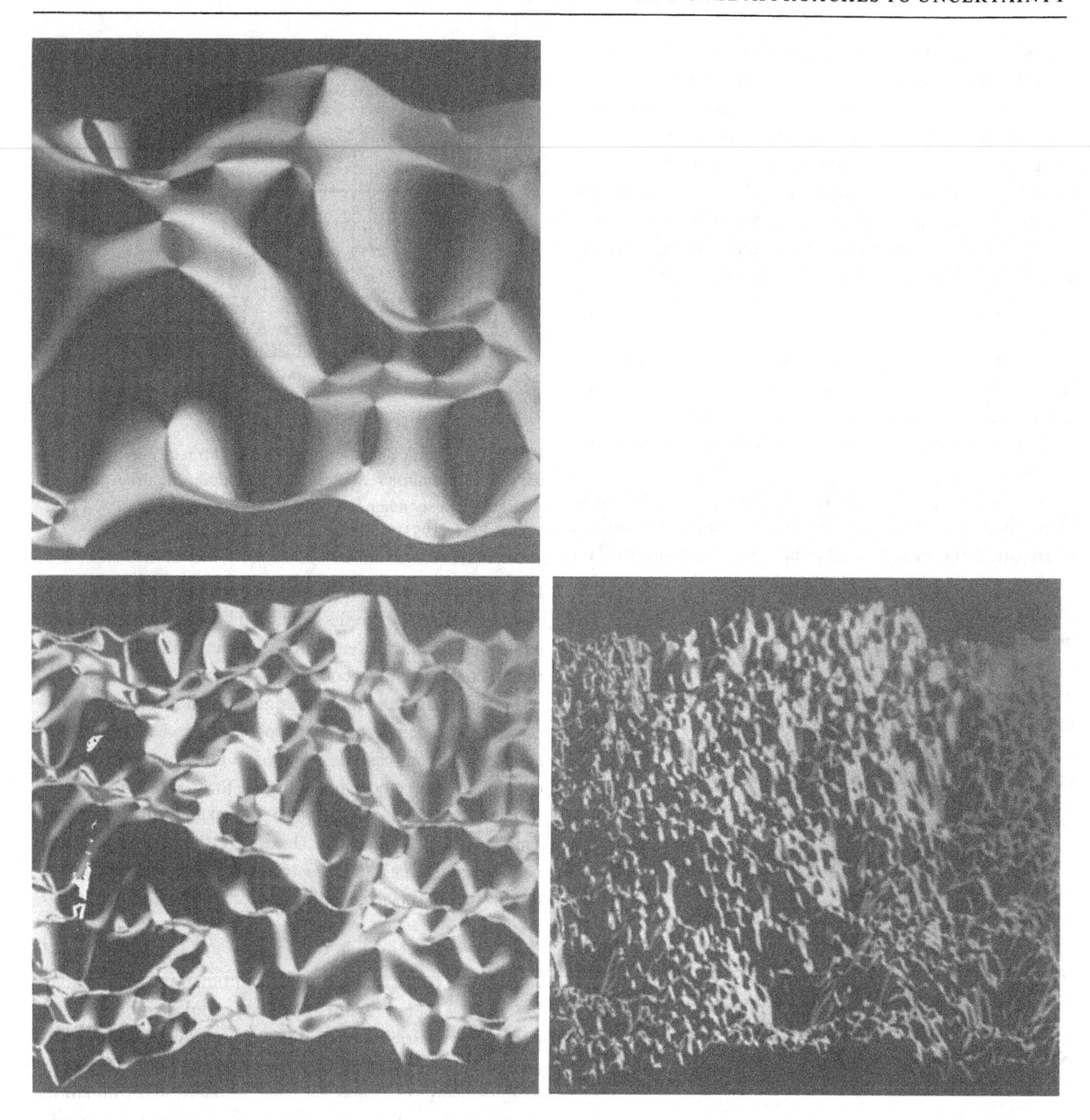

Fig. 10 Successive stages in the construction of a universal multifractal temperature field shown with resolution increasing by factors of four, counterclockwise from the upper left corner. The temperature is represented as a surface with false colours, incipient singularities (the high valued spikes) and associated 'Levy holes' are particularly evident in the low resolution image in the design. The parameters used for the simulation were those estimated from atmospheric measurements analysed in Schmitt *et al.* (1992), i.e. $\alpha = 1.3$, $C_1 = 0.5$, $H = \frac{1}{3}$. From Lovejoy & Schertzer (1991c).

$D(\gamma)$) will be too sparse to be observed (they will almost surely not be present on a given realization). This restriction on the accessible values of $c(\gamma)$ is shown in Fig. 11; to explore more of the probability space, we will require many scans. With N_s scans, the accessible range of singularities can readily be estimated. If each scan has a range of scales λ (= the ratio of the size of the picture to the smallest resolution = the number of 'pixels' on a side), then we can introduce the 'sampling dimension' (Schertzer & Lovejoy, 1989a; Lavallée, 1991; Lavallée *et al.*, 1991a): $D_s = \log N_s / \log \lambda$. It is not hard to see (Fig. 12) that the accessible range will be $\gamma < \gamma_s$, with $c(\gamma_\tau) = \Delta + \Delta_s$ (see Fig. 30 for a concrete illustration in rain).

$c(\gamma)$ has many other properties that are illustrated graphi-

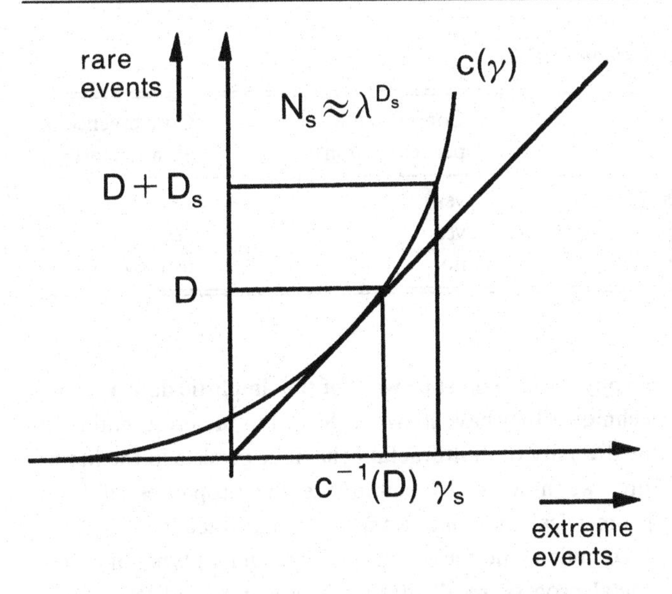

Fig. 11 A schematic diagram showing a typical codimension function for a conserved process ($H=0$). The lines $c(\gamma)=D$, $\gamma=C^{-1}(D)$ indicate the limits of the accessible range of singularities for a single realization, dimension D. The corresponding lines for $D+D_s$, where D_s is the sampling dimension, are also shown. As we analyse more and more samples, we explore a larger and larger fraction of the probability space of the process, hence finding more and more extreme (are rare) singularities. From Tessier *et al.* (1993).

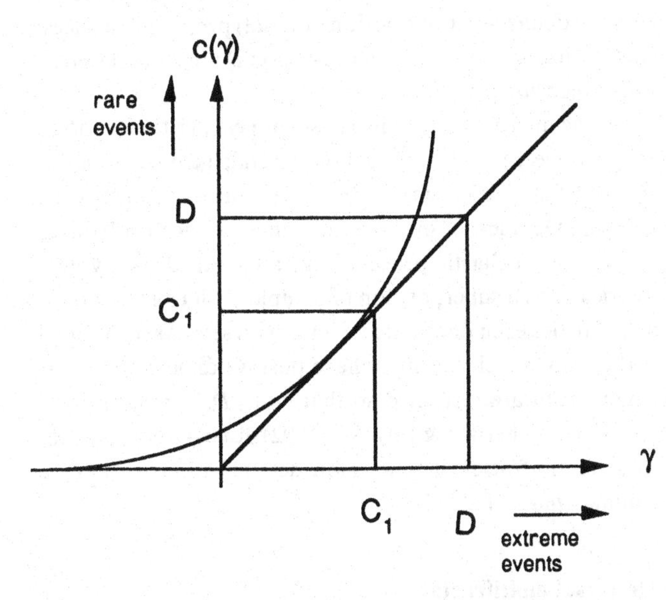

Fig. 12 Same as previous, but showing the fixed point $C_1=c(C_1)$, (with $0 \leq C_1 \leq D$) the singularity corresponding to the mean of the process. The diagonal line is the bisectrix ($\gamma=c(\gamma)$). From Tessier *et al.* (1993).

cally. A fundamental property which is readily derived by considering statistical moments (below), is that it must be convex. It must also satisfy the fixed point relation $C_1=c(C_1)$ as indicated in Fig. 12. C_1 is thus the codimension of the mean process; if the process is observed on a space of dimension D, it must satisfy $D \geq C_1$, otherwise, following the above, the

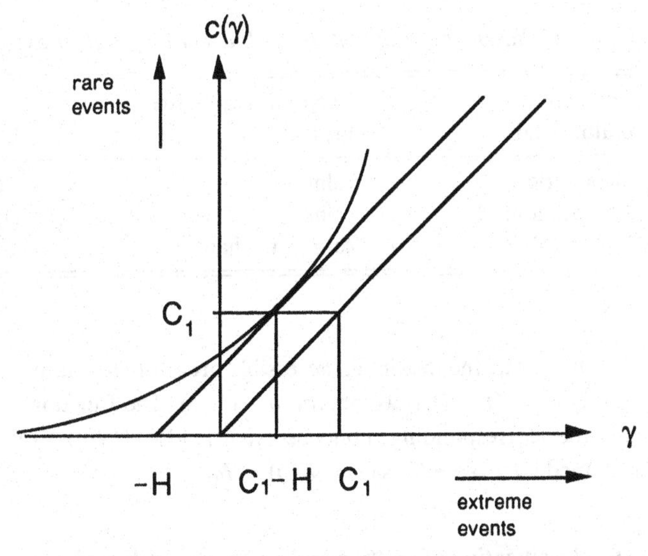

Fig. 13 Same as 11, but for a nonconserved process. All the singularities are shifted by $-H$. From Tessier *et al.* (1993).

mean will be so sparse that the process will (almost surely) be zero everywhere; it will be 'degenerate'. We can also consider the (nonconserved) ΔR_λ; it is obtained from φ_λ by multiplication by λ^{-H}, since $\varphi_\lambda=\lambda^{\gamma}$, we have $\Delta R_\lambda=\lambda^{\gamma-H}$; i.e. by the translation of singularities by $-H$ (see Fig. 13).

Rather than specifying the statistical properties via the scaling of probabilities $c(\gamma)$ can (equivalently) be specified by the scaling of the statistical moments[43]. Consider the qth order statistical moments $\langle \varphi_\lambda^q \rangle$. We can now define the multiple scaling exponent[44] $K(q)$:

$$\langle \varphi_\lambda^q \rangle = \lambda^{K(q)} \tag{8}$$

$K(q)$, $c(\gamma)$ are related by the following Legendre transformations (Parisi & Frisch, 1985):

$$K(q)=\max_\gamma (q\gamma - c(\gamma)); \; c(\gamma)=\max_q (q\gamma - K(q)) \tag{9}$$

which relate points on the $c(\gamma)$ function to tangencies on the $K(q)$ function and visa versa; $\gamma=K'(q)$, $q=c'(\gamma)$. For example, a quantity which will be useful below in estimating the multifractal parameters of radiances and reflectivities is the sampling moment q_s which is the maximum order moment that can be accurately estimated with a finite sample.

[43] Gupta & Waymire (1990) have introduced the idea of multiple scaling of random variables rather than fields/measures ('GW multiscaling'): in other words a multiple scaling without the notion of scales. In GW multiscaling there are no multifractals and there is no 'hard' behavior (see below).

[44] The turbulent codimension notation $c(\gamma)$ and $K(q)$ is related to the '$f(\alpha)$' dimension notation (Halsey *et al.*, 1986) by the following: $\alpha=(D-\gamma)$, $f(\alpha)=D-c(\gamma)$, $\tau(q)=(q-1)D-K(q)$. Because the dimension notation fundamentally depends on the dimension of the observing space D; it cannot be used in stochastic processes such as those of interest here where we deal with infinite dimensional probability spaces, $D \Rightarrow \infty$. The dimension notation is useful for multifractal probability measures; in turbulence, we deal with spatial measures ε which do not reduce to probability measures.

Table 2. *Classification of multifractals according to their extreme singularities*

Type of multifractal	Types of singularities present	Localized?	Conservation per realization?	Convergence of all moments?
Geometric	calm	yes	yes	yes
Microcanonical	calm	no	yes	yes
Canonical	calm, wild, hard	no	no	usually no

Recalling that the maximum accessible order of singularity was $\gamma_s = c^{-1}(D + D_s)$, we obtain: $q_s = c'(\gamma_s)$. The functions for the corresponding nonconserved fields ($H \neq 0$) are obtained by $\gamma \Rightarrow \gamma - H$, $K(q) \Rightarrow K(q) - Hq$.

The classification of multifractals: nonlocal, wild and hard multifractals, multifractal phase transitions

We now discuss various different types of multifractals. To this end, we must first make a distinction between the 'bare' and 'dressed' multifractal properties (Schertzer & Lovejoy, 1987a, b). The 'bare' properties are those which have been discussed above, they correspond to the construction of the process over a finite range of scales λ. In contrast, the 'dressed' quantities (see the right hand side of Fig. 8) are obtained by integrating (averaging) a completed cascade over the corresponding scale. Experimentally measured quantities are generally 'dressed' since geophysical sensors typically have resolutions which are much lower than the smallest structures in the fields they are measuring (which in the atmosphere, is typically of the order of 1 mm or less). The dressed quantities will generally display an extreme, 'hard' behavior involving divergence of high order statistical moments. Specifically, for spatial averages over observing sets with dimension D there is a critical order moment q_D (and corresponding order of singularity $\gamma_D = K'(q_D)$) such that:

$$\langle \varphi_\lambda^q \rangle = \infty \qquad q \geq q_D \tag{10}$$

where q_D is given by the following equation:

$$K(q_D) = (q_D - 1)\, D \tag{11}$$

The associated qualitative change of behaviour at q_D (or γ_D) can be considered a multifractal phase transition (Schertzer *et al.*, 1993). Unfortunately, these general multiplicative processes with their corresponding hard behaviour have received relatively little attention in the literature; it is much more usual to introduce various constraints which have the effect of severely limiting the occurence of extreme events. While these restrictions lead to simplifications in the theoretical treatment which are justified when studying strange attractors, they are too restrictive to be appropriate in geophysics; one must be wary of the simplistic data analysis techniques they have spawned. Since this underestimation of the diversity of multifractal behaviour persists in the literature, we now briefly summarize the properties of both 'geometrical' and 'microcanonical' multifractals.

To understand the corresponding different types of multifractal process, recall that we have considered 'canonical' multifractals subject only to the weak constraint of conservation of φ only over the entire statistical ensemble, individual realizations will not be conserved. If on the contrary, we impose the much stronger constraint of conservation on each realization, then large fluctuations are suppressed and we obtain a 'microcanonical' process. Specifically, we find that 'wild' singularities with $\gamma > D$ are supressed. Both canonical and microcanonical multifractals are stochastic processes, they are defined on (infinite dimensional) probability spaces: each realization in a space of dimension D must be viewed as low-dimensional cuts.

Just as microcanonical processes are calmer than canonical processes, another type of multifractal; 'geometric' multifractals (Parisi & Frisch, 1985) can be defined which are even calmer. Geometric multifractals involve no probability space, nor stochastic process; they are defined purely geometrically as a superposition of completely localized (point) singularities each distributed over fractal sets. As mentioned earlier, since such sets must have positive dimensions, their singularities are restricted so that $c(\gamma) \leq D$. Schertzer *et al.* (1991) and Schertzer & Lovejoy (1992) discuss this classification of multifractals in much more detail; their properties are summarized in Table 2.

Universal multifractals

The above discussion is quite general and at this level, it has the unpleasant consequence that an infinite number of scaling parameters (the entire $c(\gamma)$, $K(q)$ functions) will be required to fully specify the multiple scaling of our field. Fortunately, real physical processes will typically involve both nonlinear 'mixing' (Schertzer *et al.*, 1991) of different multifractal processes, as well as a 'densification' (Schertzer & Lovejoy, 1987a, b) of the process leading to the dynamical excitation of intermediate scales. Rather than just the discrete scales (factors of 2) indicated in Figs. 7 and 8, there is the continuum indicated in 10. Either mixing or densification are sufficient[45] so that we obtain the following (bare[46]) universal[47] multifractal functions[48]:

$$c(\gamma-H)=C_1[\gamma/(C_1\alpha')+1/\alpha']\alpha'; \quad \alpha\neq 1 \tag{12}$$
$$c(\gamma-H)=C_1\exp[(\gamma/C_1)-1] \quad \alpha=1$$

$$K(q)-qH=\begin{cases}\frac{C_1}{\alpha-1}(q^\alpha-q) & \alpha\neq 1\\ C_1\, q\log(q) & \alpha=1\end{cases} \quad (\text{for } \alpha<2, q\geq 0)$$
$$\frac{1}{\alpha}+\frac{1}{\alpha'}=1 \tag{13}$$

The multifractality parameter α is the Lévy index and indicates the class to which the probability distribution belongs[49]. There are actually 5 qualitatively different cases. The case $\alpha=2$ corresponds to multifractals with Gaussian generators[50], the case $1<\alpha\leq 2$ corresponds to multifractal processes with Lévy generators and unbounded singularities, $\alpha=1$ corresponds to multifractals with Cauchy generators. These three cases are all 'unconditionally hard' multifractals, since for any D, divergence of moments will occur for large enough q (q_D is always finite). When $0<\alpha<1$ we have multifractal processes with Lévy generators and bounded singularities. By integrating (smoothing) such multifractals over an observing set with large enough dimension D it is possible to tame all the divergences yielding 'soft' behavior, these multifractals are only conditionally 'hard'. Finally[51] $\alpha=0$ corresponds to the monofractal 'β model'. Universal multifractals have been empirically found in both turbulent temperature and wind data (Schertzer *et al.*, 1991a; Schmitt *et al.*, 1992; Kida, 1991). They have also have recently found applications in high energy physics (Brax & Pechanski, 1991; Ratti, 1991; Ratti *et al.*, 1994), oceanography (Lavallée *et al.*, 1991b), topography (Lavallée *et al.*, 1993), as well as the low frequency component of the human voice (Larnder *et al.*, 1992). The first empirical estimates[52] of C_1, α in cloud radiances[53] are discussed in Lovejoy & Schertzer, 1990 (see also Gabriel *et al.*, 1988 for the first test of universality in an empirical data set[54]).

It is interesting to note here that the probability distributions associated with the various (bare) universality classes are respectively lognormal ($\alpha=2$), and log-Lévy ($\alpha<2$). The latter are in turn approximately log-normal since, with the exception of their extreme tails, Lévy distributions are themselves nearly normal (this 'tail' is pushed to lower and lower probability levels as $\alpha\to 2$). The multifractal nature of rain is therefore quite in accord with the widespread hydrological, meteorological (and generally geophysical) lognormal phenomenology. Of particular relevance here are numerous studies that have claimed that rainrates, cloud and radar echo sizes, heights and lifetimes, as well as total rain output from storms over their lifetimes are either log-normal or 'truncated log-normal' distributions (Biondini, 1976; Lopez, 1976, 1977a; Drufuca, 1977; Houze & Cheng, 1977; Konrad, 1978; Warner & Austin, 1978 etc.). Furthermore, the cascade models that generate them are actually just concrete implementations of vague laws of 'proportionate effects' (see Lopez, 1977a, b for an invocation of this law in the rain context). Shifting our attention to the dressed quantites, the above statement still holds for (nonextreme) fluctuations (up to γ_D, q_D), but will (drastically) underestimate the frequency of occurence of extreme events ($\gamma>\gamma_D$, $q>q_D$).

Using the universal multifractal formulae above, some of the results discussed earlier may be expressed in simpler form. Formulae which will prove useful below are for the sampling order moment q_s (the maximum order moment that can be reliably estimated with a finite sample), and q_D, the critical order for divergence:

$$q_s=[(D+D_s)/C_1]^{1/\alpha} \tag{14a}$$
$$(\alpha-1)/C_1[(q_D^\alpha-q_D)/(q_D-1)]=D \tag{14b}$$

For $q>q_c=\min(q_s,q_D)$, $K(q)$ will be linear, for $\gamma>\gamma_D$ $c(\gamma)$ will also be linear:

[45] This applies only to canonical multifractals; there seems to be no corresponding universality for geometric or microcanonical multifractals.

[46] The corresponding dressed functions are the same only for $\gamma<\gamma_D$, and $q<q_D$; for finite sample sizes, they becoming linear for larger γ, q.

[47] The problem of universality was for some time obscured by the exclusive study of (nonuniversal) discrete cascades in which the limits of more and more random variables and smaller and smaller scale structures were confounded (both limits occured simultaneously as the number of discrete steps approached infinity). On the contrary, universality results when more and more random variables are involved within a fixed and finite range of scales. The limit of the range of scales approaching infinity (the small scale limit) is taken only later. An example of the widespread anti-universality predjudice is the recent statement by Mandelbrot (1989): '... in the strict sense there is no universality whatsoever ... this fact about multifractals is very significant in their theory and must be recognized ...' (*op cit*, p. 16).

[48] These formulae (with $H=0$) first appeared in Schertzer & Lovejoy (1987a, Appendix C). Recently, in the special case $H=0$, Kida (1991), Brax & Peschanski (1991) have obtained equivalent formulae using different notations. They use the expressions 'log stable' and 'log Lévy' multifractals respectively. These terms are somewhat inaccurate since due to the dressing problem, the distributions will only be approximately log stable or log Lévy.

[49] Similar looking formulae (but for random variables, not multifractal measures) can be obtained in GW multiple scaling, Gupta & Waymire (1990).

[50] This is nearly the same as the lognormal multiscaling model of turbulence proposed by Kolmogorov (1962), Obhukhov (1962), except that the latter missed the essential point about the divergence of high order moments, thinking in terms of pointwise processes.

[51] A more detailed discussion about these fives cases and in particular about the generators of the Lévy variables can be found in Schertzer *et al.*, 1988; Fan 1989; and Schertzer & Lovejoy 1989a; see also Lovejoy & Schertzer (1990a, b, 1991a, b) for some applications and review.

[52] Recent (greatly improved) analyses indicate that the original estimates of α were not too accurate. See Tessier *et al.*, 1992, and below.

[53] For theoretical discussion of multifractal clouds and their associated radiance fields, see Lovejoy *et al.* (1990), Gabriel *et al.* (1990), Davis *et al.* (1990, 1991a, b).

[54] Only the hypothesis $\alpha=2$ was tested.

$$K(q)=q\gamma_{d,s}-c(\gamma_{d,s}) \qquad q>q_c \tag{15a}$$

$$c(\gamma)=\gamma q_D-K(q_D) \qquad \gamma>\gamma_D \tag{15b}$$

where $\gamma_{d,s}$ is the highest order dressed singularity present in the sample.

MULTIFRACTAL ANALYSES OF RAIN

Trace moment analyses

Soon after the discovery of multifractals, it was realized that radar rain data would provide ideal testing grounds for multifractal theories as well as data analysis techniques[55]. A whole series of new multifractal analysis techniques (trace moments, functional box-counting, elliptical dimensional sampling), were developed and tested for the first time on rain data. In this section we first discuss what might be viewed as first generation multifractal analysis techniques: methods that can be applied (with various limitations) to general multifractals. These methods are the multifractal analogues of the nonparametric methods of standard statistics. Further we indicate how a second generation of techniques can be be developed which explicitly exploit the existence of universality classes. These are the analogues of parametric statistics, and just as parametric statistical methods have more statistical power than nonparametric methods, the specific (universal) multifractal analysis techniques (when applicable) will lead to much more accurate multifractal characterizations. All these techniques are essentially experimental in the sense that no proper goodness of fit statistics are known; at the moment, confidence in the results of analyses can be obtained primarily by comparing the results of different and complementary methods as well as by extensively testing the analysis on numerical simulations.

The first multifractal rain analyses were performed on radar volume scans of rain from the McGill radar weather observatory[56] (Schertzer & Lovejoy, 1985b, 1987a, see Figs. 5, 15). Volume scans are made every 5 minutes, at 200 ranges (r) and 375 azimuthal (θ) and 13 elevation angles. In the trace moment analysis described here, data were resampled in the vertical onto constant altitude projections ('CAZLORs') at 3 levels (3, 4, 5 km altitudes, z) at 30 minute intervals in time. The analysis was performed using the data to estimate the trace moments. In this technique, the data are systematically degraded in resolution, average reflectivities being calculated over grids whose resolution is successively doubled, the resulting spatial averages are then raised to a series of powers q and the result averaged over each image, and then over many realizations.

To give a formal definition, consider the conserved ($H=0$)

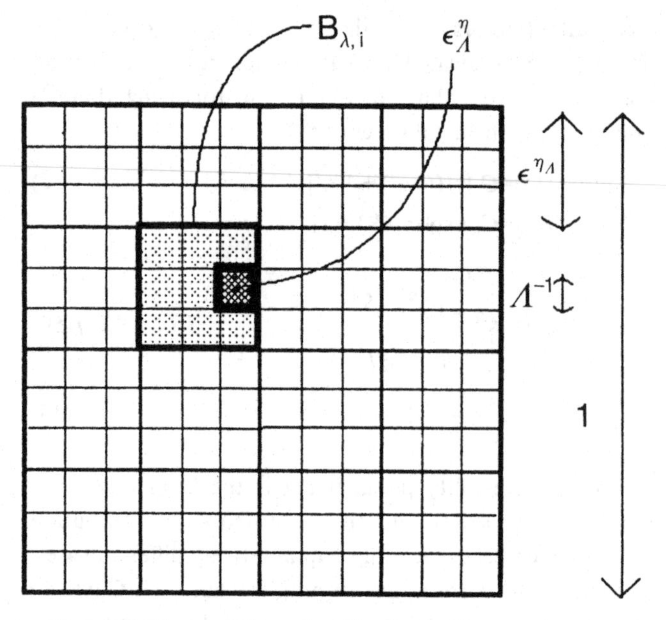

Fig. 14 A schematic diagram illustrating the different averaging scales used in the double trace moment technique, the single trace moment is obtained by taking $\eta=1$. The idea is straightforward; at the highest available resolution (λ') various powers (η) are taken. They are then degraded to an intermediate resolution (λ) by averaging, finally the qth power of the result is averaged over all the data sets. From Tessier *et al.*, 1993.

multifractal flux density at (fine) resolution Λ' (the ratio of the outer (largest) scale of interest to the smallest scale of homogeneity). The (dressed) flux over an observing set (B_λ, this corresponds to the j-th low resolution 'pixel') with dimension D, (lower) resolution λ ($\lambda<\Lambda'$) is simply an integral over the density:

$$\Pi_\Lambda(B_{\lambda,i})=\int_{B_{\lambda,i}}\varphi_\lambda\, d^Dx \tag{16}$$

We may now define the qth order 'Trace moments' (Schertzer & Lovejoy[57], 1987a) by summing $\Pi_\Lambda^q(B_{\lambda,i})$ over each individual realization[58] (each satellite picture, covering the region A has λ^D disjoint covering sets B_λ which are summed over in equation 16, see the schematic illustration, Fig. 14), and then ensemble averaging over all the realizations:

$$\mathrm{Tr}_\lambda(\varphi_\Lambda^\eta)^q=\left\langle\sum_i \Pi_\Lambda^q(B_{\lambda,i})\right\rangle\approx\lambda^{K(q)-(q-1)D} \tag{17}$$

This formula will break down for moments $q>q_D$, and (when finite samples are used to estimate the ensemble average) when $q>q_s$. Although it allows the determination of $K(q)$ (at least for small enough q), and hence in principle the

[55] Schertzer & Lovejoy (1989b) and Lovejoy & Schertzer (1990b) develop this idea and argue that it is true of many other geophysical fields.

[56] For an analogous analysis of tropical Atlantic radar data see Gupta & Waymire (1990).

[57] Although the formalism above was developed here, essentially the same method was empirically applied to rain in Schertzer & Lovejoy (1985b).

[58] Without the ensemble averaging, we have a partition function, appropriate for analyzing strange attractors.

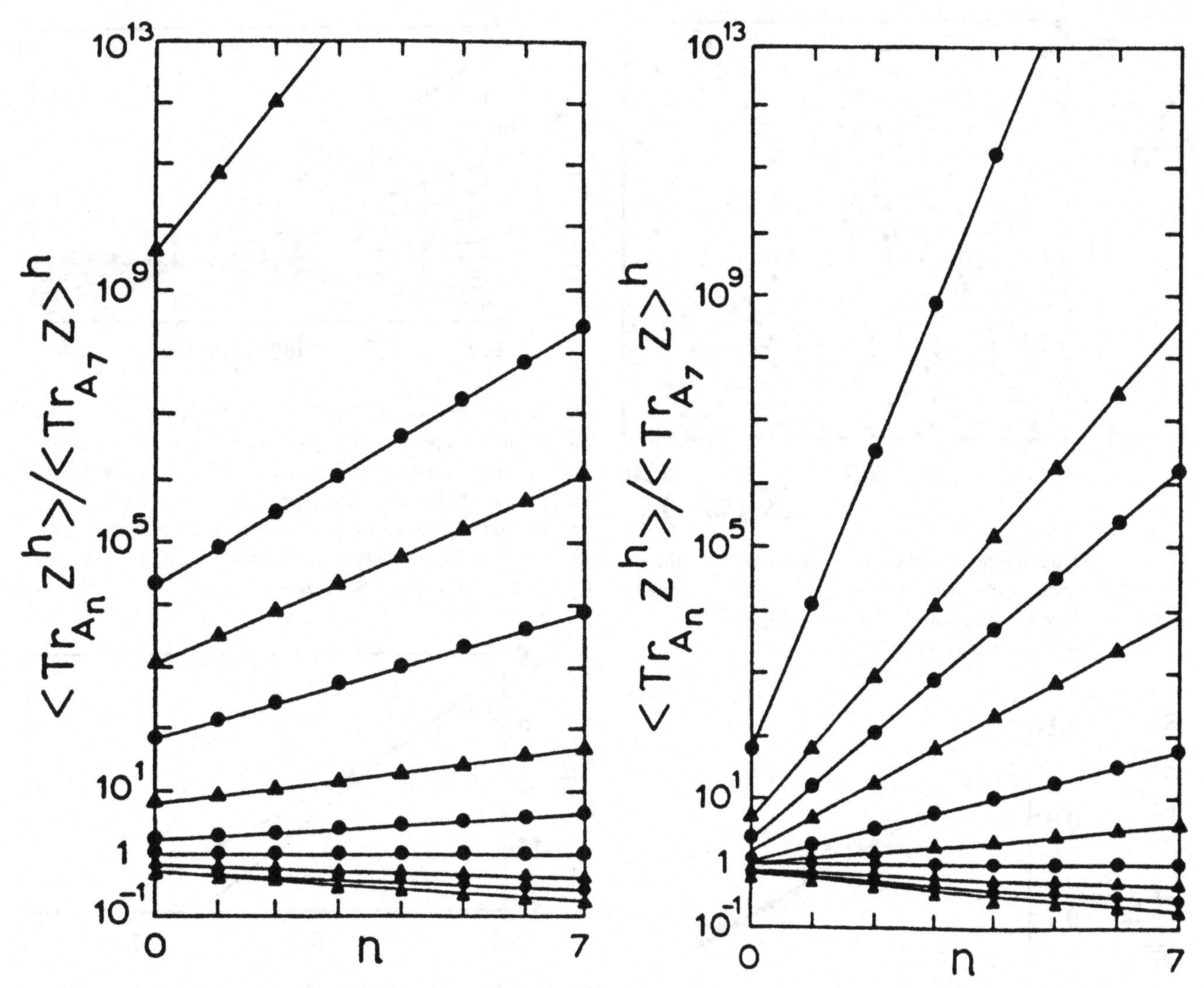

Fig. 15a The hth trace moments (in the notation in the text, $q=h$) estimated from 70 CAZLORs in the horizontal, averaging over a straight line (A is the set of downrange elements used in the averaging), the data is the 3 km altitude subset of that used in Fig. 5. The resolution is $\lambda=2^n$. Lines from top to bottom are for the following values of h: 5, 3, 2.5, 2, 1.5, 1.2, 1., 0.8, 0.6, 0.3. Note that the scaling is extremely accurately followed. From Schertzer & Lovejoy (1987a).

Fig. 15b Same as 15a but for averaging in the horizontal, A is a plane. From Schertzer & Lovejoy (1987a).

determination of C_1, α (via equation 14) this method will involve ill-conditioned nonlinear regressions ($K(q)$ vs. q).

Fig. 15 shows the result using 70 realizations, clearly showing that the multiple scaling is very well respected. The resolution can be degraded along ranges ($D=1$), (r,Θ) simultaneously ($D=2$), (r,Θ,z) or (r,Θ,z,t) simultaneously ($D=3,4$ respectively). Fig. 5 shows the resulting exponents including a 1.5 dimensional case obtained by using simulated fractal measuring networks[59]. The exponents are nearly independent of dimension for low order moments (q), but for $q\geq 1.1$ become increasingly separated, asymptotically tending to straight lines with slopes $\approx D$ for large q. It was argued (Schertzer & Lovejoy, 1987a) that this behaviour could be simply explained since for that data set $q_D\approx 1.1$ (Fig. 2b). Some recent results on the 'pseudo-scaling' (Schertzer & Lovejoy 1983a, 1984) obtained when $q>q_D$ and the relation of this to multifractal 'phase transitions' is discussed in Schertzer *et al.* (1991b).

More recently (Lovejoy & Schertzer, 1991a) trace moments were used to investigate the multiple scaling of rain at scales much smaller than the minimum (≈ 1 km) of the above radar analysis. One of the analyses (Pham & Miville, 1986) was performed on data obtained by rapidly (≈ 1 s) exposing large pieces (128×128 cm) of chemically treated blotting paper to rain, estimating the position and size of the drops. Fig. 16a shows the result of one such exposure, and Fig. 17 shows the resulting trace moment analysis and Fig. 18 the scaling exponent estimates for scales ≥ 8 cm. This analysis indicates that at least down to this scale, rain is multiscaling. The break observed at ≈ 8 cm could be due to

[59] This was close to the dimension estimated for typical gage networks; a better estimate (Lovejoy *et al.*, 1986) is 1.75.

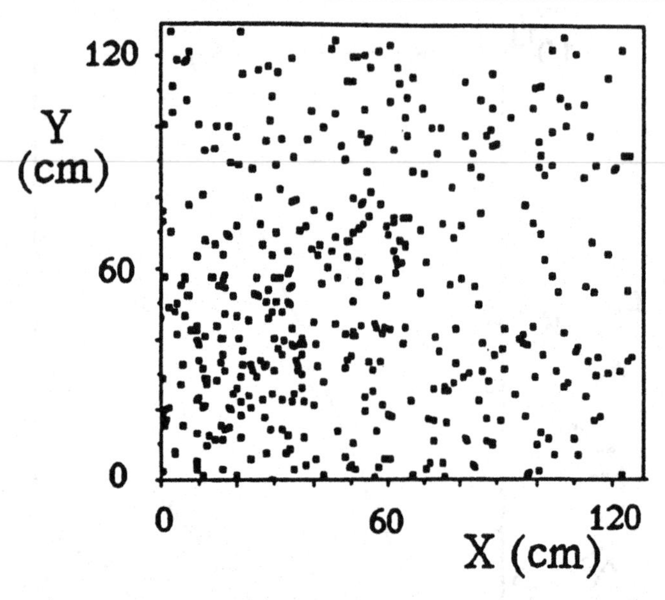

Fig. 16a Each point represents the centre of a raindrop for the 128 x 128 cm piece of chemically treated blotting paper discussed in the text. There are 452 points, the exposure was about 1 s. From Lovejoy & Schertzer (1990c).

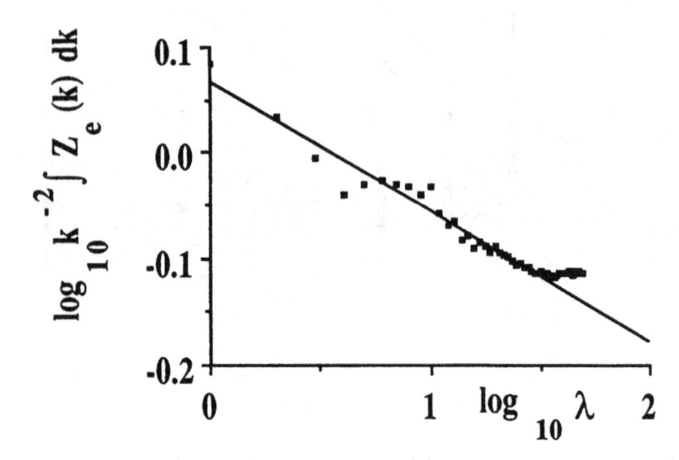

Fig. 16b Plot showing k^{-2} times the integrated energy spectrum of the radar reflectivity of the distribution in Fig. 16a plotted against $\lambda = k/k_0$ where k is the wave number, k_0 corresponds to the largest scale ($k_0 = |k_0| = 2\pi/128$. The straight line (slope -0.12) indicates a scaling power law spectrum (Gaussian white noise yields a slope 0) up to $\lambda \approx 30$ which corresponds to ≈ 4 cm. From Lovejoy & Schertzer (1991a).

finite sample effects (a single exposure was used with only 452 drops), or the break could be more fundamental; related to the inner scale at which rain can no longer be treated as a field[60], where its particulate nature must be considered[61]. Fig. 16b shows the Fourier energy density integrated over circles, confirming the breakdown by the flattening of the spectrum

[60] However, the spectrum in Fig. 4a suggests an inner dissipation scale of the order of millimeters.

[61] The proper mathematical framework is mathematical measures, associating with each drop a position r_i, and volume V_i.

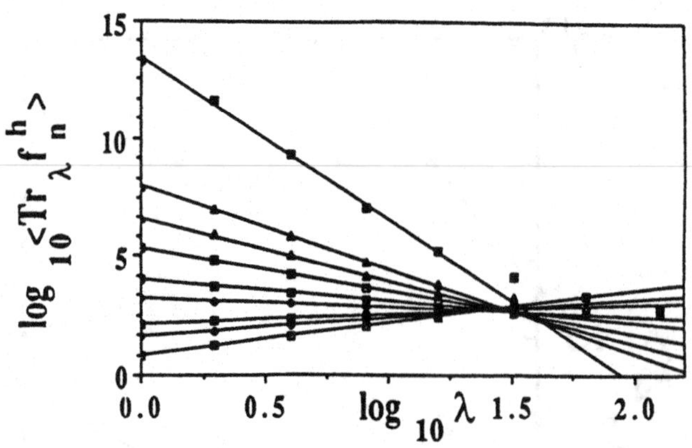

Fig. 17 A log-log plot of $\langle \mathrm{Tr}_\lambda f_n^q \rangle = \lambda - K_D(q)$ vs. λ where f_n is the number of drops per unit area at resolution λ (= the scale ratio). (The h in the figure is the same as the q used here – this is also true for Figs. 18, 19). Note that the largest scale ($\lambda = 1$) was 128 cm and that convergence to power laws occurs only for lengths ≥ 4 cm. The curves, top to bottom, are for $q = 5$, 3, 2.5, 2, 1.5, 1.2, 0.8, 0.6, 0.3. From Lovejoy & Schertzer (1991a).

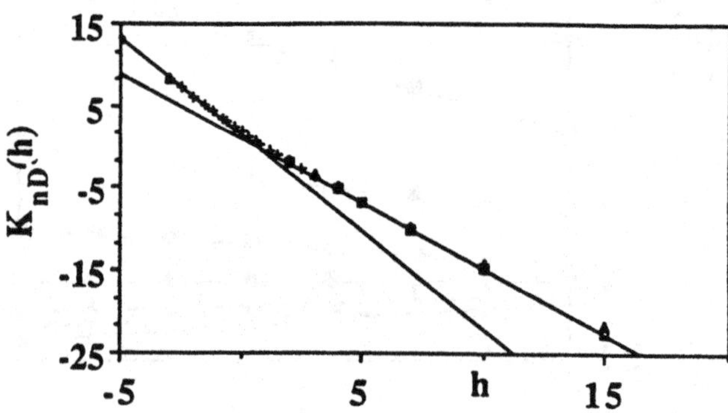

Fig. 18 $-K_{nD}(q) = K(q) - (q-1)D$ estimated from (top to bottom) a manually analysed 1293 drop case ($q > 0$ only; $q = h$), the 452 drop (digital) case, and a 339 drop manually analysed case ($q > 0$). The straight lines are asymptotic fits to the negative and positive large (absolute) q regions for the 452 drop case. The large q slope gives $\gamma_{d,s} = 0.44$ (the largest singularity present). From Lovejoy & Schertzer (1991a).

for scales less than about 4 cm. To our knowledge, this is the first attempt to study spatial heterogeneity at the individual drop level; existing empirical studies of the distribution of the drops are numerous, but consider only their relative sizes; spatial homogeneity is simply assumed on faith. Much more research at the individual drop level will be necessary to properly understand the multifractal structure of rain. We may anticipate that the results will be important in applications: Lovejoy & Schertzer (1990a) already indicate how even monofractal approximations lead to important corrections to standard radar estimates. Two low budget feasibility studies at McGill[62] point to the difficulty in accurately

[62] The blotting paper, lidar and other feasibility studies were all performed as third year physics lab projects from 1986 to present.

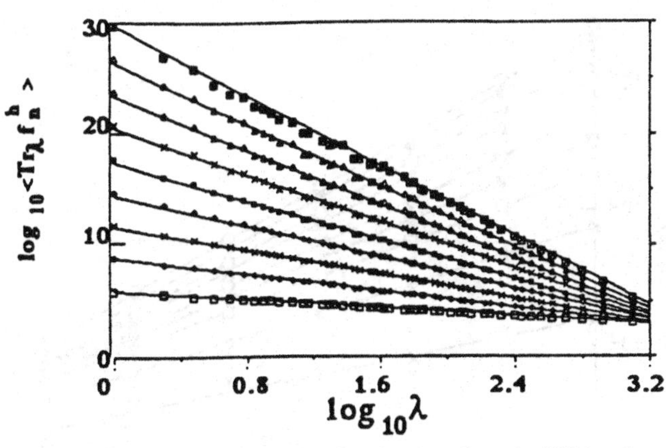

Fig. 19a Trace moment analysis for the time domain (5000 pulses $t_0 = 500$ s) for those range corrected returns that exceeded the average (this is an estimate of the drop number density f_n; it assumes either zero or one drop per pulse volume). Curves from top to bottom are for $q = 10, 9, 8, 7, 6, 5, 4, 3, 2$ respectively. Note that the scaling is extremely accurately followed. From Lovejoy & Schertzer (1991a).

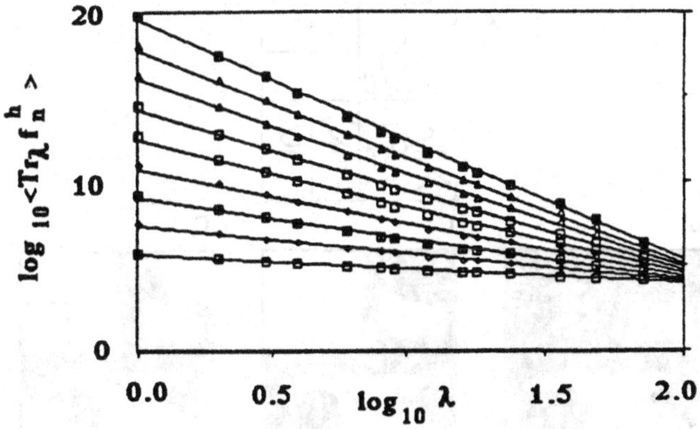

Fig. 19b Trace moment analysis for downrange domain (each pulse return is divided into 180 pulselength sections, 3 m apart, hence the largest scale is $L_0 = 540$) for those range corrected returns that exceeded the average. Curves from top to bottom same as for Fig. 19a. Note that the scaling is extremely accurately followed. From Lovejoy & Schertzer (1991a).

obtaining spatial information about large numbers of drops: stereophotography of a ≈ 1 m^3 region (Bochi Kebe & Howes, 1990), and photography of rain illuminated by sheets of laser light (to obtain horizontal rain intersections with rain, Harris & Lewis, 1991) both indicate that the relevant measurements will be quite difficult, primarily due to the very small cross-sections (at visible wavelengths) of the rain drops which makes their detection quite difficult[63].

To extend these results to slightly larger scales, high powered lidars (Weisnagel & Powell, 1987) were used to detect the optical backscattering from very small volumes[64] the sensitivity was such that individual drops 1 mm in diameter could be detected at ≈ 10 km distances. The YAG laser used had a pulse repetition frequency of 10 Hz, data were logged over 180 downrange bins, several thousand in time. Often, especially in light rain, pulse volumes were empty, and they rarely contained a large number of drops. Shorter pulse length lasers should be able to probe down to the individual drop level (this may indeed be the most promising approach for further studies). Fig. 19a, b, c shows the resulting trace moment analyses (space, time, space/time), showing not only the surprising accuracy with which the (multiple) scaling is respected, but also the possibility of using this approach for studying the space/time transformations associated with rain. A different approach currently being studied at McGill is to use stereo photography with high powered flash lamps.

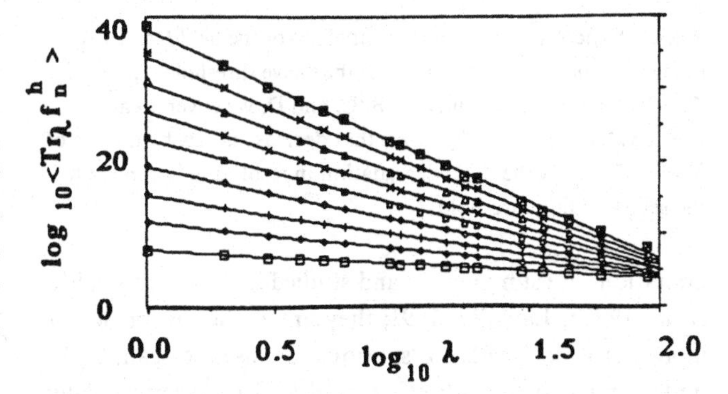

Fig. 19c Trace moment analysis for the (x, t) domain (180 pulses, 0.1s apart in time, space resolution 3 m) for those range corrected returns that exceeded the average. Curves from top to bottom same as for Fig. 19a. Note that the scaling is extremely accurately followed. This data set is the same as that shown in Fig. 19b except that analysis was performed on 'squares' in (x, t) space rather than by intervals (downrange) only. By comparing the slopes in 19a, b, c, the elliptical dimension of (x, t) space can be estimated. From Lovejoy & Schertzer (1991a).

Functional box-counting

Since the discovery of multifractal universality classes in 1986–7, a primary goal has been to test the (multi)scaling and to estimate the basic parameters H, C_1, α in rain over wide ranges of scale. While the trace moment analyses discussed above clearly established the multiple scaling nature of rain, they suffer from a number of limitations which make them difficult to use to estimate the universal parameters. These

[63] The cross-section is however significantly enhanced for both forward and back scattering.

[64] The pulse lengths were 3 m and the widths varied from 0.3 mm to 30 cm at distances of 10 and 1000 m respectively. The associated pulse volumes were thus in the range 10^{-6} to 10^{-2} m^3: 10^{15} to 10^{11} times smaller than typical radar volumes.

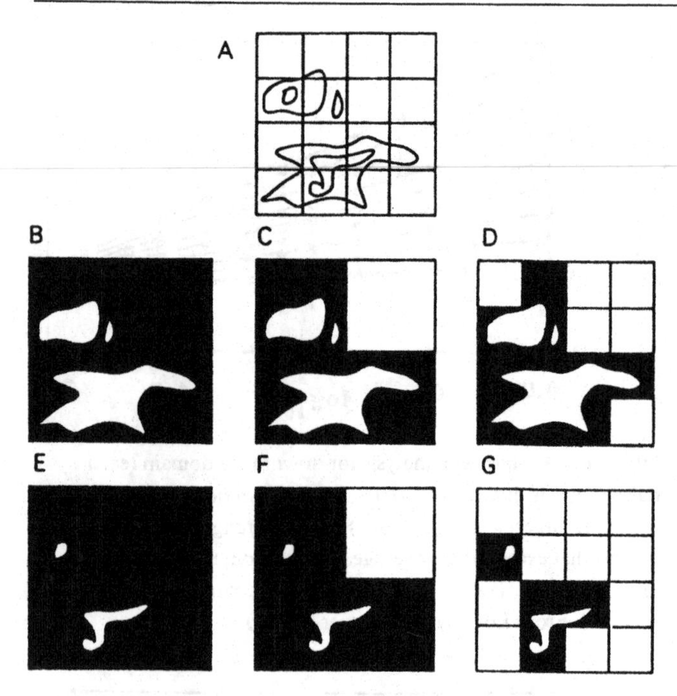

Fig. 20 Functional box-counting analysis of the field $f(r)$. In A the field is shown with two isolines that have threshold values $T_2 > T_1$; the box size is unity. In B, C, and D, we cover areas whose value exceeds T_1 by boxes that decrease in size by factors of 2. In E, F and G the same degradation in resolution is applied to the set exceeding threshold T_2.

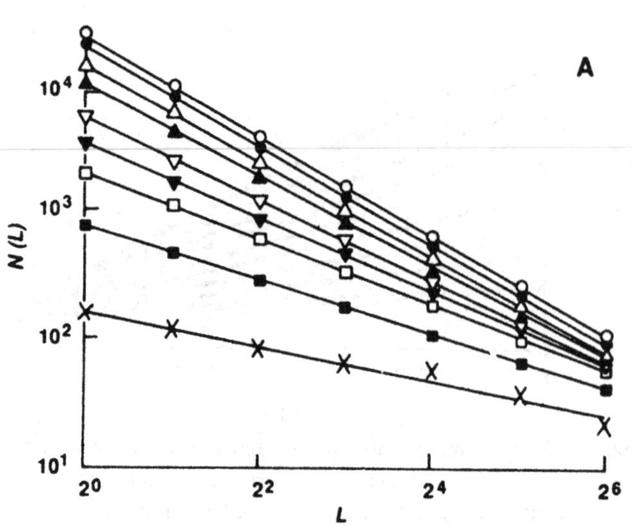

Fig. 21a A plot of $N(L)$ versus L for a single radar scan with nine radar reflectivity thresholds increasing (top to bottom) by factors of ≈ 2.5, analyzed with horizontal boxes increasing by factors of 2 in linear scale. The negative slope (dimension) decreased from 1.24 to 0.40. From Lovejoy *et al.* (1987).

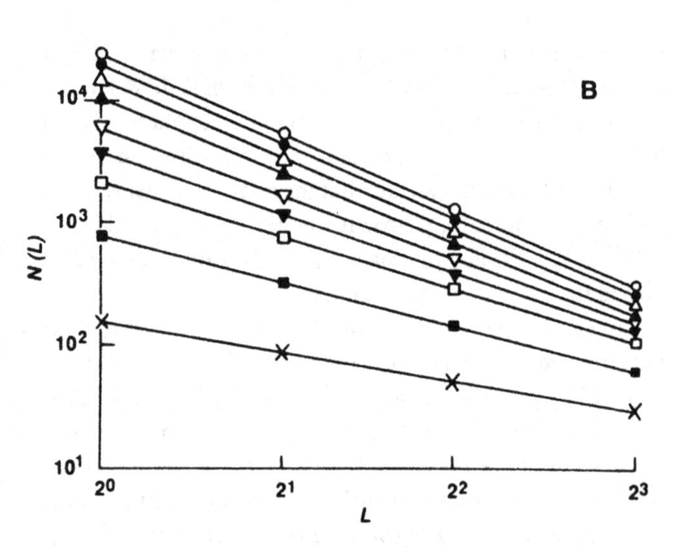

Fig. 21b Same volume scan as Fig. 21a except that the boxes used are cubical and yielded values of dimension that decreased from 2.18 to 0.81 for the same thresholds. Only eight vertical levels were available. See text for discussion of the vertical anisotropy, and 'elliptical box counting'. From Lovejoy *et al.* (1987).

limitations are summarized and studied in detail in Lavallée *et al.*, 1991a, Lavallée, 1991; they are a) the divergence of moments which leads to 'spurious' or 'pseudo-scaling', b) finite sample size. Both effects will lead to asymptotically straightline exponents (as observed in both Figs. 1 and 18); corresponding to multifractal phase transitions. In order to overcome these difficulties, other methods which avoid the use of statistical moments were developed. The first of these was 'functional box counting' (Lovejoy *et al.*, 1987). This method is straightforward: the empirical fields are first converted into finite resolution sets by using a series of thresholds; the sets of interest being defined by the regions exceeding the threshold (T), see the schematic illustration Fig. 20. In the second step, the resolution of these sets is degraded systematically by covering the sets with boxes of increasing size (the standard 'box-counting' procedure for analysing strange attractors). The dimension as a function of threshold is then obtained as the (negative) logarithmic slope of the number of boxes $N_T(L)$ as a function of the log of the box size (L). Fig. 21a, b shows the result when this method is applied to radar rain data, Fig. 22 when it is applied to the associated cloud fields (from satellite data). Again, the (multiple) scaling is well respected.

Hubert & Carbonel (1988, 1989, 1991) have used functional box-counting to study rainfall time series from Burkina Faso raingauges, finding that the multiscaling extends from one day to at least a year. For example they found that the fractal dimension of wet days was ≈ 0.8 which meant that local rule of thumb knowledge of the climate (7 wet months/year) could be extended down to at least a day since $\log 7/\log 12 \approx 0.8$.

Other related applications of functional box counting in rain can be found in Olsson *et al.* (1990), satellite cloud pictures in (Gabriel *et al.*, 1988; Baryshnikova *et al.*, 1989; Detwiller, 1990), and *in situ* measurements of cloud liquid water (Duroure & Guillemet, 1990). Although functional box-counting has the advantage of avoiding the use of statistical moments, it has the basic problem that it is not easy

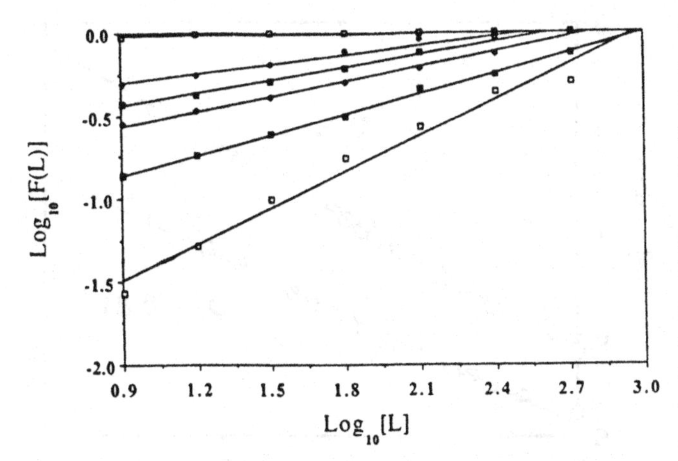

Fig. 22a A plot of the fraction $F_T(L)$ of cloud pictures exceeding a threshold T, for six radiance thresholds with L increasing for 8 to 512 km, at visible wavelengths. From a GOES (geostationary) satellite picture over the Montreal region (summer with mostly cloud cover). The minimum digital count is 24 (ground), maximum is 52 (bright cloud) corresponding to a brightness ratio of $(52/24)^2 \approx 4.7$. The fraction is estimated by using box counting to degrade the resolution of exceedance sets, and then calculating the fraction of all the boxes available at resolution L: $F_T(L) = N_T(L)/L^{-2}$. The straight lines indicate that over the range (which includes most of the meso-scale), that the scaling is accurately followed. From Gabriel *et al.* (1988).

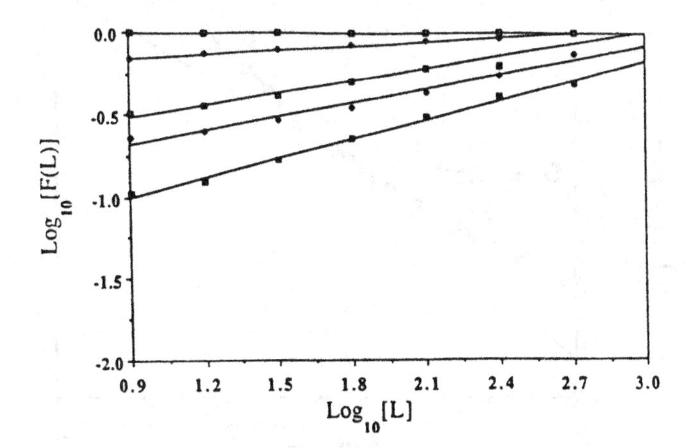

Fig. 22b Same as Fig. 22a except for the corresponding infrared image. The straight lines correspond to effective black body temperatures of (top to bottom) 17, 9, 2, −5, −23 °C respectively. Here the lowest radiances (proportional to the fourth power of the temperature) comes from the sparsest (highest) cloud tops.

to relate the threshold to the order of singularity[65] γ. Another related problem is its tendency to 'saturate' in certain situations because all the boxes larger than a given scale can be filled, a problem likely associated with statistical nonconservation ($H \neq 0$). Finally, the method does not take into account whether a given box is filled by more than one pixel (it is an all or nothing estimator). For a critique of this method, see Lovejoy & Schertzer, 1990a (Appendix A), and Lavallée (1991).

Direct application of functional box-counting requires gridded data. We now describe a variant which is useful for highly inhomogeneous raingage network data (indeed, as shown in Lovejoy *et al.* (1986), they are more nearly uniform over a fractal than over a surface[66]). Define the 'exceedance stations' as all the stations whose rain rate exceed a given threshold, and then calculate the number of pairs of exceedance stations that are closer than a distance[67] l (this is proportional to the average number of exceedance stations in a circle radius l). The resulting scaling exponent is called the 'correlation dimension'; it will less than or equal to[68] the corresponding fractal (box-counting) dimension. Fig. 23a shows[69] the result when the method is applied to daily rain accumulations for 1983 (roughly 8000 stations were used), for various exceedances levels up to 150 mm/day. Although the statistics become poor at the high thresholds, the lines are fairly straight indicating that the scaling is well respected (over the range ≈ 3 km–5000 km). Note that the zero level exceedance line is also included; this has a nontrivial fractal dimension (≈ 1.78 here) due to the fractal nature of the network.

This method of improving statistics by examining pairs of points can also be applied to time series. Tessier (1993) has used this method to effectively study the scaling of increasingly wet stations using the same global daily rainfall data set. Fig. 23b shows that the value 0.8 for wet/dry days (using a threshold of 0.1 mm/day) seems to be global (rather than specific to the Sahel, Hubert & Carbonnel 1988). However, Fig. 23c, d indicate that a definite scale break occurs at ≈ 3 weeks for higher thresholds, consistent with the existence of a 'synoptic maximum' (see Fig. 4d; i.e. a break whose duration corresponds to the lifetime of global sized rain events).

The probability distribution/multiple scaling technique (pdms)

A more successful way of estimating $c(\gamma)$ is to directly exploit the scaling of the probability distributions of the multifractals as indicated in equation 7. Methods which directly exploit this equation were developed and baptized 'prob-

[65] This can be done approximately via the relation $T \sim L^{-\gamma}$, but the normalization (which is required to nondimensionalize this relation and determine the proportionality constant), is nontrivial, and cannot be completely determined at a single averaging resolution.

[66] In fact it is even better to treat the density of stations as a multifractal.

[67] To account for the curvature of the earth, the following measure of distance should be used:

$$l = r\sqrt{(8(1-\cos\theta/2))},$$

where r is the radius of the earth, θ is the angle subtended at the centre of the earth by the two stations (Lovejoy *et al.*, 1986).

[68] In practice, the difference is usually small.

[69] We thank C. Hooge for help with this analysis.

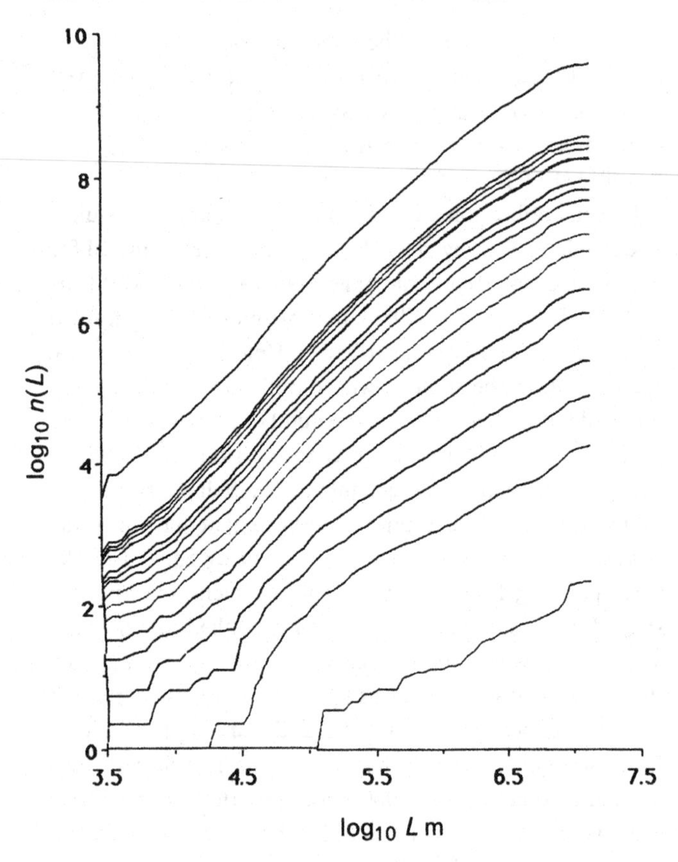

Fig. 23a The correlation function (the average number of exceedance stations in a circle) $n(L)$ as a function of radius (L), for various thresholds. From top to bottom, the thresholds (in mm for daily accumulations) are 0, 0.1, 0.2, 0.4, 0.8, 1.6, 2.5, 3.2, 5.0, 6.4, 10.0, 12.8, 17.5, 20.0, 25.0, 40.0, 52.5, 76.0, 150.0. Note that at least some of the deviations from straight line (scaling) are due to the imperfect scaling of the network itself (top line).

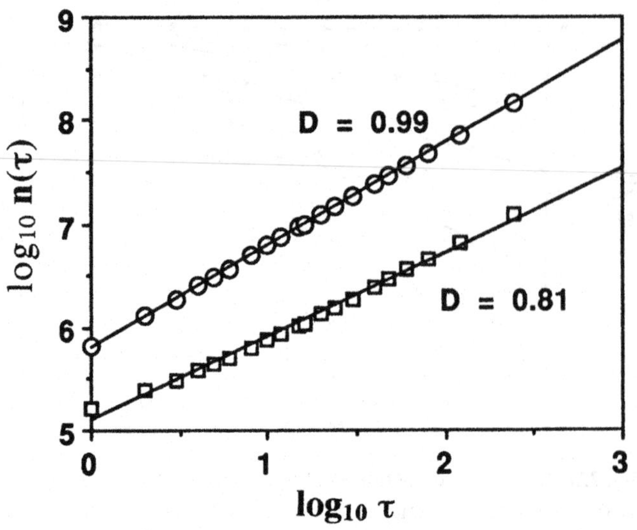

Fig. 23b The number $n(\tau)$ of exceedance pairs for daily accumulations in time τ (for a year), accumulated over all 8000 stations. Since many stations frequently had missing data, it was first confirmed (top curve) that the pattern of missing data was not itself fractal (the slope is consistent with a dimension 1 on the time axis, hence nonfractal data outages). The line below is only for those stations whose accumulation was above the minimum detectable level 0.1 mm/day. From Tessier (1993).

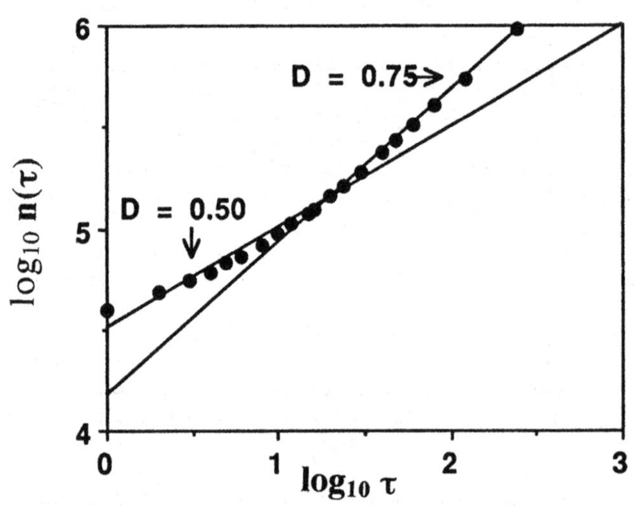

Fig. 23c Same as Fig. 23b except for a threshold of 1.28 cm/day, showing a clear break at about three weeks (the 'synoptic maximum'). From Tessier (1993).

ability distribution/multiple scaling' techniques (PDMS) in Lavallée *et al.* (1991a), Lavallée (1991). They are distinguished from other histogram based techniques (e.g. Paladin & Vulpiani, 1987; Atmanspacher *et al.*, 1989) in that they overcome the nontrivial problem of the (slowly varying) proportionality constants in equation 7 by examining the histograms over a range of scales rather than at a single scale. The drawback of these methods is that they are quite sensitive to the correct normalization of the field: the ensemble average of the large scale spatial average must satisfy $\langle R_1 \rangle = 1$ (i.e. in rain, the large scale, climatological rate must be used). An early implementation in rain is given in Seed (1989) (see e.g. Fig. 24) who studied radar reflectivities of four separate convective storms in Montreal. However, the statistical estimation of H, C_1, a from $c(\gamma)$ is a poorly conditioned nonlinear regressions and leads to low accuracies in the estimates. Nonetheless, Seed found α in the range 0.3–0.6 and C_1 in the range 0.6–1.0. Although he averaged in space, he pooled statistics into histograms involving many (~ 144) consecutive 5 minute PPIs. His estimates are in fact close to the more accurately estimated temporal parameters found here.

Another application of PDMS to rain is described in Tessier *et al.* (1994), where it is applied to the global meteorological network (Fig. 25) used to estimate global rain.

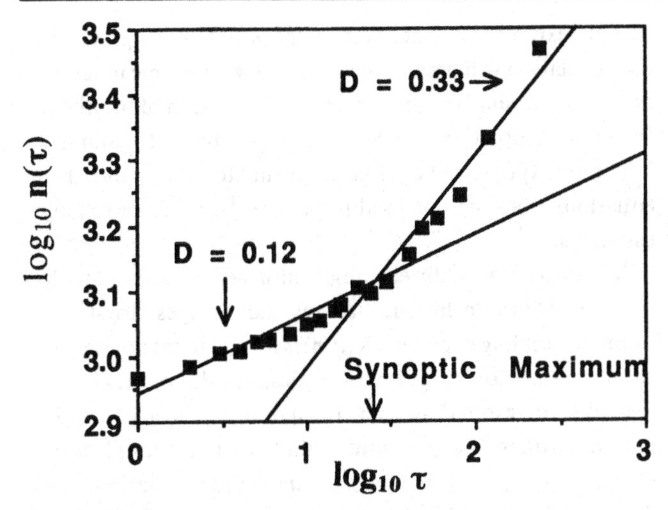

Fig. 23d Same as 23c except for a (very high) threshold of 10.24 cm/day, showing the same break at about three weeks and much lower dimensions. From Tessier (1993).

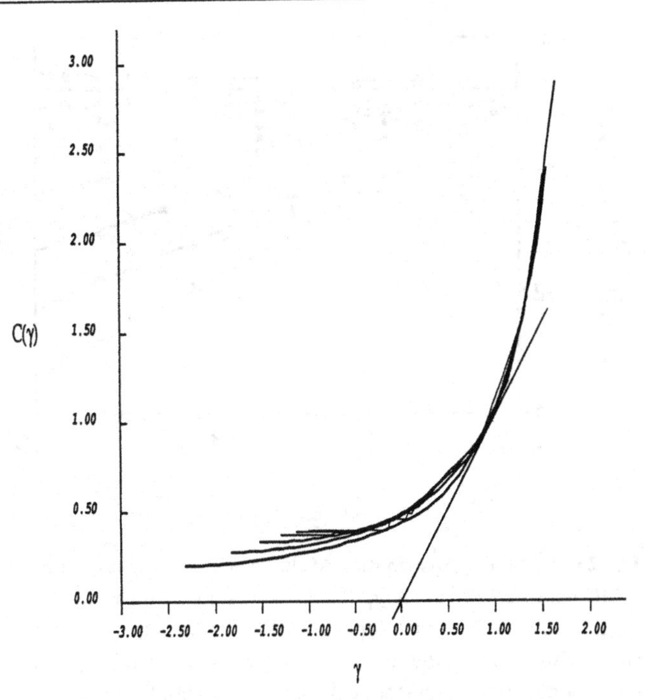

Fig. 24 An early implementation of the PDMS technique on radar reflectivities during a summer storm in Montreal (August 14th, 1988), using 144 consecutive CAPPIs (Constant Altitude Plan Position Indicators, i.e. constant altitude reflectivity maps), at five minute intervals. The codimension function was estimated at scales of 2, 4, 8, 16, 32 and 64 km from the corresponding probability distributions (of spatially degraded reflectivities) using the following approximation: $c(\gamma) \approx -\log(Z_\lambda)/\log\lambda$ (i.e. setting the proportionality coefficient in equation 7 equal to unity). The resulting estimates for each of 6 resolutions is shown, along with a smooth curve obtained by nonlinear regression using the theoretical (universal) formula equation 12. The universal parameters can be graphically estimated using the construction line (the bisectrix $x=y$) shown, which exploits the fact that for a conserved multifractal ($H=0$), $c(C_1)=C_1$, $c'(C_1)=1$, i.e. the bisectrix will be tangent to $c(\gamma)$ at the point C_1 (see Figs. 12, 13). From the graph we immediately deduce that $H \approx 0$, (from equation 12, $H \neq 0$ leads to a left/right shift of $c(\gamma)$ with respect to the bisectrix), $C_1 \approx 0.9$, and from some other point on the curve (e.g. the value of $c(0)$) we deduce $\alpha \approx 0.55$. From Seed (1989).

Direct estimates of universality parameters: double trace moments

Above, we reviewed the results of multifractal analysis techniques which in principle could be applied to arbitrary multifractals. They enjoyed the apparent advantage of making no assumptions about the type of multifractal being analyzed. In practice however, the techniques are overly ambitious: for a finite (and usually small) number of samples of a process, they attempt to deduce an entire exponent function, with the result that there is considerable uncertainty in the resulting estimates of $c(\gamma)$ or $K(q)$. With the realization that physically observable multifractals are likely to belong to universality classes, it is natural to develop specific methods to directly estimate the universality parameters (H, C_1, α). These parameters can then be used to determine $c(\gamma)$, $K(q)$ from equations 12–13.

The double trace moment (DTM) technique (Lavallée, 1991; Lavallée *et al.*, 1991d) directly exploits universality by generalizing the trace moment; it introduces a second moment η by transforming the high resolution field $\varphi_{\Lambda'} \Rightarrow \varphi^\eta_{\Lambda'}$. This transforms the flux Π into an 'η flux'

$$\Pi^{(\eta)}_\Lambda(B_\lambda) = \int_{B_\lambda} \varphi^\eta_\Lambda \, d^D x, \text{ see (16)} \quad (18)$$

The double trace moment can then be defined as:

$$\mathrm{Tr}_\lambda(\varphi^{(\eta)}_\Lambda)^q = \left\langle \sum_i [\Pi^{(\eta)}_\Lambda(B_{\lambda,i})]^q \right\rangle \approx \lambda^{K(q,\eta)-(q-1)D} \quad (19)$$

where we have introduced the (double) exponent $K(q,\eta)$, which reduces to the usual exponent when $\eta=1$: $K(q,1)=K(q)$ (the sum is over all disjoint boxes indexed by i). Note that the basic implementation of the DTM is quite straightforward[70]; the field at the highest available resolution is raised to the power q, the result is iteratively degraded in resolution and the qth moment averaged over the field and the ensemble of samples available – see Fig. 14 for a schematic illustration.

The entire transformation from single to double trace moments (i.e. taking η powers and then integrating) can be summarized in the following formulae (where the prime indicates transformed, double trace quantities, not differentiation):

[70] Note that, if $H>0$ the data will require some 'prewhitening' before the application of the DTM, i.e. power law filtering to yield a conserved field.

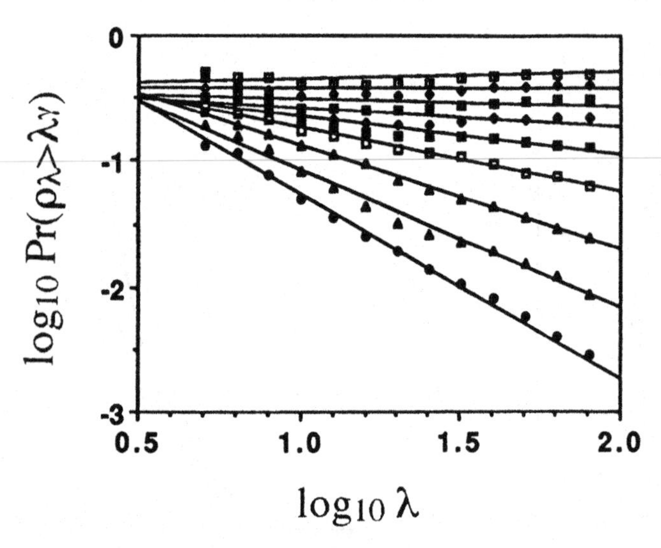

Fig. 25 Probability Distribution/Multiple Scaling analysis of the NMC network (Fig. 26), $\log_{10}\Pr(\rho_\lambda > \lambda^\gamma)$ vs. $\log_{10}\lambda$ for γ increasing in 0.1 intervals from 0 (top) to 0.8 (bottom line), the absolute slopes give $c(\gamma)$, ρ_λ is the station density at resolution λ. $\log_{10}\lambda = 0$ corresponds to the largest scale (here ≈ 15000 km), the (multiple) scaling is well followed up to $\log_{10}\lambda \approx 2$ (≈ 150 km); for smaller scales, the finite number of stations leads to a break in the scaling. Using nonlinear regressions, the universal parameters were estimates as $\alpha \cong 0.8$, $C_1 \approx 0.5$. From Tessier *et al.* (1994).

$$\gamma \Rightarrow \gamma' = \eta\gamma - K(\eta) \tag{20a}$$
$$c(\gamma) \Rightarrow c'(\gamma') = c(\gamma) \tag{20b}$$
$$q \Rightarrow q' = q/\eta \tag{20c}$$
$$K(q) \Rightarrow K'(q') = K(q,\eta) = K(\eta q') - q'K(\eta) \tag{20d}$$

Note the subtlety in the above: due to the integration in equations 18–19, we are dealing with dressed rather than bare quantities, hence the dressed singularities (equation 20a) transform with an extra term $(-K(\eta))$; necessary since the dressing operation enforces conservation of the η flux.

The real advantage of the DTM technique becomes apparent when it is applied to universal multifractals (Lavallée,1991) since we obtain the following transformations of C_1:

$$C_1 = \left.\frac{dK}{dq}\right|_{q=1} \Rightarrow C_1' = \left.\frac{dK'}{dq'}\right|_{q'=1} = C_1\eta^\alpha \tag{21}$$

Therefore, $K'(q') = K(q,\eta)$ has a particularly simple dependence on η:

$$K(q,\eta) = \eta^\alpha K(q) \tag{22}$$

α can therefore be estimated on a simple plot of $\log K(q,\eta)$ vs. $\log \eta$ for fixed q. By varying q, we improve our statistical accuracy. Finally, note that since equation 20d is only valid when the relevant statistical moments converge, and the sample size is sufficiently large to accurately estimate the scaling exponents, whenever $\max(q\eta, q) > \min(q_s, q_D)$ the above relation will break down; $K(q,\eta)$ will become independent of η. We shall see that effective exploitation of the above involves a 'bootstrap' procedure in which the well estimated low q, η exponents are used to estimate α, C_1, and then equations 14a, b can be used to predict the range of reliable estimates.

In comparison with existing multifractal analysis methods, the DTM technique has two advantages. First, the estimated scaling exponent $K(q,\eta)$ is independent not only of the normalization at the largest scales, but also of the change $\gamma \Rightarrow \gamma + b$ corresponding to a translation in γ space – in the bare quantities[71]. The second is that when a multiplicative change of γ is made $(\gamma \Rightarrow a\gamma)$ then relation for $K(q,\eta) \rightarrow K(q, a-\eta) = a^\alpha K(q,\eta)$ (when a corresponds to a contraction in the γ space, but is also equivalent to the integration of the fields φ_A at an unknown power a by the experimental apparatus). This implies that the determination of α will also be independent of the power a to which the process is raised. In other words the universality has been exploited to give a method to determine α which is invariant under the general transformation $\gamma \Rightarrow a\gamma + b$!

Estimating H

We have seen that in multiplicative processes, it is convenient to isolate an underlying conserved quantity which has basic physical significance; in turbulence it was the energy flux to smaller scales, in rainfall we denoted it by φ, and related it to the rain fluctuations via equation 3. In terms of the scaling, conservation means $\langle\varphi_\lambda\rangle$ = constant (independent of λ), hence $K(1)=0$. If we consider the energy spectrum of φ_λ, it is of the form $k^{-\beta}$ with[72] $\beta = 1 - K(2)$, i.e. the spectrum is always less steep than a $1/f$ noise[73].

The reason for dwelling on this is that it illustrates a basic point common to most geophysical fields viz, their spectra often have $\beta > 1$, hence they cannot be conservative processes, they must be[74] (fractionally) differentiated by order $-H$ (the spectra must be power law filtered by k^H) to become conservative. Lavallée (1991) analyzed simulations of conserved processes fractionally integrated and differentiated by varying amounts. As long as he differentiated (filtered by k^H with $H > 0$) using the DTM technique, he obtained stable and accurate estimates of both C_1 and α; however when he fractionally integrated $(H < 0)$, he only recovered α. C_1 was

[71] This is also true of single trace moments or partition function approaches.

[72] This formula is a consequence of the fact that the energy spectrum is the Fourier transform of the autocorrelation function which is a second order moment.

[73] The difference is often not great since $K(2)$ is usually small: $= C_1(2^\alpha - 2)/(\alpha - 1)$, and $0 \le \alpha \le 2$.

[74] See Schertzer & Lovejoy (1991, Appendix B.2) for more discussion of fractional derivatives and integrals.

not accurately determined[75]. From the C_1, α estimated this way, we can determine $K(2)$ from equation 11 and hence[76], writing β for the spectral slope of the observed process, the order of fractional integration required to go from the conserved process to the nonconserved (observed) process is given by:

$$H=\frac{\beta-1+K(2)}{2}=\frac{\beta-1}{2}+\frac{C_1(2^\alpha-2)}{2(\alpha-1)} \quad (23)$$

In many data analyses, it is possible to avoid the use of Fourier space. In $1-D$ we have already recalled that replacing the time series by its differences is approximately the same as multiplying by k in Fourier space[77]. To generalize this to two (or more) dimensions, one possibility is to use a finite difference Laplacian. This multiplies by $|k|^2$ in Fourier space, hence the spectrum by $|k|^4$; although this is quite drastic we have found that it apparently works fairly well. A method involving less smoothing which also works well, is to replace the field by the modulus of the local finite difference gradient operator.

As a final comment, it is possible to directly estimate H via (first order) structure functions (the scaling of absolute differences). However, current direct methods are designed for time series analysis, the optimum extension for fields in two or higher dimensions is not clear. Other methods such as the probability distribution and R/S analysis methods used in Lovejoy (1981) when applied to multifractals yield results which are not directly related to the multifractal parameters; the value $H\approx0.5$ quoted earlier (which assumed simple scaling) needs careful reconsideration.

ESTIMATES OF UNIVERSAL MULTIFRACTAL EXPONENTS IN RAIN —

Analyses of rain gage network in space and multifractal objective analysis

A basic problem with *in situ* geophysical measurements – such as those from rain networks – is that the networks are typically sparse, they have 'holes' at all scales. An early method for dealing with this problem involved characterizing the sparseness by fractal dimensions, for example, Lovejoy *et al.* (1986) found a fractal dimension of ~1.75 in the $\approx10\,000$ station network reporting to the World Meteorological Organization indicating that 'holes' do indeed occur over a wide range of scales. Using (generalized) intersection theorems and ordinary trace moments, Montariol & Giraud (1986), Giraud *et al.* (1986), Marquet & Piriou (1987) and Ladoy *et al.* (1987) showed how corrections to network infered rain statistics could be made by subtracting appropriate network codimensions from the corresponding measured rain codimensions. Below, we examine the daily rainfall accumulations observed by raingages at synoptic weather stations[78] covering the earth for the year 1983 (Fig. 26).

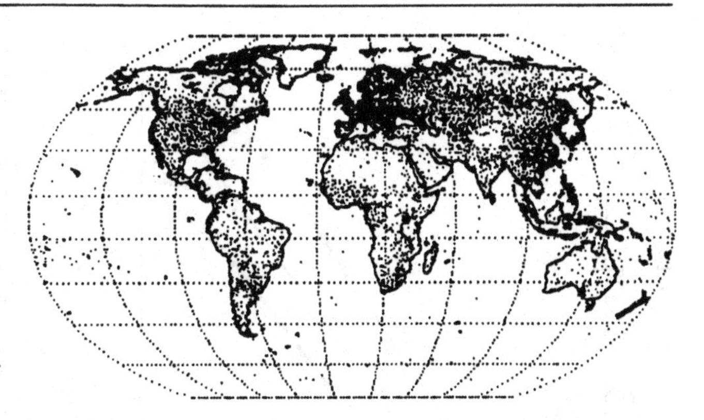

Fig. 26 Position of the stations reporting daily rainfall accumulations in 1983 that have been used in our analysis.

Actually, it is better to treat the density of stations as a multifractal measure (rather than the stations themselves as a fractal set, see Fig. 25), and then to statistically correct for the multifractal nature of the network density. In what follows we summarize the results of Tessier *et al.* (1994). Consider that the measuring stations have a multifractal density ρ_λ when measured at resolution λ. This is found (see Fig. 25) to be a reasonable approximation to the density field over the range $\sim5.0\times10^3$ km to $\sim1.5\times10^2$ km (the lower limit arises because there were only ~8000 stations which is quite small for this type of analysis).

Over the multifractal range, the station density may be estimated in a variety of ways, for example by counting the number N_λ of stations in a circle of radius λ^{-1} (taking the size of the earth = 1), and then[79] $\rho_\lambda\approx N_\lambda\lambda^2$. Consider now the product measure $M_\lambda=\rho_\lambda R_\lambda$. In the ith circle $B_{\lambda,i}$, it can be estimated as follows:

$$\lambda^2\sum_{j\in B_{\lambda,i}}R_j\cong\lambda^2N_{\lambda,i}R_{\lambda,i}=M_{\lambda,i}=\rho_{\lambda,i}R_{\lambda,i} \quad (24)$$

where the sum is over the measured rain rates (indexed by j) of the stations in the ith circle. If we now suppose statistical independence of ρ, R, by taking qth powers and ensemble averaging we immediately obtain:

$$K_R(q)=K_M(q)-K_\rho(q) \quad (25)$$

[75] This indicates that as long as the spectrum is less steep than the underlying conserved process ($\beta<1-K(2)$), that we can recover C_1.

[76] In the case of turbulence, it is not necessary to infer the relation since it is given by dimensional analysis from known dynamical quantities. In rain, we don't know the corresponding dynamical (partial differential) equations, nor their conserved quantities, so that this type of empirical inference is unavoidable.

[77] Because of the finite differencing, this will not be exactly true at the highest frequencies corresponding to the resolution the series.

[78] This data set was archived at the National Meteorological Center (NMC) of NOAA; it is not exactly the same as the WMO set.

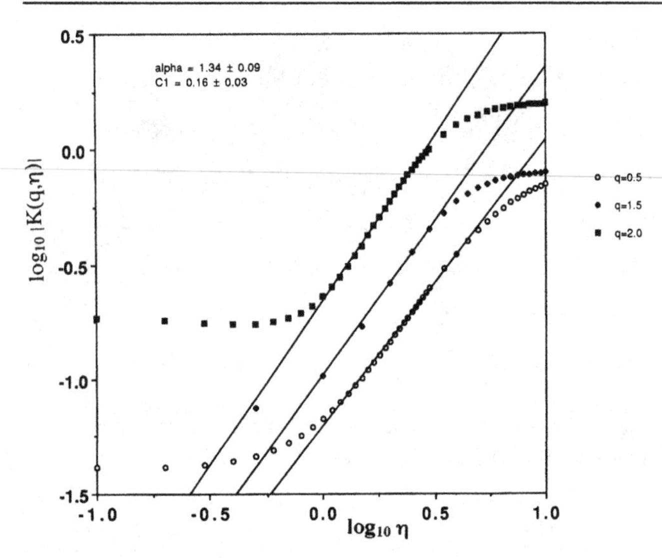

Fig. 27 log($|K(q,\eta)|$) versus log(η) for daily rainfall accumulations on a global network after the corrections explained in the text. The regression lines (for $q=0.5$, 1.5, 2, bottom to top respectively) give a value of $\alpha=1.35\pm0.1$ and $C_1=0.16\pm0.05$. From Tessier *et al.* (1993).

Tessier *et al.* (1993) indicate how to generalize this result to double trace moments ($K(q,\eta)$); the principle is the same, subtract the measured K_M, from the network[80] K_ρ: $K_R(q,\eta)=K_M(q,\eta)-K_\rho(q,1)$. From such an analysis we obtain $\alpha=1.35\pm0.1$ and $C_1=0.16\pm0.5$ as may be seen on Fig. 27 where we have plotted $\log|K(q,\eta)|$ vs $\log\eta$ for $q=0.5$, 1.5, 2. We see that for large values of η the curve $K(q,\eta)$ becomes flat, here due to limited sample size, whereas at low values of η, it also becomes flat due to the sensitivity of the low order moments to noise and/or the presence of a minimum order of singularity.

The classical radar observer's problem for multifractal reflectivity fields and estimates of C_1, α from radar

Up to now, we have discussed various fractal and multifractal analyses of radar rain reflectivity data, carefully distinguishing this from the rain rate. The exact relationship between the radar reflectivity and the rain rate (R) is an unsolved problem going back to the 1940s. Standard (non-scaling) theory already leads to power law relations between the two and we have already mentioned the monofractal (Lovejoy & Schertzer, 1990c) corrections that can be used to improve the latter. In this section we summarize some recent theoretical results (Lovejoy & Schertzer, 1990a) on this 'classical'[81] multifractal 'observer's problem'.

[79] Ignoring factors of π; this will be a good approximation when $N_\lambda \gg 1$.

[80] The double trace moment is with respect to the measure $\rho d^D x$, rather than the usual $d^D x$; hence the $K_\rho(q,1)$ rather than $K_\rho(q,\eta)$. $K_\rho(q,1)$ is easy to estimate since $K_\rho(q,1)=K_M(q,0)$.

[81] 'Classical' because we assume subresolution homogeneity; the multifractals are (unrealistically) assumed to be cutoff at this scale.

In its classical form (Marshall & Hitschfeld, 1953, Wallace, 1953), the observer's problem makes assumptions of subsensor homogeneity (specifically that the rain drops have uniform (Poisson) statistics over scales smaller than the radar 'pulse volume': typically about 1 km^3). The variability in observed 'effective[82]' radar reflectivity factor Z_{eff} is then considered to arise from two sources. The first is the natural variability of interest characterized by the 'reflectivity factor' Z (proportional to the variance of the drop volumes). The second arises as a result of the random positions of each of the drops within the pulse volume. Under certain assumptions about the homogeneity of the field and on the form of the drop size distribution (finite variance, $q_D>2$), Z can be related to the rain rate, total volume of liquid water, or other parameters of interest[83].

Figs. 2a, b, c, d, Fig. 4a, b and Figs. 16a, b already point to the inadequacy of the assumptions of homogeneity (even at subwavelength scales); even the assumption of finite variance is not trivially respected. In spite of this, based on these assumptions, much work has been done to devise sampling and averaging strategies to obtain Z from Z_{eff}. In this section, we indicate that even with these subsensor homogeneity assumptions, that correction can still arise if we allow for a multifractal Z field from the largest scales down to the radar scale; hence even in the standard theory, we must still account for multifractal effects. Introducing the natural log of the range in scales[84] ($\zeta=\ln\lambda$) and the measured codimension function $c_{\text{eff}}(\gamma)$ (for Z_{eff}) we seek to relate this to the underlying $c(\gamma)$ (for Z). The basic result of Lovejoy & Schertzer (1990a) is:

$$\Delta c(\gamma)=c(\gamma)-c_{\text{eff}}(\gamma)=\frac{c'(\gamma)[\ln c'(\gamma)-1]}{\zeta}+O\left(\frac{1}{\zeta^2}\right) \quad (26)$$

hence for λ large enough, $c_{\text{eff}}(\gamma)\Rightarrow c(\gamma)$: in the limit where the natural variability builds up over a sufficiently wide range of scales (i.e. that the radar resolution is much smaller than the outer scale of the rain producing processes), the two are equal[85]. In other words, in this limit the natural variability is so strong that it completely dominates that arising from random fluctuations due to drop phases. This answers the

[82] In this subsection we denote the 'effective reflectivity factor' by Z_{eff}, and the 'reflectivity factor' by Z; in the rest of the paper, for convenience, we drop the subscript 'eff'; Z denotes the measured 'effective' quantity, we do not require the (unobserved, theoretical) 'reflectivity factor'.

[83] The fact that individual radar echoes have long tails (e.g. Figs. 2b, c) and that the drops are highly non uniformly distributed means that in reality, Z can only be statistically estimated from Z_{eff} – ultimately it will probably be simpler to statistically relate the Z_{eff} directly to the rain rate; use of the unmeasureable Z will be unnecessary.

[84] Taking a typical radar resolution of 1 km, and an external scale for the rain processes at 1000 to 10000 km, we find ζ in the range ln(1000) to ln(10000) ~ 7–9.

[85] Using data from Seed (1989), Lovejoy & Schertzer (1990a) estimated that the largest correction to $c(\gamma)$ is ~0.14.

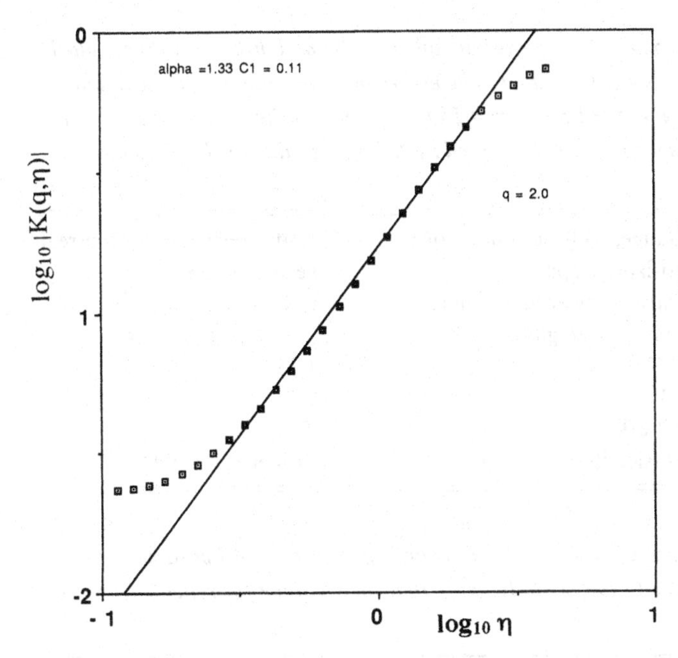

Fig. 28a $\log(|K(q,\eta)|)$ versus $\log(\eta)$ for the gradient of vertically pointing radar reflectivities in the vertical direction (128 elevations at 21 m intervals), for $q=2$, statistics accumulated over 8192 consecutive pulses at 2 second intervals. The straight line indicates $\alpha=1.35$, $C_1=0.11$. From Tessier *et al.* (1993).

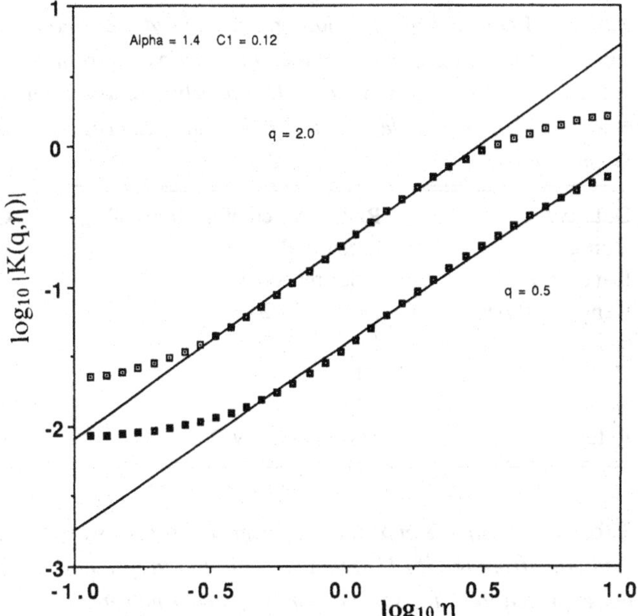

Fig. 28b $\log(|K(q,\eta)|)$ versus $\log(\eta)$ for the gradient of the horizontal radar reflectivities (same data as in Fig. 4f) at 75 m resolution. The bottom curve is for $q=0.5$ and the top for $q=2.0$. The straight line indicate $\alpha=1.40$, $C_1=0.12$. From Tessier *et al.* (1993).

question raised by Zawadzki (1987) as to which variability is strongest.

We now seek to explore the relation between the reflectivity factors and rain rates. Limiting ourselves to studying the implications of the usual semi-empirical relations, (based again on subresolution homogeneity) we find the simplest statistical relation between Z and R is a power law i.e. $Z \propto R^a$. Such power laws are frequently invoked in rain (e.g. the well known semi-empirical Marshall–Palmer (1948) law has exponent $a=1.6$). Writing $Z=\lambda^{\gamma_z}$ and $R=\lambda^{\gamma_R}$ this is equivalent to the linear transformation of singularities: $\gamma_z=a\gamma_R$ where γ_z is the singularity in Z and γ_R is the corresponding singularity in R. We have already seen (equation 20 with $\eta=a$) that under such transformations, $\alpha \Rightarrow \alpha$, $C_1 \Rightarrow C_1 a^{\alpha}$.

Fig. 28a shows the results for vertical pointing radar reflectivities yielding $\alpha_z \approx 1.35$ and $C_{1z} \approx 0.11$ (vertical direction) and Fig. 28b for a horizontal pointing radar yielding $\alpha_z \approx 1.40$ and $C_{1z} \approx 0.12$ (horizontal direction) showing remarkable agreement with α for network rain, and between the vertical and horizontal directions. To estimate H, we may use the horizontal estimate of β (≈ 1.45, Fig. 4f), to yield $H_{Zhor} \approx 0.32$, and in the vertical, using the estimate $\beta \cong 2.3$ (Tessier *et al.*, 1991a), we obtain $H_{Zhor} \approx 0.73$. The agreement between the values of α is particularly significant considering the apparently very different natures of the data sets involved. The C_1 estimates are comparable, although

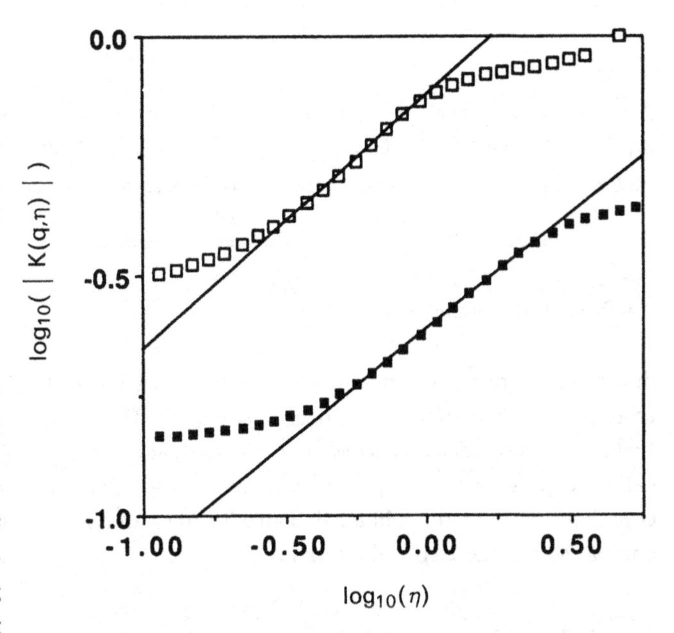

Fig. 28c $\log(|K(q,\eta)|)$ versus $\log(\eta)$ for the gradient of vertically pointing radar reflectivities in time (8192 consecutive pulses at 2 second intervals) and statistics accumulated for 128 elevations at 21 m intervals (same data set as Fig. 28a). The straight line indicates $\alpha=0.50$, $C_1=0.60$. From Tessier *et al.* (1993).

Table 3. *A comparison of various gauge and radar estimates of α, C_1, H over various space scales and directions. Particularly significant is the agreement between the a estimates from such disparate sources. The errors in α are estimated to be about ± 0.1, in C_1, ± 0.05. The value of H is poorly estimated. The value $\alpha \approx 1.35$, $C_1 \approx 0.15$ is the same as that obtained by Tessier* et al. *(1991a) for visible and infra red cloud radiances. It is also very near that found by Schmitt et al. (1991) for turbulent temperatures*

Data type	Radar reflectivity, Montreal	Gauge, daily accumulations	Radar reflectivity, Montreal
Domain	horizontal space	horizontal space	vertical space
Data type	radar reflectivity	daily gauge accumulations	radar reflectivity
Range of scales	75 m to 19.2 km	$\approx$150 km to global	21 m–2.5 km
α	1.40	1.35	1.35
C_1	0.12	0.16	0.11
H	0.32	0.2 ± 0.3	0.73
References:	Tessier *et al.* 1993	Tessier 1993	Tessier *et al.* 1993

Table 4. *A comparison of various gauge and radar estimates of α, C_1 over various time and space scales. All parameters were estimated from the DTM technique with the exception of the Seed (1989) study. Note that the C_1 value for reflectivities are not expected to be the same as for the gauge rain rates*

Data type	Gauge, daily accumulations	Gauge, daily accumulations	Gauge, daily accumulations	Gauges, daily accumulations	Radar reflectivity	Radar reflectivity
Location	Global network	Reunion island	Nimes	Germany	Montreal	Montreal
Sample characteristics	1000 stations, 1–64 days	1 station, 30 years, scales 1–64 days	1 station, 30 years, scales 1–64 days	1 station, 45 years, scales 1–32 days	4 storms, 144 PPIs each 1 km resolution every 5 minutes	1 storm, vertically pointing radar 20 m resolution, every 2 s for $5\frac{1}{2}$ hours
α	0.5	0.5	0.5	0.6	0.3–0.6	0.5
C_1	0.6	0.2	0.6	0.5	0.6–1.2	0.6
References	Tessier *et al.* 1993	Hubert *et al.* 1993	Ladoy *et al.* 1993	Larnder & Fraedrich*	Seed 1989	Tessier *et al.* 1993

Note:
* = Private communication.

differences[86] are to be expected if only because of the Z–R relation, and the horizontal/vertical anisotropy[87]. This is the first empirical agreement between any fundamental statistical rain gauge and reflectivity parameters and gives us confidence in the value obtained. Table 3 shows an overall comparison of these spatial estimates.

Double Trace Moments and the statistics of rain in time

The scaling of rain in space coupled with the scaling of the dynamic (wind) field leads to temporal scaling. Theoretically, the appropriate framework for treating the problem is via scaling space/time transformations and Generalized Scale Invariance; this is discussed in the next section, here we summarize various recent empirical results. Table 4 is a summary of six independent analyses from four different locations, different data types, ranges of scale and analysis methods, studies indicating a remarkable consistency[88] in estimates of α, C_1, especially α. Considering only the gauge estimates, we obtain $\alpha \approx 0.5 \pm 0.1$, $C_1 \approx 0.5 \pm 0.1$. Unfortunately, at present the temporal value of H is not well known. A rough estimate can be obtained from the high frequency end of Fig. 4d: $\beta \approx 0.5$. If we use the corresponding values of α, C_1 we obtain[89] (using equation 23), $H \approx 0.1 \pm 0.1$, i.e. it is

[86] Differences could also arise because of the limited sample sizes used in the various studies, and because C_1 may in fact vary climatologically: the theoretical arguments for the universality of C_1 are less convincing than for α.

[87] From the values quoted above, it seems that the most obvious effect of anisotropy is to modify the values of H.

[88] The theoretical arguments leading to the expectation of a universal C_1 are much weaker than those for a universal α.

[89] Using the mean of the individual H values also yields $H \approx 0.1 \pm 0.1$.

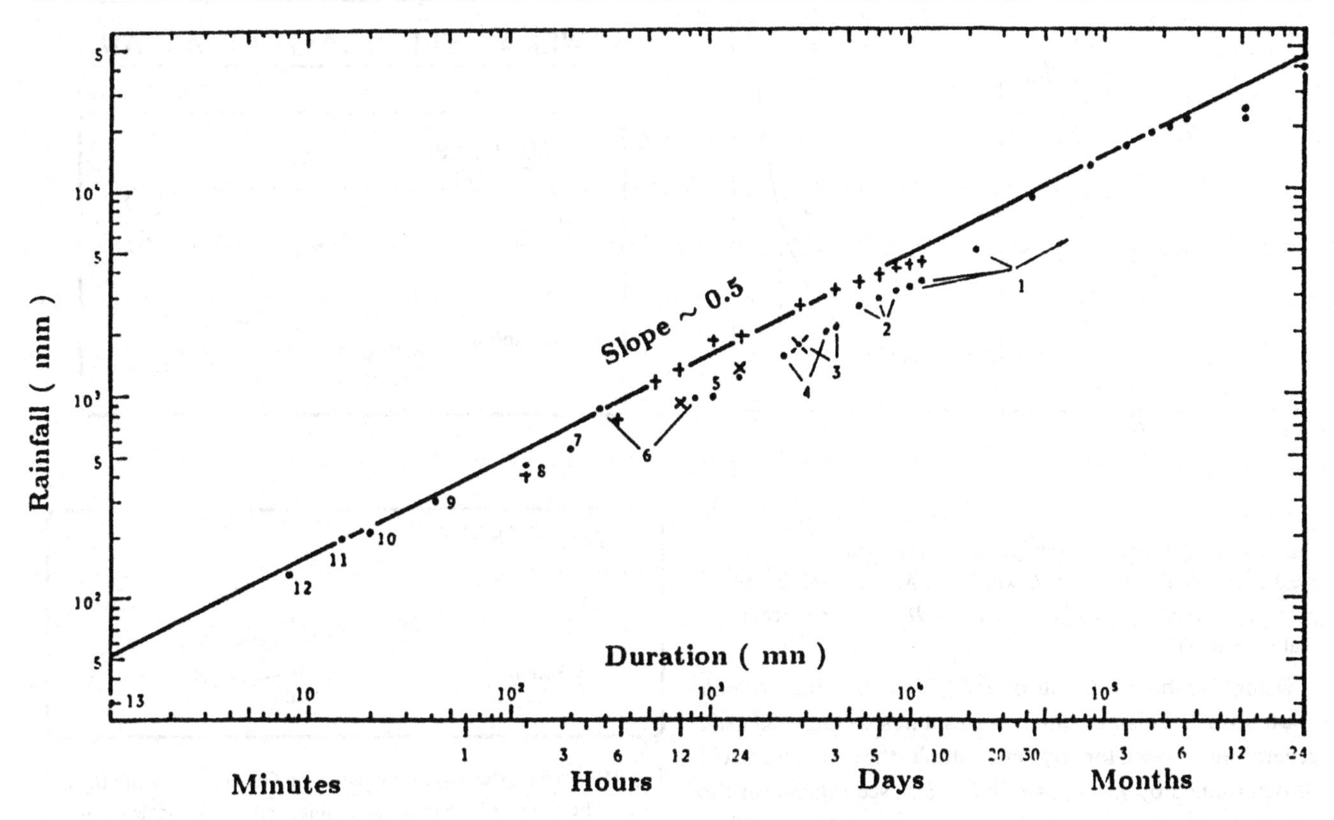

Fig. 29 The world's record point rainfall values, 1 – Cherrapunji, India; 2 – Silver Hill Plantation, Jamaica; 3 – Funkiko, Taiwan; 4 – Baguio, Philippine Is.; 5 – Thrall, Texas; 6 – Smethport, Pa; 7 – D'Hani, Texas; 8 – Rockport, W.Va; 9 – Holt, Mo.; 10 – Cutea de Arges, Romania; 11 – Plumb Point, Jamaica; 12 – Fussen, Bavaria; 13 – Unionville, Md.; values from Jennings (1950). (+) La Reunion, France; (x) Paishih, Taiwan; values from Paulhus (1965).

possible that in time, rain is a conserved process ($H=0$). In comparison, we may apply the double trace moment technique to time series from the vertical pointing radar data discussed in the previous sub-section (Fig. 28c), finding $\alpha\approx0.50$, $C_1\approx0.60$. The spectral slope yielded $\beta\approx1.1$, hence we obtain $H\approx0.4$ for the effective reflectivity.

EXTREME RAINFALL EVENTS

One of the attractive features of multifractal models of rain is that they naturally generate violent extreme events. In this subsection, we show that they are apparently of the same type as those which actually occur. The following is a summary of recent work by Hubert *et al.* (1993).

To begin with, consider a multifractal rainfall time series with maximum order of singularity γ_{max}. This maximum may arise through a variety of mechanisms: it could be due to geometrical or microcanonical constraint, the result of a cascade with bounded singularities, or – of relevance here (Table 3) – associated with universal multifractals with $0\leq\alpha<1$ (see equation 12). In any case even if the process itself has unbounded orders of singularity (as in the universal multifractals with $1\leq\alpha<2$), in any finite sample there will always be a maximum order of singularity present γ_s. Whichever way it arises, for any fixed averaging period (resolution), this maximum order of singularity places an upper bound on the extreme values that will be observed. To see this, consider the maximum rain accumulation A_τ ($=\tau R_\tau$) over time τ ($=\lambda^{-1}$):

$$A_\tau\approx\tau^{1-\gamma_{max}} \qquad (27)$$

We therefore will expect log–log plots of maximum accumulations A_τ versus duration τ to be straight, Fig. 29 (from Rémémeras & Hubert, 1990) shows a typical result showing the maximum recorded point rainfall depths for different durations going from minutes to several years. These measurements easily fit straight line with slope equal to about 0.5 (hence $\gamma_{max}\approx0.5$).

If we consider the empirically observed C_1 and α values for the temporal rain process we can readily explain this remarkable alignment. Recall from Table 3, (gauge results only) that $\alpha\approx0.5\pm0.1$, $C_1\approx0.4\pm0.1$ from series of quite disparate origins. Since $0\leq\alpha<1$, the following maximum order of singularity γ_0 of the process is obtained (equaton 12):

$$\gamma_0=\frac{C_1}{1-\alpha}+H \qquad (28)$$

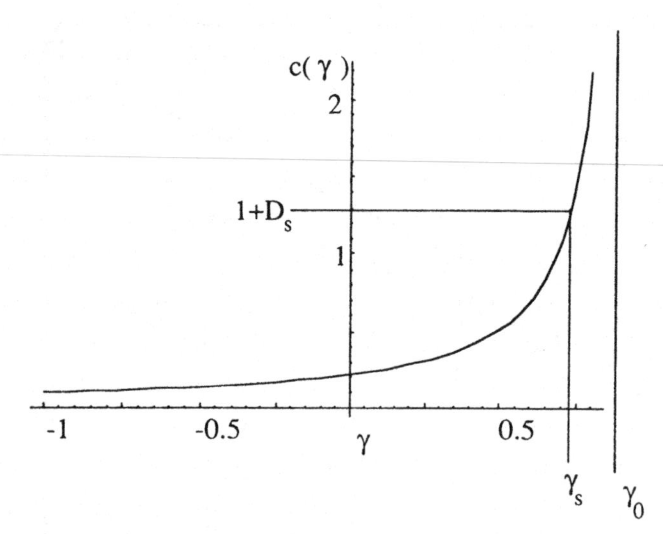

Fig. 30 The $c(\gamma)$ curve corresponding to the estimated parameters $\alpha \approx 0.51 \pm 0.05$, $C_1 \approx 0.44 \pm 0.16$, with γ_s for $N_s = \lambda^{D_s}$ samples, and γ_0 ($= \gamma_s$ for an infinite number of samples; $D_s = N_s = \infty$). From Hubert *et al.* (1993).

Using the above estimate of α, C_1, (with $H=0$) we obtain $\gamma_0 \approx 0.8 \pm 0.2$. This maximum corresponds to the stochastic generating process, for any finite sample, the actual limit will be determined by $\gamma_s = \gamma_{\max} = c^{-1}(D + D_s)$, see Fig. 30 for the corresponding illustration). However, when $\alpha < 1$, the difference $\gamma_s - \gamma_0$ is typically small:

$$\gamma_0 \left(1 - \alpha \left(\frac{C_1}{D}\right)^{-\frac{1}{\alpha}}\right) \leq \gamma_s \leq \gamma_0 \tag{29}$$

with the upper limit (implying $\gamma_s \to \gamma_0$) occuring when $D_s \to \infty$ (an infinite number of samples), and the lower limit applying for single samples ($D_s = 0$). Using the above gauge values in Table 3 for C_1, α, we obtain with $D_s = 0$, $\gamma \approx 0.7 \pm 0.2$, whereas for an infinite number of independent rain series, $\gamma \approx 0.9 \pm 0.3$. In both cases, the predicted slopes ($1 - \gamma \approx 0.3 \pm 0.2$, 0.1 ± 0.3 respectively) are close to those observed in Fig. 29 (≈ 0.5).

These results help to reconcile two opposing views on extreme precipitation, the 'extreme maximum precipitation' (PMP) and probability approaches (based on frequency analyses) since it simultaneously clarifies the role of the accumulation period and sample size in determining the observed maxima. It also provides a solid theoretical ground for the derivation of rate-duration-frequency curves.

GENERALIZED SCALE INVARIANCE, STRATIFICATION AND SPACE TIME TRANSFORMATIONS

Vertical stratification of rain

We have considered the simplest scaling system involving no prefered orientation; isotropic (self-similar) scaling whose theory has been developed over a considerable period of time, particularly in fluid turbulence. However, the atmosphere is not a simple fluid system, nor is it isotropic; gravity leads to differential stratification, the Coriolis force to differential rotation and radiative and microphysical processes lead to further complications. However even when the exact dynamical equations are unknown it can still be argued that at least over certain ranges, these phenomena are likely to be symmetric with respect to scale changing operations. This view is all the more plausible when it is realized that the requisite scale changes to transform the large scale to the small scale can be very general.

To see this, introduce a scale changing operator $\mathbf{T}_\lambda$ defined by: $\mathbf{T}_\lambda B_1 = B_\lambda$, where B_1 will be a large scale averaging set, B_λ, the corresponding set 'reduced' by factor λ. For example, considering multifractals, 'self-similar' measures will satisfy equations 7 or 9 with $\mathbf{T}_\lambda = \lambda^{-\mathbf{I}} = \lambda^{-1}\mathbf{I}$ where $\mathbf{I}$ is the identity matrix i.e. $\mathbf{T}_\lambda$ is a simple reduction by factor λ. However, much more general scaling transformations are possible; detailed analysis shows that practically the only restriction on $\mathbf{T}_\lambda$ is that it has group properties, viz.: $\mathbf{T}_\lambda = \lambda^{-\mathbf{G}}$ where $\mathbf{G}$ is the generator of the group of scale changing operations (this formalism is called 'generalized scale invariance' or GSI, (Schertzer & Lovejoy 1983b, 1985a, b, 1987a, b, 1989a,

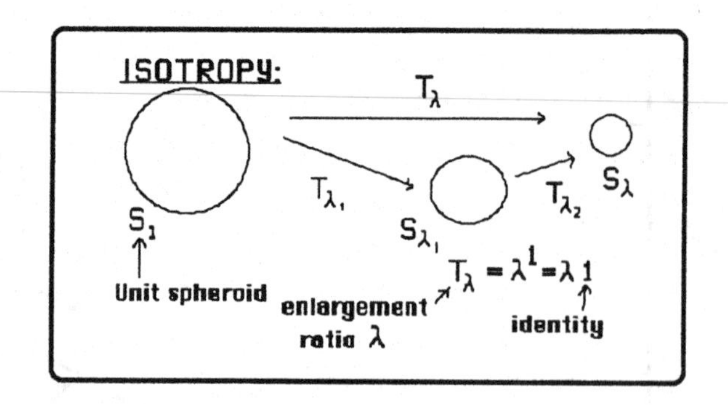

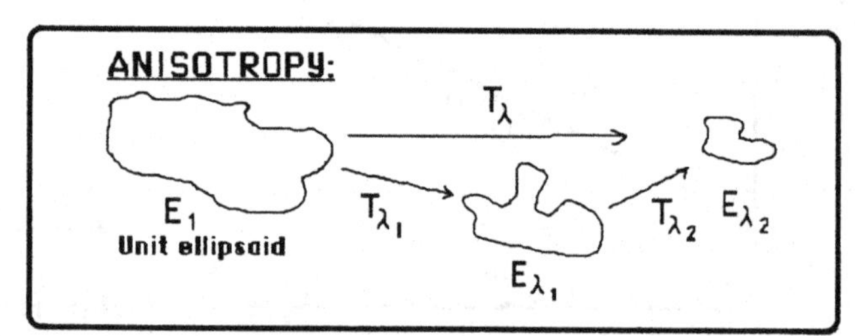

Fig. 31 A schematic diagram illustrating Generalized Scale Invariance. The top box indicates the familiar self-similar scale invariance involving ordinary 'zooms', with scale changing operator $\mathbf{T}_\lambda = \lambda^{-\mathbf{G}}$ with $\mathbf{G}$ the identity. The bottom box illustrates the more general case, the main requirement on $\mathbf{T}_\lambda$ is that it satisfy group properties. From Schertzer & Lovejoy (1989b).

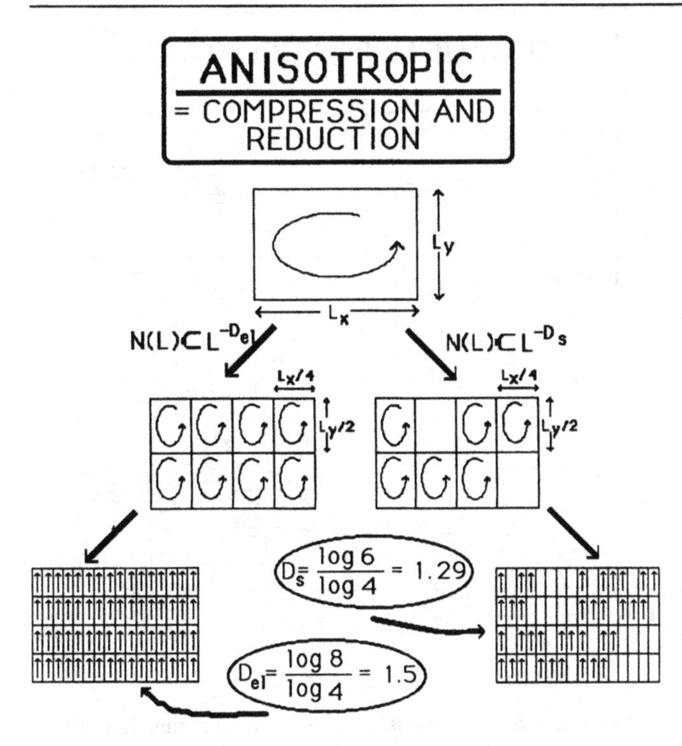

Fig. 32 A schematic diagram analogous to Fig. 7, but showing an anisotropic cascade, here with

$$\mathbf{G}=\begin{bmatrix}1 & 0\\ 0 & 1/2\end{bmatrix}.$$

At large scales eddies are flattened in the horizontal (like Hadley and Ferrel cells), whereas at small scales, they are more vertically aligned (like convective cells). The cross-sectional area clearly decreases with the 3/2 power of the horizontal scale; the elliptical dimension is 3/2. Left and right hand sides again show (stratified) homogeneous and (stratified) β model turbulence. From Lovejoy & Schertzer (1986a).

1991b), see Fig. 31 for a schematic illustration). For example 'self-affine' measures involve reductions coupled with compression along one (or more) axes; **G** is again a diagonal matrix but with not all diagonal elements equal to one (see Fig. 32 for a schematic of such a cascade, Fig. 33a for the corresponding balls). If **G** is still a matrix ('linear GSI') but has off-diagonal elements, then $\mathbf{T}_\lambda$ might compress an initial circle B_1 into an ellipsoid as well as rotate the result (see Fig. 33b). Linear and nonlinear GSI has already been used to model galaxies, clouds and rain (for examples of the corresponding balls B_λ, see Fig. 33c, d for examples with rotation (also Fig. 37), see Fig. 3d, e, for stratification only, see Fig. 3f, g, h, i). Empirically, the trace of **G** (called the 'elliptical dimension' d_{el} of the system) has been estimated in both rain and wind fields to have the values 2.22 and 2.55 respectively, indicating that the fields are neither isotropic ($d_{el}=3$), nor completely stratified ($d_{el}=2$), but are rather in between, becoming more and more stratified at larger and larger scales.

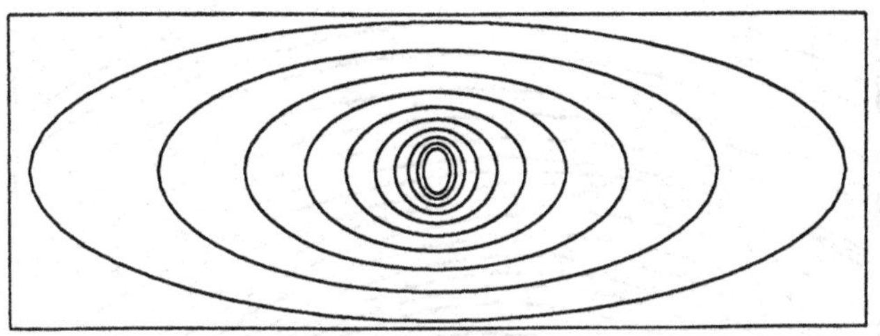

Fig. 33a The series of balls B_λ for a example of linear GSI with only diagonal elements (a 'self-affine' transformation) showing the stratification of structures that result. From Schertzer & Lovejoy (1989b).

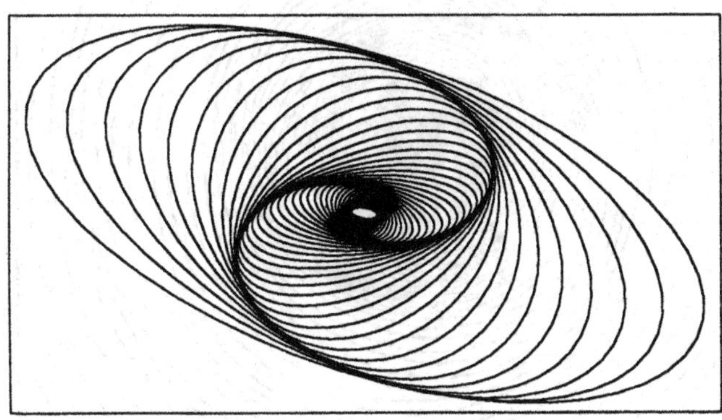

Fig. 33b Same as Fig. 33a, but with off diagonal elements showing the rotation and stratification of structures that result. From Schertzer & Lovejoy (1989b).

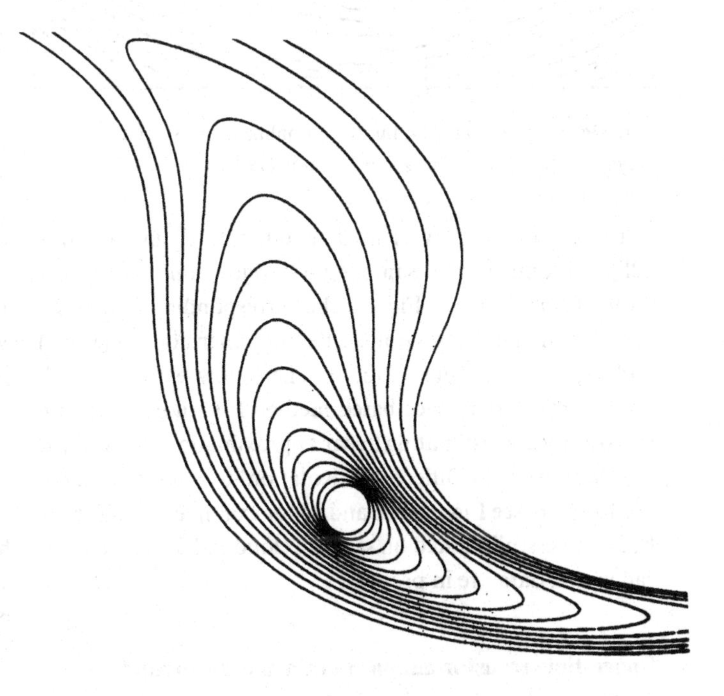

Fig. 33c Same as Fig. 33a, but for a nonlinear but deterministic generator **G**. From Schertzer & Lovejoy (1989b).

Fig. 33d Same as Fig. 33a, but for a nonlinear but stochastic generator **G**. From Schertzer & Lovejoy (1991a).

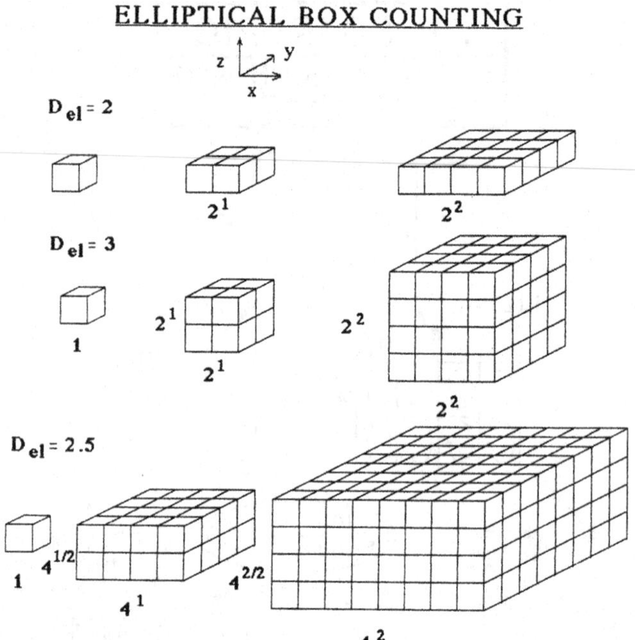

Fig. 34 The variation on usual (isotropic) box counting that can be used to estimate del when the direction but not magnitude of the stratification is known; this method applied to radar reflectivities (Figs. 13a, b) yielded $d_{el} \approx 2.22 \pm 0.07$ (Lovejoy *et al.*, 1987). This figure is from Schertzer & Lovejoy (1989b).

The method that was used to estimate d_{el} in rain was 'elliptical dimensional sampling' (Lovejoy *et al.*, 1987), the basic idea is shown in Fig. 34; the corresponding functional box-counting indicates quite different dimensions in (x, y, z) and (x, y) space, (Figs. 21a, b). There now exists a much improved Fourier space based method for studying scaling anisotropy and estimating linear approximations to **G** called the 'Monte Carlo Differential Rotation' technique (Pflug *et al.*, 1991a, b, see Figs. 35a, b and 37 for examples). It has now been successfully tested on satellite cloud radiances; tests on radar rain data are in progress.

Space-time transformations in rain and the prediction problem

In both geophysical and laboratory flows, it is generally far easier to obtain high temporal resolution velocity data at one or only a few points than to obtain detailed spatial information at a given instant. It is therefore tempting to relate time and space properties by assuming that the flow pattern is frozen and is simply blown past the sensors at a fixed velocity without appreciable evolution, and to directly use the time series information to deduce the spatial structure. This 'Taylor's hypothesis of frozen turbulence' (Taylor, 1938) can often be justified because in many experimental set ups, the flow pattern is caused by external forcing at a well defined velocity typically much larger than the fluctuations under study. However, in geophysical systems (in particular in the atmosphere and ocean) where no external forcing velocity exists, the hypothesis has often been justified by appeal to a 'meso-scale' gap separating large scale motions (two dimensional turbulence associated with 'weather') and small scale three dimensional turbulence (viewed as a kind of 'noise' superposed on the weather[90]. If such a separation existed, it might at least justify a statistical version of Taylor's hypothesis in which the large scale velocity is considered statistically constant (i.e. stationary). Various statistical properties such as spatial and temporal energy spectra would be similar even though no detailed transformation of a given

[90] Zawadzki (1973) finds that from 5 to 40 minutes this version of Taylor's hypothesis is consistent with radar rain data, but that for longer times it is inconsistent. We suspect that his data may be much more consistent with the generalizations of Taylor's hypothesis discussed here. For a discussion of conventional Taylor hypotheses in rain, see Gupta & Waymire (1987).

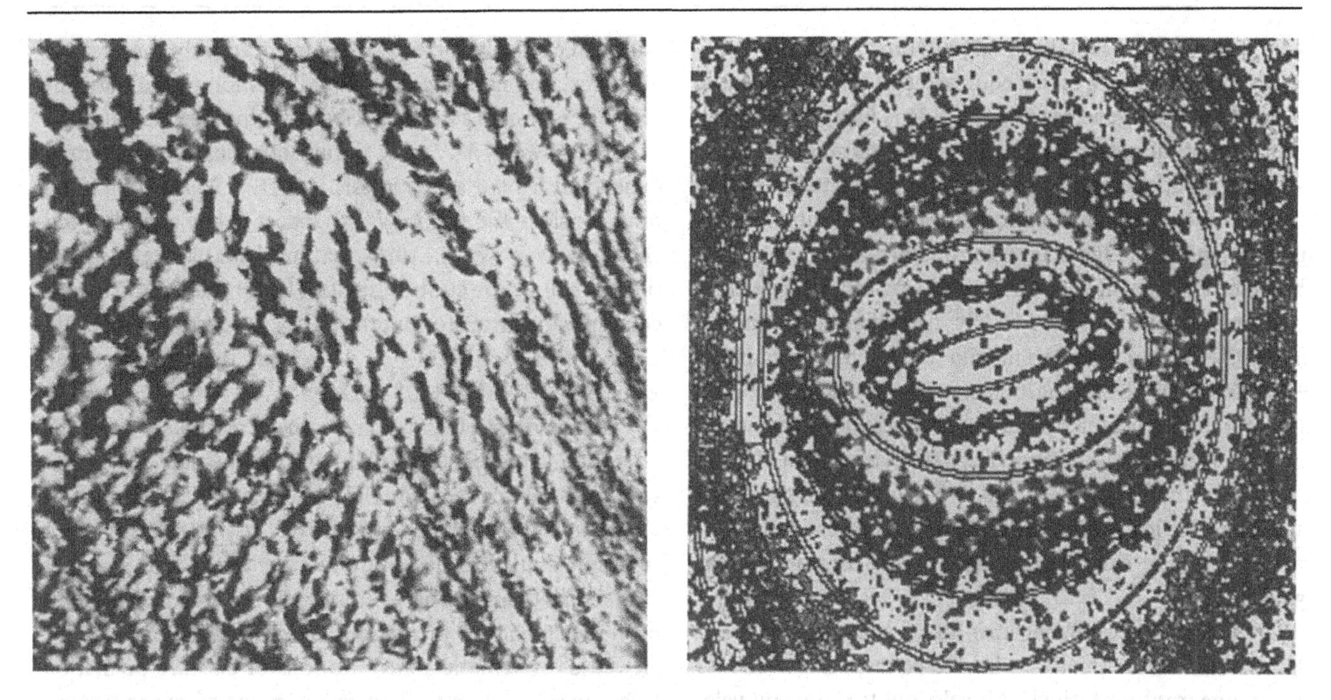

Fig. 35a Illustrations of the Monte Carlo differential rotation technique applied to NOAA satellite infra-red image of a field of Marine Stratocumulus clouds at 1.1 km resolution (256 × 256 points), the left is a grey scale rendition in real space, the right is the modulus squared of the fourier transform in Fourier space. The superposed ellipses are the best fits corresponding to a sphero-scale of 3.5 km,

$$\mathbf{G}=\begin{bmatrix}0.57 & -0.40\\ 0.40 & 1.43\end{bmatrix}$$

(in linear GSI, the Fourier space generator is the transpose of the real space generator). From Pflug *et al.* (1993).

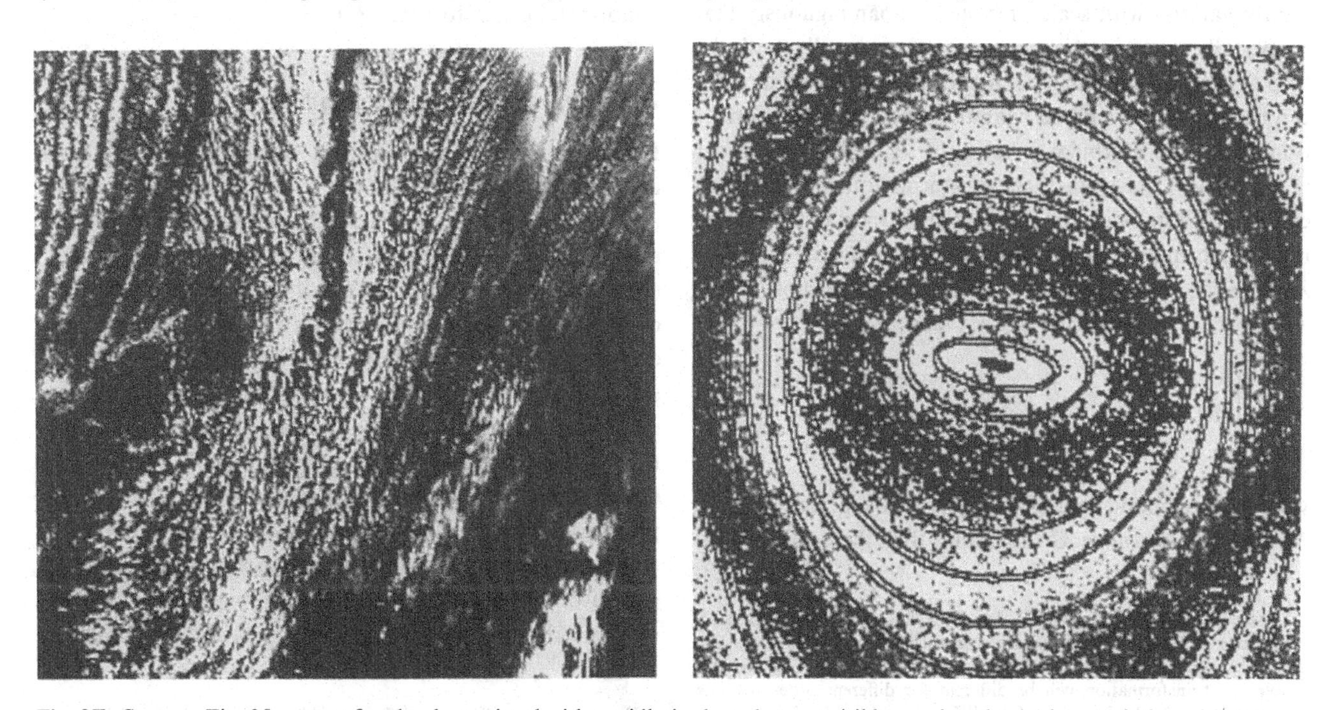

Fig. 35b Same as Fig. 35a except for cloud associated with a midlatitude cyclone, at visible wavelengths, 1.1 km resolution with 512 points on a side. The estimated generator, was

$$\mathbf{G}=\begin{bmatrix}0.68 & -0.18\\ 0.16 & 1.32\end{bmatrix}$$

the sphero-scale, 3.9 km. From Pflug *et al.* (1993).

time series to a particular spatial pattern would be possible. Only some kind of statistical equivalence would be possible.

However, as argued here, the scaling is likely to continue over most of the meteorologically significant range of scales. No large scale forcing velocity can be appealed to in order to transform from space to time; a turbulent velocity must be used (equation 5). Due to the multifractal nature of the wind, the exact scaling will depend on the $K(q)$ of the energy flux[91] (ε). For example, $<v_l> \approx l^{1/3-K(1/3)}$ according to Schmitt *et al.* (1992), in wind tunnel wind data, $C_1 \approx 0.25$, $\alpha \approx 1.3$, hence $-K(1/3) \approx 0.07$. Denoting this small intermittency correction by δ, we expect that rather than being scale independent, the space-time transformation has a scale dependent velocity $\langle v_l \rangle \approx l^H$ with $H_v = 1/3 + \delta$. The two geophysically relevant statistical Taylor's hypotheses therefore correspond to $H_v = 0$ or $H_v = 1/3 + \delta$ depending on the existence (or not) of the 'gap'.

The theoretical arguments mentioned above make it clear that the turbulent velocity is likely to be the relevant one for space-time transformations; this rules out the constant velocity ($H_v = 0$) hypothesis[92]. In fact, as discussed in Pflug *et al.* (1991, 1993), a good way to directly measure the series of 'balls' is to look for lines of constant energy density in Fourier space; Fig. 36 shows the result using vertically pointing radar data), the elliptical character of the (vertical) spatial wavenumber/frequency isolines, with eccentricity clearly varying with scale; this shows unambiguously that empirically (z, t) space is anistropic. Theoretically, with the help of in the formalism of Generalized Scale Invariance, we can understand this by deriving the space-time transformation from the (turbulent) value of $H_v (\approx 1/3)$. Consider (x, y, t) space, the space-time transformation can be simply expressed by statistical invariance with respect to the following transformation (generalized reduction in scale by factor λ): $x \Rightarrow \lambda x$, $y \Rightarrow \lambda y$, $t \Rightarrow \lambda^{1-H_v}$ or, using the notation $\mathbf{r} = (x, y, z, t)$, $\mathbf{r}_\lambda = T_\lambda \mathbf{r}_1$ with $T_\lambda = \lambda^{-G}$ and:

$$\mathbf{G} = \begin{bmatrix} 1 & 0 & 0 \\ 0 & 1 & 0 \\ 0 & 0 & 1-H_v \end{bmatrix}$$

we therefore obtain Trace $\mathbf{G} = 3 - H_v$ i.e. by measuring d_{el} or H_v we can determine $\mathbf{G}$ (assuming that there are no off-diagonal elements corresponding to rotation between space and time, and ignoring differential rotation in the horizontal). The isotropic statistical Taylor's hypothesis is therefore expressed by $d_{el} = 3$ ($H_v = 0$), the anisotropic, turbulent scale-dependent Taylor's hypothesis is $H_v \approx 1/3$, $d_{el} \approx 8/3$. If we now consider the full (x, y, z, t) space, it has already been shown (Lovejoy *et al.*, 1987) that in (x, y, z) space $d_{el} = 2.22$ (i.e. the z direction contributes 0.22 to the trace of $\mathbf{G}$) for the corresponding transformation in (x, y, z) space for radar rain data, hence for the (x, y, x, t) process, the corresponding value is $d_{el} \approx 2 + 0.22 + 2/3 \approx 2.89$.

The generator of space/time transformations defines the operation required to go from large to small space/time structures. When it is coupled with the multifractal probability generator (characterized by H, C_1, α), it provides a complete statistical description of the space/time process, and hence – in principle – all the information necessary to produce stochastic predictions. Such predictions may be viewed as systematic generalizations of existing prediction techniques based on the 'stochastic memory' of the system. Work is currently in progress at McGill and the Météorologie Nationale to use this approach to improve nowcasting methods for extrapolating radar echoes and satellite estimated rain areas[93] (e.g. Bellon *et al.*, 1980). For any given set of data, they have the potential to provide the theoretically optimum prediction: all that is required is knowledge of the multifractal generators ($\mathbf{G}$, H, C_1, α).

Dynamical simulations of rainfall

In this section we indicate briefly how to exploit the universality (and the measured H, C_1, α parameters) to perform multifractal simulations. The first 'continuous'[94] multifractal models of this type were discussed in Schertzer & Lovejoy (1987a, b), and Wilson (1991). Wilson *et al.* (1991) gives a comprehensive discussion including many practical (numerical) details[95]. In particular, the latter describes the numerical simulation of clouds and topography, including how to iteratively 'zoom' in, calculating details to arbitrary resolution in selected regions. Although we will not repeat these details here, enough information has been given in the previous sections to understand how they work. First, for a conserved (stationary) multifractal process φ_λ we define the generator $\Gamma_\lambda = \log \varphi_\lambda$. To yield a multifractal φ_λ, Γ_λ must be exactly a $1/f$ noise, i.e. its spectrum[96] is $E(k) \approx k^{-1}$ (this is

[91] *A priori*, any of the statistics $\langle v_l^q \rangle^{1/q}$ could be used in space-time transformations; all that is required is a parameter with the dimensions of velocity. It is therefore possible that (due to the multiple scaling of v_l) that the relevant transformation will be different for different orders of rain singularities γ (indeed, in view of the different a values found in time and space, this is necessary). Here for simplicity, we ignore this possible complication and consider transformations of the low order singularities corresponding to $q \sim 1$.

[92] Using lidar data, Lovejoy & Schertzer (1991a) find the value ($H_v = 0.5 \pm 0.3$) which is not accurate enough to usefully estimate δ. A more accurate radar based estimate $H_v = 0.38 + 0.05$ was announced by Tsonis *et al.* (1990).

[93] For rain forecasts for up to six hours, such techniques are already the best available.

[94] 'Continuous' since it does not involve integer ratios between eddies and sub-eddies; it is continuous in scale, avoiding the artificial straight lines of the (discrete ratio) cascades.

[95] For a recent illustration, see Fig. 10 from Lovejoy & Schertzer (1991c).

[96] For $\alpha < 2$, the generator variance diverges, we use 'generalized' spectra – see Schertzer & Lovejoy (1987a), appendix C.

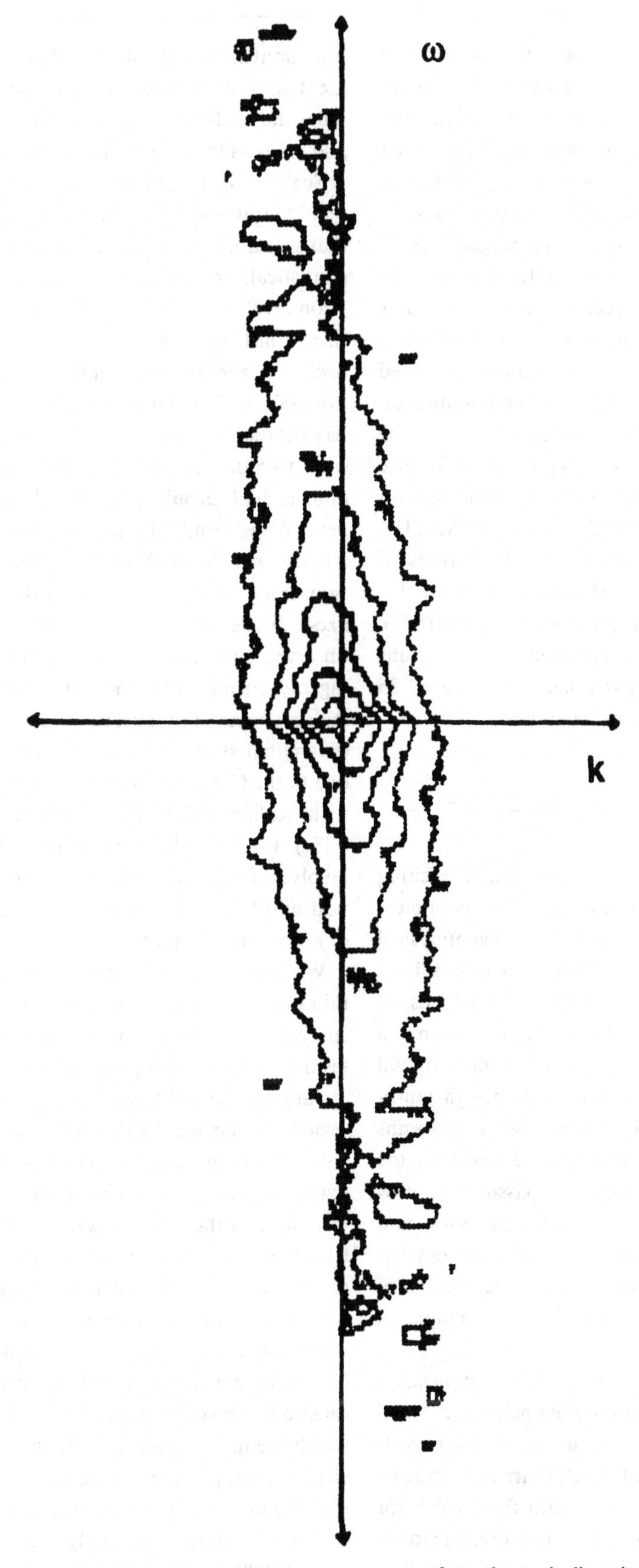

Fig. 36 We plot here the contours of the (two dimensional) energy spectrum from the vertically pointing radar reflectivities; 2 second temporal resolution, 21 m spatial resolution. The figure is the result of averaging the Fourier space energy density (modulus squared of the Fourier transform) over 20 consecutive (z, t) planes, each with 256×512 points (Fourier conjugate axes are k, ω respectively). Note the clear differential stratification. The rotation of the principle axes with respect to the (Fourier) axes seems not to be differential (hence due to a mean wind). From Tessier *et al.* (1993).

necessary to ensure the multiple scaling of the moments of φ_λ). To produce such a generator, we start with a stationary Gaussian or Lévy 'subgenerator'. The subgenerator is a noise consisting of independent random variables with either Gaussian ($\alpha=2$) or extremal Lévy distributions (characterized by the Levy index α), whose amplitude (e.g. variance in the Gaussian case) is determined by C_1. The subgenerator is then fractionally integrated (power law filtered in Fourier space) to give a (generalized) k^{-1} spectrum. This generator is then exponentiated to give the conserved φ_λ which will thus depend on both C_1 and α. Finally, to obtain a nonconserved process with spectral slope β, the result is fractionally integrated by multiplying the Fourier transform by k^{-H}. The entire process involves two fractional integrations and hence four FFTs. 512×512 fields can easily be modeled on personal computers (they take about 3 minutes on a Mac II), and $256\times256\times256$ fields (e.g. space-time simulations of dynamically evolving multifractal clouds) have been produced on a Cray 2 (Brenier[97], 1990, Brenier *et al.*, 1990). We used the multifractal parameters estimated by the various methods described above, taking $H\approx0.3$, $C_1\approx0.1$, $\alpha\approx1.35$ in space to produce the simulation shown in Fig. 37.

CONCLUSIONS

For over ten years, scaling ideas have provided an exciting new perspective for dealing with rain and other dynamical processes occuring in the atmosphere and other geophysical systems; in this paper we have attempted to give a brief review of this mushrooming field. To maintain a focus, as indicated in the title, we have restricted our attention as much as possible to an account of the necessary multifractal formalism and to specific results on rain. Although multifractal notions are also relevant in stream flows, river basins and other areas of hydrology, we have omitted these from the discussion. We have only mentioned in passing the now burgeoning literature concerning scaling analyses and modeling of clouds and their associated radiative transfer. Finally we have only given a brief outline of the relation of our results to turbulence theory and to recent empirical turbulence results.

During the period covered by this review, scaling ideas were extended far beyond the restrictive bounds of the fractal geometry of sets to directly deal with the multifractal statistics (and dynamics) of fields. Multifractals are now increasingly understood as providing the natural framework for scale-invariant nonlinear dynamics. Furthermore, due to the existence of stable attractive multifractal generators they provide attractive physical models. This implies that many of the details of the dynamics are irrelevant (universal behaviour) and leads to new and powerful multifractal simulation and analysis techniques (many of which were discussed).

Scaling ideas have also been enriched by extensions in another quite different direction: scaling anisotropy. Recall that a scaling system is one in which small and large scale (statistical) properties are related by a scale changing operation involving only the scale ratio: there is no characteristic size. Until recently, this scale change was restricted to ordinary 'zooms' or magnifications. Since the 1950s, this isotropic self-similar scaling has provided the theoretical basis of the standard model of atmospheric dynamics: a large scale two dimensional turbulence and a small scale three dimensional turbulence. The only generalization of scaling beyond self-similarity was a slight variation called 'self-affinity' which combined the zoom with a (differential) 'squashing' along certain fixed directions (e.g. coordinate axes). While this extension is necessary to account for the observed atmospheric stratification (implying a single scaling, anisotropic turbulence), it is still very special. In particular, geophysical applications generally involve not only differential stratification but also differential rotation (e.g. due to the Coriolis force). The formalism developed to deal with scaling anisotropy is Generalized Scale Invariance (GSI). GSI goes far beyond self-affinity: not only does it involve both differential rotation and stratification, it allows both effects to vary from place to place in either deterministic or even random manners.

We have argued that due to the enormous quantities of rain data spanning many orders of magnitude in time and space, that rain has and will continue to play a leading role in testing and developing new ideas in scaling and nonlinear dynamics. The rapid progress of this field makes the task of reviewing difficult. In the first part, we have attempted to concentrate on results which most clearly demonstrate the scaling of rain; by combining many different measurement techniques although the exact limits are still not clear, we have seen that it is possible that rain is scaling over most of the meteorologically significant range of scales. In general, we did not attempt detailed intercomparisons of different empirical results, largely because many were derived from essentially experimental data analysis techniques (such as functional box-counting, area-perimeter relations etc.), which are now somewhat outdated and which in any event were often applied to quite different data sets. Nevertheless, the measurements presented here cover the entire range from ≈1 mm (blotting paper analysis of rain drop distributions), to ≈10000 km (the global rain network), and give us considerable confidence that the basic (multi) scaling holds reasonably well.

Whereas early analysis and modeling techniques were

[97] This paper describes how such clouds simulations were used to produce a video called 'Multifractal Dynamics'.

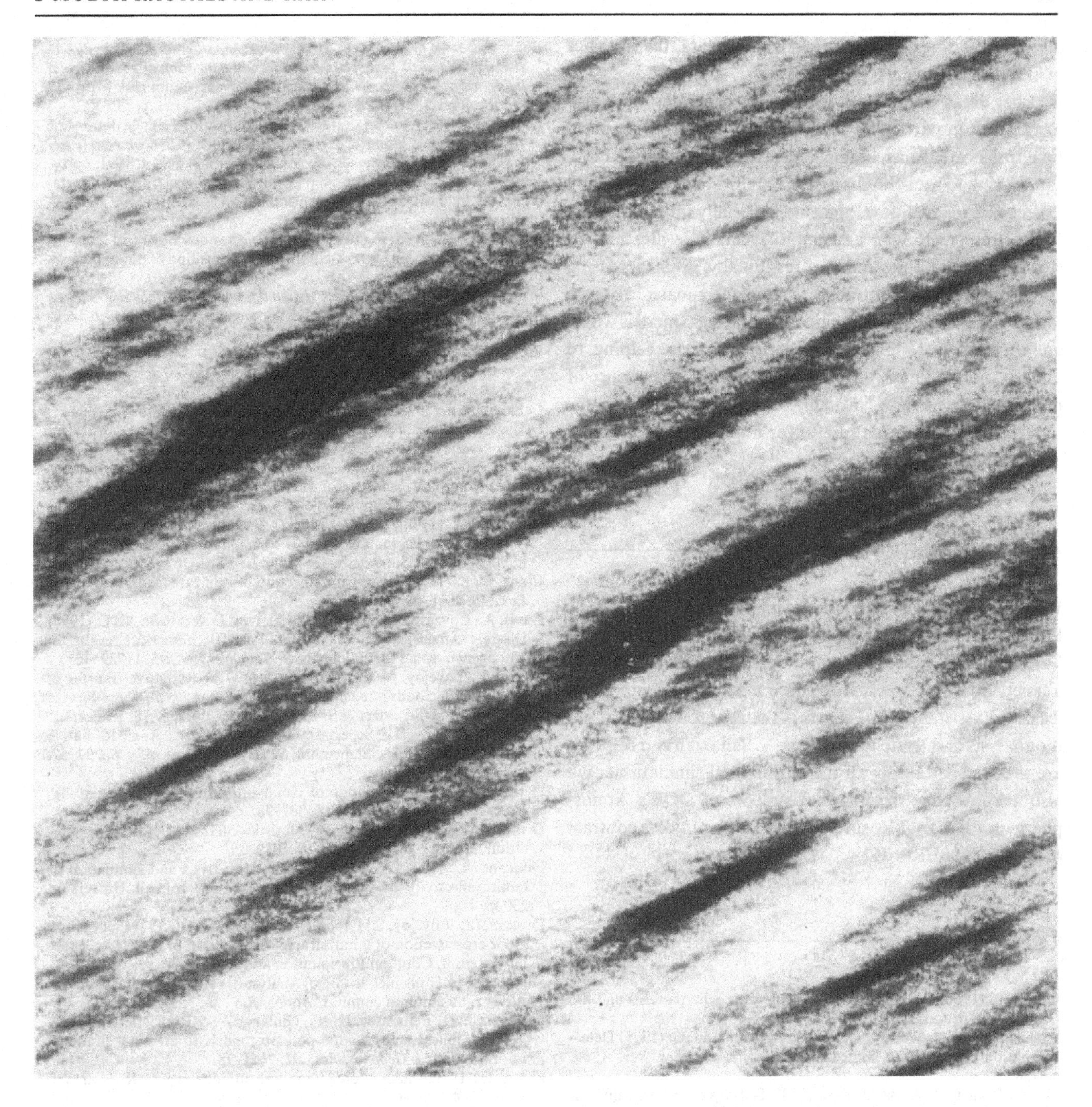

Fig. 37 Numerical simulation of a universal continuous cascade multifractal rain field on a 512×512 point gridwidth $\alpha = 1.3$, $C_1 = 0.1$, $H = 0.3$, G the same as empirically estimated in Fig. 35b. We thank S. Pechnold for help with the simulation.

based on *ad hoc* geometric notions relevant to fractal sets, the more mature multifractal framework outlined in the main part of the paper depended on two important breakthroughs. The first was the connection with the physics of rain processes via the problematic of passive scalar clouds, and the second, the continuous (cascade) modelling of the latter with the associated discovery of multifractal universality classes. Universality also provided the framework for the development of a new generation of 'specific' multifractal analysis methods that are analogous to parametric methods in standard statistics and are statistically quite robust. Indeed, the results of the Double Trace Moment technique are now providing consistent estimates of multifractal parameters in rainfall measured over wide ranges of scale in both time and space, as estimated from both rain gage and radar measurements (see Tables 3 and 4 for summaries). Although these results are quite recent, they suggest that the field is maturing, and that it is now the time for developing a variety of multifractal applications. Some have already been mentioned; the multifractal objective analysis problem, the multifractal observers problem for radar data, the statistics of extreme rain events.

Other areas where work is only just beginning were also mentioned, in particular the problem of multifractal space/time transformations, scaling anisotropy, and stratification, as well as their modeling. These are areas where we may soon expect exciting new developments, especially for multifractal forecasting methods, and multifractal classification of storms, morphologies and textures. Finally, given increasing confidence in our multifractal parameter estimates, all these ideas can be tested on (dynamical) multifractal models which are thus likely to play an important role in helping to understand the larger problem of resolution dependence of remotely sensed data, including the relation between the radiance and rain fields (useful for improving satellite rain estimating algorithms).

AKNOWLEDGEMENTS

We thank A. Davis, N. Desaulniers-Soucy, M. Duncan, F. Fabry, C. Hooge, P. Hubert, P. Guerrido, P. Ladoy, C. Larnder, D. Lavallée, S. Pecknold, K. Pflug, F. Schmitt, Y. Tessier and B. Watson for helpful comments, discussions and technical assistance. F. Francis is thanked for helping us in the analysis of the satellites images, V. Sahakian and F. Begin are thanked for help with the multifractal simulations. We also acknowledge the financial support of DOE's Atmospheric Radiation Measurement (ARM) project, contract #DE-FG03-90ER61062.

REFERENCES

Armijo, L. (1966) Statistical properties of radar echo patterns and the radar echo process. *J. Atmos. Sci.*, **23**, 560–8.

Atmanspacher, H., Scheingraber, H. & Weidenmann, G. (1989) Determination of $f(a)$ for a limited random point set. *Phys. Rev.* A, **40**, 3954.

Bak, P., Tang, C. & Weissenfeld, K. (1987) Self-organized criticality: an explanation of 1/f noise. *Phys. Rev. Lett.*, **59**, 381–4.

Baryshnikova, Y. S., Zaslavskii, G. M., Lupyan, E. A., Moiseev, S. S. & Sharkov, E. A. (1989) Fractal dimensionality of cloudiness IR images and turbulent atmospheric properties. *Akad. Nauk USSR, Issledovanii Zemli, iz Kosmos*, **1** (in Russian), 17–26.

Bell, T. L. (1987) A space-time stochastic model of rainfall for satellite remote sensing studies. *J. Geophys. Res.*, **92**, 9631–44.

Bellon, A., Lovejoy, S. & Austin, G. L. (1980) A short-term precipitation forecasting procedure using combined radar and satellite data. *Mon. Wea. Rev.*, **108**, 1554–66.

Benzi, R., Paladin, G., Parisi, G. & Vulpiani, A. (1984) *J. Phys.* A, **17**, 3521.

Bialas, A. & Peschanski, R. (1986) Moments of rapidity distributions as a measure of short-range fluctuations in high-energy collisions. *Nucl. Phys. B*, **273**, 703–18.

Biondini, R. (1976) Cloud motion and rainfall statistics, *J. Appl. Meteor.*, **15**, 205–24.

Blanchard, D. C. (1953) Raindrop size-distribution in Hawaiian rains. *J. Meteor.*, **10**, 457–73.

Bochi Kebe, G. & Howes, K. (1990) *Determination of the three dimensional spatial distribution and of the fractal dimension of passive scalar fields using stereoscopic techniques*, 3rd year honours lab project report, 46pp, McGill Physics dept.

Bouquillon, C. & Moussa, R. (1991) Caracterisation fractale d'une serie chronologique d'intensité de pluie. *Proceedings of Rencontres Hydrologiques Franco-Roumaines, Ecoles des Mines*, 3–6 Sept. 1991, 6pp.

Bras, R. L. & Rodriguez-Iturbe, I. (1976) Rainfall generation: a nonstationary time varying multidimensional model. *Wat. Resour. Res.*, **12**, 450–6.

Brax, P. & Pechanski, R. (1991) Levy stable law description of intermittent behaviour and quark-gluon phase transitions, *Phys. Lett.* B, 225–30.

Brenier, P. (1990) *Simulations dynamiques multifractale des nuages*, Master's Thesis. Ecole Normale Superieure des Sciences et Techniques Avancées, Paris, France.

Brenier, P., Schertzer, D., Sarma, G., Wilson, J. & Lovejoy S. (1990) Continuous Multiplicative cascade models of passive scalar clouds. *Annales Geophys.*, **8** (Special Issue), 320.

Cahalan, R. F. (1991) Landsat Observations of Fractal Cloud Structure. *Scaling, fractals and non-linear variability in geophysics*, D. Schertzer & S. Lovejoy (eds.), pp. 281–96, Kluwer.

Cole, J. W. (1964) *Statistics related to the shape and scale of pattern elements*. Final Report, Contract Cwb-10709, Travelers Research Center Hartford, Conn.

Come, J. M. (1988) *Caractérisation fractale des relations Périmetre/Surface des précipitations en zone soudano-sahélienne*. MSc. thesis, Ecole Nationale Supérieure des Mines de Paris, 102pp.

Crane, R. K. (1990) Space-time structure of rain rate fields. *J. Geophy. Res.*, **95**, 2011–20.

Davis, A., Lovejoy, S., Gabriel, P., Schertzer, D. & Austin, G. L. (1990) Discrete Angle Radiative Transfer. Part III: numerical results on homogeneous and fractal clouds. *J. Geophys. Res.*, **95**, 11729–42.

Davis, A., Lovejoy, S. & Schertzer, D. (1991a) Radiative transfer in multifractal clouds. *Scaling, fractals and non-linear variability in geophysics*, D. Schertzer & S. Lovejoy (eds.), pp. 303–18, Kluwer.

Davis, A., Lovejoy, S. & Schertzer, D. (1991b) Discrete Angle Radiative transfer in a multifractal medium, *SPIE proceedings 1558*, pp. 37–59, San Diego, 21–26 July.

Detwiller, A. (1990) Analysis of cloud imagery using box-counting. *Inter. J. of Remote Sensing*, **11**, 887–98.

Drufuca, G. (1977) Radar derived statistics on the structure of precipitation patterns. *J. Appl. Met.*, **16**, 1029–35.

Duncan, M. (1993) Universal multifractal analysis and simulation of radar reflectivity of rain, Ph.D. dissertation, McGill University, 230pp.

Duncan, M., Lovejoy, S., Fabry, F. & Schertzer, D. (1992) Fluctuating radar cross-section of a multifractal distribution of scatterers. Proc. 11th Internat. Conf. on Precipitation and Clouds, pp. 997–1000.

Duroure, C. & Guillemet, B. (1990) Analyse des Hétérogénéités spatiales daes stratocumuls et cumulus. *Atmos. Res.*, **25**, 331–50.

Eagleson, P. S., Fennessey, N. M., Qinlang, W. & Rodriguez-Iturbe, I. (1987) Application of spatial poisson models to air mass thunderstorm rainfall. *J. Geophys. Res.*, **92**, 9661–78.

Fan A. H. (1989) Chaos additif et multiplicatif de Levy. *C. R. Acad. Sci. Paris*, I **308**, 151–4.

Fraedrich, K. & Larnder, C. (1993) Scaling regimes of composite rainfall time series. *Tellus*, **45A**, 289–98.

Frisch U. P., Sulem P. L. & Nelkin M. (1978) A simple dynamical model of intermittency in fully developed turbulence. *J. Fluid Mech.*, **87**, 719–24.

Gabriel, P., Lovejoy, S., Davis, A., Schertzer, D. & Austin, G. L. (1990) Discrete Angle Radiative Transfer. Part II: renormalization approach to scaling clouds. *J. Geophys. Res.*, **95**, 11717–28.

Gabriel, P., Lovejoy, S., Schertzer, D. & Austin, G. L. (1988) Multifractal analysis of satellite resolution dependence. *Geophys. Res. Lett.*, pp. 1373–6.

Giraud, R., Montariol, F., Schertzer, D. & Lovejoy, S. (1986) The codimension function of sparse surface networks and intermittent fields. Astract vol., *Nonlinear Variability in Geophysics*, 35–6, McGill University, Montreal, Canada.

Ghilardi, P. (1990) A search for chaotic behaviour in storm rainfall events. *Annales Geophys.* (special issue), p. 313.

Grassberger, P. (1983) Generalized dimensions of strange attractors. *Phys. Lett.* A **97**, 227.

Gupta, V. & Waymire, E. (1987) On Taylor's hypothesis and dissipation in rainfall. *J. Geophys. Res.*, **92**, 9657–60.
Gupta, V. & Waymire, E. (1990) Multiscaling properties of spatial rainfall and river flow distributions. *J. Geophys. Res.*, **95**, 1999–2010.
Halsey, T. C., Jensen, M. H., Kadanoff, L. P., Procaccia, I. & Shraiman, B. (1986) Fractal measures and their singularities: the characterization of strange sets. *Phys. Rev.* A **33**, 1141–51.
Harris, D. & Lewis, G. (1991) *Determination of fractal distribution of precipitation using laser scattering*. 3rd year honours project report, 47pp, McGill Physics dept.
Hentschel, H. G. E. & Proccacia, I. (1983) The infinite number of generalized dimensions of fractals and strange attractors. *Physica* **8D**, 435–44.
Houze, R. A. & Chee-Pong Cheng (1977) Radar characteristics of tropical convection observed during GATE: mean properties and trends over the summer season. *Mon. Wea. Rev.*, **105**, 964–80.
Hubert, P. (1991) Analyse multifractale de champs temporels d'intensités des precipitations. *Proceedings of Rencontres Hydrologiques Franco-Roumaines, Ecoles des Mines*, 3–6 Sept. 1991, 6pp.
Hubert, P. & Carbonnel, J. P. (1988) Caractérisation fractale de la variabilité et de l'anisotropie des précipitations intertropicales. *C. R. Acad. Sci. Paris*, **307**, 909–14.
Hubert, P. & Carbonnel, J. P. (1989) Dimensions fractales de l'occurence de pluie en climat soudano-sahelien. *Hydro. Continent.*, **4**, 3–10.
Hubert, P. & Carbonnel, J. P. (1991) Fractal characterization of intertropical precipation variability. *Scaling, fractals and non-linear variability in geophysics*, D. Schertzer & S. Lovejoy (eds)., pp. 209–13, Kluwer.
Hubert P., Tessier, Y., Ladoy, P., Lovejoy, S., Schertzer, D., Carbonnel, J. P., Violette, S., Desurosne, I. & Schmitt, F. (1993) Multifractals and extreme rainfall events, *Geophs. Res. Lett.*, **20**, 931–4.
Hurst, H. E. (1951) Long-term storage capacity of reservoirs. *Trans. of the Amer. Soc. of Civil Engin.*, **116**, 770–808.
Jennings, A. H. (1950) *Mon. Wea. Rev.*, **78**, 4–5.
Kida, S. (1991) Log stable distribution and intermittency of turbulence. *J. Phys. Soc. of Japan*, **60**, 5–8.
Keddem, B. & Chiu, L. S. (1987) Are rain rate processes self-similar? *Wat. Resour. Res.*, **23**, 1816–18.
Koleshnikova, V. N. & Monin, A. S. (1965) Spectra of meteorological field fluctuations. *Izv. Atmos. Oceanic Physics*, **1**, 653–69.
Kolmogorov, A. N. (1940) Wienersche spiralen and einige andere interessante kurven in Hilbertschen Raum. *C. R. (Doklady) Acad. Sci. URSS (N.S.)*, **26**, 115–18.
Kolmogorov, A. N. (1962) A refinement of previous hypotheses concerning the local structure of turbulence in viscous incompressible fluid at high Reynolds number. *J. Fluid Mech.*, **83**, 349.
Konrad, T. G. (1978) Statistical models of summer rainshowers derived from fine-scale radar observations. *J. Appl. Meteor.*, **17**, 171–88.
Kumar, P. & Foufoula-Georgiou, E. (1993) A new look at rainfall fluctuations and scaling properties of spatial rainfall using orthogonal wavelets. *J. Appl. Meteor.*, **32**, 209–22.
Ladoy, P., Schertzer, D. & Lovejoy, S. (1986) Une étude d'invariance locale-regionale des temperatures. *La Météorologie*, **7**, 23–34.
Ladoy, P., Marquet, O., Piriou, J. M., Lovejoy, S. & Schertzer, D. (1987) Inhomogeneity of geophysical networks, calibration of remotely sensed data and multiple fractal dimensions. *Terra Cognita*, **7**, 2–3.
Ladoy, P., Lovejoy, S. & Schertzer, D. (1991) Extreme fluctuations and intermittency in climatological temperatures and precipitation. *Scaling, fractals and non-linear variability in geophysics*, D. Schertzer & S. Lovejoy (eds)., pp. 241–50, Kluwer.
Ladoy, P., Schmitt, F., Schertzer, D. & Lovejoy, S. (1993) Variabilité temporelle des observations pluviometriques à Nimes, *C. R. Acad. des Sci.*, **317**, II, 775–82.
Lamperti, J. (1962) Semi-stable stochastic processes. *Trans. Am. Math. Soc.*, **104**, 62–78.
Larnder, C., Desaulniers-Soucy, N., Lovejoy, S., Schertzer, D., Braun, C. & Lavallée, D. (1992) Evidence for universal multifractal behaviour in human speech. *Chaos and Bifurcation*, **2**, 715–19.
Lavallée, D. (1991) *Multifractal analysis and simulation techniques and turbulent fields*, 142pp, PhD. thesis, McGill University.
Lavallée, D., Schertzer, D. & Lovejoy, S. (1991a) On the determination of the co-dimension function. *Scaling, fractals and non-linear variability in geophysics*, D. Schertzer & S. Lovejoy (eds)., pp. 99–110, Kluwer.
Lavallée, D., Lovejoy, S. & Schertzer, D. (1991b) Universal multifractal theory and observations of land and ocean surfaces, and of clouds, *SPIE proceedings 1558*, San Diego, 21–26 July.
Lavallée, D., Lovejoy, S., Schertzer, D. & Ladoy, P. (1993) Nonlinear variability and Landscape topography: analysis and simulation. *Fractals in Geography*, L. De Cola & N. Lam (eds.), Prentice Hall, 158–92.
Levich, E. & Tzvetkov, E. (1985) *Phys. Rep.*, **120**, 1–45.
Lopez, R. E. (1976) Radar characteristics of the cloud populations of tropoical disturbances in the northwest Atlantic. *Mon. Wea. Rev.*, **194**, 269–83.
Lopez, R. E. (1977a) Some properties of convective plume and small fair-weather cumulus fields as measured by acoustic and lidar sounders. *J. Appl. Meteor.*, **16**, 861–5.
Lopez, R. E. (1977b) The log-normal distribution and cumulus cloud populations. *Mon. Wea. Rev.*, **105**, 865–72.
Lovejoy, S. (1981) *Analysis of rain areas in terms of fractals, 20th conf. on radar meteorology*, pp. 476–84, AMS Boston.
Lovejoy, S. (1982) The area-perimeter relations for rain and cloud areas. *Science*, **216**, 185–7.
Lovejoy, S. (1983) La geometrie fractale des regions de pluie et les simulations aleatoires. *Houille Blanche*, **516**, 431–6.
Lovejoy, S., Tardieu, J. & Monceau, G. (1983) Etude d'une situation frontale: analyse meteorologique et fractale. *La Météorologie*, **6**, 111–18.
Lovejoy, S. & Schertzer, D. (1985a) Generalised scale invariance and fractal models of rain. *Wat. Resour. Res.*, **21**, 1233–50.
Lovejoy, S. & Schertzer, D. (1985b) Rainfronts, fractals and rainfall simulations. Hydro. Appl. of Remote sensing and data trans., *Proc. of the Hamburg symposium*, IAHS publ. no. 145, pp. 323–34.
Lovejoy, S. & Mandelbrot, B. (1985) Fractal properties of rain and a fractal model. *Tellus*, **37A**, 209–32.
Lovejoy S. & Schertzer, D. (1986a) Scale invariance, symmetries, fractals and stochastic simulations of atmospheric phenomena, *Bulletin of the AMS*, **67**, 21–32.
Lovejoy, S. & Schertzer, D. (1986b) Scale invariance in climatological temperatures and the spectral plateau. *Annales Geophys.*, **4B**, 401–10.
Lovejoy, S., Schertzer, D. & Ladoy, P. (1986) Fractal characterization of inhomogeneous measuring networks. *Nature*, **319**, 43–4.
Lovejoy, S., Schertzer, D. & Tsonis, A. A. (1987) Functional box-counting and multiple elliptical dimensions in rain. *Science*, **235**, 1036–8.
Lovejoy, S. & Schertzer, D. (1988) Extreme variability, scaling and fractals in remote sensing: analysis and simulation, *Digital image processing in Remote Sensing*, J. P. Muller (ed.), pp. 177–212, Francis and Taylor.
Lovejoy, S. & Schertzer, D. (1989) Comments on 'Are rain rate processes self-similar?' *Wat. Resour. Res.*, **25**, 577–9.
Lovejoy, S. & Schertzer, D. (1990a) Multifractals, universality classes, satellite and radar measurements of clouds and rain. *J. Geophys. Res.*, **95**, 2021–34.
Lovejoy, S. & Schertzer, D. (1990b) Our multifractal atmosphere: a unique laboratory for nonlinear dynamics. *Physics in Canada*, **46**, 62–71.
Lovejoy, S. & Schertzer, D. (1990c) Fractals, rain drops and resolution dependence of rain measurements. *J. Appl. Meteor.*, **29**, 1167–70.
Lovejoy, S., Gabriel, P., Davis, A., Schertzer, D. & Austin, G. L. (1990) Discrete Angle Radiative Transfer. Part I: scaling and similarity, universality and diffusion. *J. Geophys. Res.*, **95**, 11699–715.
Lovejoy, S. & Schertzer, D. (1991a) Multifractal analysis techniques and rain and cloud fields from 10^3 to 10^6m. *Scaling, fractals and non-linear variability in geophysics*, D. Schertzer & S. Lovejoy (eds.), pp. 111–44, Kluwer.
Lovejoy, S. & Schertzer, D. (1991b) Universal Multifractal temperature simulations. *EOS*, **72**, 1–2.
Lovejoy, S. & Schertzer, D. (1993) Scale invariance and multifractals in the atmosphere. *Encyclopedia of the Environment*, Pergamon, 527–32.
Mandelbrot, B & Van Ness, J. W. (1968) Fractional Brownian motions, fractional noises and applications. *SIAM Review*, **10**, 422–50.
Mandelbrot, B. & Wallis, J. R. (1968) Noah, Joseph and operational hydrology. *Wat. Resour. Res.*, **4**, 909–18.
Mandelbrot, B. & Wallis, J. R. (1969) Some long-run properties of geophysical records. *Wat. Resour. Res.*, **5**, 228.
Mandelbrot, B. (1974) Intermittent turbulence in self-similar cascades: divergence of high moments and dimension of the carrier. *J. Fluid Mech.*, **62**, 331–50.

Mandelbrot, B. (1983) *The Fractal Geometry of Nature*. Freeman, San Francisco, 318pp.
Mandelbrot, B. (1984) Fractals in physics: squig clusters, diffusions, fractal measures and the unicity of fractal dimensionality. *J. Stat. Phys.*, **34**, 895–930.
Mandelbrot, B. (1989) Fractal geometry: what is it and what does it do? *Fractals in the Natural Sciences*, M. Fleischman, D. J. Tildesley & R. C. Ball, pp. 3–16, Princeton University Press.
Marquet, O. & Piriou, J. P. (1987) *Modélisations et analyses multifractales de la pluie*. Thèse Ingenieur des travaux, Ecole Nationale de la Météorologie Nationale, Toulouse.
Marshall, J. S. & Palmer, W. M. (1948) The distribution of raindrops with size. *J. Meteor.*, **5**, 165–6.
Marshall, J. S. & Hitschfeld, W. (1953) Interpretation of fluctuating echoes from randomly distributed scatters, part I. *Can. J. Phys.*, **31**, 962–94.
Meneveau, C. & Sreenivasan, K. R. (1987) Simple multifractal cascade model for fully developed turbulence. *Phys. Rev. Lett.*, **59** (13), 1424–7.
Meneveau, C. & Sreenivasan, K. R. (1991) The multifractal nature of turbulent energy dissipation. *J. Fluid Mech.*, **224**, 429–84.
Montariol, F. & Giraud, R. (1986) *Dimensions et multidimensions des réseaux de mesure et des précipitations*. Thèse Ingenieur des travaux, Ecole Nationale de la Météorologie Nationale, Toulouse.
Novikov, E. A. & Stewart, R. (1964) Intermittency of turbulence and spectrum of fluctuations in energy-dissipation. *Izv. Akad. Nauk. SSSR. Ser. Geofiz.*, **3**, 408–12.
Obhukhov, A. (1962) Some specific features of atmospheric turbulence. *J. of Geophys. Res.*, **67**, 3011.
Olsson, J., Niemczynowicz, J., Berndtsson, B. & Larson, M. (1990) Fractal properties of rainfall time series. *Annales Geophys.* (special issue), p. 142.
Osborne, A. R. & Provenzale, A. (1989) Finite correlation dimension for stochastic systems with power-law spectra. *Physica* D, **35**, 357–81.
Paladin, G. & Vulpiani, A. (1987) *Phys. Rev. Lett.*, **156**, 147.
Parisi, G. & Frisch, U. (1985) A multifractal model of intermittency, *Turbulence and predictability in geophysical fluid dynamics and climate dynamics*, pp. 84–8, Ghil, Benzi & Parisi (eds.), North-Holland.
Paulhaus, J. L. H. (1965) *Mon. Wea. Rev.*, **93**, 331–5.
Pflug, K., Lovejoy, S. & Schertzer, D. (1991a) Generalized Scale Invariance, Differential Rotation and Cloud Texture. *Nonlinear Dynamics of Structures*, R. Z. Sagdeev, U. Frisch, F. Hussain, S. S. Moiseev & N. S. Erokhin (eds.), pp. 71–80, World Scientific.
Pflug, K., Lovejoy, S. & Schertzer, D. (1991b) Generalized Scale Invariance, Differential Rotation and Cloud Texture: analysis and simulation. *J. Atmos. Sci.* (in press).
Phan, T. D. & Miville, B. (1986) *Fractal dimension of rain drops distribution*. 3rd year honours lab report, 116pp, McGill physics dept.
Ratti, S. (1991) Universal multifractal analysis of multiparticle production in hadron-hadron collisions ar r(s) = 16.7GeV. *Proceedings of Hadron physics conf.*, China, Sept. 1991.
Ratti, S., Salvadori, G., Lovejoy, S. & Schertzer, D. (1991) Preprint FNT (in preparation).
Richardson, L. F. (1922) *Weather prediction by numerical process*. Republished by Dover, 1965.
Rémémeras, G. & Hubert, P. (1990) Article: *'Hydrologie' de l'Encyclopedia Universalis*, Paris, XI, pp. 796–806.
Rodriguez-Iturbe, I., Gupta, V. K. & Waymire, E. (1984) Scale considerations in the modelling of temporal rainfall. *Wat. Resour. Res.*, **20**, 1611–19.
Rodriguez-Iturbe, I., de Power, B. F. & Valdés, J. B. (1987) Rectangular pulses point process models for rainfall: analysis of empirical data. *J. Geophys. Res.*, **92**, 9645–56.
Rodriguez-Iturbe, I., de Power, B. F., Sharifi, M. B. & Georgakakos, K. P. (1989) Chaos in rainfall. *Wat. Resour. Res.*, **25**, 1667–75.
Rosso, R. & Burlando, P. (1990) Scale Invariance in temporal and spatial rainfall. *Annales Geophys.* (special issue), p.145.
Rhys, F. S. & Waldvogel, A. (1986) Fractal shape of hail clouds. *Phys. Rev. Lett.*, **56**, 784–7.
Schertzer, D. & Lovejoy, S. (1983a) The dimension of atmospheric motions, Preprints, *IUTAM Symp. on turbulence and chaotic phenomena in fluids*, pp. 141–4, Kyoto, Japan.
Schertzer, D. & Lovejoy, S. (1983b) Elliptical turbulence in the atmosphere. *Proceedings of the 4th symposium on turbulent shear flows*, 11.1–11.8, Karlsrhule, West Germany.
Schertzer, D. & Lovejoy, S. (1984) *Turbulence and chaotic phenomena in fluids*, T. Tatsumi (ed.), pp. 505–8, North-Holland.
Schertzer, D. & Lovejoy, S. (1985) The dimension and intermittency of atmospheric dynamics, *Turbulent Shear Flow*, **4**, 7–33, B. Launder (ed.), Springer.
Schertzer, D. & Lovejoy, S. (1985b) Generalised scale invariance in turbulent phenomena. *Physico-Chemical Hydrodynamics Journal*, **6**, 623–35.
Schertzer, D. & Lovejoy, S. (1986) Intermittency and singularities: generalised scale invariance in multiplicative cascade processes. *Abstract vol., Non-linear variability in Geophysics Workshop*, 14–15, McGill U., Montreal, Canada.
Schertzer, D. & Lovejoy, S. (1987a) Physically based rain and cloud modeling by anisotropic, multiplicative turbulent cascades. *J. Geophys. Res.*, **92**, 9692–714.
Schertzer, D. & Lovejoy, S. (1987b) Singularités anisotropes, et divergence de moments en cascades multiplicatifs. *Annales Math. du Que.*, **11**, 139–81.
Schertzer, D., Lovejoy, S., Visvanthan, R., Lavallée, D. & Wilson, J. (1988) Multifractal analysis techniques and rain and cloud fields. *Fractal Aspects of Materials: Disordered Systems*, D. A. Weitz, L. M. Sander & B. B. Mandelbrot (eds.), pp. 267–70, Materials Research Society, Pittsburg.
Schertzer, D. & Lovejoy, S. (1989a) Nonlinear variability in geophysics: multifractal analysis and simulations. *Fractals: Their physical origins and properties*, Pietronero (ed.), pp. 49–79, Plenum Press, New York.
Schertzer, D. & Lovejoy, S. (1989b) Generalized scale invariance and multiplicative processes in the atmosphere. *Pageoph*, **130**, 57–81.
Schertzer, D. & Lovejoy, S. (1992) Hard vs. soft multifractal processes, *Physica* A, **185**, 187–94.
Schertzer, D. & Lovejoy, S. (1991b) Nonlinear geodynamical variability: Multiple singularities, universality and observables. *Scaling, fractals and non-linear variability in geophysics*, D. Schertzer & S. Lovejoy (eds.), pp. 41–82, Kluwer.
Schertzer, D., Lovejoy, S., Lavallée, D. & Schmitt, F. (1991) Universal hard multifractal turbulence, theory and observations. Nonlinear Dynamics of Structures. R. Z. Sagdeev, U. Frisch, F. Hussain, S. S. Moiseev, N. S. Erokhin (eds.), pp. 213–35, World Scientific.
Schertzer, D., Lovejoy, S. & Lavallée, D. (1993) Generic multifractal phase transitions and self-organized criticality. In: *Cellular Automata: Prospects in Astrophysical Applications*, Pendang, J. M. & Lejeune, A. (eds.), World Scientific, 216–27.
Schertzer, D., Lovejoy, S., Schmitt, F. & Lavallée, D. (1991c) Multifractal analysis and simulations of nonlinear geophysical signals and images. *Proceedings, 13th GRETSI Colloqium*, Juan-Les-Pins, 16–20 Sept. 1991, pp. 1313–25.
Schertzer, D. & Lovejoy, S. (1994) Multifractal generation of self-organized criticality. In: *Fractals in the Natural and Applied Sciences*. Novak, M. M. (ed.), Elsevier, 325–39.
Schmitt, F., Lavallée, D., Schertzer, D. & Lovejoy, S. (1992) Empirical determination of universal multifractal exponents in turbulent velocity fields. *Phys. Rev. Lett.*, **68**, 305–8.
Seed, A. (1989) *Statistical problems in measuring convective rainfall*. PhD. Thesis, McGill University, 141pp.
Segal, B (1979) High-Intensity rainfall statistics for Canada. *Comm. Res. Centre, Dept. of Communications, Canada, report CRC 1329-E*, Ottawa, Canada.
Sèze, G. & Smith, L. (1990) On the dimension of a cloud's boundary. Preprint vol., *7th conf. on Atmos. Rad.*, AMS, Boston, pp. 47–57.
Taylor, G. I. (1938) The spectrum of turbulence. *Proc. R. Soc. Lond.* A, **164**, 476–90.
Tessier, Y. (1993) Multifractal objective analysis: rain and clouds, Ph.D. thesis, McGill University, 143pp.
Tessier, Y., Lovejoy, S. & Schertzer, D. (1993) Universal multifractals in rain and clouds: theory and observations. *J. Appl. Meteor.*, **32**, 223–50.
Tessier, Y., Lovejoy, S. & Schertzer, D. (1994) The multifractal global rain gauge network: analysis and simulation. *J. Appl. Meteor.*, **33**, 1572–86.
Tsonis, A. A. & Elsner, J. B. (1989) Chaos, Strange Attractors and Weather. *Bull. Amer. Meteor. Soc.*, **70**, 14–23.
Tsonis, A. A., Elsner, J. B., Lovejoy, S. & Schertzer, D. (1990) The space/time fractal structure of rain and Taylor's hypothesis. *EOS*, 466.
Visvanathan, R., Weber, C. & Gibart, P. (1991) The stochastic coherence and the dynamics of global climate models and data. *Scaling,*

fractals and non-linear variability in geophysics, D. Schertzer & S. Lovejoy (eds.), pp. 269–78, Kluwer.

Wallace, P. R. (1953) Interpretation of fluctuating echoes from randomly distributed scatters, part II. *Can. J. Phys.*, **31**, 995–1009.

Warner, C. & Austin, G. L. (1978) Statistics of radar echoes on day 261 of GATE. *Mon. Wea. Rev.*, **106**, 983–94.

Waymire, E. (1985) Scaling limits and self-similarity in precipitation fields. *Wat. Resour. Res.*, **21**, 1271–81.

Weisnagel, J. & Powell, A. (1987) *Lidar detection of rain drops.* 3rd year honor project report, 47pp, McGill Physics Dept.

Welch, R. M., Kuo, K. S., Wielicki, B. A., Sengupta, S. K. & Parker, L. (1988) Marine stratocumulus cloud fields off the coast of Southern California observed by LANDSAT imagery, part I: Structural characteristics. *J. Appl. Meteor.*, **27**, 341–62.

Wilson, J. (1991) *Physically based stochastic modelling of rain and cloudfields.* MSc. thesis, McGill University, 97pp.

Wilson, J., Lovejoy, S. & Schertzer, D. (1986) An intermittent wave packet model of rain and clouds. *2nd conf. on satellite meteor. and remote sensing*, AMS Boston, pp. 233–6.

Wilson, J., Lovejoy, S. & Schertzer, D. (1991) Physically based cloud modelling by scaling multiplicative cascade processes. *Scaling, fractals and non-linear variability in geophysics*, D. Schertzer & S. Lovejoy (eds.), pp. 185–208, Kluwer.

Yaglom, A. M. (1966) The influence of the fluctuation in energy dissipation on the shape of turbulent characteristics in the inertial interval. *Sov. Phys. Dokl.*, **2**, 26–30.

Yano, J.-I. & Takeuchi, Y. (1988) Fractal dimension analysis of horizontal cloud pattern in the intertropical convergence zone. *Scaling, fractals and non-linear variability in geophysics*, D. Schertzer & S. Lovejoy (eds.), pp. 297–302, Kluwer.

Zawadzki, I (1973) Statistical properties of precipitation patterns. *J. Appl. Meteor.*, **12**, 459–72.

Zawadzki, I (1987) Fractal versus correlation structure in rain. *J. Geophys. Res.*, **92**, 9683–93.

3 Is rain fractal?

I. ZAWADZKI

Department of Physics, University of Québec in Montréal, Canada

ABSTRACT Some scaling properties of the distribution of rain rate in space are investigated. The scales of concern range from the radar coverage range (of the order of 400 km) down to individual raindrops. Preliminary results show that over this wide range of scales rain fields exhibit: preferential scales in the range of a few tens of kilometers; behaviour compatible with multifractal structure between scales of 0.5 to 12 km and some clustering properties of the distribution of the small raindrops in space probably related to raindrop collisions and breakup.

INTRODUCTION

Rain rate fields exhibit variability at all scales down to individual raindrops. In this sense the answer to the title question is positive. However, our interest focuses more on restrictive properties, such as scale invariance, that could serve in modelling the process. One should not expect the restrictive properties (if they exist) to extend uniformly over all scales of the rainfall fields. On scales smaller then the size of individual cumulus turbulence probably prevails in determining the distribution of water substance. On larger scales other phenomena will affect markedly the distribution of precipitation in space. For example, orography (including the distribution of humidity sources) plays undoubtedly a role in organizing convective elements.

The ability to study rain at all scales is limited. A single radar does not cover a typical precipitation system. Quantitative data from networks of radars covering extended areas are now only becoming operational. Thus, one is limited to circular areas of the radius of the order of 200 km. On the other end of the scale the nature of the radar signal limits the resolution to the order of one square kilometer (depending on the range and the time integration of the radar signal). For smaller scales little spatial information is available and one must rely on the time records of rain measuring devices.

In this work analysis of radar data and time records of raindrop detection will be made, in search for the evidence of scaling behaviour.

MESOSCALE

Several precipitation systems were analyzed using radar derived rain rate distribution in space. Basic resolution was 4×4 km, maximum range was 190 km and data were presented as constant altitude plan-position indicator (CAPPI) at 2 km height. Typical results are those of a system which occurred on June, 29th 1977, associated with the passage of a front. The distribution of rain at one time instant is shown in Fig. 1. The moment analysis was performed with these data both for quasi-instantaneous precipitation patterns and for data averaged over a number of radar scans. Results are shown in Fig. 2. The change of moment with scale does not exhibit a behaviour indicative of scaling for any time resolution. Nevertheless, for single scan data a linear fit to the values may appear as satisfactory approximation. However, multiscaling models based on such an approximation lead to a power-law form of the autocorrelation function (ACF) of the model field. The correlation analysis of the precipitation patterns shows, however, a more complex behaviour. In Fig. 3 the time averaged spatial ACF is shown (for definitions, see Zawadzki, 1973). Sections along the major and minor axis are shown in log-log coordinates in Fig. 4, where some instantaneous values of the spatial ACF are also shown.

It is clear that in this case the quite uniform N-S distribution of rain is insufficiently covered by a single radar to reveal any structure. Along the minor axis of the ACF, on the other hand, the most striking feature is the breaking point at *c.* 80 km which may be indicative of a preferred scale of organization. The scale of a few tens of kilometers is usually associated in convective rainbands with clusters of convective

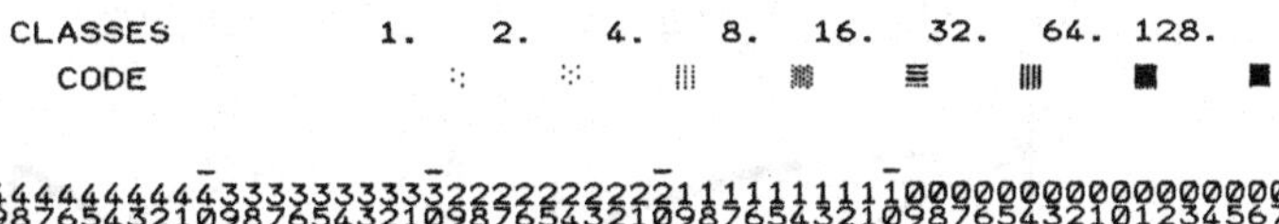

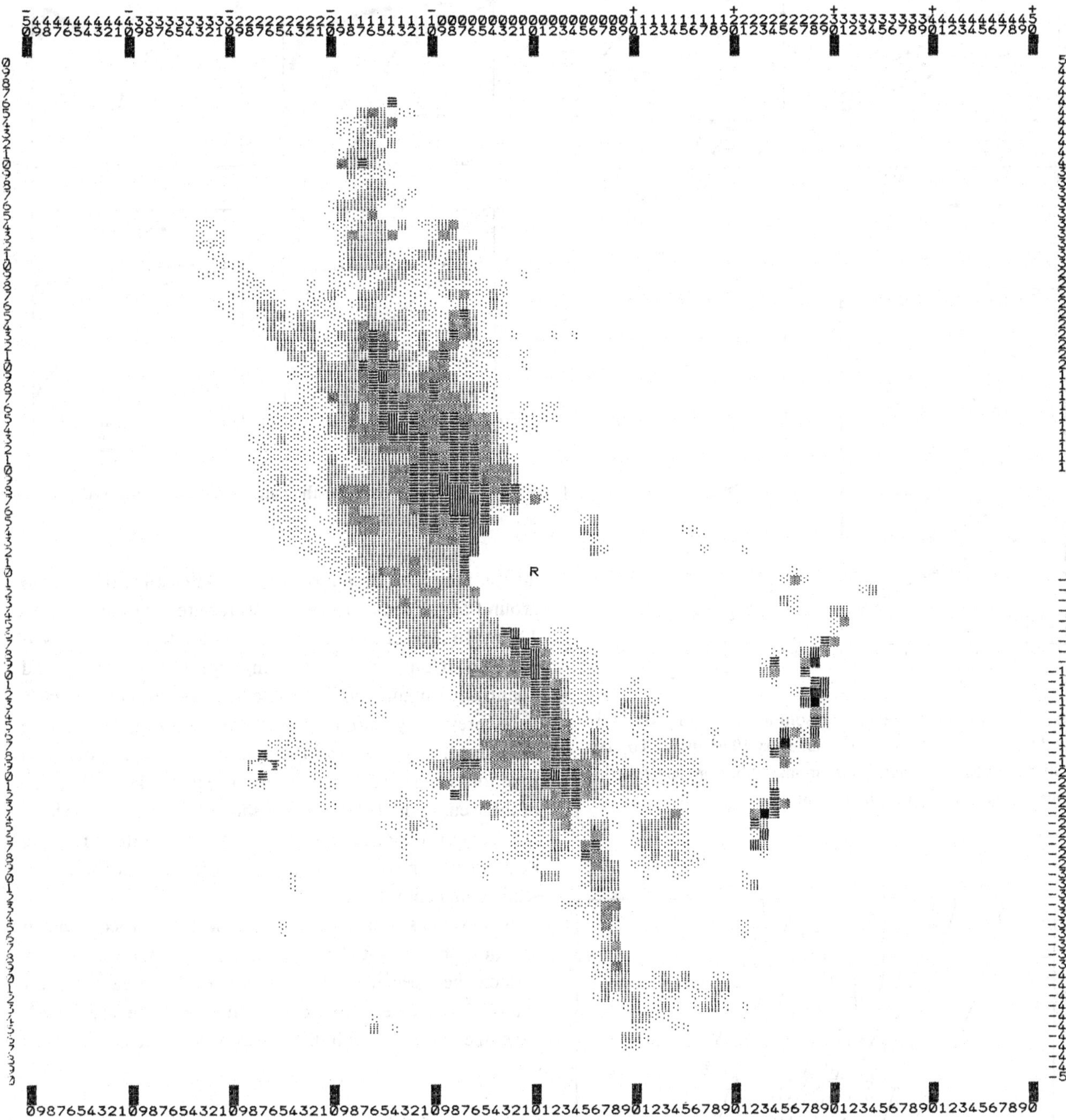

Fig. 1 Frontal rain band at 11:05 on 29/6/77. Data correspond to 2 km height and shades indicate rain rate in mm/h.

cells. On both sides of the breaking point a power law approximation to the ACF is reasonable. At any given time the curves show some secondary maxima or minima but the essential pattern is the same as for the time averaged ACF. The breaking point changes in time as the precipitation systems go through an evolution of the organization on the restricted scale of radar coverage.

CONVECTIVE SCALE

A three days record of high resolution radar data taken on 11 September, 1981 was examined in the search for the rain structure in scales below ten kilometers. The sample represents 24 hours of precipitation. The raw data were smoothed first to a resolution of 0.5 km^2 in order to eliminate signal

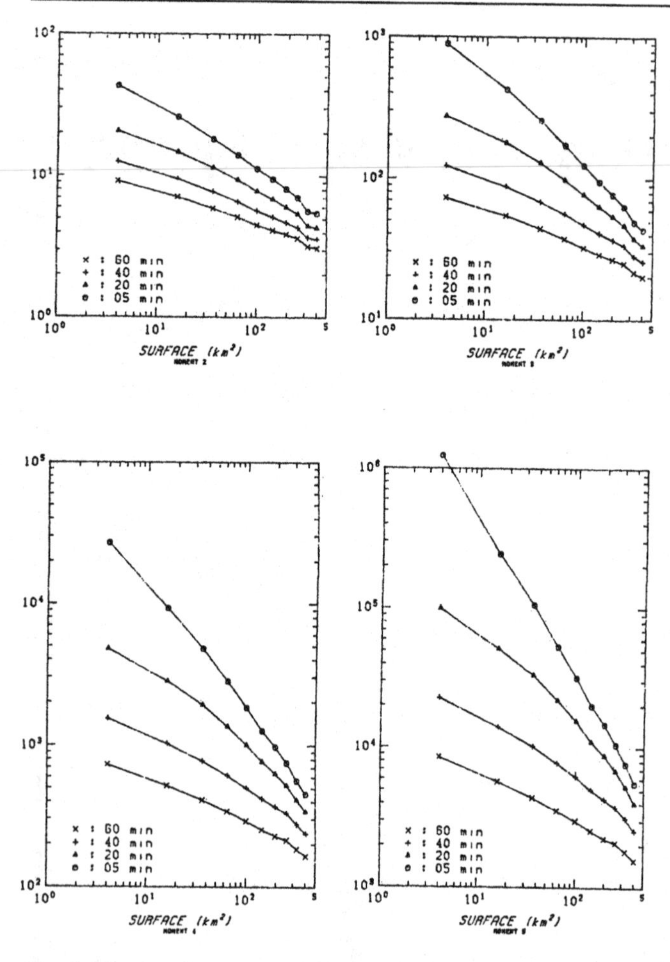

Fig. 2 Change of moments with resolution of data for the indicated number of radar scans (1 scan every 10 minutes). Top left, second moment; (top right) third moment; bottom left, fourth moment; bottom right, fifth moment.

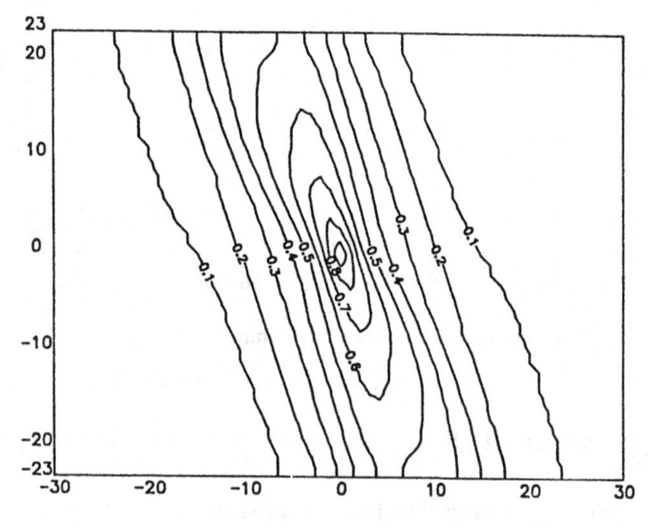

Fig. 3 Time averaged autocorrelation pattern of the rain system of 29/6/77. Space lag is in multiples of 4 km.

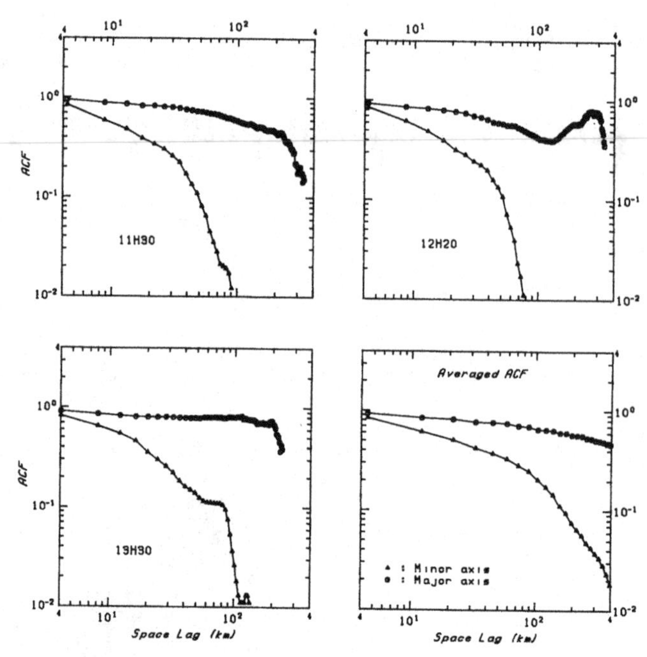

Fig. 4 Cross sections along the major and minor axes of the spatial ACFs (29/06/77).

fluctuations. This resolution together with constraints due to ground clutter limits the radar coverage to the range of 12 to 40 km. In azimuth the data were limited to 80° in the N-E quadrant. Comparison with raingauge data insures a good quantitative quality of the radar data (Zawadzki *et al.*, 1986).

As previously, moments and correlation analyses were performed. Fig. 5 shows the values of some moments as functions of the spatial resolution. The power law behaviour of these curves is striking. Fig. 6 shows the slopes of the lines in Fig. 5 as functions of the moment order. For the three time resolutions a behaviour usually associated with the multiscaling concept is seen.

However, as seen in Fig. 7, the spatial autocorrelation function (quite isotropic at these scales) shows again a broken line aspect in log-log coordinates. The breaking point is at *c.* 7 km which may be associated with the size of well developed and intense individual convective cells.

SMALL SCALE

Since raindrops are the basic elements of rain, fractal properties at small scales are better investigated by the analysis of distribution of raindrops in space. Data of this sort are not obtainable in a practical manner for scales beyond a few square meters. An alternative source of data are the time records of falling drops, such as given by the distrometer of Joss-Waldvogel (1967). For a given diameter of drops these records can be interpreted as drops distributed in space.

Two records of the temporal length of 15 minutes, com-

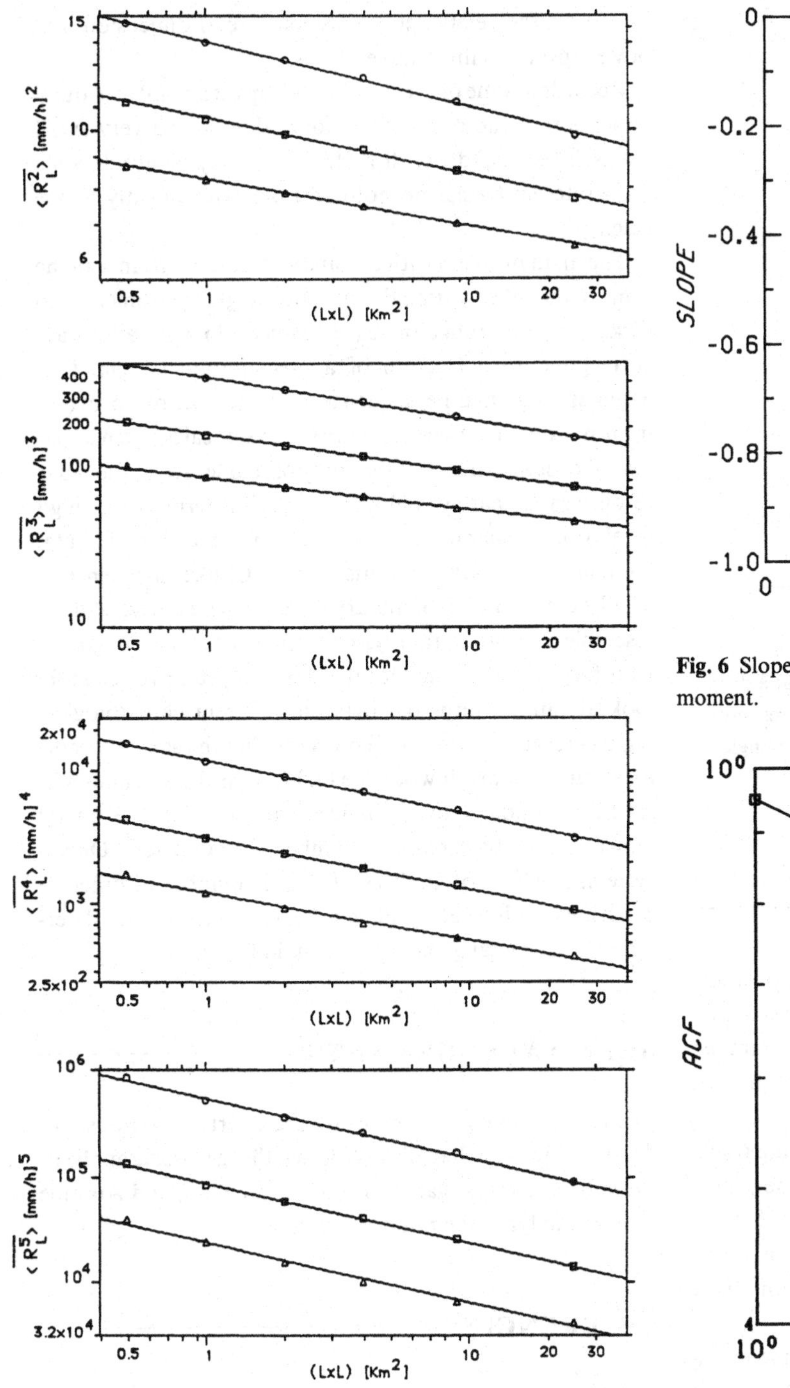

Fig. 5 Moment value as a function of resolution for the indicated moments and for data averaged in time over 1, 2, and 4 radar scans indicated by the circles, squares and triangles, respectively.

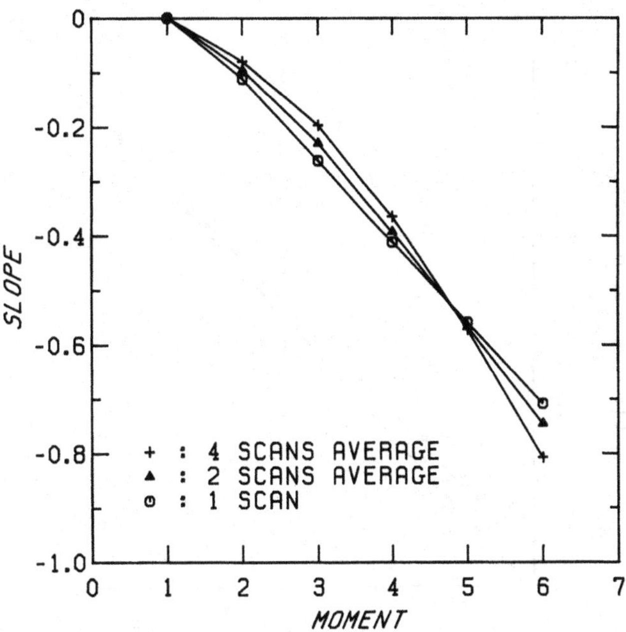

Fig. 6 Slopes of the lines in Fig. 5 as functions of the order of moment.

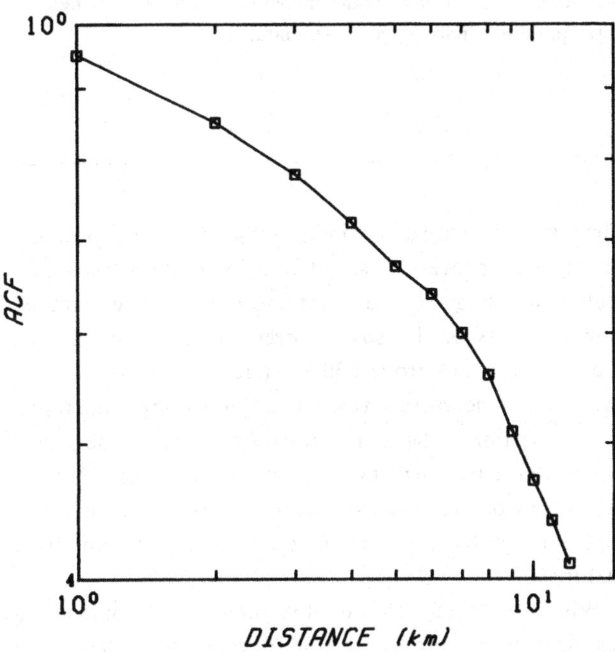

Fig. 7 Spatial autocorrelation of the rainfall events for which the moment analysis is shown in Figs. 5 and 6.

prising ca. 8000 drops were analyzed in search for the correlation dimension. The data analysis was performed for drops in various diameter categories and results were compared with random generated data of the same sample size. For drops larger than 0.5 mm in diameter no significant evidence of any structure was found. For smaller drops the analysis shown in Fig. 8 shows a correlation dimension of 0.93. This small but significant effect, if true, has a simple physical explanation in the collisional break-up that generates a cluster of small drops after each event. It is interesting to note that although the phenomenon was extensively studied in laboratory no clear evidence of its occurrence in nature was available up to now. However, it should be pointed out that for the very small diameters the distrometer may be not sufficiently accurate.

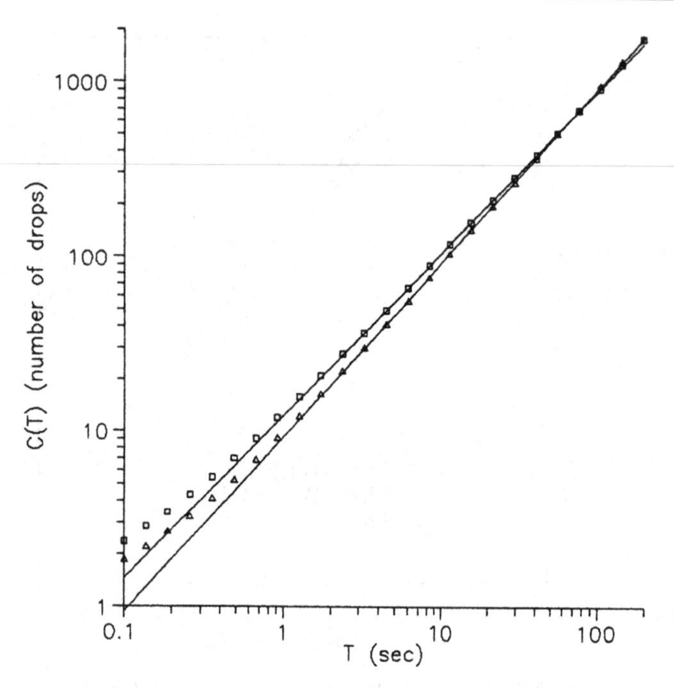

Fig. 8 Number of drops within a time interval centred at a drop arrival time as a function of the interval duration. Squares indicate actual data for drop diameters below 0.5 mm and triangles correspond to randomly generated data.

DISCUSSION

The results presented here indicate that no scaling or multiscaling properties are present in rain distribution in space for scales exceeding the size of mature precipitating cumulus. Log-log plots of the space correlation function indicate preferential scales around 30 km but varying from case to case and in time with each case. This is not a surprising result. Storm dynamics tells us that in strong systems some grouping of convective elements is the consequence of the dynamics of the process. Analysis, such as presented here, represents an objective and unambiguous way of revealing these features.

Within scales of the order of a cumulus size, some multiscaling properties appear to hold for rain. This conclusion is based on moment analysis and it is not sustained by the characteristics of the autocorrelation function. This illustrates that fulfilling one necessary condition for scaling does not suffice for concluding about the scaling nature of the process. However, the apparent contradiction may be due to limitations in the theoretical framework within which this contradiction arises. The empirical study presented by Fox (1989) on the relationship between fractal dimension and power spectra is illustrative of this.

Records of time of arrival of raindrops at a point at ground indicate that the rain rate is not scaling at the very small scales. The small drops, for which the analysis reveals some moderate clustering, do not contribute significantly to the rate.

The main problem with the study of scale invariance in the rain is the observational one. No single instrument can observe a wide range of scales. Meteorological radar data correspond to a fraction of a precipitation system. One assumes, that a time sequence of these records provides information on the spatial distribution on scales much larger than the radar range. For the study of smaller scales the radar resolution is barely sufficient, and furthermore, if high resolution is maintained, the noise contaminates the data. Sample size of data is a critical issue. Unlike turbulence at sm ll scales in the boundary layer, where all situations resemble each other, rain systems come in distinct categories: stratiform, frontal band, local convection, etc. Even a causal look at rain patterns occurring in different climatological regions reveals obvious differences within the same categories of rain systems. It would be highly desirable to characterize this variability by a single parameter, like the fractal dimension, or by a reduced number of parameters. Only a systematic effort of analysis of a large number of different situations, with an adequate *a priori* classification will determine if the fractal geometry can fulfill this goal.

ACKNOWLEDGEMENTS

The results presented summarize the efforts of various students. Jacques Lachapelle performed the mesoscale analysis, Marc Besner investigated the convective data, and Antoine Saucier analyzed the raindrop data.

REFERENCES

Fox, C. G. (1989) Empirically derived relationship between fractal dimension and power law form frequency spectra. *PAGEOPH*, **131**, 211–39.

Joss, J. & Waldvogel, A. (1967) Ein Spectrograph für Niederschlagstropfen mit automatischer Auswertung. *Pure Appl. Geophys.*, **68**, 240–6.

Zawadzki, I. (1973) Statistical properties of precipitation patterns. *J. of Appl. Met.*, **3**, 459–72.

Zawadzki, I., Desrochers, C., Torlaschi, E. & Bellon, A. (1986) A radar-raingauge comparison, *Proc. 23rd Conf. on Radar Met., Amer. Met. Soc.*, Boston, Mass., pp. 121–4.

4 Multifractal structure of rainfall occurrence in West Africa

P. HUBERT
CIG, Ecole des Mines de Paris, Fontainebleau, France

F. FRIGGIT
EIIER, Ouagadougou, Burkina Faso

J. P. CARBONNEL
CNRS, Université P. & M. Curie, Paris, France

ABSTRACT Rainfall occurrence related to a particular location, defined as the set of rainy periods observed, can be regarded as a fractal object belonging to the 1-D space of time. The dimension of this object, which is bounded by 0 and 1, is estimated via the functional box counting method. A large number of West African rainfall time series has been analysed. The resulting dimension is a function of the time scale and of the accepted threshold of rainfall intensity. In all cases under study, for a given time scale, a decreasing fractal dimension of rainfall occurence with increasing rainfall intensity threshold was observed. A main time scale range of practical interest was found to be from some days to some months. It is possible to attribute a multifractal structure to the process of rainfall occurrence. It can be used for simulation and/or estimation purposes. Attempts to find regional patterns and trends, and to compare them to those of inter annual rainfall means were undertaken.

INTRODUCTION

In a given location, rainfall is an intermittent process. That means that, for this location, one can observe a succession of wet and dry states. These states must be carefully defined, with areal, time interval and threshold references. A time period would be defined as wet if a given area receives during a given time interval an amount of water greater than the given threshold.

A raingauge defines accurately an observed area, being its collection surface (generally 400 cm^2). Raingauge measurements are typically performed at seven or eight o'clock every morning. Then, one can qualify successive daily periods as wet or dry. A period is considered wet if the amount of rainfall gathered during this period is greater than or equal to the given threshold. It is considered dry if the amount of rainfall is less than the threshold. A recording raingauge has also a well defined collection area and enables one to reach a better time resolution. However, it is an expensive equipment, requiring a more skilled manpower.

METHODOLOGY

Daily observation records were used in this study. They were extracted from the rainfall data base the CIEH (Comité Inter Africain d'Etudes Hydrauliques, Ouagadougou, Burkina Faso) kindly made available.

For each daily series, and for different rainfall thresholds, the observation period was divided into dry and wet periods. Rainfall occurrence then appears as a disconnected set supported by the time axis. From a geometrical point of view this set resembled the result obtained in course of generation of a Cantor dust from a segment (Fig. 1). Such objects can now be considered classical in the light of the works of Mandelbrot (1975, 1977), who introduced the concepts of fractal objects in geosciences.

The fractal dimension of rainfall occurrence has been estimated with the help of the box counting method (Hentschel and Proccacia, 1983a; Hentschel & Proccacia, 1983b; Lovejoy, Schertzer & Tsonis, 1987). Given a fractal object of the dimension D, included in a space of Euclidian dimension

Fig. 1 Daily rainfall records in Dedougou (Burkina Faso) covering the time span of 45 years. The first solid line shows the length of a whole year. In the lower rows, representing consecutive years from 1922 to 1966, the rainy days, during which precipitation greater than 0.1 mm was observed, were marked.

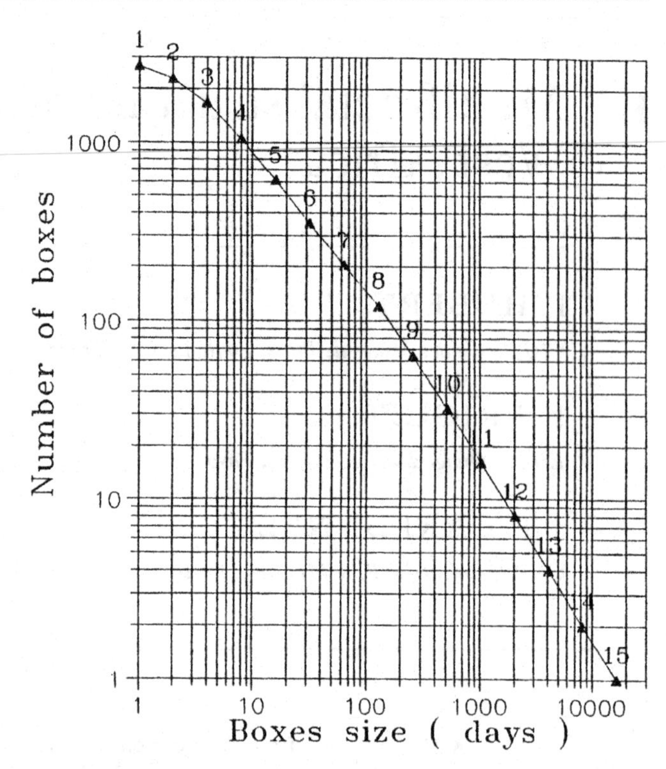

Fig. 2 Log-log diagram of the box counting method applied to the daily rainfall series of Dedougou (Burkina Faso).

E, the number N of a-sided boxes (respectively segments, squares or cubes if E is equal to 1, 2 or 3) necessary to cover the fractal object reads:

$$\log [N(a)] = -D \log (a) + K; \qquad K = \text{const.} \tag{1}$$

So, on a log-log diagram, the points whose coordinates are a and $N(a)$ would fall around a straight line with slope $-D$. In the present work, the set under study is included in a space of Euclidian dimension 1, the boxes are segments and the fractal dimension of rainfall occurrence ranges between 0 and 1.

STUDY OF DAILY TIME SERIES

The first series studied, with a 0.1 mm threshold, was that of Dedougou (Burkina Faso, latitude 12.28 N, longitude 3.29 W), the length of which is about 45 years (Hubert & Carbonnel, 1989). The results of the box counting are shown in Fig. 2, where one can see the alignment of points 4 to 8 on the one hand, and the alignment of points 9 to 15 on the other hand.

The alignment of points 9 to 15 along a line with the slope of -1 is trivial. The time scale of point 9 is 256 days and for such a scale, a rainy spell has always been observed. Then for time scales equal to or greater than 256 days the rainfall occurrence is merged with the time axis and its fractal dimension is obviously equal to 1.

The alignment of points 4 to 8 is more relevant. One can see on the log-log diagram that these points lay along a line with the slope of -0.79. This gives the rainfall occurrence a fractal dimension of 0.79 for time scales ranging from 8 to 128 days. This structure and this dimension may be related to the duration of the rainy season in West Africa (about 7 months, from April to October). A Cantor dust generator with such an initiator would yield a dimension equal to 0.783 [log(7)/log(12)] very close to our empirical computation.

The Soudano-Sahelian region experienced an alternation of dry and wet periods during this century, here conceived as groups of years with relatively low or high precipitations (Nicholson, 1983; Hubert, Carbonnel & Chaouche, 1989). It seemed interesting to look for possible variations of the fractal dimension of rainfall occurrence during the century. In fact, no significant differences could be seen between four contrasted long periods, of the length of 11 years each, beginning respectively in 1924 (wet), 1937 (dry), 1953 (wet) and 1969 (dry). The fractal dimension remained quite constant and equal to the previously estimated value of .79. This constancy is in agreement with the results of Chaouche (1988), who showed that the limits of the rainy season have not changed in this century while annual rainfall heights were subject to very large variations. In addition, from the self similarity observed from 8 to 256 days, one may infer that the internal structure of the rainy season has not changed.

This study has been repeated for different stations of the Soudano-Sahelian region and for different daily rainfall thresholds. 44 stations have been chosen from the daily rainfall data base of the CIEH in order to constitute a network as dense as possible over the Soudano-Sahelian climatic region during a continuous period as long as possible. The choice of the 44 series is the empirical result of a compromise between these conflicting demands. Sorting the series by countries we find 8 series in Mauritania, 6 in Senegal, 11 in Mali, 6 in Burkina Faso, 7 in Niger and 6 in Chad. Their location can be seen on the different maps quoted thereafter. All these daily series embrace a common period of the length of 41 years, from 1936 to 1976, which will be under study here. An original procedure, including the box counting method, but especially an automatic estimation of the slope and of the time range of the daily rainfall occurrence fractal behaviour for a given threshold has been devised.

The results of the Dedougou station can be generalized. In all cases a fractal behaviour was found for time scales ranging from some days to some months (about two orders of magnitude).

New conclusions resulted from this analysis. The value of the fractal dimension depends on the location of the raingauge and on the choosen daily rainfall threshold. The illustration of this latter point can be seen in Fig. 3, for the Mopti station. The dimension of the rainfall occurrence decreases as the threshold defining this occurence increases, what gives rise to a multifractal structure (Schertzer & Lovejoy, 1988). As far as the influence of the location is concerned, three maps were drawn in Figs. 4, 5 and 6, for thresholds 0.1, 10 and 40 mm, respectively. A strong dependence of the fractal dimension of rainfall occurence on latitude was observed. It was quite similar to that of mean annual rainfall, although more regular.

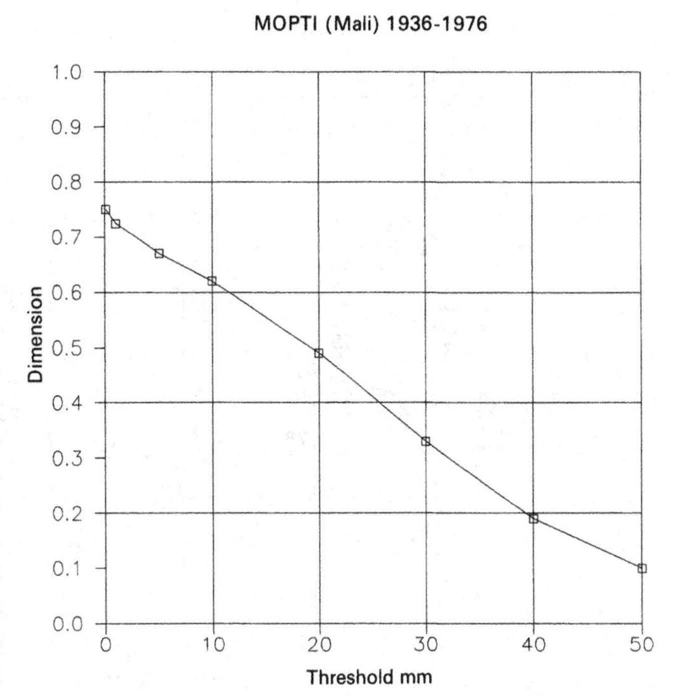

Fig. 3 Fractal dimension of the rainfall occurrence at Mopti (Mali) vs the daily rainfall threshold.

CONCLUSIONS

The main result of this study of daily rainfall series of the Soudano-Sahelian region is the evidence of the multifractal structure of the rainfall occurrence in this region for times scales ranging from some days to some months. This preliminary result opens some new perspective, and makes the analyst raise several questions.

Intensive studies, often using Markov chains (Masson, 1977), have been done in the field of synthetic rainfall series generation. Since the reviews of Buishand (1978) and Waymire & Gupta (1981), new developments have to be noticed (Smith & Karr, 1985; Foufoula-Georgiou & Lettenmaier, 1987; Tsakiris, 1988). Some attempts of simulation of fractal (Lovejoy & Schertzer, 1986; Wilson, Lovejoy & Schertzer, 1986; Chils, 1988) or multifractal fields (Schertzer & Lovejoy, 1987; 1988; Wilson, Lovejoy & Schertzer, 1988) have been made. Such simulations can be applied to rainfall occurrence and give a new insight into synthetic rainfall series generation.

A large field of research about the structure of the rainfall occurrence using the fractal language and tools is now open. Beside our work, Tessier *et al.* (1989) and Olsson *et al.* (1990) have also dealt with this area, giving valuable contributions. However, the amount of data analyzed is still small, while the shorter time scales, less than a day, have to be explored. The computation algorithms must be assessed, improved or even renewed (Lavalle *et al.*, 1990). At last, these investigations must be managed in close linkage with those regarding atmospheric phenomena and sharing the same theoretical background.

REFERENCES

Buishand, T. A. (1978) Some remarks on the use of daily rainfall models. *J. Hydrol.*, **36**, 295–308.
Chaouche, A. (1988) *Structure de la saison des pluies en Afrique Soudano-Sahélienne*, Thèse de doctorat, Ecole des Mines de Paris.
Chilès, J. P. (1988) Fractal and geostatistical methods for modeling of a fracture network. *Mathematical Geology*, **20**, 631–54.

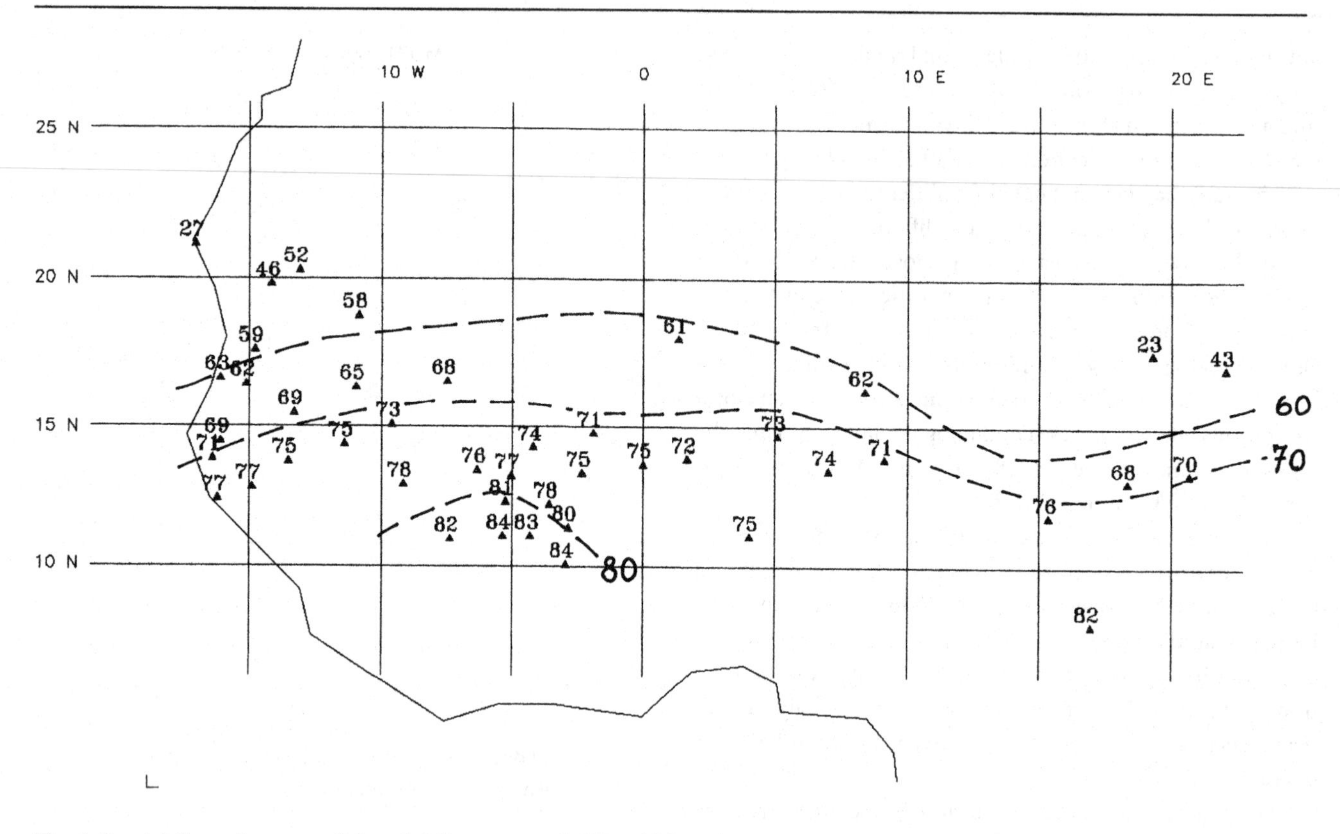

Fig. 4 Fractal dimension map of the rainfall occurrence in West Africa (threshold = 0.1 mm).

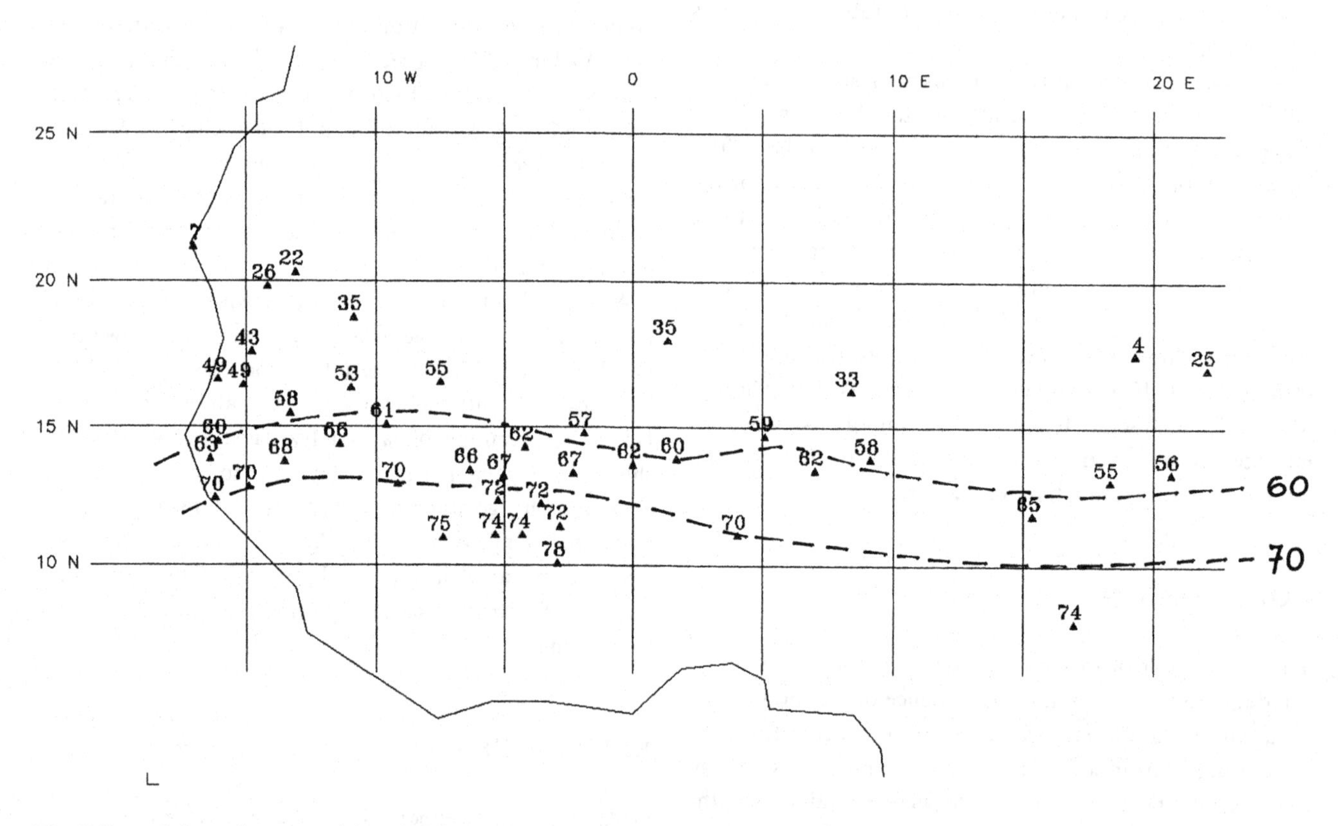

Fig. 5 Fractal dimension map of the rainfall occurrence in West Africa (threshold = 10 mm).

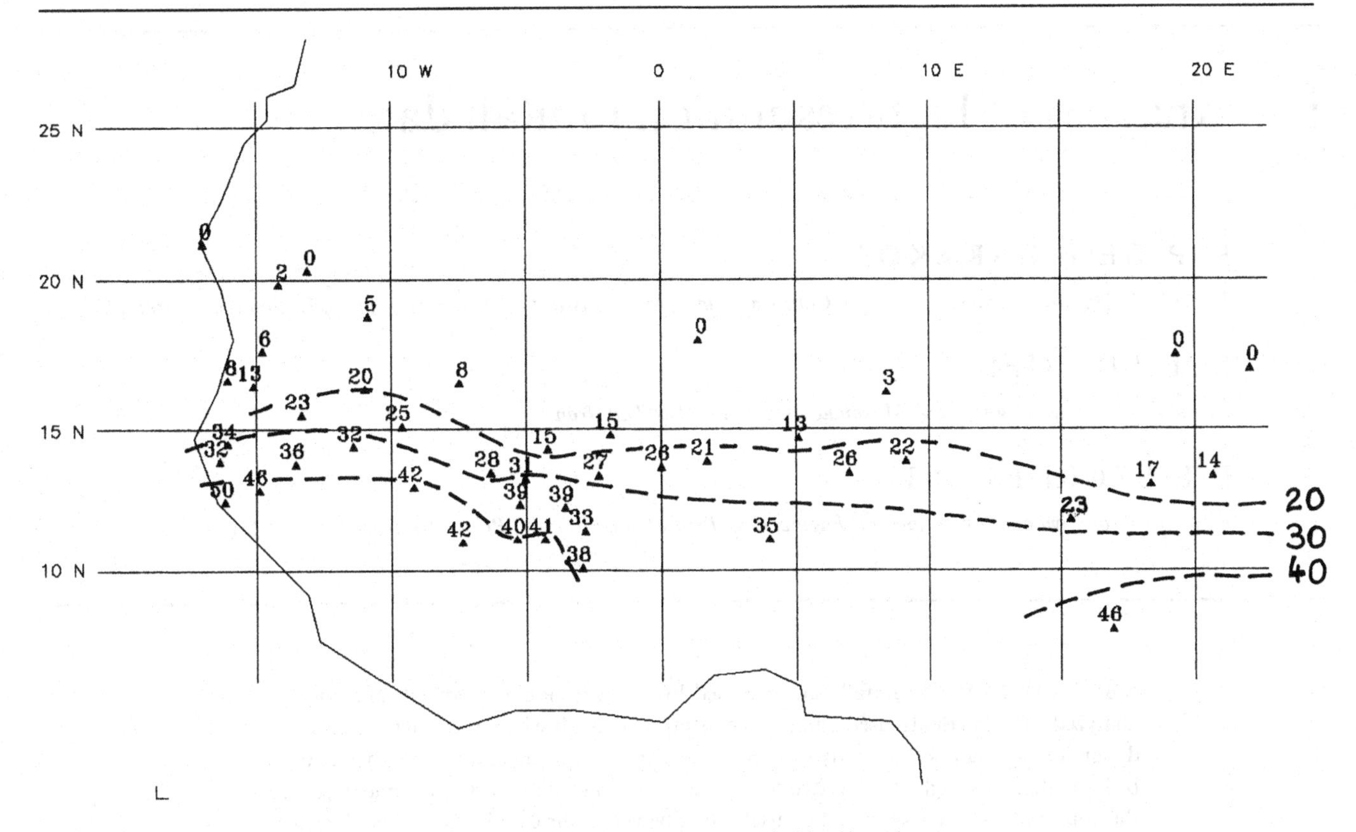

Fig. 6 Fractal dimension map of the rainfall occurrence in West Africa (threshold = 40 mm).

Foufoula-Georgiou, E. & Lettenmaier, D. P. (1987) A Markov renewal model for rainfall occurences. *Water Resour. Res.*, **23**, 875–84.

Hentschel, H. G. E. & Proccacia, I. (1983a) The infinite number of generalised dimensions of fractals and strange attractors. *Physica*, **8D**, 435–44.

Hentschel, H. G. E. & Proccacia, I. (1983b) Relative diffusion in turbulent media: the fractal dimension of clouds. *Phys. Rev.*, **A29**, 1461–70.

Hubert, P. & Carbonnel, J. P. (1989) Dimensions fractales de l'occurrence de pluie en climat soudano-sahélien. *Hydrologie Continentale*, **4**, 2–10.

Hubert, P., Carbonnel, J. P. & Chaouche A. (1989) Segmentation des séries hydrométéorologiques, application à des séries de précipitations et de débits de l'Afrique de l'Ouest. *J. Hydrology*, **110**, 349–67.

Lavallée, D., Schertzer, D. & Lovejoy, S. (1991) On the determination of the codimension function. In: Schertzer, D. & Lovejoy, S. (Eds.), *Nonlinear Variability in Geophysics 1*, Kluver Academic Press, Amsterdam (in press).

Lovejoy, S. & Schertzer D. (1986) Scale invariance, symmetries, fractals and stochastic simulations of atmospheric phenomena. *Bulletin of the AMS*, **67**, 21–32.

Lovejoy, S., Schertzer, D. & Tsonis, A. A. (1987) Functional box-counting and multiple elliptical dimensions in rain. *Science*, **235**, 1036–8.

Mandelbrot, B. B. (1975) *Les objets fractals, forme, hasard et dimension*, Paris.

Mandelbrot, B. B. (1977) *The fractal geometry of nature*, San Francisco, p. 461.

Masson, J. M. (1977) Persistance des états pluvieux en fonction de leur durée, analyse de 52 années d'enregistrements pluviométriques a Montpellier Bel Air, *Cahiers de l'ORSTOM, série Hydrologie*, **14**, 173–89.

Nicholson, S. E. (1983) Sub-saharian rainfall in the years 1976–1980: evidence of continued drought. *Mon. Weather Rev.*, **111**, 1646–54.

Olsson, J., Niemczynowicz, J., Berndtsson, R. & Larson, M. (1990) Some fractal properties of rainfall, Draft paper, Department of Water Resources Engineering, University of Lund, Sweden.

Schertzer, D. & Lovejoy, S. (1987) Physical modeling and analysis of rain and clouds by anisotropic scaling and multiplicative processes. *J. Geophys. Res.*, **92**, D8, 9693–714.

Schertzer, D. & Lovejoy, S. (1988) Multifractal simulations and analysis of clouds by multiplicative processes. *Atmospheric Research*, **21**, 337–61.

Smith, J. A. & Karr, A. F. (1985) Statistical inferences for point processes models of rainfall. *Water Resour. Res.*, **21**, 73–9.

Tessier, Y., Lovejoy, S. & Schertzer, D. (1989) Multifractal analysis of global rainfall from 1 day to 1 year, Abstract Volume, *European Geophysical Society*, *XIV General Assembly*, Barcelona.

Tsakiris, G. (1988) Stochastic modelling of rainfall occurences in continuous time. *Hydrol. Sci. J.*, **33**, 437–47.

Waymire, E. & Gupta, V. K. (1981) The mathematical structure of rainfall representations: 1. A review of the stochastic rainfall models, 2. A review of the theory of point processes, 3. Some applications of the point process theory to rainfall processes. *Water Resour. Res.*, **17**, 1261–94.

Wilson, J., Lovejoy, S. & Schertzer, D. (1986) An intermittent wave packet model of rain and clouds, *Proc. 2nd Conf. on Satellite Meteorology and Remote Sensing*, AMS, Boston, pp. 233–6.

Wilson, J., Lovejoy, S. & Schertzer, D. (1988) Physically based cloud modeling by multiplicative processes. In: Schertzer, D. & Lovejoy, S. (Eds.), *Nonlinear Variability in Geophysics 1*, Kluver, Amsterdam (in press).

5 Analysis of high-resolution rainfall data

K. P. GEORGAKAKOS

Hydrologic Research Center, San Diego, California, and Scripps Institution of Oceanography, La Jolla, California, USA

M. B. SHARIFI

Department of Civil Engineering, Mashhad University, Mashhad, Iran

P. L. STURDEVANT

Department of Civil and Environmental Engineering, Princeton University, Princeton, New Jersey, USA

ABSTRACT Point-rainfall data recorded by a fast-responding optical raingauge were analyzed. The methods used range from statistical analysis to the fractal and chaotic dynamics approaches. The study showed the evidence of scaling and chaotic dynamics. It is believed that the insight into the dynamics of rainfall data with very fine increment, gained in the course of the exercise, could be useful in advancing our capability to reliably estimate probable maximum rainfall for design purposes.

INTRODUCTION AND BACKGROUND

The realization that it is possible to have a temporal natural process that has a random appearance but which is generated by a deterministic set of ordinary differential equations, triggered by Lorenz (1963) in his now well known example of the dynamics of a convecting fluid, has initiated a wealth of attempts to re-investigate natural phenomena thought to be inherently random. Rainfall rate is one such natural variable and a few investigations of its nature and dynamics have already appeared in the literature that provide some evidence for the existence of a deterministic generating mechanism in the rainfall process at small spatial scales (Rodriguez-Iturbe *et al.*, 1989, and Sharifi *et al.*, 1990). The mathematical methods for the investigation of this 'new' dynamics (called chaotic dynamics) require samples with very fine temporal resolution, that goes beyond the resolution available with conventional *in situ* raingauges. The work presented herein reports results obtained using very-fine increment convective-rainfall data recorded by a specially-calibrated optical raingauge in Iowa City, Iowa, USA, during the summer of 1989. Results of both conventional statistical analysis and modern chaotic-dynamics analysis are reported.

The theory, vocabulary and methods of investigation of chaotic-dynamics are new to the field of hydrology/hydrometeorology and we devote a few paragraphs for their outline in the next section before we describe the experimental facility. Further, results are presented and discussed, and finally conclusions and prospects are outlined. The reader interested in gaining a more in-depth understanding of the field of chaotic-dynamics, including methods of data analysis, is referred to texts such as Berge *et al.* (1984), Ruelle (1989), Schuster (1988), Moon (1987) and to the descriptive but insightful book by Gleick (1987).

ELEMENTS OF ANALYSIS METHODS FOR CHAOTIC DYNAMICS

Continuous spectrum constructed from samples of a natural process was usually taken as a sure sign of randomness. It has now been established, however, that certain class of dissipative deterministic systems are capable of generating such a spectrum, too. In particular, systems of ordinary nonlinear differential and difference equations have been constructed that generate continuous spectra and motion that is chaotic (e.g., Moon, 1987). That is, given two nearby trajectories in a region of the space of state variables of the system under consideration (phase or state space), system evolution in time forces exponential divergence of the aforementioned trajectories, resulting in the loss of predictability; and this for a perfectly deterministic system. Furthermore, in the presence of chaotic dynamics, system trajectories in phase space remain on a phase-space object whose dimension is less (and in some cases considerably less) than the dimension of the embedding phase space. Such an object is called a strange attractor. In most cases of chaotic dynamics, the dimension of the embedded attractor is not an integer but it is a fraction, greater than its topological dimension, and it is called a fractal dimension (Mandelbrot, 1983).

Inherent in the notion of fractality are the notions of self-similarity and scaling (Mandelbrot, 1983), which, if present, advocate the absence of a measure for scale for an object or a process. For random variables, scaling is defined asymptotically through the probability distribution (usually the exceedance probability distribution). That is, there exists an exponent $D>0$ such that the $\mathrm{Prob}(X>x)$ scales as x^{-D} for large x. The probability distributions with such a property are called asymptotically hyperbolic distributions.

The theory of chaotic dynamics has received a strong impetus by the findings of several investigators that have reported evidence of chaotic dynamics in experimental data and observations of natural processes (e.g., Berge *et al.*, 1984, Tsonis & Elsner, 1988, Nicolis & Nicolis, 1987, Rodriguez-Iturbe *et al.*, 1989, Sharifi *et al.*, 1990). Instrumental in such studies has been the design of methods for detecting the presence of a strange attractor and of trajectory divergence (and, therefore, of chaotic dynamics) for a natural process based on a time series sampled from the process. Grassberger & Procaccia (1983a, b) have proposed a method of time-delays that helps determine whether there is a strange attractor in the observed time series with a dimension that is less than the embedding dimension of the phase space. Mayer-Kress (1987) has, in addition, determined error bounds for the dimension computations that depend on the sample size and the range of scaling present in the sample data. To complement inference studies, Wolf *et al.* (1985) have developed an algorithm to confirm and measure exponential divergence of nearby sampled trajectories from a natural process. In the following, we describe briefly the basis of the aforementioned methods of inference for sampled processes that exhibit chaotic dynamics.

The methodology of Grassberger & Procaccia (1983a, b) (see also Berge *et al.*, 1984, for a comprehensive exposition) is based on the concept of the correlation integral and correlation dimension. The latter dimension is obtained from the correlation between random points on the presumed strange attractor and bounds the fractal dimension from below. Given a sample time series: $X(t_i)$, $i=1,2,\ldots,$ of a natural process, a trajectory can be constructed in a p-dimensional phase space by taking as coordinates for the ith point of the trajectory: $X(t_i)$, $X(t_i+\tau)$, $X(t_i+2\tau),\ldots,X(t_i+(p-1)\tau)$, where τ is an appropriate time delay. Then, to measure spatial correlation, the correlation integral $C(r)$ is determined:

$$C(r)=\lim_{N\Rightarrow\infty} 1/N^2 \{\text{number of pairs } i,j \text{ whose distance } |X_i-X_j|<r\} \tag{1}$$

where X_i is the ith coordinate vector. For many attractors this function has been found to exhibit a power law dependence on r as $r\Rightarrow 0$. The exponent in the power law dependence is the correlation dimension ν, which is found as the slope of the $\log[C(r)]$ versus $\log(r)$ curve. The dimension ν is determined for various embedding dimensions p, and if the values of ν remain constant after a certain embedding dimension (saturation of ν occurs), there is an evidence of the presence of a strange attractor with a correlation dimension equal to the saturation value of ν.

A measure of the time-averaged exponential rate of divergence of trajectories initiated from two nearby initial conditions is provided by the Lyapunov exponents. An hypersphere in the n-dimensional phase space of a continuous dynamical system evolves to an hyperellipsoid due to the locally deforming nature of the system. The ith one-dimensional Lyapunov exponent is defined in terms of the length of the ellipsoid principal axis $p_i(t)$ as:

$$\lambda_i=\lim_{t\Rightarrow\infty}\frac{1}{t}\log_2[p_i(t)/p_i(0)] \tag{2}$$

where the λ_i's are ordered from largest to the smallest, and t denotes time. Since the orientation of the ellipsoid changes continuously as it evolves, the direction associated with a given exponent varies in a complicated way through the attractor, and one cannot associate a direction with a given attractor. If d_0 is a measure of the initial distance between two nearby starting points, in a short time the distance is

$$d(t)=d_0\,2^{\lambda_1 t},$$

where the largest Lyapunov exponent, λ_1, controls the linear extent of the ellipsoid growth. The existence of a positive Lyapunov exponent implies the divergence of nearby trajectories and, thus, the presence of chaotic dynamics. The numerical algorithm of Wolf *et al.* (1985) has been used in this work for the computation of the largest positive Lyapunov exponent from the sample time series.

EXPERIMENTAL APPARATUS AND DATA COLLECTION

Through the four months of the summer of 1989, a project was run for the purpose of collecting rainfall data. The first two months were spent in acquiring background information, setting up the equipment and data acquisition system, and testing. During the last two months, actual

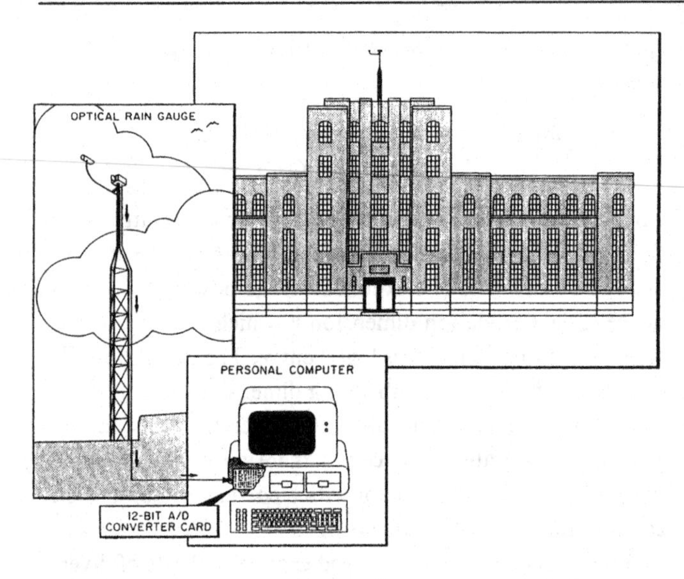

Fig. 1 Optical raingauge and recording apparatus at the Iowa Institute of Hydraulic Research in Iowa City, Iowa.

rainfall data was collected and analyzed. The objective of this project was to collect high resolution rainfall data, to create a prototype storm directory, and to perform preliminary analysis on the rain data. This project is part of an ongoing rainfall analysis program designed to provide insight into the dynamics of very-fine increment rainfall data.

The device used for measuring and recording rainfall was an optical raingauge, placed on a fifteen-foot tower located on the roof of the building that houses the Iowa Institute of Hydraulic Research of The University of Iowa, and connected to a micro computer (Fig. 1). The infrared beam emitted from the gauge is sensitive to the light scattered by the falling raindrops and the optical raingauge records these scintillations on a path- and time-averaged basis. This process is effective for both light and heavy rainfall and the amount of light scattered is converted to voltage for recording. The raingauge has a DC voltage range of −1.0 to 5.0 Volts and was manufactured by Scientific Technology Incorporated (STI). Actual measurement range is from −.35 Volts, corresponding to .1 mm/h rainfall, to 3.65 Volts, corresponding to 1000 mm/h rainfall. In terms of accuracy, from 10 mm/h to 100 mm/h, the raingauge is accurate to within 1% and from 1 mm/h to 500 mm/h accuracy is to within 4% outside the previous range (Scientific Technology, Inc.). A series of three articles (Wang *et al.*, 1978, 1979, 1980) discussed the development of obtaining path averaged rain rates from a divergent laser beam. These articles present the advantages of this method and the effects of the following conditions: rain-drop size, updrafts and downdrafts along the laser beam, path length, and rain-drop terminal velocity.

The signal is read by output circuitry electronics found within the gauge itself. Voltage is then interfaced directly to the microcomputer via a fifty foot straight wire through the roof of the building. A 12-bit analog to digital converter card was contained within the hardware, giving us an accuracy of plus or minus .0025 Volts (Fig. 1).

The optical raingauge implements precipitation measurements on a 40-inch beam path. It was used to sample rainfall once every 5 seconds. The noise level of the measurements was established in days of calm weather to be equal to −0.35 Volts. The raingauge was set up and operating as of July 7, 1989, with the first recorded rainfall occurring on July 15. Data were taken twenty-four hours a day with data stored in files only for the times for which voltage exceeded −0.35 Volts. Quality control consisted of scanning the files manually for obvious outliers (none were detected), and the data were divided in storm time series. Intervals greater than 1 hour between recordings were used to separate different storms. In that way an inventory of fine-increment rainfall data corresponding to eleven individual rainfall events was created. For more details on data collection and processing the interested reader is referred to Sturdevant *et al.* (1990).

RESULTS AND DISCUSSION

Analysis of the fine-increment rainfall data was based on the computation of first and second moment statistics of the time series corresponding to each individual rainfall event, computation of relative frequency and exceedance histograms, and correlation dimension analysis. Sturdevant *et al.* (1990) present all the results in detail. In this paper we only comment on important findings and present selected results. Table 1 presents selected sample statistics for the eleven rainfall events.

Characteristic of the time series plots is their wavelike appearance dominated by irregular (in terms of length, frequency and magnitude) bursts of rain (e.g., Fig. 2). The average duration of a rain burst is about twenty minutes ranging from as low as five minutes to as long as forty minutes. Another characteristic of the time series plots is an apparent scale invariance. Storms with similar durations and with maximum rainfall amounts that differ by an order of magnitude(e.g., 7.9 mm/hr and 71.9 mm/hr) can not be distinguished based on the overall appearance of the time series plots. Thus, if the rainfall amount axis was not indicated on the plot it would be impossible to recognize the high from the low intensity storm.

The possibility of correlation between the length of a rainfall event and several statistics computed from its observed intensity was examined based on the values of Table 1. The sample mean and sample variance vs. duration plots show significant scatter with the variability in the mean and variance decreasing for longer duration rainfall events. The correlation coefficient vs. duration plots indicate that the

Table 1. *First and second moment statistics for the eleven recorded rainfall events during the summer of 1989 in Iowa City*

Storm Number	Storm Date	Duration (hrs)	Mean (mm/hr)	Standard Deviation (mm/hr)	Lag-12	Lag-60	Lag-180
1	7.15.89	4.406	.240	.295	.92	.572	.13
2	7.18.89	3.957	1.455	1.931	.849	.387	.303
3	7.19.89	1.426	1.022	.433	.78	.46	.02
4	7.23.89	.589	16.54	16.713	.773	.048	.016
5	8.03.89	.925	34.78	32.259	.88	.5	.2
6	8.13.89	.982	.534	.392	.863	.256	−.34
7	8.14.89	1.597	2.35	4.31	.788	.341	−.076
8	8.23.89	3.14	2.39	1.46	.884	.447	.544
9	8.23.89	0.726	1.964	1.614	.795	.242	−.216
10	8.26.89	2.779	5.09	12.38	.889	.282	2.779
11	8.28.89	0.3	1.058	.848	.844	.133	−1.51

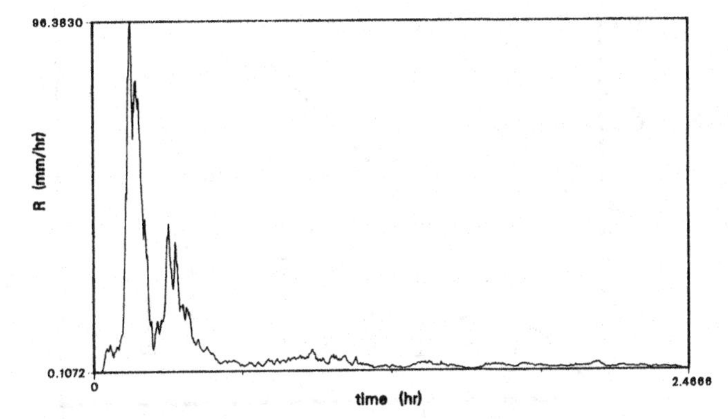

Fig. 2 Five-seconds rainfall data during the August 26, 1989 rainfall event in Iowa City. Rainfall is in mm/hr.

Table 2. *Scaling regions and associated scaling exponents for the eleven recorded rainfall events during the summer of 1989 in Iowa City*

Storm No.	Storm Date	Scaling Region (mm/hr)	Scaling Exponent
1	7.15.89	.317–1.2	2.0
		1.2–2.0	3.8
2	7.18.89	2.2–10.5	2.1
		10.5–13	5.6
3	7.19.89	1.32–2.2	6.9
4	7.23.89	8–40	1.0
		40–63	3.6
5	8.03.89	12.3–69	0.5
		69–124	5.8
6	8.13.89	.52–2	1.6
		2–2.85	7.9
7	8.14.89	3–15.5	0.9
		15.5–26	6.0
8	8.23.89	no scaling exponent	
9	8.23.89	3.03–7	3.3
10	8.26.89	13.2–70.1	1.0
		70.2–83	7.0
11	8.28.89	.324–1.33	1.0
		1.33–3	2.8

one- and five-minute correlation coefficients have a mild dependence on storm duration. For increasing duration, the correlation coefficient increases by the rate of 0.1 per 4 to 5 hours.

Regarding dependence of the sample mean and variance statistics on the sample size, Sturdevant *et al.* (1990) show that due to the sudden jumps that are present in the relevant plots, a sample size of about 1000 of five-second data points is necessary for stable estimates of both sample statistics. A sample size of about 2000 five-second data points is required for stable estimates of the sample five-minute-lag correlation coefficient.

Exponential shape was predominant in the relative frequency histograms. The log-log plots of the relative frequency of rainfall magnitudes greater than a certain magnitude showed two scaling regions for most rainfall events. Table 2 shows the slope of the log-log plot (referred to as scaling exponent) and the scaling region for all eleven storms. Fig. 3 presents a typical example of a log-log plot. It can be seen that the estimates of the scaling exponent in the high intensity scaling region range from 2.8 to 7.9, while in the low intensity scaling region they ranged from 0.5 to 1.6. In most cases the estimates in the low scaling region are more reliable due to the presence of more data points. These results give a preliminary indication of self similarity and asymptotic scaling (e.g., Mandelbrot, 1983). Analysis of many more time series is necessary, however, before a definite statement to this effect can be made.

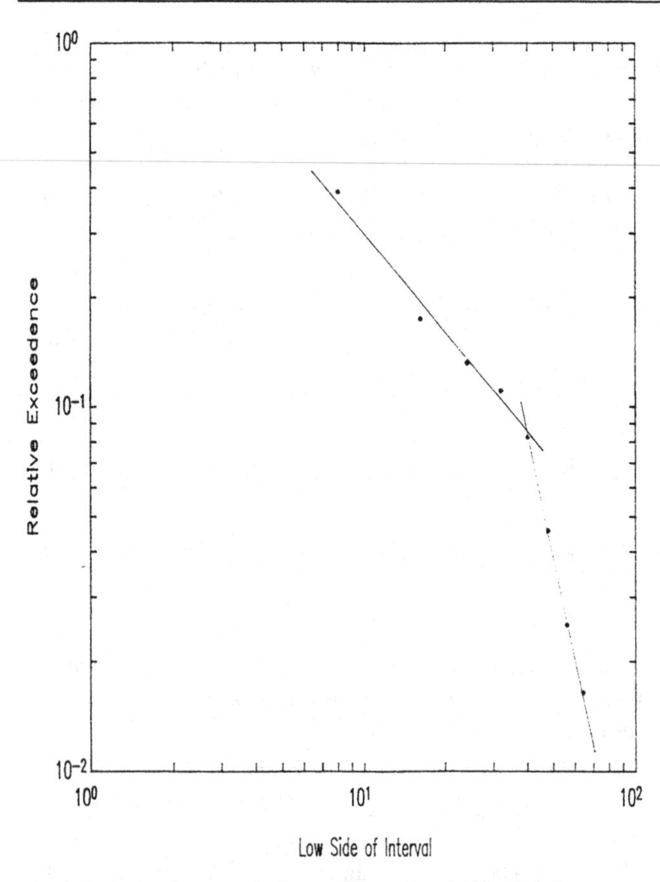

Fig. 3 Relative frequency of exceeding a threshold rainfall rate vs. the rainfall rate threshold in a log-log plot. Rainfall event of July 23, 1989, in Iowa City. Shown are the fitted lines that determine the scaling exponent D.

Sturdevant *et al.* (1990) show that the spectra of the eleven rainfall events studied have continuous regions over a wide range of frequencies. Such a finding warranted the use of correlation dimension analysis in search of chaotic dynamics in the time series of rainfall events. Correlation dimension analysis followed the work of Sharifi et al. (1990). They used the Grassberger & Procaccia (1983a, b) algorithm to compute the correlation dimension from the correlation integral. In this work, error bounds on the correlation dimension were also computed using an algorithm similar to that proposed by Mayer-Kress (1987). Only rainfall events with more than 2000 data points were used. Table 3 presents the estimates of correlation dimension for each of the rainfall events studied together with their sample size. The results are in good agreement with those of Sharifi *et al.* (1990) and lend credence to the conjecture that a low dimensional strange attractor is in the heart of storm rainfall. As an example, Fig. 4 presents the correlation dimension v vs. embedding dimension p for the rainfall event of August 26, 1989 in Iowa City. The estimate of v is 2.5 with an error bound that implies a maximum v of about 3.5. Lyapunov exponent analysis according to Wolf's algorithm (Wolf *et al.*, 1985) was also undertaken. The results appear in Table 3 for each of the three storms and range in value from 10^{-3} to 10^{-5}. Fig. 5 gives the estimate of the Lyapunov exponent for the August 26, 1989 storm as a function of sample size.

Table 3. *Correlation-dimension and largest positive Lyapunov-exponent estimates for rainfall events with sample size greater than 2000 data points*

		Correlation Dimension and Largest Lyapunov Exponent Estimates		
Storm No.	Date	No. of Data Points	Correlation Dimension	Lyapunov Exponent
1	8.14.89	2236	2.1	3.7×10^{-3}
2	8.23.89	2265	3.7	1.7×10^{-5}
3	8.26.89	2421	2.5	4.3×10^{-3}

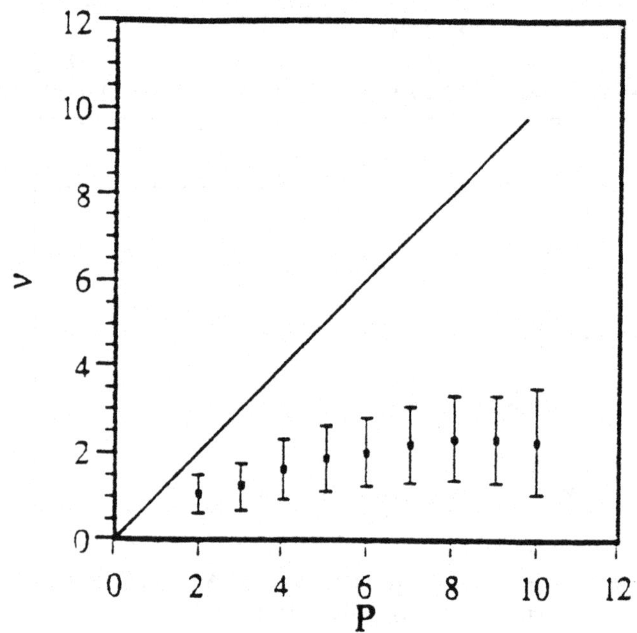

Fig. 4 Correlation dimension v vs. embedding dimension p for the rainfall event of August 26, 1989, in Iowa City. Error bars are also indicated on the figure.

Given that for some of the storms the correlation dimension was less than 3, the strange attractor can be reconstructed in a three dimensional space using the method of time-lagged coordinates (Grassberger & Procaccia, 1983). Fig. 6 is an example of a three-dimensional phase space for the August 26, 1989 rainfall event. Several folding and stretching regions (characteristic of a fractal object in phase space) can be identified on the sample trajectories.

CONCLUSIONS AND PROSPECT

Data recorded by a fast-responding optical raingauge have been analyzed using classical statistical analysis, and modern

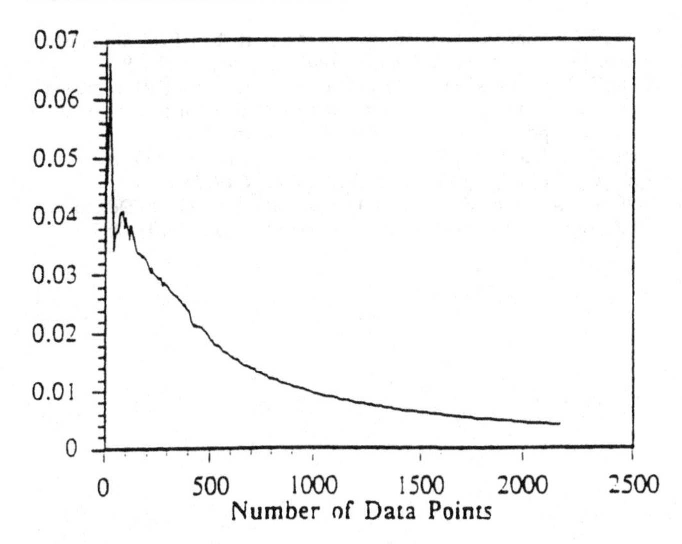

Fig. 5 Largest positive Lyapunov exponent vs. sample size for the rainfall event of August 26, 1989, in Iowa City.

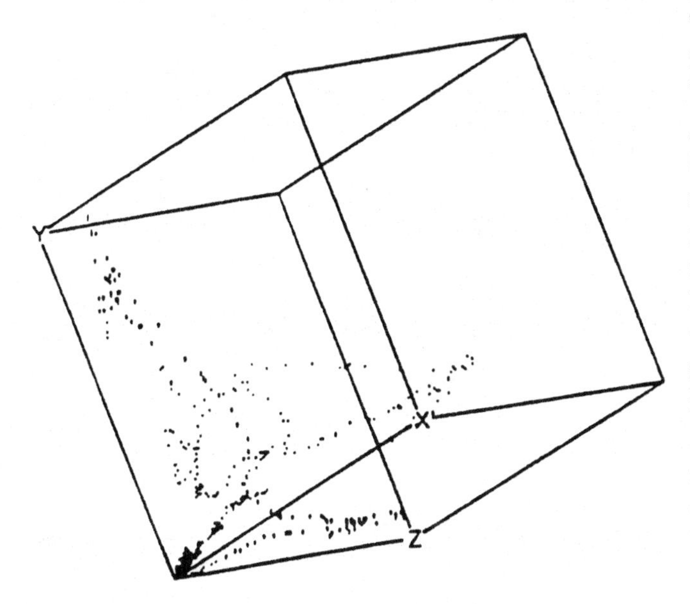

Fig. 6 Strange attractor projection in three dimensional space for the rainfall event of August 26, 1989, in Iowa City.

fractal and chaotic-dynamics analysis methods. Eleven convective rainfall events over the Iowa City, Iowa, area have been used in the analysis. Evidence of scaling and chaotic dynamics was found during the analysis. The accurate, high-resolution rainfall data supporting the aforementioned findings are rare. There is certainly a need for more work in this area before conclusive evidence is obtained. On a basic research level, high resolution data provide the only hope for an experimental verification of the conjectures put forth in this work. From a utilitarian point of view, the insight into the dynamics of very-fine increment rainfall data can be very useful in advancing our capability to reliably estimate probable maximum rainfall for hydraulic and hydrologic design.

For future studies of surface storm dynamics, the optical raingauge has been complemented by *in situ*, fast-responding, accurate sensors of air temperature, humidity, wind direction and speed and barometric pressure. A basic research effort has begun to reconstruct the governing ordinary differential equations that presumably generate chaotic dynamics and a strange attractor in phase space. The effort is based on the reduction of the partial differential equations that describe the known physics of the rainfall process to a set of a few ordinary differential equations, aided by the high resolution samples of the observed meteorological variables. If successful, the research effort would confirm and describe in quantitative terms the nature of chaotic dynamics in rainfall. As of the beginning of the summer of 1990 all the sensors were operational.

ACKNOWLEDGEMENTS

The work reported in this paper was sponsored by the National Science Foundation through the Presidential Young Investigator (PYI) Award to K.P. Georgakakos with Award No. CES-8657526. Additional support was provided by the Research Experience for Undergraduates (REU) supplement to the PYI Award. The computational work was performed at the Computational Laboratory for Hydrometeorology and Water Resources (Hydromet Lab) of the University of Iowa. Particular thanks go to the Iowa Institute of Hydraulic Research Administration for their permission to use the roof of the IIHR building for the instrumentation. The invaluable assistance of Jim Cramer, Data Systems Coordinator of the Hydromet Lab, in setting up and maintaining the meteorological station and the associated data acquisition system is gratefully acknowledged.

REFERENCES

Berge, P., Pomeau, Y. & Vidal, C. (1984) *Order within Chaos: Towards a Deterministic Approach to Turbulence*, John Wiley, New York.
Gleick, J. (1987) *Chaos, Making a New Science*, Viking, New York.
Grassberger, P. & Procaccia, I. (1983a) Characterization of strange attractors, *Phys. Rev. Lett.*, **50**, 722–8.
Grassberger, P. & Procaccia, I. (1983b) Measuring the strangeness of strange attractors. *Physica*, **9D**, 189–208.
Lorenz, E. A. (1963) Deterministic non-periodic flow. *J. Atmos. Sci.*, **20**, 130–41.
Mandelbrot, B. B. (1983) *The Fractal Geometry of Nature*, Freeman, New York.
Mayer-Kress, G. (1987) Application of dimension algorithms to experimental chaos. In: Hao Bai-Lin (Ed.), *Directions in Chaos*, World Scientific, New Jersey.
Moon, F. C. (1987) *Chaotic Vibrations, An Introduction for Applied Scientists and Engineers*, John Wiley, New York.
Nicolis, C. & Nicolis, G. (1987) Evidence for a climatic attractor. *Nature*, **326**, 523–4.
Rodriguez-Iturbe, I., De Power, F. B., Sharifi, M. B. & Georgakakos, K. P. (1989) Chaos in rainfall. *Water Resour. Res.*, **25**, 1667–76.
Ruelle, D. (1989) *Chaotic Evolution and Strange Attractors*, Cambridge University Press.

Sharifi, M. B., Georgakakos, K. P. & Rodriguez-Iturbe, I. (1990) Evidence of deterministic chaos in the pulse of storm rainfall. *J. of the Atmos. Sci.*, **47**(7), 888–93.
Schuster, H. G. (1988) *Deterministic Chaos*, VCH Weinheim, Germany.
Sturdevant, P. L., Chambers, V. A., & Georgakakos, K. P. (1990) *Measurement and analysis of fine-increment storm rainfall*, First Year Report to NSF, Research Experience for Undergraduates Program, Award No. CES-8657526, Iowa Institute of Hydraulic Research, The University of Iowa, Iowa City, Iowa, 146 pages, June 1990.
Tsonis, A. A. & Elsner, J. B. (1988) The weather attractor over very short timescales. *Nature*, **333**, 545–7.
Wang, T. I., Earnshaw, K. B. & Lawrence, R. S. (1978) Simplified optical path-averaged rain gauge. *Applied Optics*, **17**, 384–90.
Wang, T. I., Earnshaw, K. B. & Lawrence, R. S. (1979) Path-averaged measurements of rain rate and raindrop size distribution using a fast response optical sensor. *J. of Appl. Meteor.*, **18**, 654–60.
Wang, T. I., Earnshaw, K. B. & Tsay, M.-K. Tsay (1980) Optical rain gauge using a divergent beam. *Applied Optics*, **19**, 3617–21.
Wolf, A., Swift, J. B., Swinney, H. L. & Vestano, J. A. (1985) Determining Lyapunov exponents from a time series. *Physica* D, **16**, 285–317.

6 Application of fuzzy theory to snowmelt runoff

K. MIZUMURA

Civil Engineering Department, Kanazawa Institute of Technology, Ishikawa, Japan

ABSTRACT Fuzzy theory (logic) is introduced to reduce uncertainty in the prediction of snowmelt runoff. It has been used to control plants, traffic junctions, subway systems, etc. The tanks model of Sugawara seems to be the most reliable method enabling computation of the snowmelt runoff in Japanese conditions. However, it is difficult to identify the parameters of this model and much data are needed for calibration. Fuzzy logic is the tool that gives the best prediction while it does not require the optimal parameters of the prediction model (tanks model). If the fuzzy logic is employed, the deviation between the observed values and the predicted ones is automatically minimized step by step. The prediction by the fuzzy logic is based on the value of the membership functions used. The effect of different membership functions on the prediction is tested by changing coefficients in time. As a result, despite the complexity of the phenomenon of snowmelt runoff, the prediction is in a good agreement with observation.

INTRODUCTION

A fuzzy set theory developed by Zadeh (1965) is presently being applied in many fields. For example, Mamdani (1974, 1981) used a fuzzy algorithm to control a plant (laboratory-built steam engine). Further, Pappis & Mamdani (1977) used the fuzzy logic for a traffic-junction control. Recent use of fuzzy methods can be found in the field of complex industrial processes (Tong, 1977) and feedback analysis (Cumani, 1982, Tanaka *et al.*, 1982, and Tong, 1980). Fujita (1985) predicted runoff from rainfall by adopting a fuzzy logic. The main purpose of this study is to predict the snowmelt runoff by the combined system of the tanks model (Sugawara, 1979) and the fuzzy logic controller.

A part of Japan facing the Sea of Japan is known to have heavy snowfalls. The main cause is the monsoon, blowing from a high pressure system over Siberia to a low pressure system over the North Pacific Ocean during the winter season. The monsoon winds pick up the moisture while passing over the Sea of Japan and deliver a heavy snowfall when rising along the high mountains on Honshu Island of Japan. Therefore, snowmelt runoff is an important source of water supply in this region, where it is used for hydropower, rice cultivation, and as drinking water. Snowmelt runoff in the spring is stored in reservoirs and, thus, an accurate runoff prediction is necessary for water level controls in the reservoirs. Rainfall–runoff and snowfall–snowmelt processes were studied in the Tadami River watershed northwest of Tokyo in Japan (Fig. 1). The watershed area is 478.6 km^2 and the altitudes of several mountains are about 2000 m.

FUZZY SET THEORY

A fuzzy set A in X is characterized by a membership function $f_A(x)$ which associates with each point in X a real number in the interval [0,1], with the value of $f_A(x)$ representing the grade of membership of x in A (Zadeh, 1965). Therefore, the closer to unity the value of $f_A(x)$ is, the higher the grade of membership of x in A. If B is a fuzzy set, then the union, the intersection, and the complement, are defined as follows:

$$f_A \vee f_B = \max\{f_A, f_B\} \quad (1)$$

$$f_A \wedge f_B = \max\{f_A, f_B\} \quad (2)$$

$$f_{A'} = 1 - f_A \qquad f_{B'} = 1 - f_B \quad (3)$$

where the prime ′ denotes the complement of the set. A fuzzy reasoning based on the fuzzy set theory is given by the following fuzzy conditional statements:

If x is A then y is B. (4)

If x is A then y is B else y is C. (5)

Defining the above conditional statements as $R_{A\ B}$ and $R_{A\ BC'}$

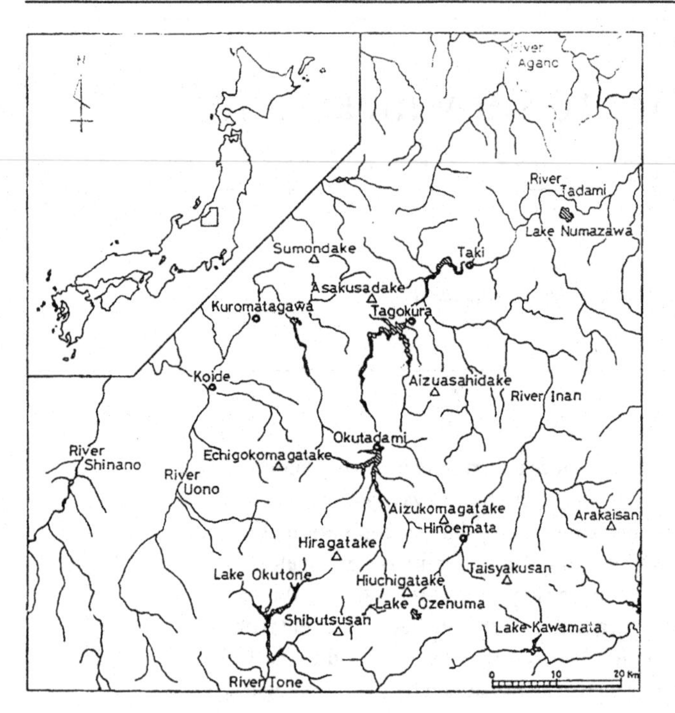

Fig. 1 Study area.

respectively, Mamdani *et al.* (1974) proposed the following equations:

$$R_{A\to B}(i,j)=f_A(x)*f_B(y) \tag{6}$$

$$R_{A\to BC}(i,j)=f_A(x)*f_B(y)+f_{A'}(x)*f_C(y) \tag{7}$$

in which '*' and '+' denote intersection and union, respectively. The operation '*' is defined in the product space of x and y as follows:

$$R_{A\to B}(i,j)=\min\{f_A(x_i),f_B(y_j)\} \tag{8}$$

$$R_{A\to BC}(i,j)=\max\{R_{A\to B}(i.j).\ R_{\bar{A}\to C}(i,j)\} \tag{9}$$

Next, consider a method of application of the fuzzy reasoning to control. This is based on the fuzzy reasoning such that 'y is B'' is inferred from 'If x is A then y is B' and 'x is A''. Expression that (x,y) is R by 'If x is A then y is B', 'y is B'' is reasoned by 'B' = A'R', if 'x is A'' is given. That is

$$f_{B'}(y)=f_{A'}(x)\ R_{A\to B} \tag{10}$$

Equation (10) can be transformed into

$$f_{B'}(y_j)=\max[\min\{f_{A'}(x_i),\ R(i,j)\}] \tag{11}$$

Next, consider a method of fuzzy reasoning in the control process. This is based on the fuzzy reasoning such that 'y is B' is inferred from 'If x is A then y is B' and 'x is A''. Expressing that '(x,y) is R' by 'If x is A then y is B', 'y is B'' is stated as 'B' = A'R', if 'x is A'' is given. That is:

$$f_{B'}(y)=f_{A'}(x)\ R_{A\to B} \tag{12}$$

Equation (10) can be transformed into

$$f_{B'}(y_j)=\max[\min\{f_{A'}(x_i),\ R_{A\to B}(i,j)\}] \tag{13}$$

The following examples illustrate the fuzzy computations. Assume that the membership functions are expressed by

$$f_A(x_i)=[0.3,\ 1.,\ 0.5,\ 0.4] \tag{14a}$$

$$f_{A'}(x_i)=[1.,\ 0.9,\ 0.5,\ 0.2] \tag{14b}$$

$$f_B(y_j)=[0.2,\ 0.8,\ 0.9,\ 0.4,\ 0.1] \tag{14c}$$

Then:

$$R_{A\to B}(i,j)=\begin{bmatrix}0.3\\1.0\\0.5\\0.4\end{bmatrix}[0.2,\ 0.8,\ 0.9,\ 0.4,\ 0.1]$$

$$=\begin{bmatrix}0.2&0.3&0.3&0.3&0.1\\0.2&0.8&0.9&0.4&0.1\\0.2&0.5&0.5&0.4&0.1\\0.2&0.4&0.4&0.4&0.1\end{bmatrix} \tag{15}$$

$$f_{B'}(y_j)=f_{A'}(x_i)\ R_{A\to B}(i,j)$$

$$=[1.,\ 0.9,\ 0.5,\ 0.2]\begin{bmatrix}0.2&0.3&0.3&0.3&0.1\\0.2&0.8&0.9&0.4&0.1\\0.2&0.5&0.5&0.4&0.1\\0.2&0.4&0.4&0.4&0.1\end{bmatrix}$$

$$=[0.2,\ 0.8,\ 0.9,\ 0.4,\ 0.1] \tag{16}$$

If $f_{A'}=f_A,\ f_{B'}=f_B$.

TANKS MODEL SYSTEM

The model, referred to as the 'tanks model' hereafter, is represented by a cascade series of conceptual reservoirs, as shown in Fig. 2. The number of tanks used is five. This is because of the inclusion of the snowmelt runoff having an important effect on the groundwater flow. Although the study area is not wide, snowmelt gradually influences runoff from this watershed for a long time. Sugawara (1979) classified the structure of the tanks model and suggested appropriate values for its parameters. A tank for snow accumulation is located in the upper position of the series of five tanks. The snowfall is stored in this upper tank and melts when the air temperature becomes higher than 0 °C. The snowmelt and rainfall move into the second tank together. Most of the snowmelt and rainfall stored in the second tank is discharged through side outlets, and the remainder infiltrates to the third tank through a bottom outlet. It is assumed for the sake of simplicity, that there is no interaction between the rainfall and the snowfall in the model; that is, that rain does not melt snow. Sugawara suggests that this assumption is reasonable

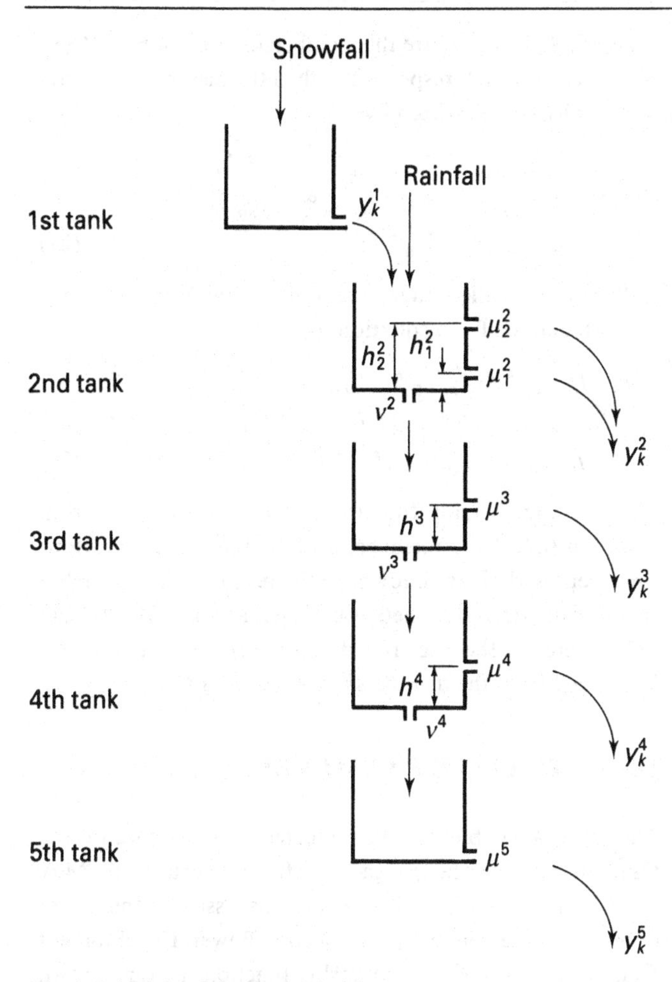

Fig. 2 Principle of the Tank Model.

to describe the snowmelt–runoff process. As the elevation of this watershed is higher than the elevation of the measuring stations, the average snowfall on the watershed may be several times that of the measuring station.

Accordingly, if S_i is the snow-pack depth at the measuring station at time step i, then λS_i represents the average snowfall in this watershed, where λ is greater than 1. Next, to simplify the snowmelt process, snowmelt can be assumed to be proportional to the air temperature T_k in the kth time step, and is expressed by mT_k if T_k is greater than 0 °C. Further, considering that the snowmelt does not occur when the air temperature is lower than 0 °C, or when the present snowfall accumulation X_k^1 in the first tank is zero, the snowmelt y_k^1 from the first tank can be expressed by the following equation:

$$y_k^1 = \begin{cases} 0 & \text{if } X_k^1 = 0 \quad \text{or} \quad T_k < 0 \\ X_k^1 & \text{if } X_k^1 \leq mT_k \text{ and } T_k \geq 0 \\ mT_k & \text{if } X_k^1 > mT_k \text{ and } T_k \geq 0 \end{cases} \tag{17}$$

where $X_k^1 = \sum_{i=1}^{k-1} (\lambda S_i - y_i^1)$; m and λ are parameters; and T_k is the air temperature at the time step k. The inflow to the second tank becomes $x_k = y_k^1 + r_k$; where r_k is the rainfall at the time step k. The superscript and the subscript indicate the tank number and the time step, respectively. Runoff through two side outlets from the second tank can be obtained as:

$$y_k^2 = \begin{cases} 0 & \text{if } X_k^2 \leq h_1^1 \\ \mu_1^2(X_k^2 - h_1^2) & \text{if } h_1^2 < X_k^2 \leq h_2^2 \\ \mu_2^2(X_k^2 - h_2^2) + \mu_1^2(X_k^2 - h_1^2) & \text{if } h_2^2 < X_k^2 \end{cases} \tag{18}$$

where X_k^2 is the storage in the second tank; μ_1^2 and μ_2^2 are the discharge coefficients; h_1^2 and h_2^2 are the elevations of the side outlets from the tank bottom; and the superscript refers to the second tank. Discharge via the bottom outlet of the second tank into the third tank is given by multiplying the discharge coefficient, v^2, and storage in the second tank, X_k^2:

$$z_k^2 = v^2 X_k^2 \tag{19}$$

The storage at the $(k+1)$th time step is represented by the following equation:

$$X_{k+1}^2 = X_k^2 - y_k^2 - z_k^2 + x_{k+1} \tag{20}$$

in which z_k^2 is the discharge from the bottom of the second tank into the third tank. Runoff from the third tank is formulated as follows:

$$y_k^3 = \begin{matrix} 0 & \text{if } X_k^3 \leq h^3 \\ \mu^3(X_k^3 - h^3) & \text{if } X_k^3 > h^3 \end{matrix} \tag{21}$$

$$z_k^3 = v^3 X_k^3 \tag{22}$$

$$X_{k+1}^3 = X_k^3 - y_k^3 - z_k^3 + z_{k+1}^2 \tag{23}$$

where z_k^3 is the discharge from the tank bottom into the fourth tank; X_k^3 is the storage in the tank; μ^3 and v^3 are the discharge coefficients, respectively, and h^3 is the elevation of the side outlet from the tank bottom. The calculations in the fourth and fifth tanks are the same as those in the third tank. The total runoff $y(k)$ at the time step k can be expressed as:

$$y_k^2 + y_k^3 + y_k^4 + y_k^5.$$

The data used for this procedure are rainfall, snowfall, snow-pack accumulation, runoff, and air temperature at 9 p.m. at the measuring station. Rainfall, snowfall, and runoff are averaged daily from 9 a.m. to 9 a.m. The air temperature is greatly influenced by the Sea of Japan and the minimum during a year appears in February. It increases in the snowmelt period from March to May, while the air temperatures higher than 20 °C in April are caused by the foehn phenomenon. Fig. 3 shows the meteorological data, i.e. air temperature, rainfall and snow data.

APPLICATION OF FUZZY SET THEORY –

Fuzzy reasoning can be used to predict the deviation of the runoff computed by the tank model from the observed data.

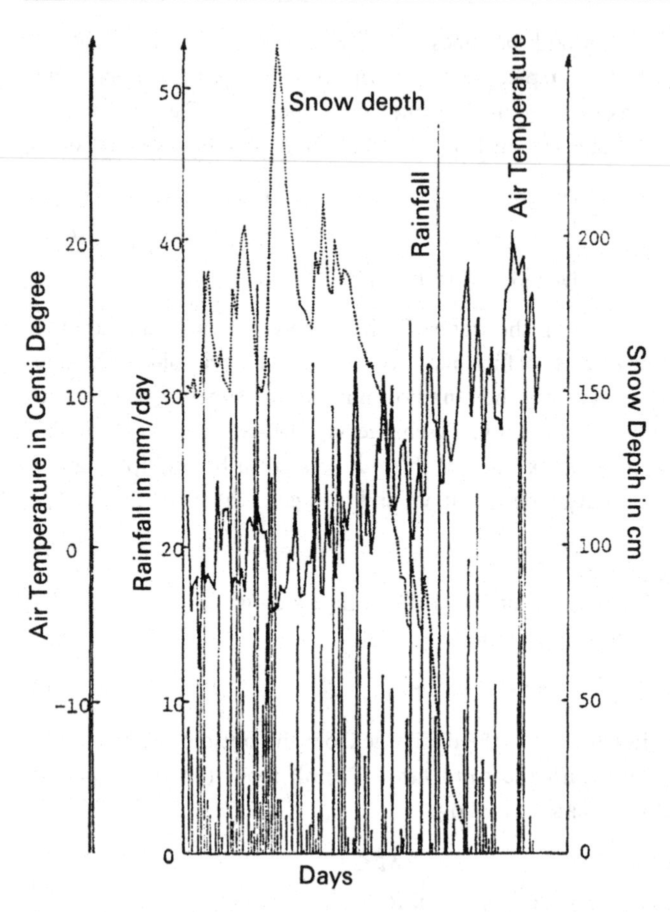

Fig. 3 Meteorological data.

The discharge deviation $Q(t)$ during the snowmelt season is most likely dependent on the discharge deviation, rainfall, air temperature, and snow deposit at the previous time steps. Therefore, the discharge $Q(t)$ may be expressed by the following function:

$$Q(t) = \text{Function of } [r(t-1), r(t-2), \ldots, r(t-m_r), T(t-1), T(t-2), \ldots, T(t-m_T), S(t-1), S(t-2), \ldots, S(t-m_S), Q(t-1), Q(t-2), \ldots, Q(t-m_Q)] \tag{24}$$

where r is the rainfall, T is the air temperature, S is the snow deposit, and m_r, m_T, m_S, m_Q are the memory lengths of rainfall, air temperature, snow deposit, and discharge deviation, respectively. One of the possible conditional statements of equation (24) is explained by the following form:

$$R_t = \text{If } r(t-1) \text{ then } r(t-2) \text{ then} \ldots \text{then } r(t-m_r) \text{ then } T(t-1) \text{ then } T(t-2) \text{ then} \ldots \text{then } T(t-m_T) \text{ then } S(t-1) \text{ then } S(t-2) \text{ then} \ldots \text{then } S(t-m_S) \text{ then } Q(t-1) \text{ then } Q(t-2) \text{ then} \ldots \text{then } Q(t-m_Q). \tag{25}$$

This statement is time dependent. Since the set of conditional statements $R_1, R_2, \ldots, R_t$ until the time t is obtained, the whole fuzzy relation becomes:

$$\Pi_t = R_1 \text{ oe } R_2 \text{ or} \ldots \text{or } R_t \tag{26}$$

When f_Q, f_r, f_T and f_S are the membership functions of $Q(t)$, $r(t)$, $T(t)$, and $S(t)$, respectively, then the membership function of $Q(t+1)$ is written by:

$$f_{Q(t+1)} = f_{r(t-m_r+1)} \circ f_{r(t-m_r+2)} \circ \quad \circ f_{r(t)} \circ f_{T(t-m_T+1)} \circ \circ f_{T(t)} \circ f_{S(t-m_S+1)} \circ f_{S(t-m_S+2)} \circ \quad \circ f_{S(t)} \circ f_{Q(t-m_Q+1)} \circ \circ f_{Q(t)} \circ \Pi_t \tag{27}$$

This is so called fuzzy reasoning. Further, the fuzzy relation derived from equation (6) reads:

$$R_t = f_{r(t-m_r)} * f_{r(t-m_r+1)} * \quad * f_{r(t-1)} * f_{T(t-m_T)} * f_{T(t-m_T+1)} * \quad * f_{T(t-1)} * f_{S(t-m_S)} * f_{S(t-m_S+1)} * \quad * f_{S(t-1)} * f_{Q(t-m_Q)} * f_{Q(t-m_Q+1)} * \quad * f_{Q(t)} \tag{28}$$

Thus the membership function of $Q(t+1)$ is derived from equation (27) if the membership functions in the previous time steps and Π_t are known. In the result, the membership function of $Q(t)$ is obtained by using equations (28), (26), and (27). Therefore the discharge difference $Q(t+1)$ at time $t+1$ is inferred from the membership function of $Q(t+1)$.

ILLUSTRATIVE EXAMPLES

The set of available data is the meteorological data on the Okutadami watershed, measured from February to May 1979, as shown in Fig. 3. The detailed discussion of these data is given in the reference by Electric Power Development Company (1980). The membership functions assumed read:

$$f_{r(t)} = \begin{cases} \exp(-p|x-r(t)|); & x \geq 0 \\ 0 & x < 0 \end{cases} \tag{29a}$$

$$f_{Q(t)} = \begin{cases} \exp(-p|y-Q(t)|); & y+Q_T(t) \geq 0 \\ 0 & y+Q_T(t) < 0 \end{cases} \tag{29b}$$

$$f_{T(t)} = \begin{cases} \exp(-p|z-T(t)|); & z \geq 0 \\ 0 & z < 0 \end{cases} \tag{29c}$$

$$f_{S(t)} = \begin{cases} \exp(-p|w-S(t)|); & S \geq 0 \\ 0 & x < 0 \end{cases} \tag{29d}$$

where $Q_T(t)$ is the runoff computed by the tanks model, and p is a parameter determining the shape of the membership functions.

The result computed for $p=1$ is plotted in Fig. 4. The predicted discharge is obtained by taking the fuzzy mean as if it does not have a peak.

$$Q(t) = \frac{\int f_{Q(t)}(\xi)\,\xi\,d\xi}{\int f_{Q(t)}(\xi)\,d\xi} \tag{30}$$

The comparison of the error variance between the predicted and the observed runoff for different values of p in the membership functions is represented in Fig. 5. It is not sensitive to variations in values of p.

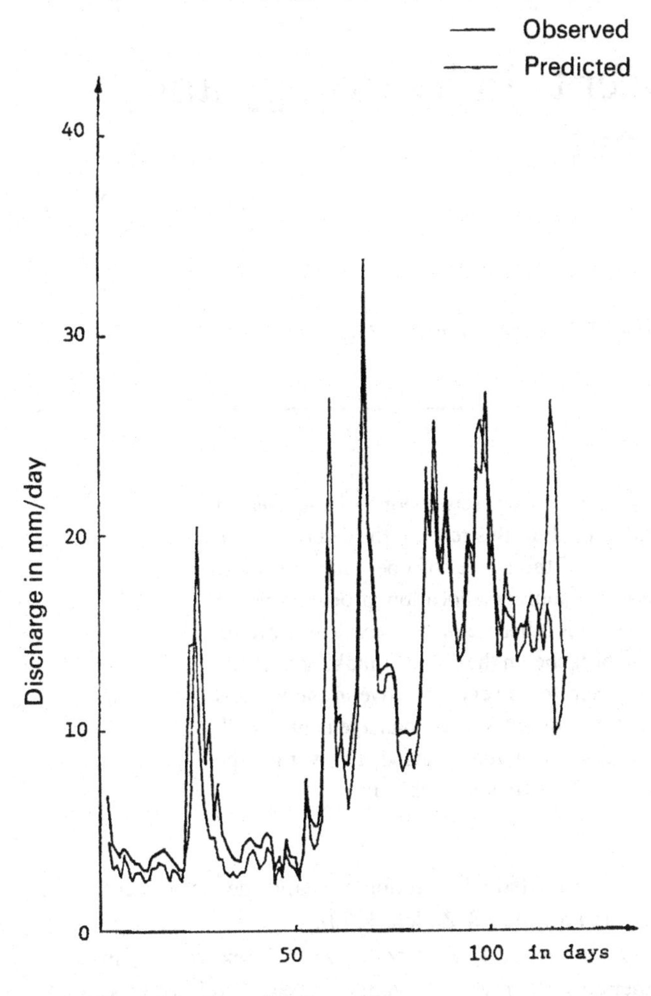

Fig. 4 Observed and predicted runoff.

Fig. 5 Error variance for different values of p.

CONCLUDING REMARKS

The deviation of the runoff predicted by the tanks model from the observed one is estimated by using the fuzzy logic. The rainfall, the air temperature, and the snow deposit at the previous time step are employed as the membership functions. The combined system of the tanks model and the fuzzy logic model was found to predict the runoff during the snowmelt period very well.

REFERENCES

Asai, K. & Negoita, C. V., Eds. (1984) *Introduction to Fuzzy System Theory*, Ohm Book Co., Tokyo (in Japanese).

Cumani, A. (1982) On a possibilistic approach to the analysis of fuzzy feedback systems. *IEEE Trans. on Systems, Man and Cybernetics*, SMC-**12**, 417–22.

Electric Power Development Company (1980) *Data Report of Snow, Weather and Discharge in Okutadami Region* (in Japanese).

Fujita, M. (1985) An application of fuzzy set theory to runoff prediction, Proc. 29th Japanese Conference on Hydraulics. *Jap. Soc. Civil Eng.*, pp. 263–8 (in Japanese).

Mamdani, E. H. (1974) Application of fuzzy algorithms for control of simple dynamic plant. *Proc. IEEE*, **121**, 121–12, 1585–8.

Mamdani, E. H. & Gaines, B. R., Eds. (1981) *Fuzzy Reasoning and Its Applications*, Academic Press, New York.

Mizumura, K. (1981) A combined snowmelt and rainfall runoff, *Proc. of the 1st International Conference on Time Series Methods in Hydrosciences*, Burlington, Ont.

Mizumura, K. & Chiu, Chao-Lin (1984) Application of autoregressive model and Kalman filtering in prediction of runoff from combined snowmelt and rainfall, *Proc. of the 4th International Symposium on Stochastic Hydraulics*, Urbana-Champaign, Ill.

Mizumura, K. & Chiu Chao-Lin (1985) Prediction of combined snowmelt and rainfall runoff. *J. Hydraul. Engineering, Amer. Soc. Civil Engineers*, **111**, 179–93.

Pappis, C.P. & Mamdani, E.H. (1977) A fuzzy logic controller for a traffic junction. *IEEE Trans. on Systems, Man, and Cybernetics*, SMC-**7**, 707–17.

Sugawara, M. (1979) *Runoff Analysis*, Kyoritsu Shuppan Co., Tokyo (in Japanese).

Tanaka, H., Uejima, S. & Asai, K. (1982) near regression analysis with fuzzy model. *IEEE Trans. on Systems, Man, and Cybernetics*, SMC-**12**, 903–7.

Tong, R.M. (1977) A control engineering review of fuzzy systems, *Automatica*, **13**, 559–69.

Tong, R.M. (1980) Some properties of fuzzy feedback systems. *IEEE Trans. on Systems, Man, and Cybernetics*, SMC-**10**, 327–30.

Zadeh, L.A. (1965) Fuzzy sets. *Information Control*, **8**, 338–53.

7 On the value of fuzzy concepts in hydrology and water resources management

J. KINDLER and S. TYSZEWSKI

Institute of Environmental Engineering, Warsaw University of Technology, Warsaw, Poland

ABSTRACT Evaluation of the applicability of the fuzzy sets theory in the area of hydrology and water resources management is attempted. In this respect, the determination of the membership functions and the interpretation of the results of operations on these functions are of crucial significance. Using water resources allocation problems as an example, the advantages of fuzzy set approaches vis à vis other techniques are demonstrated and discussed. Although the advantages of fuzzy approaches in the decision-making contexts are not always straightforward, these approaches seem to be very attractive in the various diagnostic and classification problems in hydrology and water resources management as well. This is illustrated by the application of some elements of fuzzy sets theory within the framework of a decision support system for a choice of an analog catchment.

INTRODUCTION

Water resources systems include a number of physical, economical, social, and environmental factors that must be considered in making choices among alternative options for resource use and control. The development and application of planning, management, and policy-oriented models for helping water resources managers have been taking place for several decades throughout the world. Most of them deal in one or another way with the uncertainty issue – uncertainty due to the random character of natural processes governing water supply (precipitation, streamflow, etc.), uncertainty concerning management objectives and evaluation criteria, and uncertainty about the future embedded above all in future demand projections. To deal quantitatively with uncertainty, the techniques and tools provided by the probability, decision, control, and information theories have been for a long time employed. There is no doubt about the usefulness of these techniques and tools for the solution of many problems in water resources management.

It should be acknowledged, however, that much of water resources management takes place in an environment in which the basic input information, the goals, the constraints, and the consequences of possible actions are not known precisely. In other words, water resources managers and modellers are bound to deal with imprecision – mostly due to insufficient data and imperfect knowledge – which should not be equated with randomness and the consequent uncertainty (Bellman & Zadeh, 1970).

As discussed by Kacprzyk (1983), the existence of management problems with relevant imprecision-related aspects was early recognized by many decision scientists, however, for a long time, there was no appropriate formal apparatus for handling imprecision. The situation changed in 1965 when L. A. Zadeh introduced the concept of the fuzzy sets theory. The new concept attracted attention of many analysts and modellers in many fields and the already numerous literature on this subject is growing every year.

The interest in the fuzzy sets theory has not bypassed water resources systems. The possibilities of application of the fuzzy concepts and techniques in this field have been discussed in an early paper by Hipel (1982). Several applications were attempted during the past eight years, including some by the authors of this paper (e.g. Kindler *et al.*, 1985; Kindler, 1990). It seems to be the right time, therefore, to undertake some evaluation of the applicability of the fuzzy sets theory in the field of hydrology and water resources management.

This paper is a modest attempt towards such evaluation. It uses two examples of a choice of the best analog catchment and a simple water resources allocation problem to illustrate the essence of the fuzzy concepts as applied in hydrology and water resources. The paper is written under assumption that basic definitions, operations on the fuzzy sets, and the extension principle are known to the reader. These examples are followed by a brief discussion of a few rather fundamental problems related to:

(a) identification of a membership function;
(b) aggregation operator; and
(c) interpretation of the calculation results.

THE CHOICE OF THE BEST ANALOG CATCHMENT

Assume that n controlled catchments (i.e. catchments with hydrological observation records) are given and each of them is characterized by m physiographic and climatic characteristics $\{x_{i,1}, x_{i,2}, \ldots, x_{i,m}\}$; $i=1,2,\ldots,n$. The problem is to choose the catchment which is most similar (from the point of view of these m characteristics) to the uncontrolled catchment described with the characteristics $(x_{0,1}, x_{0,2}, \ldots, x_{0,m})$.

The identification of the best analog catchment is based on the following asumptions:

(a) the ith controlled catchment $(x_{i,1}, x_{i,2}, \ldots, x_{i,m})$ is similar to the analyzed catchment $(x_{0,1}, x_{0,2}, \ldots, x_{0,m})$ in a sense of similarity in runoff generation mechanism, if each one of their m characteristics is similar;
(b) the ith controlled catchment is similar to the analyzed one from the point of view of kth characteristic, if the value of $x_{i,k}$ is equal or close to the value of $x_{0,k}$;
(c) as a measure of the degree of similarity, the value $\mu_k(x_{i,k})$ of the triangular membership function will be adopted, describing the degree of membership of the value $x_{i,k}$ to the fuzzy set 'value close to $x_{0,k}$' (Kindler *et al.*, 1985). See Fig. 1.

$$\mu_k(x_{i,k}) = \begin{cases} 0 & \text{for } x_{i,k} < (1-\alpha_k)\cdot x_{0,k} \\ \dfrac{x_{i,k} - (1-\alpha_k)\cdot x_{0,k}}{\alpha_k \cdot x_{0,k}} & \text{for } x_{i,k} \in \langle (1-\alpha_k)\cdot x_{0,k}, x_{0,k} \rangle \\ \dfrac{x_{i,k} - x_{0,k}}{\alpha_k \cdot x_{0,k}} & \text{for } x_{i,k} \in \langle x_{0,k}, (1+\alpha_k)\cdot x_{0,k} \rangle \\ 0 & \text{for } x_{i,k} > (1+\alpha_k)\cdot x_{0,k} \end{cases} \tag{1}$$

where α_k is the parameter of the membership function.

The values of $\mu_k(x_{i,k})$; $k=1,2,\ldots,m$ which define 'partial similarities' of the ith controlled catchment to the uncontrolled one, must now be aggregated into one synthetic value $\mu(i)$ describing global similarity (from the point of view of all m characteristics). Since the analog catchment must be similar to the analyzed one from the point of view of all characteristics, it seems to be most natural that the aggregating function $\mu = G(\mu_1, \mu_2, \ldots, \mu_m)$ should correspond to the aggregation operator 'and' which corresponds to the intersection (Byczkowski *et al.*, 1988):

$$\mu(i) = \min\{\mu_1(x_{i,1}), \mu_2(x_{i,2}), \ldots, \mu_m(x_{i,m})\} \tag{2}$$

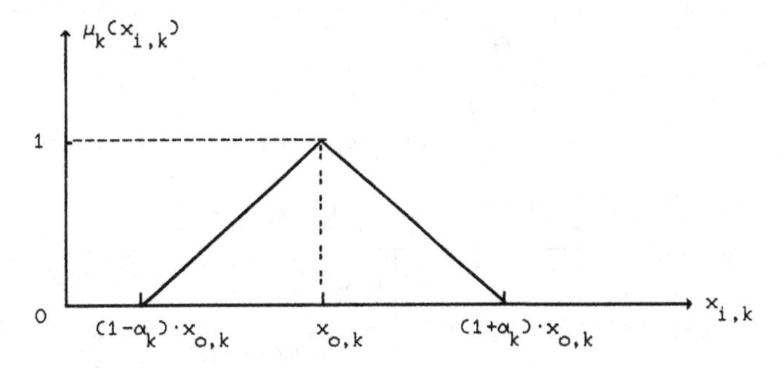

Fig. 1 Triangular membership function defining 'partial similarity' from the point of view of kth characteristics between ith controlled (gauged) catchment and the catchment brought under analysis.

The best analog catchment is the one for which the aggregated measure of similarity attains its maximum:

$$\arg \max_{i=1,2,\ldots,n} \{\mu(i)\} \tag{3}$$

WATER RESOURCES ALLOCATION BY LINEAR PROGRAMMING (FLP)

The classical problem of allocating limited resources among competing activities is often formulated as a linear programing (LP) problem of a general form:

$$\max Z = \sum_{j=1}^{n} c_j \cdot x_j$$

subject to: (4)

$$\sum_{j=1}^{n} a_{i,j} \cdot x_j \leq b_i; \quad i=1,2,\ldots,m;$$

$$x_j \geq 0; \quad j=1,2,\ldots,n$$

where: $x_1, \ldots, x_n$ are the decision variables; Z is the objective function; $c_j, a_{i,j}, b_i$ are parameters.

Utilization of such a linear model for allocation of water resources requires that the numerical values of model parameters $c_j, a_{i,j}$ and b_i are given by the decision-maker (DM) concerned with a problem at hand.

Assume that in the system shown in Fig. 2 it is necessary to allocate available water resources to water users U_1 and U_2, i.e. it is necessary to define such values of x_1 and x_2 that the total economic returns due to water use are as high as possible, and at the same time minimum flow requirements QN_1 and QN_2 in certain parts of the system are satisfied.

The symbols used in Fig. 2 have the following meaning:

– Q_1, Q_2 are discharges available in two river profiles; $Q_1 = 60$ m^3/s, $Q_2 = 50$ m^3/s,

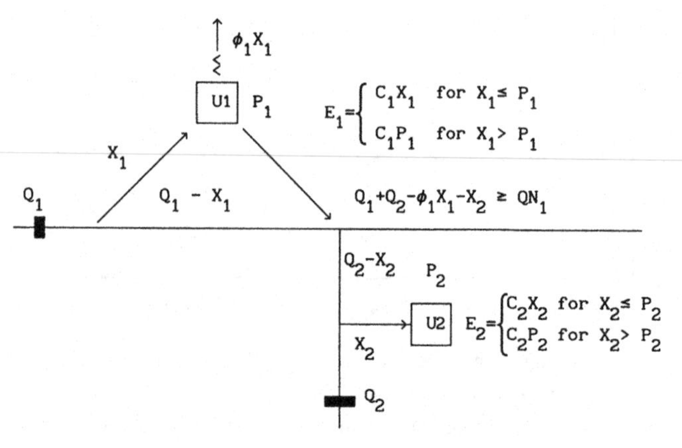

Fig. 2 General scheme of a water resources system considered.

- P_1, P_2 are water requirements of users U_1 and U_2; $P_1 = 50$ m³/s, $P_2 = 42$ m³/s,
- ϕ_1 is the consumptive use coefficient of user U_1; $\phi_1 = 0.8$,
- QN_1, QN_2 are minimum flow requirements; $QN_1 = 45$ m³/s; $QN_2 = 10$ m³/s,
- E_1, E_2 are economic effects resulting from water supply to users,
- c_1, c_2 are unit effect ceofficients; $c_1 = 1$ \$/m³/s, $c_2 = 1.6$ \$/m³/s,
- x_1, x_2 are the amounts of water actually supplied to water users U_1 and U_2 (decision variables – m³/s).

Mathematically, the problem stated above is:

$$\max Z = c_1 \cdot x_1 + c_2 \cdot x_2 \tag{5}$$

s.t.

$$x_1 \leq P_1;\ x_2 \leq P_2;\ x_1 \leq Q_1;\ x_2 \leq Q_2;\ x_1 \geq 0;\ x_2 \geq 0;$$
$$\phi_1 \cdot x_1 + x_2 \leq Q_1 + Q_2 - QN_1;\ x_2 \leq Q_2 - QN_2$$

Taking into account numerical values of system (model) parameters, the problem (5) can be writen as:

$$\max Z = x_1 + 1.6 \cdot x_2 \tag{6}$$

s.t.

$$x_1 \leq 50;\ x_2 \leq 42;\ x_1 \leq 60;\ x_2 \leq 50;\ x_1 \geq 0;\ x_2 \geq 0;$$
$$0.8 \cdot x_1 + x_2 \leq 65;\ x_2 \leq 40$$

Graphical representation of problem (6) is shown in Fig. 3. The optimal allocation decision is defined by point C with coordinates $\hat{x}_1 = 31.25$ and $x_2 = 40.00$. Such allocation pattern generates a profit of:

$$Z_{\max} = c_1 \hat{x}_1 + c_2 \hat{x}_2 = 95.25$$

APPLICATION OF ZIMMERMANN'S FLP FORMULATION

The fuzzy LP formulated by Zimmermann (1976) belongs to a certain class of decision problems analyzed in terms of the so called 'satisficing' decisions. Following notation intro-

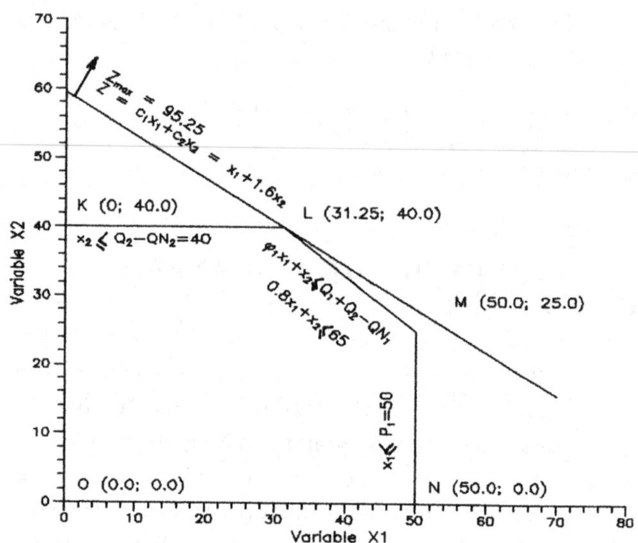

Fig. 3 Representation and solution of a water resources allocation problem (classical LP).

duced by Zimmermann (1976) and Hamacher *et al.* (1978), this problem is represented in form:

$$\max Z = \sum_{j=1}^{n} c_j \cdot x_j$$

s.t. (7)

$$\sum_{j=1}^{n} a_{i,j} \cdot x_j \leq b_i; \quad i = 1, \ldots, m$$

$$x \in X$$

where: $x = [x_1, \ldots, x_n]$

$$X = \left\{ x\colon \sum_{j=1}^{n} d_{k,j} \cdot x_j \leq f_k;\ k = 1, \ldots, m_1;\ x_j \geq 0;\ j = 1, \ldots, n \right\}$$

This type of a decision problem requires from the DM that he defines aspiration levels concerning accomplishment of individual goals (objectives, constraints) to be statisfied by the system under consideration (e.g. water resources system). Moreover, the DM has to define the criteria of accomplishment of the goals. For example, if one of the goals of a given water resources system is water quality control, the DM states that all decisions (solutions, alternatives) $x = [x_1, \ldots, x_n]$, which ensure that concentrations of a certain pollutant in the control profile of the river which are less than b [mg/l], are satisfactory from the point of view of water quality control. In such case pollutant concentration b is an aspiration level of DM.

In the fuzzy linear programming problem (FLP) formulated by Zimmermann it is assumed, that aspiration levels concerning the goal Z_0 and constraints $b_1, \ldots, b_m$ are not ordinary numbers but fuzzy numbers $\boldsymbol{Z}_0, \boldsymbol{b}_1, \ldots, \boldsymbol{b}_m$ characterized by the triangular membership function of the following form:

$$\mu_{Z_0}(\gamma) = \langle \mathbf{Z}_0, Z_0, \bar{Z}_0 \rangle; \quad \mu_{b_i}(\beta_1) = \langle \mathbf{b}_i, b_i, \bar{b}_i \rangle; \quad i = 1, \ldots, m \tag{8}$$

The criteria of satisfying the goal and constraints, Zimmermann determines in form of the membership functions of a fuzzy goal:

$$\mu_0\left(\sum_{j=1}^{n} c_j \cdot x_j\right)$$

and fuzzy constraints $\mu_i\left(\sum^{n} a_{i,j} \cdot x_j\right)$ determined in the following way (see Figs. 4 and 5):

$$\mu_0(x) = \mu_0\left(\sum_{j=1}^{n} c_j \cdot x_j\right) = \begin{cases} 0 & \text{for} \quad \sum_{j=1}^{n} c_j \cdot x_j < \mathbf{Z}_0 \\ \dfrac{\sum_{j=1}^{n} c_j \cdot x_j - \mathbf{Z}_0}{Z - \mathbf{Z}_0} & \text{for } \mathbf{Z}_0 \leq \sum_{j=1}^{n} c_j \cdot x_j \leq \bar{Z}_0 \\ 1 & \text{for} \quad \sum_{j=1}^{n} c_j \cdot x_j > 0 \end{cases} \tag{9}$$

And per analogy:

$$\mu_i(x) = \mu_i \sum^{n} a_{i,j} \cdot x_j) \tag{10}$$

Following the definition of fuzzy optimality introduced by Bellman & Zadeh (1970), solution of Zimmermann's problem is such vector $\hat{x} = [\hat{x}_1, \ldots, \hat{x}_n]$ for which the degree of satisfaction due to the simultaneous satisfaction of the constraints and accomplishment of the goal attains the highest possible value:

$$\hat{x} = \arg \max_{x \in X} \mu_D(x) \tag{11}$$

where $\mu_D(x)$ is a membership function of fuzzy decision:

$$\mu_D(x) = \min \{\mu_0(x), \mu_1(x), \ldots, \mu(x)\} \tag{12}$$

Problem (11) can be written in the following equivalent form:

$$\hat{x} = \arg \max_{x \in X} \lambda \tag{13}$$

s.t.

$$\mu_i(x) \geq \lambda; \quad i = 1, \ldots, m$$

In other words, solution of Zimmermann's problem is obtained by determination of the highest possible value $\hat{\lambda}$ (which measures the degree of satisfaction related to the solution of the problem) for which the set of satisficing decisions:

$$X_{\hat{\lambda}} = \{x: \mu_1(x) \geq \hat{\lambda}; \quad i = 0, 1, \ldots, m\} \tag{14}$$

is nonempty.

It should be noticed that set $X_{\hat{\lambda}}$ may include more than a single element. In case of card $(X_{\hat{\lambda}}) > 1$, we encounter another decision problem, namely how to chose $\hat{x} = [\hat{x}_1, \ldots, \hat{x}_n]$ from the set $X_{\hat{\lambda}}$. Solution of this problem is beyond the scope of the fuzzy set theory ($X_{\hat{\lambda}}$ is not a fuzzy one) and it needs additional criteria unconsidered so far.

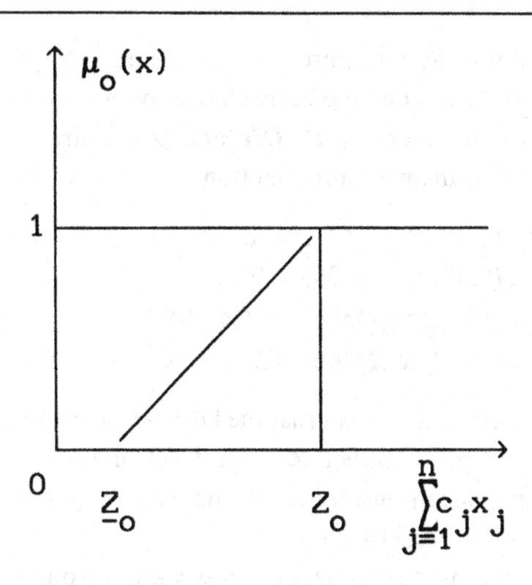

Fig. 4 Membership function of the fuzzy goal.

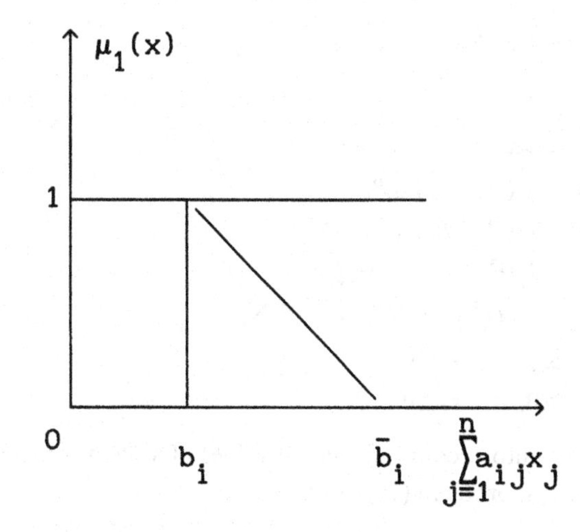

Fig. 5 Membership function of the fuzzy constraint.

Following the original Zimmermann's formulation, it is proposed to replace problem (13) by a classical LP problem of the following form:

$$\hat{x} = \arg \max_{x \in X} \lambda \tag{15}$$

s.t.

$$\sum_{j=1}^{n} c_j \cdot x_j - \lambda \cdot (Z_0 - \mathbf{Z}) \geq \mathbf{Z}$$

$$\sum_{j=1}^{n} a_{i,j} \cdot x_j - (\bar{b}_i - b_i) \leq \bar{b}_i; \quad i = 1, \ldots, m$$

APPLICATION OF FUZZY LP TO ALLOCATION OF WATER RESOURCES —

Consider the same water resources system as the one introduced before (see Fig. 2), but now assume that the numerical values of water requirements P_1 and P_2 of users U_1 and U_2

and minimum flow requirements QN_1 and QN_2 cannot be precisely defined. These parameters are only known in form of the fuzzy numbers $\boldsymbol{P}_1$, $\boldsymbol{P}_2$, $\boldsymbol{QN}_1$ and $\boldsymbol{QN}_2$ characterized by the triangular membership functions:

$$\mu_{P_1} = \langle \underline{P}_1, P_1, \bar{P}_1 \rangle = \langle 45,50,58 \rangle;$$
$$\mu_{P_2} = \langle \underline{P}_2, P_2, \bar{P}_2 \rangle = \langle 35,42,48 \rangle$$
$$\mu_{QN_1} = \langle \underline{QN}_1, QN_1, \bar{QN}_1 \rangle = \langle 35,45,50 \rangle;$$
$$\mu_{QN_2} = \langle \underline{QN}_2, QN_2, \bar{QN}_2 \rangle = \langle 3,10,15 \rangle$$

Moreover it is assumed, that the DM can define fuzzily – in form of a fuzzy number $\boldsymbol{Z}$ – the level of his aspirations concerning accomplishment of the system goals, i.e. $\mu_Z = \langle \underline{Z}, Z, \bar{Z} \rangle = \langle 80,110.140 \rangle$.

Under this assumptions, the fuzzy water resources alocation problem can be stated as:

$$\hat{x} = \arg\max_{x \in X} \lambda$$

s.t. (16)

$$c_1 x_1 + c_2 x_2 - \lambda \cdot (Z - \underline{Z}) \geq \underline{Z}$$
$$x_1 + \lambda \cdot (\bar{P}_1 - P_1) \leq \bar{P}_1$$
$$x_2 + \lambda \cdot (\bar{P}_2 - P_2) \leq \bar{P}_2$$
$$\phi_1 \cdot x_1 + \lambda \cdot (\bar{QN}_1 - QN_1) \leq Q_1 + Q_2 - \mathrm{QN}_1$$
$$x_2 + \lambda \cdot (\bar{QN}_2 - QN_2) \leq Q_2 - \mathrm{QN}_2$$
$$x_1 \leq Q_1; \quad x_2 \leq Q_2$$
$$x_1 \geq 0; \quad x_2 \geq 0$$

Taking into account numerical values of system (model) parameters, problem (16) can be written as:

$$\hat{x} = \arg\max_{x \in X} \lambda$$

s.t. (17)

$$x_1 + 1.6 \cdot x_2 - 30 \cdot \lambda \geq 80$$
$$x_1 + 8 \cdot \lambda \leq 58$$
$$x_2 + 6 \cdot \lambda \leq 48$$
$$0.8 \cdot x_1 + 10 \cdot \lambda \leq 75$$
$$x_2 + 7 \cdot \lambda \leq 47$$
$$x_1 \geq 60; \quad x_2 \leq 50$$
$$x_1 \geq 0; \quad x_2 \geq 0$$

The solution of problem (17) shown in Fig. 6 has the following interpretation. For the fuzzy parameters of the problem and DM's fuzzy aspirations, the optimal decision is defined by point S with coordinates $\hat{x}_1 = 32.49$ and $\hat{x}_2 = 42.31$. At the same time, the degree of truth $\hat{\lambda}$ in the assertion:

allocation of $\hat{x}_1 = 32.49$ m³/s of water to user U_1 and $\hat{x}_2 = 42.31$ m³/s to user U_2 generates a profit of $Z_{\max} = \underline{Z} + \hat{\lambda} \cdot (Z - \underline{Z}) = 80 + 0.671 \cdot (110 - 80) = 100.13$ \$

is 0.671.

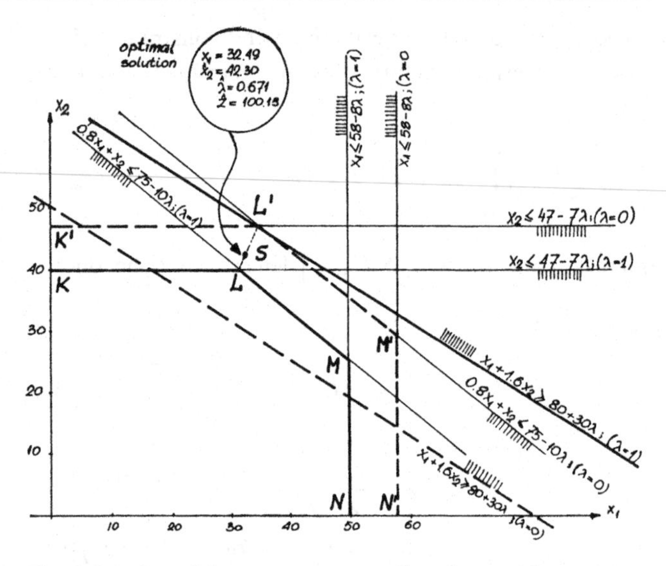

Fig. 6 Solution of the water resources allocation problem.

EVALUATION OF THE FUZZY SETS APPLICATIONS IN HYDROLOGY AND WATER RESOURCES MANAGEMENT

Problems related to identification of the membership function

In the problem concerning the choice of the best analog catchment, the fundamental and most critical issue is determination of the membership function $\mu_k(x_{i,k})$. If a given physiographic or climatic characteristic is being explicitly and numerically used in the hydrological catchment model, the sensitivity analysis may be used for determination of the parameters of the membership function. Those parameters can also be determined by an expert hydrologist (subjective judgement). Another possibility presented by Tyszewski *et al.* (1990) is to determine the parameters of the membership function on the basis of the so called 'teaching sample'.

The essence of this approach is that for each one of the n catchments, an expert hydrologist (or a group of experts) – based on his (or their) experience and/or detailed hydrological studies – has to evaluate how good analogues are each of the remaining $n-1$ catchments. In this process, the p-degree scale may be used, describing how analogous one catchment to the other is. This way the so called 'teaching sample' is created:

$$\begin{bmatrix} \omega_{1,2} \; \omega_{1,3} \; \omega_{1,4} \; \cdots\cdots \; \omega_{1,n} \\ \omega_{2,1} \; \omega_{2,3} \; \omega_{2,4} \; \qquad \omega_{2,n} \\ \cdots\cdots\cdots\cdots\cdots\cdots \\ \omega_{n,1} \; \omega_{n,2} \; \omega_{n,3} \; \qquad \omega_{n,n-1} \end{bmatrix} \tag{18}$$

where: ω is the degree to which the jth catchment is analogous to the ith catchment.

Having built the 'teaching sample', now the problem of

determining membership function (1) is reduced to identification of such values of parameters α_k, for which the correspondence in the choice of the best analog catchment made by fuzzy sets method and by an expert is as large as possible.

This can be mathematically formulated as the following optimization problem:

$$\max F = \sum_{i=1}^{n} \omega_{i,NA_i} \qquad (19)$$

s.t.

$$\alpha_{\min_k} < \alpha_k < \alpha_{\max_k} k = 1,2,\ldots,m$$

where: n is the number of catchments in the 'teaching sample'; m is the number of physiographic and climatic characteristics; α_k is the parameter of the membership function of the kth characteristics; and ω_{i,NA_i} is the evaluation how good the catchment NA_i is as an analog for the ith catchment.

The problems related to aggregation of partial goals and interpretation problems

Consider a simple decision problem as to determine the value of a decision variable x which satisfies:

(a) goal: 'x greater than $\boldsymbol{Z}$';
(b) constraint: 'x less than $\boldsymbol{b}$';
where $\boldsymbol{Z}$ and $\boldsymbol{b}$ are fuzzy aspiration levels of the goal and constraint respectively.

Assume that fuzzy numbers $\boldsymbol{Z}$ and $\boldsymbol{b}$ are characterized by triangular membership functions:

$$\mu_Z(x) = \langle \underline{Z}, Z, \bar{Z} \rangle = \langle 2,6,9 \rangle$$
$$\mu_b(x) = \langle \underline{b}, b, \bar{b} \rangle = <5,8,12>$$

With the aid of expressions (9) and (10) it is possible to construct membership functions of the fuzzy goal $\mu_0(x)$ and fuzzy constraint $\mu_1(x)$; (see Fig. 7).

This simple example provides a good illustration of a problem how to choose the most appropriate aggregation operator. By the analogy to the conventional set theory in which the intersection contains the elements belonging to one set 'and' to the other one, likewise for fuzzy sets the intersection corresponds to 'and' operator. The appropriateness of this operator is therefore very much connected to the semantical meaning of of the connective 'and'. However, the fuzzy sets literature considers several other aggregation operators (min-type, weighted-sum-type, max-type, and others) and the problem is still fairly open to debate.

According to Bellman & Zadeh (1970), fuzzy decision set $\mu_D(x)$ is obtained by intersection:

$$\mu_D(x) = \min\{\mu_0(x), \mu_1(x)\}$$

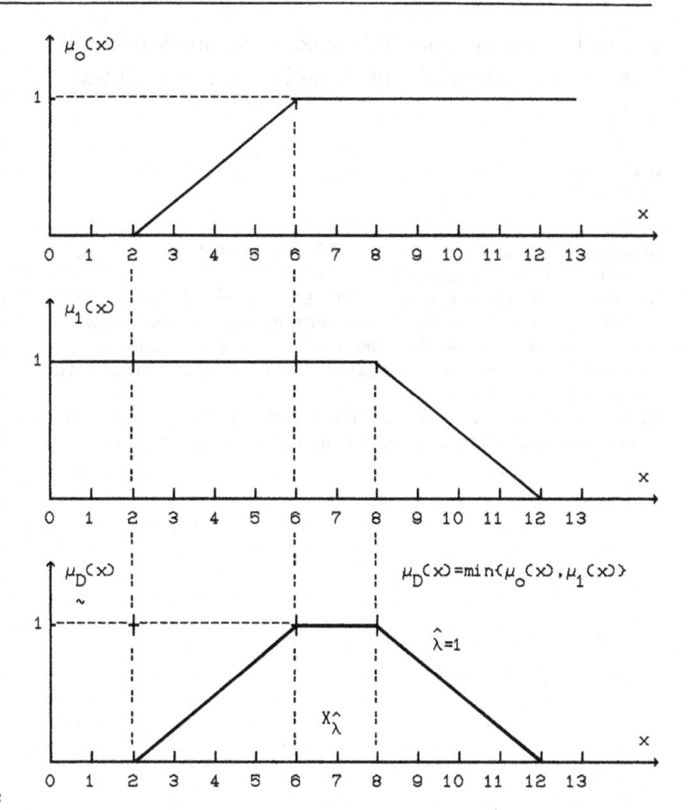

Fig. 7 Membership functions of the fuzzy goal, fuzzy constraint, and fuzzy decision.

and the fuzzy optimal decision is defined as such for which the membership function $\mu_D(x)$ attains its maximum. As shown in Fig. 7, fuzzy optimal decision defined this way is not univocal. Thus, the problem is what is to be done in such situations?

CONCLUSIONS

Although the fuzzy sets theory and the associated concepts are truly compatible with the type of information normally available, their practical application in hydrology and water resources raises some doubts. The fundamental problem is how to identify membership functions. One of the possible ways of attaining this goal has been demonstrated in this paper. The 'teaching sample' approach has been used in the diagnostic problem of the analog catchment but its eventual use in the allocation problem would be much more difficult. The aggregation of goals and constraints in the decision-type problems is also open to debate. Finally interpretation of the fuzzy results is not an easy task. Although it cannot be the other way round and 'fuzzy-in' must lead to 'fuzzy-out', sometimes one may be wondering whether the computational effort related to the use of fuzzy sets apparatus is worth trying. The theory seems to be much more applicable in the diagnostic problems than in the decision-making contexts.

REFERENCES

Bellman, R. E. & Zadeh, L. A. (1970) Decision-making in a fuzzy environment. *Management Science*, **17**, 4.

Byczkowski, A., Tyszewski, S., Okruszko, T. & Rogowski, R. (1988) *Opracowanie koncepcji i struktury systemu ekspertowego dla potrzeb okreslania przeplywow charakterystycznych rzek w profilach niekontrolowanych*, Opracowanie wykonane dla Polskiego Towarzystwa Gleboznawczego, Warsaw.

Hamacher, H., Leberling, H. & Zimmermann, H.-J. (1978) Sensitivity analysis in fuzzy linear programming. *Fuzzy Sets and Systems*, **1**(1).

Hipel, K. W. (1982) *Fuzzy Set Techniques in Decision Making, IFAC Theory and Application of Digital Control*, Pergamon Press.

Kacprzyk, J. (1983) *Multistage Decision-making under Fuzziness, Theory and Applications*, Verlag TUV Rheinland, Koln.

Kindler, J., Tyszewski, S., Wierzchon, S. & Rogowski, R. (1985) *The Fuzzy sets and multiobjective methods in analysis of regional water resources systems.* Warsaw University of Technology, (manuscript in English).

Kindler, J., Tyszewski, S. & Zielinski, P. (1984) Water resources allocation with imprecise demand estimates, *Proceedings of the 9th World Congress of IFAC*, Budapest, Hungary.

Kindler, J. (1990) Water resources planning in the 90s; the case of resources allocation with unprecise demand estimates, *Proceedings of the International Symposium on Water Resource Systems Applications, Winnipeg, Canada, June 12–15, 1990.*

Zadeh, L. A. (1965) Fuzzy sets. *Information and Control*, **8**.

8 Application of neural network in groundwater remediation under conditions of uncertainty

S. RANJITHAN, J. W. EHEART and J. H. GARRETT, JR.

Department of Civil Engineering, University of Illinois at Urbana-Champaign, USA

ABSTRACT The design of groundwater contamination remediation based on hydraulic head gradient control method determines the locations of the pumping wells and their pumping rates. In a heterogeneous medium such a design will be sensitive to the spatial characteristics of the underlying geological parameters. The geological uncertainty is due to the heterogeneity of the hydraulic conductivity of the porous medium. Under conditions of uncertainty, incorporation of transmissivity fields with spatial characteristics that most influence the design will reduce the sensitivity of the design. A new class of artificial intelligence technique known as neural networks has been identified as appropriate for pattern association tasks. A neural network based screening tool is being developed to identify transmissivity fields with such spatial characteristics. The ongoing research embraces training a neural network to learn the association between transmissivity fields and their impact on the design, and using the trained network to classify randomly generated feasible transmissivity fields according to their level of impact on the design.

INTRODUCTION

Safe and effective designs for groundwater remediation is a topic that is currently gaining increased worldwide attention. There are many alternative techniques available for groundwater contaminant containment and restoration. The hydraulic gradient control for containment and removal of groundwater contamination is one of the techniques under investigation among the researchers in the field of groundwater management (Gorelick *et al.* (1984); Atwood & Gorelick (1985); Keely (1984); Colarullo (1984); Wagner & Gorelick (1987); Valocchi & Eheart (1987); Gorelick (1987); Wagner & Gorelick (1989); Morgan (1990)). The hydraulic gradient control technique uses a series of extraction/injection wells to control plume migration, extract and possibly treat the contaminated groundwater, and re-inject treated or fresh water. A brief description of this method is given further. Performance of a reclamation design based on this technique is highly influenced by the accuracy of the aquifer parameters used to develop the design. The hidden nature of aquifer geology and insufficient data and knowledge about them lead to uncertainty in the estimation of hydrogeologic parameters. Furthermore, the accuracy of estimated parameters cannot be easily verified since the response times of aquifers are very long. Therefore design of groundwater pollution control under conditions of parameter uncertainty plays an important role in reliable aquifer protection programs.

A major source of uncertainty is due to insufficient knowledge about the spatial variability of the hydraulic conductivity (or the transmissivity in a two dimensional case) of the aquifer medium. In general the uncertainty in geology is assumed to be due to the heterogeneity of the porous medium as manifested in the hydraulic conductivity (or transmissivity) parameter. Although uncertain parameters are not completely known, they could be assumed to lie within a range estimated from partially known information. Most of the methods reported in the literature adopt the stochastic approach of considering many realizations of the random hydraulic conductivity field. A design that satisfies many realizations will, in general, tend to be more reliable than one that satisfies fewer realizations. However, all the realizations included in a design do not equally constrain the design. The realization that most constrains the design will be called the most pessimistic realization. Generation of pessimistic realizations, and only considering these realizations, is important in that it would lead to reliable designs with the least amount of computational effort. Development of a tool that could generate pessimistic realizations for hydraulic gradient control design is the focus of this work. Some preliminary results were reported in Ranjithan *et al.* (1991).

The characteristics that make a realization to be pessimistic are essentially spatial in nature, i.e., they depend on the spatial distribution of the hydraulic conductivity values with respect to the pump location and the plume boundary. However these spatial characteristics are currently not well defined. An initial attempt was made by Eheart *et al.* (1990) to identify some spatial characteristics that would constitute a pessimistic realization, but the authors acknowledge that these characteristics are not complete. It would be ideal if a random field generator could generate hydraulic conductivity fields which contain these specific spatial characteristics. Even under the optimistic assumption that such spatial characteristics are *a priori* known, numerically incorporating a spatial characteristic along with the estimated statistics in the generation phase of the random hydraulic conductivity fields makes the generation task intractable. For example, Eheart *et al.* (1990) proposed an optimization approach for parameter configuration: first they identified three spatial characteristics in a hydraulic conductivity field that would render it pessimistic and then used appropriate weights in the objective function to obtain a feasible hydraulic conductivity field with those characteristics. This approach suffers from computational complexities due to the non-linear constraints in the parameter configuration technique.

It may be possible to approach this design problem using rule-based heuristic methods. Although no such application to this problem is reported in the literature, the complexities of capturing spatial relations and reasoning through a purely rule-based approach to the problem of sampling design for plume delineation were reported in an unpublished report (Ranjithan & Morgan, 1988). In that report, it was concluded that

(a) eliciting the knowledge behind spatial reasoning was very difficult and subject to incompleteness;
(b) the number of rules became incredibly large to represent even rather trivial spatial reasoning tasks; and
(c) it became impossible to handle a situation involving simultaneous interactions of multiple spatial reasoning tasks.

In light of these limitations in both analytical and symbolic approaches, an alternative method must be developed.

An alternate approach, described further, is to use a tool that could pick out from many realizations only those that have a spatial distribution of hydraulic conductivity values commonly found among pessimistic realizations. This entails complex pattern recognition. Neural networks have been found to be good at recognizing patterns after being trained to do so. A neural network-based tool for recognizing particular patterns of spatial distribution of hydraulic conductivity values that constrain a design is described in this paper.

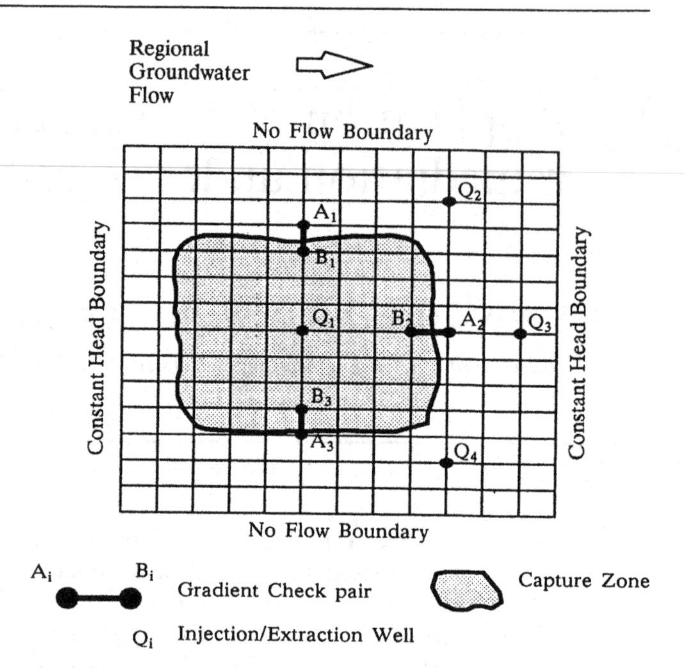

Fig. 1 Schematic diagram of a typical problem.

DESCRIPTION OF THE PROBLEM AND MANAGEMENT MODEL

This section briefly describes the groundwater contamination problem and a management model for remediation that are used in the subsequent sections. A typical problem scenario is as shown in Fig. 1. In the absence of complete knowledge of the location of the contaminant plume, the plume is assumed to lie within a capture zone that has to be cleaned up. The domain is discretized into rectangular grids and the properties within each grid are assumed to be uniform. This discretization approach is required to compute the head at a location (using a standard finite-difference or finite-elements based numerical method) for a problem with a heterogeneous hydraulic conductivity field.

One approach to groundwater remediation is through hydraulic head gradient control. The goal of the hydraulic gradient control design is to create a hydraulic head field that captures the contaminant and flushes the contaminant to extraction wells which carry it to the surface; i.e., minimize the cost of pumping strategy that induces inwardly directed hydraulic gradients at specific check points along the boundary of the capture zone. An optimization model could be written as shown in Fig. 2 (see Atwood & Gorelick, 1985, for details). The first set of constraints (equation (2) in Fig. 2) ensures that the hydraulic gradient along the boundary of the plume are inwardly directed for successful plume contain-

$$\text{minimize} \quad \sum_{j=1}^{NW} (u_j + v_j) \qquad (1)$$

$$\text{subject to} \quad \sum_{j=1}^{NW} R_{ij}(u_j - v_j) \geq g_i \qquad i = 1, 2, \ldots, NGP \qquad (2)$$

$$\sum_{j=1}^{NW} (u_j - v_j) \geq 0 \qquad (3)$$

$$0 \leq u_j \leq Q_{max} \qquad (4)$$
$$0 \leq v_j \leq Q_{max} \qquad j = 1, 2, \ldots, NW \qquad (5)$$

where

u_j - extraction rate of well j

v_j - injection rate of well j

g_i - target hydraulic gradient at check point i

R_{ij} - gradient response coefficient; impact of unit extraction at well j upon gradient at check point i

NGP - # of gradient check points

NW - # of wells

Q_{max} - maximum possible pumping rate for a single well

Fig. 2 Optimization model for well design.

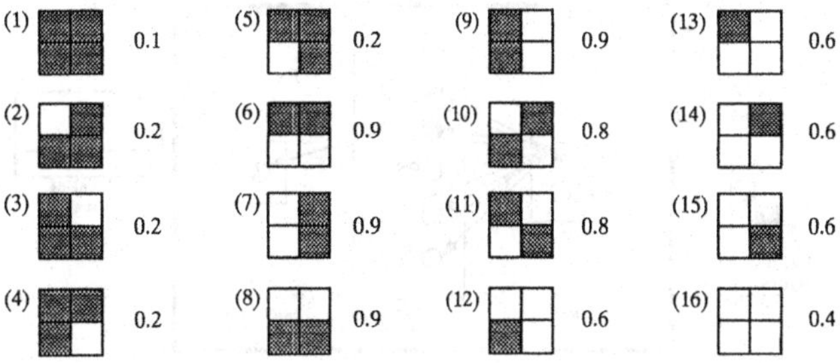

ment. The second constraint (equation (3) in Fig. 2) ensures that the total withdrawal rate is larger than the total injection rate. The constraint set (equation (2) in Fig. 2) represents one single hydraulic conductivity field. For each hydraulic conductivity field incorporated into the design, its corresponding constraint set must be included in the model described above. Under conditions of uncertainty, a typical stochastic approach would be to include as many realizations of the uncertain hydraulic conductivity field as possible in the model given in Fig. 2. However, only a small set of (pessimistic) realizations will influence the final design. The neural network-based tool proposed here will generate the pessimistic realizations which alone could then be considered in the optimization model to obtain a reliable design thus reducing the size of the optimization model.

NEURAL NETWORK-BASED PATTERN RECOGNIZER

This section presents the over-all framework for building the neural network-based pattern recognition tool for identification of pessimistic realizations. The development of the neural network-based screening tool is first illustrated using a simple example. The architecture of the pro-type neural network and the required input-output to this network are described. The generation of training examples and the performance of the neural network are also presented.

Approach

The method presented here uses a 'generate and screen' approach. A standard random field generator is used to generate many realizations of the hydraulic conductivity. Then a neural network-based pattern recognition tool scans

(1) 0.1 (5) 0.2 (9) 0.9 (13) 0.6
(2) 0.2 (6) 0.9 (10) 0.8 (14) 0.6
(3) 0.2 (7) 0.9 (11) 0.8 (15) 0.6
(4) 0.2 (8) 0.9 (12) 0.6 (16) 0.4

Fig. 3 Illustrative example: spatial distributions of hydraulic conductivity and corresponding level of criticalness. (Dark regions represent low conductivity regions and light regions represent high conductivity regions.)

each realization and identifies those possessing the characteristics of pessimistic realizations. Lack of knowledge about the actual constitution of these pessimistic realizations requires the tool to be able to learn the relationship between a hydraulic conductivity field and its being or not being a pessimistic realization. In the absence of a known closed-form relationship, it is imperative that this relationship be captured through observations, i.e., the proposed tool should learn the association from examples. Backpropagation learning in feed-forward type neural networks (Rumelhart *et al.*, 1986) has been found to be effective at learning and recognizing the association between a set of input and output signals. Neural networks technology is a rapidly emerging field of research within artificial intelligence that has the advantage of being able to capture knowledge that is vague, complex, and not explicitly expressed by mathematical or symbolic (i.e., rule-based, frame-based) means. An illustrative example is used in the following section to describe the implementation of the proposed tool.

Neural network: an illustrative example

A simple example is used to illustrate the function of backpropagation learning in feed-forward type neural networks. The illustrative example has a 2 grid × 2 grid pattern where each grid represents the average hydraulic conductivity within a region. The average hydraulic conductivity is represented by a continuous scale varying from very low to very high, 0.0 representing very low and 1.0 representing very high. All the possible spatial distributions of hydraulic conductivity in this 2 × 2 grid representation are shown in Fig. 3 (the dark area represent low hydraulic conductivity regions, 0.1 on the continuous scale, and the light area represent high hydraulic conductivity regions, 0.9 on the continuous scale). The value assigned beside each pattern represents how critical that spatial distribution of hydraulic conductivity is towards the total pumping rate (0.0 representing least critical and 1.0 representing most critical).

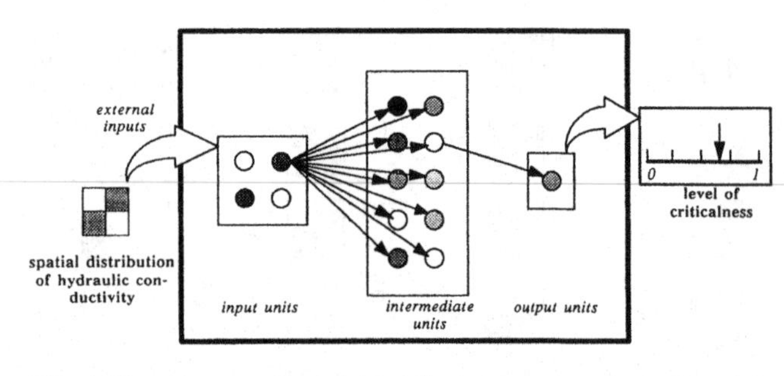

Fig. 4 Neural network model for illustrative example (each unit in a layer is connected to all the units in the subsequent layer).

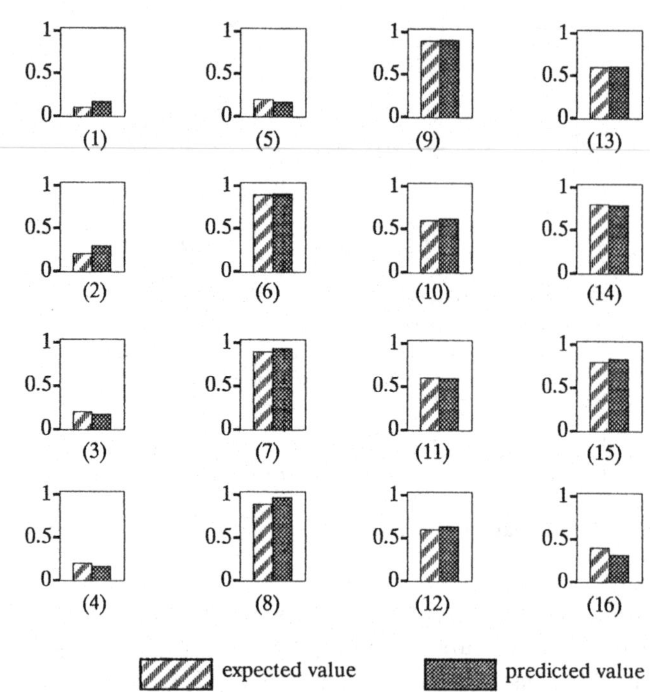

Fig. 5 Results comparing the training performance of neural network model for the illustrative example (vertical axis represents level of criticalness).

Although these values were assigned arbitrarily, in the real problem a measurable quantity will be used as a surrogate for criticalness. The pattern association task at hand is to learn the association between each spatial distribution of hydraulic conductivity values and the corresponding level of criticalness.

A neural network model capable of performing this pattern recognition task is shown in Fig. 4. This is a collection of highly interconnected units. Each unit in the network is capable of a limited amount of processing involving integrating all incoming input signals (a product of activation and connection strength), computing an output signal, and sending the output signal to other units to which a unit is connected. Activations at a unit generally vary continuously between 0 and 1. In the input layer, each grid is represented by a unit. There are four input units in this example. The activations, each of whose magnitude represent the average hydraulic conductivity in the region represented by that unit, are externally input at the input units to represent the different hydraulic conductivity patterns shown in Fig. 3. The activation at the output unit represents the level of criticalness of a given input pattern. The intermediate units are what the neural network model uses for internal representation of the mapping between the input and output signals. In this neural network model each input unit is connected to all the intermediate units and each intermediate unit is connected to the output unit as shown in Fig. 4. The association between the input patterns and the output values will be captured in the connection strengths in the network.

Initially the connection strengths are set randomly and therefore the association between the input and output patterns is not properly captured. That means if an input pattern, say the second pattern in Fig. 3, is input to the neural network model, the activation of output unit will not likely be close to the expected value of 0.2. The backpropagation training procedure modifies the connection strengths in the network until the input-output association is properly captured. The modifications are made by propagating backwards through the network the error at the output node for all the pairs of input-output patterns. During the training phase, the input-output patterns are presented to the network and the training procedure continues until the error at the output unit is reduced to an acceptable level for all training cases presented. The details of connection strength updating schemes are discussed in Chapter 8 of Rumelhart *et al.* (1986).

Testing of the trained neural network model involves two different aspects. One is the capability of the network to predict correctly the output for the input patterns used in the training; this is a test of accuracy. The other is the capability of the network to correctly predict the output for input patterns that were not in the training set; this is the test of generalization capability. The results of the first test are presented in Fig. 5. Here the actual output, i.e. the level of criticalness given in Fig. 3, is compared with the predicted output when each of the 16 patterns in Fig. 3 was presented to the trained network. These results indicate that the network has been successfully trained to closely predict the actual outputs for the 16 different input patterns. In order to test the generalization capabilities some noise was introduced into the input patterns and compared the output with the output corresponding to the input pattern without noise. The input patterns with noisy data can be assumed to represent distributions of hydraulic conductivity that are similar to but not exactly the same as the input patterns shown in Fig. 3. The results (see Fig. 6) indicate that the neural network model has fairly good generalization capabilities. Here again the actual

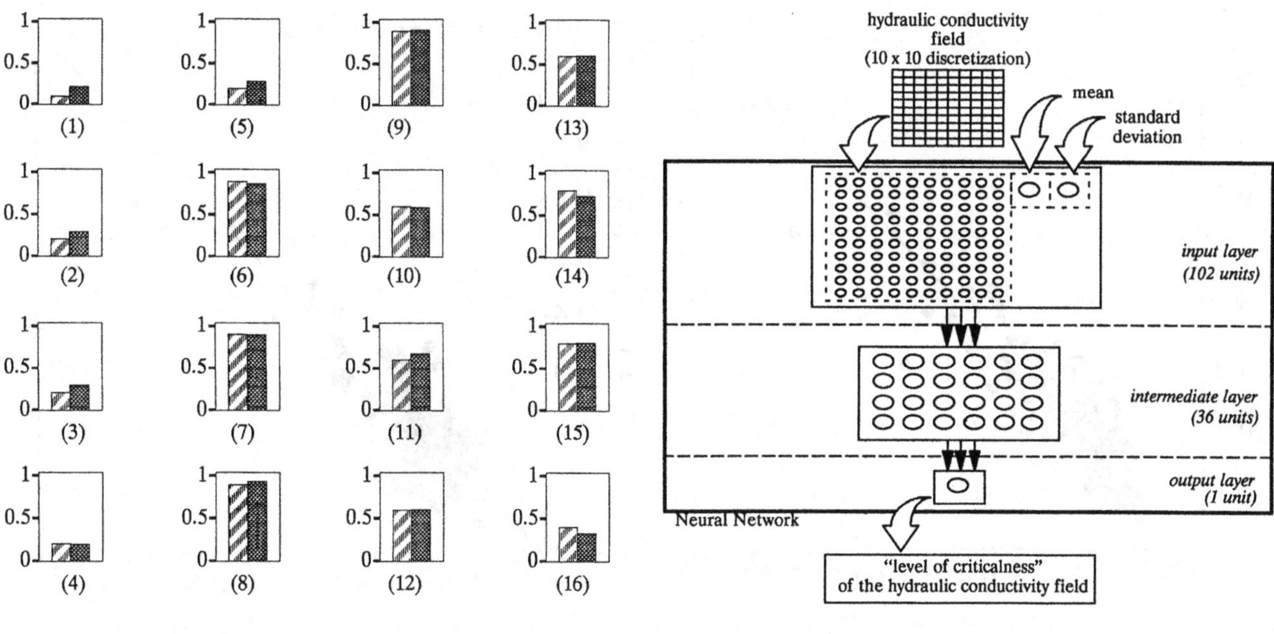

Fig. 6 Results comparing the generalization capabilities of neural network model for the illustrative example (vertical axis represent level of criticalness).

Fig. 7 Schematic diagram of the prototype neural network.

output, i.e. the level of criticalness given in Fig. 3, is compared with the predicted output when each of the 16 patterns in Fig. 3 was presented with noisy data.

A prototype neural network model

The concept of neural networks was explained and illustrated by a simple example in the previous section. This section describes the components of a prototype neural network model for screening out critical realizations and then discusses the overall architecture of the network and the training performance. Fig. 7 shows a schematic diagram of the prototype neural network.

Input patterns: a 10 grid × 10 grid discretization of the physical domain was used to represent the distribution of hydraulic conductivity values. (The dimensions of discretization is not limited by the method.) This required 100 input units that represented the hydraulic conductivity values. Further, it was found that inclusion of the mean and standard deviation of the hydraulic conductivity values in a realization as inputs improved the training performance of the neural network model. Therefore two input units, whose activations represent the normalized values of these two parameters, were included in the input level. All the input activations at the input units were normalized to a range between 0 and 1.

Output value: our research indicates that the total pumping rate required for optimal pumping strategy to contain a plume in a realization could be used as a surrogate for the level of criticalness of that realization. A set of realizations that require high total pumping rates (evaluated by solving a groundwater management model for each realization) comprise the set of critical realizations. First the total pumping rate required for a realization is obtained by solving the management model for that realization. Rank ordering a large set of realizations according to the total pumping rates gives a relative measure of criticalness of each realization. However, the total pumping rate will vary depending on the number of potential reclamation wells in the management model. Our investigations also indicate that a realization requires high total pumping rate independent of the number of potential wells in the management model. This useful discovery provides the means by which one could develop a single neural network model, for identifying critical realizations, that is independent of the number of potential reclamation wells.

The output layer in the neural network model has a single unit representing the level of criticalness. The total pumping rate required by the optimal pumping strategy for a management model (with an arbitrary number of potential reclamation wells) is used as a surrogate for the level of criticalness of a realization. The activation at the output unit is again normalized between 0 and 1.

Neural network architecture: the architecture of the prototype neural network model consist of 102 input units, one output unit, and 36 intermediate units. The number of intermediate units was obtained through a trial-and-error procedure. The state-of-the-art theory does not provide any

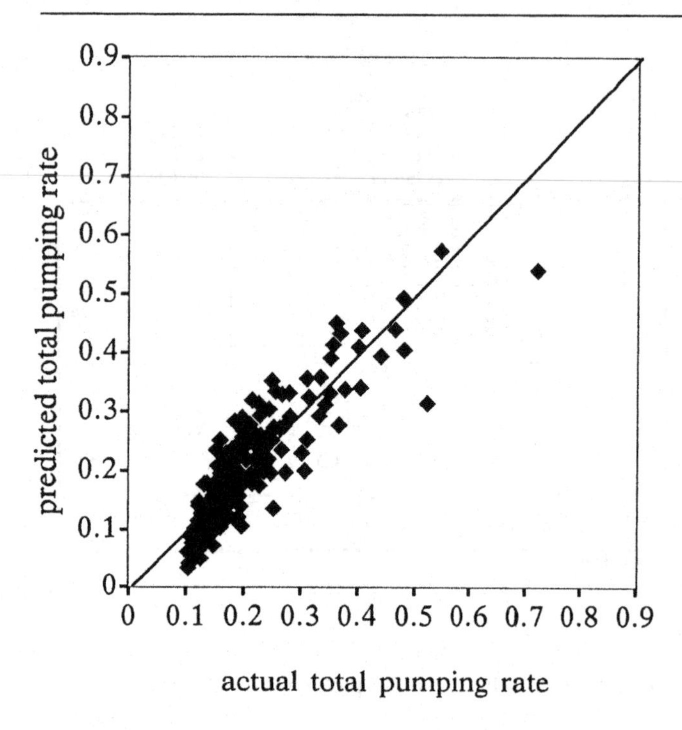

Fig. 8 Performance of a neural network model. Training: No. of training cases = 100, No. of potential reclamation wells = 9. Testing: No. of testing cases = 200, No. of potential reclamation wells = 9.

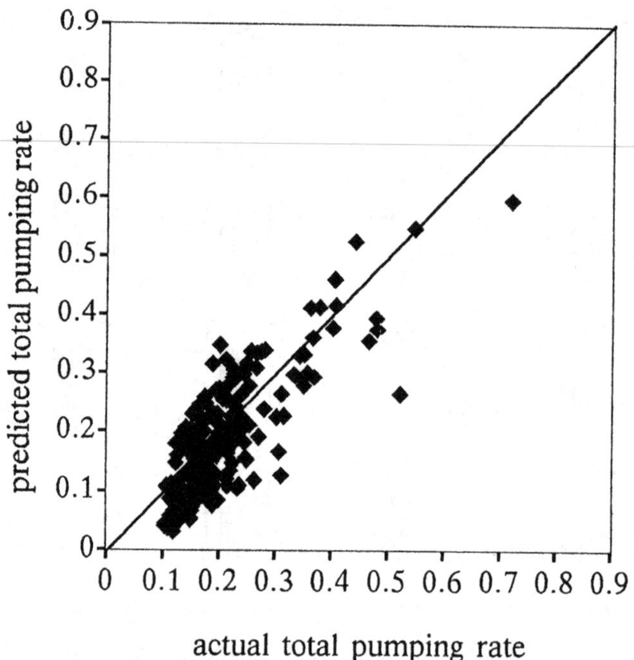

Fig. 9 Performance of a neural network model. Training: No. of training cases = 100, No. of potential reclamation wells = 6. Testing: No. of testing cases = 200, No. of potential reclamation wells = 9.

analytical method to determine the best number of intermediate units, however, empirical results show that too many or too few intermediate units are not the best choice and there exists an optimal number in between.

Performance of the neural network models

The trained neural network model was tested using a different set of 200 realizations. For each of these realizations the required total pumping rate was computed and the set of 200 realizations were then rank ordered based on the total pumping rates. The input activations for each of the realizations were computed and fed into the neural network model. The realizations were then rank ordered according to the predicted output values from the neural network model. The two rank orders were then compared in order to evaluate the performance of the neural network model.

The results presented here are based on a neural network model that was trained on 100 training cases. The trained neural network model was then tested on another set of 200 realizations. Fig. 8 shows the performance of the neural network model that was trained based on total pumping rates obtained for 9 potential reclamation wells; the predictions are compared against the rankings based on the actual total pumping rates obtained by solving the management model with 9 potential reclamation wells. Perfect prediction would be illustrated by a straight line at 45 degrees to the horizontal; note that the predictions are closely distributed around this line. Fig. 9 shows the performance of the neural network model that was trained based on total pumping rates obtained for 6 potential reclamation wells; the predictions are compared against the rankings based on the actual total pumping rates obtained by solving the management model with 9 potential reclamation wells. Again a very close prediction is observable although two different numbers of potential reclamation wells were involved.

These results indicate that the neural network model does provide a tool for quickly screening a set of realizations to identify the critical realizations. The neural network approach improves on computational efficiency by cutting down on the computational effort required for solving the steady state flow model (to obtain the coefficients required in the formulation of the management model) and the solution of the management model. It could be argued that considerable computational effort is spent on the preparation of the training cases and the training process itself. On the other hand, once the neural network is trained it could be used in a wide range of situations repeatedly with very little computational effort. Furthermore, considerable improvements in computational effort could be achieved through parallel processing due to the inherent parallel nature of the neural network models. Computational effort for solving a nonlinear management model with only one realization to obtain the total pumping rate will be tremendously smaller than that required for a non-linear management model with multiple realizations. Therefore, the use of a screening tool could save large computational effort when a management model has a non-linear formulation.

The following section shows the usefulness of the neural network-based screening tool through an application. This illustration shows how pumping strategies with high reliability levels, for aquifer remediation under uncertainty, could be achieved by considering only a few critical realizations instead of a very large number.

AN APPLICATION

This section presents an application of the neural network-based screening tool to illustrate the usefulness of this tool. For this application we used the management model as described previously. Spatial variability in hydraulic conductivity was considered to be the uncertain parameter in this model. A stochastic approach to hydraulic head gradient control was adopted to obtain the pumping strategies. In this case multiple realizations were considered simultaneously in the model (1)–(5). The reliabilities of pumping strategies with and without the use of the screening tool were compared. The reliability of a pumping strategy was evaluated using the approach described in Wagner & Gorelick (1989); a post optimality Monte Carlo analysis was performed and the reliability of a pumping strategy was computed based on the percentage of realizations for which successful containment was achieved.

Design without the screening tool

In this case, 100 realizations were generated using COVAR (William & El-Kadi, 1986). These realizations were simultaneously considered in the model (1)–(5) and solved to obtain the optimal pumping strategy. The reliability level of the optimal pumping strategy was computed based on the number of realizations (from a different set of 100 realizations) for which successful containment was achieved. This procedure was carried out for various cases of potential number of reclamation wells and different statistical distributions of the uncertain parameter. The results are summarized in column (3) of Table 1.

Design with the screening tool

In this case, 100 realizations were first generated using COVAR (William & El-Kadi, 1986) and then they were screened by the neural network-based screening tool. After screening, the 10 most critical realizations were picked out. These critical realizations were then simultaneously considered in the model (1)–(5) and solved to obtain the optimal pumping strategy. Again the reliability levels were evaluated as described previously. The results are summarized in column (4) of Table 1.

Table 1. *Comparison of reliability levels of optimal pumping strategies with and without the use of the neural network-based screening tool*

(1) number of potential reclamation wells	(2) standard deviation of log hydraulic conductivity	(3) reliability level of pumping strategy for multiple (100) realizations	(4) reliability level of optimal pumping strategy for 10 most critical realizations
6	0.5	100%	100%
6	1.0	90%	90%
100	0.5	96%	96%
100	1.0	96%	94%
100	2.0	N/A	95%

A comparison of the reliability levels shown in columns (3) and (4) of Table 1 indicates that high reliability levels could be achieved with only as few as 10 critical realizations instead of 100 unscreened realizations. In order to prove that this behavior is not a random occurrence the following test was done; the reliability levels of pumping strategies obtained by simultaneously considering only 10 random realizations in the management model were compared with the results shown in column (4) of Table 1. These results are compared in Table 2. Results presented here suggest that critical realizations do possess specific characteristics that are not present in any realization.

CONCLUSION AND SUMMARY

A stochastic approach to design under conditions of uncertainty is to ensure that the final design satisfies many realizations of the uncertain parameter field. However, only few of the realizations will constrain the final design. Identification of these pessimistic realizations, which is something a human familiar with the problem will be able to do manually, will ensure a reliable design with reduced computational effort. This paper presented a neural network-based technique for recognizing pessimistic realizations from a large set of realizations. A feed-forward type neural network is first trained, through presentation of examples, to identify the pessimistic realizations. The network learns the association between the spatial distribution of the hydraulic conductivity values in a realization and its impact upon the final groundwater remediation design. These pessimistic realizations

Table 2. *Comparison of reliability levels of optimal pumping strategies obtained when simultaneously considering (a) 10 random realizations, (b) 10 most critical realizations in the management model*

standard deviation of log hydraulic conductivity	reliability level of optimal pumping strategy for 10 random realizations	reliability level of optimal pumping strategy for 10 most critical realizations
0.5	74%	96%
1.0	76%	94%
2.0	74%	95%

could then be incorporated into a groundwater management model for design under conditions of uncertainty in the hydraulic conductivity parameter.

An application of the screening tool was also illustrated through a simple example. This example showed that pumping strategies with high reliability levels could be achieved with only as few as 10 critical realizations. The critical realizations were obtained by screening, using the neural network-based tool, a large set of realizations. The reliability levels of pumping strategies obtained using only the 10 most critical realizations were comparable to the reliability levels of pumping strategies obtained using 100 unscreened realizations. These results indicate that the critical realizations do have specific characteristics that make those realizations to constrain the design the most. Results also show that a neural network based pattern recognition tool could identify the critical realizations.

A trained neural network could screen a set of realizations with very little computational effort. However, considerable computational effort is required in the training process; for generation of the training cases and error backpropagation in the network. The proposed approach consumes the bulk of the computational time up-front and very little at run-time. This approach could therefore be used very effectively in an interactive design procedure where critical realizations need to be generated at run-time. The current implementation of the neural network is carried out using a network simulator on a serial computer. This does not harness the inherent parallel nature of neural networks. The overall computational time could be reduced by implementing the neural network on a parallel machine.

Some management models have non-linear formulations. Solving such a model with multiple realizations could lead to computational complexities, so much in some cases that no solution could be found. In such cases it would be very efficient to screen the realizations and use only a few of the most critical realizations. An extension of the current research would be to investigate the applicability and usefulness of a neural network-based screening tool in a case with a non-linear management model.

The work presented here pertains to a small hypothetical scenario. Further research is warranted to examine the general applicability of the proposed screening tool An imminent question is whether the same neural network could be used for different values of ensemble statistics (i.e., mean, standard deviation, and correlation scale) of the hydraulic conductivity field. An important extension would be to examine the performance of the neural network-based screening tool for different domain sizes of the groundwater problem and different ensemble statistics of the hydraulic conductivity parameter.

REFERENCES

Atwood, D. F. & Gorelick, S. M. (1985) Hydraulic gradient control for groundwater contaminant removal. *J. Hydrol.*, **76**, 85–108.

Colarullo, S. J., Heidari, M. & Maddock, T. III (1984) Identification of an optimal groundwater management strategy in a contaminated aquifer. *Water Resour. Bull.*, **20**(5), 747–60.

Eheart, J. W., Rahman, R. M., Keith, S. M. & Valocchi, A. J. (1990) A game-theoretic parameter configuration technique for aquifer restoration design. *J. of Contaminant Hydrology*, **6**, 205–26.

Gorelick, S. M., Voss, C. I., Gill, P. E., Murray, W., Saunders, M. A. & Wright, M. H. (1984). Aquifer reclamation design: The use of contaminant transport simulation combined with nonlinear programming. *Water Resour. Res.*, **20**(4), 415–27.

Gorelick, S. M. (1987) Sensitivity analysis of optimal groundwater contaminant capture curves: spatial variability and robust solutions, *Proceedings of the Conference and Exposition: Solving Groundwater Problems with Models, National Water Well Association*, Denver, February 10–12.

Keely, J. (1984) Optimizing pumping strategies for contaminant studies and remedial actions. *Groundwater Monitoring Review*, Summer 1984.

Morgan, D. R. (1990) *Decision Making Under Uncertainty Using a New Chance Constraint Programming Technique: A Groundwater Reclamation Application*, Ph.D. dissertation, Department of Civil Engineering, University of Illinois at Urbana-Champaign.

Ranjithan, S. & Morgan, D. R. (1988) *A rule-based expert system for groundwater contamination assessment*, Report: CE398 Class Project, Dept of Civil Engineering, University of Illinois at Urbana-Champaign.

Ranjithan, S., Garrett, J. H. Jr. & Eheart, J. W. (1991) Application of Neural Network to Groundwater Remediation. In: Allen, R. (Ed.), *Expert Systems in Civil Engineering – Knowledge Representation*, Amer. Soc. of Civil Eng., New York (in review).

Rumelhart, D. E., McClelland, J. L. & the PDP Research Group, (1986) *Parallel Distributed Processing – Volume 1: Foundations*, MIT Press, Cambridge, Massachusetts, USA.

Valocchi, A. J., & Eheart, J. W. (1987) Incorporating parameter uncertainty into groundwater quality management models. In: Beck, M. B. (Ed.), *Systems Analysis in Water Quality Management*, Proceedings of a Symposium held in London, UK, 30 June–2 July, Pergamon.

Wagner, B. J., & Gorelick, S. M. (1987) Optimal groundwater management under parameter uncertainty. *Water Resour. Res.*, **23**(7), 1162–74.

Wagner, B. J., & Gorelick, S. M. (1989) Reliable aquifer remediation in the presence of spatially variable hydraulic conductivity: from data to design. *Water Resour. Res.*, **25**(10), 2211–25.

Williams, S. A., & El-Kadi, A. I. (1986) *COVAR – A computer program for generating two-dimensional fields of autocorrelated parameters by matrix decomposition*, International Groundwater Modeling Center, Holcomb Research Institute, Butler University, Indianapolis.

9 Application of pattern recognition to rainfall–runoff analysis

K. MIZUMURA

Civil Engineering Department, Kanazawa Institute of Technology, Ishikawa, Japan

ABSTRACT Traditionally human beings predict future runoffs from present rainfalls. One of the recent methodologies of prediction stemming from the pattern recognition technique is presented. The possible range of values of the predicted runoff is estimated by the discriminant functions. The discriminant functions are derived from data sets on several events of rainfall and runoff in the same watershed. The predicted runoff is in good agreement with the observed one.

INTRODUCTION

Forecasting the runoff resulting from a rainfall belongs to the classical basic issues of hydrology. It is shown that the pattern recognition method, which is used in as diverse fields as medical diagnosis, mail problems, banking processes, coastal changes and cybernetics (Mizumura, 1988) is useful also in hydrological forecasting. The method dwells on the obvious statement that much and little rainfall correspond to much and little runoff, respectively.

RAINFALL–RUNOFF PROCESS

The physical system of transformation of rainfall into runoff is very complex. Moreover the runoff consists of three components such as surface flow, interflow, and groundwater flow. Therefore, even if the model strictly described the underlying physical phenomena, it would be difficult to solve the governing equations. The rainfall–runoff process is heavily dependent upon many characteristics of each watershed. For the sake of runoff prediction the rainfall–runoff process is treated here as a black box. Thus, one can employ either of such methods as differential equations, integral equations, least square methods, Wiener–Hopf equation, Kalman filtering etc. Yet another approach originating from the pattern recognition methodology will be tackled here. The forecasted range of runoff will be estimated using rainfall and runoff in the previous time steps. This forecast is based on the process similar to the analogue form of reasoning of human beings. The forecast equation can be expressed as:

$$[\text{Range of } \hat{Q}_n] = \text{Function of } \{R_{n-1}, R_{n-2}, \ldots, R_{n-i}, Q_{n-1}, Q_{n-2}, \ldots, Q_{n-j}\} \tag{1}$$

where $\hat{Q}_n$ is the forecasted runoff at the n time step, R_{n-i} is the rainfall at the $(n-i)$th time step, Q_{n-j} is the runoff at the $(n-j)$th time step, and i and j are the memory lengths of rainfall and runoff, respectively. Equation (1) states that in the process of runoff forecast one makes use of rainfall and runoff data during the previous i, and j, days, respectively. The function in equation (1) is derived by combination pattern between the range of forecasted runoff, and observed rainfall and runoff in the previous time steps. Therefore, this combination pattern is determined before the runoff forecast is made. That is, the decision functions necessary to classify patterns are computed. In the forecast phase, the range of the forecasted runoff is defined as the range where the discriminant function takes its maximum value.

FUNCTIONAL APPROXIMATION OF PROBABILITY DENSITY FUNCTIONS

Let $p(\mathbf{x})$ define an estimate of $p(\mathbf{x})$ in which $p(\mathbf{x})$ denotes $p(\mathbf{x}|\omega_i)$. The estimate is defined by minimizing the following equation:

$$R = \int_{\mathbf{x}} u(\mathbf{x})[p(\mathbf{x}) - \hat{p}(\mathbf{x})]^2 \, \mathrm{d}\mathbf{x} \tag{2}$$

in which $u(\mathbf{x})$ is a weighting function and $\mathbf{x}$ is a pattern vector. Expand the estimate $p(\mathbf{x})$ in the series

$$\hat{p}(\boldsymbol{x})=\sum_{j=1}^{m} c_j\phi_j(\boldsymbol{x}) \tag{3}$$

in which C_j are the coefficients to be determined and $\{\phi_j(\boldsymbol{x})\}$ are a set of specified basis functions. Substituting equation (3) into equation (2) one gets:

$$R=\int_{\boldsymbol{x}} u(\boldsymbol{x})[p(\boldsymbol{x})-\sum_{j=1}^{m} c_j\phi_j(\boldsymbol{x})]^2\,\mathrm{d}\boldsymbol{x} \tag{4}$$

A necessary condition for the minimum of R is:

$$\partial R/\partial C_k=0; \quad k=1,2,\ldots,m \tag{5}$$

The result is

$$\sum_{j=1}^{m} c_j \int_{\boldsymbol{x}} u(\boldsymbol{x})\phi_j(\boldsymbol{x})\phi_k(\boldsymbol{x})\,\mathrm{d}\boldsymbol{x}=\int_{\boldsymbol{x}} u(\boldsymbol{x})\phi_k(\boldsymbol{x})p(\boldsymbol{x})\,\mathrm{d}\boldsymbol{x} \tag{6}$$

The right hand side of equation (6) is approximated by the sample average. If the basis functions $\{\phi_k(\boldsymbol{x})\}$ are orthonormal, then $A_k=1$ for all k. Thus

$$C_k=\frac{1}{NA_k}\sum_{j=1}^{m} u(\boldsymbol{x}_i)\phi_k(\boldsymbol{x}_i); \quad k=1,2,\ldots,m \tag{7}$$

When the terms $u(\boldsymbol{x}_i)$ are independent of k and are common to all coefficients, then:

$$C_k=\frac{1}{N}\sum_{j=1}^{m} \phi_k(\boldsymbol{x}_i); \quad k=1,2,\ldots,m \tag{8}$$

Next, design a Bayes classifier by employing probability density functions which have been directly estimated from the simulated samples. One may approximate these functions by an expansion of the following form:

$$\hat{p}(\boldsymbol{x}|\omega_i)=\sum_{j=1}^{m} C_{ij}\phi_j(\boldsymbol{x}) \tag{9}$$

in which the first subscript of the coefficients denotes the class ω_i. The Hermite polynomial functions are applied as the functions $\{\phi_j(x)\}$. In the one-dimensional case these functions are given by the following recursive relation:

$$H_{l+1}(x)-2x\,H_l(x)+2l\,H_{l-1}(x)=0 \tag{10}$$

The first few terms of $H(x)$ are as follows:

$$\begin{aligned} &H_0(x)=1;\\ &H_1(x)=2x;\\ &H_2(x)=4x^2-2\\ &H_3(x)=8x^3-12x\\ &H_4(x)=16x^4-48x^2+12 \end{aligned} \tag{11}$$

A n-dimensional orthogonal set of functions is easily obtained by forming arbitrary pairwise combinations of the one-dimensional functions.

$$\begin{aligned} &\phi_1(\boldsymbol{x})=H_0(x_1)H_0(x_2)\ldots H_0(x_n)=1\\ &\phi_2(\boldsymbol{x})=H_1(x_1)H_0(x_2)\ldots H_0(x_n)=2x_1\\ &\phi_3(\boldsymbol{x})=H_0(x_1)H_1(x_2)\ldots H_0(x_n)=2x_2\\ &\ldots\\ &\phi_{n+1}(\boldsymbol{x})=H_0(x_1)H_0(x_2)\ldots H_1(x_n)=2x_n\\ &\phi_{n+2}(\boldsymbol{x})=H_1(x_1)H_1(x_2)\ldots H_0(x_n)=4x_1x_2\\ &\phi_{n+3}(\boldsymbol{x})=H_1(x_1)H_0(x_2)\ldots H_0(x_n)=4x_1x_3\\ &\ldots\\ &\phi_{2n}(\boldsymbol{x})=H_1(x_1)H_0(x_2)\ldots H_1(x_n)=4x_1x_n\\ &\phi_{2n+1}(\boldsymbol{x})=H_0(x_1)H_1(x_2)\ldots H_0(x_n)=4x_2x_3 \end{aligned} \tag{12}$$

The next problem is to determine the coefficients C_{ij} for use in the expansion of $p(\boldsymbol{x}|\omega_i)$. For class ω_i:

$$C_{ik}=\frac{1}{N_i}\sum_{j=1}^{N_i} \phi_k(\boldsymbol{x}_{ij}) \tag{13}$$

in which N_i is the number of patterns in the class ω_i and k ranges from 1 to m. The decision functions for this problem are given by:

$$d_i(\boldsymbol{x})=\hat{p}(\boldsymbol{x}|\omega_i)\,p(\omega_i); \quad i=1,2,\ldots,n \tag{14}$$

in which

$$\hat{p}(\boldsymbol{x}|\omega_i)=\sum_{j=1}^{m} C_{ij}\phi_j(\boldsymbol{x}).$$

A pattern $\boldsymbol{x}$ is assigned to the class ω_i if for that pattern $d_i(x)>d_j(x)$ for all $j\neq i$.

ILLUSTRATIVE EXAMPLES

Fig. 1 represents the hyetograph and the hydrograph at the watershed of the river Onga on Kyushu Island. The units used are milimeters per day for the rainfall and cubic meters per second for the runoff. The total number of the data points used is 100. Fig. 2 shows the comparison of the observed data and the runoff forecasted with the help of the pattern recognition method. The result is in a relatively good agreement with the observed data.

Further, to improve the forecast of the runoff, the tanks model (Fig. 3) is employed and the deviation of the computed runoff from the observed data is predicted by the pattern recognition method. Rainfall is stored in the first tank with two side outlets, active after some thresholds of stored volumes are exceeded, and one bottom outlet draining into the second tank. The discharge from the first tank is given by

$$y_k^1=\begin{cases}0 & \text{if } X_k^1\leq h_1^1\\ \mu_1^1(X_k^1-h_1^1) & \text{if } h_1^1<X_k^1\leq h_2^1\\ \mu_2^1(X_k^1-h_2^1)+\mu_1^1(X_k^1-h_1^1) & \text{if } h_1^1<X_k^1\end{cases} \tag{15}$$

where μ_1^1 and μ_2^1 are the discharge coefficients, X_k^1 is the water level in the first tank, and h_1^1 and h_2^1 are two threshold

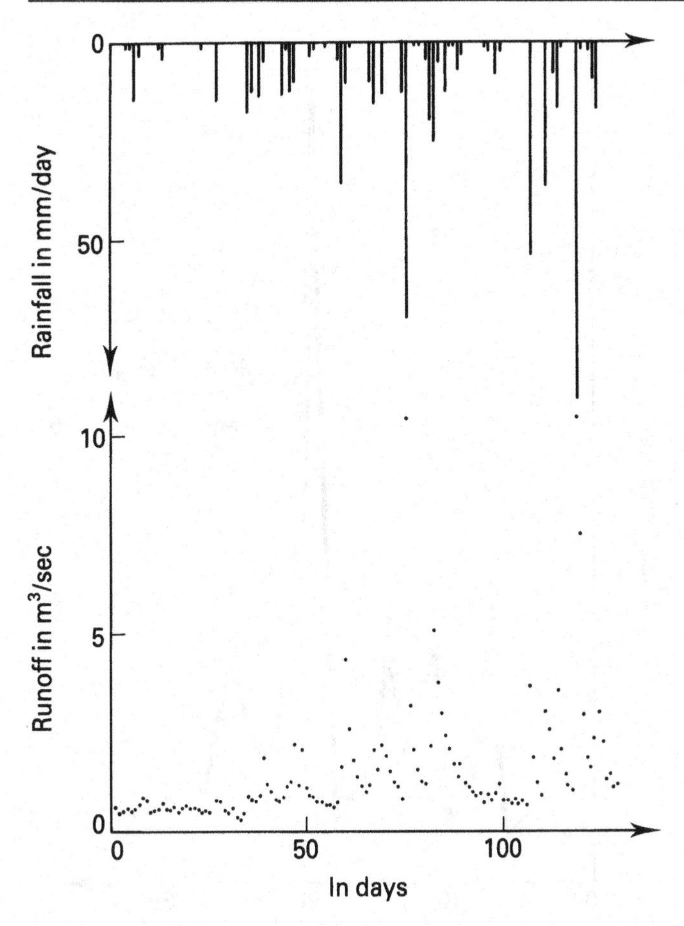

Fig. 1 Hyetograph and hydrograph at the watershed of the river Onga.

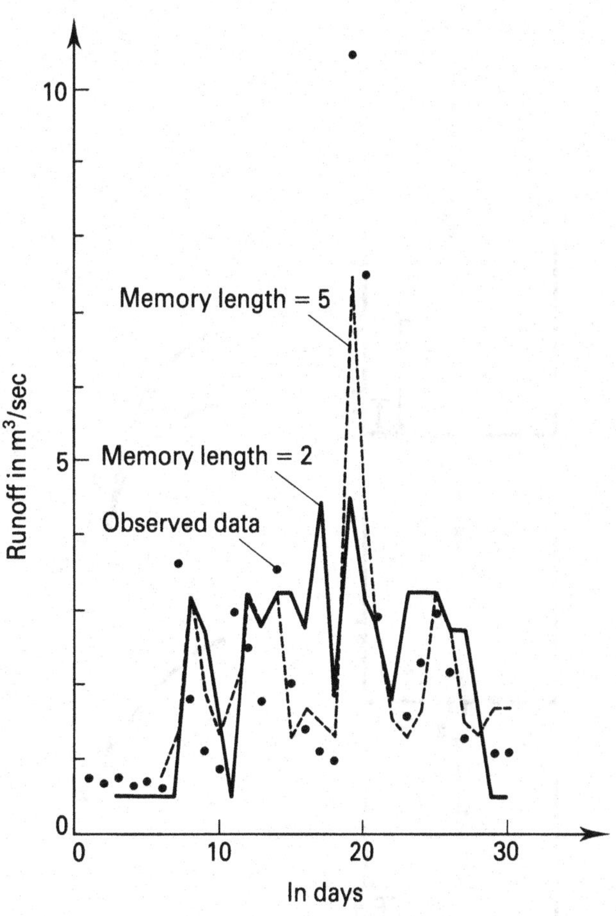

Fig. 2 Prediction of runoff via pattern recognition method without the Tanks Model.

elevations above the tank bottom for side outlets. Further, the infiltrated discharge z^1 to the second tank is given by:

$$z_k^1 = v^1 X_k^1 \tag{16}$$

where v^1 is the discharge coefficient. For the other tanks the relationships are analogous. The total discharge is given by $y_k^1 + y_k^2 + y_k^3 + y_k^4$. The superscript denotes the tank number. The difference between the observed runoff and the runoff computed by the tanks model is forecasted by the method of pattern recognition discussed previously. The parameters used for computing the discharge by the tanks model are given as follows:

$$\begin{aligned} &\mu_1^1 = 0.08;\ \mu_2^1 = 0.2;\ \mu_1^2 = 0.013;\ \mu_1^3 = 0.013;\ \mu_1^4 = 0.008;\ \mu_2^2 = 0.008; \\ &\mu_2^3 = 0.008;\ \mu_2^4 = 0.004;\ v^1 = 0.3;\ v^2 = 0.16;\ v^3 = 0.013;\ h_1^1 = 1.2; \\ &h_2^1 = 20;\ h_1^2 = 1.5;\ h_2^2 = 2.0;\ h_1^3 = 5.0;\ h_1^4 = 0.001;\ h_2^3 = 6;\ h_2^4 = 0.02 \end{aligned} \tag{17}$$

To check the relevance of the memory length, the value of the AIC (Akaike Information Criterion) is employed as follows:

$$\mathrm{AIC} = N_0[1 + \ln 2\pi + \ln 1/N_0 \sum_{i=1}^{N_0} (y_i - \hat{y}_i)^2] + 4N \tag{18}$$

where N_0 is the total number of data points, y_i are the observed data, $\hat{y}_i$ are the forecasted values, N is the number of parameters involved in the model. Comparison of the values of the AIC shows that the model of the memory length of $m = 5$ is the best.

The results of the prediction of runoff achieved with the help of the pattern recognition methodology with the Tanks Model are shown in Fig. 4. The simulation is significantly better than achieved without the Tanks Model (Fig. 2).

CONCLUDING REMARKS

Application of the method of pattern recognition to the process of transformation of rainfall into runoff yielded satisfactory results. Since this forecast is based on a Bayesian approach, the detailed physical information on the watershed is not necessary. The eventual new data do not play such an important role as they do in forecasting via the Kalman filtering technique. The use of the tanks model improves the forecast.

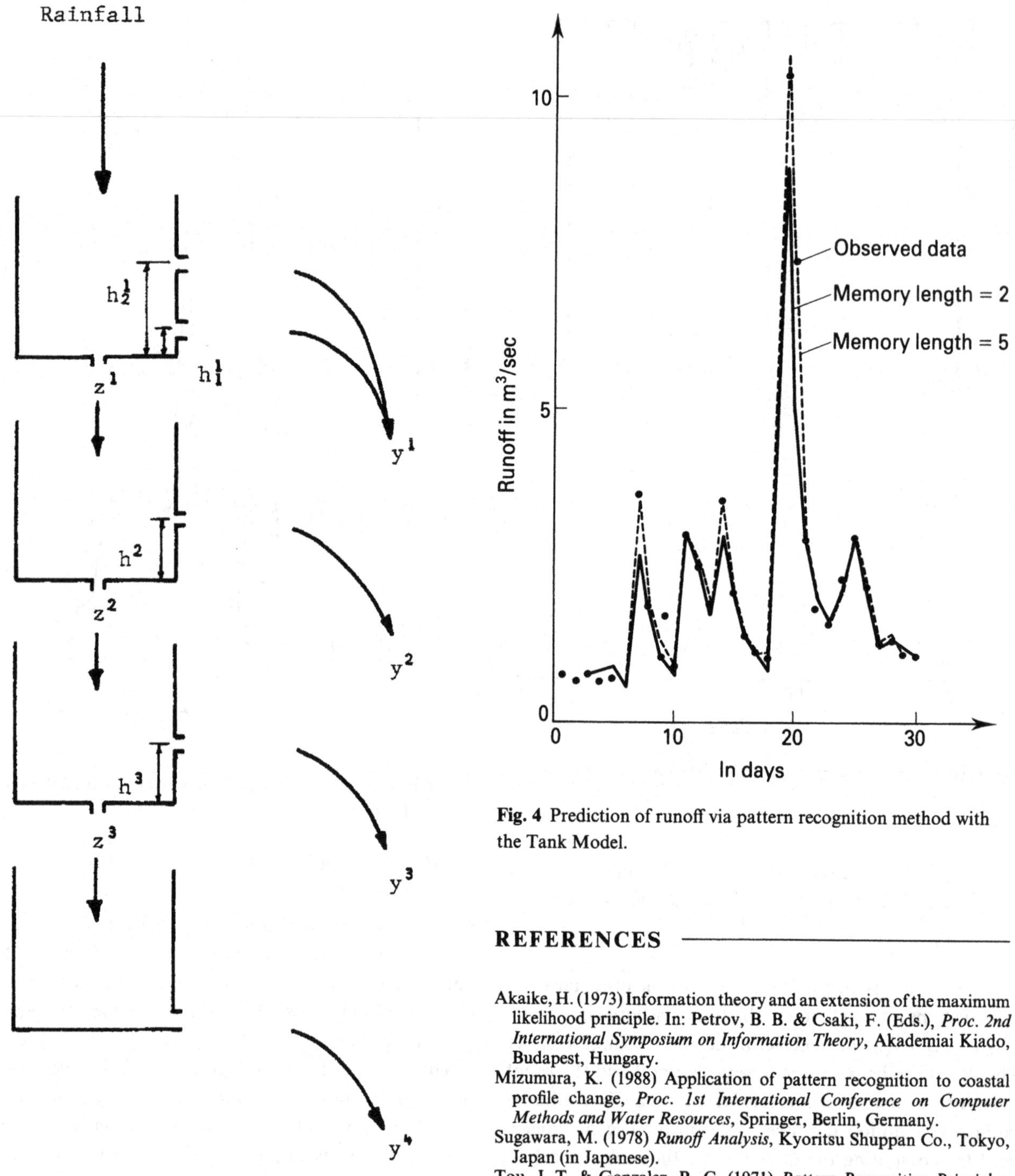

Fig. 3 The principle of the Tank Model.

Fig. 4 Prediction of runoff via pattern recognition method with the Tank Model.

REFERENCES

Akaike, H. (1973) Information theory and an extension of the maximum likelihood principle. In: Petrov, B. B. & Csaki, F. (Eds.), *Proc. 2nd International Symposium on Information Theory*, Akademiai Kiado, Budapest, Hungary.

Mizumura, K. (1988) Application of pattern recognition to coastal profile change, *Proc. 1st International Conference on Computer Methods and Water Resources*, Springer, Berlin, Germany.

Sugawara, M. (1978) *Runoff Analysis*, Kyoritsu Shuppan Co., Tokyo, Japan (in Japanese).

Tou, J. T. & Gonzalez, R. C. (1971) *Pattern Recognition Principles*, Addison-Wesley, New York.

10 Nonparametric estimation of multivariate density and nonparametric regression

W. FELUCH

Technical University of Warsaw, Institute of Environmental Engineering, Warsaw, Poland

ABSTRACT The p.d.f.s typically used in hydrology for determination of exceedance probability (e.g. design floods) are typically based on the parametric approach. In the two- or three-dimensional cases and in the case of regression problems the multivariate normal distribution is in common use. Nonparametric density estimators in multivariate random variables are a new approach to estimation and regression. As an alternative to the standard parametric estimators, the nonparametric multivariate Parzen estimator has been used in the analysis. The results of the analysis indicate that the parametric and nonparametric estimators are performing comparatively well. Some conclusions are offered concerning the applications of the nonparametric approaches.

INTRODUCTION

Various probability distributions are used in hydrology for determination of exceedance probability. Flood frequency analysis is an example, where typically only one-dimensional random variables are considered (Flood Frequency and Risk Analyses, 1986; Kaczmarek, 1970). Sometimes models involving two- or three-dimensional random variables are investigated. For example, multivariate models for low or high water stages were developed by Zielińska (1963, 1964), Yevjevich (1967) and Strupczewski (1967), under the assumption of normality of the underlying probability distribution.

Another parametric estimation approach in hydrology is a classical regression problem also based on multivariable normal distribution (Kaczmarek, 1970).

Recently investigations based on the nonparametric approach (nonparametric method of estimation (NME)) have been initiated in hydrology (Adamowski, 1985, Feluch, 1987, Adamowski & Feluch, 1988, Schuster & Yakowitz, 1985). The nonparametric method of estimation enables one to estimate unknown probability densities without any prior assumptions concerning the shape of the density function.

A well-known method of nonparametric estimation is the kernel method (for the one-dimensional case, cf. Feluch (1987) and Adamowski & Feluch (1988)).

The kernel estimator of unknown density $f(x)$ based on random sample $\{X_i\}_{i=1}^n$ of independent and identically distributed real-valued random variables is given by

$$\hat{f}(x)=\frac{1}{nh}\sum_{i=1}^{n} K\left(\frac{x-X_i}{h}\right) \tag{1}$$

where h is the bandwidth or smoothing factor related to n, and $K(.)$ is the kernel function which satisfies the conditions

$$\sup_{-\infty<y<\infty} |K(y)|<\infty \tag{2}$$

$$K(y)>0 \tag{3}$$

$$\int_{-\infty}^{\infty} K(y)dy=1 \tag{4}$$

The kernel estimator (1) is a sum of 'bumps' placed at the observations. The kernel function $K(.)$ determines the shape of these bumps, while the bandwidth h determines their width. An illustration is given in Fig. 1, where the individual bumps $n^{-1}h^{-1}K\left[\frac{x-X_i}{h}\right]$ are shown and the estimate $\hat{f}$ is constructed by adding them up.

In practical applications, the kernel function $K(.)$ is assumed to be a symmetrical function, that is $K(y)=K(-y)$. Usually, one of the following two kernels is used:

– the optimal kernel function (Epanechnikov, 1969):

$$K(y)=\begin{cases}\frac{3}{4\sqrt{5}}\left(1-\frac{y^2}{5}\right) & \text{for} \quad |y|\le\sqrt{5}\\ 0 & \text{for} \quad |y|>\sqrt{5}\end{cases} \tag{5}$$

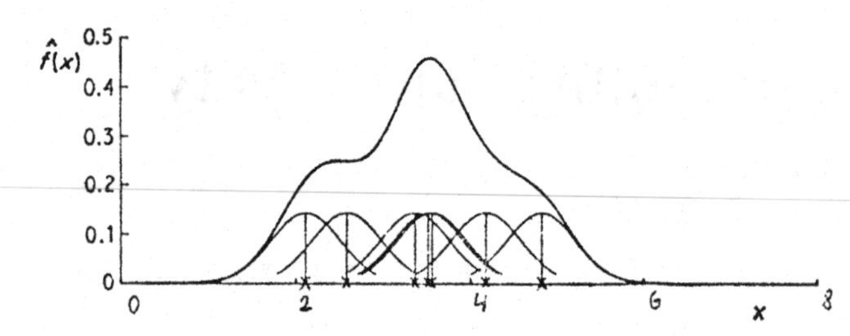

Fig. 1 Kernel estimates showing individual kernels.

– the Gauss kernel function (Tapia & Thompson, 1878; Scott & Terrel, 1987)

$$K(y)=\frac{1}{\sqrt{2\pi}}\exp\left(-\frac{\xi^2}{2}\right)\quad \text{for } -\infty<y<\infty \tag{6}$$

The optimal kernel function is the one which minimizes the mean integrated square error of the form

$$\text{MISE}(\hat{f},f)=\text{E}\int_{-\infty}^{\infty}[\hat{f}(x)-f(x)]^2\,\text{d}x \tag{7}$$

MISE for the Gaussian kernel function is asymptotically 4% larger then that for the optimal kernel (Tapia & Thompson, 1978).

The smoothing factor h is computed from the data, and is based on minimizing the MISE as well. For Gaussian kernel the smoothing factor is given by Rudemo (1982) and developed in hydrology by Feluch (1994).

The estimator (1) was applied to flood frequency analysis (Feluch, 1987, Adamowski & Feluch, 1988) and to minimum yearly discharge frequency analysis (Feluch, 1988).

MULTIVARIATE NONPARAMETRIC DENSITY FUNCTION

Assume, that a random sample of size n of k-dimensional random variables is given

$$X_i=\langle X_1^i,X_2^i,\ldots,X_k^i\rangle \qquad i=1,2,\ldots n \tag{8}$$

with an unknown density function $f(x)$, $x=\langle x_1,x_2,\ldots x_k\rangle$, which can be estimated nonparametrically as (Cacoullos, 1966, Epanechnikov, 1969; Feluch, 1989):

$$\hat{f}(x)=\frac{1}{n}\sum_{i=1}^{n}\prod_{l=1}^{k}\frac{1}{h_l}K\left(\frac{x_l-X_l^i}{h_l}\right) \tag{9}$$

where x is a k-vector of variables $x=\langle x_1,x_2,\ldots x_k\rangle$, h_l is a smoothing factor of the lth variable, that is lth coordinate of the random vector X.

Using the Gauss kernel function one gets the exceedance probability

$$\hat{p}(x)=\frac{1}{n}\sum_{i=1}^{n}\prod_{l=1}^{k}\varphi\left(\frac{x_l-X_l^i}{h_l}\right) \tag{10}$$

where

$$\varphi(y)=\frac{1}{\sqrt{2\pi}}\int_{y}^{\infty}\exp\left(-\frac{\xi}{2}\right)^2\text{d}\xi \tag{11}$$

In the above estimators there are k smoothing factors. Rudemo (1982) has proposed a procedure for estimation of an optimal value of the smoothing factor using a cross-validation (CV) technique for minimizing MISE (7).

In the multivariate case the MISE can be written in the form

$$\text{E}\int[\hat{f}(x)-f(x)]^2\,dx=\text{E}\int\hat{f}^2(x)\,dx-2\int f(x)\,\text{E}[\hat{f}(x)]\,\text{d}x + \int f^2(x)\,\text{d}x, \tag{12}$$

where the integration is over the region of the values of the function arguments, that is

$$\int\ldots\equiv\underbrace{\int_{-\infty}^{\infty}\ldots\int_{-\infty}^{\infty}}_{k\text{-times}}\ldots$$

The last term of equation (12) does not depend on $\hat{f}$. In the sense of minimizing MISE the ideal choice of smoothing factors $\{h_l\}_{l=1}^{k}$ is the one corresponding to the choice which minimizes the quantity R defined by

$$R(\hat{f},f)=\text{E}\int\hat{f}^2(x)\,\text{d}x-2\int f(x)]\,\text{E}[\hat{f}(x)]\,\text{d}x \tag{13}$$

Using the CV technique one gets the estimator of $R(\hat{f},f)$ (Silverman, 1986) in the form:

$$\hat{R}(\hat{f})=\int\hat{f}^2(x)\text{d}x-\frac{2}{n(n-1)}\sum_{\substack{i,j=1\\ i\neq j}}^{n}\prod_{l=1}^{k}\frac{1}{h_l}K\left[\frac{X_l^i-X_l^j}{h_l}\right] \tag{14}$$

Using the Gaussian kernel function in equation (14) has two advantages:

- the extrapolation and interpolation ranges may be better derived than when using the Epanechnikov kernel,
- the minimization of equation (14) is simple because derivatives of this estimator exist everywhere.

Using the Gaussian kernel yields

$$\hat{R}(\hat{f})=\frac{1}{n\pi^{k/2}2^k}\prod_{l=1}^{k}\frac{1}{h_l}\left[1+2\sum_{\substack{i,j=1\\ i<j}}^{n}\left(\frac{d_{ij}}{n}-\frac{2^{1+k/2}}{n-1}\right)d_{ij}\right] \tag{15}$$

where

$$d_{ij}=\exp\left[\frac{1}{2}\sum_{l=1}^{n}\left(\frac{X_l^i-X_l^j}{h_l}\right)^2\right] \tag{16}$$

The problem of finding the smoothing factors $\{h_l\}_{l=1}^k$ can be solved minimizing the above expression (Feluch, 1994).

Required sample sizes for given accuracy

This problem was considered by Silverman (1986). He assumed, that the true density f is standard multivariate normal, and that the kernel is normal too. At first he assumed f at the point **0**, and he found the smoothing factor h which minimized the mean square error at this point. He ensured that the relative mean square error ($\tilde{M}SE$) $E\{\hat{f}(\mathbf{0})-f(\mathbf{0})\}^2/f^2(\mathbf{0})$ is less than 0.1. Table 1 (columnn 1) gives the sample size required to achieve this objective as a function of dimension. Table 1 shows, how the required sample size increases with dimension. Furthemore Silverman suggested that in all dimensions up to 10, the sample sizes required to yield a value of 0.1 for the relative mean integrated square error ($\tilde{M}ISE$), $E\int(\hat{f}(\boldsymbol{x})-f(\boldsymbol{x}))^2d\boldsymbol{x}/\int f^2(\boldsymbol{x})d\boldsymbol{x}$ are approximately 1.7 times those given in Table 1, column (2). If a used measure of the global fit were more sensitive to tail behaviour, then the sample sizes required would be still larger.

The above results justify the following conclusion important in hydrological applications: the dimension of the random vector has to be strictly related to the sample size. As hydrological data are typically scarce, the value of the dimension of the random vector is not larger than three.

Table 1. *The minimum sample size required to achieve that the required relative errors $\tilde{M}SE$ and $\tilde{M}ISE$ are less than 0.1*

Dimensionality	$\tilde{M}SE<0.1$	$\tilde{M}ISE<0.1$
1	4	7
2	19	33
3	67	110
4	223	380
5	768	1300
6	2790	4800
7	10700	18000
8	43700	74000
9	187000	320000
10	842000	1430000

Estimation of the exceedance probability of the high/low water discharge

According to the above conclusions concerning the required sample size, the value of the dimension of the random vector assumed in this example is two. It is very important for small rivers, when the period of observation is short.

The flood discharge can be characterized from practical engineering view point as a random vector

$$X=\langle Q_{\max}^{\text{high}}, V_{\text{high}}\rangle \tag{17}$$

where $Q_{\max}^{\text{high}}$ [cms] is the maximum (culmination) discharge in the high-flow period and V_{high} [cm] is the volume of the flood wave.

Analogically, the low flow coordinates of the random vector are

$$X=\langle Q_{\min}^{\text{low}}, T_{\text{low}}\rangle \tag{18}$$

where $Q_{\min}^{\text{low}}$ [cms] is the minimum discharge in the low flow period and T_{low} [days] is a low water flow period. The exceedance probability is given by the estimator (10).

The high and low flow periods occur few times a year. The annual probability of this phenomena is applied in engineering. There are relations between the annual probability and the probability in the sample (Zielińska, 1963; Strupczewski, 1967).

The results of the multivariate estimation of the annual maximum and minimum flow in the river Vistula at the gauge Sandomierz and in the river Warta at the gauge Skwierzyna are presented in Figs. 2 and 3. The shapes of the curves of exceedance probability for both rivers do largely differ. They both feature stabilization of the values in the tail parts. This is the effect of the property of the density estimator (9), giving values close to zero behind the observed range. The results show that the behaviour of the above estimator limited extrapolation of density function and exceedance probability functions. Of course there are other estimators which have different properties – the smoothing factor is not constant. These estimates give good results in the extrapolation range (Feluch, 1994).

ESTIMATOR OF REGRESSION FUNCTION

Regression estimator assumes that a random variable Y is related to a random vector X of dimension k. The random sample (8) is extended to the form of

$$X_i, Y_i=\langle X_1^i, X_2^i, \ldots X_k^i, Y_i\rangle \qquad i=1,2,\ldots n \tag{19}$$

Assume that the above sample originates from a multivariate distribution with joint density $f(x,y)$. The marginal density of X is (Rao, 1983)

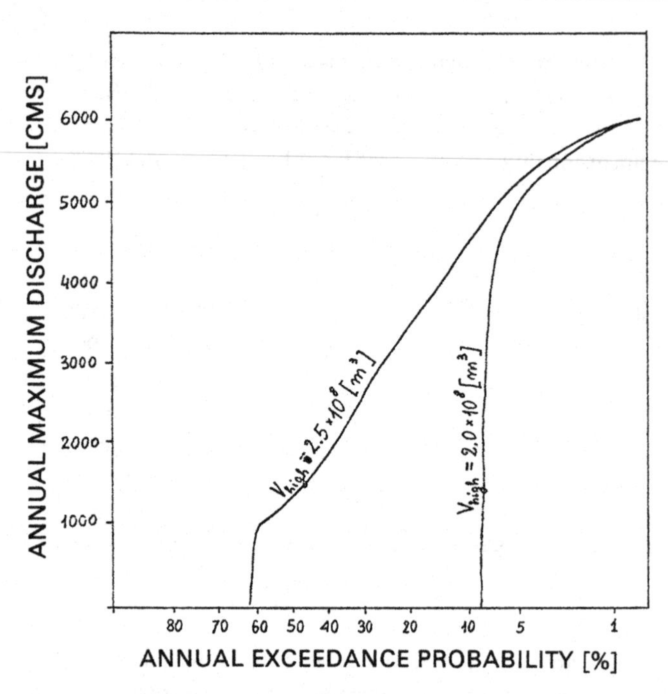

Fig. 2 The multivariate estimation of the annual maximum discharges based on the high flow in the river Vistula at the gauge Sandomierz (observation period: 1920–63).

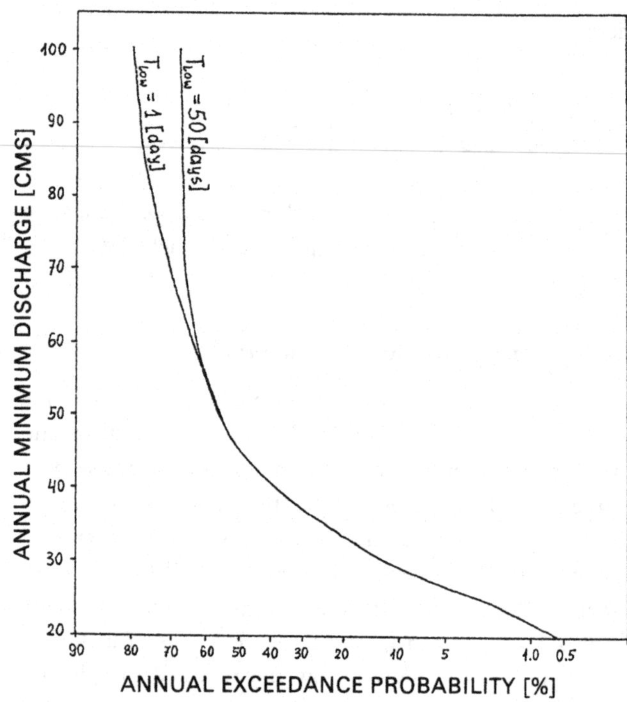

Fig. 3 The multivariate estimation of the annual minimum discharges based on low flow in the river Warta at the gauge Skwierzyna (observation period 1901–59).

$$g(\boldsymbol{x}) = \int_{-\infty}^{\infty} f(\boldsymbol{x}, y)\mathrm{d}y \tag{20}$$

and the conditional density of Y given $\boldsymbol{X}=\boldsymbol{x}$ is

$$f(y/\boldsymbol{x}) = \frac{f(\boldsymbol{x}, y)}{g(\boldsymbol{x})} \tag{21}$$

The conditional mean or regression of Y on $\boldsymbol{X}$ is

$$r(\boldsymbol{x}) = \mathrm{E}(Y|\boldsymbol{X}=\boldsymbol{x}) \tag{22}$$

$$r(\boldsymbol{x}) = \frac{\int_{-\infty}^{\infty} y f(\boldsymbol{x}, y)\mathrm{d}y}{g(\boldsymbol{x})} \tag{23}$$

The nonparametric estimator of the unknown joint density $f(\mathbf{x}, y)$ can be expressed as

$$f(\boldsymbol{x}, y) = \frac{1}{n}\sum_{i=1}^{n} \frac{1}{h_y} K\left(\frac{y - Y_i}{h_y}\right) \prod_{l=1}^{k} \frac{1}{h_l} K\left(\frac{x_l - X_l^i}{h_l}\right) \tag{24}$$

where h_y is a smoothing factor corresponding to the realization of the random variable Y.

The nonparametric estimator of the marginal density (23) is given by

$$\hat{g}(\boldsymbol{x}) = \hat{f}(\boldsymbol{x}) \tag{25}$$

where $\hat{f}(\boldsymbol{x})$ is a multivariate estimator of an unknown density function (9). Based on (23)–(25) the nonparametric estimator of the regression function can be expressed as

$$\hat{r}(\boldsymbol{x}) = \frac{\sum_{i=1}^{n} Y_i \prod_{l=1}^{k} \frac{1}{h_l} K\left(\frac{x_l - X_l^i}{h_l}\right)}{\sum_{i=1}^{n} \prod_{l=1}^{k} \frac{1}{h_l} K\left(\frac{x_l - X_l^i}{h_l}\right)} \tag{26}$$

The above estimator is a local weighted average of the Y_i given a random vector $\boldsymbol{x}$.

Nonparametric regression as applied to the classical relation: river discharge – water stage

Based on the nonparametric regression estimator (26), the relationship between river discharge and water stage was developed.

Regression analysis of the relationship was conducted, whose results are presented in Fig. 4. In engineering applications the shape of this relationship is not known theoreticaly and the choice of the shape is arbitrary. The nonparametric method gives the possibility to choose better and more objective shape of this relationship.

Based on the results presented in Fig. 4 it can be concluded that the nonparametric regression algorithm (26) gives excellent results especially in areas with a high density of observation data. When the density of the data is low, the curve is oscillating (cf. the high part of relationship presented in Fig. 4). This is suggesting, that sometimes, for very low values of the density of the observed data, interpolation with this algorithm can be defective.

Outside the observed data range, the shape of the curve is stabilizing. It is natural for this algorithm; if there are no observations, the values of the density are equal to zero.

From practical engineering view point it is known, that the interpolation and extrapolation in this relationship have to be very smooth and the extent of a reasonable extrapolation is limited.

Based on these results it can be concluded that the nonparametric regression estimator (26) has to be modifited for this application by using, for example, the variable kernel in the estimator. The variable kernel can yield a smoother curve in the interpolation and extrapolation areas (Adamowski, 1989; Feluch, 1994).

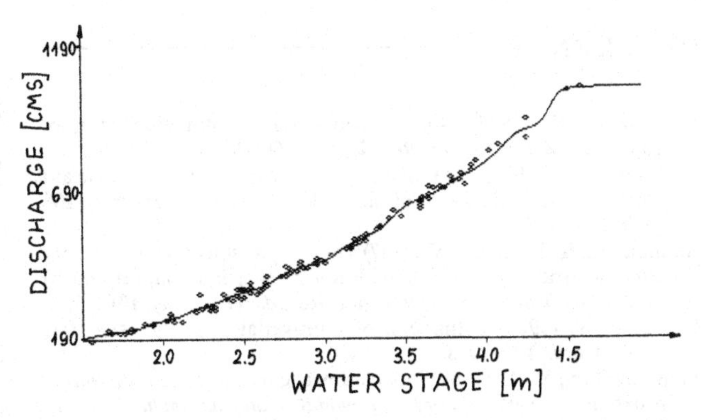

Fig. 4 The relationship between river discharge and water stage for the river Odra at Gozdowice.

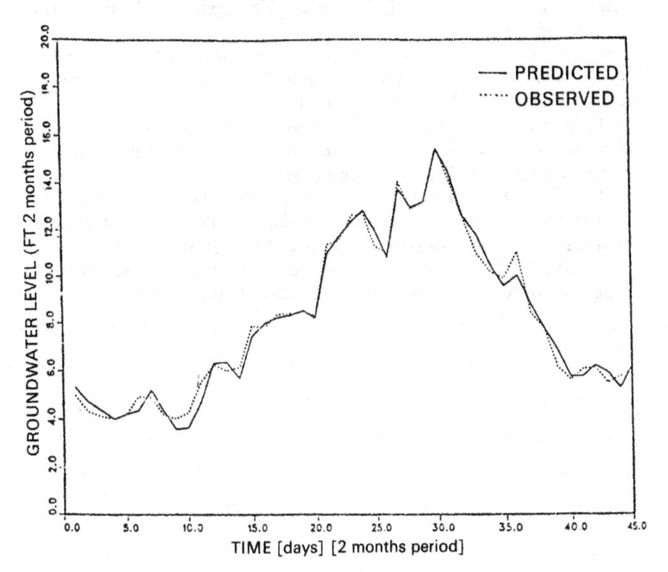

Fig. 5 Observed vs predicted groundwater levels for the data from Kuczera (1982).

Nonparametric regression in the relation of concurrent time series of runoff and groundwater elevation

The data set analysed is a concurrent time series of runoff and groundwater level in Silver Springs in Florida taken from Kuczera (1982). A split-sample procedure was employed in using the nonparametric regression program, whereby the first half of the data set was used to estimate the smoothing parameters (exactly two) and the second half was used as a basis for verifying the model's predictive capabilities. The plot of the observed versus predicted values of groundwater levels time series is shown in Fig. 5; for the observed data (used to estimate smoothing factors); and in Fig. 6 for extrapolation (i.e. split-sample experiment).

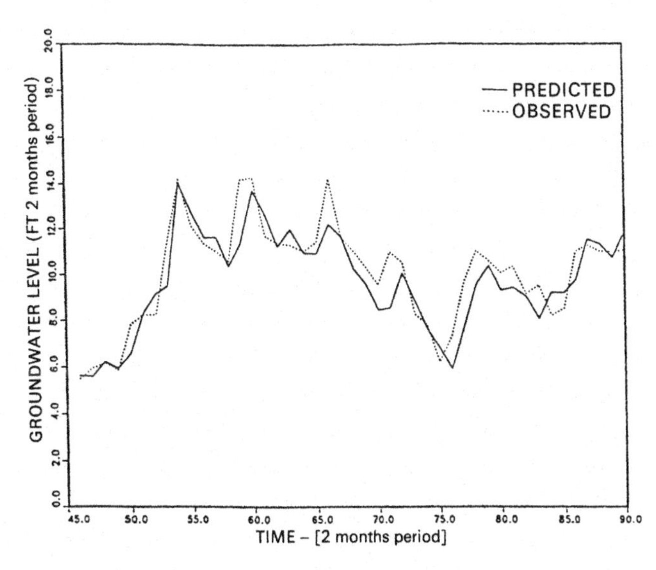

Fig. 6 Observed vs extrapolated groundwater levels for the data from Kuczera (1982).

CONCLUSIONS

Methods of nonparametric estimation of density and of regression function can be used in hydrologic applications because they require few and mild assumptions, and are capable of inferring complicated densities or relationships (in regression case). These properties are very important in practical hydrology.

When applying nonparametric approach to the estimation of density function and regression one has to be particularly careful about:

(a) the relationship between the sample size required for achieving the prescribed accuracy of estimation and the dimensionality of the problem;
(b) the accuracy of interpolation and extrapolation in the region where there are no data or where only few data are available.

REFERENCES

Adamowski, K. (1985) Nonparametric kernel estimation of flood frequencies. *Water Resour. Res.*, **21**(11), 1585–90.

Adamowski, K. (1989) A Monte Carlo comparison of parametric and nonparametric estimation of flood frequencies. *J. Hydrol*, **108**, 205–308.

Adamowski, K. & Feluch, W. (1987) The comparison of parametric and nonparametric methods of flood frequencies estimation, (in Polish). *Wiadomości Instytutu Meteorologii i Gospodarki Wodnej*, **10**(2–3).

Cacoullos, T. (1966) Estimation of multivariate density. *Ann. Inst. Statist. Math.*, **18**, 179–89.

Epanechnikov, V. A. (1969) Nonparametric estimates of multivariate probability density. *Theory of probability and its applications*, **14**, 153–8.

Feluch, W. (1987) Nonparametric estimation of density function applied to yearly maximum discharge, *Proc. Seminar of Institute of Environmental Engineering*, Technical University of Warsaw (in Polish), pp. 21–37.

Feluch, W. (1988) *Nonparametric estimation methods of characteristic flows in small catchments – one-dimensional case*, Project Report under contract with IMUZ, Warsaw (in Polish).

Feluch, W. (1989) *Nonparametric estimation methods of characteristic flows in small catchments – multivariate case*, Project Report under contract with IMUZ, Warsaw (in Polish).

Feluch, W. (1994) *The choice methods of kernel estimation of probability density function and regression in hydrology* (in Polish), Politechnika Warszawska, Prace Naukowe, Inżyniera Śroclowiska, z. 15.

Kaczmarek, Z. (1970) Statistical Methods in Hydrology and Meteorology. *Wyd. Komunikacji i Łaczności*, Warsaw (in Polish).

Kuczera, G. (1982) On the relationship between the reliability of parameter estimates and hydrologic time series data used in calibration. *Water Resour. Res.*, **18**(1), 146–54.

Praska Rao, B. L. (1983) *Nonparametric Functional Estimation*, New York, Academic Press.

Rudemo, M. (1982) Empirical choice of histograms and kernel density estimators. *Scand. J. Statist.*, **9**, 65–78.

Schuster, E. & Yakowitz, S. (1985) Parametric/nonparametric mixture density estimation with application to flood-frequency analysis. *Water Resour. Bull.*, **21**(5), 797–804.

Scott, D. W. & Terrell, G. R. (1987) Biased and unbiased cross-validation in density estimation. *Journal of the American Statistical Association*, **82**, 1131–46.

Silverman, B. W. (1986) *Density Estimation for Statistics and Data Analysis*, Chapman and Hall, London.

Strupczewski, W. (1967a) Determination of the probability of repeating phenomena. *Acta Geoph. Pol.*, **15**(2) (in Polish).

Strupczewski, W. (1967b) *Statistical analysis of swell wave shapes*, Ph. D. thesis, Technical University of Warsaw.

Tapia, R. & Thompson, J. (1978) *Nonparametric Probability Density Estimation*, The Johns Hopkins University Press, Baltimore.

United States Water Resources Council Hydrology Committee (USWRC) (1982) *Guidelines for determining flood flow frequency*, Bull. 17b (revised), U.S. Gov. Print. Office, Washington, D.C.

Yevjevich, V. (1967) *An objective approach to definition and investigation of continental hydrologic drought*. Colorado State University, Fort Collins, Hydrol. Pap. No. 23.

Zielińska, M. (1963) Statistical methods of defining low water stages – i. *Przegląd Geofizyczny*, **8**(1–2), 75–87.

Zielińska, M. (1964) Statistical methods of defining low water stages – ii. *Przegląd Geofizyczny*, **9**(2), 109–20.

11 Nonparametric approach for design flood estimation with pre-gauging data and information

GUO SHENG LIAN

Department of Engineering Hydrology, University College Galway, Ireland, on leave from Wuhan University of Hydraulic and Electric Engineering, Wuhan, People's Republic of China

ABSTRACT The main task of flood frequency analysis is to obtain design flood magnitudes from a streamflow record. The gauged record is rarely long enough to yield an estimate of an extreme flood which is sufficiently accurate to be applied with confidence in hydraulic engineering. Therefore extending a data record back in time using historical or palaeoflood data has the potential to provide a considerable amount of additional information on very large floods. Parametric estimation methods are readily applicable to flood frequency analysis when pre-gauging data is available. However, all parametric approaches need an assumption about the underlying parent distribution which is never known in hydrologic processes. A new nonparametric kernel estimation model is proposed and developed. With limited real data and simulation experiments, results show that quantiles estimated by nonparametric methods are better than those obtained by some selected parametric estimators both in terms of the descriptive ability and predictive ability. The choice of the optimum kernel function, the uncertainty of the threshold of perception value and the difference between fixed kernel and variable kernel estimators are also discussed. It is expected that the nonparametric approach will be widely used in practice as it is free of serious limitations of classical parametric models.

INTRODUCTION

Statistical methods of flood frequency estimation in current use are mainly based on the assumption that observed flood series comes from a population whose probability density function is known. Several estimation methods exist which may be used in these circumstances to obtain estimates of parameters and quantiles. Such methods are said to be 'parametric'. However, in hydrological practice, there is no compelling evidence in favour of any one parametric distribution or fitting procedure. Based on goodness-of-fit tests, several different distributions can appear to fit the data equally well, but each distribution gives quite different estimates of a given quantile, especially in the tail of the distributions (Guo, 1986). To achieve some degree of uniformity in the determination of flood quantiles, some countries have imposed a certain probability distribution coupled with a certain parameter estimation procedure, for example, the log Pearson type 3 (LP3) distribution was recommended by USWRC, (1968, 1981), the general extreme value (GEV) distribution was suggested for use in the UK and Ireland (NERC, 1975). A uniform method is desirable, but is of questionable validity with currently available knowledge. It is evident that the parametric method, which depends on the assumption of a parent distribution, has its disadvantage and limitations. Dooge (1986) pointed out that '*no amount of statistical refinement can overcome the disadvantage of not knowing the frequency distribution involved*'.

Currently used flood frequency analysis methods were seriously criticized by Klemes (1986, 1987). Due to recent developments in nonparametric statistics theory, another possible approach to flood frequency analysis is to estimate the density function nonparametrically. Adamowski (1985, 1989) has suggested that the nonparametric approach provides an alternative to making unjustified assumptions concerning the unknown distribution in flood frequency. He pointed out (1989) that: '*The nonparametric method is therefore intuitively very attractive in the absence of a clear indication of which parametric family should be selected*'. All work that has been done in this area is based on two sets of relatively long recorded data and limited Monte Carlo experiments.

The gauged records are rarely long enough to yield an estimate of a design flood which is accurate enough to be applied in hydraulic engineering. Therefore, there is a considerable interest in augmenting the flow record with information from other sources. One of these techniques of data extension is to consider historical information and to include extraordinarily large records in flood frequency analysis. Another source of information regarding presample floods is palaeohydrology, in which the data and magnitudes of flood events are deduced from physical evidence such as deposited mud layers and age of vegetation at the flood site (Baker, 1987). The methods of incorporation of historical floods and palaeologic information into parametric models and the usefulness of doing so have been discussed by many hydrologists (USWRC, 1981; Stedinger & Cohn, 1986; Hirsch, 1987; Guo, 1990). The main problem of this study is how to incorporate this valuable pre-gauging information into nonparametric kernel density estimation process. Before the widespread application of nonparametric methods in flood frequency analysis is possible, several questions also should be answered and discussed in detail, such as the estimation of the smoothing factor, the choice between fixed kernel and variable kernel estimators and the reliability of extrapolation of the frequency curve beyond the largest observed data.

NONPARAMETRIC DENSITY ESTIMATION

Many different types of nonparametric density estimators are available, including the histogram, kernel, penalized-likelihood estimator, etc. The histogram is probably the oldest probability density estimator, and it is the classical nonparametric statistical tool for the graphical display of data. The most popular and well developed theoretically is the kernel estimator (Rao, 1983; Adamowski, 1989).

The fixed kernel estimator (FKE)

For a given kernel function $K(.)$, which is a probability density function symmetric about zero, a positive smoothing parameter h and a sample $x_1, x_2, \ldots, x_n$, the kernel estimate of probability density function at each fixed point x is:

$$f(x/h)=\frac{1}{nh}\sum_{i=1}^{n} K\left(\frac{x-x_i}{h}\right) \quad (1)$$

The kernel estimate is nonnegative and integrates to one.

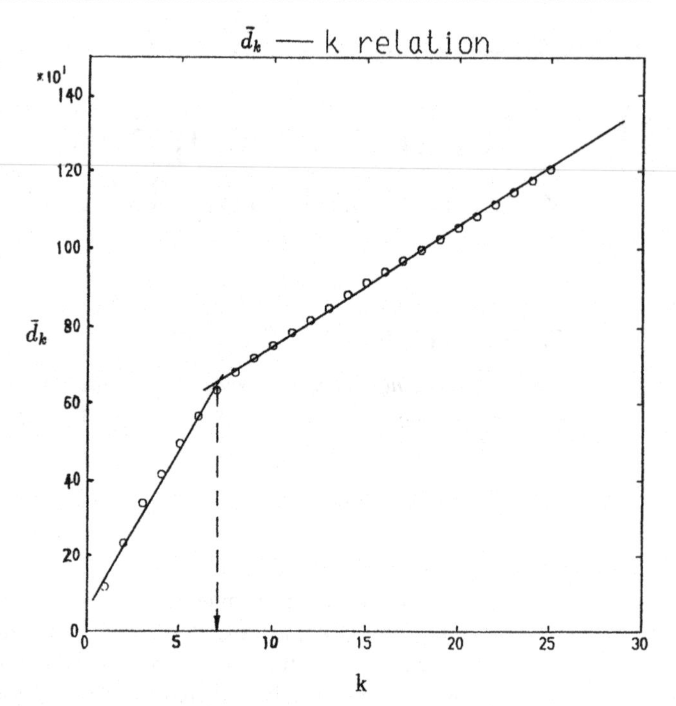

Fig. 1 Plots of $\bar{d}_k$ vs k at Guan Ting catchment, China, where sample size $n=50$, $\bar{x}=1121$ m^3/s, $C_v=0.86$ and $C_s=2.25$.

The variable kernel estimator (VKE)

Breiman *et al.* (1977) defined VKE as

$$f(x/a_k)=\frac{1}{n}\sum_{i=1}^{n}\frac{1}{a_k d_{i,k}} K\left(\frac{x-x_i}{a_k d_{i,k}}\right) \quad (2)$$

where d_{ik} is the interpoint distance between x_i and its kth nearest neighbour among $x_1, x_2, \ldots, x_n$ data points, a_k is a constant smoothing factor and $K(.)$ is the smoothing kernel function whose functional form has to be obtained. In the VKE method, it is necessary to assume a kernel and determine the values of k, a_k and $d_{i,k}$. Determination of the values of k is critical. The value of k should range from 1 to $n/2$. Breiman *et al.* (1977) established empirically that the optimal value of k corresponds to a knee (a sudden change in slope) in the relationship of $\bar{d}_k$ to k, where $\bar{d}_k$ is the arithmetic mean value of $d_{i,k}$. For example, the relationship of $\bar{d}_k$ and k in Guan Ting catchment (China) is plotted in Fig. 1, in which the value of k can be determined ($k=7$).

The choice of kernel function

The choice of a kernel has been investigated extensively by many researchers (Rao, 1983; Adamowski, 1985). It is now known that many symmetric unimodal kernel functions are nearly optimal. This implies that the choice of kernel is not critical as even a suboptimal kernel leads to only a small loss of accuracy with respect to integrated mean square error criterion (Rao, 1983).

An optimal kernel given by Adamowski (1989)

$$K(x)=\begin{cases}0 & x<-1\\ \frac{3}{4}(1-x^2) & -1\leq x\leq 1\\ 0 & x>1\end{cases} \tag{3}$$

is circular in shape, bounded, and not dependent on the sample size or underlying density function.

A CHOICE BETWEEN FKE AND VKE ESTIMATOR

Two catchments whose sample skewness read 2.25 and 1.07, respectively, were chosen for the purpose of comparing the performance of FKE and VKE. The nonparametric kernel estimates as well as the recorded data were plotted on the same EV1 probability paper. Fig. 2a indicated that nonparametric variable kernel estimate gives more reasonable extrapolation than that of fixed kernel estimate for Guan Ting catchment ($C_s=2.25$). The differences between the variable and fixed kernel estimates will reduce as the sample skewness decreases. It is also shown that the quantiles estimated by fixed kernel are closer to the real data points than those estimated by variable kernel in the case of small sample skewness (e.g. Bu Xi catchment, $C_s=1.07$), see Fig. 2(b).

It is now known that the choice between fixed and variable kernel estimator depends on sample skewness. Each of these kernels may have its own advantages and disadvantages. For example, the fixed kernel is suitable for small skewness sample and easy calculation of h, but it may obtain unreasonable extrapolation curve in high skewness sample. The question naturally arises as to what kind of kernel should be selected for any particular sample.

In order to answer this question, a Monte Carlo experiment was carried out. For simulation purposes the GEV distribution was selected as a parent population because it fits a wide range of flood observations and is mandated in the United Kingdom and Ireland (NERC, 1975). The probability density function of GEV can be expressed as

$$f(x)=\begin{cases}\frac{1}{\alpha}\exp\left[-\frac{x-u}{\alpha}-\exp\left(-\frac{x-u}{\alpha}\right)\right]; & \beta=0\\ \frac{1}{\alpha}\left[1-\frac{\beta(x-u)}{\alpha}\right]^{\frac{1}{\beta}-1}\exp\left[-\left(1-\frac{\beta(x-u)}{\alpha}\right)^{\frac{1}{\beta}}\right]; & \beta\neq 0\end{cases} \tag{4}$$

where u, α and β are location, scale and shape parameters, respectively.

For a given sample size n and parent parameters, 10000 traces of data are generated by Monte Carlo method and each simulated series is ranked in ascending order. The ranked values are then averaged over the 10000 to obtain $\bar{x}_{(i)}$ values which represent the parent population $E[x_{(i)}]$ values.

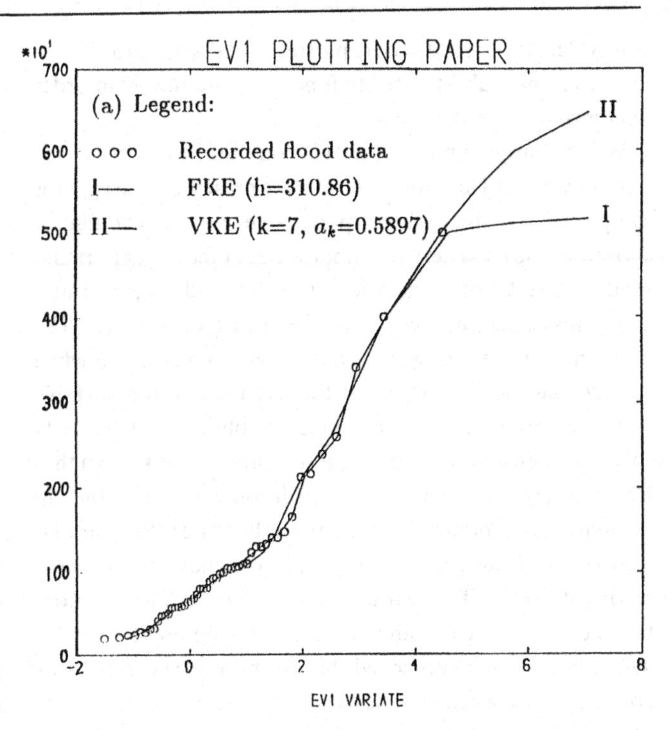

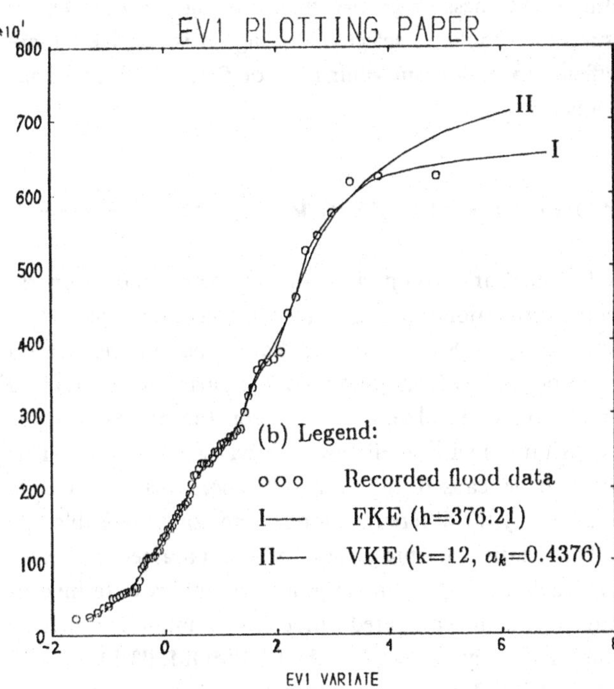

Fig. 2 Flood quantiles estimated by nonparametric FKE and VKE methods; (a) Guan Ting catchment ($C_s=2.25$); (b) Bu Xi catchment ($C_s=1.07$).

$$\bar{x}_{(i)}=\frac{1}{10000}\sum_{i=1}^{10000}x_{(i),j} \tag{5}$$

The nonparametric fixed and variable kernel probability density estimators were calculated by equations 1 and 2, respectively for this representative sample $x_{(i)}$. These probability density estimates were plotted on the same graph with the parent probability density function (equation 4) in order to compare the differences.

While a large number of simulation tests has been carried out, only a small amount of results are presented in Fig. 3 for sample size 50 and different parent skewnesses. The figures show that the fixed kernel estimator can fit the rising part and mode of the density function quite well for all parents, but it gives unsatisfactory results in the fitting of the tail. The differences in the fitting of the tail become more evident when sample skewness is large. On the other hand, the variable kernel estimator can fit the tail reasonably well, particularly for large skewness samples, see Fig. 3c and d. It is known that hydrologists and engineers are much concerned with the tail behaviour in flood frequency analysis, because the extrapolation of the frequency curve mostly depends on the large recorded floods. Those results show that the VKE is more flexible than the FKE and is suitable for different samples. Therefore it is recommended that the fixed kernel estimator could be used when sample skewness is less than 1.0; if the sample skewness is greater than 1.0, the variable kernel estimator should be selected although it is relatively more difficult to obtain smoothing factor than the fixed kernel estimator.

MONTE CARLO EXPERIMENTS

A Monte Carlo comparison of parametric and nonparametric estimation of flood frequencies has been reported by Adamowski (1989). He showed that nonparametric method is competitive with parametric counterparts. However, in his simulation work, all samples came from the log Pearson type 3 distribution with small parent skewness ($C_s = 0.444$). This could have caused that the differences between results obtained by the different methods were not remarkable.

The statistical properties of nonparametric kernel approach used for estimating flood quantiles were investigated again and compared with other common 'parametric' flood frequency models, namely, LP3/MOM, P3/MOM and GEV/PWM. The relative bias and root mean square error (RMSE) are chosen as criteria for comparison:

$$\mathrm{Bias}(T) = \frac{1}{N}\sum_{i=1}^{N}\frac{\hat{Q}_{i,T} - Q_T}{Q_T} \tag{6}$$

$$\mathrm{RMSE}(T) = \left\{\frac{1}{N}\sum_{i=1}^{N}\left[\frac{\hat{Q}_{i,T} - Q_T}{Q_T}\right]^2\right\}^{0.5} \tag{7}$$

where T is return period; N is the number of Monte Carlo repetitions; $\hat{Q}_{i,T}$ and Q_T represent the calculated and theoretical quantile values, respectively.

Hundred samples with size 50 were simulated from Pearson type 3 distribution (Whittaker, 1973) with parent parameters $\bar{Q} = 100$, $C_v = 0.5$ and $C_s = 1.0$, 1.5 and 2.0. The relative bias and RMSE of quantile estimates were calculated for each sample by each algorithm. In the case of nonparametric VKE, the same order of k was fixed in advance for all the samples. Only the results for 100-year design flood estimates were given in Table 1.

Table 1. *Relative bias and RMSE in the estimation of 100-year design floods. 100 samples with size 50 were simulated from Pearson type 3 distribution with parameters $\bar{Q} = 100$, $C_v = 0.5$ and $C_s = 1.0$, 1.5 and 2.0*

	$C_s = 1.0$		$C_s = 1.5$		$C_s = 2.0$	
model	BIAS	RMSE	BIAS	RMSE	BIAS	RMSE
VKE ($k = 10$)	0.013	0.156	−0.007	0.172	−0.034	0.164
VKE ($k = 20$)	0.001	0.145	−0.013	0.177	−0.046	0.164
P3/MOM	−0.021	0.126	−0.044	0.144	−0.060	0.145
LP3/MOM	0.476	0.581	0.076	0.173	0.063	0.174
GEV/PWM	0.043	0.159	0.050	0.179	0.164	0.269

Table 1 shows that the quantiles estimated by the nonparametric VKE method are less biased than those estimated by the parametric models. As regards the RMSE, the nonparametric estimates are better than LP3/MOM and GEV/PWM and competitive with P3/MOM. The P3/MOM algorithm performed well due to the fact that the samples originated from the Pearson type 3 parent distribution. The differences between VKE estimates are quite small when the order k changes from 10 to 20. It is expected that the accuracy of nonparametric estimates will be improved if the choice of k in VKE is made for each sample rather than being fixed in advance for all examples.

NONPARAMETRIC MODEL WITH HISTORICAL DATA AND PALEOFLOOD INFORMATION

Consider the form of the annual maximum flood series presented in Fig. 4. There are a total of g known floods, m of which are known to be the m largest in the period of n years. The n year period contains within it the systematic record period of s years ($s \leq n$). The number e of the m largest floods occurred during the systematic record ($e \leq m$ and $e < s$). Note also that $g = s + m - e$. Assume that there is a threshold of perception X_0 such that the m largest floods are larger than or equal to it and the remainder are smaller than it. We know about the $m - e$ floods in the pre-gauging period ($n - s$) specifically because they were greater than or equal to X_0.

Arrange the g known floods in ascending order, i.e.,

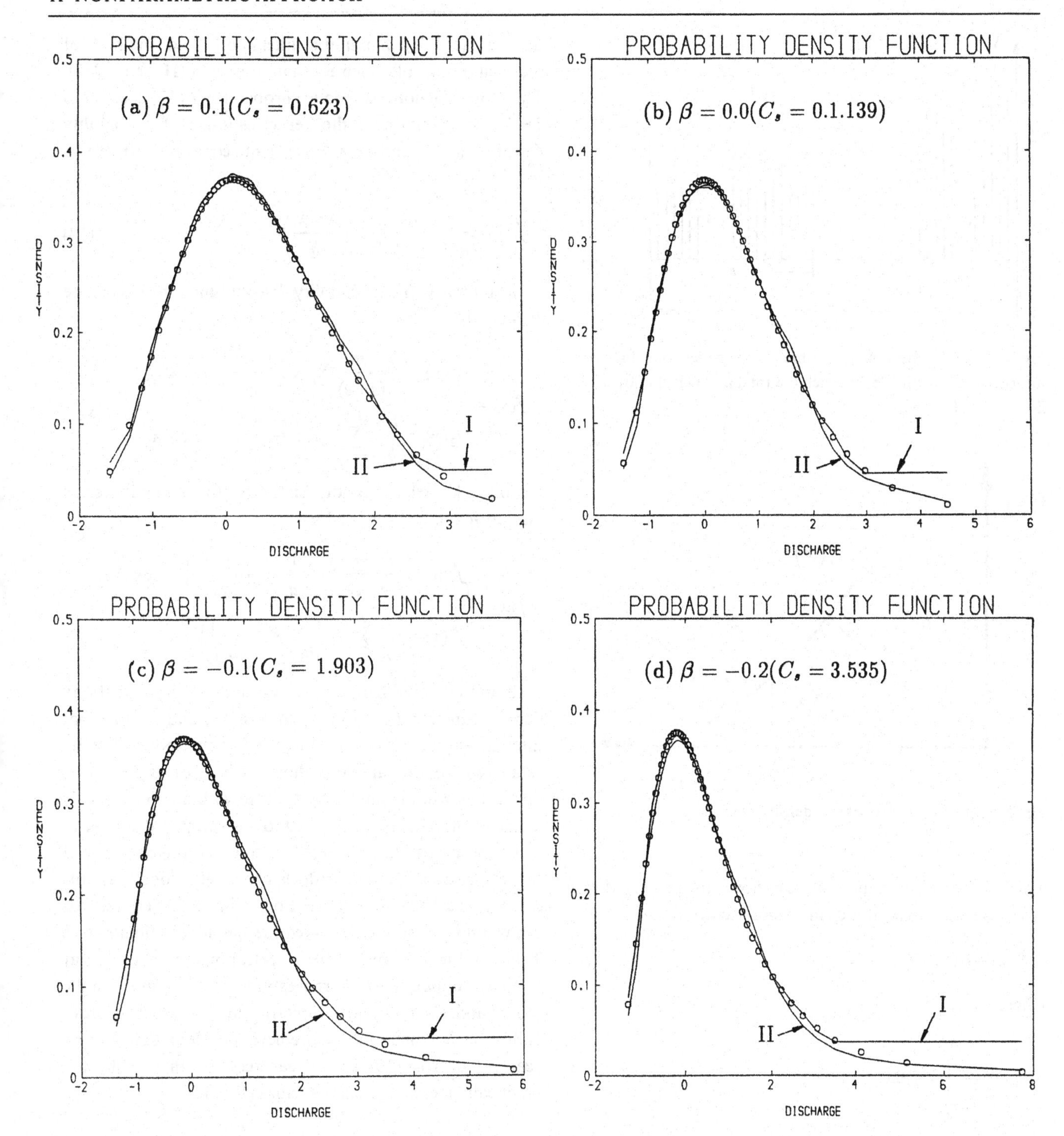

Fig. 3 Comparison between fixed and variable kernel estimator for the GEV representing sample with size $n=50$, $u=0$, $\alpha=1$ and different β values. Solid line: I – fixed kernel estimator, II – variable kernel estimator, circle – recorded AM flood data.

$x_1<x_2<\ldots<x_g$. The probability density function $f(x)$, shown in Fig. 5, can be expressed as:

$$f(x)=\begin{cases} f_1(x) & 0<x<X_0 \\ f_2(x) & x\geq X_0 \end{cases} \tag{8}$$

For $f_2(x)$, it is known that there are m floods larger than or equal to the threshold value X_0 in the period of n years. Therefore, the nonparametric fixed kernel estimation can be obtained directly as follows

$$\hat{f}_2(x)=\frac{1}{nh}\sum_{i=s-e+1}^{g} K\left(\frac{x-x_i}{h}\right) \tag{9}$$

As regards the estimation of $f_1(x)$, $n-s-m+e$ floods

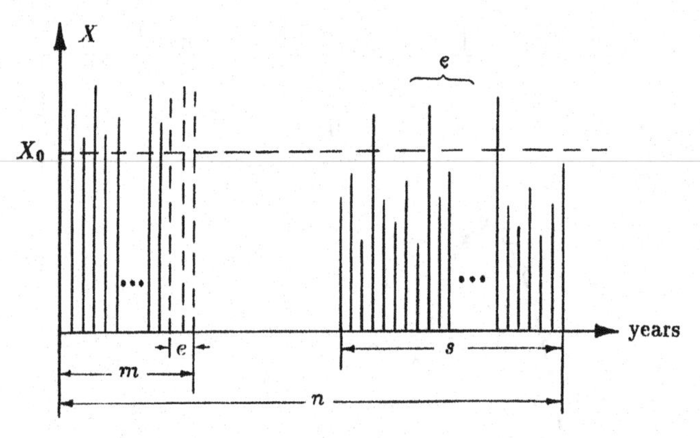

Fig. 4 Sketch of the annual maximum flood series when historical information is available, in which the total number of known floods $g=s+m-e$.

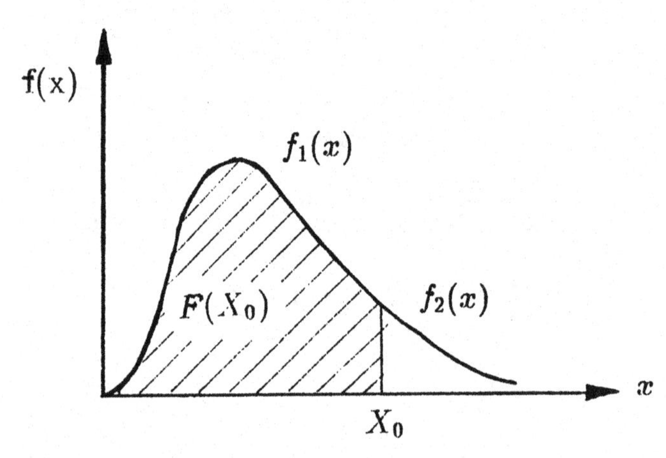

Fig. 5 Sketch of the probability density function.

which are smaller than X_0 are unknown in the period of n years. Because these floods must not exceed X_0, one gets:

$$f_1(x)=\hat{f}_1(x)/F(X_0) \tag{10}$$

where

$$F(X_0)=\int_0^{X_0} f(x)\mathrm{d}x$$

The non-exceedance probability distribution function $F(X_0)$ can be estimated empirically by

$$F(X_0)=\frac{n-m}{n} \tag{11}$$

which depends only on the values of m and n. This result is based on the fact that the probability $P(x)$ is an order statistic of a uniformly distributed random variable. These order statistics follow a beta distribution regardless of the distribution of discharge. The exceedance probability (P_e) for X_0 can be estimated by maximum likelihood (and method of moments) both of which estimate it as m/n (Hirsch, 1987). Therefore, the non-exceedance probability $F(X_0)$ is equal to $1-P_e$. Substitution of the kernel estimate of probability density function and equation (11) into equation (10), results in

$$\hat{f}_1(x)=\frac{n-m}{nh(s-e)}\sum_{i=1}^{s-e} K\left(\frac{x-x_i}{h}\right) \tag{12}$$

In summary, the probability density function $f(x)$ can be estimated by

$$\hat{f}(x)=\begin{cases}\hat{f}_1(x)=\dfrac{n-m}{nh(s-e)}\displaystyle\sum_{i=1}^{s-e} K\left(\frac{x-x_i}{h}\right) & x<X_0\\ \hat{f}_2(x)=\dfrac{1}{nh}\displaystyle\sum_{i=s-e+1}^{gs-e} K\left(\frac{x-x_i}{h}\right) & x\geq X_0\end{cases} \tag{13}$$

for fixed kernel estimator. Similarly, the variable kernel estimator can be expressed by

$$\hat{f}(x)=\begin{cases}\hat{f}_1(x)=\dfrac{n-m}{n(s-e)}\displaystyle\sum_{i=1}^{s-e} \frac{1}{a_k d_{i,k}} K\left(\frac{x-x_i}{a_k d_{i,k}}\right) & x<X\\ \hat{f}_2(x)=\dfrac{1}{n}\displaystyle\sum_{i=s-e+1}^{g} \frac{1}{a_k d_{i,k}} K\left(\frac{x-x_i}{a_k d_{i,k}}\right) & x\geq X\end{cases} \tag{14}$$

It should be noted here that the estimate of the probability density function (pdf) in equations (13) and (14) do not integrate to one. This is due to the fact that pdf is directly estimated from a sample without such a constraint on its integral as would usually be the case with an algebraically defined pdf. This is a deficiency of the proposed nonparametric kernel estimation method. However, this property of the estimation procedure does not affect flood quantile estimation scheme in practice. For example, 50-year records can result in reasonable 100-year design flood estimates (see Table 1), but it is impossible to extrapolate to a 1000-year flood by nonparametric kernel estimation procedure. On the other hand, even though parametric models (assuming probability density function) can predict a thousand- or a million-year flood from 20- to 50-year records, the results are practically meaningless, cf. Klemes (1987).

METHOD OF ESTIMATING THE SMOOTHING FACTOR

The maximum likelihood method is proposed and developed for estimation of the smoothing factor h in equation (13). Since FKE and VKE are of the same form, only the derivation and formulae of fixed kernel estimator will be shown in the following. Consider the following problem with a maximum likelihood criterion for choosing h

$$\text{maximize } L(h) = \prod_{j=1}^{g} \hat{f}(x_j); \qquad h \geq 0 \tag{15}$$

It may be seen from equation (15), that $h=0$ maximizes $L(h)$, corresponding to an estimate with a Dirac function at each sample point. Thus a slightly modified maximum likelihood criterion is used

$$\text{maximize } L'(h) = \prod_{j=1}^{g} \hat{f}^{*}(x_j)$$

$$= \prod_{j=1}^{s-e} \hat{f}_1(x_j) \prod_{j=s-e+1}^{g} \hat{f}_2(x_j) \tag{16}$$

where

$$\hat{f}_1(x_j) = \frac{1}{(s-e)h} \sum_{\substack{i=1 \\ i \neq j}}^{s-e} K\left(\frac{x_j - x_i}{h}\right) \tag{17}$$

$$\hat{f}_2(x_j) = \frac{1}{nh} \sum_{\substack{i=s-e+1 \\ i \neq j}}^{s-e} K\left(\frac{x_j - x_i}{h}\right) \tag{18}$$

and in the terms of log likelihood

$$LL'(h) = \sum_{j=1}^{s-e} \log|\hat{f}_i(x_j)| + \sum_{j=s-e+1}^{g} \log|\hat{f}_2(x_j)| \tag{19}$$

Substituting equations (17) and (18) into equation (19), taking partial derivative with respect to h, equating to zero, and rearranging gives:

$$\sum_{j=1}^{s-e} \left\{ \frac{\sum_{\substack{i=1 \\ i \neq j}}^{s-e} \hat{K}\left(\frac{x_j - x_i}{h}\right)}{\sum_{\substack{i=1 \\ i \neq j}}^{s-e} \hat{K}\left(\frac{x_j - x_i}{h}\right)} \right\} + \sum_{j=s-e+1}^{g} \left\{ \frac{\sum_{\substack{i=s-e+1 \\ i \neq j}}^{g} \hat{K}\left(\frac{x_j - x_i}{h}\right)}{\sum_{\substack{i=s-e+1 \\ i \neq j}}^{g} \hat{K}\left(\frac{x_j - x_i}{h}\right)} \right\} - n = 0 \tag{20}$$

where $\hat{K}(x) = 1.5x$ for $|x| \leq 1$ and zero elsewhere.

Equation (20) only has one unknown parameter h, and can be solved by using simple optimization techniques for given kernel and sample. It was found that the choice of tolerance (defined as ε) in estimating the smoothing factor h in equation (20) is also important. For example, in Yi Chang catchment whose sample size and statistics are $n=820$, $s=101, m=8, \bar{x}=51\,612\,\text{m}^3/\text{s}, C_v=0.19$ and $C_s=0.43$, a large tolerance ($\varepsilon=0.1$) value results an irregular density function (Fig. 6a, $h=3351$). Vice versa, a small tolerance value ($\varepsilon=0.0001$) produces oversmoothing density estimates (Fig. 6d, $h=13\,324$). Fig. 6b and c show that density estimates are more reasonable when ε equals 0.03 ($h=5397$) and 0.001 ($h=7957$) respectively. Unfortunately, the tolerance or criterion use to choose the smoothing factor varies with the sample and depends on personal experience and judgement. However, the effect on the tail of a distribution is relatively small and can be neglected, see Fig. 6.

FLOOD QUANTILE ESTIMATION

In flood frequency analysis, the main task is to estimate a design quantile value corresponding to a given return period. It is, of course, sufficient to estimate only the exceedance probability $P(x)$ corresponding to ordered values of observations. The estimation of exceedance probability $P(x)$ does not only depend on the density function, but also depends on the value of threshold of perception, X_0, of historical floods. We consider the following two situations:

(a) if $x \geq X_0$, then

$$\hat{P}(x) = \int_x^{\infty} \hat{f}_2(x)\mathrm{d}x = \frac{1}{nh} \sum_{i=s-e+1}^{g} C_i(x) \tag{21}$$

where

$$C_i(x) = \int_x^{\infty} K\left(\frac{x - x_i}{h}\right) \mathrm{d}x$$

(b) if $x < X_0$, then

$$\hat{P}(x) = \int_x^{X_0} \hat{f}_1(x)\mathrm{d}x + \int_{X_0}^{\infty} \hat{f}_2(x)\mathrm{d}x$$

$$= \frac{n-m}{n(s-e)h} \sum_{i=1}^{s-e} C_i'(x) + \frac{1}{nh} \sum_{i=s-e+1}^{g} C_i''(x) \tag{22}$$

where

$$C_i'(x) = \int_x^{X_0} K\left(\frac{x - x_i}{h}\right) \mathrm{d}x$$

and

$$C_i''(x) = \int_{X_0}^{\infty} K\left(\frac{x - x_i}{h}\right) \mathrm{d}x$$

The estimation of $C_i'(x)$ and $C_i''(x)$, will depend on the location of X_0 and the kernel which have been discussed by Guo (1990). Generally speaking, accurate determination of X_0 does not significantly influence the effectivity of the proposed nonparametric kernel estimation models.

COMPARISON OF PROPOSED NONPARAMETRIC MODEL WITH PARAMETRIC METHODS

The performance of the proposed nonparametric model was compared with some parametric models, namely, LP3/

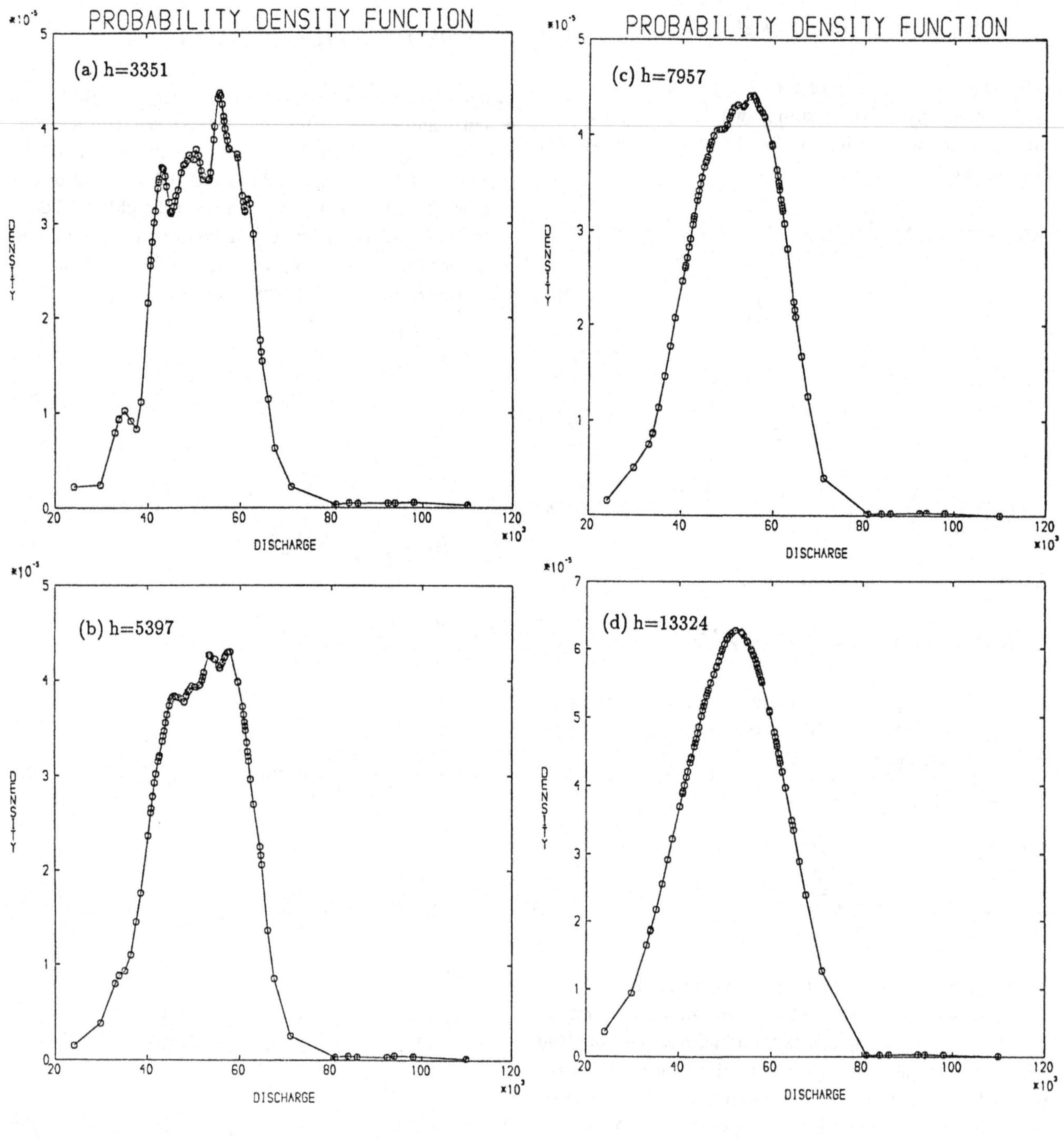

Fig. 6 Fixed kernel probability density function estimates with different tolerance values at Yi Chang catchment, where $n = 820$, $s = 101$, $m = 8$, and $X_0 = 75000\ \mathrm{m^3/s}$.

Table 2. *List of sample sizes and statistics for two catchments in the UK (the units of $\bar{X}$ and X_0 are m^3/s)*

catchment	n	s	m	e	$\bar{X}$	X_0	C_v	C_s
Avon at Bath	90	32	16	6	200	155	0.43	1.56
Adelphi Weir	73	31	25	15	240	215	0.40	0.48

Table 3. *Comparison of 100-year flood quantile estimates for different models [m^3/s]*

catchment	P3/HMOM	LP3/HMOM	ML/EV1	Proposed
Avon at Bath	378	412	344	379
Adelphi Weir	451	558	490	471

HMOM, P3/HMOM and EV1/ML, in which the EV1/ML procedure is that proposed by Leese (1973) and HMOM is the method of historically weighted moments (USWRC, 1982). Two catchments in the England were selected for the analysis. The annual maximum flood series of these catchments were used as examples in the Flood Studies Report (NERC, 1975, pp. 216–17) and their sample size and statistics are listed in Table 2.

Table 3 only gives the quantile estimates for the return periods equal to 100 at Avon at Bath and at Irwell at Adelphi Weir catchments. For the convenience of visual comparison, the flood quantiles estimated by different models were plotted on EV1 probability paper, in which exceedance Gringorten formula is used (Hirsch, 1987).

$$P_i = \begin{cases} \dfrac{i-0.44}{m+0.12}\,\dfrac{m}{n}; & i=1,\ldots,m \\ \dfrac{m}{n}+\dfrac{n-m}{n}\,\dfrac{i-m-0.44}{s-e+0.12}; & i=m+1,\ldots,g \end{cases} \tag{23}$$

It is shown that the proposed model can fit the real data much more closely than can P3/HMOM, LP3/HMOM and EV1/ML, see Figs. 7 and 8. Another large difference is that the flood quantiles estimated by parametric models increase rapidly with T in the high return periods, particularly for large skewness samples; while the extrapolation of the nonparametric kernel estimator suggests that a possible upper bound exists for a given catchment.

SUMMARY AND CONCLUSIONS

Historical flood peaks and palaeologic information reflect the frequency of large floods and thus should be incorporated into flood frequency analysis. They also can help to judge the adequacy of an estimated flood frequency relation-

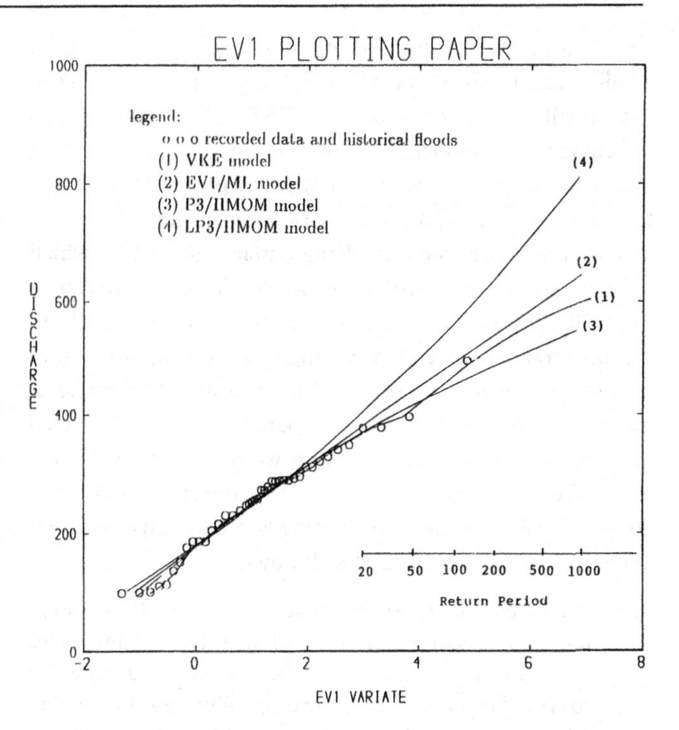

Fig. 7 Comparison of nonparametric quantile estimates with parametric models at Avon at Bath catchment, UK, where $n=90$, $s=32$, $m=16$, $e=6$ and $X_0=200$ m^3/s.

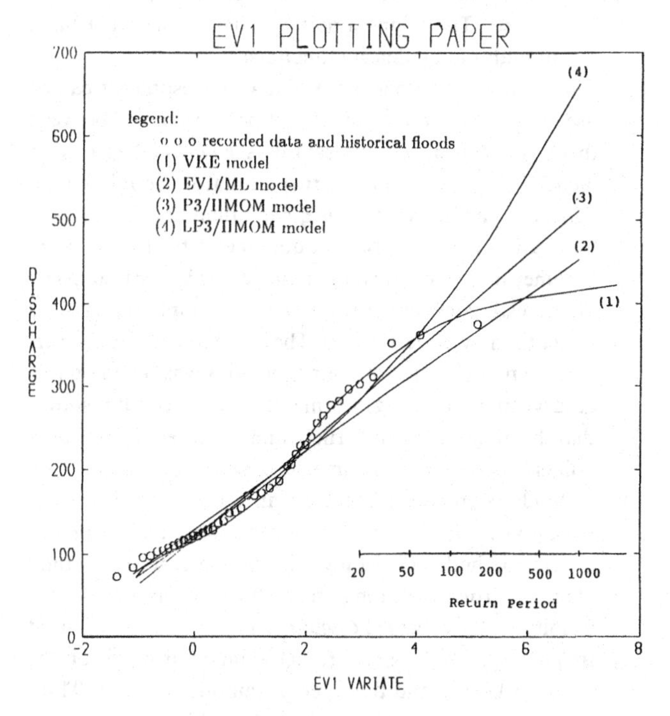

Fig. 8 Comparison of nonparametric quantile estimates with parametric models at Irwell at Adelphi catchment, UK, where $n=73$, $s=31$, $m=25$, $e=15$ and $X_0=240$ m^3/s.

ship. Parametric estimation methods, such as maximum likelihood estimators (Leese, 1973; Stedinger & Cohn, 1986), historically weighted moments (USWRC, 1982) and graphical curve fitting (Hirsch, 1987), are readily applicable to flood frequency estimation when historical or palaeoflood data is available. However, all parametric approaches do need an assumption about the underlying parent distribution which is never known in hydrologic processes. To overcome some of the limitations of the 'parametric' method, there has been a recent trend to develop new nonparametric approaches. In this study, a new nonparametric kernel estimation model is proposed, investigated and compared with some selected parametric methods. The choice between FKE and VKE, the method of estimating smoothing factor as well as the optimal kernel were also discussed. Some useful discussions and conclusions are summarized as follows.

(a) The nonparametric approach does not require the assumption of any particular form of density function. This procedure allows the annual maximum flood series to 'speak for themselves', i.e., the probability density function is directly estimated from the sample. Most of the common 'parametric' models are unimodal, thus representing homogeneous phenomena which is why parametric models sometimes fail to fit annual maximum flood series well especially if the actual histogram is multimodal as can happen in some catchments. The nonparametric model on the other hand can fit multimodal density functions.
(b) Either the fixed kernel or variable kernel estimator can be used to estimate the density function. The choice between them will depend on the sample skewness. It is recommended that the variable kernel estimator should be used when the sample skewness is greater than one.
(c) Simulation results show that quantiles estimated by nonparametric method are more reliable and better than parametric estimates, especially when the sample skewness is larger than two (see Table 1). The comparison of a nonparametric model with parametric methods was based on real data when pre-gauging information is available. It is shown that the nonparametric kernel estimator fitted the real data points closer than its parametric counterparts.
(d) It should be stressed that there is an important difference in the application philosophy of parametric and nonparametric methods. Generally, the nonparametric method places a strong weighting on the flood magnitudes in the vicinity of the specified discharge x. The extrapolation of frequency curve by nonparametric model is based on the shape of kernel, and the value of smoothing factor. Thus, only a few large observations contained in the interval of smoothing factor will influence the extrapolation. On the contrary, the parametric approach uses both historical and systematic information about the annual maxima distribution as a whole. This gives rise to results which run counter to hydrologic intuition. For example, it might be questioned why the addition of a few very large historical floods should cause an adjustment in the LP3 or GEV estimates of 5-year magnitudes. Comments in a similar vein were made by Klemes (1986, Fig. 5).

In summary, the proposed nonparametric kernel estimation model provides an alternative way in flood quantile estimation when historical data or palaeofloods are available.

ACKNOWLEDGEMENTS

The author wishes to record his thanks to Professor C. Cunnane for his supervision. Grateful acknowledgements for general encouragement and help are due to Professors J. E. Nash and Ye Shouze.

REFERENCES

Adamowski, K. (1985) Nonparametric kernel estimation of flood frequency, *Water Resour. Res.*, **21**(11), 1585–90.
Adamowski, K. (1989) A Monte Carlo comparison of parametric and nonparametric estimation of flood frequencies. *J. Hydrol.*, **108**, 295–308.
Baker, V. R. (1987) Paleoflood hydrology and extraordinary flood events. *J. Hydrol.*, **96**, 79–99.
Breiman, L., Meisel, W. & Purcell, E. (1977) Variable kernel estimates of multivariate densities. *Technometrics*, **19**(2), 135–44.
Dooge, J. C. I. (1986) Looking for hydrologic laws. *Water Resour. Res.*, **22**(9), 465–85.
Guo, S. L. (1986) *Flood frequency analysis in Hubei Province, China*, M.Sc. Thesis, National University of Ireland, Galway, Ireland.
Guo, S. L. (1990) *Flood frequency analysis based on parametric and nonparametric statistics*, Ph.D. Thesis, National University of Ireland, Galway, Ireland.
Hirsch, R. M. (1987) Probability plotting position formulas for flood records with historical information. *J. Hydrol.*, **96**, 185–99.
Klemes, V. (1986) Dilettantism in hydrology: – transition or destiny? *Water Resour. Res.*, **22**(9), 177–88.
Klemes, V. (1987) Hydrological and engineering relevance of flood frequency analysis. In Singh, V. P. (Ed.), *Hydrologic Frequency Modelling*, Reidel, Dordrecht.
Leese, M. N. (1973) Use of censored data in the estimation of Gumbel distribution parameters for annual maximum flood series. *Water. Resour. Res.*, **9**(6),1534–42.
Natural Environmental Research Council (NERC) (1975) *Flood Studies Report*, London, England.
Rao, B. L. S. P. (1983) *Nonparametric Function Estimation*, Academic Press, New York, N.Y.
Stedinger, J. R. & Cohn, T. A. (1986) The value of historical and paleoflood information in flood frequency analysis. *Water Resour. Res.*, **22**(5), 785–93.
U.S.W.R.C. (1976, 1977 & 1982) *Guidelines for determining flood flow frequency*, Bulletin 17, Hydrology Committee, Water Resources Council, Washington, D.C.
Whittaker, J. (1973) A note on the generation of gamma random variables with non-integral shape parameter, In: *Floods and Droughts, Proceedings of the Second Symposium on Hydrology*, Water Resources Publication, Fort Collins, Colo., 591–4.

IV

Random fields

1 Analysis of regional drought characteristics with empirical orthogonal functions

I. KRASOVSKAIA

HYDROCONSULT AB, Uppsala, Sweden

L. GOTTSCHALK

Department of Geophysics, University of Oslo, Norway

ABSTRACT An approach to quantify regional meteorological drought is presented. Different definitions and types of drought are discussed and the importance of problem and place related drought criteria is emphasized. The method of empirical orthogonal functions is used for interpolation of drought characteristics: drought area and areal deficit. The method allows quantification of drought characteristics in respect to any chosen drought criteria. The approach is illustrated with one example from Kerala, south-western India.

INTRODUCTION

It is difficult to define precisely what drought is, but in general terms it can be regarded as a condition of 'lack of sufficient water to meet requirements which are dependent on the distribution of plant, animal and human populations, their lifestyle and their use of the land' (Gibbs, 1975). It is obvious from this general definition that an universal quantitative measure of drought does not exist. Traditionally three types of drought are distinguished: meteorological, hydrological and agricultural. Meteorological drought can be defined as a prolonged and abnormal moisture deficiency. Hydrological drought can be thought of as a period during which the actual water supply is less than the minimal water supply necessary for normal operation in a particular region. Agricultural drought is usually described in terms of crop failure and is said to exist when soil moisture is depleted so that the yield of plants is reduced considerably (after Thomas, 1965).

As drought (contrary to flood or rainfall) is a 'non-event', it is impossible to precisely state the date of its onset or end. It is possible to precisely determine the existence of drought when it has already begun. Similarly, the drought does not cease with the first rainfall but disappears gradually, with the rate dependent on drought severity for the particular local conditions. As the effect of a drought is dependent on the concrete problem and local natural and economic conditions, studies of any drought type make sense only in regard to a particular problem and place.

A study of regional drought characteristics has been carried out in connection to a project on water resources assessment in Kerala, Western India. More precisely, the task was to quantify the demand in groundwater to compensate rainfall deficit during meteorological drought in different parts of the region of study. For this specific problem, the regional drought, evaluated from a regional set of data, is more important than point drought, analysed from individual precipitation time series. The former one is directly related to the regional water shortages due to rain deficit and their impacts. Different theoretical approaches characterizing regional drought have been developed and reviewed by Sen (1980), Santos (1983) and Correia (1987). Regional patterns of precipitation can be conveniently described with the help of empirical orthogonal functions (EOF) and this approach is developed and applied here for the determination of regional drought characteristics in Kerala for seasonal data.

METEOROLOGICAL DROUGHT

There are different ways of characterizing meteorological drought conditions. Precipitation analysis with respect to amounts and distribution can be an effective tool to define meteorological drought quantitatively. Meteorological drought can be, for example, subdivided into absolute drought, partial drought and dry spell (Rodda, 1965; Raju *et al.*, 1983), basing on the length of the period during which precipitation is less than a given amount.

Another way to describe meteorological drought is to regard rainfall 'surplus' or 'deficit' with respect to 'normal' rainfall. The latter is assumed to be the mean rainfall for a certain period (month, season or year). For Western India, for example, the even rainfall distribution throughout the year prevails. That is, as the approximation, 1/12th of the annual precipitation value (8.3% of the total annual rainfall) can be assumed as the average monthly value (Raju *et al.*, 1983). The value 8.3% has been chosen on the basis of agroclimatic considerations, as this amount creates identical conditions for the growth of the perennial crops and also provides for adequate water supply for other purposes. Negative deviations from the 'normal' precipitation give the 'deficit'. As stated before, drought criteria should be related to the local conditions. In this study the following drought criteria, suggested by the Indian Meteorological Department, have been used along with long term means:

(a) deficient rainfall (negative deviation of 20–60% from the mean);
(b) scanty rainfall (negative deviation of more than 60% from the mean).

Meteorological droughts can also be described in terms of an 'effective rainfall', which is usually defined as a rainfall sufficient to counteract evaporation and runoff and to maintain soil moisture above the wilting point. According to Indian Meteorological Department the amount of 'effective rain' equals 2.5 mm/day, which gives about 76 mm/month. This criterion has also been used in the study. It is also possible to quantify the concept of meteorological drought in terms of probabilities of lengths of rainless periods and their sequences and a of certain rainfall amount, defined for a particular problem and place.

It can be assumed that life and farming in any region are adapted to the prevailing climatic pattern, so that maximum advantage is taken of the months with high average rainfall. However, variations in precipitation during these 'wet' months can cause drought conditions. On the other hand, almost totally 'dry' months during a certain time of the year are not harmful as agricultural activities are usually well adapted to this pattern. That is why the study puts the stress on the deviations from the normal climatic situation with respect to harmful droughts.

EMPIRICAL ORTHOGONAL FUNCTIONS

The general idea of using empirical orthogonal functions (EOF) is to make a linear transformation of the original data and produce a new, orthogonal set of functions. The method simplifies, and excludes redundant information. The type of analysis used to derive EOFs has much in common with such methods as the principal component analysis or the eigenvector analysis. The theory of expansion into empirical orthogonal functions has been treated by, for example, Holmström (1963) and Obled & Creutin (1986). A short summary of the method is presented below.

Consider a set of time series $Z_i(t), i=1,\ldots,N$ over a time interval (a,b). $Z'_i(t)$ are the corresponding series with respective time averages subtracted. An expansion into EOF has the form:

$$Z'_i(t)=\sum_{n=1}^{M} h_{ni}\beta_n(t), \qquad i=1,\ldots,N \tag{1}$$

where: h_{ni} are weight coefficients (summing to M over $n=1,\ldots,M$) varying between the series but constant in time, and β_n are sets of functions common to all series. These functions will be called amplitude functions.

Requiring fastest possible convergence of the series expansion and adding a normalizing condition to the weight coefficients, we get orthogonal sets of weight coefficients and amplitude functions with the properties:

$$\sum_{i=1}^{M} h_{ni}h_{mi}=\delta_{nm}M \tag{2}$$

and

$$\frac{1}{T}\int_a^b \beta_n(t)\beta_m(t)\,\mathrm{d}t=\delta_{nm}\frac{1}{T}\int_a^b (\beta_n(t))^2\,\mathrm{d}t=\delta_{nm}\lambda_n \tag{3}$$

where: δ_{nm} is the Kronecker delta and λ_n are the eigenvalues of the covariance matrix.

The weight coefficients, h_{ni}, are the elements of the eigenvectors of the covariance matrix. The new set of functions created by this expansion is empirical in the sense that they are based on the series themselves and not restricted to any predetermined polynomial form. Normally the sets of EOFs are arranged in descending order according to the proportion of variance explained by each function. An important property is that a small number of functions reproduce a great part of the total variance. An EOF representation using $M=N$ linearly independent functions is a complete description of the original data.

REGIONAL DROUGHT CHARACTERISTICS

Drought conditions are described by some critical level, say D_0, in accordance with the drought criteria referenced above. The deficit at the point (x,y) at time t is defined by the following non-negative function:

$$D(x,y,t)=\begin{cases} D_0 - Z(x,y,t) \text{ when } Z(x,y,t) \leq D_0 \\ 0 \text{ when } Z(x,y,t) > D_0 \end{cases} \tag{4}$$

Important regional drought characteristics are the total area with deficit $A_D(t)$ and the total areal deficit $D_T(t)$. The deficit area can be simply defined as a portion of the total area determined from the quotient between the number of stations with deficit and the total number (Sen, 1980). Santos (1983) on the other hand did not give equal weight to all stations but in proportion to an estimated effective area for each station.

In the present study the deficit area is determined from the regionally interpolated deficit function as:

$$A_D(t) = \iint_A I_D(x,y,t)\,\mathrm{d}x\,\mathrm{d}y \tag{5}$$

Where $I(x,y)$ is an indicator function defined as:

$$I(x,y,t)=\begin{cases} 1 \text{ when } Z(x,y,t) \leq D_0 \\ 0 \text{ when } Z(x,y,t) > D_0 \end{cases} \tag{6}$$

The total areal deficit is determined from the expression:

$$D_T(t) = \iint_A D(x,y,t)\,\mathrm{d}x\,\mathrm{d}y \tag{7}$$

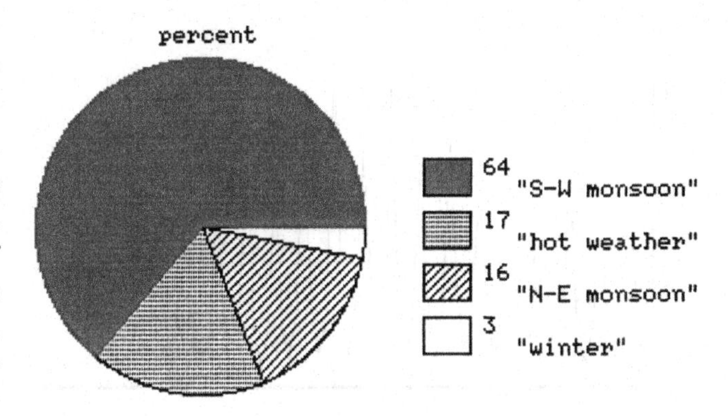

Fig. 1 Seasonal distribution of precipitation.

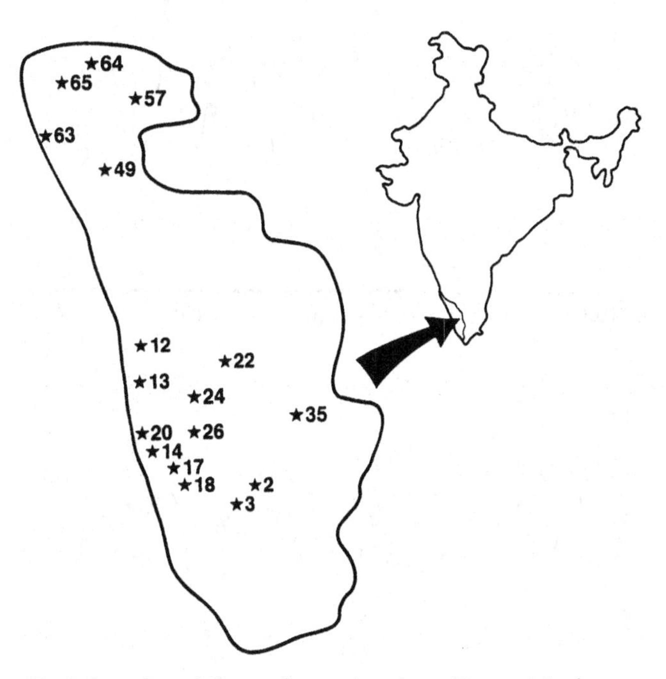

Fig. 2 Location of observation stations (monthly precipitation values for 1951–70).

CASE STUDY

The area of study is situated in the zone of monsoon climate. The concept of monsoon contains two key ideas – regularity and reversal. Strong seasonal contrasts of weather are associated with the mean seasonal circulation. The monsoon period of four months is out of step with conventional three months seasons. Several alternatives can be used instead of the traditional concept of season. In this study we followed a general classification, suggested by Rao (1981):

(a) Season 1 'Winter' – January–February;
(b) Season 2 'Hot weather season' ('spring') – March–May;
(c) Season 3 'Southwest monsoon period' ('summer') – June–September;
(d) Season 4 'Northeast monsoon period' ('autumn') – October–December.

There is a clear dominance of monsoon in the climatic pattern of the region. The amount of rain bound to the southwest monsoon (see Fig. 1) is one order of magnitude more than during the rest of the year, which makes both annual and monsoon isohyets look rather similar. The diagram in Fig. 1 is based on the analysis of data for 17 precipitation stations, whose location is given in Fig. 2.

The mean annual amount of precipitation in the region varies between 2000 mm and 4000 mm, with values decreasing in the direction to the north. The average amount of rainy days is 125. The annual precipitation sums can vary up to 30%, with the average about 20% (based on the coefficient of variation). The variability of rain amounts depends to a great extent on the variation in the duration and continuity of the southwest monsoon. The monsoon rains vary considerably not only from year to year, but also in the intensity and distribution within one event. The dry spells that interrupt the rainy weather vary in duration and frequencies. These variations have an important role due to the great effect of the southwest monsoon on human life and economy.

The EOF-analysis of seasonal precipitation data showed that 95% of variation is already explained by the first three amplitude functions. The first amplitude function describes the mean variation pattern for the whole region, while the second seems to describe a delay in the onset of the monsoon period (Fig. 3). The weight coefficients for seasonal data have

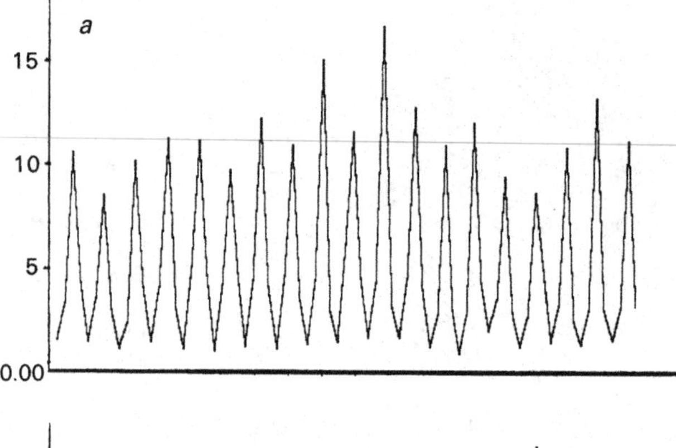

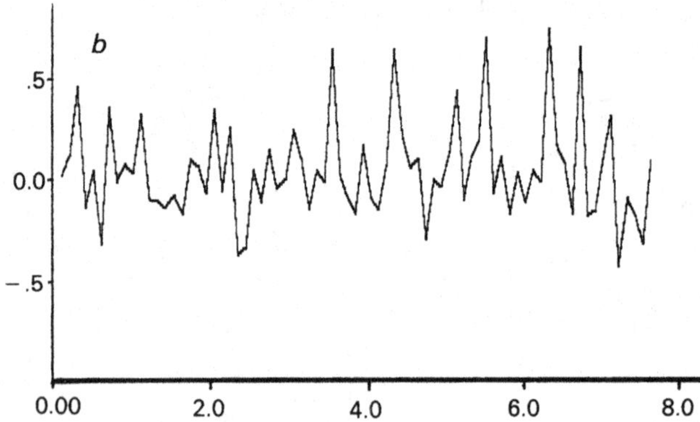

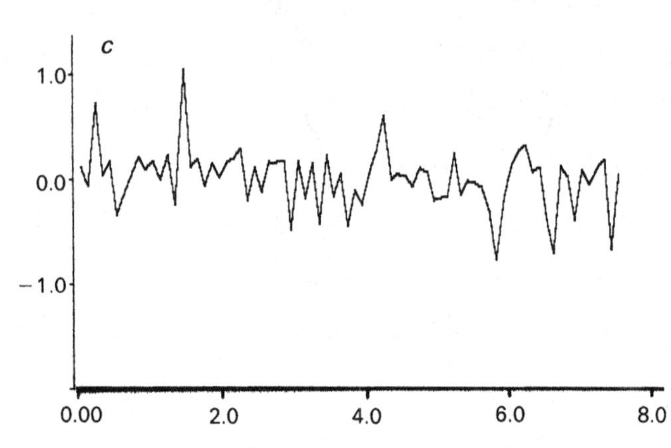

Fig. 3 The first three amplitude functions for seasonal data; (a), (b), (c).

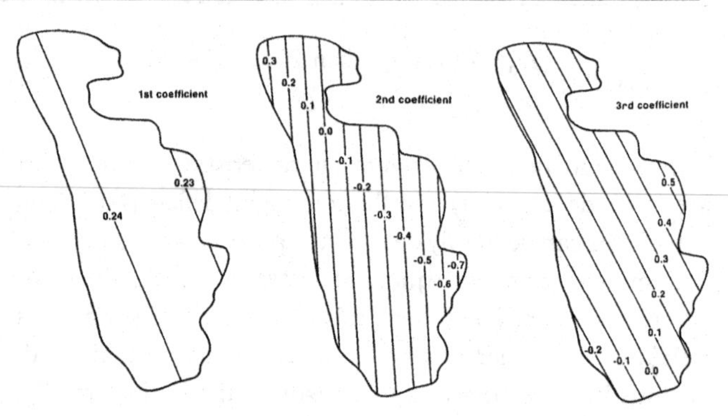

Fig. 4 Trend surfaces for the first three weight coefficients.

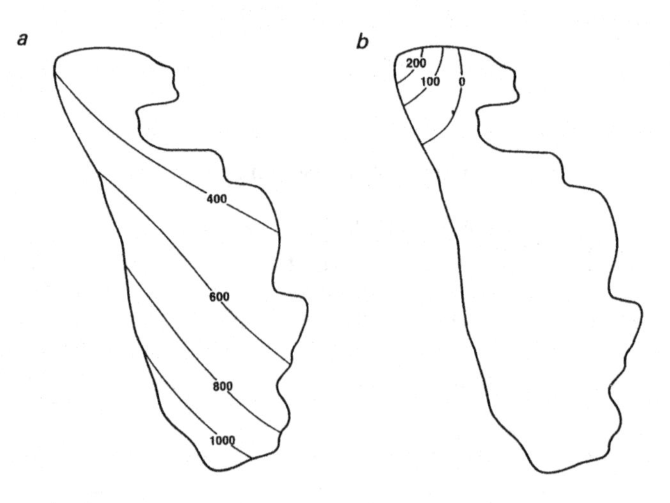

Fig. 5 Area with deficit and mean deficit for south-west monsoon season for: (a) dry year (1966), area with deficit 24 000 km^2, mean deficit 612 mm; and (b) wet year (1968), area with deficit 1336 km^2, mean deficit 121 mm. Drought criterion: long term mean ($D_0 = m(x,y)$).

been subject to trend surface analysis in the search for the trend pattern in the region (Fig. 4). They vary smoothly over the region of study, hence they can be considered to be functions of spatial coordinates $h = h(x,y)$. In absolute values the second coefficient gives the dominant influence. The regional pattern reflects the approach of the monsoon from south-west towards north-east.

Inserting equation (1) of the EOF-model into equation (4), one gets a model for a description of time-space variation of regional drought deficit:

$$\hat{D}(x,y,t) = \begin{cases} D_0 - \hat{Z}(x,y,t) \text{ when } \hat{Z}(x,y,t) \leq D_0 \\ 0 \text{ when } \hat{Z}(x,y,t) > D_0 \end{cases} \quad (8)$$

where

$$\hat{Z}(x,y,t) = m(x,y) + \sum_{n=1}^{M} h_n(x,y)\beta_n(t) + \varepsilon \quad (9)$$

and $m(x,y)$ is the mean value at the point (x,y), and ε is a random error caused by the truncation of the series expansion ($M < N$). Three amplitude functions will be used in the sequel ($M = 3$). The individual precipitation series can be considered to be independent in time. This fact allows us to further simplify the model equation (9) to the following expression:

$$\hat{Z}(x,y) = m(x,y) + \sum_{n=1}^{M} h_n(x,y)\beta_n + \varepsilon \quad (10)$$

where $\beta, n = 1, \ldots, M$ are independent stochastic variables in the time domain. $h_n(x,y)$ and $m(x,y)$ are deterministic functions in space. In the present study point kriging is applied for spatial interpolation of these functions. Figs. 5a and 5b show examples of calculations of the regional deficit for two

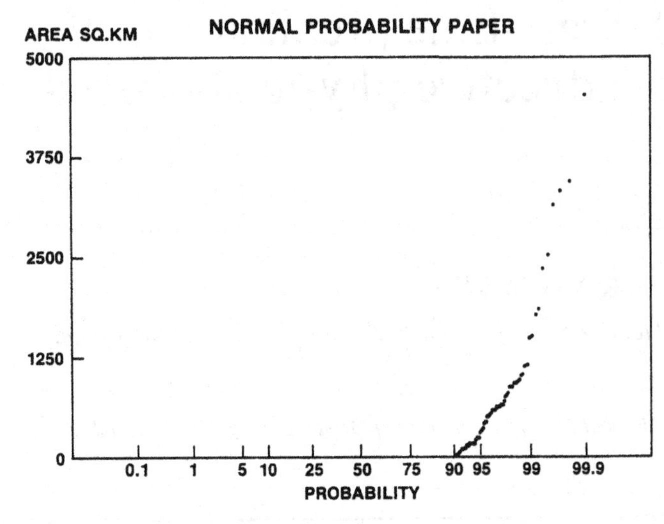

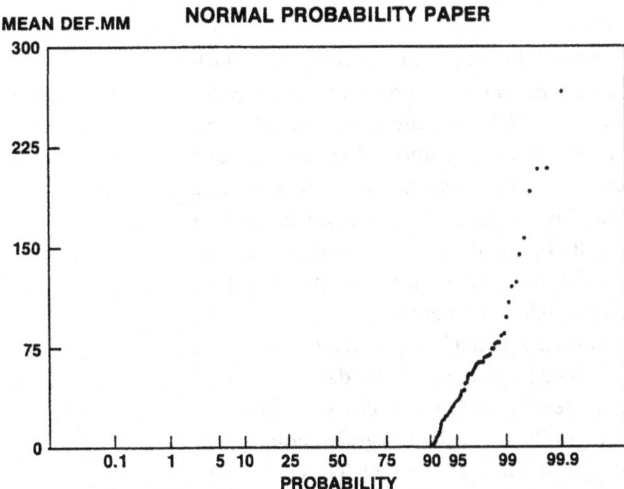

Fig. 6 Probability distribution function for the area with deficit and mean deficit for this area for season 3 (June–Sept.). Drought criterion: deficient rain ($D_0 = 1563$ mm).

drought criteria: the long term mean value for a season ($D_0 = m(x, y)$) and the 'effective rain' ($D_0 = 76$ mm $*$ length of the season in months). For comparison two years have been chosen: a very dry, shown as (a) in the figure, and a very wet, shown as (b).

The elements of individual amplitude functions β_n are independent and therefore uniquely described by one dimensional distribution functions as well as the error term ε. This allows us to derive the probability distribution functions for the drought characteristics A_D and D_T defined from equations (5) and (7) by means of Monte Carlo methods. (For details see Gottschalk & Krasovskaia, 1987).

The calculated probability distribution functions (see an example in Fig.6) show different patterns. The spatial variations (mainly a north-south trend) for seasons 2 and 4 are of the same order of magnitude as the variability between the years. In this case there are always parts of coastal Kerala that receive less precipitation than the critical level (those used in the study), but the probability that the whole territory is below the critical level is negligible. Only for the 'scanty rain' criterion there is a 3% probability that some part receives less precipitation during the fourth season than this critical amount. For southwest monsoon season the fluctuations between years are much bigger than the spatial variability. There is as much as 30% probability that none of the territory has precipitation below the long term seasonal mean. For 'deficient rain'-criterion this probability is 90%.

CONCLUSIONS

An approach to quantification of regional meteorological drought based on the concept of Empirical Orthogonal Functions (EOF) is presented. The EOF technique allows one to interpolate the precipitation data in the region and also the values of the deficit function, based on the precipitation amounts. This gives a possibility to quantify the drought characteristics used for the region of study, that is the amount of deficit D_T (in mm) and the area affected by drought (experiencing this deficit) A_D (in km^2). The choice of the drought characteristics was guided by the local climatological conditions, agricultural techniques and the task to quantify the demand for supplementary use of ground water in case of drought. An example is given of the use of the amplitude functions obtained by means of EOF techniques for derivation of the probability distribution functions for the drought characteristics D_T and A_D with the help of Monte Carlo methods.

REFERENCES

Correia, F. N. (1987) Engineering risk in regional drought studies, In: Duckstein, L. & Plate, E. (Eds.), *Engineering Reliability and Risk in Water Resources*, NATO ASI Series.

Gibbs, W. J. (1975) *Drought – its definition, delineation and effects*, WMO, Special Environmental Report No 5, WMO No 403.

Gottschalk, L. & Krasovskaia, I. (1987) *Drought conditions in coastal Kerala, Coastal Kerala Ground Water Project*, Government of India, Central Ground Water Board/Sida.

Holmström, I. (1963) On a method for parametric representation of the state of the atmosphere. *Tellus*, **15**, 127–49.

Obled, Ch. & Creutin, J. D. (1986) Some developments in the use of empirical orthogonal functions for mapping meteorological fields. *Journal of Climate and Applied Meteorology*, **25**(9), 1189–204.

Raju, K. C. B. *et al.* (1983) *Ground water resources of Noyil Ponnani and Vattamalai Karai basins*, Government of India, Central Ground Water Board, Ministry of Irrigation

Rao, Y. P. (1981) The climate of the Indian subcontinent in *World survey of climatology, Vol. 9: Climates of Southern and Western Asia*, Elsevier, Amsterdam.

Rodda, J. C. (1965) A drought study in southeast England. *Wat. & Wat. Engr.*, **69**, 316–21.

Santos, M. A. (1983) Regional droughts: A stochastic characterization. *J. Hydrol.*, **66**, 183–211.

Sen, Z. (1980) Regional drought and flood frequency analysis: theoretical consideration. *J. Hydrol.*, **46**, 265–79.

Thomas, H. E. (1965) Reality of drought is always with us. *Natural History*, **74**, 50–62.

2 Worth of radar data in the real-time prediction of mean areal rainfall by nonadvective physically-based models

K. P. GEORGAKAKOS* and W. F. KRAJEWSKI

Department of Civil and Environmental Engineering and Iowa Institute of Hydraulic Research, The University of Iowa, Iowa City, Iowa 52242–1585, USA

**Now: Hydrologic Research Center, San Diego, California, and Scripps Institution of Oceanography, UCSD, La Jolla, California, USA*

ABSTRACT The utilization of operationally available radar data for improved short-term predictions of mean areal rainfall on hydrologic scales can be accomplished by the use of a physically-based spatially-lumped rainfall prediction model. The state-space form of such a model admits covariance estimation algorithms for the determination of rainfall forecast variance. In particular, when the model is linear in the state, covariance analysis can be performed without the use of radar reflectivity data. Covariance analysis of a particular linear physically-based model indicates that the utility of the radar reflectivity data of various elevation angles is limited in mean areal rainfall predictions, even when a very small density of rain gauges exists over the region of interest and good quality radar data are used. This applies to both raw reflectivity and radar-rainfall data converted through a *Z–R* relationship. The ratio of mean areal rainfall prediction variances, defined as variance with radar data divided by variance without radar data, was found to be greater than 0.8 in most cases. On the other hand, the radar data reduced the estimated variance of the vertically-integrated liquid water content considerably, even when high density rain gauge data were present. The conclusions of this study are representative of covariance analyses procedures that require linear or linearized rainfall prediction models and, for such procedures, are independent of the particular model used. On the other hand, the model used is a spatially-lumped model and can not utilize information on storm velocity offered by the radar data time series. Extensions to two-dimensional stochastic-dynamic formulations are proposed.

INTRODUCTION

Forecasting mean areal rainfall accurately and reliably on a basin scale and in real time has been one of the pressing needs of hydrology. Recently, it has been addressed in review papers of Georgakakos & Hudlow (1984) and Georgakakos & Kavvas (1987). They discuss the suitability of spatially-lumped quantitative rainfall prediction models developed by Georgakakos & Bras (1984a, b) and Georgakakos (1984, 1986c) for use in real-time flood forecasting. The models were developed in an effort to improve short-term precipitation predictions on the scale of small- and medium-size hydrologic basins (100 to 1000 km^2). They have already been coupled to hydrologic models (Georgakakos, 1986a, b) to form integrated hydrometeorological forecast systems for the real-time prediction of floods and flash floods (Georgakakos, 1987). The models, as originally formulated, use as input point data of readily available variables such as surface air temperature, dew point temperature and pressure. Real-time observations of rain gauge rainfall are used in model state updating. It seems reasonable to expect improvements in model predictions if additional information on the relevant processes could be included. In particular, further improvements are expected if real time observations of radar reflectivity were used.

The rainfall prediction models are based on the principle of the conservation of liquid water mass and utilize adiabatic and pseudo-adiabatic air-parcel ascent processes for the determination of the condensation and deposition source term. Computations of cloud precipitation rate are based on parameterizations of cloud micro-physics. Evaporation of cloud drops in the sub-cloud layer of unsaturated air is also modeled. Those models predict hourly precipitation rates given input in the form of surface air temperature, pressure and dew-point temperature. The models have been formulated in state-space form and state estimators have been designed to process mean areal rainfall observations in real time for: 1) state updating and 2) determination of forecast uncertainty. Tests of the models with field data showed considerable skill in predicting rainfall with hourly forecast lead times.

This work is an attempt to quantify the utility of radar data in the real time prediction of mean areal rainfall on hydrologic scales down to 100 km^2. The spatially-lumped stochastic-dynamical rainfall prediction model of Georgakakos & Bras (1984a, b) is used in the following to integrate rainfall microphysics and dynamics with uncertainty measures of observations from various sensors with the purpose to determine the reduction of rainfall prediction variance attained by the real-time utilization of radar data. Linear covariance analysis is applied that does not require observed radar rainfall. Such a type of analysis is attractive since good radar reflectivity data of sufficient quantity for covariance analysis are difficult to obtain. This situation will change soon with deployment in the 1990s of a network of more than 150 NEXRAD (Next Generation Weather Radar) radars in the United States and Western Europe (Leone *et al.*, 1989).

In this paper both steady state and unsteady covariance measures are presented, with the unsteady measures being applicable to particular storms and the steady state measures being more generally applicable as long-term covariance averages. Both vertically-averaged reflectivity data and low-elevation angle reflectivity data converted to rainfall through a standard *Z–R* relationship are considered. The covariance analysis performed assumes only random errors in the observed data and does not account for potential systematic errors in radar data (e.g., due to solid precipitation effects on reflectivity). Therefore, a preprocessing of the radar data is assumed (for NEXRAD precipitation processing system see Ahnert *et al.*, 1983, and Hudlow *et al.*, 1983) such as the one planned for NEXRAD. Analyses such as the one presented in this paper are necessary if the deployment of the NEXRAD radars is to bring significant improvements to hydrologic and hydrometeorological forecasting practices.

Following the mathematical formulation of the next section, the steady state assessment is presented later, followed by the unsteady covariance analysis. Conclusions and recommendations for further research are also given.

MATHEMATICAL FORMULATION

Covariance Analysis

Consider the discrete form of the dynamic equation of the Georgakakos & Bras (1984 a, b) precipitation forecasting model for sampled input data:

$$x_{k+1} = x_k + f_k - h_k x_k + w_k \quad \text{for } k = 0,1,2,\ldots \tag{1}$$

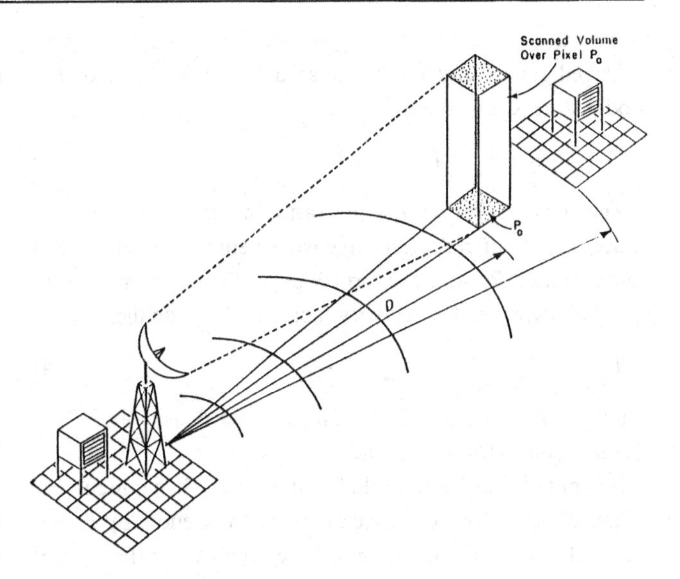

Fig. 1 Schematic of radar coverage of the atmospheric column above the area where mean areal surface rainfall predictions are sought.

where the state x_k represents the condensed water equivalent mass in a cloud column composed of both cloud and rain water at time t_k, f_k is the mass input to the column due to condensation over the interval $\Delta t = t_{k+1} - t_k$, and $h_k x_k$ is the mass output due to precipitation at cloud base over the same interval. The stochastic process w_k represents random errors in model structure, input and parameter estimates. It is a zero mean process with variance Q. It is noted that f_k and h_k differ from the analogous quantities in Georgakakos & Bras (1984 a, b) in that they represent the total mass input and output over the period Δt rather than instantaneous input and output mass rates. For the purposes of this work $\Delta t = 1$ hour.

The corresponding observation equation for mean areal surface rainfall is

$$R_k = \Phi_k x_k + v_{r_k} \quad \text{for } k = 1,2,\ldots \tag{2}$$

where R_k is the rainfall amount observed at ground level over the period $(t_k - t_{k-1})$, $\Phi_k x_k$ is the surface rainfall accumulation predicted by the Georgakakos and Bras model for the same period, and v_{r_k} represents the random sequence of observation errors, with zero mean and variance R_r. It is noted that the mean areal surface rainfall observation R_k could be based only on rain-gauge observations, on base scan (low-tilt angle) radar rainfall observations, or on both of the above sensor data. Such situations should be reflected in the choice of a value for R_r.

For the locations under a meteorological-radar coverage an additional observation may be available. This observation is the radar-measured raindrop reflectivity. For illustration purposes, Fig. 1 presents a schematic of the radar coverage of the atmospheric column above the area where mean areal rainfall predictions are sought. Appendix A (see also Georgakakos & Krajewski, 1989) derives a second observation equation that gives the observed vertically-averaged reflectivity factor Z_k as a linear function of the model state x_k at time t_k:

$$Z_k = \Psi_k x_k + v_{z_k} \quad \text{for } k=1,2,\ldots \tag{3}$$

The term v_{z_k} represents the random sequence of observation errors in reflectivity measurements with zero mean and variance R_z, and the term $\Psi_k x_k$ is the model-predicted vertically-averaged reflectivity factor with Ψ_k defined as

$$\Psi_k = (720\,\varepsilon_4^3 \alpha^7)/(\pi \rho_w Z_{c_k}) \tag{4}$$

where ε_4 and α are parameters, ρ_w is water density and Z_{c_k} is cloud depth in [m] at time t_k.

It is noted that there could be situations when concurrent values of v_{r_k} and v_{z_k} are correlated, and we denote by ρ_{rz} the relevant correlation coefficient. One such case is the case of utilizing base scan radar data for the determination of R_k.

Given the dynamics, parameters and the parameters of the noise statistics, as well as initial conditions on x_k, it is possible (e.g. Gelb, 1974, Jazwinski, 1970) to use state estimators to obtain the state prediction variance and the state estimation (or updating when observations become available) variance of the stochastic-dynamical model in real time. The state covariance propagation and updating for the above formulation is accomplished by the following set of recursive equations.

Propagation

$$P_{k+1}^- = P_k^-(1-h_k)^2 + Q \tag{5}$$

Updating

$$P_{k+1}^+ = P_{k+1}^- - K_{k+1} H_k P_{k+1}^- \tag{6}$$

where P_k^+ is the estimated state variance at time t_k given observations up to and including time t_k; P_{k+1}^- is the estimated state variance at time t_{k+1}, given observations before time t_{k+1}; K_{k+1} is the estimator gain matrix at time t_{k+1}; and

$$H_k = \begin{bmatrix} \Phi_k \\ \Psi_k \end{bmatrix} \tag{7}$$

Gelb (1974) gives:

$$K_{k+1} = P_{k+1}^- H_k^T [H_k P_{k+1}^- H_k^T + R_v]^{-1} \tag{8}$$

where R_v is defined by

$$R_v = \begin{bmatrix} R_r & \rho_{rz}(R_r R_z)^{1/2} \\ \rho_{rz}(R_r R_z)^{1/2} & R_z \end{bmatrix} \tag{9}$$

In case vertically averaged reflectivity data are not available, the matrices H_k and R_v become scalars equal to Φ_k and R_r, respectively. Then, R_r is the variance of observation error associated with rain gauge data alone, with base scan radar data, or with both types of data.

Substituting for the elements of H_k, R_{k+1}^-, and R_v in equations (6) and (8) yields:

$$P_{k+1}^+ = \frac{P_{k+1}^- R_r R_z (1-\rho_{rz}^2)}{\psi_k^2 R_r P_{k+1}^- + \Phi_k^2 R_z P_{k+1}^- + R_r R_z (1-\rho_{rz}^2) - 2\rho_{rz}\psi_k \Phi_k P_{k+1}^- (R_r R_z)^{1/2}} \tag{10}$$

Dividing both numerator and denominator by $R_r R_z P_{k+1}^-$ one obtains

$$P_{k+1}^+ = (1-\rho_{rz}^2)\left[\frac{\psi_k^2}{R_z} + \frac{\Phi_k^2}{R_r} + \frac{1-\rho_{rz}^2}{P_{k+1}^-} + \frac{2\rho_{rz}\Psi_k\Phi_k}{(R_r R_z)^{1/2}}\right]^{-1} \tag{11}$$

When vertically averaged reflectivity data are not available, Equation (11) still holds true but with the terms containing ρ_{rz} and Ψ_k set to zero. Given expressions for h_k, Φ_k and Ψ_k, estimates of the model-error variance Q and estimates of the observation-error variances R_r and R_z, one can use equations (5) and (11) to obtain predicted and updated estimates, P_{k+1}^- and P_{k+1}^+, of the state variance for all times t_{k+1} during a storm. The reader is reminded that in the present model the state is the condensed water equivalent mass in the cloud, and it consists of both rainwater and cloudwater. Using the observation Equation (2) we can obtain an expression for the error variance in the surface rainfall predictions, Σ_{k+1}, for all k by

$$\Sigma_{k+1} = \Phi_k^2 P_{k+1}^- \tag{12}$$

Rainfall variance sensitivity

One way to examine the sensitivity of the rainfall prediction error variance Σ_k with respect to R_r and R_z is to study the time variation of the normalized derivatives $S_R = (R_r/\Sigma_k)(\partial\Sigma_k/\partial R_r)$ and $S_z = (R_z/\Sigma_k)(\partial\Sigma_k/\partial R_z)$ (e.g., Rabitz, 1989). In the following, expressions for the derivatives $(\partial\Sigma_k/\partial R_r)$ and $(\partial\Sigma_k/\partial R_z)$ are obtained using the covariance propagation and updating equations.

Taking derivatives of both sides of Equation (12) with respect to R_r and R_z yields:

$$\frac{\partial\Sigma_{k+1}}{\partial R_r} = \Phi_k^2 \frac{\partial P_{k+1}^-}{\partial R_r} \tag{13}$$

$$\frac{\partial\Sigma_{k+1}}{\partial R_z} = \Phi_k^2 \frac{\partial P_{k+1}^-}{\partial R_z} \tag{14}$$

Then, using Equation (5) we can obtain the following forms for the right-hand side derivatives:

$$\frac{\partial P_{k+1}^-}{\partial R_r} = (1-h_k)^2 \frac{\partial P_k^+}{\partial R_r} \tag{15}$$

$$\frac{\partial P_{k+1}^-}{\partial R_z} = (1-h_k)^2 \frac{\partial P_k^+}{\partial R_z} \tag{16}$$

Finally, the derivatives in the right-hand side of (15) and (16) can be obtained from equation (11) as:

$$\frac{\partial P_k^+}{\partial R_r} = (1-\rho_{rz}^2)\left[\frac{\Phi_{k-1}^2}{R_r^2} + \frac{(1-\rho_{rz}^2)}{(P_k^-)^2}\frac{\partial P_k^-}{\partial R_r} - \frac{\rho_{rz}\Psi_{k-1}\Phi_{k-1}}{R_r(R_rR_z)^{1/2}}\right]$$
$$\left[\frac{\Psi_{k-1}^2}{R_z} + \frac{\Phi_{k-1}^2}{R_r} + \frac{1-\rho_{rz}^2}{P_k^-} - \frac{2\rho_{rz}\Psi_{k-1}\Phi_{k-1}}{(R_rR_z)^{1/2}}\right]^{-2} \quad (17)$$

$$\frac{\partial P_k^+}{\partial R_z} = (1-\rho_{rz}^2)\left[\frac{\Psi_{k-1}^2}{R_z^2} + \frac{(1-\rho_{rz}^2)}{(P_k^-)^2}\frac{\partial P_k^-}{\partial R_z} - \frac{\rho_{rz}\Psi_{k-1}\Phi_{k-1}}{R_r(R_rR_z)^{1/2}}\right]$$
$$\left[\frac{\Psi_{k-1}^2}{R_z} + \frac{\Phi_{k-1}^2}{R_r} + \frac{1-\rho_{rz}^2}{P_k^-} - \frac{2\rho_{rz}\Psi_{k-1}\Phi_{k-1}}{(R_rR_z)^{1/2}}\right]^{-2} \quad (18)$$

The set of equations (15) and (17) form a recursive set which can be used to obtain $\partial P_{k+1}^-/\partial R_r$ for all k. Similarly, equations (16) and (18) can be used to obtain $\partial P_{k+1}^-/\partial R_z$. Notice that Equations (17) and (18) depend on P_k^-. They should be simulated together with the covariance propagation and updating equations. As for initial conditions to start the iterations, one can use

$$\partial P_0^+/\partial R_r = 0 \quad (19)$$

and

$$\partial P_0^+/\partial R_z = 0 \quad (20)$$

Steady state covariance analysis

An approximate assessment of the contribution of radar data to the prediction and estimation of rainfall can be made by specializing the covariance propagation and updating equations for a hypothetical steady state. The advantage of the simplification is that the results are applicable generally and do not pertain to individual storms. Thus, they offer added insight into the question of radar utility for forecasting, in real time, mean areal rainfall. Assuming characteristic values h, Φ, and Ψ for h_k, Φ_k and Ψ_k for an extended period of time, and $\rho_{rz}=0$, one can rewrite equations (5) and (11) as:

$$P_{k+1}^- = P_k^+(1-h)^2 + Q \quad (21)$$

and

$$P_{k+1}^+ = (\psi^2/R_z + \Phi^2 R_r + 1/P_{k+1}^-)^{-1} \quad (22)$$

For the given assumptions and for time invariant Q, R_z and R_r, it is known (e.g., Gelb, 1974) that the system of equations (21) and (22) reaches a steady state with $P_k^+ \Rightarrow P_\infty^+$ and $P_k^- \Rightarrow P_\infty^-$, for k large enough. Using equations (21) and (22) and denoting by 1/B the first two terms in the denominator of equation (22) we obtain:

$$P_\infty^- = (1-h)^2(1/B + 1/P_\infty^-)^{-1} + Q \quad (23)$$

which can be solved for the positive root:

$$P_\infty^- = [(B-AB-Q)^2 + 4BQ)^{1/2} - (B-AB-Q)]/2 \quad (24)$$

with A denoting $(1-h)^2$. The steady state value of the estimation (or updating) covariance P_∞^+ is obtained by substitution of P_∞^- from equation (24) in equation (22).

In the absence of vertically-averaged reflectivity data, Z_k, equation (22) reduces to:

$$P_{k+1}^+ = (\Phi^2/R_r + 1/P_{k+1}^-)^{-1} \quad (25)$$

and the steady state value $P_{\infty,r}^-$ for this case is given by

$$P_{\infty,r}^- = [(B_r - AB_r - Q)^2 + 4B_rQ)^{1/2} - (B_r - AB_r - Q)]/2 \quad (26)$$

with $1/B_r$ denoting the first term in the denominator of equation (25).

The ratio $R_P = P_\infty^-/P_{\infty,r}^-$ is a measure of the improvement in state and, via Equation (12), rainfall prediction accuracy gained by using vertically averaged reflectivity data over the rainfall model updated with only surface rainfall data. The ratio $R_P = P_\infty^+/P_{\infty,r}^+$ is a measure of the improvement in state estimation (or updating) accuracy gained by using reflectivity data. Such ratios can be computed for various values of h, Φ, Ψ, Q, R_r and R_z and an assessment of the worth of reflectivity data for real-time mean areal rainfall prediction and estimation of rain and cloud water content can be made.

Using dimensionless quantities, one can denote by F_r the ratio of the standard deviation of the measurement noise v_r to the average value of the observations of surface rainfall $\bar{R}_k$ during the steady state period; F_z the ratio of the standard deviation of v_z to the average value of the observations of vertically-averaged reflectivity $\bar{Z}_k$ during the steady state period; F_q the ratio of the standard deviation of the model error to the average value of the model state $\bar{x}_k$ during the steady state period. It follows that

$$R_r = (F_r \phi \bar{x}_k)^2 \quad (27)$$
$$R_z = (F_z \Psi \bar{x}_k)^2 \quad (28)$$
$$Q = (F_q \bar{x}_k)^2 \quad (29)$$

Also,

$$1/B = (1/F_z^2 + 1/F_r^2)/\bar{x}_k^2 \quad (30)$$

and

$$1/B_r = 1/(F_r^2 \bar{x}_k^2) \quad (31)$$

Then, the ratios $P_\infty^-/P_{\infty,r}^-$ and $P_\infty^+/P_{\infty,r}^+$ become functions of only F_r, F_z, F_q and h. Note that F_r and F_z can be interpreted as coefficients of variation of the rainfall and vertically-averaged reflectivity observations.

As an example, we present the limiting case of $h=1$ which corresponds to a 100% depletion of the cloud water. In such a case,

$$P_\infty^- = P_{\infty,r}^- = Q \quad (32)$$

with

$$P_{\infty}^{-}=P_{\infty,r}^{-}=1 \tag{33}$$

Also

$$P_{\infty}^{+}=\frac{\bar{x}_k^2}{(F_r^2)^{-1}+(F_z^2)^{-1}+(F_q^2)^{-1}} \tag{34}$$

$$P_{\infty,r}^{+}=\frac{\bar{x}_k^2}{(F_r^2)^{-1}+(F_q^2)^{-1}} \tag{35}$$

$$P_{\infty}^{+}/P_{\infty,r}^{+}=\left[\frac{(F_z^2)^{-1}}{(F_r^2)^{-1}+(F_q^2)^{-1}}\right]^{-1} \tag{36}$$

It can be seen that for a given F_q, as F_z increases for a given F_r, the variance ratio tends to 1 indicating no utility for the vertically-averaged reflectivity data. However, as F_r increases for a given F_z, the utility of the radar data increases. Also, notice that as F_q increases so is the potential utility of the radar data since there is a larger number added to 1 in the denominator of the last relationship. In a following section, various cases of steady state variance ratios are presented and discussed.

The formulation of equations (24), (26), (27) and (29) can also be used to assess the utility of base scan radar data in obtaining measurements of surface mean areal rainfall as a substitute for or in conjunction with rain gauge data. In such a case, B in equation (24) would be computed as in equation (30) with the first term neglected and F_r representing errors in mean areal rainfall observations due to errors in base scan radar data. B_r would be computed as in Equation (31) with F_r representing errors in mean areal rainfall as computed from point rain gauge data.

STEADY STATE ASSESSMENT

Several useful results can be obtained utilizing the steady state concepts of the last section. It is noted that even though a true steady state is infrequently reached during a storm period (for evidence of the existence of steady rain periods see Kessler, 1969), steady state covariance analysis may be successfully employed to determine relative magnitudes of covariances for periods characterized by certain quasi-steady state behavior (e.g., intense rainfall periods, light rainfall periods). Also, steady state analysis is a general analysis in that it does not depend on particular time series of meteorological variables that correspond to only a few storms.

Low-Tilt Angle Radar Data

At first, we examine the worth of the low-tilt angle radar data in the prediction of mean areal surface rainfall. In this case, the state space form of the model consists of equations (1) and (2), where v_{r_k} represents the random error in the measurement of mean areal surface rainfall. We distinguish two situations.

(a) Only rain gauges have been used to obtain the measurement of mean areal surface rainfall. (This situation is referred to as situation (a) in the following and will be denoted with an appropriate index.)

(b) Only low-tilt angle radar data have been used to obtain the measurement of mean areal rainfall, utilizing a standard Z–R relationship of the type (e.g. Burgess & Ray, 1986):

$$Z=c_1R^{c_2} \tag{37}$$

with c_1 and c_2 being known constants, Z denoting the radar reflectivity factor in $\mathrm{mm}^6/\mathrm{m}^3$ and R denoting the mean areal rainfall rate in mm/h. (This situation is referred to as situation (b) in the following and will be denoted with an appropriate index.)

In both situations, v_{r_k} in equation (2) represents random errors (vs. biases) in the measurement of mean areal surface rainfall.

Steady state analysis of the type developed in the previous section yields equation (26) for the steady state predicted state variance with B_r denoting R_{r_a}/Φ^2 for situation (a), and R_{r_b}/Φ^2 for situation (b). R_{r_a} is the observation error variance associated with the determination of mean areal rainfall by rain gauge data. R_{r_b} is the analogous quantity for the situation (b) involving low-tilt angle radar data. The rest of the quantities in equation (26) are as previously defined.

Denote by R_P the ratio of the steady state rainfall prediction variances with and without low-tilt angle radar data. Equation (12) suggests that R_P is also equal to the ratio of the steady-state state prediction variances:

$$R_P=P_{\infty_b}^{-}/P_{\infty_a}^{-} \tag{38}$$

where subscripts a and b distinguish the aforementioned two situations. By letting r_a approach infinity, we can derive a lower bound, R_P^*, for R_P. This limiting case corresponds to non-existent or very bad quality rain gauge data and, thus, in most cases, values of R_P are expected to be considerably greater than R_P^*.

Using equations (25) and (21) for $1/R_{r_a}=0$, gives

$$P_{\infty_a}^{-}=Q/(1-A) \tag{39}$$

Substituting equations (29) and (31) in equation (26) for situation (b), and equation (29) in equation (39) one obtains for R_P^*:

$$R_P^*=\frac{1-A}{2}\,\frac{\{[1-A-F_q^2/(F_{r_b}^2)]^2+4F_q^2/(F_{r_b}^2)\}-[1-A-F_q^2/(F_{r_b}^2)]}{F_q^2/(F_{r_b}^2)} \tag{40}$$

Thus, the limiting ratio R_P^* depends on h (through A) and on the ratio F_q/F_{r_b}. Fig. 2 presents the ratio R_P^* as a function of

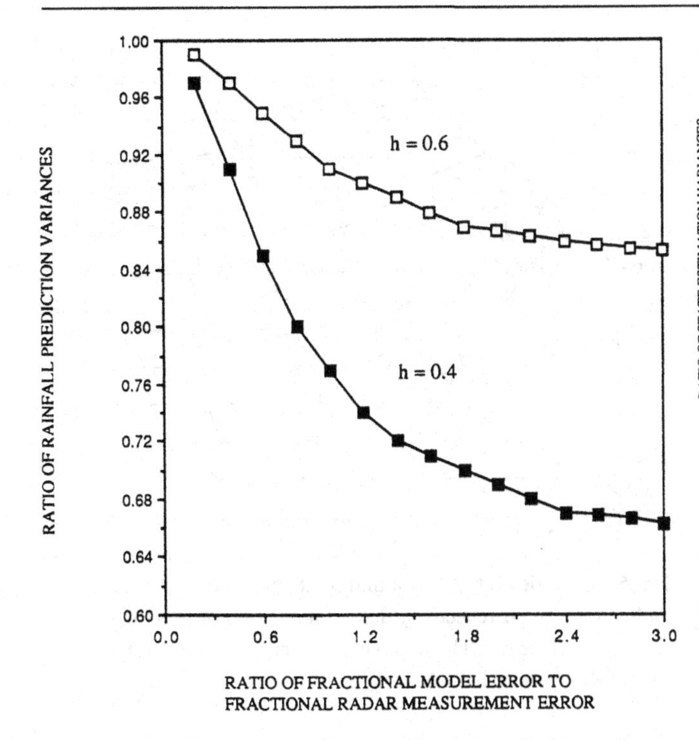

Fig. 2 Minimum values of the ratio of rainfall prediction variances with and without low-tilt angle radar reflectivity data as function of the ratio of fractional model error to fractional radar measurement error for moderate ($h=0.6$) and light ($h=0.4$) rainfall.

F_q/F_{r_b} for several values of h. For $h=1$, rainfall fully depletes the cloud water x_k during each time step: t_{k-1} to t_k. Such a situation occurs under conditions of fully developed intense rainfall (e.g. Kessler, 1986). The case $h<1$, corresponds to rainfall that does not deplete cloud water completely during each time step, thus allowing for memory in the rainfall production process. This case is referred to as the case of moderate to light mean areal rainfall since lower rainfall rates result as compared to the case with $h=1$ for the same input meteorological conditions.

The curves in Fig. 2 reveal that only in cases of light mean areal rainfall and large ratios F_q/F_{r_b}, do low-tilt angle reflectivity data improve mean areal rainfall predictions significantly. For $h=0.6$, even for ratios F_q/F_{r_b} up to 3, (implying highly uncertain model and highly accurate radar data), R_P^* remains above 0.85. For intense rainfall ($h=1$), there is no memory in the rainfall process and the ratio R_P is identically equal to 1 with both numerator and denominator equal to Q in equation (38).

Vertically integrated radar reflectivity data

CASE A: INTENSE RAINFALL

At first, we examine the character of equation (36), that gives the ratio of state estimation (or updating) variances with and

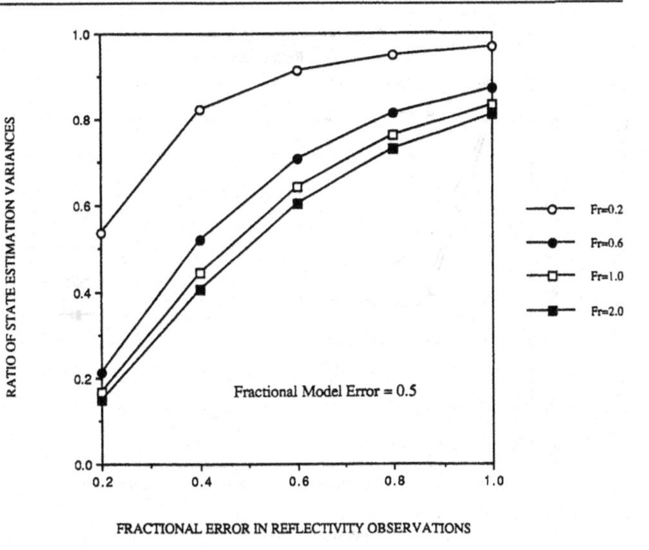

Fig. 3 Ratio of estimation variances with and without radar reflectivity data as a function of the fractional error in reflectivity observations. Intense rainfall.

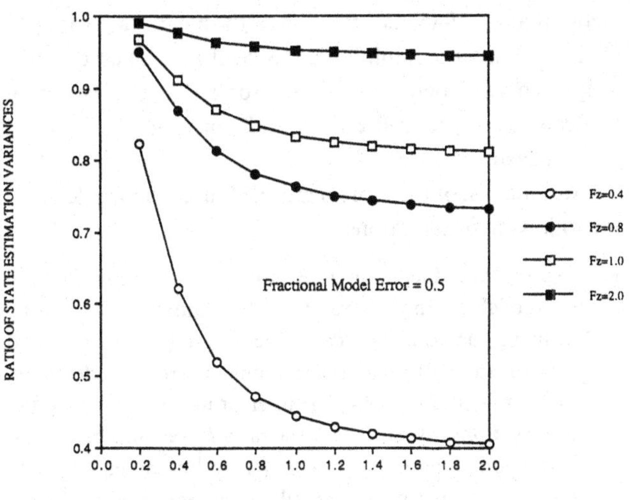

Fig. 4 Ratio of estimation variances with and without radar reflectivity data as a function of the fractional error in surface rainfall observations. Intense rainfall.

without vertically-averaged reflectivity observations (reflectivity observations in short) as a function of the uncertainty in reflectivity measurements, surface-rainfall measurements and in the model (model error variance). The situation under study is for $h=1$. For such a case, the ratio of prediction variances is equal to 1, and independent of the observation errors. Figs. 3 and 4 present the ratio of estimation variances, $R_E=P_\infty^+/P_{\infty,r}^+$ for various values of the fractional observation error in reflectivity, F_z, and in surface rainfall, F_r, using the fractional model error, F_q, as a parameter. Fig. 5 shows dependence of R_E on F_q for several values of F_r and for $F_z=0.4$. The value of $F_q=0.5$ used in Figs. 3 and 4 represents the best estimated value of $\sqrt{Q}$ by previous studies (i.e.,

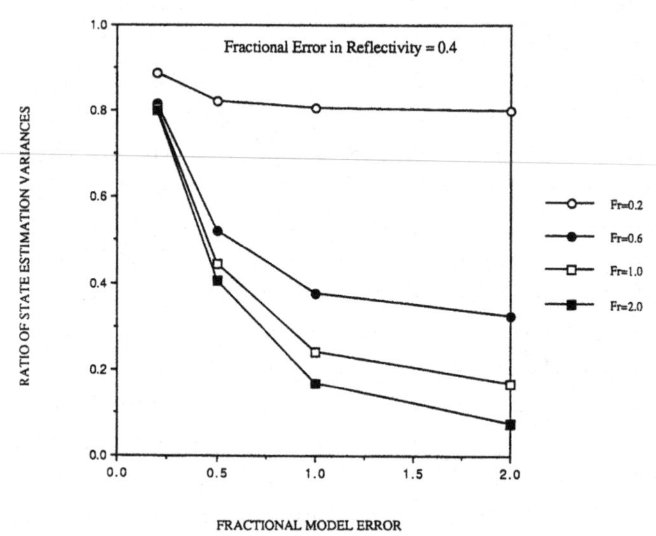

Fig. 5 Ratio of estimation variances with and without radar reflectivity data as a function of fractional model error. Intense rainfall.

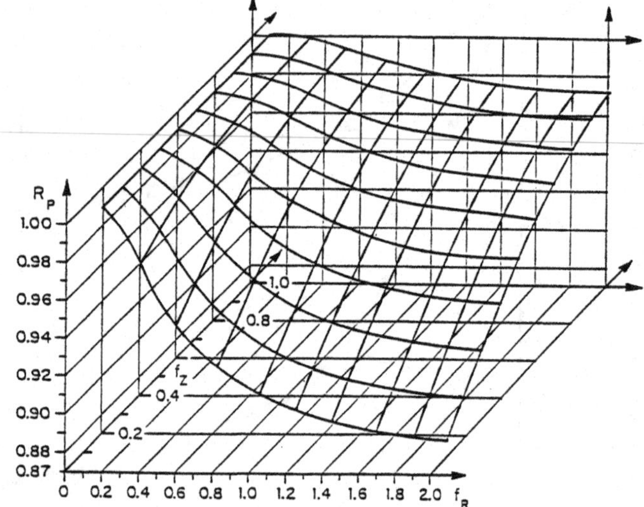

Fig. 6 Isometric plot of the rainfall-prediction variance ratio with and without radar reflectivity data as a function of fractional errors in surface rainfall and reflectivity measurements, for $F_q = 0.5$. Moderate to light rainfall.

Georgakakos, 1984) as a fraction of the mean state value $\bar{x}_k$, for cases of intense rainfall. The value of $F_z = 0.4$ used in Fig. 5 represents a 40 percent random error in the observations of vertically-averaged reflectivity and corresponds to good radar measurements.

Examination of the aforementioned three Figures leads to the following observations:

1. Significant reduction in the estimation variance of the total mass of cloud and rainwater can be attained in real-time by utilizing good quality radar reflectivity data (see Figs. 3 and 4). For up to 80 percent measurement error in reflectivity, and for greater than 60 percent error in the surface rainfall observations (or estimates), the ratio R_E remains below 0.8. For good quality radar data (error less or equal to 40 percent), even for very small measurement error in the observations of surface rainfall (down to 20 percent), the radar data reduce R_E below the 0.8 level.
2. For good quality radar data (of about 40 percent measurement error) the ratio R_E drops sharply with increasing error in the measurement of surface rainfall (i.e., Fig. 4, curve labeled $F_z = 0.4$).
3. For bad quality measurements of surface rainfall (more than 100 percent measurement error), even fair-quality radar measurements (of an 80 to 100 percent measurement error) are useful in reducing the ratio R_E to the 0.8 level or below. It is noted that such measurements of surface rainfall are typical in areas with sparse raingauges.
4. The greater the model uncertainty, the smaller the ratio R_E is and, therefore, the higher the potential utility of radar reflectivity data for the real-time estimation of vertically integrated cloud and rain water (see Fig. 5).

CASE B: MODERATE TO LIGHT RAINFALL

This case corresponds to $h < 1$. Now attention is directed toward equations (24) and (26) for the ratio of state and rainfall prediction variances: $R_P = P_\infty^- / P_{\infty,r}^-$. It is noted that in contrast to the previous case examined, radar data can reduce the state and rainfall prediction variance ($R_P < 1$). In terms of the right-hand side of equations (24) and (26), R_P is given as:

$$R_P = \frac{(F_r^2)^{-1}}{(F_r^2)^{-1} + (F_z^2)^{-1}} \cdot \frac{[(1-A-C)^2 + 4C]^{1/2} - (1-A-C)}{[(1-A-C_r)^2 + 4C_r]^{1/2} - (1-A-C_r)} \tag{41}$$

with

$$A = (1-h)^2 \tag{42}$$

$$C = F_q^2(1/F_r^2 + 1/F_z^2) \tag{43}$$

$$C_r = F_q^2/F_r^2 \tag{44}$$

A value of h equal to 0.60 was used for the tests of this case which represents an average value for moderate to light rainfall. Fig. 6 presents an isometric plot of the ratio of prediction variances with and without radar reflectivity data for $F_q = 0.5$. It can be seen that the ratio R_P can take on values less than 1 when poor quality measurements for the estimation of mean areal surface rainfall are available together with high quality radar reflectivity data. For example, values of R_P less than 0.9 are observed for errors greater than 120 percent in surface rainfall observations and less than 40 percent in radar reflectivity data.

Figs. 7 and 8 present the ratio R_P as a function of F_q for various values of F_r and for two cases of radar reflectivity data. A case of good quality reflectivity data ($F_z = 0.4$, Fig. 7) and a case of very good quality reflectivity data ($F_z = 0.2$, Fig. 8) are presented. Both figures show marginal utility of vertically-averaged radar reflectivity data (even of a very good quality) when surface rainfall data, even of a bad

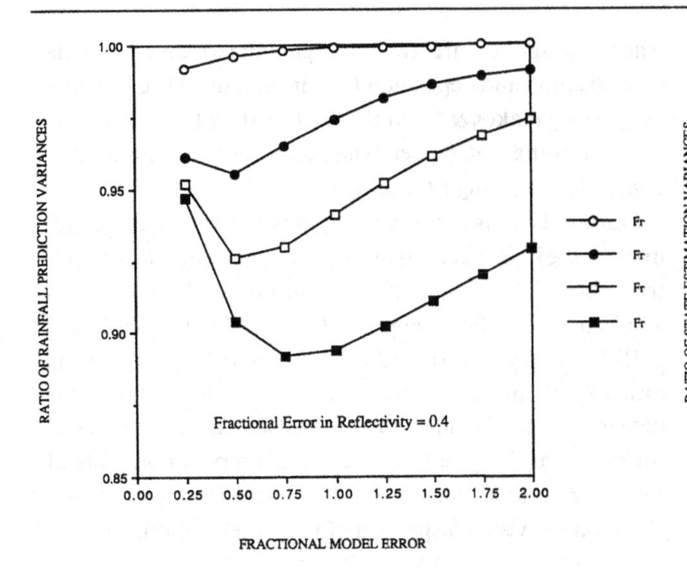

Fig. 7 Ratio of prediction variances with and without radar reflectivity data as a function of fractional model error, for $F_z = 0.4$. Moderate to light rainfall.

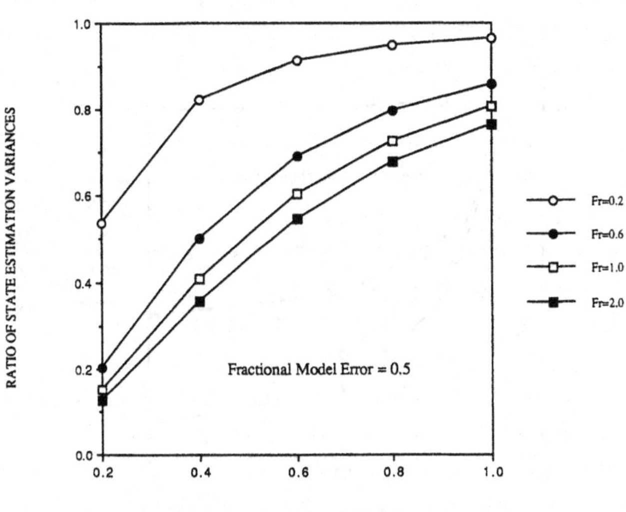

Fig. 9 Ratio of rainfall estimation variances with and without radar reflectivity data as a function of the fractional error in reflectivity observations. Moderate to light rainfall.

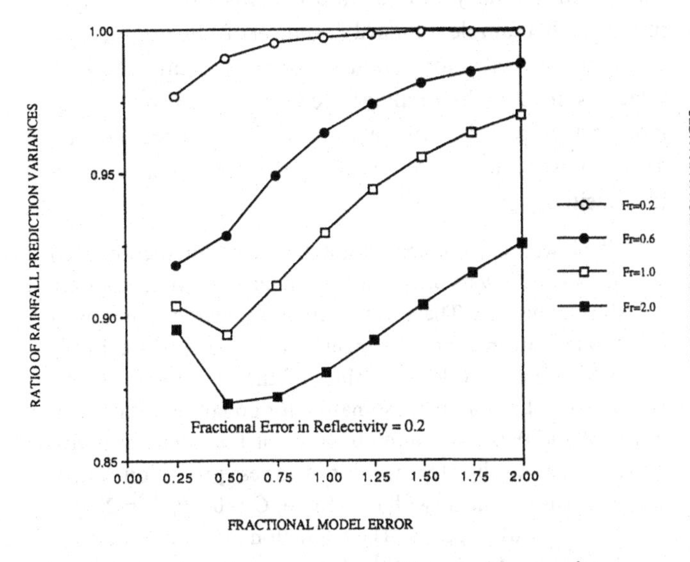

Fig. 8 Ratio of prediction variances with and without radar reflectivity data as a function of fractional model error, for $F_z = 0.2$. Moderate to light rainfall.

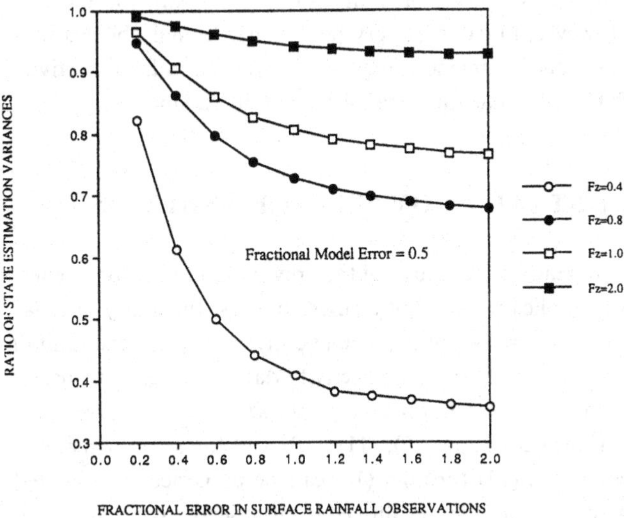

Fig. 10 Ratio of rainfall estimation variances with and without radar reflectivity data as a function of the fractional error in surface rainfall observations. Moderate to light rainfall.

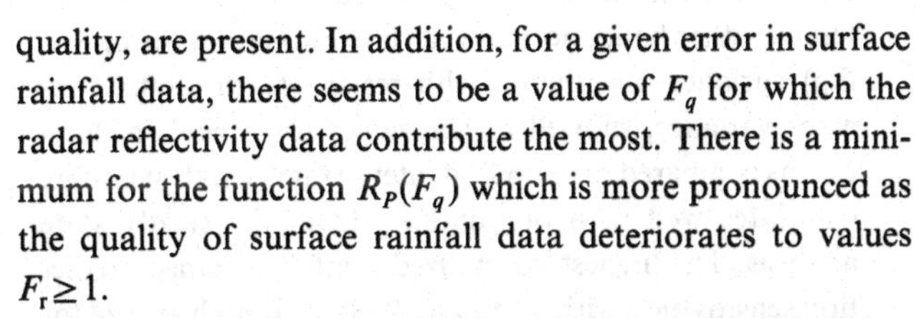

quality, are present. In addition, for a given error in surface rainfall data, there seems to be a value of F_q for which the radar reflectivity data contribute the most. There is a minimum for the function $R_P(F_q)$ which is more pronounced as the quality of surface rainfall data deteriorates to values $F_r \geq 1$.

Once the steady state rainfall prediction variance is determined by equations (24) and (26) with and without radar reflectivity data, equations (22) and (25) can be used to obtain the corresponding steady-state state estimation variances. Then, the ratio $R_E = P_\infty^+ / P_{\infty,r}^+$ can be determined for various values of F_r, F_z and F_q, given $h = 0.6$. Figs. 9, 10, and 11 are analogous to Fig. 3, 4, and 5 and display dependence of the ratio in state (rain and cloud water) estimation (or updating) variances with and without reflectivity data on fractional errors F_z, F_r and F_q, respectively. The comments made previously for the case $h = 1$ apply here too. The difference is that for $h = 0.6$, the ratio of state estimation variances is somewhat lower than for $h = 1$, for given values of F_r, F_z and F_q, especially for large values of F_r and F_z. The reduction in R_P is small, however, with decreasing h and the assessment of the worth of radar reflectivity data made for $h = 1$ applies for moderate to light rainfall as well. Clearly, the contribution of vertically-averaged radar reflectivity data to the real time prediction of mean areal surface rainfall is significantly less than their contribution to the real time

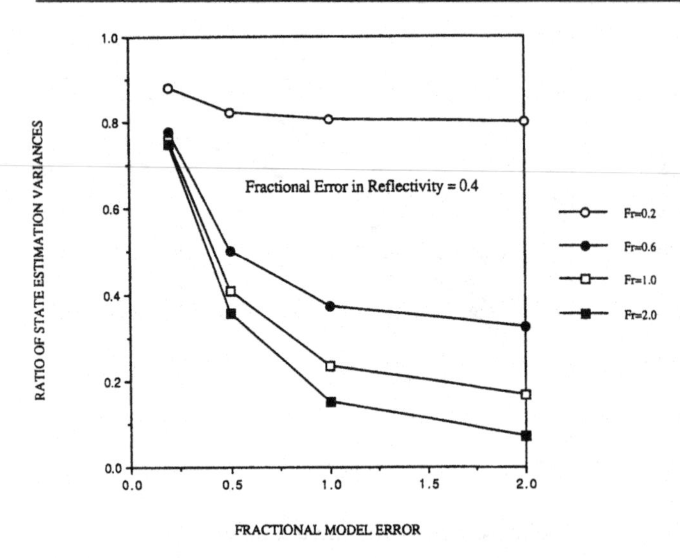

Fig. 11 Ratio of rainfall estimation variances with and without radar reflectivity data as a function of the fractional model error. Moderate to light rainfall.

estimation of vertically integrated cloud and rain water. However, in cases of very poor conventional observations (e.g., due to sparse raingauge networks), radar reflectivity data will reduce the rainfall prediction variance.

UNSTEADY COVARIANCE ANALYSIS

The steady state results of the previous section, albeit generally applicable, can not be used to assess the utility of radar data (either low-tilt angle data converted to surface rainfall or vertically-averaged reflectivity data) in cases of transient model behavior. In those cases, the set of unsteady, recursive variance equations (5), (11), (12) and the set of sensitivity equations (13) through (18) can be numerically simulated with h_k, Φ_k and Ψ_k, evaluated from observed meteorological storm data. In the following, we report on the results of an extensive sensitivity analysis that was performed on the variance and sensitivity equations utilizing meteorological data from two long-lasting storms. The storms have been previously used for the calibration of the rainfall model. The first storm occurred in September, 1962, and was recorded at Logan Airport in Boston, Massachusetts. It spanned a period of 43 hours. The second storm occurred in May, 1950, and was recorded at Tulsa International Airport in Oklahoma. It spanned a period of 41 hours. Hourly data of surface air temperature, pressure and dew-point temperature were used for the computation of h_k, Φ_k and Ψ_k. Sensitivity analysis was performed with respect to Q, R_r, R_z and ρ_{rz}. Q was further parameterized as

$$Q=(C_0+C_1/f_k)^2 \quad f_k>0$$
$$Q=C_0^2 \quad f_k=0 \tag{45}$$

where C_0 and C_1 are two new parameters and f_k is the condensation and deposition term in equation (1) computed as in Georgakakos & Bras (1984a). Equation (45) reflects the hypothesis that the model dynamics is more accurate when convection is strong (f_k is large).

During the sensitivity runs, C_0 took the values 5, 2 and 0 mm; C_1 took the values 0, 0.1, and 0.5 mm^2/hr; $(R_r)^{1/2}$ took the values 100, 150 and 200 percent of the observed mean areal surface rainfall in mm/hr; $(R_z)^{1/2}$ took the values 1, and 2 dBZ; and ρ_{rz} took the values 0, 0.5, and 0.8. Plots of the ratios R_P, R_E and of the normalized sensitivities S_R and S_Z (as defined earlier), as functions of time step k, were produced for each sensitivity run, and were superimposed on plots of rainfall rate as a function of k. Conclusions were drawn based on the visual inspection of the aforementioned plots.

Storm 1

The average hourly rainfall rate for this storm was 1.91 mm/hr with a sample standard deviation of 1.86 mm/hr. The storm rainfall lasted for 43 hours. When a standardized Z–R relationship (Marshall-Palmer, see Battan, 1973) was used it gave an average reflectivity factor of 27.5 dBZ. The following are the most significant observations that we made, based on the sensitivity analysis runs.

1. The lowest instantaneous rainfall variance prediction ratio R_P was as low as 0.75 and that value was obtained rarely during the runs. The lowest average prediction ratio was as low as 0.80, in good agreement with the steady state theory predictions. The lowest values of the instantaneous and average state variance estimation (or updating) ratios were 0.18 and 0.22, and were obtained for the same sensitivity run as were the aforementioned lowest prediction ratios. That run was made with $C_0=5$ mm, $C_1=0$, $(R_r)^{1/2}=200\%$, $(R_z)^{1/2}=1$ dBZ, $\rho_{rz}=0$. The prediction ratio instantaneous (average) value for $(R_r)^{1/2}=100\%$ was 0.75 (0.85).
2. The time variation of the normalized rainfall prediction variance sensitivities with respect to the reflectivity error variance, was found to bear no significant relationship to concurrent variations of the observed hourly rainfall rates. On the other hand, normalized state estimation (or updating) variance sensitivities with respect to the reflectivity error variance, showed higher values for periods of light rain as compared to periods of intense rain. This last finding correlates well with our findings from the steady state analysis. The highest normalized rainfall variance prediction sensitivities, with respect to R_z, were as high as 0.24 for instantaneous values and 0.12 for average values during a run. The run parameters that gave the aforementioned sensitivity values were: $C_0=5$ mm/hr, $C_1=0$, $(R_r)^{1/2}=100\%$, $(R_z)^{1/2}=1$ dBZ, $\rho_{rz}=0.8$. The highest normalized rainfall variance estimation (or updating) sensitivities with respect to R_z were as high as 3.0 for instantaneous values

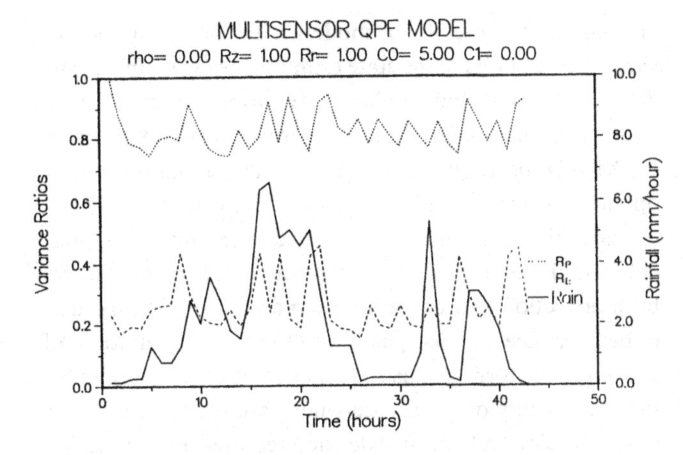

Fig. 12 Hourly rainfall prediction variance ratio and hourly state estimation (or updating) variance ratio for the Boston, Massachusetts, storm. Also, shown on the right scale are the concurrent rainfall rates in mm/hr. The parameter values are: $\rho_{RZ}=0$, $(R_r)^{1/2}=100\%$, $R_z=1$dBZ, $C_0=5.00$, and $C_1=0.0$.

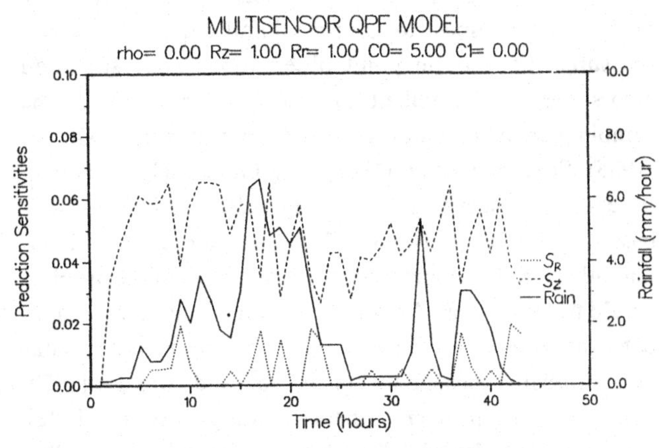

Fig. 13 Hourly normalized prediction sensitivities of the rainfall prediction variance with respect to R_z and R_r for the Boston, Massachusetts, storm. Also, shown on the right scale are the concurrent rainfall rates in mm/hr. The parameters are as in Fig. 12.

and 2.6 for average values during a run. The same run gave the highest prediction and estimation sensitivities.

3. The effect of reducing C_0 while the other run parameters were being held constant, was to smooth out temporal fluctuations in the plots of all the variance ratios and normalized sensitivities. The opposite effect was observed when ρ_{rz} and C_1 were increased to values higher than 0, and when $(R_z)^{1/2}$ was increased from 1 dBZ to 2 dBZ. An increase in $(R_r)^{1/2}$ resulted in a decrease of the average variance ratios.

Figs. 12 and 13 are presented to illustrate the character of the sensitivity plots. The variance ratios (Fig. 12) and the normalized prediction sensitivities with respect to R_z and R_r (Fig. 13) are shown for the case with run parameters: $C_0=5$ mm/hr, $C_1=0$, $(R_r)^{1/2}=100\%$ of observed surface rainfall, $(R_z)^{1/2}=1$ dBZ and $\rho_{rz}=0$.

Storm 2

The average rainfall rate for this storm was 3.27 mm/hr with a sample standard deviation of 3.87 mm/hr. The rainfall duration was 41 hours. Using a standardized Z–R relationship, an average reflectivity factor of 31.25 dBZ was obtained for this storm. Compared to the Boston Storm 1, the Tulsa storm had much higher hourly rainfall rates (a maximum of about 16 mm/hr vs a maximum of 7 mm/hr for Storm 1), and it exhibited a more pulse-like nature characteristic of Oklahoma convective storms. The general comments made for Storm 1 regarding the behavior of variance ratios and normalized prediction sensitivities are applicable to the results obtained from the sensitivity runs of Storm 2. The main difference was that the minimum instantaneous rainfall prediction variance ratio was obtained for a lower value of Q ($C_0=2$, $C_1=0$) and its value, 0.90, was higher than for Storm 1. The minimum instantaneous state estimation (or updating) variance was also higher at 0.25 and it was obtained for the aforementioned value of C_0 and C_1. Characteristic of the runs with Storm 2 data was that the rainfall prediction variance and state estimation variance ratios exhibited smaller fluctuations than those observed in the case of Storm 1. The values of the normalized prediction sensitivity, with respect to R_z, were about the same magnitude in Storm 2 as in Storm 1. A significant increase (100 percent) of the prediction sensitivities, with respect to R_r, was observed in the runs of Storm 2, as compared to those in the runs of Storm 1. Figs. 14 and 15 are analogous to Figs. 12 and 13, respectively, and correspond to a Storm 2 run with parameters: $C_0=5$, $C_1=0$, $\rho_{rz}=0$, $(R_r)=100$ percent of observed surface rainfall, and $(R_z)^{1/2}=1$ dBZ.

CONCLUSIONS AND FUTURE DIRECTIONS

This work has employed methods of modern estimation theory applied to physically-based spatially-lumped rainfall prediction models to quantify the potential improvement offered by radar data in real-time short-term mean areal rainfall predictions. Utility of both base scan and volume scan radar data has been examined. Improvement of rainfall predictions was measured by the reduction in the rainfall prediction variance attained when radar data was used. The results show that in the presence of raingauges (even for a small number of them), the ratio of predicted mean areal rainfall variance with radar data to the same quantity without radar data approaches 0.75 as a low value with more typical values in the range of 0.85–0.95. In addition, data from storms in Boston, Massachusetts, and Tulsa, Oklahoma, indicated greater improvement in mean areal rainfall

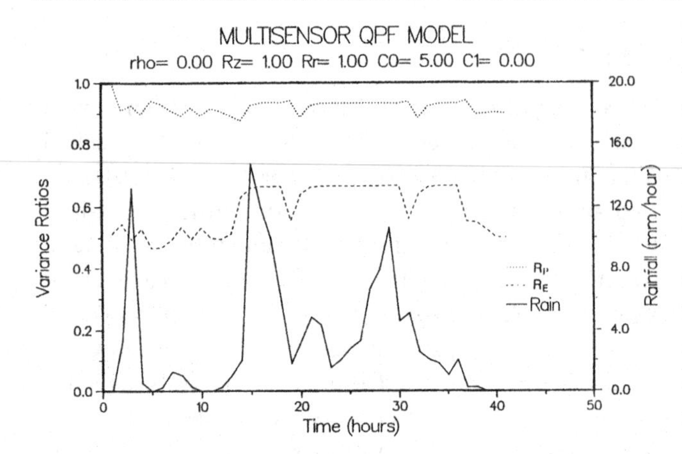

Fig. 14 As in Fig. 12, only for the Tulsa, Oklahoma, storm.

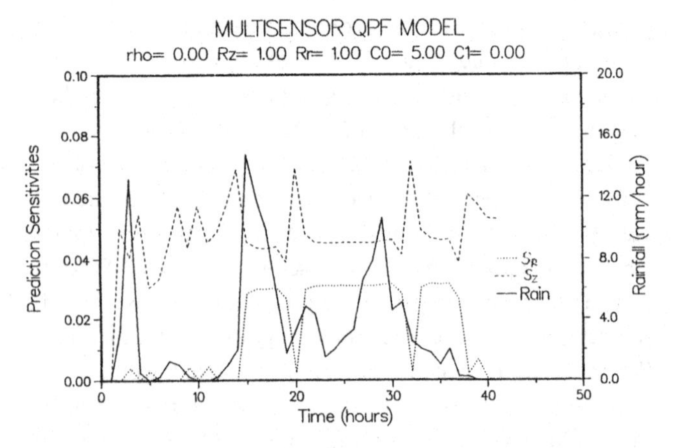

Fig. 15 As in Fig. 13, only for the Tulsa, Oklahoma, storm.

predictions for Boston when radar data were used. The radar data, however, offered significant reduction of the uncertainty associated with the real-time estimation of the model state: vertically integrated condensed water equivalent mass.

The lack of a continuous record of good quality radar data that would cover several storms in various hydroclimatic regimes prevented the use of actual radar data for the covariance analysis. Instead, linear estimation theory methods were used that decouple the mean from the covariance equation and allow covariance analysis in the absence of actual data. To satisfy this requirement, a linear (in the state) rainfall prediction model was used which has been previously used in the real-time prediction of areal rainfall. However, given the good performance of the model in the short term prediction of rainfall and the published evidence that its optimal parameters remain reasonably constant in various regimes (e.g. Georgakakos, 1984), the results reported herein should be representative of other linear or linearized rainfall prediction models.

Several case studies using actual radar and raingage data together with the input data are underway. These will allow for covariance analysis of nonlinear (in the state) spatially-lumped rainfall prediction models with the mean and covariance equations of the state estimator being coupled. The data involved include radar observations from Darwin, Australia; Champaign, Illinois; Norman, Oklahoma (from NEXRAD prototype facility); and the Japanese radars operated by the Japan Ministry of Construction.

The authors are currently involved in extending the present formulation into a two-dimensional model that includes both advection and convection dynamics, of the type described by Lee & Georgakakos (1990). The addition of advection of storm clouds in such models offers promise for increased utility of radar data when used for the estimation of storm velocity. Such an extension requires reliable models for the statistical description of the spatial structure of radar measurement errors.

The described study underscores the importance of the investigations of errors in mean areal precipitation estimates. In the case of using raingage data, the sources of errors include sampling errors due to sparseness of raingage network with respect to hourly rainfall spatial variability, measurement error in point observations and estimation errors due to a particulant interpolation method used. The sampling errors are addressed by Bras & Rodriguez-Iturbe (1985), Silverman *et al.* (1981), and Gabriel (1981) among others. Point accuracy of raingage observations was discussed by Sevruk & Hamon (1984), Larson & Peck (1974) and others. Still, a comprehensive assessment of the problem would be very useful. A similar, but somewhat more complicated situation exists in the evaluation of the errors in radar rainfall-related measurements. Wilson and Brandes (1979) summarized the problem, others including Zawadzki (1982), Sachidanada & Zrnic (1987), Chandrasekhar & Bringi (1987) and Jameson (1989) addressed in a comprehensive way the problem of reflectivity measurements and radar-rainfall estimates error in a simple sampling volume. However, the problems of spatial structure of radar errors, clearly relevant for the discussed problem (see also Krajewski, 1987), have not been properly addressed.

Finally, the recent radar technologies of multiparameter radar rainfall observations (Jameson, 1989) can be readily incorporated into the presented framework for analysis of the prediction error variance. This is because the described framework is based on physical models and data uncertainty measures, and thus, it is suitable for optimal combining of multisensor observations relevant for rainfall prediction and estimation.

APPENDIX A

Given reflectivity-factor measurements Z_i^j $(i=1,\ldots,n)$ at various radar-antenna elevation angles Φ_j with $j=1,\ldots,m$, assumed to be taken at instances t_1 through t_n during the time interval Δt, the following reflectivity-factor vs. size-distribution relationship exists (e.g., Burgess & Ray, 1986):

$$Z_i^j = \int_0^\infty D^6 n_i^j(D)\, \mathrm{d}D \tag{A1}$$

where the size distribution of hydrometeors $n(D)$ has been localized as $n_i^j(D)$ for the time instant t_i and the tilt angle Φ_j. The size distribution $n(D)$ is assumed exponential of the type:

$$n(D) = N_0 \exp(-D/\varepsilon_4) \tag{A2}$$

Localization of the distribution is through localization of its parameters N_0 and ε_4. Because of the absence of upper-air meteorological data for the temporal scales involved (hourly), utilization of reflectivity factor for each tilt angle is not recommended. Instead, as a compromise, we propose to use an average reflectivity factor $Z(t_i)$, defined by:

$$Z(t_i) = \frac{1}{m} \sum_{j=1}^{m} Z_i^j \tag{A3}$$

as a characteristic measure of non-zero reflectivity factors over the height of the storm clouds and for time instant t_i. In addition, we assume that an 'equivalent' size distribution, $n_i'(D)$, which is uniform over the height of the storm clouds exists at time t_i such that:

$$Z(t_i) = \int_0^\infty D^6\, n_i'(D)\, \mathrm{d}D \tag{A4}$$

It is noted that due to the high power of D use of ∞ as an upper bound might introduce appreciable errors in the computations for relatively flat size distributions. Also, for diameters of hydrometeors near zero no return signal (and hence no reflectivity factor) is expected. However, for simplicity we retain the previous equation with the understanding that it might be improved.

It also holds that for time instant t_i,

$$X(t_i) = Z_{cl}[u(t_i)]\, \pi/6\, \rho_w \int_0^\infty D^3\, n_i(D)\, \mathrm{d}D \tag{A5}$$

where ρ_w represents the density of liquid water, and $Z_{cl}[u(t_i)]$ is the average depth of the storm clouds over the hydrologic area of interest and for time t_i. The cloud height is a function of the model vector input of meteorological indices $u(t_i)$ at time t_i. The size distribution $n_i(D)$ is an average size distribution over the depth of the clouds over the domain of interest. It is noted that the two distributions: $n_i'(D)$ and $n_i(D)$ need not be the same since the functional form of the right-hand side of equations (A4) and (A5) differs. It is hypothesized that the difference between the two size distributions lies in the average hydrometeor size ε_4, with $\varepsilon_4' = \alpha\varepsilon_4$, and ε_4' corresponding to $n_i'(D)$. Given that the larger hydrometeors produce the bulk of the radar return signal, it is expected that parameter α will satisfy: $\alpha > 1$.

The last two equations (after substitution of the exponential size distribution relationships for $n_i(D)$ and $n_i'(D)$ have four unknowns: X, Z, N_0, ε_4. One, then, can eliminate two of them by solving the equations for those two unknowns. Since $X(t_i)$ is the model state variable and $Z(t_i)$ is observed, it is convenient to solve for the parameters of the hydrometeor size distribution. This way, if reflectivity factor measurements are available, they could be used in a second observation equation within the model state space form. The second observation equation would be:

$$Z(t_i) = \Phi'[u(t_i)]X(t_i) + v'(t_i) \tag{A6}$$

with

$$\Phi'[u(t_i)] = \frac{720\, \varepsilon_4^3 \alpha^7}{\pi \rho_w Z_{cl}[u(t_i)]} \tag{A7}$$

and with $v'(t_i)$ accounting for random measurement errors in reflectivity-factor observations.

ACKNOWLEDGEMENTS

The research work was partially supported by The Hydrologic Research Laboratory of the U.S. National Weather Service. A slightly modified form of this paper was published in *Water Resources Research*, **27**(2), 185–7 (1991).

REFERENCES

Ahnert, P., Hudlow, M., Johnson, E., Greene, D. & Dias, M. (1983) *Proposed 'on-site' precipitation processing system for NEXRAD*, Preprints of the 21st Conference on Radar Meteorology, pp. 378–85.

Battan, L. (1973) *Radar Observations of the Atmosphere*, Chicago Press.

Bras, R. L. & Rodriguez-Iturbe, I. (1985) *Random Functions and Hydrology*, Addison-Wesley Publishing Co., Reading, MA.

Burges, D. & Ray, P. S. (1986) *Principles of radar, in mesoscale meteorology and forecasting*. In: Ray, P. S. (Ed.), *American Meteorological Society*, Boston, Massachusetts, pp. 85–117.

Chandrasekar, V. & Bringi, V. N. (1987) Simulation of radar reflectivity and surface measurements of rainfall. *Journal of Atmospheric and Oceanic Technology*, **4**(6), 464–78.

Gabriel, K. R. (1981) Gage density and variability of rainfall estimates: a complement to Silverman, Rogers, and Dahl's study. *Journal of Applied Meteorology*, **20**, 1537–42.

Gelb, A. (Ed.) (1974) *Applied Optimal Estimation*, MIT Press, Cambridge, Massachusetts.

Georgakakos, K. P. & Bras, R. L. (1984a) A hydrologically useful station precipitation model. Part I – Formulation. *Water Resour. Res.*, **20**(11), 1585–96.

Georgakakos, K. P. & Bras, R. L. (1984b) A Hydrologically Useful Station Precipitation Model. Part II – Applications. *Water Resour. Res.*, **20**(11), 1597–610.

Georgakakos, K. P. & Hudlow, M. D. (1984) Quantitative precipitation forecast techniques for use in hydrologic forecasting. *Bulletin of the American Meteorological Society*, **65**(11), 1186–200.

Georgakakos, K. P. & Kavvas, M. L. (1987) Precipitation analysis, modelling and prediction in hydrology. *Reviews of Geophysics*, **25**(2), 163–78.

Georgakakos, K. P. & Krajewski, W. F. (1989) *Short-term rainfall and flood forecasting using radar data and hydrometeorological models*,

Preprints of the International Symposium on Hydrological Applications of Weather Radar, University of Salford, United Kingdom, August 1989.

Georgakakos, K. P. (1984) Model error adaptive parameter determination of a conceptual rainfall prediction model, *IEEE 1984 Proceedings of the 16th Southeastern Symposium on System Theory*, IEEE Computer Society Press, Silver Spring, Maryland, p. 111.

Georgakakos, K. P. (1986a) A generalized stochastic hydrometeorological model for flood and flash-flood forecasting. Part I – Formulation. *Water Resour. Res.*, **22**(13), 2083–95.

Georgakakos, K. P. (1986b) A generalized stochastic hydrometeorological model for flood and flash-flood forecasting. Part II – Case Studies. *Water Resour. Res.*, **22**(13), 2096–106.

Georgakakos, K. P. (1986c) State estimation of a scalar dynamic precipitation model from time-aggregate observations. *Water Resour. Res.*, **22**(5), 744–8.

Georgakakos, K. P. (1987) Real-time flash flood prediction. *Journal of Geophysical Research*, **92**(D8), 9615–29.

Hudlow M. D., Greene, D., Ahnert, P., Krajewski, W., Sivaramakrishnan, T., Johnson, E. & Dias, M. (1983) *Proposed 'off-site' precipitation processing system for NEXRAD*, Preprints of the 21st Conference on Radar Meteorology, pp. 394–403.

Jameson, A. R. (1989) Comparison of microwave techniques for measuring rainfall, submitted to *Journal of Applied Meteorology*.

Jazwinski, A. H. (1970) *Stochastic Processes and Filtering Theory*, Academic Press, New York, New York.

Kessler, E. (1969) On the Continuity of Water Substance in Atmospheric Circulations. *Meteorological Monographs* **10**(33), 84 pp.

Kessler, E. (1986) Model relationships among storm cloudiness, precipitation, and airflow, in: *Thunderstorm Morphology and Dynamics*, Ed. E. Kessler, Univ. of Oklahoma Press, Norman, Oklahoma, pp. 297–312, 2nd edition.

Krajewski, W. F. (1987) Cokriging radar-rainfall and rain gage data, *J. Geophysical Research*, **92**(D8), 9571–80.

Larson, L. W. & Peck, E. L. (1974) Accuracy of precipitation measurements for hydrological modeling. *Water Resour. Res.*, **10**, 857–63.

Lee, T. H. & Georgakakos, K. P. (1991) Two-dimensional stochastic-dynamical quantitative precipitation forecasting model, *Journal of Geophysical Research*.

Leone, D. A., Endlich, R. M., Petriceks, J., Collis, R. T. H. & Porter, J. R. (1989) Meteorological considerations used in planning the NEXRAD Network. *Bulletin of the American Meteorological Society*, **70**(1), pp. 4–13.

Marshall, J. S. & Palmer, W. M. (1948) The distribution of raindrops with size. *J. of Meteorology*, **5**, 165–6.

Rabitz, H. (1989) Systems analysis at the molecular scale. *Science*, **246**, 221–46.

Sachidananda, M. & Zrnic, D. S. (1987) Rain rate estimates from differential polarization measurements. *Journal of Atmospheric and Oceanic Technology*, **4**, 588–98.

Sevruk, B. & Hamon, W. R. (1984) International comparison of national precipitation gauges with reference pit gauge. *WMO Instruments and Observing Methods Report No. 17*, pp. 140.

Silverman, B. A., Rogers, L. K. & Dahl, D. (1981) On the sampling variance of raingage networks. *Journal of Applied Meteorology*, **20**, 1468–78.

Wilson, J. W. & Brandes, E. A. (1979) Radar measurement of rainfall – a summary. *Bull. Am. Meteorol. Soc.*, **60**(9), 1048–58.

Zawadzki, I. (1982) The quantitative interpretation of weather radar measurements. *Atmos. Ocean*, **20**, 158–80.

3 Uncertainty analysis in radar-rainfall estimation

W. F. KRAJEWSKI

Department of Civil and Environmental Engineering and Iowa Institute of Hydraulic Research, The University of Iowa, Iowa City, Iowa, USA

J. A. SMITH

Department of Civil Engineering and Operations Research, Princeton University, Princeton, New Jersey, USA

ABSTRACT Two Monte Carlo simulation experiments which address the problem of radar-rainfall estimation are presented. One of the problems associated with hydrologic use of radar-rainfall data is the need to adjust radar rainfall estimates to raingage estimates. The adjustment, which is performed in real time, can be done in the mean field sense. The problem of development of such an adjustment scheme is difficult due to largely unknown statistical structure of radar errors and the fundamental sampling differences between these two sensors. To investigate the problem, mean field bias is modeled as a random process that varies not only from storm to storm but also over the course of a storm. State estimates of mean field bias are based on hourly rain gage data and hourly accumulations of radar rainfall estimates. The procedures are developed for the precipitation processing system to be used with products of the Next Generation Weather Radar (NEXRAD) system. To implement the state estimation procedure parameters of the bias model must be specified. The performance of the state estimation is investigated within a Monte Carlo simulation framework. The results highlight the dependence of the state estimation problem on the parameter estimation problem. The second experiment addresses the problem of converting radar-measured reflectivity into rainfall rate. This is typically done using a *Z–R* relationship. The parameters of such relationship can be estimated using climatological data and nonparametric estimation framework. In the paper the effects of thresholds imposed on the observations included in the estimation are investigated.

INTRODUCTION

Modern weather radars are capable of providing detailed quantitative information on spatial and temporal distribution of rainfall. The utility of this information can be significantly increased if radar-rainfall estimates are complemented by their uncertainty bounds. However, in order to produce these bounds in a reliable and consistent manner a statistical framework needs to be developed which would account for all the uncertainty sources. Among the many errors affecting radar-rainfall estimates are those associated with *Z–R* relationship effect and the mean field bias effect. Other error sources are discussed in detail by Wilson & Brandes (1979), Zawadzki (1984), Austin (1987), Chandrasekar & Bringi (1987), Joss & Waldvogel (1990) and others.

Mean field bias is modeled as a random process that varies from storm to storm and, during a storm, on an hourly time scale. The estimation procedures are designed for the precipitation processing systems used for the Next Generation Weather Radar (NEXRAD) system. Discussions of general design features of the precipitation processing systems are given in Hudlow *et al.* (1984) and Shedd *et al.* (1989) among others.

This paper describes an attempt to quantify these errors using mathematical models and numerical simulation. Smith & Krajewski (1991) describe estimation of the mean field bias of radar-rainfall estimates using a recursive algorithm which could be applied in real time. The statistical model proposed by Smith & Krajewski (1991) has four parameters and their study focuses on a scheme to estimate these parameters from radar and raingage data. In this paper the effects of the propagation of the parameter errors on the mean field bias estimates are studied. Since the mean field bias can be interpreted as randomizing the multiplicative parameter in the Z–R relationship, it is of interest to study the errors in Z–R parameter estimation. This was the subject of work by Krajewski & Smith (1991) and is continued here.

BIAS MODEL FORMULATION

To estimate the mean field bias a statistical model is developed that relates radar measurements of rainfall to the true rainfall field. A component of the model is a time-varying random process representing mean field bias. Specification of the model serves two purposes. First, it provides a precise interpretation of the mean field bias. It also serves as the basis for development of an observation equation that relates radar and rain gage measurements to the mean field bias. The observation equation and model of mean field bias are the principal tools required for development of procedures to estimate the mean field bias. The notation to be used is introduced below. The presentation follows that by Smith & Krajewski (1991).

Precipitation rate at time t and spatial location x is denoted $\xi_\tau(x)$. The index τ represents time, in hours, since the last period of no rainfall. The precipitation rate process for radar scan t of hour s averaged spatially over the bin specified by azimuth i and range j is denoted

$$R_{s,t}(i,j) = |D_{ij}|^{-1} \int_{D_{ij}} \xi_\tau(x)\,dx; \quad \tau = (s-1) + t\Delta t \tag{1}$$

where D_{ij} is the land area beneath the radar sample volume with azimuth i and range j, $|D_{ij}|$ is the surface area associated with D_{ij}, and Δt is the time resolution of radar observations (in hours). The number of scans γ during an hour is $1/\Delta t$. For the US NEXRAD system the time resolution is approximately 6 minutes ($\Delta t = 0.1$ hour) during precipitation periods so the number of scans during an hour will be 10. As noted in (1) the times τ for which observations are available, relative to the start of the storm, are given by $(s-1) + t\Delta t$, with t ranging from 1 to γ. The equivalent radar reflectivity factor (Battan, 1973) for scan t in hour s at azimuth i and range j is denoted $z_{s,t}(i,j)$.

The most popular method of estimating rainfall using radar is to convert the equivalent radar reflectivity factor to rainfall rate by a Z–R relationship. Typically, it has a power law form. The statistical model presented below specifies that rainfall rate can be represented as the product of two terms. The first term is a power function of equivalent reflectivity factor with range-dependent parameters. The second term specifies a multiplicative error model for radar rainfall estimates. The statistical model is expressed as follows:

$$R_{s,t}(i,j) = [a(j)\, Z_{s,t}(i,j)^{b(j)}]\,[B(s)\varepsilon_{s,t}(i,j)] \tag{2}$$

In the formulation given above the model can be interpreted as a regression model for log rainfall rate versus log reflectivity factor. The error field ε has, for each s, t, i, and j, a log-normal distribution with mean 1 and range-dependent standard deviation. The mean field bias $B(s)$ is a Markov chain with median 1. Its complete distribution is specified below. Unlike the mean field bias, the error field e is spatially varying over the radar field and varies from scan to scan. Both error processes are assumed to be mutually independent and to be independent of the reflectivity process. Underlying the distributional assumptions on the error processes is the assumption that rainfall rate and reflectivity factor follow a log-normal distribution. An alternative representation of the model is

$$R_{s,t}(i,j) = [A_s(j)\, Z_{s,t}(i,j)^{b(j)}]\,[\varepsilon_{s,t}(i,j)] \tag{3}$$

where

$$A_s(j) = a(j)\, B(s) \tag{4}$$

This formulation leads to the interpretation of the bias process as producing a randomized Z–R relationship. In this case the randomized multiplicative coefficient is given by the process $\{A_s(j)\}$. The formulation of (3) also leads to an interpretation of the statistical model as a 'random coefficient' regression model. This interpretation is useful in development of parameter and state estimation procedures.

Denote the natural logarithm of the mean field bias for hour s by $\beta(s)$, that is,

$$\beta(s) = \ln[B(s)] \tag{5}$$

We assume that $\beta(s)$ is a stationary Markov chain over the integers $(1, \ldots, T)$ satisfying

$$\beta(s) = a_1\beta(s-1) + W(s); \qquad W(s) \sim N(0, v) \tag{6}$$

where $0 \leq a_1 \leq 1$, v is nonnegative, T is the storm duration in hours, and $W(s)$ is a sequence of independent normally distributed random variables with mean 0 and variance v. The log bias process has mean 0. We denote the stationary variance of the log bias process by the parameter a_2 that is,

$$a_2 = \mathrm{var}[\beta(s)]; \qquad s = 0, \ldots, T \tag{7}$$

The correlation function of the log bias process is given by

$$\mathrm{Cor}\{\beta(s), \beta(s+k)\} = a_1^k \tag{8}$$

Two very distinct conceptual models of radar bias have simple representations in terms of model parameters. It follows from the assumption of stationarity that

$$v = a_2(1 - a_1^2) \tag{9}$$

Note that if a_1 equals 1, v must equal 0. If the correlation parameter a_1 equals 1, the bias process can vary randomly from storm to storm, but is fixed over the duration of a storm. The log bias, which applies over the duration of a storm, has a normal distribution with mean 0 and variance a_2. Conversely, if the correlation parameter is less than 1, the bias varies not only from storm to storm but also over the course of a storm.

To develop procedures for estimating the bias process $B(s)$ it is necessary to specify the relationship between radar and rain gage observations and the mean field bias. The number of rain gages reporting measurable rainfall for hour s of the storm is denoted $\eta(s)$. The accumulated rainfall measured by gage k during hour s is denoted $G_s(k)$. Gage locations are specified in terms of the radar grid coordinates; the location of the kth gage is denoted $(i(k),j(k))$. It follows from (2) that

$$B(s)=\frac{\sum_{k=1}^{\eta(s)}\sum_{t=1}^{\gamma} R_{s,t}[i(k),j(k)]}{\sum_{k=1}^{\eta(s)}\sum_{t=1}^{\gamma} a[j(k)]\, Z_{s,t}[i(k),j(k)]^{b[j(k)]}\varepsilon_{s,t}[i(k),j(k)]} \tag{10}$$

Recall that γ is the number of radar samples in an hour. Based on (10) and the log-normality of $B(s)$ the following approximation is used for the observation equation:

$$Y(s)=\beta(s)+M(s); \qquad M(s)\sim N\{0,\sigma[\eta(s)]^2\} \tag{11}$$

where

$$Y(s)=\ln\left\{\frac{\sum_{k=1}^{\eta(s)} G_w(k)}{\sum_{k=1}^{\eta(s)}\sum_{t=1}^{\gamma} a[j(k)]\, Z_{s,t}[i(k),j(k)]^{b[j(k)]}}\right\} \tag{12}$$

$\sigma(n)$ is a nonnegative function representing the observation error given that the number of gages with measurable rainfall is η, and $M(s)$ is a sequence of independent normally distributed random variables with mean 0 and variance $\sigma(n)$. It is assumed that the error function is a power function of the number of gages, that is,

$$\sigma(n)^2=a_3\eta^{a_4} \tag{13}$$

The power law form of (13) allows the error model to account for correlation of gauge observations.

The observation $Y(s)$ is the log ratio of mean gage rainfall to mean radar rainfall at gage locations. The error process $M(s)$ accounts for the approximation of gage observations for the numerator in (10) and for the approximation of radar observations to the random variable in the denominator of (10). Therefore, both discrete approximation and measurement errors are taken into account.

STATE ESTIMATION OF MEAN FIELD BIAS

The formulations of the bias model and observation model in the previous section provide a model system in which procedures can be developed for correcting radar rainfall estimates for mean field bias. The procedures developed below will be applied in the NEXRAD precipitation processing algorithms. The resulting precipitation estimates serve multiple purposes. They are used for flash flood forecasting, main stem flood forecasting, routine river flow and stage forecasting, and in preparation of long-term hydrologic forecasts. Automated rain gage data are required, along with radar data, for implementation of the bias estimation procedures. For the majority of radar umbrellas in the US it will be possible to obtain more than 30 gages with automated hourly data. For more than one-third of the radar umbrellas 100 or more automated gages are currently available.

In the flash flood application interest focuses on the most recent rainfall estimates. To correct these estimates for mean field bias we must estimate the bias for the most recent hour given observations prior to and including the most recent hour. This type of state estimation problem is referred to as a filtering problem. For main stem river forecasting, with the longer response times of catchments to precipitation, precipitation estimates prior to the most recent hour are of significant interest. Consequently in some situations we will want to correct for bias in preceding hours given observations prior to, including, and following a given hour. This state estimation problem is referred to as a smoothing problem. The final type of problem we may face is one in which inadequate gage data are available for the current hour to make a bias computation. In this case we will want to estimate the current bias from observations preceding the current hour. This problem is one of prediction. We define our state estimation problems more formally below.

State estimation is distinguished from parameter estimation by virtue of the fact that the objects to be estimated are random variables rather than unknown real-valued parameters. In the problem at hand the mean field bias $\{B(s)\}$ has been modeled as a random process with distributional law specified by (5)–(7). The observations related to the bias process are specified by the observation equation (12). State estimators are derived, whenever possible, as the conditional expectation of the process given the observations. We define our state estimation problems below.

Let $Y(1),\ldots, Y(T)$ be observations of log radar bias, as defined in (12). The state estimation problem is to compute the conditional expectation of the bias $B(s)$ given observations $Y(1),\ldots, Y(u)$ for u less than or equal to T, that is,

$$\hat{B}(s|u)=\mathrm{E}[B(s)|\,Y(u),\ldots, Y(1)] \tag{14}$$

If $u\leq s$, the problem is one of prediction; if $u=s$, the problem is one of filtering; and if $u\geq s$, the problem is one of smoothing.

To evaluate accuracy of the state estimators we want to compute the conditional error variance

$$V(s|u)=\mathrm{E}\{[\hat{B}(s|u)-B(s)]^2|\,Y(u),\ldots, Y(1)\} \tag{15}$$

The bias model is defined in terms of the log bias process $\beta(s)$. State estimators for $B(s)$ will consequently be derived in terms of state estimators for $\beta(s)$. The following argument shows how this is done. From (6) and (11) it can be seen that the random variables $\beta(s)$, $Y(1),\ldots,Y(u)$ have a multivariate normal distribution. It follows from Theorem 2.5.1 in Anderson (1958) that the conditional distribution of $\beta(s)$ given $Y(1),\ldots,Y(u)$ is normal, we will write

$$[\beta(s)|Y(u),\ldots,Y(1)] \sim N[\hat{\beta}(s|u),\Sigma(s|u)] \quad (16)$$

where

$$\hat{\beta}(s|u) = \mathrm{E}[\beta(s)|Y(u),\ldots,Y(1)] \quad (17)$$

and

$$\Sigma(s|u) = \mathrm{E}\{[\hat{\beta}(s|u)-\beta(s)]^2|Y(u),\ldots,Y(1)\} \quad (18)$$

It follows that the conditional distribution of $B(s)$ given $Y(1),\ldots,Y(u)$ is log-normal with parameters $\beta(s|u)$ and $\Sigma(s|u)$. Consequently,

$$\hat{B}(s|u) = \exp\{\hat{\beta}(s|u) + (\tfrac{1}{2})\Sigma(s|u)\} \quad (19)$$

and

$$V(s|u) = \hat{B}(s|u)^2\{\exp[\Sigma(s|u)] - 1\} \quad (20)$$

The remaining problem is to compute state estimates for log bias. We begin with the filtering problem for $\beta(s)$. It can be shown (Smith & Krajewski, (1991)) that the smoothing and prediction problems simplify to filtering problems. We suppress the conditioning argument u for filtering problems, so $\hat{\beta}(s)$ equals $\hat{\beta}(s|s)$ and $\Sigma(s)$ equals $\Sigma(s|s)$.

The Kalman filter can be used to recursively compute $\hat{\beta}(s)$ and $\Sigma(s)$. To initiate the procedure note that

$$\hat{\beta}(0) = \mathrm{E}[\beta(0)] = 0 \quad (21)$$

and

$$\Sigma(0) = \mathrm{E}\{[\hat{\beta}(0)-\beta(0)]^2\} = a_2 \quad (22)$$

For $s>1$, the conditional expectations can be computed recursively using the following relations, developed originally by Kalman (1960):

$$\hat{\beta}(s) = a_1\hat{\beta}(s-1) + \{H(s)/[a_3\eta(s)^{a_4} + H(s)]\}\, e(s) \quad (23)$$

$$\Sigma(s) = H(s)\,\{1 - H(s)/[a_3\eta(s)^{a_4} + H(s)]\} \quad (24)$$

where

$$e(s) = Y(s) - a_1\hat{\beta}(s-1) \quad (25)$$

is the 'innovation' and

$$H(s) = a_1^2\Sigma(s-1) + a_2(1-a_1^2) \quad (26)$$

The prediction problem is easily solved by noting that

$$\beta(s+k) = a_1^k\,\beta(s) + \sum_{j=0}^{k-1} a_1^j\, W(s+k-j) \quad (27)$$

It follows that

$$\hat{\beta}(s+k|s) = a_1^k\hat{\beta}(s) \quad (28)$$

Similarly,

$$\Sigma(s+k|s) = a_1^{2k}\,\Sigma(s) + \sum_{j=0}^{k-1} a_1^{2j}a_2(1-a_1^2) \quad (29)$$

Discussion on the smoothing procedure can be found in Smith & Krajewski (1991). In this section it has been implicitly assumed that parameters of the observation equation and bias model are known. Generally, this will not be the case. In order to effectively implement the state estimation procedures we must develop procedures to estimate unknown parameters of the model. Smith & Krajewski (1991) presented a recursive algorithm for parameters estimation. In the next section we will describe a Monte Carlo simulation experiment demonstrating the effects of parameter uncertainty on the bias estimation.

EFFECT OF PARAMETER UNCERTAINTY

The values of parameters of the bias model are never known exactly even if the model structure is perfect. The two main sources of parameter uncertainty are 1) the measurement errors; and 2) the sampling errors. Smith & Krajewski (1991) presented the results of extensive Monte Carlo simulation experiments of the bias model parameter estimation. They focused on the effects of sampling errors and concluded that, if the record is large enough the parameters can be estimated quite accurately. The most important finding was that the estimates were unbiased. In this section we extend their experiment to investigate the effects of parameter estimator errors propagation on the radar rainfall field bias estimators.

In our experiment the raingage and radar observations are generated according to the bias model presented earlier. In the generation step assumed (true) values of the parameters are used. Then, the state (bias) estimation takes place. Two scenarios were investigated. In the first one the true values of the parameters were used while in the second one the perturbed values were used. That way the bias errors in the first scenario are the consequence of the sampling error only. This sampling error is due to limited number of storms used in the bias estimation. The second scenario accounts for both the sampling errors and the errors in the model parameters. The perturbations in the second scenarios were generated randomly from a Gaussian distribution with the mean and variance corresponding to the distribution of the parameter values obtained in the experiments by Smith & Krajewski (1991). Values corresponding to several sample sizes are used. The sample size ranges from 10 to 500 storms. Storms are defined as in Smith and Krajewski (1991). The results of the simulation are presented in Figs. 1 and 2 for both scenarios, respectively. The statistics shown in the Figures are the mean, the variance and the lag-1 correlation of the bias errors. For both scenarios the results are unbiased, but there is considerably more variability in the second scenario which takes into account the parameter errors. The residual correlation is low in both cases.

It should be emphasized that the presented results establish the upper limit of bias estimation performance (using the presented model) since the simulation assumed the perfect model structure. The presented framework, however, offers a way to estimate radar-rainfall bias in real time using a limited number of raingages (10 in the simulation experiment) and producing an accuracy measure associated with the bias estimates.

Z–R PARAMETER ESTIMATION

The discussion in the proceeding sections was based on the assumption that the parameters of a *Z–R* relationship, $a(j)$ and $b(j)$ are known. In this section we will investigate the estimation process of these parameters since accurate specification of *Z–R* parameters provides the fundamental building block for constructing high quality radar rainfall estimates. We will focus on a particular method of *Z–R* estimation, the so-called climatological method.

The method was developed in response to the need for accurate *Z–R* estimation from non-synchronous radar and rain gage observations. The concept was originally suggested by Miller (1972), and has been followed by the work of Calheiros & Zawadzki (1987), Atlas *et al.* (1990), Smith *et al.* (1989), and Rosenfeld *et al.* (1990). In general terms, the method establishes *Z–R* relations by relating reflectivity and rainfall rate values corresponding to the same probability of exceedence. These values can be obtained not only from the concurrent pairs, but from all the available historical radar and raingage data as well. The approach could be particularly attractive if used in conjunction with methods such as the Area-Time Integral (ATI) (Doneaud *et al.*, 1984) and the Height-Area Rainfall Threshold (HART) technique (Atlas, *et al.*, 1990) to estimate areally averaged rainfall. Krajewski & Smith (1991) discussed many statistical aspects of the method and presented the results of a Monte Carlo experiment of *Z–R* parameter estimation. In this paper we will continue the framework of their experiment focusing on the single issue of the effects of rainfall thresholds often used in the climatological method.

Climatological method

The key feature of the climatological method, which in spirit is a nonparametric procedure, is its reliance on order statis-

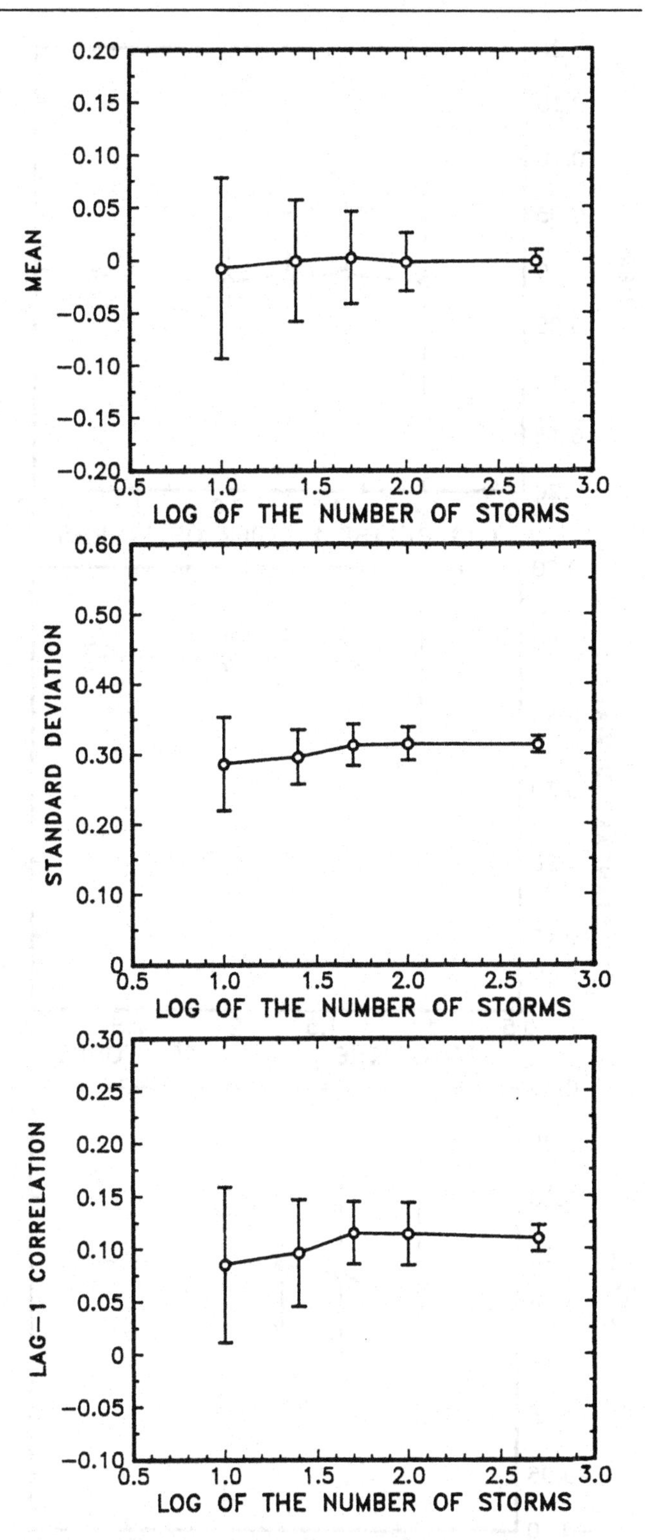

Fig. 1 Results of the Monte Carlo simulation experiment using perfect parameters of the bias model. The mean, variance, and Lag-1 correlation of the log-bias errors are shown. The true values of the model parameters were $a_1 = 0.8$, $a_2 = 0.1$, $a_3 = 1.0$, and $a_4 = -1.0$. The data points correspond to 10, 25, 50, 100, and 500 storms each with average duration of 5 hours. One hundred realizations were performed. The vertical bars denote a range of one standard deviation. The average number of rain-reporting gages per hour was 10.

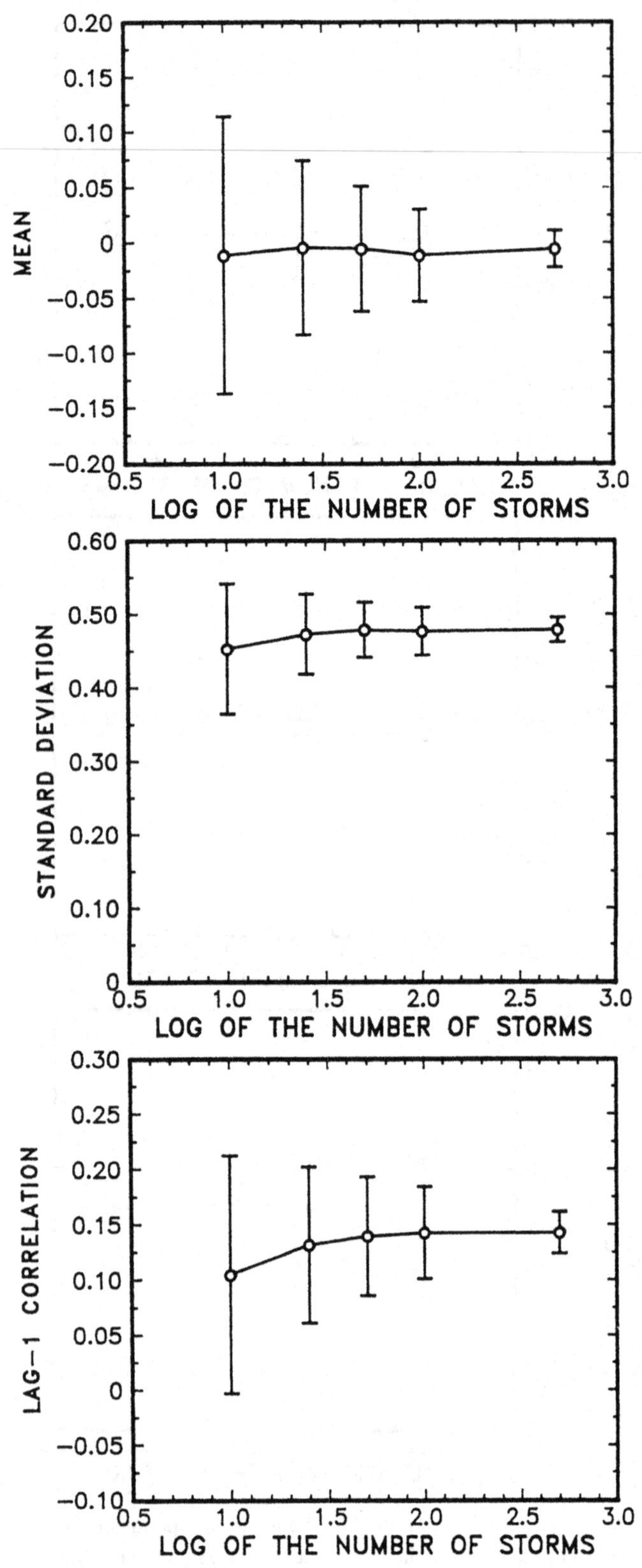

Fig. 2 Results of the Monte Carlo simulation experiment. The true parameters of the bias model are the same as those in Fig. 1. In bias calculations the randomly perturbed parameters were used.

tics, that is, the ordered values of the sample. We denote the order statistics of the rain rate sample by

$$R_{(1)} \leq \ldots \leq R_{(n)} \tag{30}$$

Similarly the order statistics of the reflectivity sample are denoted

$$Z_{(1)} \leq \ldots \leq Z_{(m)} \tag{31}$$

The quantile function of the reflectivity distribution is defined by

$$Q_z(p) = F_z^{-1}(p) \tag{32}$$

where $F_z(\cdot)$ is the probability distribution function of the radar reflectivity. In words, the quantile $Q_z(p)$ provides the reflectivity value that is exceeded with probability $1-p$. Similarly, $Q_\mathrm{R}(p)$ will denote the quantile function of the rainfall rate distribution. Nonparametric procedures for estimating climatological Z–R relationships can be based on the observation that

$$Q_z(p) = aQ_\mathrm{R}(p)^b \tag{33}$$

which follows directly from equation of the power law type relationship typically adopted for Z–R estimation. Climatological Z–R relations can be obtained (Calheiros & Zawadzki (1987)) by relating $Q_\mathrm{R}(p_i)$ and $Q_z(p_i)$ for k values of exceedance probability, $p_1, \ldots, p_k$. We can choose Z–R parameters to minimize the sum of squared differences

$$H(a,b) = \sum_{i=1}^{k} \{\ln[Q_z(p_i)] - \ln[a\, Q_R(p_i)^b]\}^2 \tag{34}$$

To implement estimators based on equation (34) the quantile functions must be replaced by estimators obtained from samples of reflectivity and rainfall rate. Sample quantile estimators are of the form

$$\hat{Q}_\mathrm{R}(p) = \mathrm{R}_{(i)} \quad \text{if} \quad \frac{i}{n+1} \leq p \leq \frac{i+1}{n+1} \tag{35}$$

More sophisticated quantile estimators based on order statistics are discussed in Serfling (1980).

It can be shown (see Cramer, 1946) that the probability density function of the mth order statistic of a random sample with probability density function f_R is given by

$$f_{R,m,N}(R) = \frac{N!}{(m-1)!\,(N-m)!} p(R)^{m-1}\,[1-p(R)]^{N-m} f_R(R) \tag{36}$$

In some cases this result can be used to evaluate sampling properties of quantile estimators. Even for parametric models, however, equation (36) will often be too complicated to provide useful sampling properties of quantile estimators. An attractive feature of quantile estimators, including the sample quantile estimator of equation (35), is asymptotic normality (see Serfling, 1980) represented as follows:

$$n^{1/2}\{[\hat{Q}(p_i),\ldots,\hat{Q}(p_k)]-[Q(p_i),\ldots,Q(p_k)]\} \overset{\mathscr{D}}{\Rightarrow} N(0,\Sigma) \quad (37)$$

where Σ is a k by k covariance matrix. This result will provide a guide for the simulation experiments carried out in the following section.

Nonparametric procedures have two major advantages: parametric distributional assumptions are not required, and estimators can be derived which are more robust to common sources of error in radar and rain gage samples. A disadvantage of nonparametric procedures, compared to parametric models using maximum likelihood estimates, is that it is more difficult to quantitatively assess the accuracy of the estimators. This can be achieved only by using simulation. Krajewski & Smith (1991) investigated the effects of several factors on the accuracy of Z–R parameters estimated within the climatological framework. Factors they considered included: the sample size, radar and raingage observations errors, and the thresholds applied to eliminate the low intensity data which are typically very noisy. The problem of appropriate threshold selection is important in the climatological context as it affects the quantile estimators used in the estimation. This was recognized by Rosenfeld *et al.* (1993) who recommend that the thresholds should correspond to each other. In the next section we present the effects of mismatched thresholds.

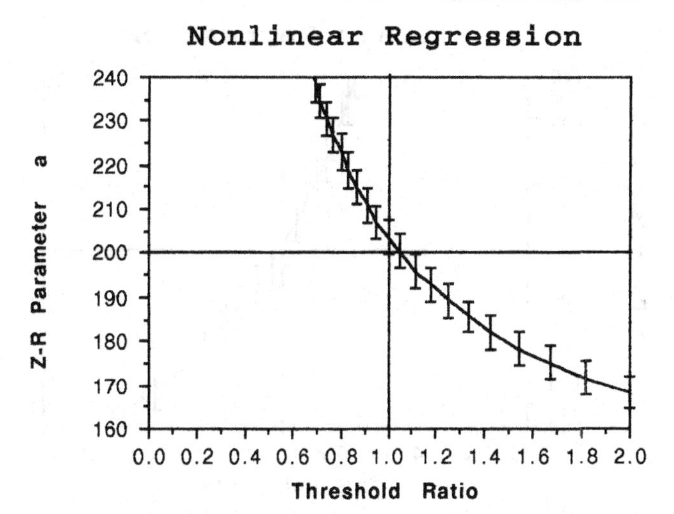

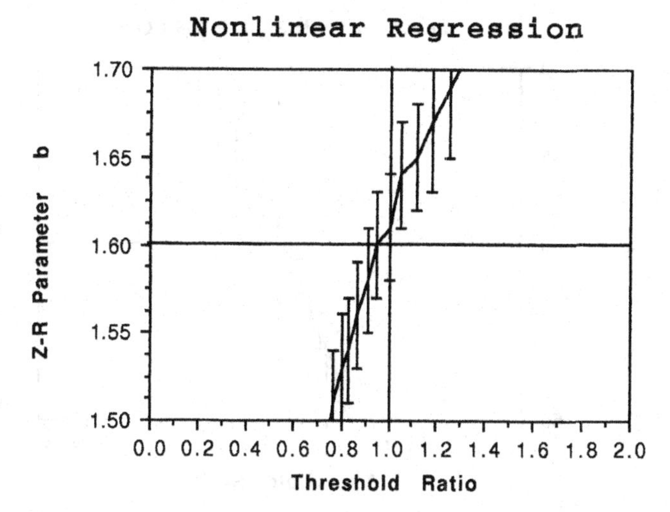

Fig. 3 The Z–R parameter estimators as a function of threshold ratio. The rainfall threshold was kept constant at 0.2 mm/hr while the radar reflectivity threshold was adjusted according to the indicated ratio. For ratio equal to 1.0 both thresholds match through the true Z–R relationship. The top plot shows the parameter a, the bottom plot shows the parameter b. The true values of the parameters are indicated. The number of realizations was 250. The vertical bars indicate one standard deviation range. The estimation was performed using a nonlinear regression.

Monte Carlo experiment

In order to investigate the effect of selecting radar and raingage observations threshold on the performance of the climatological Z–R estimation an extensive Monte Carlo experiment was designed and conducted. The experiment parallels that by Krajewski & Smith (1991). In the experiment the observations were generated from a known (Marshall–Palmer) Z–R relationship for which $a=200$ and $b=1.6$. Independent samples of varied size were generated from an exponential distribution for rainfall and a corresponding (through the true Z–R relationship) distribution for radar reflectivity. The parameter of the exponential distribution was selected to give the mean rainfall rate of 2 mm/h. The size of the generated sample was 10000. The observations in both radar and raingage samples were contaminated by the simulated measurement errors. For the raingage observations the error was random from Gaussian distribution with zero mean and the standard deviation of 10% of the true value. For the radar the value of 1 dBZ was used. This value corresponds to a good quality radar. For each sample nine quantiles corresponding to the probabilities 0.031, 0.063, 0.125, 0.250, 0.500, 0.750, 0.875, 0.938, and 0.967, were calculated. Then, both a nonlinear regression and a linear regression were performed based on the model presented earlier. The process was repeated for 250 independent realizations and statistics of the parameter estimators were calculated.

To eliminate the noisy low intensity rainfall observations (which are of low importance for total rainfall volume) thresholds are often imposed. It was noticed by Krajewski & Smith (1991) that the climatological parameter estimators of a and b were sensitive to the selection of these thresholds. The experiment performed in this work shows these effects systematically. In Fig. 3 and Fig. 4 the results of the Z–R

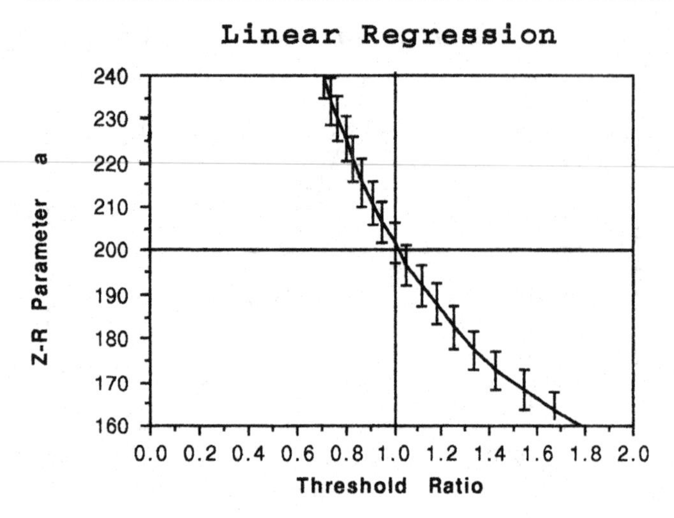

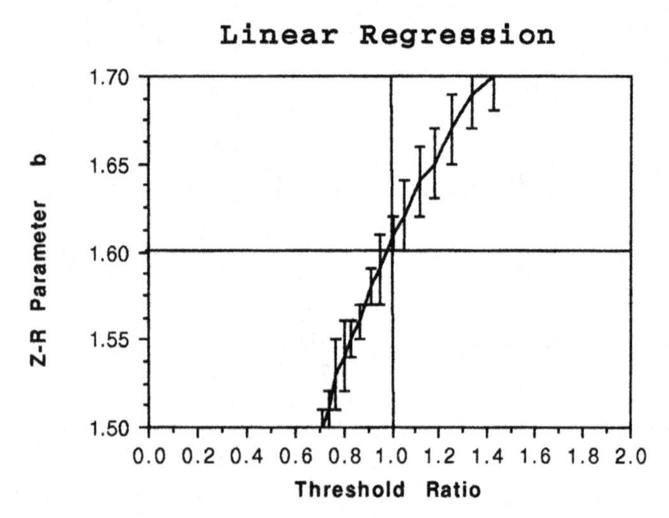

Fig. 4 The same as in Fig. 3, but the estimation was performed using a linear regression.

parameter estimation are presented for nonlinear, and linear regression, respectively. The curves shown correspond to the mean of results obtained based on 10000 element samples. It is quite striking that both parameters obtained using either method show significant sensitivity to the right specification of the thresholds. Since in dealing with actual data we usually do not have enough information about the probability distribution of rainfall one should expect difficulties in estimating Z–R relationship using the climatological data.

CONCLUSIONS

Two problems regarding radar rainfall estimation were investigated using a Monte Carlo simulation approach. The first problem concerns real-time estimation of the mean field bias. It was found that the uncertainty in the parameters of the bias model considerably increase the uncertainty in the mean field bias estimates. Even with the uncertain parameters, however, the bias estimates remain free of systematic errors. The second problem concerns the use of a climatological method of estimating the Z–R relationship. In practical implementations the method, based on matching quantiles of equal probability, disregards values below certain thresholds (both for radar reflectivity and raingage data). It was shown that the parameters of the Z–R relationship are very sensitive to the correct specification of the thresholds. The estimates of the parameters of the Z–R relationship are unbiased only if the thresholds correspond to each other through a correct Z–R relationship.

Monte Carlo simulation should be used more extensively in rainfall research to complement real data analysis. The results of idealized scenarios, which can be studied within a Monte Carlo simulation framework, can provide bounds for the performance of various methods to be expected in reality. The biggest advantage of using a Monte Carlo approach to study rainfall measurement and prediction problems is that it allows for isolation of various effects – something clearly impossible while working with actual data.

ACKNOWLEDGEMENTS

The authors would like to acknowledge the support of federal agencies. The work of W. F. Krajewski was supported by NASA grant No. NAG-5-1126, NOAA Climate and Global Change Program grant NA89AA-D-AC195, and by the National Weather Service under Cooperative Agreement NA86-AA-H-HY126 between the Hydrologic Research Laboratory and the Iowa Institute of Hydraulic Research. The work of J. A. Smith was supported in part by the NEXRAD Joint System Program Office.

REFERENCES

Anderson, T. W. (1958) *An Introduction to Multivariate Statistical Analysis.* John Wiley and Sons Inc., New York.
Atlas, D., Rosenfeld, D. & Short, D. (1990) The estimation of convective rainfall by area integrals, 1. The theoretical and empirical basis. *J. Geophys. Res.*, **95**, 2153–60.
Austin, P. M. (1987) Relation between measured radar reflectivity and surface rainfall. *Mon. Wea. Rev.*, **115**, 1053–69.
Battan, L. J. (1973) *Radar Observation of the Atmosphere*, The University of Chicago Press.
Calheiros, R. V. & Zawadzki, I. (1987) Reflectivity-rain rate relationships for radar hydrology in Brazil. *J. Clim. Appl. Meteor.*, **26**, 118–32.
Chandrasekar, V. & Bringi, V. N. (1987) Simulation of radar reflectivity and surface measurements of rainfall. *Journal of Atmospheric and Oceanic Technology*, **4**(6), 464–78.
Cramer, H. (1946) *Mathematical Methods of Statistics*, Princeton University Press, Princeton, NJ.
Doneaud, A. A., Niscov, S. I., Priegnitz, D. L. & Smith, P. L. (1984) The area-time integral as an indicator for convective rain volumes. *J. Clim. Appl. Meteor.*, **23**, 555–61.
Hudlow, M. D., Farnsworth, R. K. & Ahnert, P. R. (1984) *NEXRAD Technical Requirements for Precipitation Estimation and Accompany-*

ing Economic Benefits. Hydro Tech. Note 4, Office of Hydrology, Nat. Weather Service, NOAA, Silver Spring, Maryland.

Joss, J. & Waldvogel, A. (1990) Precipitation measurement and hydrology, In: Atlas, D. (Ed.) *Radar in Meteorology*, American Meteorological Society, Boston.

Kalman, R. E. (1960) A new approach to linear filtering and prediction problems. *Journal of Basic Engineering (ASME)*, **82D**, 35–45.

Krajewski, W. F. & Smith, J. A. (1991) On the estimation of climatological Z–R relationships. *Journal of Applied Meteorology*, **30**, 1436–45.

Miller, J. R. (1972) *A climatological Z–R relationship for convective storms in the northern Great Plains*. Preprints of the 15th Conference on Radar Meteorology, AMS, Boston.

Rosenfeld, D., Wolff, D. B. & Atlas, D. (1993) General probability-matched relations between radar reflectivity and rain rate. *Journal of Applied Meteorology*, **30**, 50–72.

Rosenfeld, D., Atlas, D. & Short, D. (1990) The estimation of convective rainfall by area integrals, 2. The height-area rainfall threshold (HART) method. *J. Geophys. Res.*, **95**, 2161–76.

Serfling, R. (1980) *Approximation Theorems of Mathematical Statistics*, J. Wiley, New York.

Shedd, R., Smith, J. A. & Walton, M. L. (1989) The sectorized hybrid scan strategy for NEXRAD, *Proceedings of the International Symposium on Hydrological Applications of Weather Radar*, Salford, England.

Smith, J. A., Shedd, R. & Walton, M. L. (1989). *Parameter estimation for the NEXRAD Hydrology Sequence*, Preprints, 24 Conference on Radar Meteorology, 259–63, AMS, Boston.

Smith, J. A. & Krajewski, W. F. (1991) Estimation of the mean field bias of radar rainfall estimates. *Journal of Applied Meteorology*, **30**, 397–412.

Wilson, J. W. & Brandes, E. A. (1979) Radar measurement of rainfall – a summary. *Bull. Am. Meteorol. Soc.*, **60**(9), 1048–58.

Zawadzki, I. (1984) Factors affecting the precision of radar measurements of rain, Preprints. *22nd Conference on Radar Meteorology*, Boston, Amer. Meteor. Soc.

4 Design of groundwater monitoring networks for landfills

P. D. MEYER, J. W. EHEART, S. RANJITHAN and A. J. VALOCCHI
Department of Civil Engineering, University of Illinois at Urbana-Champaign, USA

ABSTRACT Designing a system to monitor groundwater for contamination from landfills involves a tradeoff between cost, time of detection, and probability of detection. As monitoring wells are spaced more closely together, the probability of detecting a leak improves. Locating monitoring wells further downgradient of the landfill also improves detection probability because the plume disperses more and is less likely to move undetected between two monitoring wells. However, closer spacing costs more, and location further away from the source implies a greater time of detection and a greater probability that a water supply well will become contaminated.

An important problem in designing a monitoring system is that the hydraulic conductivity properties of the aquifer around and under the landfill are often poorly understood. It is possible to test for these properties only at points and interpolation between them may be done only with some uncertainty.

A method is discussed for designing a monitoring system under parameter uncertainty. This method places a given number of wells (the number selected by the analyst or user) in locations that maximize the probability of detection of a plume. The method requires some prior knowledge of the statistical properties of the conductivity parameters of the aquifer and some knowledge of the probability of a leak occurring at any given point in the landfill. The method is microcomputer based and currently runs on an advanced microcomputer workstation. The possibility for its adaptation to a more readily available microcomputer is discussed.

INTRODUCTION

The response of the public to certain environmental issues has recently been what might be called reactionary; the Alar scare of 1989 comes to mind. The so-called NIMBY response of the public to local government attempts to site municipal solid waste landfills (MSWLFs) is a further example. Fear of groundwater contamination from a leaky landfill has been a major reason for the difficulty in locating MSWLFs. When faced with a skeptical, and even reactionary public, the ability of technical experts to present designs that minimize risk becomes very important (not to mention the ability to communicate those risks to the public). In the case of a MSWLF, concern often centers on the risk of exposure to contaminated groundwater.

The risk of exposure to groundwater contamination from a landfill can be reduced in several ways. The landfill can be located in a hydrogeologic environment that inhibits the transport of contaminants into potential groundwater sources. The landfill can also be designed to minimize the chance of leakage. The risk of contamination, however, cannot be completely eliminated. Even a well-located and well-designed landfill may release contaminants. Risk of exposure can be further reduced by monitoring the quality of the groundwater. In the event of a release, it is the purpose of groundwater monitoring to detect the contaminant early enough that appropriate action can be taken to prevent exposure.

For the purposes of this discussion, a groundwater monitoring network is a series of wells located around a landfill and sampled periodically for contaminants. Such a monitoring network, unfortunately, may not completely eliminate the risk of exposure to contaminants released from the landfill. Since it is impossible to predict with certainty the path of a contaminant plume in groundwater, a plume may travel between monitoring wells and go undetected by the network. Uncertainty in the contaminant path arises because the natural variability of subsurface properties limits our understanding of site specific contaminant transport. Cost constraints limit the amount of hydrogeologic information that can be gathered. In addition, it may also be impossible to predict important factors such as the location of a contaminant release. The net effect of our incomplete information is that the success of a monitoring network cannot be predicted absolutely. It is thus appropriate to design a monitoring network that has a high probability of detecting groundwater contamination.

This paper presents a method for groundwater monitoring network design that explicitly incorporates uncertainty in the description of contaminant transport. The method uses a numerical model of groundwater flow and contaminant transport coupled to an optimization model. The method provides network alternatives that have a high probability of detection and can also be used to develop tradeoffs between the number of wells in the monitoring network, the position of the compliance boundary, and the probability of contaminant detection. An application to a field site is also presented.

The 1984 amendments to the Resource Conservation and Recovery Act (RCRA) required the Environmental Protection Agency (EPA) to establish criteria for municipal solid waste landfills. These criteria, as stated in RCRA, should provide that 'no reasonable probability of adverse affects on health or the environment' arise from the disposal of waste in the landfill. EPA has proposed extensive revisions to the regulations for MSWLFs. Among the proposed rules is a requirement that groundwater monitoring be carried out to determine both the background (upgradient) water quality and the quality of groundwater passing a specified (downgradient) compliance boundary. The compliance boundary is either the landfill boundary, or an alternative boundary specified by the State that can be up to 150 meters from the landfill boundary. The number, location, and depth of monitoring wells is to be proposed by the landfill owner or operator and approved by the State. The State's evaluation is to be based upon site specific hydrogeologic information.

The network design method presented here can be used for site specific evaluation of decisions such as the appropriate position of the compliance boundary and the number and location of monitoring wells. The method can also be used to evaluate the performance of existing monitoring networks. In addition, the method can be used in a more generic manner by examining the effect on network design of changes in parameter values. These results can be used to develop rules-of-thumb that may be appropriate for screening network design alternatives.

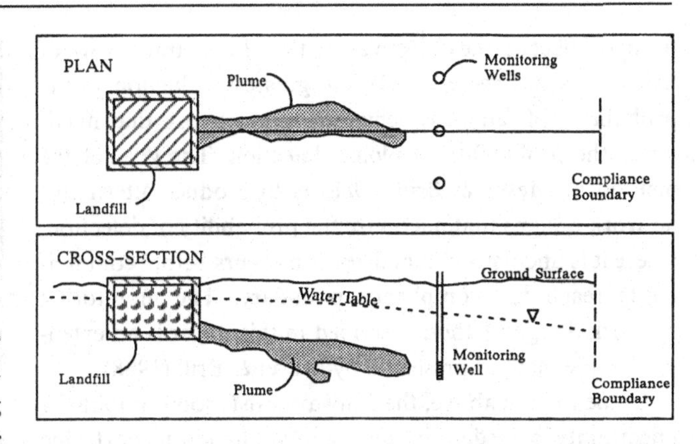

Fig. 1 Groundwater contamination scenario (after Massmann & Freeze, 1987).

A REVIEW OF THE MONITORING NETWORK DESIGN METHOD

A comprehensive framework for design of a landfill operation was presented by Massmann & Freeze (1987). Their design objective maximizes a sum of benefits, costs, and risks. Risk is present in the operation of a landfill because of the possibility that the landfill will fail to contain its contents. In the framework of Massmann & Freeze (1987), failure occurs when a contaminant leaks from the landfill and reaches a prespecified compliance boundary with a concentration in excess of a predetermined level (e.g. the detection limit). If a landfill fails in this manner, the consequences of that failure, such as increased regulation and possible remedial action, have costs associated with them. These costs are the components of risk in the objective function of Massmann & Freeze (1987). Fig. 1 illustrates the components of the problem. Groundwater monitoring enters into the landfill design process because it is able to influence the objective function by providing a means to detect a leak from the landfill, thus reducing the probability of failure. Monitoring is able to play this role because the monitoring wells are located between the landfill and the compliance boundary, as is indicated in Fig. 1. Upon detection of a contaminant at a monitoring well, a landfill owner/operator is thus able to take action to prevent failure (i.e. to prevent the contaminant from reaching the compliance boundary). Note that detection, as used here, means detection before the contaminant reaches the compliance boundary.

Meyer & Brill (1988) used the framework of Massman & Freeze (1987) to develop a method for the optimal design of groundwater monitoring networks. Meyer & Brill (1988) reasoned that, since monitoring can directly influence the landfill owner/operator's objective function, the most desirable monitoring network is that which has the greatest positive effect on the objective function. In the framework of Massmann & Freeze (1987), the greatest reduction in the probability of failure is achieved by a network that maximizes the probability of plume detection. The goal of the method of Meyer & Brill (1988) is to produce alternative network designs that maximize the probability of detection, where it is understood that detection occurs before contaminants reach the compliance boundary. The monitoring network design method described in this paper is an extension of the method presented by Meyer & Brill (1988).

As mentioned above, there always exists some amount of uncertainty in the description of contaminant transport. One of the most important parameters, in terms of its contribution to uncertainty, is the hydraulic conductivity. Hydraulic conductivity is a measure of how easily water can move through a porous material. In particular, the transport of contaminants in groundwater is greatly affected by the manner in which the hydraulic conductivity varies in space. For the landfill monitoring problem, the location of the contaminant source constitutes another source of uncertainty. That is, the contaminant will leak from within the boundary of the landfill, but the precise location of the leak cannot be predicted. Although the hydraulic conductivity and the contaminant source location are here assumed to be the major contributors to contaminant transport uncertainty, there may be additional parameters that are also important.

The uncertainty in parameter values is incorporated into the monitoring network design using a Monte Carlo simulation. The Monte Carlo procedure involves the generation of random contaminant plumes using a numerical model. The two parameters discussed above, the hydraulic conductivity and the location of the contaminant source, are modeled as random variables. Each random plume results from the random selection of both the contaminant source location and the spatially varying hydraulic conductivity. Figs. 2A, B and C represent three random plumes that might result from a Monte Carlo simulation. The small circles in Fig. 2 represent potential well locations, the set of points at which a monitoring well may be installed.

In the monitoring network design method, the movement of each of the random plumes is simulated until it just reaches the compliance boundary, as shown in Fig. 2. This is the point of failure. If a plume can be detected before reaching this point, it is possible that failure can be averted. A plume is detectable if the concentration at a potential well location is above the detection limit at the point of failure. The shaded well locations in Figures 2A, B and C, are the locations at which each plume is detectable. The Monte Carlo simulation for network design consists of generating a large number of random plumes and keeping track of the potential well locations at which each plume is detectable.

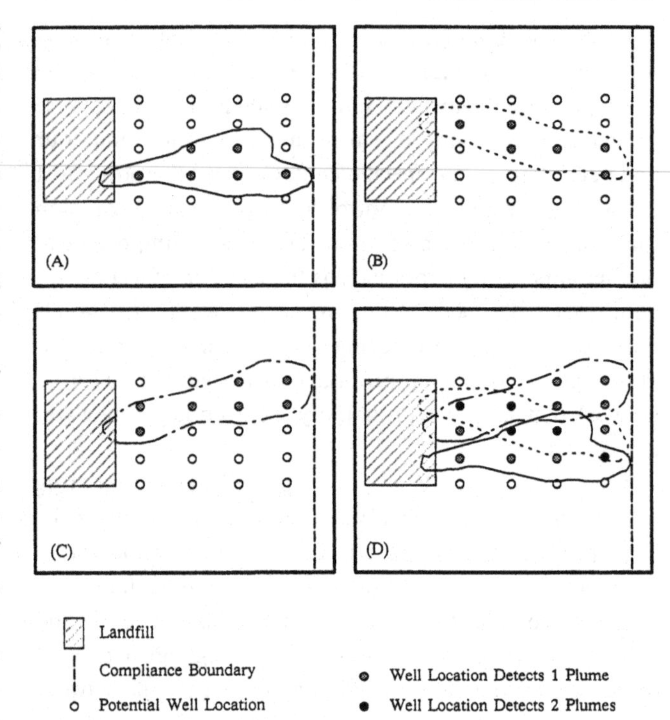

Fig. 2 An illustration of the monitoring network design method: (A), (B) and (C) show random plumes from a Monte Carlo simulation and the potential well locations indicated at which each plume is detectable before reaching the compliance boundary. (D) shows how the plumes can be combined to indicate potential well locations that maximize the number of plumes detected.

The goal of the network design is to find the well locations that maximize the probability that a future (unknown) plume is detected. For a given monitoring network configuration, this probability can be estimated by the fraction of simulated, random plumes that the network detects. This quantity can be determined since the Monte Carlo simulation keeps track of the potential well locations at which each plume is detectable. The problem thus becomes to identify the well locations that maximize the number of simulated plumes detected. Fig. 2D illustrates the concept. There are five locations at which two of the three plumes are detectable. One well can thus detect a maximum of two plumes. If the network consists of two properly located wells, all three plumes can be detected.

The optimal location of a single well can always be determined in a straightforward manner by counting the number of detectable plumes at each potential well location. The optimal location of a single well will be that location at which the greatest number of plumes is detectable. It is likely, however, that the network will be composed of more than one monitoring well. In a practical Monte Carlo simulation, consisting of many more than three random plumes and twenty potential well locations, the determination of the optimal locations for two or more wells is no longer trivial, since there will be a large number of network configurations to consider. In this case, an efficient means to search the many possible configurations is required.

Church & ReVelle (1974) presented a facility location model called the Maximal Covering Location Problem (MCLP) that Meyer & Brill (1988) showed could be used in groundwater monitoring network design. Facility location is a broad subject that, in general, deals with the problem of locating facilities to optimally satisfy a demand. Meyer & Brill (1988) drew an analogy between groundwater monitoring network design and facility location by observing that a monitoring well could be thought of as a facility and a contaminant plume could be viewed as a demand. Detection of a plume is analogous to satisfying demand. The MCLP can be used to search the many possible network configurations to find a given number of well locations such that a maximum number of plumes from the Monte Carlo simulation are detected. The MCLP formulation is as follows:

$$\text{maximize } Q=\sum_{i=1}^{I} y_i \qquad (1)$$

$$\text{subject to } \sum_{j\in N_i} x_j \geq y_i \qquad i=1,\ldots,I \qquad (2)$$

$$\sum_{j=1}^{J} x_j = P \qquad (3)$$

$$x_j = 0 \text{ or } 1 \qquad j=1,\ldots,J \qquad (4)$$

$$y_i = 0 \text{ or } 1 \qquad i=1,\ldots,J \qquad (5)$$

where Q is the number of plumes detected before reaching the compliance boundary,
I is the number of plumes generated,
J is the number of potential well locations,
$y_i = 1$ if plume i is detected before reaching the compliance boundary; $=0$ otherwise,
$x_j = 1$ if a well is located at j; $=0$ otherwise,
P is the number of wells in the network, and
N_i is the set of well locations at which plume i is detectable before reaching the compliance boundary.

Constraints of type (2) stipulate that a plume cannot be detected unless a well is included in the network at a location at which the plume is detectable. Constraint (3) requires that P wells be included in the monitoring network solution. Constraints of type (4) and (5) prevent the unrealistic case of a fraction of a well being installed, or a fraction of a plume being detected.

For a fixed number of monitoring wells, the MCLP determines the locations of the wells that maximize the number of simulated plumes detected. Further increasing the number of detected plumes can only be accomplished at the expense of using additional wells. There is thus a tradeoff between the number of simulated plumes detected and the number of monitoring wells. Groundwater monitoring network design is a multiobjective problem in which it is desired to both maximize the probability of detection and minimize the cost of the network. The number of random, simulated plumes detected is used as a surrogate for the probability of detection and the number of wells in the network is the surrogate for cost. The tradeoff between these two conflicting objectives can be found by solving the MCLP several times for different values of P. Such tradeoffs are calculated for the application discussed below.

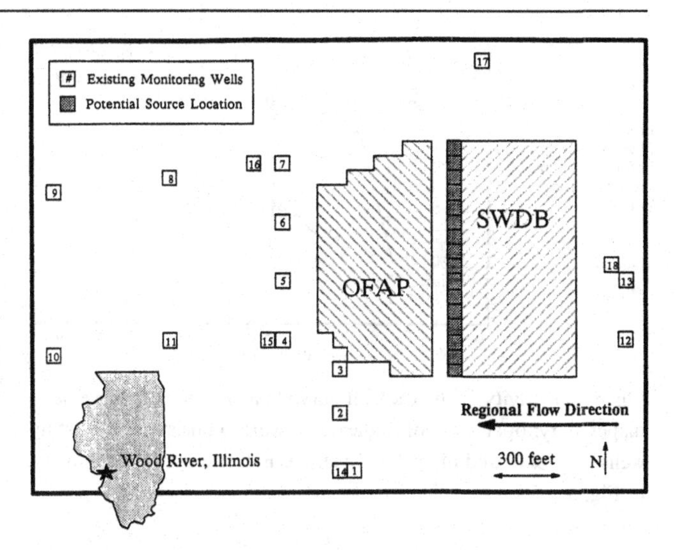

Fig. 3 Shell Oil model showing the location of existing monitoring wells, the solid waste disposal basin, and the old fly ash pond.

AN APPLICATION OF THE MONITORING NETWORK DESIGN METHOD

The application presented here is to a hazardous waste disposal site at the Shell Oil Company–Wood River Manufacturing Complex located in Wood River, Illinois. Fig. 3 shows a layout of the waste disposal area.

Hazardous waste has been disposed of in the Solid Waste Disposal Basin (SWDB). Potential contamination from the SWDB is the concern. There is also a buried disposal area referred to as the Old Fly Ash Pond (OFAP) that is important because it restricts the location of monitoring wells – wells cannot be located in the OFAP. Fig. 3 also shows the locations of existing monitoring wells. The model of the area near the SWDB was based upon available hydrogeological information on the Shell site, obtained primarily from Dames and Moore (1981).

The potential source area was taken to be the downstream edge of the SWDB as indicated in Fig. 3. The actual source for each random plume covered a single model grid cell with its location chosen at random according to a uniform probability distribution. The Monte Carlo simulation con-

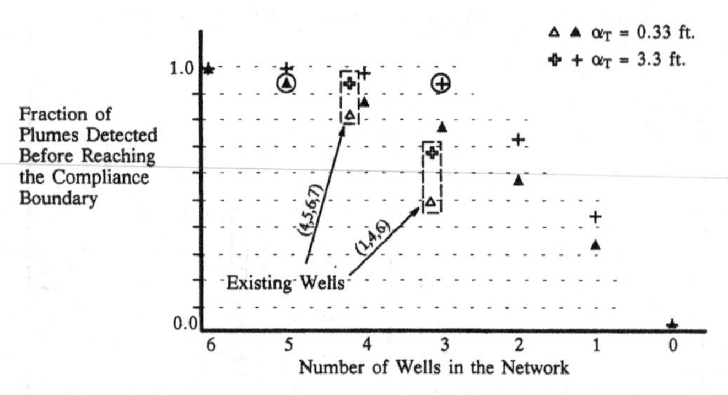

Fig. 4 Sensitivity of the tradeoff curve to a change in transverse dispersivity; open symbols indicate networks consisting of existing wells (1, 4 & 6) and (4, 5, 6 & 7); circles indicate networks shown in Fig. 5.

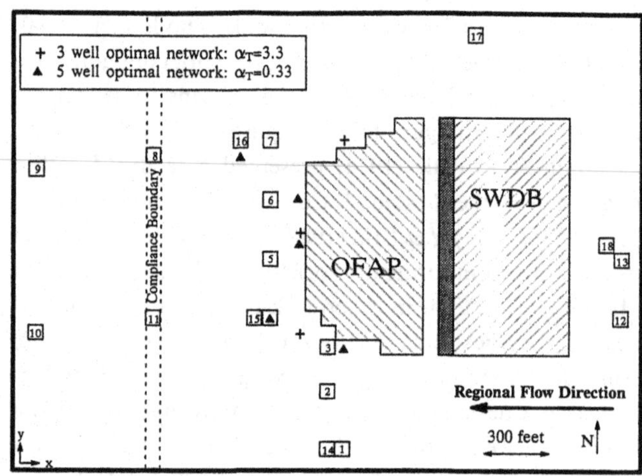

Fig. 5 Effect on optimal well locations of a change in transverse dispersivity. Each network detects 94% of the simulated plumes.

sisted of 160 randomly generated plumes and 290 potential well locations.

A graphical summary of monitoring network performance is the tradeoff between the number of monitoring wells in the network and the maximum fraction of simulated plumes detected before reaching the compliance boundary. This tradeoff is shown in Fig. 4 for the Shell site (refer to the solid triangle or solid plus symbols).

As the number of wells in the network increases, the fraction of plumes detected also increases. As Fig. 4 indicates, five or six wells, optimally located, can detect nearly all the plumes. Fig. 4 presents results for two problems whose only difference is the value of transverse dispersivity. The transverse dispersivity is a groundwater transport parameter that reflects the spread of a contaminant in the direction perpendicular to the direction of regional flow. As transverse dispersivity increases, the simulated plumes tend to spread out and cover a larger area. Consequently, they are easier to detect. Fig. 4 clearly shows that the fraction of plumes detected is sensitive to the transverse dispersivity.

Fig. 4 also shows the modeled performance of two networks consisting of existing wells. The detection monitoring network currently in use at the Shell site consists of existing wells 1, 4, and 6. The fraction of plumes detected by this network (for both transverse dispersivity values) are indicated in Fig. 4 with the open symbols at $P = 3$. The performance of this network is quite sensitive to the transverse dispersivity. This network is also clearly inferior. The network design method predicts that the network currently in use is about 30% below optimal, in terms of the probability of detecting a contaminant plume. It is only fair to note, however, that the boundary conditions of the groundwater flow model impose a regional gradient parallel to the x-axis of the model domain. It is likely that the operators of the Shell site believe the regional gradient is directed slightly toward existing well 1. The groundwater flow data available to us, however, is limited. Herzog *et al.* (1988) note in an evaluation of the Shell site detection monitoring network that wells 1, 4, and 6 are not located to detect all probable contaminant pathways. In any case, this example illustrates how the network design method can be used to evaluate existing monitoring networks. A network made up of four existing wells (4, 5, 6, and 7) is also included in Fig. 4. This network is much closer to optimal.

Fig. 5 illustrates the optimal well locations for two of the networks of Fig. 4 (these two networks are circled in Fig. 4). The networks shown in Fig. 5 are the three well network for the large value of transverse dispersivity and the five well network for the small value of transverse dispersivity. Each of the networks detects 94% of the simulated plumes. As can be seen, the value of transverse dispersivity is related to the average distance between monitoring wells.

The network design method requires the specification of the compliance boundary. Because the method simulates contaminant plumes only until they reach the compliance boundary, this boundary can be interpreted as a limit on the acceptable area of contamination. That is, if the compliance boundary is set farther from the contaminant source, the implication is that it is acceptable to contaminate a larger part of the aquifer. The effect on network design of different compliance boundary positions is examined below.

Tradeoff curves for three compliance boundary positions are presented in Fig. 6. Fig. 7 shows the locations of these boundaries. As Fig. 6 indicates, there is little difference between the fraction of plumes detected for the mid-position and the far-position compliance boundaries. When the compliance point is set at the near-position, however, the optimal detection percentages are significantly reduced. Remember that monitoring wells cannot be placed within the old fly ash pond. The potential locations for wells are thus limited to the small area between the OFAP and the near-position compliance boundary. The results of Fig. 6 suggest that the

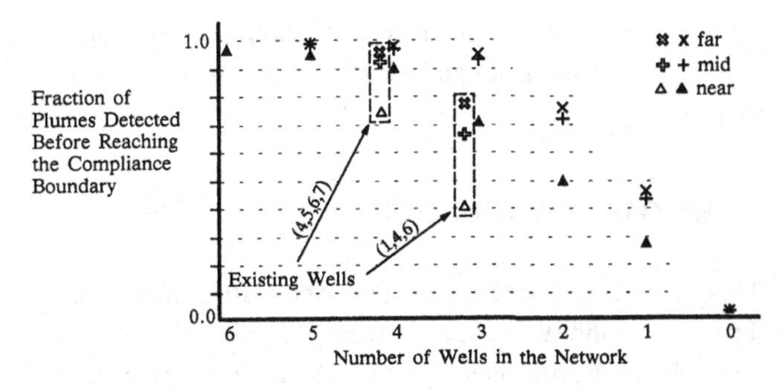

Fig. 6 Tradeoff curves for several compliance boundary locations ($\alpha_T = 3.3$). Open symbols indicate networks consisting of existing wells (1, 4 & 6) and (4, 5, 6 & 7).

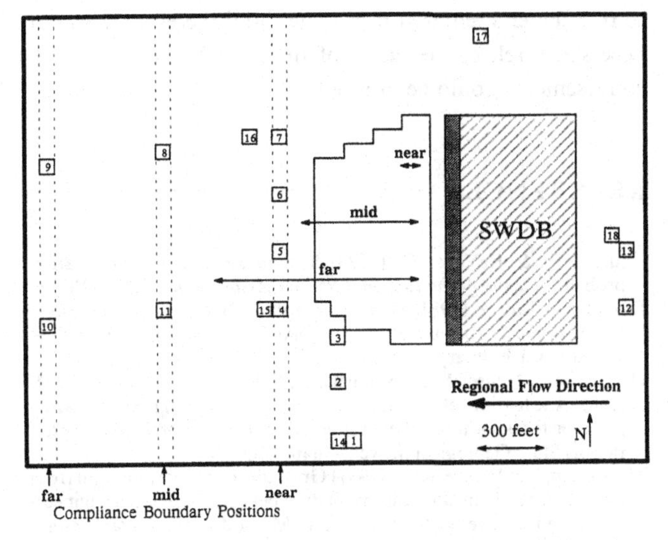

Fig. 7 Three locations of the compliance boundary and the associated range within which a single row, four well solution can be found detecting at least 97% of the simulated plumes.

existence of the OFAP prevents the monitoring wells from being optimally located when the near-position compliance boundary is used.

The question then arises, how does the best location for monitoring depend upon the compliance boundary position? Fig. 6 indicates the effect of compliance boundary position on the fraction of plumes detected, but not the effect on the monitoring well locations. The following specific question was posed for the Shell site. If a monitoring network consisting of four wells located in a single row (i.e. in a line pointing north) were to be used, what is a good location for the network? A network was said to be in a 'good' location if it detected at least 97% of the plumes (an arbitrary value). The answer to this question depends on the location of the compliance boundary.

Fig. 7 shows the range in which a four well network (with all wells located in the same row) that detects at least 97% of the plumes can be found. The range is quite large when the compliance boundary is in the far position, and becomes progressively smaller as the compliance boundary is placed closer to the contaminant source. Note that the range of 'good' locations stops short of the edge of the SWDB. For the near-position compliance boundary, the only 'good' location is near the source, entirely within the old fly ash pond area. Because of the existence of the OFAP, a single row, four well network cannot be found that will detect a high percentage of plumes *and* limit the extent of contamination to the near-position boundary. According to Fig. 7 the best that can be done is to limit the contamination to the mid-position boundary. This can be accomplished by locating four wells just outside the OFAP. To emphasize this point, the performance of the single row, four well network consisting of existing wells 4, 5, 6, and 7 is included in Fig. 6. The detection percentage for this network is quite high when the compliance boundary is in the mid or the far position, but declines considerably for the near-position compliance boundary. The performance of the network consisting of existing wells 1, 4, and 6 is included in Fig. 6 for comparison.

An important implication of Fig. 7 is that if it is allowable for a large portion of the aquifer to be contaminated (i.e. if the distance between the contaminant source and the compliance boundary can be large), then less care is needed in locating monitoring wells because there are many good locations. If, however, the goal is to limit strictly the contamination of the aquifer, then the monitoring wells must be precisely located.

It is important to understand the difference between the model of detection monitoring embodied in the design method presented here and the conceptual model under which conventional detection monitoring at MSWLFs is carried out. In conventional practice, the monitoring wells define the compliance boundary. That is, failure occurs when contamination is detected in the *monitoring* wells. At the time of initial detection, however, there is a significant chance that the contamination will already extend beyond the location of the monitoring wells. The implicit assumption in conventional practice is that the extent of contamination can be minimized by locating the monitoring wells as close to the waste disposal facility as possible. No explicit limit on the extent of contamination enters the design, however. By distinguishing between the location of monitoring wells and the position of the compliance boundary, the design method presented in this paper is able to incorporate explicitly a limit on the extent of contamination. The probability of detection can then be measured with respect to this limit and the number and location of monitoring wells can be determined within the context of this previously defined contamination limit.

Fig. 7 illustrates two points concerning well location that are relevant to the discussion of conventional monitoring network design. One, there may be a large area in which monitoring wells can be placed and still achieve a high probability of detection. The size of this area apparently depends upon the distance between the source and the compliance boundary. Two, in order to achieve the highest probability of detection with a fixed number of wells, these wells should not be located very close to either the compliance boundary or the contaminant source. This observation suggests that current monitoring practice is not optimal.

CONCLUDING REMARKS

Monitoring in the vicinity of landfills is an important means of reducing the risk of exposure to contaminants. The inherent uncertainty in the transport of contaminants in groundwater makes monitoring network design an uncertain process. The method for groundwater monitoring network design presented in this paper explicitly incorporates groundwater transport uncertainty. The location of wells is based upon maximizing the probability of contaminant detection. By measuring detection with respect to a fixed compliance boundary the method takes into account the goal of limiting the contaminated area. The application presented above illustrates the use of the method both to design a monitoring network for a specific site and to assess the performance of an existing monitoring network. The method can also be used to evaluate the relationship between problem parameters and network design. The application illustrated this for two parameters: transverse dispersivity and compliance boundary location. With further work, it may be possible to quantify these relationships to develop generic rules-of-thumb. Such rules-of-thumb could be used in screening network design alternatives at poorly characterized sites.

ACKNOWLEDGEMENTS

This paper was also presented at the Association of State and Territorial Solid Waste Management Officials 1990 National Solid Waste Forum held July 16–18, 1990 in Milwaukee, Wisconsin. Although the information described in this paper has been funded wholly or in part by the Hazardous Waste Research and Information Center of the Illinois Department of Energy and Natural Resources, it has not been subjected to the Center's required peer review and therefore does not necessarily reflect the views of the Center and no official endorsement should be inferred.

REFERENCES

Church, R. & ReVelle, C. (1974) The maximal covering location problem. *Papers of the Regional Science Association*, **32**, 101–18.

Dames and Moore (1981) *Site Investigation And Conceptual Design, Ultimate Waste Disposal Facility, Wood River, Illinois*, for Shell Oil–Wood River Refinery.

Herzog, B. L., Hensel, B. R., Mehnert, E., Miller, J. R. & Johnson, T. M. (1988) Evaluation of ground water monitoring programs at hazardous waste disposal facilities in Illinois, *Environmental Geology Notes*, Illinois State Geological Survey, Champaign, IL.

Massmann, J. & Freeze, R. A. (1987) Groundwater contamination from waste management sites: the interaction between risk-based engineering design and regulatory policy, 1, Methodology. *Water Resour. Res.*, **23**(2), 351–67.

Meyer, P. D. & Brill Jr., E. D. (1988) A method for locating wells in a groundwater monitoring network under conditions of uncertainty. *Water Resour. Res.*, **24**(8), 1277–82.

5 Spatial variability of evaporation from the land surface – random initial conditions

R. ROMANOWICZ

Institute of Environmental and Biological Sciences, Lancaster University, Lancaster LA1 4YQ, UK

J. C. I. DOOGE and J. P. O'KANE

Centre for Water Resources Research, University College Dublin, Ireland

ABSTRACT The effect of spatial variation of the initial moisture contents on the distribution of soil moisture and the evaporation rate from the land surface is evaluated. The process of drying is described by a lumped, nonlinear model representing two stages of evaporation using the thermodynamic equation.

STAGES OF SOIL DRYING

The evapotranspiration process over the catchment considered in this work, can be subdivided into two stages. If there is enough water at the surface of the soil, the evapotranspiration proceeds at the potential rate and the process is determined by the atmospheric conditions above the soil surface and the evapotranspiration rate does not depend on the state of the soil. Once the soil moisture at the surface layer is below some value prescribed by soil and vegetation conditions, it is these conditions, that control the evapotranspiration rate, independently of the atmospheric conditions. The controlling factor is usually taken as a threshold average root zone water contents below which the transport of water to the plant leaves limits the transpiration process (Gardner *et al.*, 1975; Cordova & Bras, 1981). In the literature, these two stages of evapotranspiration were tackled using switching boundary conditions (e.g. Entekhabi & Eagleson, 1989; and Kuhnel, 1989). In the first stage the condition of constant moisture flux at the surface was used and in the second stage this condition was replaced by the condition of the constant moisture contents at the surface. A third stage of evaporation is reached for very low moisture contents when the moisture movement is largely in vapour form and is usually taken as a low constant value. This process can be described by the non-isothermal theory (Philip & de Vries, 1957) and is outside the scope of this paper.

When considering the evapotranspiration process at the catchment scale, the existence of the 'switching' phenomenon makes it difficult to average the moisture contents over the catchment area because of discontinuities present in the model description. In this paper a model is proposed, describing the two stages of evapotranspiration process in a concise way without discontinuities in its structure.

In order to overcome the problem of spatial variability of parameters, the idea of scales over which they are defined, has been introduced (e.g. Tillotson & Nielsen, 1984). The larger the spatial variability of a parameter is, the more measurements of this parameter are needed for its identification. That is, the data acquisition costs and measurement errors are likely to increase. On the other hand, the increase of the length scale used decreases the spatial variation of the parameters while increasing the uncertainty of the parameter values. Hence spatial variation and uncertainty are strongly related. The uncertainties of soil moisture contents can be also caused by the inadequate knowledge of the relief, the rates of precipitation and evaporation, their duration and frequency of occurrence. All the lack of human knowledge regarding these factors can be expressed by treating them as random variables. This leads to the necessity of introducing stochastic techniques to obtain the solution of flow problems in the unsaturated zone. Several investigations of stochastic processes of flow in the unsaturated zone have been presented in the hydrological literature over the last few years. The general idea of most of these studies was to evaluate the stochastic properties of an output, assuming a complete knowledge of the stochastic properties of the input data.

STOCHASTIC APPROACHES TO EVAPORATION

A general review of the role of soil moisture in climate modelling together with the present practices of soil moisture modelling in general circulation models was given by Dooge (1986). The first and very substantial work towards the transformation of physical hydrology through the catchment scale to the GCM scale was due to Eagleson and coauthors (Eagleson, 1978; Eagleson, 1982; Milly & Eagleson, 1982; and Entekhabi & Eagleson, 1989). In the first of these works (Eagleson, 1978) a statistical dynamic approach was used to incorporate the short term dynamics of the soil moisture contents into the long-term water balance. The probability density functions of the actual infiltration during storms and actual evaporation between storms were derived, using representative probability density functions for the independent climatic variables and assuming the soil moisture contents and the corresponding rate of deep moisture percolation at the beginning and at the end of each event. Then the average volumes of infiltration and evaporation in the average long-term water balance were evaluated using the average number of storms in the rainy season. Milly & Eagleson (1982) studied the effects of spatial variability of the storm inputs and of the soil on the infiltration process using the statistical techniques of temporal averageing. They found that increased spatial variability nearly always leads to decreased infiltration and to increased surface runoff.

In one of the first approaches to modelling dynamic unsaturated flow in spatially variable soils, the case of unsaturated flow in a series of one-dimensional noninteracting soil columns was considered. The soil properties were assumed random, but uniform over depth. This type of approach can be found in the works of Dagan & Bresler (1983), Koch (1985) and Mtundu & Koch (1987). In the work of Dagan & Bresler (1983) the expectations and variances of a few water flow variables and of effective hydraulic properties have been computed. They were obtained by the statistical averageing procedure and probability density function (pdf) of saturated hydraulic conductivity, under the stationary hypothesis. Mtundu & Koch (1987) derived a model of stochastic ordinary differential equations after having introduced assumptions on the stochastic character of the variables describing precipitation, cumulative infiltration, and evapotranspiration. These equations are equivalent to stochastic Ito integrals (Ito, 1951). Their solutions have been obtained via the sample function approach. The moments of the solutions have been derived as well. This approach was further extended by Serrano & Unny (1987) and Serano (1990) to the cases when the stochastic partial differential equations for horizontal and vertical infiltration are considered. In the model developed by Koch (1985) the form of equations describing the soil moisture flow in the unsaturated zone is similar to that in the previous models. The first and second moments of excess precipitation and drainage have been derived analytically using the standard expectation formulae (Parzen, 1962).

The approach in which the soil is modelled as an assemblage of a series of vertical columns might be justified in the case of infiltration caused by a high rate of surface recharge. In the case when lateral inflow is also important, an alternative approach based on partial differential equations is preferrable. The evaluation of stochastic properties of a three-dimensional process is possible using the linearized perturbation method. For the first time this method was used to explore the effects of spatial variability on unsaturated flow by Anderson & Shapiro (1983), in a one-dimensional steady state model. Yeh *et al.* (1985a, b, c) examined the effects of spatial variability on the steady unsaturated flow using three-dimensional stochastic approach and a linearized perturbation method. This work was continued by Mantoglou & Gelhar (1987a, b, c) towards modelling large scale transient unsaturated systems in spatially variable soils. They assumed that local soil properties are realizations of three-dimensional zero-mean second-order stationary random fields. The large scale model structure has been derived by averageing the local governing flow equation. The resulting mean model representation is in the form of a partial differential equation in which the averaged or flow effective parameters occur. These effective model parameters (i.e. effective hydraulic conductivity and effective specific moisture capacity) have been evaluated using a quasi-linearized fluctuation equation and spectral representation of stationary processes. From the study of analytical relationships obtained it was concluded that the variability of soil properties produces large-scale hysteresis and anisotropy of the effective parameters. The effective parameters of the large-scale model do not depend on the actual local soil properties but rather depend on a few parameters describing the statistics of local variability (i.e. mean, variances, correlation lengths). Unlike the local parameters which are infinite in number and not identifiable, the effective parameters depend on a few parameters that should be identifiable using a finite observation set and/or prior information about the characteristics of the soil property variability.

Although the latter approach seems advantageous for future research, it cannot be used to analyze the influence of initial and boundary conditions (e.g. initial soil moisture contents at the surface) on the soil moisture distribution as well as the impact of the threshold effect (time to desaturation). This follows from the necessary assumption needed for the linearization of the moisture flow equation in the vicinity of mean effective values of parameters far from the boundaries.

MODEL DESCRIPTION

It is assumed herein that the soil moisture flow in unsaturated zone can be described by a one-dimensional model, averaged over depth, where the switching from the atmosphere-controlled to the soil-controlled evaporation is modelled with the help of a thermodynamic equation. Hence the soil–plant–atmosphere system can be considered as a physically based conceptual model, in which the preceding input to the system, i.e. the water infiltrated from the storm event is represented by the initial moisture contents. The output from the model is the actual evapotranspiration from the surface layer. The state variable will be $\Theta(t)$, the soil moisture contents of the root zone. The changes in the state of the system are then described by (Philip, 1957):

$$L\frac{d\Theta}{dt} = -E_p\frac{h_p - h_a}{1-h_a} \tag{1}$$

where L is the depth of the root zone, E_p is the potential evapotranspiration from the upper layer of the soil, and h_a and h_p are the relative humidities of the atmosphere and the soil, respectively ($h_a < h_p < 1$). It is assumed that the relative humidity of the surface soil layer can be described by the thermodynamic equation (cf. Philip, 1957; O'Kane, 1990):

$$h_p = \exp\left(\frac{M_w}{RT}(\Phi - \Phi_e)\right) \tag{2}$$

where M_w is the molar mass of water (0.018 kg/mol), R is the molar gas constant (8.314 J/mol K) and T is the constant temperature (293 K i.e. 20 °C). The initial condition for this process is:

$$\Theta = \Theta_0, \quad t = t_0 \tag{3}$$

where t_0 denotes the beginning of the considered time horizon taken as the time at which infiltration ceases.

Θ is constrained:

$$\Theta_{min} < \Theta_0 < \Theta_s$$

where

$$\Theta_{min} = \max\,(\Theta_{res}, (h_a^{-\alpha/n\Phi_e}\exp(\alpha)\Theta_s) \tag{3a}$$

Θ_{res} denotes residual moisture content, Θ_s is the soil moisture at saturation; n is a soil water parameter depending only on temperature:

$$n = M_w/RT \tag{3b}$$

where Φ_e is an air entry potential.

In order to solve the problem described by equations (1) and (3), the form of the soil moisture characteristic curve $\Phi(\Theta)$ must be assumed. In the present study a generalized version of the model developed by Gardner (1958) is used. As a first step the original exponential Gardner model for $\Phi(\Theta)$ can be considered, with an exponential relationship between the unsaturated hydraulic conductivity and the soil moisture potential. In the present paper this is written as:

$$K(\Theta) = K_s \exp\left(-\alpha_1\frac{\Phi - \Phi_e}{\Phi_e}\right) \tag{4}$$

where α_1 is a dimensionless parameter depending on soil properties and Φ_e is an air entry potential. A similar relationship will be assumed for the soil moisture characteristic curve:

$$\Theta = \Theta_s \exp\left(-\alpha_2\frac{\Phi - \Phi_e}{\Phi_e}\right) \tag{5}$$

where α_2 is a soil dependent dimensionless parameter, in general case different from α_1.

The relations (4) and (5) give the following soil moisture–hydraulic conductivity relationship:

$$K(\Theta) = K_s\left(\frac{\Theta}{\Theta_s}\right)^{\frac{\alpha_1}{\alpha_2}} \tag{6}$$

If α_1 is equal to α_2, the relation (6) becomes linear and it corresponds to a constant soil moisture diffusivity ($D = K\,d\Phi/d\Theta$), cf. Philip (1967). For the general case of α_1 not equal to α_2:

$$D(\Theta) = -\left(\frac{\Theta}{\Theta_s}\right)^{(\alpha_1-\alpha_2)/\alpha_2}\frac{K_s\Phi_e}{\alpha_2} \tag{7}$$

It can be easily seen that the relations given by equations (5), (6), and (7) fulfill the general requirements concerning the variation of diffusivity, the unsaturated hydraulic conductivity and the soil moisture retention curves for real soils. In order to use the above model one has to evaluate its parameters using some known identification techniques (Eykhoff, 1974). The adjustment of the parameter α_2 for the model can be made by fitting the analytical soil moisture retention curve to the data for some specified soils (e.g. sand and loam). The bigger the value of α_2 is, the more nonlinear the model becomes, i.e. the switching between the soil- and atmosphere-controlled evapotranspiration is more sharp (it becomes a step-like function). For the values of α_2 estimated from the conductivity–soil moisture relation for the Gardner model ($\alpha_1 = \alpha_2 = \alpha$) the analyzed model gives a very sharp switching with the actual evapotranspiration equal to the potential one, before the switch, and sudden fall of evapotranspiration after the switch.

AN ANALYTICAL SOLUTION OF THE PROBLEM

Consider the soil moisture retention curve given by equation (5). Substituting this relation to the model equations (1) and (2) and simplifying the notation ($\alpha = \alpha_2$) one gets:

$$L\frac{\mathrm{d}\Theta}{\mathrm{d}t}=-\frac{E_{\mathrm{p}}}{1-h_{\mathrm{a}}}\exp(n\Phi_{\mathrm{e}})\left(\frac{\Theta}{\Theta_{\mathrm{s}}}\right)^{-n\Phi_{\mathrm{e}}/\alpha}+\frac{E_{\mathrm{p}}h_{\mathrm{a}}}{1-h_{\mathrm{a}}} \tag{8}$$

The solution of this equation can be obtained analytically:

$$t=\frac{L(1-h_{\mathrm{a}})}{E_{\mathrm{p}}}[F(\Theta_0)-F(\Theta)] \tag{9}$$

where

$$F(\Theta_0)-F(\Theta)=-\int_{\Theta}^{\Theta_0}\frac{\mathrm{d}\Theta}{h_{\mathrm{a}}-\exp(n\Phi_{\mathrm{e}})\left(\frac{\Theta}{\Theta_{\mathrm{s}}}\right)^{-n\Phi_{\mathrm{e}}/\alpha}} \tag{10}$$

and

$$F(\Theta)=\Theta_{\mathrm{s}}\frac{m}{b}\left[\sum_{k=1}^{m-1}\frac{1}{m-k}\left(\frac{a}{b}\right)^{k-1}\left(\frac{\Theta}{\Theta_{\mathrm{s}}}\right)^{\frac{m-k}{m}}+\left(\frac{a}{b}\right)^{m-1}\ln\left[-\frac{a}{b}+\left(\frac{\Theta}{\Theta_{\mathrm{s}}}\right)^{1/m}\right]\right]$$

where $a=h_a$; $b=\exp(n\Phi_{\mathrm{e}})$ and $m=\alpha/n\Phi_{\mathrm{e}}$, $\Phi_{\mathrm{e}}<0$. (11)

The number of terms of the series in (11) is determined by the physical parameters of the soil (α and Φ_{e}), as well as the thermodynamic parameter n (equation 3b). For the light, sandy soils this parameter is bigger than for the loam and clay soils. Hence for the clay the number of terms in the series will be the smallest.

The relation (9) allows us to evaluate the time to desaturation, t_{d}, defined as the time in which the soil moisture at the surface layer decreases below some value, Θ^*, conditioned by soil and vegetation status, i.e. a threshold average root zone water content below which transport of water to plant leaves limits the transpiration process (Gardner *et al.*, 1975):

$$t_{\mathrm{d}}=\frac{L(1-h_{\mathrm{a}})}{E_{\mathrm{p}}}[F(\Theta^*)-F(\Theta_0)] \tag{12}$$

The performance of the model depends also on the values of the atmospheric humidity which stands as the lower constraint of the evapotranspiration (see equation 3a). The influence of the values of α and of the atmospheric humidity on the model performance are given in Figs. 1 and 2, respectively. It results from Fig. 1 that for the values of α around 0.15×10^{-3}, i.e. in the range corresponding to several terms of the series in (11), the relation for soil moisture content is nonlinear but invertible. For bigger values of α (over 2.5×10^{-3}) which correspond to hundred or more terms in the series in (11), the moisture content curve becomes distinctly divided into a linear pre-desaturation part and nearly constant post-desaturation part. That curve, though not invertible, models better the evaporation process. A similar effect is obtained by means of increasing the atmospheric humidity. Decreasing the potential evaporation one gets longer intervals before the desaturation occurs. For

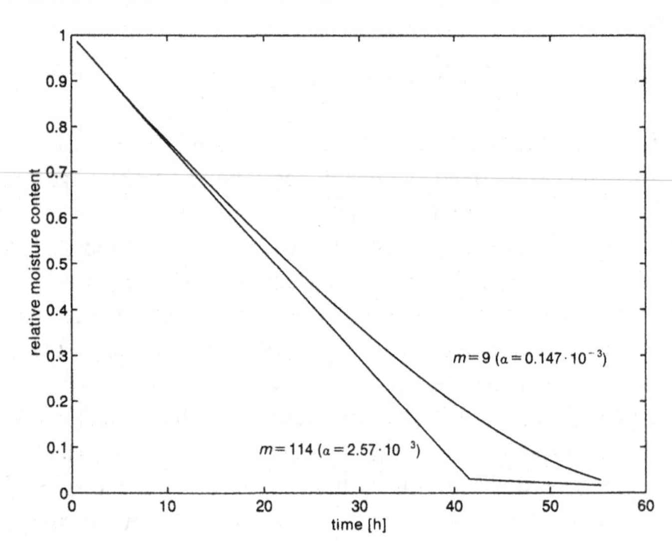

Fig. 1 Influence of the coefficient α on soil moisture deterministic model; $E_{\mathrm{p}}=0.1\cdot10^{-3}$ [m/d].

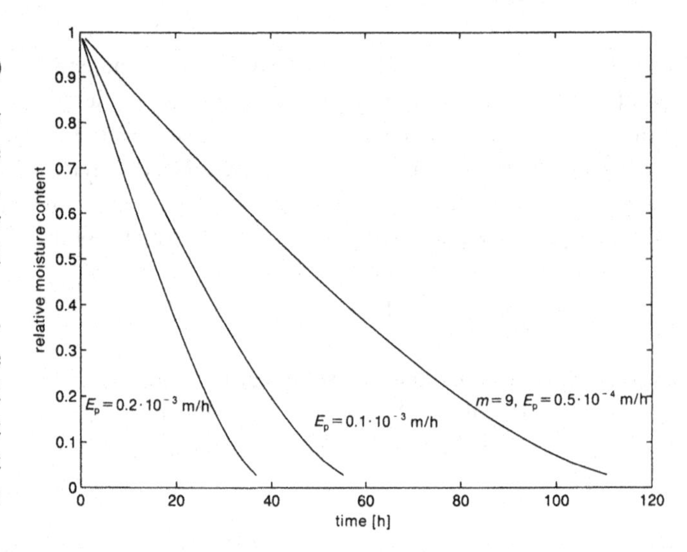

Fig. 2 Influence of potential evaporation on soil moisture content (deterministic model).

the purpose of illustrative examples in this paper the two values of α were chosen: 0.15×10^{-3} and 2.5×10^{-3}.

The purpose of this research is to solve equation (1) subject to (2)–(3) and the given soil moisture characteristic for a field whose hydraulic properties vary in space. In the next section the problem of evapotranspiration will be considered under assumption that uncertainties following from the spatial variability of soil parameters are enclosed in the random initial soil moisture content, Θ_0.

SOLUTION FOR RANDOM INITIAL CONDITIONS

The concept of a stochastic process is used to model the properties which vary in space, as well as the uncertainties following from both errors in modelling the soil moisture flow in unsaturated zone and measurement errors. The models of unsaturated flow process with stochastic inputs are usually assumed to be relatively simple in order for the analytical relationships for the moments of the probability density function to be obtained. Under the assumption that the initial soil moisture contents is a random variable, equation (8) will be treated as a random differential equation (see Soong, 1973; Bobrowski, 1987). This type of equations follows the formalism of stochastic equations with Lebesque or Lebesque–Stieltjes integrals:

$$\int_a^b X_t \mathrm{d}f(t)$$

rather than the stochastic differential equations formulated for example in the work of Mtundu & Koch (1987), cf. also Serrano (1990), following the formalism of stochastic Ito integrals:

$$\int_a^b f(t)\,\mathrm{d}X_t$$

where X_t; $t \in (a,b)$ is a stochastic process and $f(\ )$ is a real function.

The advantage of the former approach consists in the fact that the solution of the random differential equations requires only well known standard mathematical tools and there is no need of introducing special assumptions concerning the input stochastic processes. However, using this technique one has to be careful about the risk connected with the uncritical transfer of properties of the real functions solutions of deterministic differential equations, such as continuity and differentiability to the stochastic processes. Also new type of problems arise related to the probabilistic sense of the obtained solutions and to the relationships between the solutions in the sense of different definitions.

The proposed technique allows one to determine the probabilistic characteristics of the solution for the nonlinear initial condition problems as well as for the random forcing function and random parameters of the model. The algorithm used here consists in:

(a) the solution of the stated problem for deterministic initial moisture content Θ_o; and
(b) subsequent determination of the probability density function of the solution, assuming that Θ_0 is a random variable with known distribution function.

DETERMINATION OF PDF FOR SOIL MOISTURE CONTENTS

The deterministic solution of the model of interest is given by equation (9). The next step of the method consists in the evaluation of the distribution of the solution obtained assuming that the distribution of initial soil moisture content is known. Following the discussion given in the work of Bobrowski (1987), the general formula for the pdf of an absolutely continuous function $Y=h(X)$ of an absolutely continuous random vector X, assuming $h(X)$ and $X=h^{-1}(Y)$ have continuous derivatives, has the form:

$$f_Y(y)=f_X[h^{-1}(y)]\,\det J \qquad (13)$$

where $J=\dfrac{\partial h^{-1}(y)}{\partial y}$

and $f_x(\ )$ is the pdf of a random vector, X.

The distribution function of soil moisture content can be evaluated from the formula (13) provided that the inverse function of the equation (9) exists.

Let us assume that Θ_0 has a lognormal distribution given by

$$f(\Theta_0)=\frac{1}{\Theta_0\sqrt{2\pi}\sigma}\exp\left[\frac{-(\ln\Theta_0-\mu)^2}{2\sigma^2}\right] \qquad (14)$$

where μ and σ are the mean and the variance of $\ln(\Theta_0)$, respectively.

As the value of Θ_0 is constrained according to the condition (3a), the assumption of its lognormal distribution is not strictly correct and should be accompanied by the assumption of a very small variance.

The transformations enabling us to obtain the pdf of the soil moisture content pdf were developed in Romanowicz *et al.* (1990) The formulae obtained allow one to calculate the soil moisture distribution function for any assumed initial moisture content distribution.

As indicated earlier, soil moisture content evaluated from the model given by equation (8) is linear in most of its region, excluding only the values close to the lower constraint determined by the atmospheric humidity, h_a. Hence the solution can be approximated by the linear relation:

$$t=\frac{L}{E_\mathrm{p}}(\Theta_0-\Theta) \qquad (15)$$

which is valid for $\Theta>\Theta_{\min}$, where $\Theta_{\min}$ is given by equation (3a) and $t \ll L\Theta_s/E_\mathrm{p}$. It corresponds to the assumption of constant evapotranspiration with the potential rate in equation (1).

For this case and under the assumption that the distribution of the initial moisture content is given by equation (14), the distribution of the soil moisture content takes the following form:

$$f(\Theta)=\frac{1}{\left(\dfrac{E_\mathrm{p}}{L}t+\Theta\right)\sqrt{2\pi}\sigma}\exp\left(\frac{-\left[\ln\left(\dfrac{E_\mathrm{p}}{L}t+\Theta\right)-\mu\right]^2}{2\sigma^2}\right) \qquad (16)$$

It is interesting to analyze the changes of the soil moisture content with time. Obviously, it tends to zero with the rate inversely proportional to the value of potential evapotranspiration, E_p. Its expected value is equal to:

$$E(\Theta) = -\frac{E_p}{L}t + \exp(\mu_{\Theta_0} + \sigma_{\Theta_0}^2/2) \quad (17)$$

and the variance reads:

$$\sigma_\Theta = \exp(\mu_{\Theta_0} + \sigma_{\Theta_0}^2/2)\sqrt{\exp(\sigma_0^2) - 1} \quad (18)$$

Hence the coefficient of skewness of this approximated model has the form:

$$C_v = \frac{\sigma_0}{\mu_0} = \frac{\exp(\mu_{\Theta_0} + \sigma_{\Theta_0}^2/2)\sqrt{\exp(\sigma_{\Theta_0}^2) - 1}}{-\frac{E_p}{L}t + \exp(\mu_{\Theta_0} + \sigma_{\Theta_0}^2/2)} \quad (19)$$

As the expected value of the soil moisture depends on time, the process is nonstationary. The variance depends only on the parameters of the initial moisture content distribution. The distributions of the initial surface soil moisture content and of the one evaluated after certain period of time since the infiltration ceases are given in Fig. 3. The temporal changes of expected values and variances of soil moisture content for different potential evaporation rates are plotted in Fig. 4. In order to investigate the effect of nonlinearity of the evaporation equation (1) the soil moisture distribution was evaluated for the approximate solution (15) with the lognormal initial surface soil moisture content.

Under the assumption of linearity shown by the approximate solution of equation (15) we can determine the autocorrelation of the soil moisture process for two arbitrary time instants, t_1 and t_2.

$$\mathrm{E}[\Theta(t_1)\Theta(t_2)] = m_{\Theta_0}^2 - \frac{E_p}{L}(t_1 + t_2)m_{\Theta_0} + \frac{E_p}{L^2}t_1t_2 \quad (20)$$

where $t_1, t_2 < \frac{L\Theta_s}{E_p}$

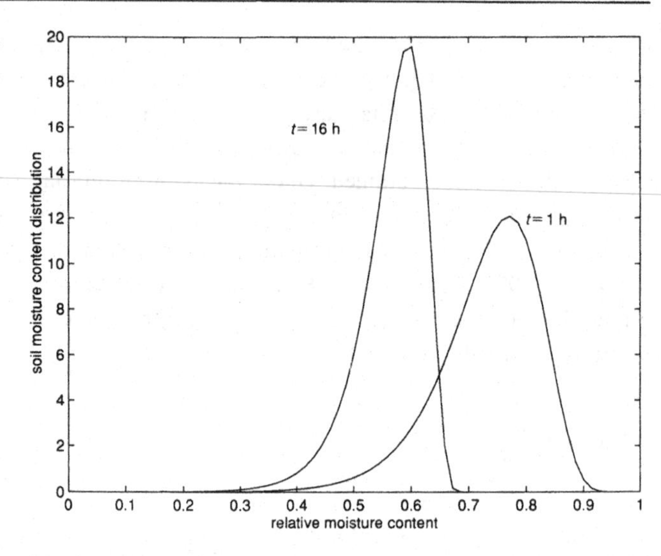

Fig. 3 Pdf of soil moisture contents, $\mathrm{E}(\Theta_0) = 0.32$, $\Theta_s = 0.42$ and $E_p = 0.1 \cdot 10^{-3}$ [m/h].

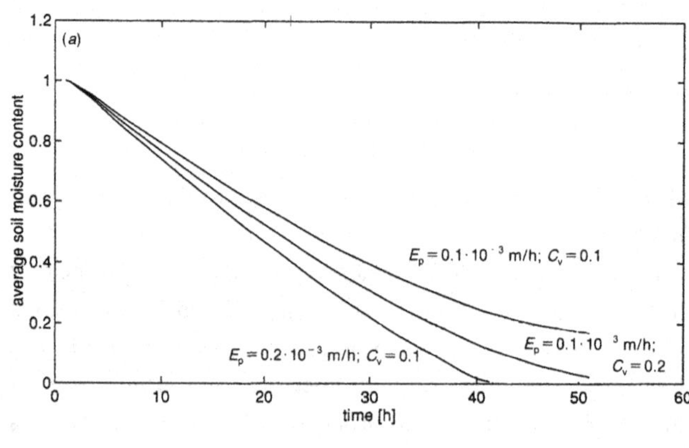

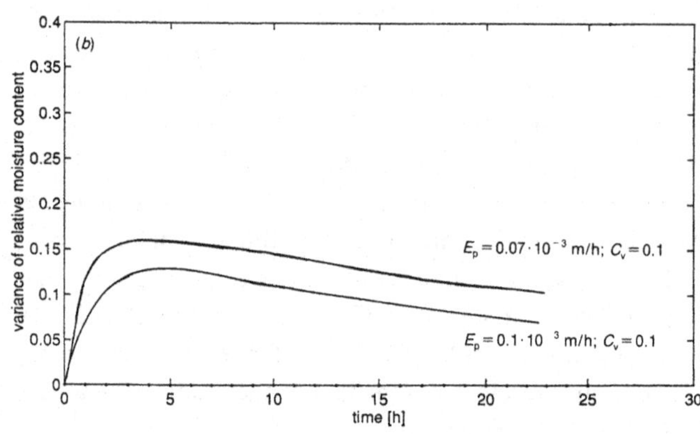

Fig. 4 Expected value (a) and variance (b) of soil moisture contents for different potential evaporation rates.

DETERMINATION OF PDF FOR TIME TO DESATURATION

Using the same equation (13) as in the last paragraph, the probability distribution of the time to desaturation can be obtained from equation (9), assuming that the initial soil moisture content has a distribution given by equation (14). The formula for the time to desaturation developed in Romanowicz et al. (1990) describes the spatial variability of the time to desaturation depending on the spatial variability of soil moisture characteristic enclosed in the initial soil moisture contents. The mean value of time to desaturation and its variance can be evaluated using the derived p.d.f function.

Assume that the process prior to desaturation can be approximated by a linear relation (15). Then the distribution of the time to desaturation will have the form:

$$f_{t_d}(t_d) = f_{\Theta_0}\left(\frac{E_p}{L}t_d + \Theta^*\right)\frac{E_p}{L} \quad (21)$$

If one assumes that the initial moisture contents has a lognormal distribution, then the approximate time to desaturation will also be lognormal with the mean equal to:

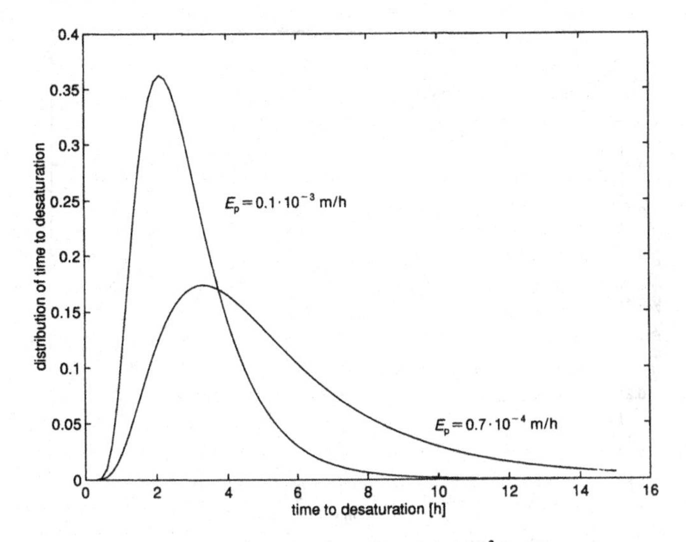

Fig. 5 Pdf of time to desaturation; $E_p = 0.1 \cdot 10^{-3}$ [m/d].

$$E(t_d) = -\Theta^* + \frac{L}{E_p} \exp(\mu_{\Theta_0} + \sigma^2_{\Theta_0}/2) \tag{22}$$

and the variance unchanged. The pdf of the time to desaturation for different values of potential evapotranspiration rates is plotted in Fig. 5, for both linear and nonlinear cases. Using the formulae given by Romanowicz *et al.* (1990), the expected values of time to desaturation for the nonlinear case were evaluated.

Following the reasoning from the work of Sivapalan & Wood (1986) we shall evaluate the portion of the catchment that is desaturated at time t, denoting it as $\gamma(t)$. The desaturated portion of the catchment can be calculated as (Romanowicz *et al.*, 1990)

$$\gamma(t) = \int_{\Theta_{min}}^{F^{-1}(t)} f_{\Theta_0}(\Theta_0) d\Theta_0 \tag{23}$$

If we assume that Θ_0 is lognormally distributed, the function $F^{-1}(t)$ is also lognormally distributed. After integration we get:

$$\gamma(t) = \tfrac{1}{2} \operatorname{erf}\left(\frac{(\ln F^{-1}(t) - \mu)}{\sqrt{2}\,\sigma}\right) + \tfrac{1}{2} \tag{24}$$

Using the above formula one can determine the influence of the values of the potential evapotranspiration E_p and the coefficient of variation of the initial moisture content on the cumulative distribution of time to desaturation as shown in Fig. 6.

EVALUATION OF THE ACTUAL EVAPOTRANSPIRATION AND ITS DISTRIBUTIONS

According to equation (1) the actual evapotranspiration was assumed to have the form:

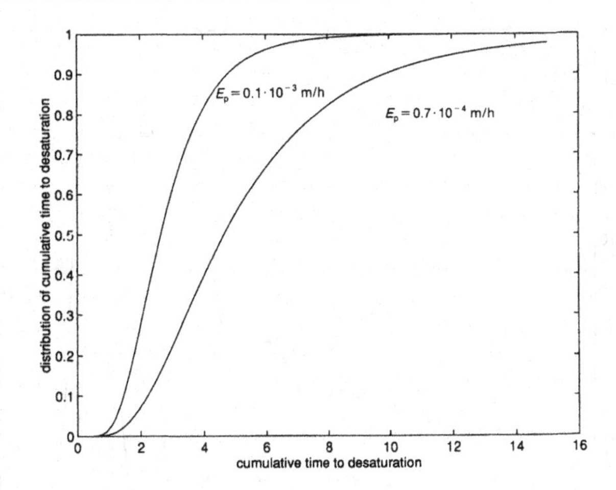

Fig. 6 Distribution of cumulative time to desaturation.

$$E_a = E_p \frac{h_p - h_a}{1 - h_a} \tag{25}$$

where h_p is given by equation (2).

Hence, for the generalized Gardner model the actual evapotranspiration takes the form:

$$E_a = E_p \frac{\exp(n\Phi_e)\left(\dfrac{\Theta}{\Theta_s}\right)^{-n\Phi_e/\alpha} - h_a}{1 - h_a} \tag{26}$$

Let us assume that the initial moisture content is a random variable and that the rest of parameters of this model are deterministic. In that case the soil moisture content, Θ, is also a random variable. The evaluation of the distribution of actual evapotranspiration can be done only numerically, using the soil moisture probability distribution function derived earlier, unless some approximation is made. The mean and the covariance of this function can be computed according to Romanowicz *et al.* (1990).

As it can be seen from Figs. 1 and 2, the soil moisture contents in the light soils remains linear in nearly whole range of its variation, while the actual evapotranspiration varies more strongly with time. This is due to the fact that the time dependence of the soil moisture contents emerges from the integration of the power function of the soil moisture contents (actual evapotranspiration). The process of integration exerts a kind of smoothing effect. Hence one can assume that the soil moisture contents is a linear function of time and its distribution is given by equation (16) for the case of lognormal initial moisture contents. This assumption allows one to determine the approximate analytical p.d.f. of actual evapotranspiration. Substituting the linear relation for the soil moisture contents (15) into equation (26) one gets

$$E_a = E_p \frac{\exp(n\Phi_e)\left(\dfrac{\Theta_0 - \dfrac{E_p t}{L}}{\Theta_s}\right)^{-n\Phi_e/\alpha} - h_a}{1 - h_a} \tag{27}$$

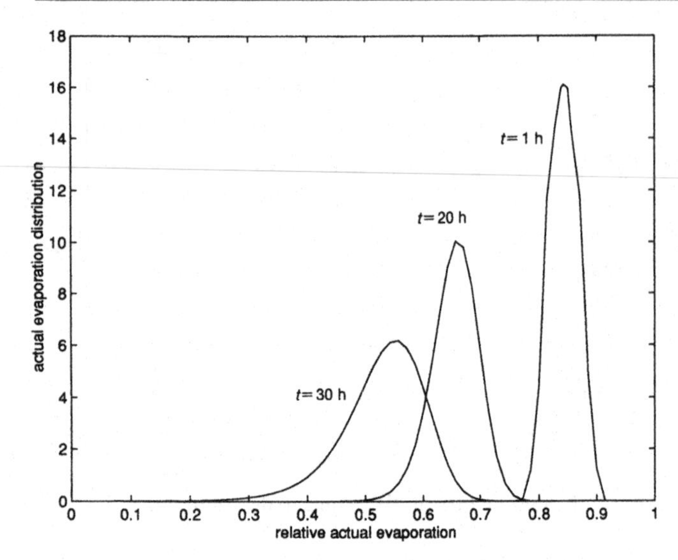

Fig. 7 Pdf of actual evaporation, $E_p=0.07\cdot10^{-3}$ [m/h], $E(\Theta_0)=0.32$.

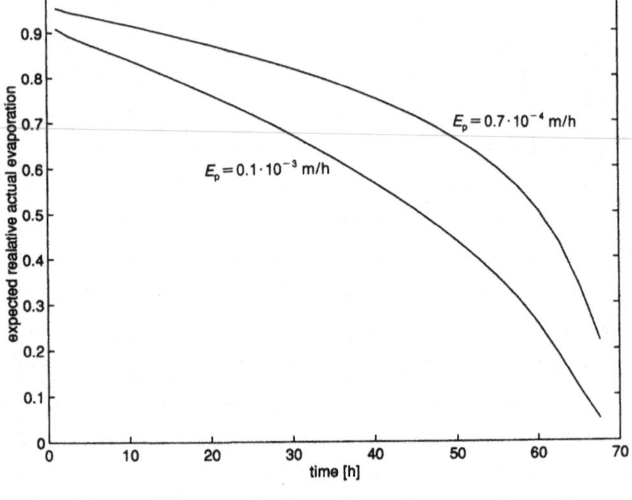

Fig. 8 Changes of expected values of actual evaporation with time.

Using the relations given by Romanowicz *et al.* (1990) one can obtain the following equation for the pdf of an approximate analytical formula for evapotranspiration:

$$f_{E_a}(E_a)=f_{\Theta_0}(\Theta_s\exp(-\alpha)[(1-h_a)E_a/E_p+h_a]^{-\alpha/n\Phi_e} + E_pt/L)\frac{\partial\Theta_0}{\partial E_a} \quad (28)$$

This relation is valid only as long as the assumption of a linear soil moisture contents distribution holds, for times shorter than $L(\Theta_s-\Theta_{\min})/E_p$. The distribution of actual evapotranspiration as well as its expected value for different values of potential evaporation, E_p, are plotted in Figs. 7 and 8.

CONCLUSIONS

Spatially variable soils with deterministic constant evapotranspiration are considered. The spatial variability of soil is enclosed in the random distribution of initial soil moisture contents. The two stages of evaporation from the land surface are described by a lumped nonlinear model using the thermodynamic equation. The distributions, means and variances of the soil moisture contents, time to desaturation, and actual evapotranspiration are evaluated.

The case of the lognormal initial soil moisture distributions is examined. In order for this assumption to be used, the variance must be small enough assuring that the values of the soil moisture contents neither exceed the saturation moisture contents nor are negative. The assumption of lognormal initial soil moisture distribution allows the analytical relations for the mean and for the variance of the soil moisture contents, and the approximate relation for the mean and the variance of the time to desaturation to be obtained. The cumulative distribution of the time to desaturation is also evaluated.

The above derivations have implicitly made use of the ergodic assumption, that means that the averageing area is infinitely large, so that the areal mean is equal to the ensemble mean. In practice, the actual catchments are probably not large enough to justify the ergodic assumption.

The theoretical results are illustrated with typical field data. This approach can be further extended for the case of the random potential evaporation, random forcing function, and random parameters of the model as well as for the two-dimensional random fields.

REFERENCES

Andersson, J. & Shapiro, A. M. (1983) Stochastic analysis of one-dimensional steady state unsaturated flow: A comparison of Monte Carlo and perturbation methods. *Water Resour. Res.*, **19**(1), 121–33.

Bobrowski, D. (1987) *Introduction to Random Ordinary Differential Equations*, PWN, Warsaw, Poland (in Polish).

Bodo, B. A., Thompson, M. E. & Unny, T. E. (1987) A review on stochastic differential equations for applications in hydrology. *Stoch. Hydrol. and Hydraul.*, **1**, 81–100.

Cordova, J. R. & Bras, R. L. (1981) Physically based probabilistic models of infiltration, soil moisture, and actual evapotranspiration. *Water Resour. Res.*, **17**(1), 93–106.

Dagan, G. E. Breslev (1983) Unsaturated flow in spatially variable fields in derivation of models of infiltration and redistribution. *Water Resour. Res.*, **9**, 413–20.

Dooge, J. C. I. (1986) *Hydrologic models of land surface processes*, Proc. ISL SCP Conf., Rome, Italy.

Eagleson, P. S. (1978), Climate, soil and vegetation, Parts 1–7. *Water Resour. Res.*, **14**(5), 705–76.

Eagleson, P. S. (1982) *Land Surface Processes in Atmospheric Circulation Models*, Cambridge University Press.

Entekhabi, D. & Eagleson, P. S. (1989) Land surface hydrology parameterization for atmospheric general circulation models including sub-grid scale variability. *J. of Climate*, **2**(8), 816–31.

Eykhoff, P. (1974) *System identification: Parameter and state estimation*, Wiley.

Gardner, W. R. (1958) Some steady-state solutions of the unsaturated moisture flow equation with application to evaporation from a water table. *Soil Sci.*, **85**, 228–32.

Gardner, W. R., Jury, W. A. & Knight, J. (1975) Water uptake by vegetation. In: *Transfer Processes in the Environment*, Vol. 1, Wiley, New York, NY, pp. 443–56.

Gupta, V. K., Rodriguez-Iturbe, I. & Wood, E. F. (1986) *Scale Problems in Hydrology*, Kluwer.

Ito, K. (1951) On stochastic differential equations. *Mem. Amer. Math. Soc.*, **4**, 1–51.

Kaczmarek, Z. (1970) *Statistical Methods in Hydrology and Meteorology*, WKiŁ, Warsaw, Poland (in Polish).

Koch, R. W. (1985) A stochastic streamflow model based on physical principles. *Water Resour. Res.*, **24**(4), 545–53.

Kuhnel, V. (1989) *Scale problems in soil moisture flow*, Ph. D. Thesis, Department of Civil Engineering, University College Dublin, Ireland.

Mantoglou, A. & Gelhar, L. W. (1987a) Stochastic modelling of large-scale transient unsaturated flow systems. *Water Resour. Res.*, **23**(1), 37–46.

Mantoglou, A. & Gelhar, L. W. (1987b) Capillary tension head variance, mean soil moisture content, and effective specific soil moisture capacity of transient unsaturated flow in stratified soils. *Water Resour. Res.*, **23**(1), 47–56.

Mantoglou, A. & Gelhar, L. W. (1987c) Effective hydraulic conductivities of transient unsaturated flow in stratified soils. *Water Resour. Res.*, **23**(1), 57–67.

Milly, P. C. & Eagleson, P. S. (1982) Infiltration, and evaporation at inhomogeneous land surfaces. *Hydrology and water resource systems, Report No. 278*, Department of Civil Engineering, MIT, Cambridge, Mass., USA.

Mtundu, N. D. & Koch, R. W. (1987) A stochastic differential equation approach to soil moisture. *Stoch. Hydrol. and Hydraul.*, **1**, 101–16.

O'Kane, J. P. (1990) *A theory of drainage, drying and rewetting of peat under natural conditions*, Report, CWRR, University College Dublin, Ireland.

Papoulis, A. (1965) *Probability, Random Variables, and Stochastic Processes*, McGraw-Hill.

Parzen, E. (1962) *Stochastic Processes*, Holden-Day, San Francisco, Calif., USA.

Philip, J. R. (1957) Evaporation, and moisture and heat fields in the soil. *J. Meteorology*, **14**, 354–66.

Philip, J. R. & de Vries, D. A. (1957) Moisture movement in porous materials under temperature gradients. *Trans. Amer. Geoph. Union*, **38**, 222–32.

Romanowicz, R. J., O'Kane, J. P. & Dooge, J. C. I. (1990) *Spatial variability of Land Surface Processes, Stochastic analysis of surface fluxes*, Report CWRR/LSP 190/17, CWRR, UCD, Dublin.

Serrano, S. E. (1990) Stochastic differential equation models of erratic infiltration. *Water Resour. Res.*, **26**, 703–12.

Serrano, S. E. & Unny, T. E. (1987) Stochastic partial differential equations in hydrology. In: Mac Neill, I. B. & Umphrey, G. J. (Eds.), *Stochastic Hydrology*.

Soong, T. T. (1973) *Random Differential Equations in Science and Engineering*, Academic Press, New York.

Tillotson, P. M. & Nielsen, D. R. (1984) Scale factors in soil science. *Soil Sci. Soc. of Amer. J.*, **48**, 953–9.

Yeh, T.-C. J., Gelhar, L. W. & Gutjahr, A. L. (1985a) Stochastic analysis of unsaturated flow in heterogeneous soils, 1. Statistically isotropic media. *Water Resour. Res.*, **21**(4), 447–56.

Yeh, T.-C. J., Gelhar, L. W. & Gutjahr, A. L. (1985b) Stochastic analysis of unsaturated flow in heterogeneous soils, 2. Statistically anisotropic media. *Water Resour. Res.*, **21**(4), 457–64.

Yeh, T.-C. J., Gelhar, L. W. & Gutjahr, A. L. (1985c) Stochastic analysis of unsaturated flow in heterogeneous soils, 3. Observations and applications. *Water Resour. Res.*, **21**(4), 465–71.

Yevjevich, V. (1972) *Probability and Statistics in Hydrology*, Water Resour. Publ., Fort Collins, Colo. USA.

6 Detecting outliers in flood data with geostatistical methods

L. GOTTSCHALK

Department of Geophysics, University of Oslo, Norway

I. KRASOVSKAIA

Hydroconsult AB, Uppsala, Sweden

Z. W. KUNDZEWICZ

Research Centre for Agricultural and Forest Environment Studies, Pol. Acad. Sci., Poznan and Institute of Geophysics, Pol. Acad. Sci., Warsaw, Poland

ABSTRACT The plausibility analysis of the regional flood data of Southern Norway is performed with the help of geostatistical methods. As the data at a number of sites are analyzed, one can account the spatial relationships in the outliers detection problem. That is the results of outliers detection may differ in comparison to the non-structured (univariate) case. The geostatistical methods applied are block kriging and Switzer's location-specific covariance analysis, with catchment areas accounted.

INTRODUCTION

An intuitive definition of an outlier can be 'an observation which deviates so much from other observations as to arise suspicions that it was generated by a different mechanism' (Hawkins, 1980). An outlying observation can be interpreted in several ways. It may represent an event of extreme magnitude (e.g. due to rare natural causes) that has unexpectedly happened in the system. In flood frequency analysis such extreme events are of outmost importance, indicating a heavy right tail of the parent distribution. On the other hand, a value differing from the remainder of the data set may be an erroneous observation. This could have been caused by instrument malfunctioning or human mistakes (e.g. at the stage of interpretation of the rating curve for high flows). In this latter case outliers may contaminate the data and reduce the useful information about the natural process.

Detection of outliers in hydrological data can be performed in a number of ways. Kottegoda (1984) considered approaches based either on distributional, mixture, or slippage alternatives for investigation of a series of maximum annual flows. Gottschalk & Kundzewicz went along similar lines in the companion paper (1991). Gottschalk (1989) showed that values of maximum annual flow in one site of the data set studied are mutually independent. Therefore the maximum annual flow data for particular sites were analyzed by Gottschalk & Kundzewicz (1989) in the way typical for non-structured (univariate) observational material, rather than for time series.

If the regional flood data (at a number of sites) are available, it is advantageous to consider spatial relationships in the outliers detection problem. Such an analysis, performed in the partial duration series context is a welcome extension to the analysis of a series records at individual sites. Assume that one observation departs considerably from the remainder of the data for an individual gauge and it could be marked an outlier by the procedure outlined in Gottschalk & Kundzewicz (1995). However, similarly high flows may have occured in adjacent catchments. That is, the observation questioned in analysis of the data from one gauge (time series) may seem quite regular in the context of regional flood data. The other way round – a plausible observation in the univariate approach may seem rogue in the context of values in the neighbourhood. Geostatistical methods have proven to be useful for outliers detection in regional groundwater quality data (Kundzewicz *et al.*, 1989) and are also applied here.

GEOSTATISTICAL ANALYSIS OF FLOODS

One of standard objectives of geostatistical analyses is to estimate the value of the variable under study, z in the point x_0, as a linear combination of regional observations $z(x_1)$; $i=1,\ldots,n$. The estimator is therefore:

$$z^*(x_0)=\sum_{i=1}^{n} \lambda_i z(x_i)=\Lambda^{\mathrm{T}}\mathrm{Z} \tag{1}$$

where Z is the column vector of observations and Λ^{T} is the transposed column vector of weights $\lambda, i=1,\ldots,n$ associated to the observations. Optimal weights can be found, based on the intrinsic assumption and under the conditions of unbiasedness, by minimizing the estimation variance. This gives rise to the system of kriging equations that read in the matrix notation:

$$\Gamma\Lambda=\Gamma_0 \tag{2}$$

where

$$\Lambda=\begin{bmatrix}\lambda_1\\ \vdots\\ \lambda_n\\ \mu\end{bmatrix}\quad \Gamma=\begin{bmatrix}\gamma(0) & \gamma(|x_1-x_2|) & \ldots & \gamma(|x_1-x_n|) & 1\\ \gamma(|x_2-x_1|) & \gamma(0) & \ldots & \gamma(|x_2-x_n|) & 1\\ \vdots & \vdots & & \vdots & \vdots\\ \gamma(|x_n-x_1|) & \gamma(|x_n-x_2|) & \ldots & \gamma(0) & 1\\ 1 & 1 & & 1 & 0\end{bmatrix}$$

$$\Gamma_0=\begin{bmatrix}\gamma(|x_1-x_0|)\\ \gamma(|x_2-x_0|)\\ \vdots\\ \gamma(|x_n-x_0|)\\ 1\end{bmatrix},$$

$\gamma(h)$ denotes the (point) semivariogram for the distance h and μ is the Lagrange multiplier. If instead of the intrinsic assumption the second order homogeneity of the stochastic field $z(x)$ is assumed, the semivariogram $\gamma(h)$ is related to the covariance $c(h)$ as

$$\gamma(h)=c(0)-c(h) \tag{3}$$

In this case the kriging equations have the form

$$\mathrm{C}\Lambda=\mathrm{C}_0 \tag{4}$$

where

$$\Lambda=\begin{bmatrix}\lambda_1\\ \vdots\\ \lambda_n\\ -\mu\end{bmatrix}\quad \Gamma=\begin{bmatrix}c(0) & c(|x_1-x_2|) & \ldots & c(|x_1-x_n|) & 1\\ c(|x_2-x_1|) & c(0) & \ldots & c(|x_2-x_n|) & 1\\ \vdots & \vdots & & \vdots & \vdots\\ c(|x_n-x_1|) & c(|x_n-x_2|) & \ldots & c(0) & 1\\ 1 & 1 & & 1 & 0\end{bmatrix}$$

$$\mathrm{C}_0=\begin{bmatrix}c(|x_1-x_0|)\\ c(|x_2-x_0|)\\ \vdots\\ c(|x_n-x_0|)\\ 1\end{bmatrix},$$

The optimal weights, i.e. formal solution to equations (1) and (4), and the estimation variance read, respectively

$$\Lambda=\mathrm{C}^{-1}\mathrm{C}_0 \tag{5}$$

$$\sigma^2(x_0)=c(0)-\Lambda^{\mathrm{T}}\mathrm{C}_0 \tag{6}$$

From the definition of the covariance matrix it results that:

$$\begin{aligned}\mathrm{E}[z(x_i)z(x_j)]&=m(x_i)m(x_j)+\mathrm{C}[\mathrm{Z}(x_i),\mathrm{Z}(x_j)]\\ &=m(x_i)m(x_j)+s(x_i)s(x_j)r(|x_i-x_j|)\end{aligned} \tag{7}$$

where

$$r(|x_i-x_j|)=\mathrm{C}[\mathrm{Z}(x_i),\mathrm{Z}(x_j)]/[s(x_i)s(x_j)]$$

If the random field is homogeneous (spatially stationary), that is $m(x_i)=m(x_j)=m$ and $s(x_i)=s(x_j)=s$, one can simplify equation (7) to the form:

$$\mathrm{E}[z(x_i)z(x_j)]=m^2+s^2r(|x_i-x_j|) \tag{8}$$

In the above derivations only one realization of the spatial stochastic field, $z(x_i)$, was considered. On the other hand, regional flood data $z(x_i, t_k)$ contain several events in time t_k; $k=1,\ldots,K$, and sites in space x_i; $i=1,\ldots,n$. Typically the number of observed space points n is less than the number of observed time points K. Existence of several observed values at each site allows site specific mean values, standard deviations, and covariances to be determined.

Once data for several events in time and for several spatial locations are available, one can estimate spatial covariances separately for each event and then average them over time. The spatial covariances for an individual case are not estimated well when the number of sites n is small. Therefore the time averaged covariances may be used. If the number of temporal events K is large, the time averaged covariances are quite stable. Let these time-averaged translation-invariant spatial covariances be denoted by:

$$\tilde{c}(|x_i-x_j|)=\mathrm{average}_k[c_k(|x_i-x_j|)] \tag{9}$$

The values calculated after equation (9) may replace the

elements in the covariance matrix C and in the covariance vector C_0 (equation 4). They will be denoted $\tilde{C}$ and $\tilde{C}_0$, respectively, in the sequel.

Switzer (1989) points out that in case of space-time data there may not be a compelling reason to use translation-invariant spatial covariances. A covariance $c(|x_i - x_j|)$ between any pair of locations x_i and x_j can be directly calculated from data. It is clear that two pairs of stations, separated by the same spatial distance, might still be observed to have quite different covariances. That is, spatial interpolation would be more precise in the vicinity of a pair of stations with higher covariance and less precise closer to a stations pair with lower covariance. Using the time-averaged translation-invariant covariance one is forced to assign the same precision to both these cases.

To overcome this Switzer (1989) suggests to calculate a location-specific and not translation-invariant covariance vector C'_0 between the location x_0 and the observation sites $x_1, \ldots, x_n$ from the formula

$$C'_0 = C(\tilde{C})^{-1}\tilde{C}_0 \tag{10}$$

where C is the $n \times n$ matrix of observed station pair covariances $c(|x_i - x_j|)$. The model is proposed using mean-centered covariances. Applying the time averaged translation-invariant spatial covariances one gets the kriging weights as $\Lambda = (\tilde{C})^{-1}\tilde{C}'_0$. Equation (10) can thus be written in the form:

$$C'_0 = C\tilde{\Lambda} \tag{11}$$

After having introduced the location-specific covariance vector into equation (5) one gets:

$$\Lambda' = C^{-1}C\,\tilde{\Lambda} = \tilde{\Lambda} = \Lambda \tag{12}$$

That is, the weights are identical with those for the translation-invariant case. However, the estimation variance is different and reads:

$$\sigma_2(x_0) = s^2(x_0) - \tilde{\Lambda}^T C'_0 = s^2(x_0) - \Lambda^T C \Lambda \tag{13}$$

where $s^2(x_0)$ is the variance assigned to the unobserved location x_0.

It is necessary to note that regional flood (runoff) data do not represent a single point at the outlet of the catchment but rather the total area A of the catchment i.e.

$$z(A_i) = \int_{A_i} z(x)\,dx \tag{14}$$

For this random field with areal support the semivariogram is related to the point semivariogram by the equation

$$\gamma(A_i, A_j) = \bar{\gamma}(A_i, A_j) - 0.5\bar{\gamma}(A_i, A_j) - 0.5\bar{\gamma}(A_j, A_j) \tag{15}$$

where

$$\bar{\gamma}(A_i, A_j) = \frac{1}{A_i A_j} \int_{A_i}\int_{A_j} \gamma(|x_i - x_j|)\,dx_i\,dx_j$$

Similarly, one can link the covariances in the form:

$$c(A_i, A_j) = \frac{1}{A_i A_j} \int_{A_i}\int_{A_j} c(|x_i - x_j|)\,dx_i\,dx_j = \frac{1}{A_i A_j} \int_{A_i}\int_{A_j} s(x_i)s(x_j)r(|x_i - x_j|)\,dx_i\,dx_j \tag{16}$$

Under assumption that $s(x)$ is a smoothly varying function in space, i.e. in accordance with the covariance model equation (8) one gets the following equation for the correlation function

$$r(A_i, A_j) = \frac{\int_{A_i}\int_{A_j} c(|x_i - x_j|)\,dx_i dx_j}{\sqrt{\int_{A_i}\int_{A_i} c(|x_i - x_j|)\,dx_i dx_j \int_{A_j}\int_{A_j} c(|x_i - x_j|)\,dx_i dx_j}} = \frac{\bar{r}(A_i, A_j)}{\sqrt{\bar{r}(A_i, A_i)\,\bar{r}(A_j, A_j)}} \tag{17}$$

As the flood data represent areas rather than points, the point covariances in the kriging equations given above should be replaced by the covariances calculated from equation (16). Using equation (5), with suggested modification to support the nature of flood data, one can calculate the estimated values of a flood at a certain site and its estimation variance based on observations at other sites. The values of absolute normalized deviations (AND) defined as the differences between the estimated (kriged) and the observed values, related to the kriging standard variations can now be subject to analysis. If the value of AND for a particular site is greater than three, the observation at this site is labelled an outlier. The above procedure is based on the well established theory of normal variables and holds if the normalized deviations are approximately normally distributed with the mean 0 and the standard deviation 1.

EXPERIMENTAL SEMIVARIOGRAMS AND CORRELATION FUNCTIONS

There are at least two ways of detecting outliers in a set of flood data with the help of geostatistics. This can be achieved either already at the stage of determination of an experimental semivariogram or at the stage of cross-validation of the theoretical semivariogram where all the data are compared with values obtained with the help of kriging. In the former case the analysis, to be successful, has to be performed at the level of each flood event and calls for examination of individual data points for the distance classes of the semi-variogram (Kundzewicz *et al.*, 1989). The latter approach is more suitable for operational applications. A theoretical semivariogram (or correlation function) determined for all data is subsequently used to estimate parameter values and kriging variances for every observational point. Linear, spherical, exponential and Gaussian models (Clark, 1979) are used in the present study as theoretical semivariograms. Once a kriged value in a point is much different from the observed value, seen in relation to the estimation variance, the observation at this point is marked an outlier.

The empirical semivariogram is determined from the experimental data set for the event t_k by the equation

$$\hat{\gamma}_k(h)=\frac{1}{2N(h)}\sum_{(i,j)\in R(h)}\left(z(x_i,t_k)-z(x_j,t_k)\right)^2 \qquad (18)$$

where

$$R(h)=\{(i,j);h-\varepsilon\le|x_i-x_j|\le h+\varepsilon\}$$

that is, all pairs of sites separated by a distance close to h are considered, and $N(h)$ is the number of elements in the distance class $R(h)$.

Equation (18) is sensitive to outliers in data and thereby a more robust estimator is needed. Trimmed means in particular distance classes have been chosen here among several candidates for robust semivariogram estimators proposed in literature. The average experimental semivariogram for all K events is determined as the simple mean:

$$\bar{\gamma}(h)=\frac{1}{2[N(h)-2N_0]K}\sum_{l=N_0+1}^{N(h)-N_0}\sum_{k=1}^{K}\left(z_l(x_i,t_k)-z(x_j,t_k)\right)^2 \qquad (19)$$

where trimming is performed by removing the N_0 highest and N_0 lowest values from the sample of $N(h)$ values.

Pairwise correlation coefficients between observation points are determined from

$$\hat{r}(h)=\frac{\frac{1}{K}\sum_{k=1}^{K}z(x_i,t_k)z(x_j,t_k)-\hat{m}^2(x_i)\hat{m}^2(x_j)}{\sqrt{\left(\frac{1}{K}\sum_{k=1}^{K}z(x_i,t_k)^2-\hat{m}^2(x_i)\right)\left(\frac{1}{K}\sum_{k=1}^{K}z(x_j,t_k)^2-\hat{m}^2(x_j)\right)}} \qquad (20)$$

where $h=|x_i-x_j|$ and $\hat{m}(x_i)$ is the estimated mean value. In case of time-space data a theoretical semivariogram or correlation function can either be fitted to the time-averaged experimental semi-variogram (equation 19) or to the scatter of correlation coefficients (equation 20) plotted against the distance h.

Equations (18), (19) and (20) are developed for point values. However, as a runoff value is representative for a catchment, i.e. area rather than a single point, some way of incorporating the areal support in the algorithm is required. The iterative procedure devised by Bardossy (1989, personal communication) reads:

- Each catchment is represented by its centre point x_i.
- An experimental semivariogram $\hat{\gamma}_k^0(h)$ is calculated using the flood observations as point values at x_i.
- A theoretical semivariogram $\gamma_k^0(h)$ is fitted to $\hat{\gamma}_k^0(h)$. Set $l=0$.
- A new experimental semivariogram is calculated from (compare equation (15)):

$$\hat{\gamma}_k^{l+1}(h)=\frac{1}{2N(h)}\sum_{(i,j)\in R(h)}\left(z(x_i,t_k)-z(x_j,t_k)\right)^2 + 0.5\bar{\gamma}^l(A_i,A_i)+0.5\bar{\gamma}^l(A_j,A_j) \qquad (21)$$

- A theoretical semivariogram $\gamma_k^{l+1}(h)$ is now fitted to $\hat{\gamma}_k^{l+1}(h)$.
- $\gamma_k^l(h)$ and $\gamma_k^{l+1}(h)$ are compared. In case of convergence it is accepted as a point semivariogram. In the other case l is set to $l=l+1$, a new experimental semivariogram is calculated from equation (21) and the iteration is continued.

In case of time-space data the experimental semivariogram $\hat{\gamma}_k(h)$ calculated from equation (18) for a single event is replaced with the averaged semivariogram $\bar{\gamma}(h)$ from equation (19). The alternative for the time-space data is to fit a theoretical function to the pairwise correlations $\hat{r}(h)$ determined from observational data with the help of equation (20). Similarly, one can account areal support and find the point correlation coefficients by an iterative procedure (cf. equation (17)):

$$\bar{r}^{l+1}(A_i,A_j)=\hat{r}(h)\sqrt{\bar{r}^l(A_i,A_i)\bar{r}^l(A_j,A_j)} \qquad (22)$$

CASE STUDY

The data base used in this study consists of a set of 42 registered flood events at 22 observation stations in southern Norway (Fig. 1) during 20 years (1970–89). The data do not embrace the observation stations at the West Coast, where strong anisotropy can be observed. Variability of runoff in the West Coast is mainly controlled by the distance from the North Sea. For each site a partial duration series of daily flood peaks is constructed. A regional flood event is defined so that a peak should have been registered at at least one station. Typically, such an event embraces peak flows in several catchments, either in the same day, or in one of the

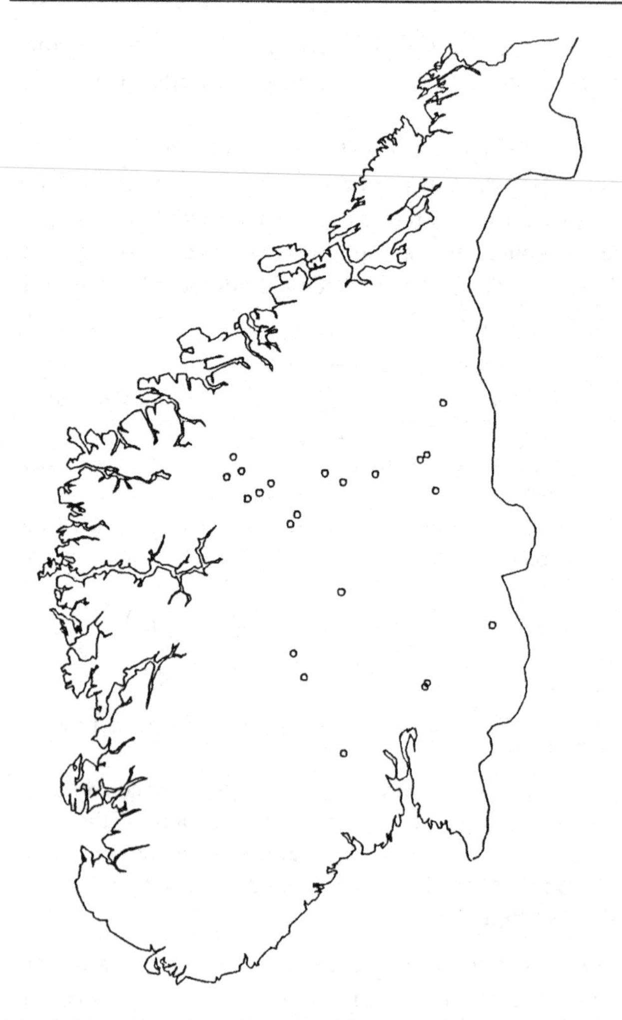

Fig. 1 Map of southern Norway with observation stations used in the study.

neighbouring days. It is also controlled that rainfall events occurred in the region at the onset of the floods.

Characteristics for the observation stations used are shown in Table 1. Each catchment is schematized in the form of a rectangle, whose axis of symmetry is drawn by the outlet and the centre of the catchment. The catchment length is taken as the distance along the symmetry axis of the rectangle, from the outlet to the orthogonal projection of the source point. The width is adjusted so as to give correct area. The angle between the axis and the North is also calculated. The approximated catchments are displayed in Figs. 2 a and 2 b together with an example flood event. The region of study is dominantly mountainous with well pronounced large valleys. This explains a rather large variation in statistical parameters of flood events among stations. In order to smoothen the variation pattern observation series are normalized with respect to their mean values.

Individual outlying observations may have a drastic influence on the experimental semivariogram calculated for single flood events. An illustration is given in Fig. 3, where experimental semivariograms for event 25 (presented in Fig. 2 a) are shown for several cases (non-trimmed/trimmed and with/without outlying observation at station 437). It is noteworthy that removal of the single outlying point flattens the semivariogram much stronger than the trimming process (removal of ten per cent highest and ten per cent lowest points from the semivariogram cloud). One of explanations is the observation that the contaminating effect of the outlier is lumped in several distance classes (see Fig. 2 a) and may locally exceed the trimmed ten per cent. It is found that the outlier detection on the event basis may be preferably performed at the stage of analysing an empirical semivariogram, rather than at the stage of cross-validation. In this latter case fitting of a theoretical semivariogram is a prerequisite.

The experimental semivariograms differ considerably in form and parameters (sill, nugget, range) from one event to another. These parameters are evaluated via optical fitting for each event. Their mean values and variability are illustrated in Table 2, both for the case of point and block approaches (equations 18 and 21). The differences in parameters of theoretical semivariograms are also reflected in the estimation variances for the events and thereby influence the decision whether or not to give the label 'outlier' to an observation. When the areal support is taken into account in the cross-validation, estimated values are almost identical to the point representation of catchment. Estimation variances are also very close but as a rule a few percent higher.

The average experimental semivariograms calculated from equations (19) and (21) are less sensitive to individual outlying observations. As an example, average empirical semivariograms estimated with and without observations from station 437 are shown in Fig. 4. The record in this site contains the two most extreme values of the total data set. The comparison of semivariograms given in Fig. 4 shows that the process of elimination of two largest outliers has a relatively smaller effect, whereas trimming influences considerably the empirical semivariograms obtained. Therefore the question emerges of the suitability of the trimming process in the analysis of the flood data. It seems, in general, that the effect of trimming on the form of an experimental semivariogram can be two-fold. If the semivariogram is very irregular, trimming is likely to regularize it by eliminating the most outlying points in particular distance classes of the semivariogram. However, if the semivariogram calculated according to eq. (18) or (19) is regular, trimming may not be necessary. This is because trimming always results in reducing the values of the semivariogram, that may be unwelcome for analyses of regional flood data.

Scatter diagram of pairwise correlation coefficients for the same data set (with and without the station 437) is shown in Fig. 5. Some few extreme points off the main concentration

Table 1. *Characteristics of observation stations from Southern Norway used in the study*

Station No.	Area km²	Length km	Width km	Angle	Maximum yearly flood runoff (in l/km²/s) Mean l/s/km²	Standard dev.	Variation coefficient
400	470	32.5	14.5	245	47.4	29.0	0.61
410	1618	45.0	36.0	148	23.4	19.7	0.84
411	1755	52.5	33.4	218	44.7	36.2	0.81
437	565	36.0	15.7	27	37.5	31.8	0.85
478	155	18.0	26.3	257	89.0	104.5	1.17
867	473	32.5	14.6	122	130.0	92.5	0.71
887	121	16.5	7.3	35	24.6	17.0	0.69
1354	248	33.5	7.4	27	13.9	10.3	0.74
1355	620	39.5	15.7	173	19.8	11.5	0.58
1357	549	28.0	19.6	152	34.8	27.4	0.79
1364	791	31.5	25.1	152	114.2	76.3	0.67
1380	25	6.5	3.8	276	11.8	10.7	0.91
1476	235	17.5	13.4	311	160.9	120.9	0.75
1489	68	7.5	9.0	190	27.4	22.8	0.83
1555	101	14.5	7.0	187	31.5	33.8	1.07
1573	262	24.5	10.7	232	39.8	37.3	0.94
1591	56	9.5	5.9	261	67.1	81.3	1.21
1603	78	9.0	8.6	77	125.7	95.2	0.76
1604	251	24.4	10.2	114	137.5	109.2	0.79
1605	260	14.0	18.6	352	142.6	128.5	0.90
1607	371	26.0	14.3	106	151.6	115.9	0.76
1609	135	19.5	6.9	152	124.8	98.4	0.79
Average					72.7	72.3	0.99

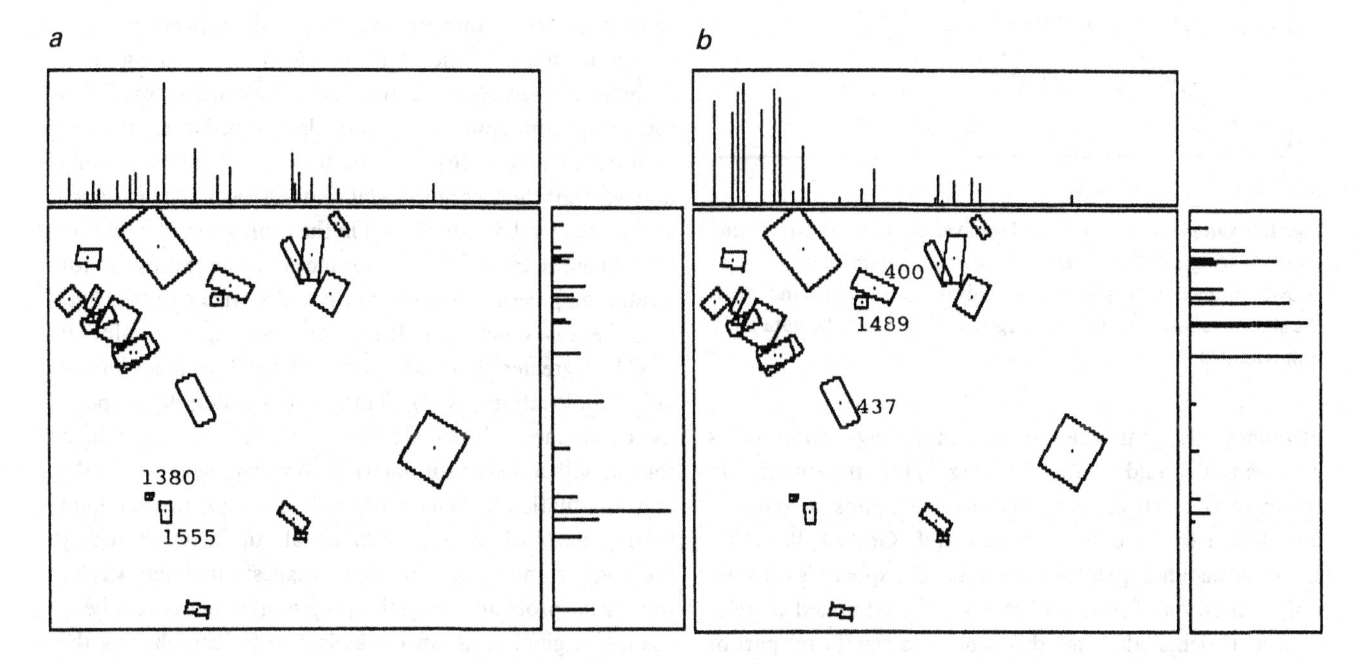

Fig. 2 Schematized map of study area displaying position of the schematized catchments and observed values of two example events; (a) event No. 25, (b) event No. 1.

Table 2. *Characteristics of interactively evaluated theoretical semivariogram parameters for the 42 flood events*

	Point approach			Block approach		
	Nugget	Sill	Range	Nugget	Sill	Range
Mean	0.056	0.612	78	0.140	0.748	78
Variability	0.0–0.27	0.10–2.21	70–126	0.05–0.35	0.15–2.30	70–126

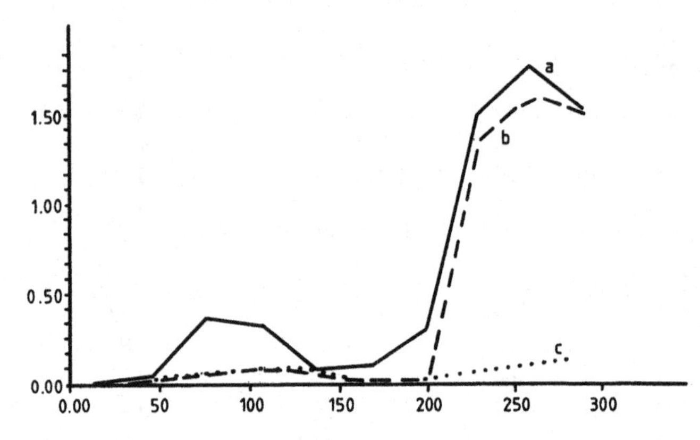

Fig. 3 Experimental semivariograms for event 25; line (a) non-trimmed, with one outlying observation; line (b) trimmed, with one outlying observation; line (c) non-trimmed, one outlying observation removed.

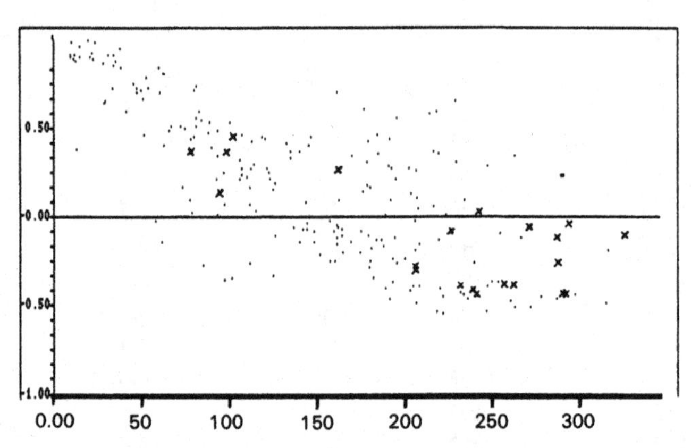

Fig. 5 Example of plots of pairwise correlation vs distance. Crosses mark corellation coefficients of the station, containing two largest outliers from the data set.

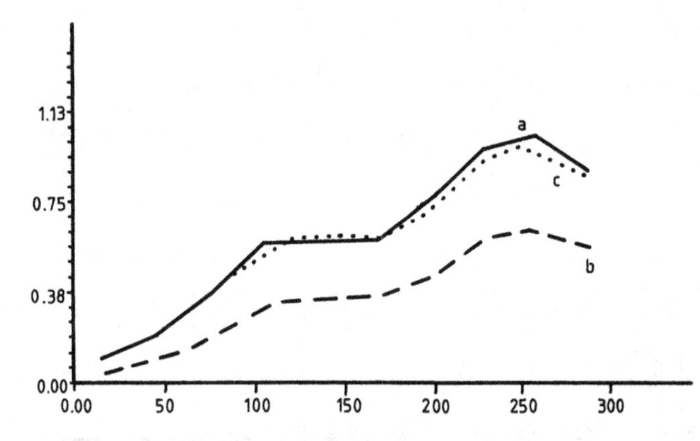

Fig. 4. Example of time averaged semivariograms; (a) experimental semivariogram with several outliers, non-trimmed; (b) as above, trimmed; (c) experimental semivariogram after removal of the station, containing the two largest outliers from the data set, non-trimmed.

of points disappear when the contaminating station 437 is removed. It should be noted (cf. Fig. 5) that the runoff data typically yield stronger scatter of data points in plots of correlation coefficients vs distance (cf. Gottschalk, 1989) than, for instance, precipitation data. The spread is thus not only a question of statistical errors in the estimated correlation coefficients, although this explains a significant part of the scatter. It can be stated that individual stations (of bad quality?) are often the offspring to groups of 'off points'.

Cross-validation is performed for each individual event both with the help of point and block kriging (with areal support accounted). In both cases it is based on theoretical curves fitted to the averaged semivariograms over all events (Fig. 6) or on curves fitted to the correlation scatter. Fig. 7 shows the correlation corresponding to the theoretical semivariogram fitted as shown in Fig. 6. It is also possible to derive a semivariogram corresponding to a fitted correlation function. Estimation variances are determined in accordance with both equation (6) and equation (13), the latter taking into account the location-specific covariances. The variance at the site of validation is used in the evaluation of equation (13). There is, however, no assurance that the expression for estimation precision is non-negative (cf. Switzer, 1989). This is so, because equation (13) may produce negative values if $\Lambda^T C \Lambda$ is greater than the estimated local variance in the unknown location ($s^2(x_0)$). This can occur despite the positive definiteness of the matrix involved. The case considered (blocks with different supports) is more complicated than the regular situation for which the method was proposed (point kriging or block kriging with equal supports). Bardossy (personal communication, 1990) suggests that the way of solving the problem of negative precision estimates may be to use the original covariance matrix also for establishing the local variance.

Analysis of the results of cross-validation illustrates the

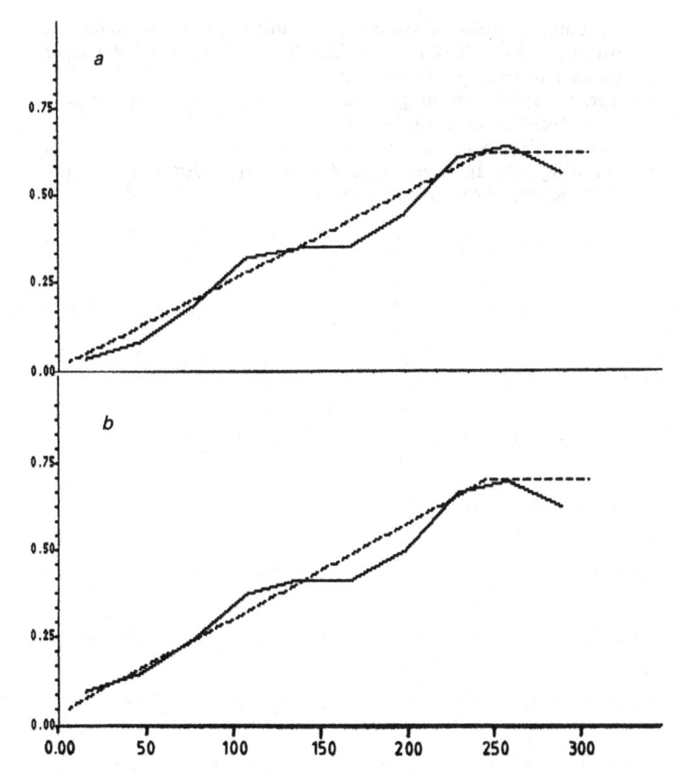

Fig. 6 Averaged experimental semivariograms for all events with fitted theoretical curves; (a) point semivariograms; (b) block semivariograms. Solid and dashed lines denote experimental, and theoretical semivariograms, respectively.

contaminating effect of outliers. There may be several observation points in the data, where observed values largely differ from the estimated values. This is caused by the fact that an outlier may influence estimated values in its neighbourhood. Removal of one or few extreme outliers may liquidate differences between observed and estimated values at a number of sites.

An example of another contaminating effect can be seen at observation sites close to the boundary between the areas with high and low runoff values. In one event high flows occured only in the upper left quadrant of the map given in Fig. 2, whereas the other quadrants contained significantly smaller flows. In such a situation of partial coverage of the study area with extreme runoff (or precipitation), that is not uncommon in the data analyzed, behaviour of the estimation process close to the boundary of distinct areas is not regular.

CONCLUSIONS

The event-based case showed, that it is advisable to perform outlier detection at the stage of analysing semivariograms. As the total variance (sill) differs considerably from one event to the other an observation can be labelled outlier in

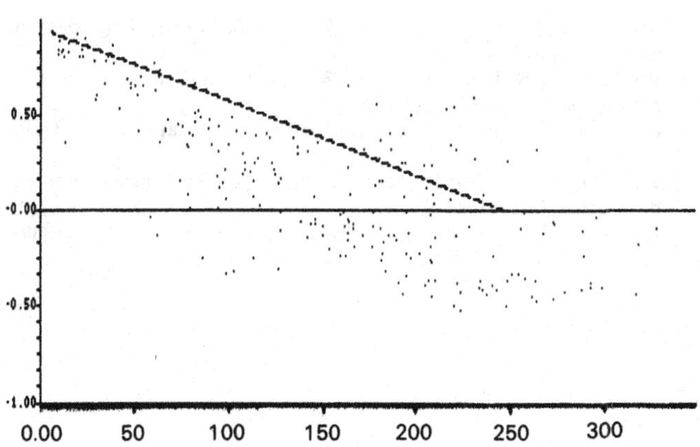

Fig. 7 Spatial correlation scatter diagram with fitted theoretical curve.

one event, whereas when compared to the full series of events would not be considered so.

Detection of outliers with cross-validation based on the time averaged semivariograms and/or pairwise correlations is well fitted for operational use. The approach seems little sensitive to such elements as, for instance, areal support and local variance. The methodology needs to be elaborated further in respect to the quality of data, statistical errors, data smoothening etc.

The philosophy of the geostatistical approach for outlier detection resembles the application of tests of Student ratios and the RST-statistics to outliers detection in non-structured data (Roesner, 1977). However, in comparison to these methods that do not take into account the regional information, cross-validation with time-averaged semivariograms gives higher precision in the determination of outlying observations. The supplementary spatial information in this latter case is reflected in the standard deviation of the estimation. This is, in general, significantly lower than the standard deviation for the time series of annual maximum flow at a site, which is, in turn, the basis of the Student ratio and RST-statistics.

ACKNOWLEDGEMENTS

The study was financially supported by the Norwegian National Council for Hydrology. The support is gratefully acknowledged. Constructive comments from Dr Andras Bardossy are appreciated.

REFERENCES

Clark, I. (1979) *Practical Geostatistics*, Elsevier.
Gottschalk, L. (1985) Hydrological regionalization of Sweden, *Hydrol. Sci. J.*, **30**(1), 65–83.

Gottschalk, L. (1989) *Analysis of rainfall floods in Norway*. Unpublished memorandum.
Gottschalk, L. & Kundzewicz, Z. W. (1995) *Analysis of outliers in Norwegian flood data*, this volume.
Hawkins, D. M. (1980) *Identification of Outliers*, Chapman and Hall, London.
Kottegoda, N. T. (1984) Investigation of outliers in annual maximum flow series, *J. Hydrol.*, **72**, 105–37.
Kundzewicz, Z. W., Bardossy, A., Plate, E. J. & Grimm-Strele, J. (1989) Plausibility analysis of structured groundwater quality data via geostatistics. In: Kobus, H. E. & Kinzelbach, W. (Eds), *Contaminant Transport in Groundwater*, Balkema, Rotterdam.
Roesner, B. (1977) Percentage points for the RST many outlier procedure. *Technometrics*, **19**(3), 307–12.
Switzer, P. (1989) Non-stationary spatial covariances estimated from monitoring data. In Armstrong, M. (Ed.), *Geostatistics*, Vol. 1, pp. 127–38, Kluwer Academic Publishers.

V

Time series and stochastic processes

1 Prediction uncertainty in seasonal partial duration series

P. F. RASMUSSEN and D. ROSBJERG

Institute of Hydrodynamics and Hydraulic Engineering, Technical University of Denmark, Lyngby, Denmark

ABSTRACT In order to obtain a good description of the exceedances in a partial duration series it is often necessary to divide the year into a number (2–4) of seasons. Hereby a stationary exceedance distribution can be maintained within each season. This type of seasonal model may, however, not be suitable for prediction purposes due to the large number of parameters required. In the particular case with exponentially distributed exceedances and Poissonian occurrence times the precision of the T-year event estimator has been thoroughly examined considering both seasonal and non-seasonal models. The two-seasonal probability density function of the T-year event estimator has been deduced and used in the assessment of the precision of approximate moments. The non-seasonal approach covered both a total omission of seasonality by pooling data from different flood seasons and a discarding of non-significant season(s) before the analysis of extremes. Mean square error approximations (bias second order, variance first and second order) were employed as measures for prediction uncertainty. It was found that optimal estimates can usually be obtained with a non-seasonal approach.

INTRODUCTION

Since its introduction into flood frequency analysis, the partial duration series (PDS) method has gained increased acceptance as an appealing alternative to the annual maximum series (AMS) method. PDS models were introduced in hydrology by Shane & Lynn (1964), and Todorovic & Zelenhasic (1970). They assumed independent and identically distributed exceedances occurring according to a Poisson process with time-dependent intensity. The efficiency of the PDS-model was demonstrated by Cunnane (1973) and Taesombut & Yevjevich (1978). Todorovic & Rousselle (1971) suggested a division of the year into distinct seasons with different, time-independent distributions of exceedances, while North (1980) and Konecny & Nachtnebel (1985) considered a continuous variation of parameters. Trigger-type cluster models were explored by Kavvas (1982a, b) and correlated exceedances by Rosbjerg (1985, 1987a). Contrary to the standard procedure, which is based on a fixed threshold level, Buishand (1989) studied PDS-models with a fixed number of peaks. Alternatives to the exponential distribution of exceedances include the Gamma-distribution (Zelenhasic, 1970), the log-normal distribution (Rosbjerg, 1987b) and the generalized Pareto-distribution (Fitzgerald, 1989). Refined studies of the prediction uncertainty of PDS-models have been carried out by Ashkar & Rousselle (1981) and Rasmussen & Rosbjerg (1989). Risk and uncertainty measures in a Bayesian framework have been developed by Rousselle & Hindie (1976) and Rasmussen & Rosbjerg (1991a).

In this paper the seasonal approach introduced by Todorovic & Rousselle (1971) is critically reviewed. When exceedances of the base level have different statistical properties, for instance different mean value (maintaining the exponential distribution assumption), a better description of the process is clearly possible by means of a seasonal model. However, it is a well known fact that the gain obtained from a sophisticated description must be balanced against the increased sampling uncertainty due to a larger number of parameters. In many cases a simple model with only a few parameters is preferable to a more complex model with regard to prediction uncertainty. The problem of estimating the annual maximum distribution from samples of maxima in separate

An extended version of this contribution (with Appendices) was published in *Water Resour. Res.*, **27**(11), Nov. 1991, copyright by the American Geophysical Union.

seasons has recently been examined by Buishand & Demarée (1990). In the present paper, however, the approach is somewhat different, since the estimator of the T-year event is obtained directly from the partial duration series without consideration of the annual maximum distribution.

The model to be used here assumes stationary distributed exceedances within each season. The occurrence of exceedances takes place according to a Poisson process which may have continuously varying intensity. When the threshold is chosen sufficiently high, the seasonality becomes more or less negligible, while a low threshold often implies a pronounced seasonality. Therefore a high threshold might be desirable. On the other hand, information is discarded when the threshold is increased. With realistically chosen thresholds some kind of seasonality will almost always be present in PDS-data, which suggests the use of seasonal models. Better results, in terms of prediction accuracy, may, however, be achieved if seasonality is simply disregarded. The mean square error of the T-year event estimator is used here to compare different model candidates. For that purpose approximate formulas for bias and variance are deduced both in the case of seasonal models and models where seasonality is present but disregarded. A third approach consists of using data only from the dominant season. This may in some cases provide satisfactory results. The three methods are compared, and some guidelines for choosing the appropriate model are given.

MODEL DESCRIPTION

The stochastic model, which we shall use in this study, assumes that the year can be divided into a number of seasons, M, and that within each season the exceedances of the threshold level are independent and identically distributed. Exceedances appear according to a Poisson process with possibly time-dependent intensity. Let λ_j be the expected number of exceedances per year in season no. j. If t years of records are available, the distribution of the total number of exceedances in season no. j, N_j, is given by

$$P\{N_j=n\}=\frac{e^{-\lambda_j t}(\lambda_j t)^n}{n!}; \quad n=0, 1, 2 \tag{1}$$

where $E(N_j)=Var(N_j)=\lambda_j t$.

Let the chosen threshold level be denoted q_0, and the magnitude of an exceedance be defined as $X=Q-q_0$, where Q is a flood peak exceeding q_0. In a particular season, say j, these magnitudes are assumed exponentially distributed with parameter α_j. The cumulative distribution function (CDF) of a single event is thus

$$F_j(x)=1-\exp(x/\alpha_j); \quad x>0 \tag{2}$$

where $E\{X\}=\alpha_j$ and $Var\{X\}=\alpha_j^2$.

The T-year flood, Q_T, is defined as the flood magnitude that is exceeded on the average once per T years. In order to derive the relationship between $x_T(=Q_T-q_0)$, T, and the parameters $\alpha_1,\dots,\alpha_M$, $\lambda_1,\dots,\lambda_M$, we first notice that the annual mean number of floods larger than q_0+x in season no. j is

$$E(N_{x,j})=\lambda_j[1-F_j(x)]=\lambda_j\exp(-x/\alpha_j) \tag{3}$$

Hence the mean value of the total number of exceedances of q_0+x in t years, N_x^t, is

$$E(N_x^t)=t\sum_{j=1}^{M}E(N_{x,j})=t\sum_{j=1}^{M}\lambda_j\exp(-x/\alpha_j) \tag{4}$$

Inserting $t=T$, $x=x_T$ and $E\{N_x^t\}=1$ and rearranging yields:

$$h(x_T,\alpha_1,\dots,\alpha_M,\lambda_1,\dots,\lambda_M,T)=\sum_{j=1}^{M}\lambda_j\exp(-x/\alpha_j) \tag{5}$$

whereby the T-year event, x_T, is given implicitly as a function of the distribution parameters and the return period, T. For given values of these parameters an iterative technique, e.g. the Newton–Raphson iteration, must be applied to obtain x_T. The N–R method is very robust in this particular case and converges to a satisfactory precision in a few steps. For practical application, estimates of the parameters can be obtained from the series of historical flow records. Maximum likelihood estimators read

$$\hat{\alpha}_j=\frac{1}{N_j}\sum_{i=1}^{N_j}X_{ji} \tag{6}$$

where X_{ji}, $i=1,2,..N_j$ are the exceedance magnitudes observed in season no. j and

$$\hat{\lambda}_j=\frac{N_j}{t} \tag{7}$$

where t is the number of observation years.

PDF OF THE T-YEAR EVENT ESTIMATOR

The non-seasonal probability density function (PDF) of $\hat{x}_T$, i.e. the special case $M=1$, was deduced by the authors (Rasmussen & Rosbjerg, 1989). By means of this PDF explicit, exact expressions for the mean value, $E\{\hat{x}_T\}$, and the standard deviation, $S\{\hat{x}_T\}$, were derived. The case of seasonality was considered by Ashkar & Rousselle (1981), but they used a definition of the T-year event based on the annual maximum series. In this section we shall briefly show how the PDF of $\hat{x}_T$, defined directly from the PDS, and corresponding moments can be deduced when seasonality is present.

For simplicity we shall confine ourselves to the case $M=2$, yet the following results are general and can be extended to comprise an arbitrary number of seasons. The sampling distributions of N_j and of $\hat{\alpha}_j$ conditional upon N_j are given by (1) and

$$f_{\hat{\alpha}_j}(a|N_j=n)=\frac{n^n e^{\frac{-na}{\alpha_j}}}{\alpha_j\Gamma(n)}\left(\frac{a}{\alpha_j}\right)^{n-1} \tag{8}$$

respectively (see for instance Ashkar & Rousselle, 1981). Inserting estimated parameters in (5) and rearranging allow us to express α_1 as a function, k, of $\hat{x}_T$ and the remaining parameter estimates

$$\hat{\alpha}_1=k(\hat{x}_T,\hat{\alpha}_2,\hat{\lambda}_1,\hat{\lambda}_2)=-\frac{\hat{x}_T}{\ln\left(\frac{1}{\hat{\lambda}_1}\left[\frac{1}{T}-\hat{\lambda}_2\exp\left(-\frac{\hat{x}_T}{\hat{\alpha}_2}\right)\right]\right)} \tag{9}$$

This arrangement is, of course, only permissible if $\hat{\lambda}_1>0$. Another basic requirement can be expressed in the inequality

$$1/T-\hat{\lambda}_2\exp(-\hat{x}T/\hat{\alpha}_2)>0 \tag{10}$$

If the left side of (10) is equal to zero, the total contribution to T will be explained by the second season, and 1 should be zero, which only happens with probability zero. For fixed values of $\hat{\alpha}_2$, $\hat{\lambda}_1$ and $\hat{\lambda}_2^2$ consider the transform of variables $\hat{\alpha}_1\rightarrow\hat{x}_T$. The PDF of $\hat{x}_T$, conditional upon $\hat{\alpha}_2$, $\hat{\lambda}_1$ and $\hat{\lambda}_1$, is

$$f_{\hat{x}_T}(x|\hat{\alpha}_2=a_2,\hat{\lambda}_1=n_1/t,\hat{\lambda}_2=n_2/t)=f_{\hat{\alpha}_1}(a_1|N_1=n_1)\frac{1}{\left|\frac{\partial\hat{x}_T}{\partial\hat{\alpha}_1}\right|_{a_1=k}} \tag{11}$$

where $k=k(x,a_2,n_1/t,n_2/t)$

By appropriate summation the conditioning upon $\hat{\lambda}_1$ can be eliminated

$$f_{\hat{x}_T}(x|\hat{\alpha}_2=a_2,\hat{\lambda}_2=n_2/t)=\sum_{n_1=1}^{\infty}f_{\hat{\alpha}_1}[k(x,a_2,n_1/t,n_2/t)|N_1=n_1]\cdot k_x'(x,a_2,n_1/t,n_2/t)\,\mathrm{P}(N_1=n_1) \tag{12}$$

where k_x' is the partial derivative of k with respect to x. The unconditional PDF of $\hat{x}_T$ may now be obtained by integration of the probability of $\hat{\alpha}_2$ conditional on N_2 followed by summation of the probability of N_2. For a given value of $\hat{\lambda}_2$ the estimator $\hat{\alpha}_2$ is restricted by (10). The final expression reads:

$$f_{\hat{x}_T}(x)=\sum_{n_2=1}^{\infty}\int_0^{x/\ln(n_2/t)T}\sum_{n_1=1}^{\infty}f_{\hat{\alpha}_1}[k(x,a_2,n_1/t,n_2/t)|N_1=n_1]\cdot k_x'(x,a_2,n_1/t,n_2/t)\cdot\mathrm{P}(N_1=n_1)\,f_{\hat{\alpha}_2}(a_2|N_2=n_2)\mathrm{d}a_2\cdot\mathrm{P}(N_2=n_2) \tag{13}$$

Observe that $\mathrm{P}\{N_j=\mathrm{n}_j\}$, $j=1,2$, as given in (1), must be adjusted by division with the probability of $N_j>0$ in order to constitute a proper probability function (note that the summations start from one). The PDF of $\hat{x}_T$ should thus be regarded as a conditional probability, namely conditional upon at least one exceedance in each season during t observation years. No further simplification of the above expression seems possible. Numerical integration must be used to determine the PDF as well as the moments of $\hat{x}_T$. Clearly, this procedure is not very tractable, and when more than two seasons are involved, the calculations become almost overwhelming, even with strong computer support. In order to obtain the bias and the standard deviation of $\hat{x}_T$ it is therefore necessary to have approximate formula available that do not involve multiple integrations and summations.

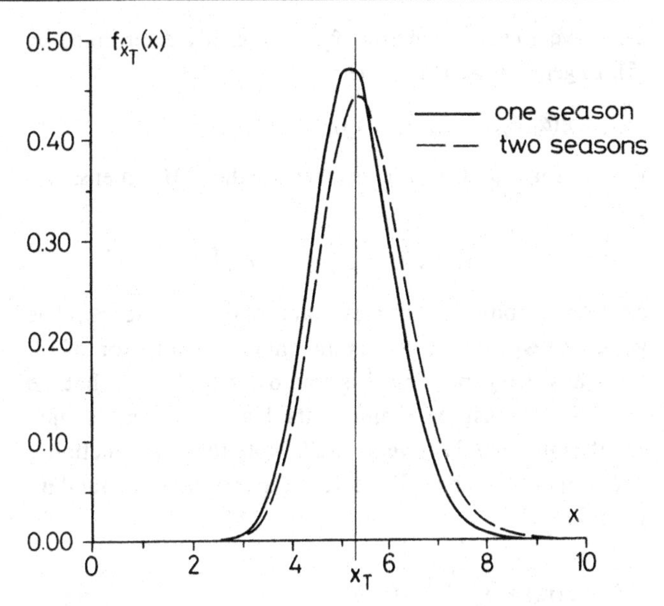

Fig. 1 Pdf of $\hat{x}_{50}$ corresponding to $t=10$ years. Non-seasonal: $\alpha=1$ and $\lambda=4$. Two-seasonal: $\alpha_1=1$, $\alpha_2=1$, $\lambda_1=1.33$ and $\lambda_2=2.67$.

Fig. 1 shows the PDF of $\hat{x}_T$ according to two distinct models, namely a non-seasonal and a two-seasonal approach. In order to make a comparison possible, the same mean exceedance has been chosen in either season, whereas the intensities are distinct: $\lambda_1=1.33$ and $\lambda_2=2.67$. Clearly, this case does not require a seasonal approach. Yet both models are correct and yield the same theoretical T-year event. As expected the two PDFs are quite similar, though the figure reveals some decrease in precision, when the seasonal approach is used.

APPROXIMATE MOMENTS OF THE T-YEAR EVENT ESTIMATOR

Approximate moments of a stochastic property, which is a function of random variables with known sampling properties, can be obtained by means of a Taylor series expansion.

The T-year event estimator, $\hat{x}_T$, is implicitly determined by (5). In general, we may write

$$\hat{x}_T = g(\hat{\alpha}_1, \hat{\alpha}_2, \ldots, \hat{\alpha}_M, \hat{\lambda}_1, \ldots, \hat{\lambda}_M) = g(\hat{\boldsymbol{\Theta}}) \quad (14)$$

where, for notational convenience, the $2M$-dimensional vector

$$\hat{\boldsymbol{\Theta}} = (\hat{\Theta}_1, \hat{\Theta}_2, \ldots, \Theta_{2M})^{\mathrm{T}} = (\hat{\alpha}_1, \hat{\alpha}_2, \ldots, \hat{\alpha}_M, \hat{\lambda}_1, \ldots, \hat{\lambda}_M)^T$$

has been introduced. The mean value of $\hat{\boldsymbol{\Theta}}$ is denoted $\boldsymbol{\mu}_{\Theta}$. The vector $\hat{\boldsymbol{\Theta}}$ will be treated as an unbiased estimator of $\boldsymbol{\Theta}$, though, strictly speaking, $\hat{\lambda}_j$ is biased due to the fact that the case $N_j = 0$ must be omitted. Its bias, however, is only significant when $\lambda_j t$ is very small. Note that the function g does not exist in an explicit form. Expansion of g around $\boldsymbol{\mu}_{\Theta}$ yields

$$\begin{aligned} \hat{x}_T = g(\boldsymbol{\mu}_\theta) &+ \sum_{i=1}^{2M} \left.\frac{\partial g}{\partial \theta_i}\right|_m (\hat{\theta}_i - \mu_i) \\ &+ 1/2! \sum_{j=1}^{2M} \sum_{i=1}^{2M} \left.\frac{\partial^2 g}{\partial \theta_j \partial \theta_i}\right|_m (\hat{\theta}_j - \mu_j)(\hat{\theta}_i - \mu_i) \\ &+ 1/3! \sum_{k=1}^{2M} \sum_{j=1}^{2M} \sum_{i=1}^{2M} \left.\frac{\partial^3 g}{\partial \theta_k \partial \theta_j \partial \theta_i}\right|_m (\hat{\theta}_k - \mu_k)(\hat{\theta}_j - \mu_j)(\hat{\theta}_i - \mu_i) + .. \end{aligned} \quad (15)$$

where index m means that the derivatives must be evaluated at $\boldsymbol{\mu}_{\Theta}$. The expectation of $\hat{x}_T$ is therefore approximately (neglecting terms higher than second order)

$$\mathrm{E}(\hat{x}_T) = g(\boldsymbol{\mu}_\theta) + \sum_{i=1}^{2M} \left.\frac{\partial^2 g}{\partial \theta_j^2}\right|_m \mathrm{Var}(\hat{\theta}_j) \quad (16)$$

where it has been used that all parameter estimators are unbiased and uncorrelated. The approximate bias of $\hat{x}_T$ is

$$\mathrm{B}(\hat{x}_T) = \mathrm{E}(\hat{x}_T) - x_T = g(\boldsymbol{\mu}_\theta) + \sum_{i=1}^{2M} \left.\frac{\partial^2 g}{\partial \theta_i^2}\right|_m \mathrm{Var}(\hat{\theta}_i) \quad (17)$$

Derivatives of g and $\mathrm{Var}\{\hat{\Theta}_i\}$ are given in Appendices A, and C, respectively, in Rasmussen & Rosbjerg (1991b). An example of the use of this formula is presented in Fig. 2. The approximation is in good agreement with the analytical, exact curve obtained by means of the theoretical PDF (13). Simulated results are given for verification. When one of the seasons is dominant over the other, in terms of mean exceedance, the bias is small, whereas there is a certain bias, when the α values are close to each other.

Equations (16) and (17) and following expressions for approximate moments depend on the population parameters, which are unknown in all practical cases. The problem is circumvented by inserting the parameter estimates $\hat{\boldsymbol{\mu}}_{\Theta} = (\hat{\alpha}_1, \hat{\alpha}_2, .., \hat{\alpha}_M, \hat{\lambda}_1, .., \hat{\lambda}_M)^T$ for $\boldsymbol{\mu}_{\Theta}$. Using this procedure on a large number of generated samples leads to a range of estimated values for the bias, but the variability is usually found to be relatively small, thus justifying the procedure. Corresponding remarks apply to the following variance approximation.

From (15) it can be seen that a first order approximation to the variance of $\hat{x}_T$ reads:

$$\mathrm{Var}(\hat{x}_T) \simeq \sum_{i=1}^{2M} \left(\frac{\partial g}{\partial \theta_i}\right)_m^2 \mathrm{Var}(\hat{\theta}_i) \quad (18)$$

In the non-seasonal case this formula is very accurate as demonstrated by Rasmussen & Rosbjerg (1989). However, in the case of two seasons the method tends to underestimate the variance when the exponential distributions have similar parameters, while overestimating it when the seasonality becomes more pronounced, see Fig. 3. When $M = 1$, the satisfactory results of (18) are mainly owing to the linearity between $\hat{x}_T$ and $\hat{\alpha}_T = \hat{\theta}_1$. When $M > 1$, the relationship between $\hat{x}_T$ and $\hat{\alpha}j$ is non-linear, which may partly explain the deviation from the simulated results. An improvement of the fit could possibly be obtained by introducing a second order approximation which reads:

$$\begin{aligned} \mathrm{Var}\,\{\hat{x}_T\} = &\sum_{i=1}^{2M} \left(\frac{\partial g}{\partial \theta_i}\right)_m^2 \mathrm{Var}(\hat{\theta}_i) \\ &+ \sum_{i=1}^{2M} \tfrac{1}{4} \left(\frac{\partial^2 g}{\partial \theta_i^2}\right)_m^2 \mathrm{Var}\,[(\hat{\theta}_i - \mu_i)^2] \\ &+ \sum_{i=1}^{2M} \sum_{j=i+1}^{2M} \left(\frac{\partial^2 g}{\partial \theta_i \partial \theta_j}\right)_m^2 \mathrm{Var}[(\hat{\theta}_i - \mu_i)(\hat{\theta}_j - \mu_j)] \\ &+ \sum_{i=1}^{2M} \left(\frac{\partial g}{\partial \theta_i}\right)_m \left(\frac{\partial^2 g}{\partial \theta_i^2}\right)_m \mathrm{Cov}\,[\hat{\theta}_i - \mu_i, (\hat{\theta}_i - \mu_i)^2] \\ &+ \sum_{i=1}^{2M} \tfrac{1}{3} \left(\frac{\partial g}{\partial \theta_i}\right)_m \left(\frac{\partial^3 g}{\partial \theta_i^3}\right)_m \mathrm{Cov}\,[\hat{\theta}_i - \mu_i, (\hat{\theta}_i - \mu_i)^3] \\ &+ \sum_{i=1}^{2M} \sum_{i=1}^{2M} 2 \left(\frac{\partial g}{\partial \theta_i}\right)_m \left(\frac{\partial^3 g}{\partial \theta_i \partial \theta_j^2}\right)_m \mathrm{Cov}\,[\hat{\theta}_i - \mu_i, (\hat{\theta}_i - \mu_i)(\hat{\theta}_j - \mu_j)^2] \end{aligned} \quad (19)$$

where

$$\mathrm{Var}[(\hat{\theta}_i - \mu_i)(\hat{\theta}_j - \mu_j)] = \begin{cases} \mathrm{Var}(\hat{\theta}_i)\,\mathrm{Var}(\hat{\theta}_j) & i \neq j \\ [\mathrm{Var}(\hat{\theta}_i)]^2\,(\gamma_{2i} - 1) & i = j \end{cases} \quad (20)$$

$$\mathrm{Cov}[\hat{\theta}_i - \mu_i, (\hat{\theta}_j - \mu_j)^2] = [\mathrm{Var}(\hat{\theta}_i)]^{3/2} \gamma_{1i} \quad (21)$$

$$\mathrm{Cov}[\hat{\theta}_i - \mu_i, (\hat{\theta}_i - \mu_i)^3] = [\mathrm{Var}(\hat{\theta}_i)]^2 \gamma_{2i} \quad (22)$$

$$\mathrm{Cov}[\hat{\theta}_i - \mu_i, (\hat{\theta}_i - \mu_i)(\hat{\theta}_j - \mu_j)^2] = \mathrm{Var}(\hat{\theta}_i)\,\mathrm{Var}(\hat{\theta}_j) \quad (23)$$

Here γ_{1i} and γ_{2i} are the coefficient of skewness and the coefficient of kurtosis, respectively, in the distribution of $\hat{\Theta}_i$.

Fig. 3 shows that for some parameter combinations the agreement with simulated results can be improved by using a second order approximation, but a general conclusion is somewhat unclear. Due to the complexity of the second order approach, along with the general uncertainty inherent in the analysis, the first order formula will, in practical applications, provide sufficient precision. In the forthcoming

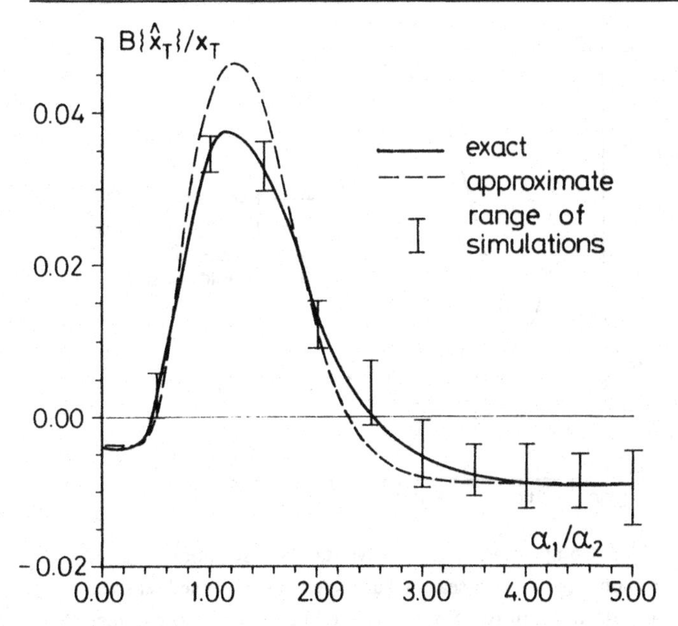

Fig. 2 Relative bias of $\hat{x}_{50}$ according to a two-seasonal model. $t=10$ years, $\alpha=1$, $\lambda_1=1.33$ and $\lambda_2=2.67$. Number of simulations = 5000.

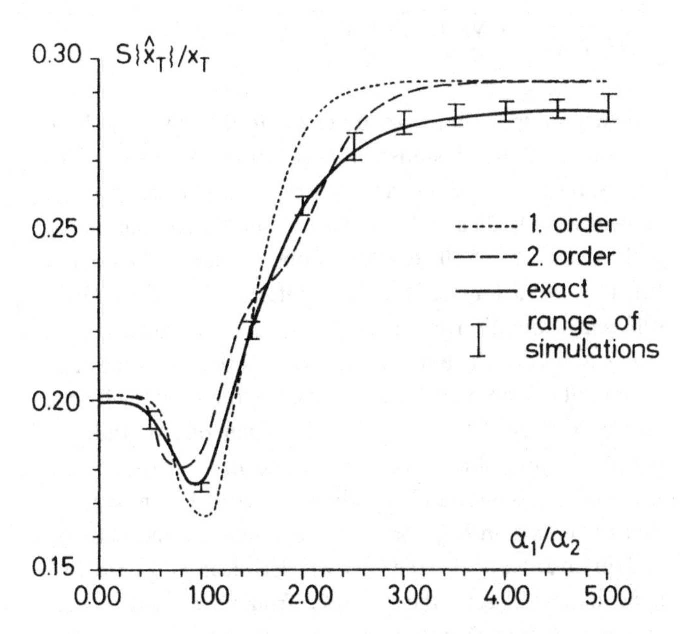

Fig. 3 Relative standard deviation of $\hat{x}_{50}$ according to a two-seasonal model. $t=10$ years, $\alpha_2=1$, $\lambda_1=1.33$ and $\lambda_2=2.67$. Number of simulations = 5000.

analysis we will therefore use (18) as an approximation of the variance.

APPROXIMATE MOMENTS WHEN SEASONS ARE POOLED

So far it has been assumed that the division of the year into seasons is given. It is, however, not obvious how this division should be made. A possible approach is to consider the year as having four seasons, spring, summer, fall and winter. In most cases floods only occur in two or maybe three seasons, being caused for example by snowmelt or heavy rainstorms. An improvement in the analysis can be obtained if floods can be recognized by their causal origin (Waylen & Woo, 1982), but this will usually require more information such as temperature and precipitation.

A flood frequency model containing many parameters is known to be less robust to sampling errors than a model with fewer parameters. On the other hand, the latter may introduce systematical errors in quantile estimates due to an incapability to describe the true statistical nature of floods. It is intuitively obvious that one should not choose a model that is more complex (in terms of the number of parameters) than absolutely required. Thus it might be appropriate to accept a certain systematical error (i.e. bias) in order to make the quantile estimates more robust. A simplified model can be achieved by pooling data from different seasons (with possibly different statistical properties) into one sample and estimating the parameter of an assumed single exponential distribution for all the exceedances. In order to compare models (in terms of mean square error), approximate formulas for bias and variance will be developed for the case where seasonality is disregarded.

If x_T is estimated from pooled seasons, the following result for the bias of $\hat{x}_T$ is obtained. Assume, as an example, that the first two seasons are pooled. Now $\hat{x}_T$ is given by $g^*(\hat{\mathbf{\Theta}}^*)$. The new parameter vector, $\mathbf{\Theta}^*$, contains $2M^*$ elements, where $M^*=M-1$. Note that

$$\hat{\Theta}_1^*=\hat{\alpha}_1^* \text{ and } \hat{\Theta}_{M^*+1}^*=\hat{\lambda}_1^*$$

are related to the original parameter estimators through

$$\hat{\alpha}_1^*=\frac{\hat{\lambda}_1\hat{\alpha}_1+\hat{\lambda}_2\hat{\alpha}_2}{\hat{\lambda}_1+\hat{\lambda}_2} \tag{24}$$

and

$$\hat{\lambda}_1^*=\hat{\lambda}_1+\hat{\lambda}_2 \tag{25}$$

The variance of $\hat{\lambda}_1^*$ is readily obtained

$$\text{Var}(\hat{\lambda}_1^*)=\text{Var}(\hat{\lambda}_1)+\text{Var}(\hat{\lambda}_2) \tag{26}$$

Some kind of approximation is required in order to determine the variance of $\hat{\alpha}_1^*$. Let

$\hat{\alpha}_1^*=h(\hat{\lambda}_1,\hat{\lambda}_2,\hat{\alpha}_1,\hat{\alpha}_2)$. Then

$$\text{Var}(\hat{\alpha}_1^*)=\left(\frac{\partial h}{\partial\hat{\lambda}_1}\right)_m^2\text{Var}(\hat{\lambda}_1)+\left(\frac{\partial h}{\partial\hat{\lambda}_2}\right)_m^2\text{Var}(\hat{\lambda}_2)$$
$$+\left(\frac{\partial h}{\partial\hat{\alpha}_1}\right)_m^2\text{Var}(\hat{\alpha}_1)+\left(\frac{\partial h}{\partial\hat{\alpha}_2}\right)_m^2\text{Var}(\hat{\alpha}_2)$$
$$=\left[\frac{\alpha_1\lambda_2-\alpha_2\lambda_2}{(\lambda_1+\lambda_2)^2}\right]^2\text{Var}(\hat{\lambda}_1)+\left[\frac{\alpha_2\lambda_1-\alpha_1\lambda_1}{(\lambda_1+\lambda_2)^2}\right]^2\text{Var}(\hat{\lambda}_2)$$
$$+\frac{\lambda_1^2}{(\lambda_1+\lambda_2)^2}\text{Var}(\hat{\alpha}_1)+\frac{\lambda_2^2}{(\lambda_1+\lambda_2)^2}\text{Var}(\hat{\alpha}_2) \tag{27}$$

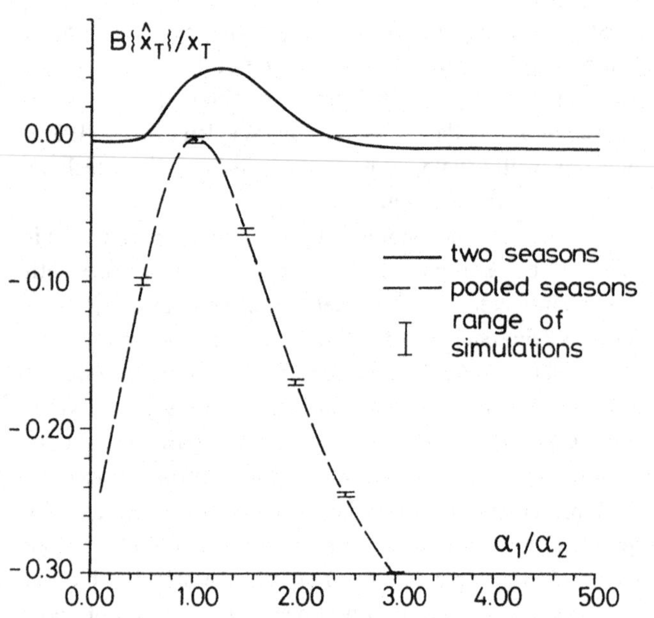

Fig. 4 Comparison of the relative bias of $\hat{x}_{50}$ corresponding to a two-seasonal approach and a pooled seasons approach. $t=10$ years, $\alpha_2=1$, $\lambda_1=1.33$ and $\lambda_2=2.67$. Number of simulations $=5000$.

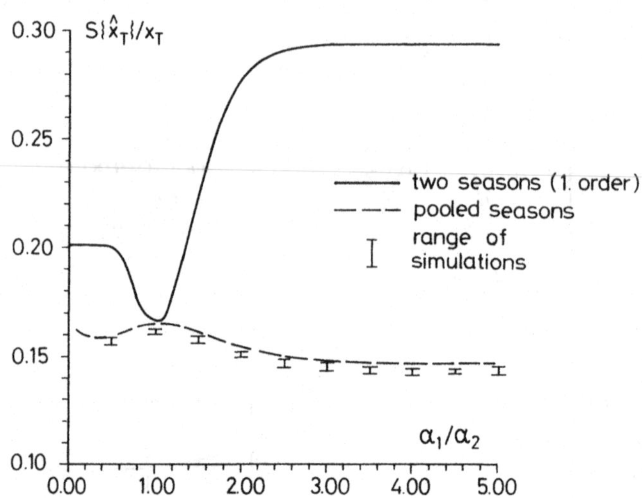

Fig. 5 Comparison of the relative standard deviation of $\hat{x}_{50}$ corresponding to a two-seasonal approach and a pooled seasons approach. $t=10$ years, $\alpha_2=1$, $\lambda_1=1.33$ and $\lambda_2=2.67$. Number of simulations $=5000$.

The bias estimate is eventually obtained as

$$B(\hat{x}_T)=g^*(\boldsymbol{\mu}_\Theta^*)+\sum_{j=1}^{2M^*}\frac{1}{2}\left(\frac{\partial^2 g^*}{\partial\theta_j^{*2}}\right)_m \operatorname{Var}(\hat{\Theta}_j^*)-g(\boldsymbol{\mu}_\Theta) \tag{28}$$

Fig. 4 shows how this approach affects the relative bias of $\hat{x}_T$. There seems to be a satisfactory agreement between calculated and simulated values. It is evident from the two curves that ignoring seasonal variations introduces a systematic error whose magnitude depends on the significance of the seasonality. In a small band around $\alpha_1/\alpha_2=1$, the absolute bias is smaller when the non-seasonal method is used, reflecting the negligible seasonal behaviour in this case.

The first order variance approximation is obtained by inserting the modified parameter vector and variance expressions, (26) and (27), into (18). A comparison of the two methods is given in Fig. 5. For all parameter combinations the approximate relative standard deviation is smallest when seasonality is disregarded. It should be noted that the variance is sensitive to the number of observation years.

COMPARISON OF MODEL PERFORMANCE

In order to illustrate the implications of different design procedures, a numerical example was designed. The results from preceeding sections have been used to construct Figs. 6a, b, which depict the relative root mean square error of $\hat{x}_T$, defined as

$$\operatorname{rrmse}(\hat{x}_T)=\frac{\sqrt{\operatorname{Var}(\hat{x}_T)+[B(\hat{x}_T)]^2}}{x_T} \tag{29}$$

The year is assumed to be characterized by two flood seasons with known statistical population parameters. The annual mean number of exceedances in the two seasons are, respectively, 1.333 and 2.667. The rrmse$(\hat{x}_{50})$ corresponding to 10 years (a) and 50 years (b) of observation is shown as a function of various relationships between α_1 and α_2. Three different estimation methods are considered, namely (i) a seasonal model, (ii) a model with pooled seasons and (iii) a dominant season model (here defined as the season with the largest α-value). It is evident that, when one of the two seasons is pronounced dominant, one need only use this season for estimating of x_T. Hence, it must be anticipated that the curves in Figs. 6a, b corresponding to methods (i) and (iii) are identical in the tails, which is clearly the case. In a band around $\alpha_1/\alpha_2=1$ the rrmse$(\hat{x}_T)$ from the pooled seasons model is less than that for the seasonal model, indicating that it would be more efficient to use the former method. Outside the band the bias dominates the rrmse of the pooled seasons, making it virtually useless. Adopting the rrmse as a criterion for model selection, one can make the following conclusions from the figures. When the seasonality in the mean exceedance is weak, it is possible to obtain more robust quantile estimators by pooling seasons together, although this violates some of the basic distribution assumptions. When the seasonality is pronounced, one need only consider the dominant season. The size of the region at the α_1/α_2-axis where a seasonal approach is justified depends on the amount of data available, i.e. on t. As t increases, the bias due to model errors constitutes a growing part of the total rrmse, but even with

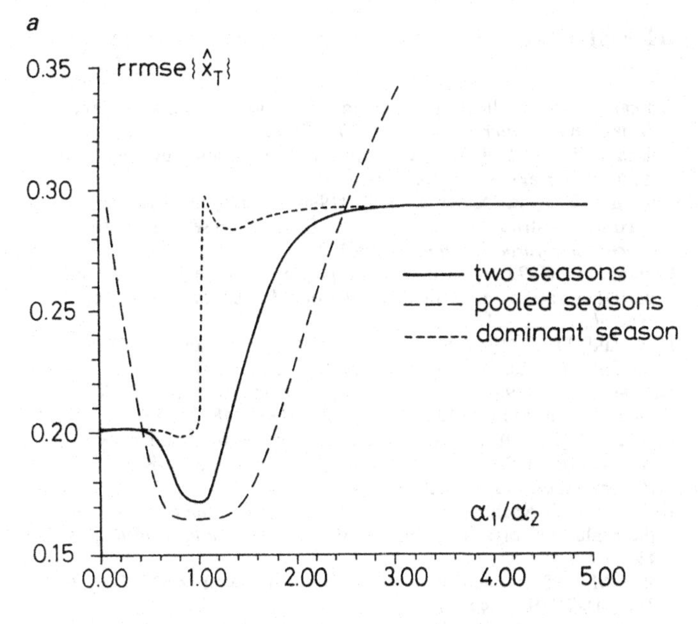

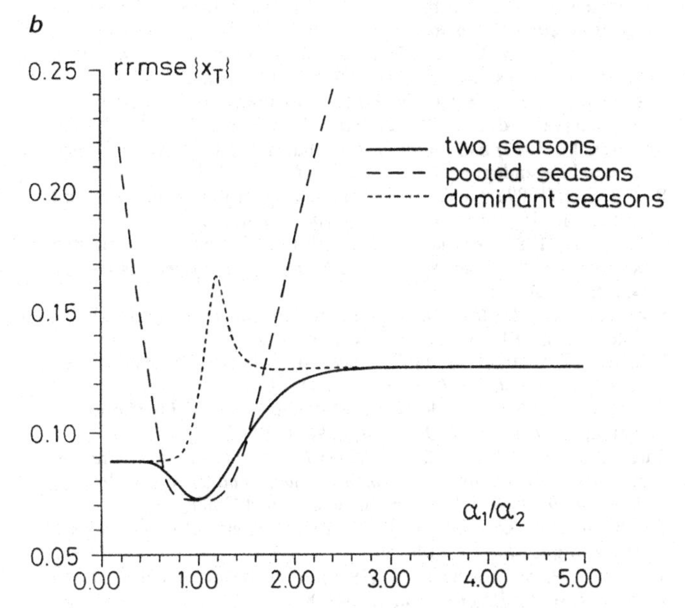

Fig. 6 Comparison of the relative root mean square error of $\hat{x}_{50}$ corresponding to a two-seasonal approach and a pooled seasons approach. $\alpha_2=1$, $\lambda_1=1.33$, $\lambda_2=2.67$, and (a) $t=10$ years, (b) $t=50$ years.

$t(\lambda_1+\lambda_2)$ as large as 200 (Fig. 6b) a seasonal procedure seems only optimal in a limited region.

Finally, a case-study was performed based on 90 years of observed discharge of the Mississippi River at St. Paul. Details of the analysis can be found in the work of Rousselle (1972). Essentially, he found that, if the year is divided into four seasons, then the assumptions concerning occurrence rates and distributions of exceedances can easily be accepted. He calculated the parameter estimates given in Table 1 (threshold level and units omitted).

An analysis was effectuated to see how different groupings

Table 1. *Mississippi River at St. Paul. Estimated mean value of exceedances and average number of exceedances in four seasons (from Rousselle, 1972)*

	1	2	3	4
α_i	1.017	1.080	2.048	1.611
λ_i	0.256	0.056	1.667	0.633

of seasons affect the mean square error (mse). Fig. 7 shows the main result of this analysis. The calculated values are obtained by means of the developed approximate formulas for bias and variance, while the simulated values are sample estimates obtained from computer-generated time series assuming the values in the table to be the population parameters. Axis-labels, such as (1)(234), mean that seasons 2, 3 and 4 have been pooled. It is clear that the smallest mse is obtained for the combinations (12)(34), (1)(2)(34), or simply (34), the latter meaning that only seasons 3 and 4 are considered when $\hat{x}_{1000}$ is estimated. Seasons 3 and 4 are the dominant ones, while seasons 1 and 2 only add noise to the estimation of x_T. Clearly, information is lost, if only season 3 is considered. Though this is the most dominant single season, the information in season 4 is not without significance and should be used to increase the robustness of $\hat{x}_{1000}$. Therefore, our recommendation is that the data from season 3 and 4 be pooled and used to estimate $\hat{x}_{1000}$ as if no seasonality was present.

CONCLUSIONS

Seasonal models can be applied in order to obtain a more precise description of the processes governing extreme hydrologic events. However, when such models are used also for predictive purposes, the prediction uncertainty should be considered in order to ensure the most reliable estimation procedure. Thus it is possible that a simplified procedure with only a few parameters involved performs better than a more complete and systematically correct model with a larger number of parameters.

For hydrological time series where exceedances of a given threshold level exhibit seasonality it has been found that the simple non-seasonal model usually outranks more sophisticated seasonal models as far as estimation precision is concerned. This statement, however, presupposes that the selection of the data for the one-seasonal approach is correct. The following guidelines may be used.

When the seasonality is weak, the seasons should be pooled, and the entire data series be used as a basis for the analysis. Seasonal approaches are deprecated in case of weak seasonality, especially when only few data are available. In

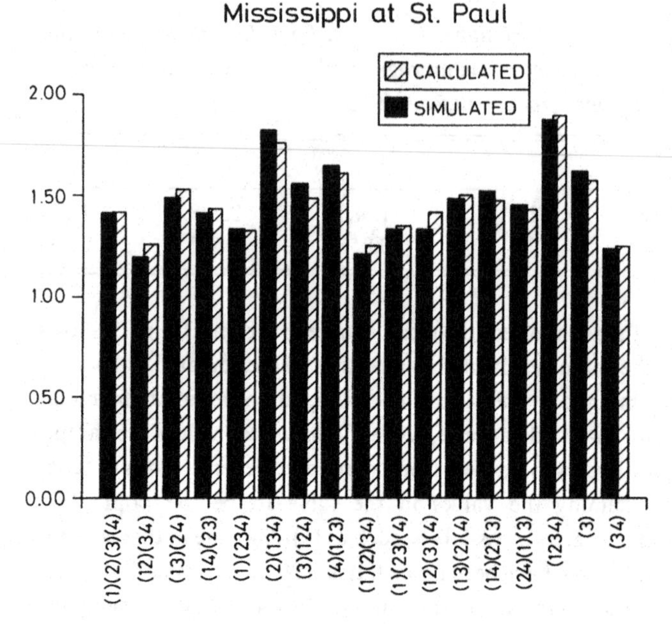

Fig. 7 Mississippi River at St. Paul. Mean square error of $\hat{x}_{1000}$ corresponding to different groupings of seasons. The parameter values are given in Table 1.

the opposite case, where the seasonality is pronounced, the most dominant season(s) should be selected for the one-seasonal approach and the remaining seasons be ignored. In this case a seasonal model might be used as well, but since no gain is obtained by introducing a more complicated estimation procedure, the simple one-seasonal approach is also recommended in this case. When the total number of peaks above the threshold is large there may be justification for the use of a seasonal approach, at least for some parameter combinations. However, most data series found in practice are too short to permit extraction of the number of peaks needed for the seasonal method to perform better than the pooled seasons method.

The above results have been developed on the basis of a particular model (non-clustered Poissonian occurrence times; exponentially distributed exceedances). Hence, the specific conclusions, of course, only apply to this particular model. In principle, however, the same kind of analysis should always be performed no matter what underlying model is assumed.

ACKNOWLEDGEMENTS

The authors want to express their appreciation to Dr T.A. Buishand for valuable comments on the calculation of the second order variance approximation.

REFERENCES

Ashkar, F. & Rousselle, J. (1981) Design discharge as a random variable: A risk study. *Water Resour. Res.*, **17**, 577–91.

Buishand, T. A. (1989) The partial duration series method with a fixed number of peaks. *J. Hydrol.*, **109**, 1–9.

Buishand, T. A. & Demarée, G. R. (1990) Estimation of the annual maximum distribution from samples of maxima in separate seasons. *Stochastic Hydrol. Hydraul.*, **4**, 89–103.

Cunnane, C. (1973) A particular comparison of annual maxima and partial duration series method of flood frequency prediction. *J. Hydrol.*, **18**, 257–71.

Fitzgerald, D. L. (1989) Single station and regional analysis of daily rainfall extremes. *Stochastic Hydrol. Hydraul.*, **3**, 281–92.

Kavvas, M. L. (1982a) Stochastic trigger model for flood peaks. 1. Development of the model. *Water Resour. Res.*, **18**, 383–98.

Kavvas, M. L. (1982b) Stochastic trigger model for flood peaks. 2. Application of the model to the flood peaks of Goksu-Karahacili. *Water Resour. Res.*, **18**, 399–411.

Konecny, F. & Nachtnebel, H. P. (1985) Extreme value processes and the evaluation of risk in flood analysis. *Appl. Math. Modelling*, **9**, 11–15.

North, M. (1980) Time-dependent stochastic model of floods. *J. Hydr. Div., ASCE*, **106**, 649–65.

Rasmussen, P. F. & Rosbjerg, D. (1989) Risk estimation in partial duration series. *Water Resour. Res.*, **25**, 2319–30.

Rasmussen, P. F. & Rosbjerg, D. (1991a) Evaluation of risk concepts in partial duration series. *Stochastic Hydrol. Hydraul.*, **5**(1), 1–16.

Rasmussen, P. F. & Rosbjerg, D. (1991b) Prediction uncertainty in seasonal partial duration series. *Water Resour. Res.*, **27**(11), 2875–83.

Rosbjerg, D. (1985) Estimation in partial duration series with independent and dependent peak values. *J. Hydrol.*, **76**, 183–95.

Rosbjerg, D. (1987a) On the annual maximum distribution in dependent partial duration series. *Stochastic Hydrol. Hydraul.*, **1**, 3–16.

Rosbjerg, D. (1987b) Partial duration series with log-normal distributed peak values. In: Singh, V. P. (Ed.), *Hydrologic Frequency Modeling*, D. Reidel Publ., pp. 117–29.

Rousselle, J. (1972) *On some problems of flood analysis*, Ph.D. thesis, Colorado State University, Fort Collins, CO.

Rousselle, J. & Hindie, F. (1976) Incertitude dans les debits de crues: approche Bayesienne. *J. Hydrol.*, **30**, 341–9.

Shane, R. M. & Lynn, W. R. (1964) Mathematical model for flood risk evaluation. *J. Hydraul. Div., ASCE*, **90**, 1–20.

Taesombut, V. & Yevjevich, V. (1978) *Use of partial flood series for estimating distribution of maximum annual flood peaks*, Hydrology Paper no. 97, Colorado State University, Fort Collin, CO.

Todorovic, P. & Rousselle, J. (1971) Some problems of flood analysis. *Water Resour. Res.*, **7**, 1144–50.

Todorovic, P. & Zelenhasic, E. (1970) A stochastic model for flood analysis. *Water Resour. Res.*, **6**, 1641–8.

Waylen, P. & Woo, M. K. (1982) Prediction of annual floods generated by mixed processes. *Water Resour. Res.*, **18**, 1283–6.

Zelenhasic, E. (1970) *Theoretical probability distributions for flood peaks*, Hydrology Paper No. 42, Engineering Research Center, Colorado State University, Fort Collins, CO.

2 A daily streamflow model based on a jump-diffusion process

F. KONECNY

Institute of Mathematics and Applied Statistics, Universität für Bodenkultur, Vienna, Austria

H.-P. NACHTNEBEL

Institute of Water Resources Management, Hydrology and Hydraulic Construction, Universität für Bodenkultur, Vienna, Austria

ABSTRACT The objective of this paper is to describe daily discharge series by a stochastic differential equation which is based on the mass balance of a linear reservoir. The input consisting of a series of jumps reflects the rainfall while the output refers to the discharges of a river basin. To account for random phenoma such as evaporation during the transformation process a perturbation term was introduced. The point process describing the shots (jumps) is based on an intensity function alternating randomly between two levels. Thus clustering of shots can be incorporated into the model.

INTRODUCTION

Numerous stochastic models have been applied to streamflow series. They can be grouped, for instance, into ARMA-type models (Fiering, 1967; Hipel *et al.*, 1977; Noakes *et al.* 1985; Kottegoda & Horder, 1980; Salas & Smith, 1981), long term memory models such as fractional Gaussian noise models (Mandelbrot & van Ness, 1968; Mandelbrot & Wallis, 1969) and related broken line models (Meija *et al.*, 1972). The third class of models refers to the transformation of an intermittent rainfall process into a continuous discharge series (Treiber & Plate, 1975; Weiss, 1973, 1977; Miller *et al.*, 1981; Kavvas & Delleur, 1984; Koch, 1985; Bodo & Unny, 1987). In this paper daily streamflow series (Beard, 1967; Quimpo, 1967; Valencia & Schaake, 1973; Mejia & Rousselle, 1976; Yakowitz, 1979; Morris, 1984; Miller *et al.* 1981; Weiss, 1977; O' Connell, 1977) are being modelled. An appropriate model should reproduce

- the distinct increase of the discharge at the beginning of floods;
- the slow decay during the falling limb of floods;
- the clustering of flood events;
- nonlinear characteristics of the system;
- seasonality in the discharge series.

Essentially the same paper was published in the *Proceedings of Hydrology Days*, 1991, pp. 185–94, H. J. Morel Seytoux (editor).

MODEL DESCRIPTION

In this paper the streamflow at the catchment outlet is described by the discharge from a linear reservoir. This conceptual reservoir (Nash, 1957; Kelman, 1980; Bodo & Unny, 1987) provides the time delay between instantaneous rainfall events and the resulting discharge. The mass balance equation is given in equation (1)

$$\frac{dS}{dt} = I - Q \tag{1}$$

where S, I and Q are the storage volume, the input and the reservoir outflow considered in continuous time. Assuming a linear reservoir, for which $Q = \alpha S$, we obtain:

$$\frac{dQ}{dt} = \alpha(I - Q) \tag{2}$$

Considering the randomness in the input by modeling the rainfall by a compound 'Poisson type' process the respective cumulative form is expressed by

$$I_t = \sum_{n=1}^{N_t} U_n \tag{3}$$

where N_t counts the shots in the time period $(0, t]$ and the shots U_n are independent and identically distributed random variables (i.i.d.r.v.).

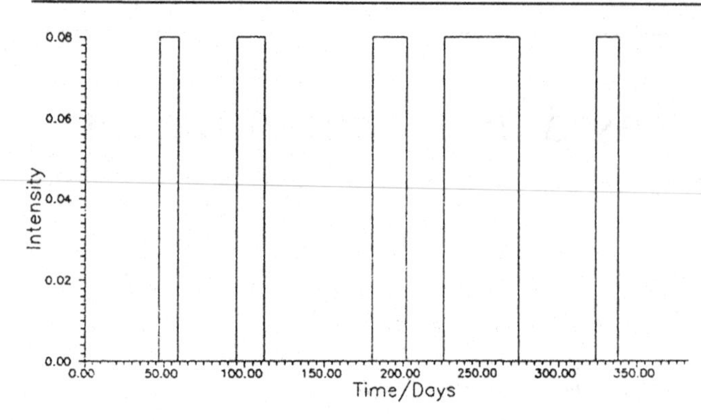

Fig. 1 Sample intensity of the RCM-process.

Often, it is observed that the counting process exhibits clusters and cannot be modeled as a Poisson process. Therefore, a point process with an intensity alternating randomly between zero and an intensity β (Fig. 1) was applied. This type of process was investigated by Kingman (1964) and applied to rainfall modeling by Smith & Karr (1983). Introducing a time-space 'Poisson type' process $v(.,.)$ the input can be rewritten as the Stieltjes integral

$$I_t = \int_0^t \int_0^\infty u v(\mathrm{d}t, \mathrm{d}u) \quad (4)$$

This representation is used in the formalism of stochastic differential equations with jump terms (cf. Snyder, 1975).

To take account for phenomena such as evaporation, time dependent and spatially distributed retention capacity in the basin a perturbation term

$$a + \sigma(Q)w \quad (5)$$

is additionally included in equation (2). a is a constant, σ a function of the outflow and w a wideband noise. Finally, the nonlinear stochastic differential equation

$$\frac{\mathrm{d}Q}{\mathrm{d}t} = \alpha(I - Q) + \sigma(Q)w + a \quad (6)$$

is obtained.

For the sake of mathematical convenience the wideband noise is substituted by white noise, which is a stationary process with constant spectral density. Formally spoken, white noise is the derivative of a standard Wiener process W. Expressing equation (6) in its integral form we obtain

$$Q_t = Q_0 + \int_0^t (AQ_s + a)\mathrm{d}s + \int_0^t \sigma(Q_s)\mathrm{d}w_s + \int_0^t \int_0^\infty u v(\mathrm{d}s, \mathrm{d}u) \quad (7)$$

In the drift term $A = -\alpha$ has been substituted and in the jump term α was included in u. Usually, equation (7) is given in the form of an Ito-differential equation

$$\mathrm{d}Q_t = (AQ_t + a)\mathrm{d}t + \sigma(Q_t)\mathrm{d}W_t + \int u v(\mathrm{d}t, \mathrm{d}u) \quad (8)$$

The solution process of this equation is called a jump-diffusion process, where $\sigma(Q_t)$ is the diffusion coefficient and $\int uv(\mathrm{d}t, \mathrm{d}u)$ stands for the jumps (shots). The diffussion coefficient $\sigma(.)$ expresses the infinitesimal variance between the shots and is obtained from (9).

$$\mathrm{E}((\mathrm{d}Q_t)^2 | Q_t = x) = \sigma^2(x)\mathrm{d}t \quad (9)$$

Three cases have been considered:

(a) $\sigma(x) = \sigma_0 = \mathrm{const}$
(b) $\sigma(x) = \sqrt{2Cx}$
(c) $\sigma(x) = Bx$

In the first case the perturbation is independent of the outflow, in the second case the infinitesimal variance is linear in x and in the third case a quadratic function is assumed. Cases (a) and (c) lead to special cases of a general linear stochastic differential equation

$$\mathrm{d}Q_t = (AQ_t + a)\mathrm{d}t + (BQ_t + b)\mathrm{d}W_t + \int u v(\mathrm{d}t, \mathrm{d}u) \quad (10)$$

The solution of this equation is given by (cf. Gikhman & Skorokhod, 1972)

$$Q_t = Q_0 Y_t + (a - Bb)\int_0^t Y_s \mathrm{d}s + b\int_0^t Y_s \mathrm{d}W_s + \int_0^t Y_s u v(\mathrm{d}s, \mathrm{d}u) \quad (11)$$

with

$$Y_t = \exp(A - B^2/2)t + BW_t + \int_0^t \int \ln(1+u) v(\mathrm{d}s, \mathrm{d}u) \quad (12)$$

IDENTIFICATION OF THE JUMP DIFFUSION PROCESS

In this section the characteristics of the process which refer to the frequency and distribution of jumps and the behaviour of the recession curve are investigated. Often, local maxima of the hydrograph occur in groups. In this model the rising limb in a hydrograph is represented by a shot of the magnitude U_n (see equation 3). The interarrival times between shots are modeled by a mixture distribution with the density

$$g(t) = \frac{p_1}{\phi_1}\exp - \left(\frac{t}{\phi_1}\right) + \frac{p_2}{\phi_2}\exp - \left(\frac{t}{\phi_2}\right) \quad (13)$$

$$p_1 + p_2 = 1 \text{ and } p_i > 0 \quad (14)$$

ϕ_1 and ϕ_2 denote the mean interarrival times in wet and dry periods and p_1 and p_2 assign weights to these periods. The corresponding point process describing the occurrences of the shots was called RCM-process (cf. Smith & Karr, 1983). It can be described by three parameters: the positive intensity level β and the transition characteristics γ_0 and γ_1 governing the transition between two intensity levels β and zero. Roughly speaking, these two levels indicate wet and dry periods. The estimation of the parameters was done by the so-called EM-algorithm which is an iterative estimation procedure to find parameter estimates in incomplete data problems. For a detailed description of this estimation method we refer to Redner & Walker (1984) and Konecny & Nachtnebel (1990).

The magnitude of the shots is represented by a sequence of positive i.i.d. random variables. In this paper a two parametric log-normal distribution was applied. Between two shots equation (8) reduces to equation (15) which describes the discharge from a reservoir superimposed by some random fluctuations.

$$dQ_t = (AQ_t + a)dt + \sigma(Q_t)dW_t \qquad (15)$$

The identification of the perturbation function $\sigma(.)$ and the drift parameters A and a is performed by utilizing daily discharge series and additional observations with a higher time resolution (cf. next section). A two-step identification procedure has been applied. First, the diffusion term is identified by the use of the quadratic variation of the observations and second, the drift parameters are estimated by the maximum likelihood method.

APPLICATION OF THE MODEL

The model was applied to an Austrian river basin located in the south-eastern part of Austria. The gauging station controls a rather flat area of 440 km^2. Due to climatic conditions runoff from snowmelt can be neglected and therefore the discharge does not exhibit a pronounced seasonality in comparison with alpine regions.

In this paper the nonlinear case (b) will be described. A daily discharge series from 1961 to 1971 has been analysed. The discharge series from 1961 is given in Fig. 2. Typical parameters obtained from the daily discharges are given in Table 1. Additionally, the hydrograph was also investigated at a time resolution of half an hour to obtain an estimate of the infinitesimal variance.

In a first step jumps (shots) had to be selected. A jump was identified as an increase in the discharge within a maximum of five days and at least 2.5 m^3/s increase per day and a peak exceeding 8 m^3/s. Minor fluctuations in the increasing limb were allowed. Finally 84 jumps were selected with a mean value of 14.2 m^3/s, a variance of 196.2 (m^3/s)^2 and a mean interarrival time of 43 days. Clustering of the peaks was obvious as can be seen from Fig. 2. Table 2 gives the estimates of the model parameters.

Table 1. *Hydrological parameters of the observation series (1961–71) and corresponding values obtained from a generated 10 years series*

	obs.	sim.
mean daily discharge (m^3/s)	4.5	4.4
standard deviation (m^3/s)	4.6	4.4
skewness (m^3/s)	5.6	6.3
minimum discharge (m^3/s)	0.93	0.9
maximum discharge (m^3/s)	56.8	74.4

Table 2. *Estimates of the model parameters*

$A=$	-0.352
$a=$	1.142
$C=$	.10
$\phi_1=$	74.8
$\phi_2=$	47.4
$p_1=$	.747

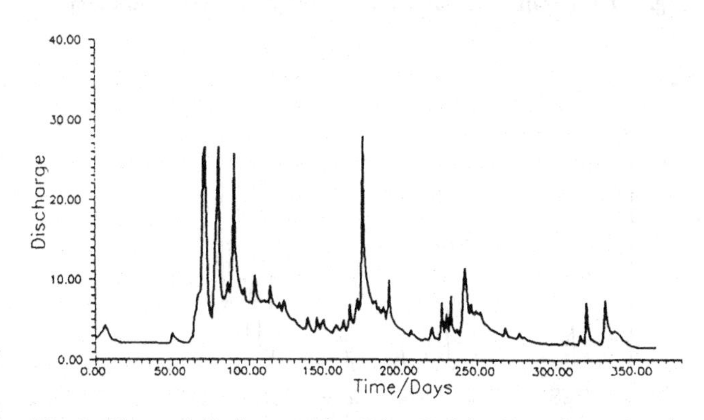

Fig. 2 Observed discharge series. River Lafnitz (Austria) 1961.

The lognormal distribution which was fitted to the observed shots which is given in Fig. 3. Based on these characteristics a daily discharge series were generated from which an example is given in Fig. 4. The corresponding parameters of the generated series are also included in Table 1.

The autocorrelation function is given in Fig. 5 which exhibits a slightly faster decrease of the autocorrelation coefficient of the generated series compared to the empirical values.

Summarizing, the presented model describes typical characteristics of a daily discharge series. The fast increase of the hydrograph and the slow decay in the falling limb are considered in the model. Also, it accounts for clustering in the shots and the nonlinearity is introduced by the diffusion

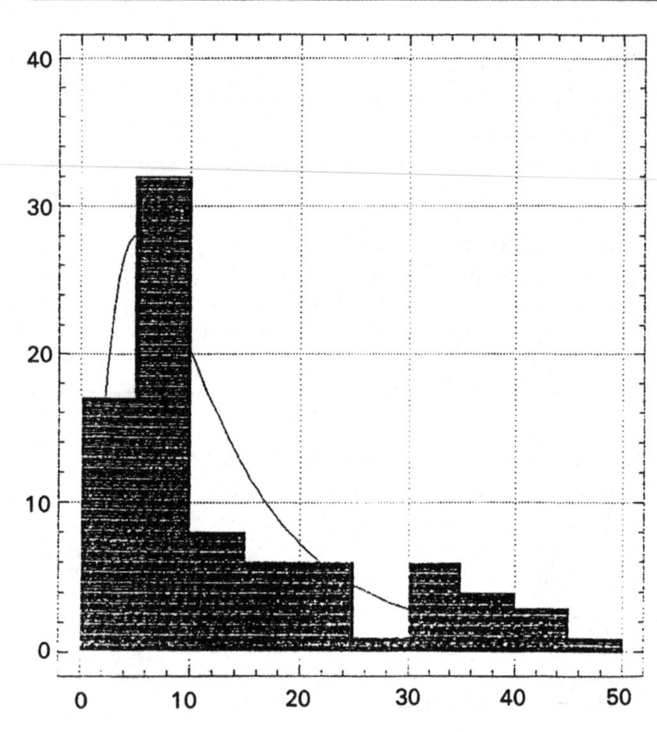

Fig. 3 Histogram of observed shots and fitted density function.

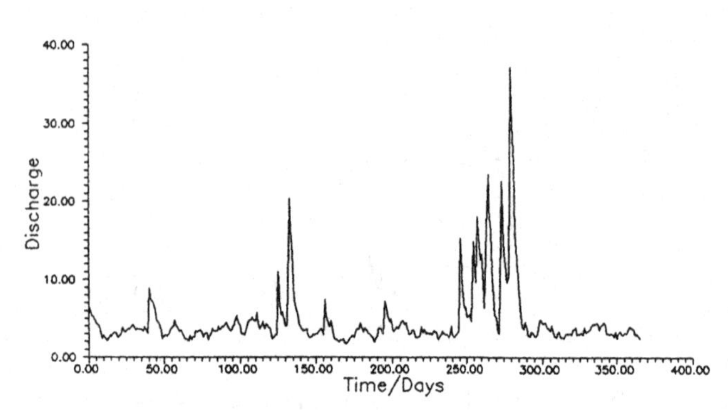

Fig. 4 Simulated discharge series.

term. In the present stage of model development a seasonality in parameters has not been considered. Comparing parameters from generated series and empirical values it can be concluded that the first three moments are sufficiently reproduced and also the autocorrelation function fits in good agreement to empirical data.

REFERENCES

Beard, L. R. (1967) Simulation of daily streamflow. *Internat. Hydrology Symposium*, Ft. Collins, Part 1, pp. 625–31.

Bodo, B. A. & Unny, T. E. (1987) On the outputs of the stochasticized Nash-Dooge linear reservoir cascade. In: Mac Neill, I. B. & Umphrey, G. J. (Eds.), *Stochastic Hydrology*, Dortrecht, Reidel Publ. Company.

Fiering, T. (1967) *Streamflow synthesis*; Harvard Univ. Press, Cambridge, Mass.

Gikhman, I. I. & Skorokhod, A. V. (1972) *Stochastic differential equations*; Springer.

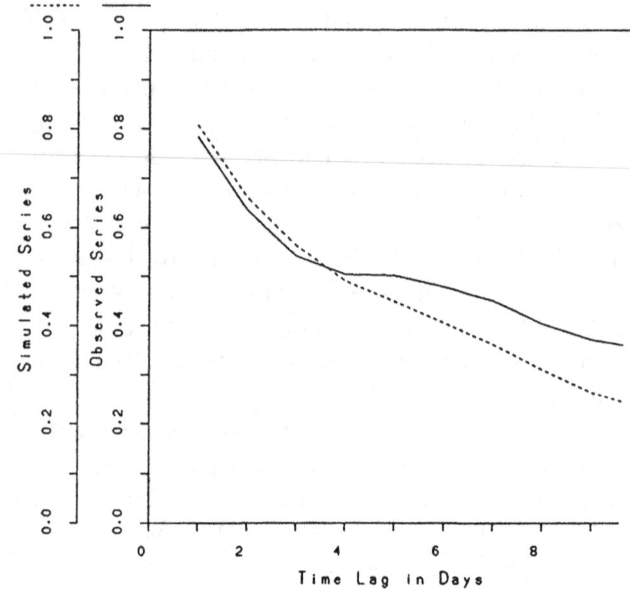

Fig. 5 Autocorrelation function (observed —— and simulated ---).

Hipel, K. W., Mc Leod, A. I. & Lennox, W. C. (1977) Advances in Box-Jenkins modeling 1, model construction. *Water Resour. Res.*, **13**(3), 567–75.

Kavvas, M. L. & Delleur, J. W. (1984) A statistical analysis of the daily streamflow hydrograph. *J. Hydrology*, **71**, 253–75.

Kelman, J. (1980) A stochastic model of daily streamflow. *J. Hydrology*, **47**, 235–49.

Kingman, J. F. C. (1964) On double stochastic processes. *Proc. Cambridge Philos. Soc.*, **60**, 923–32.

Koch, R. W. (1985) A stochastic streamflow model based on physical principles. *Water Resour. Res.*, **21**(4), 545–53.

Konecny, F. & Nachtnebel, H.-P. (1991) *A streamflow model based on a jump-diffusion process*. Working Paper, Univ. of Bodenkultur, A-1180 Vienna, Austria.

Kottegoda, N. T. & Horder, M. A. (1980) Daily flow model based on rainfall occurences using pulses and transfer function. *J. Hydrology*, **47**, 215–34.

Mandelbrot, B. B. & van Nes, J. W. (1968) Fractional Brownian motion, fractional noises and applications. *SIAM Soc. Ind. Appl. Math. Rev.*, **10**(4), 422–37.

Mandelbrot, B. B. & Wallis, J. R. (1969) Computer experiments with fractional Gaussian noise, 1, averages and variances. *Water Resour. Res.*, **5**(1), 228–67.

Meija, J. M. & Rousselle, J. (1976) Disaggregation models in hydrology revisited. *Water Resour. Res.*, **12**(2), 185–6.

Meija, J. M., Rodriguez-Iturbe, I. & Dawdy, D. R. (1972) Streamflow simulation 2, the broken line process as a potential model for hydrologic simulation. *Water Resour. Res.*, **8**(9), 931–41.

Miller, R. B., Bell, W., Ferreiro, O & Yng-Yuh Wang, R. (1981) Modeling daily river flow with precipitation input. *Water Resour. Res.*, **17**(1), 209–15.

Morris, C. D. (1984) A stochastic model for a small time interval intermittent hydrologic process. *J. Hydrology*, **68**, 247–72.

Nash, J. E. (1957) The form of the instantaneous unit hydrograph; *IAHS Gen. Ass. Toronto, Ontario, Canada*, Vol. III.

Noakes, D. J., McLeod, A. I. & Hipel, K. W. (1985) Forecasting monthly riverflow time series. *Int. J. Forecast.*, **1**, 179–90.

O'Conell, P. E. (1977) Shot noise models in synthetic hydrology. In: Ciriani, T. A., Maione, V. & Wallis, J. R. (Eds.) *Mathematical Models for Surface Water Hydrology*; J. Wiley, NY.

Quimpo, R. G. (1967) *Stochastic model of daily riverflow sequences*; Hydrology Papers No. 18, Colorado State Univ., Ft. Collins, CO.

Redner, R. A & Walker, H. F. (1984) Mixture densities, maximum likelihood and the EM-algorithm. *SIAM Review*, **26**(2), 195–239.

Salas, J. D. & Smith, R. A. (1981) Physical basis of stochastic models of annual flows. *Water Resour. Res.*, **17**(2), 429–30.

Smith, J. A. & Karr, A. F. (1983) A point process model of summer seasonal rainfall occurrences. *Water Resour. Res.*, **19**, 95–103.
Snyder, D. L. (1975) *Random Point Processes*; Wiley Interscience, NY.
Treiber, B. & Plate, E. J. (1975) A stochastic model for the simulation of daily flows; *Symposium and Workshop on the Application of Mathematical Models in Hydrology and Water Resources*, Bratislava, CSR.
Valencia, R. D. & Schaake, J. C. Jr. (1973) Disaggregation processes in stochastic hydrology. *Water Resour. Res.*, **9**(3), 580–5.
Weiss, G. (1973) *Filtered Poisson processes as models for daily streamflow data*; Ph.D. Thesis, Imperial College, London.
Weiss, G. (1977) Shot noise models for the generation of synthetic streamflow data. *Water Resour. Res.*, **13**(1), 101–8.
Yakowitz, S. J. (1979) A nonparametric Markov model for daily river flow. *Water Resour. Res.*, **15**(5), 1035–43.

3 The influence of time discretization on inferred stochastic properties of point rainfall

S. WĘGLARCZYK

Institute of Water Engineering and Water Management, Cracow Technical University, Cracow, Poland

ABSTRACT The influence of time discretization on inferred stochastic properties of the point rainfall process was investigated through analysis of its eight characteristics (number of rain spells in a given time interval, ΔN; duration of dry spell, *Td*; time interval between the beginnings of successive rainfalls, *Tb*; total depth of rainfall, *H*; duration of rainfall, *T*; average, *Ia*, and maximum rainfall intensity, *Im*; relative duration of rainfall, $Tr = T/Tb$) at time scale Δt ranging from 5 minutes to 24 hours. The analysis, based on 25-year continuous records from a daily pluviograph, showed that the process ΔN can be described by the negative binomial distribution (NBD) for $\Delta t \leq 60$ min and by the Poisson distribution for $\Delta t > 60$ min. All remaining processes considered were found independent with the log-normal probability distribution function rendering the best fit at each Δt. Rainfall and dry spell durations behave differently. For $\Delta t \geq 180$ min they have to be treated as discrete ones with the NBD. About one third of all correlations for the seven processes is strongly affected by Δt causing change of their significance.

INTRODUCTION

The uncertainty accompanying all hydrologic processes is particularly visible in the case of the atmospheric precipitation. The genesis of the process takes place within a large volume of space-time domain where the leading role is played both by strong random factors (related, among others, to air turbulence, heat transfer between ground and atmosphere, etc.) as well as by deterministic periodic factors, as the astronomical cycles, particularly the annual cycle of solar radiation. In general, the precipitation process is a nonstationary stochastic process within a year. Additionally, it is an intermittent process, i.e. the periods of time with instantaneous rainfall intensity $\xi_t > 0$ are alternated with the periods with $\xi_t = 0$ so that the probability density function (PDF) of ξ_t is of mixed, discrete-continuous, type.

The information about the process under consideration is mostly limited to the point rainfall recorded at a determined time scale (called henceforth also the time discretization level, and denoted as TDL or Δt). The natural stochastic process of the point rainfall is investigated through its representation, e.g. daily rainfall. Therefore the randomness of the process founded on the basis of this representation is 'local' and refers to the daily rainfall only. The question arises – how a change of the time scale influences the inferred properties of the investigated rainfall. This is particularly important if the TDL of the data available is different from the one needed (Committee on Precipitation, AGU Hydrology Section, 1984; Valdes & Rodriguez-Iturbe, 1985; and Woolhiser & Osborn, 1985).

The second essential factor strongly affecting the inferred properties of rainfall is the assumed definition of the rainfall event. This arbitrariness is possible because of the intermittency of the rainfall process. Many different definitions of rainfall event exist, depending on the goal of the investigations. This variety is particularly spectacular in design rainfalls (Marsalek & Watt, 1984) and rainfalls with TDL less than one day.

The objective of the paper is to show the influence which time discretization exerts on some chosen characteristics of the rainfall events. Basing on the data covering the time span of 25 years, it is accomplished by the analysis of one-dimensional rainfall characteristics treated as nonstationary stochastic processes, and by the analysis of correlations between them, made for each assumed TDL.

BACKGROUND

The basic process

The basic process, the knowledge of which would make it possible to find all the searched characteristics of rainfall, is the continuous stochastic process of rainfall intensity ξ_t, $t \geq 0$. It is an intermittent process, i.e. for every positive t_1, t_2, t

$$\begin{matrix} \text{Prob}\{\xi_t=0,\ t\in\Delta\tau\}>0 \\ \text{Prob}\{\xi_t>0,\ t\in\Delta\tau\}>0 \end{matrix} \text{ and } \begin{cases} \lim_{\Delta\tau\to\infty} \text{Prob}\{\xi_t=0,\ t\in\Delta\tau\}=0 \\ \lim_{\Delta\tau\to\infty} \text{Prob}\{\xi_t=0,\ t\in\Delta\tau\}=0 \end{cases} \quad (1)$$

where $\Delta\tau=(t_1,t_2)$ and Prob{.} denotes probability. Thus, in a given time interval there may be a rain ($\xi_t>0$) or no rain ($\xi_t=0$) with positive probability, and the probability of no alternation between two states ($\xi_t>0$ and $\xi_t=0$) in a given interval $\Delta\tau$ tends to zero with its increasing length.

If an arbitrary set of time points $\{t_1, t_2, \ldots, t_n\}$ is chosen, then, in general, the PDFs of the ξ_t-process, $F_{t_i}(x) = \text{Prob}(\xi_{t_i} \leq x)$, will be different for different t_i. So different will be also the moments of the process at given time points. In particular it refers to the expected value $E(\xi_t)$ and the variance var(ξ_t). If the ξ_t-process is nonstationary then this fact should be reflected in dependence of $E(\xi_t)$ and/or $D^2(\xi_t)$ on time.

Because of the basic property of ξ_t-process, that is its intermittency, a certain conceptual structure may be introduced, namely a rainfall event, defined as continuous rainfall between two successive non-rainy time intervals. For each rainfall event one can define many characteristics (derived processes or derived variables) which can be divided into external ones, such as the rainfall duration, the total depth, the average intensity, the beginning time of rainfall, etc., and internal characteristics of rainfall describing the function $\xi_t>0$ and its stochastic properties.

The ξ_t-process as such is rarely investigated. Usually the information about it is given in series of depth increments $\{\Delta h_i\}$, $i=1,2,\ldots,n$, taken in determined (constant or variable) time intervals:

$$\Delta h_i = \int_{t_{i-1}}^{t_i} \xi_t \mathrm{d}t \quad (2)$$

Rainfall event would be now an uninterrupted sequence of nonzero rainfall depth increments $\{\Delta h_i\}$, $i=1,2,\ldots,n$, limited by two successive zero-rainfall increments: $\Delta h_0=0$ and $\Delta h_{n+1}=0$.

Data

The data basis consisted of all rain gauge daily charts recorded at the Kraków Botanical Garden meteorological station during 5-month period (May 1–September 30) of each year from 1961 until 1985.

Each chart with total daily rainfall amount of more than 0.5 mm was digitized into a broken line resembling the recorded continuous one as closely as possible, and then corrected if necessary and possible. Daily hyetographs were chronologically linked into yearly hyetographs containing all rainfalls from successive years, beginning at 00^{00} hour of May 1 and ending not later than at 24^{00} of September 30. The set of all 'metahyetographs' ('broken-line' set) was used for making 9 derived sets of rainfall depth increments at a given time discretization level Δt which was assumed to equal 5, 10, 15, 30, 60, 180, 360, 720, and 1440 minutes. Then for each of those 9 sets and the 'broken-line' one, the sets of rainfall events (rainfalls) were determined.

For the use of the following part of the paper a rainfall event is defined as a noninterrupted sequence of Δt-minute time intervals with non-zero rainfall increments preceded and followed by at least one time interval with 'zero' rainfall increment. This last term means a threshold value beneath which all is but the noise and was set equal to 0.1 mm/h unless the recorded rainfall amount exceeded 0.05 mm. The latter condition was of value for longer discretization intervals Δt in order not to omit small rainfalls. Very small rainfalls with total depth not greater than 0.1 mm were rejected. Apart from those 9 sets of rainfalls another set was made (denoted as $\Delta t=0$) one, containing all rainfalls with variable discretization interval (made on the basis of the 'broken-line' set).

The investigated derived processes

In the present paper the following one-dimensional external characteristics of rainfall are investigated at each assumed Δt:

(1) the number of rainfalls which began in a given time interval, $\Delta N(t)$;
(2) the time interval between the end of a given rainfall and the beginning of the next one (the dry period duration), $Td(t)$ [h];
(3) the time interval between the beginning of a given rainfall and the beginning of the next one, $Tb(t)$ [days];
(4) the total depth of rainfall, $H(t)$ [mm];
(5) the duration of rainfall, $T(t)$ [h];
(6) the average rainfall intensity, $Ia(t)$ [mm/h], $Ia(t)=H(t)/T(t)$;
(7) the maximum rainfall intensity, $Im(t)$ [mm/h];
(8) the relative duration of rainfall, $Tr(t)$ [–]: $Tr(t)=T(t)/Tb(t)$, investigated mostly in its transformed form $Tr'=Tr/(1-Tr)$.

In all the cases the variable t denotes the beginning time of a rainfall event, and $t=0$ at 00^{00} hrs of May 1. Some of the above-defined characteristics are illustrated in Fig. 1.

Each characteristic X (with exception of the ΔN-process) is by definition a nonstationary stochastic process $X_t, t\in$ (May 1, 00^{00}, September 30, 24^{00}) $\equiv (0,D)$, $D=153$ days, given at a determined Δt by 25 yearly realizations $X_i^{\Delta t}(t_j)\equiv X_{ij}^{\Delta t}$, $j=1,2,\ldots,n_i^{\Delta t}$, $i=1,2,\ldots,25$, $\Delta t=0,5,\ldots,1440$ min, of size $n_i^{\Delta t}$ each, which are the basis of the analysis. Each of the 25

Table 1. *Lengths of stationarity intervals ΔD, their number $m_{\Delta t}$ per 153-day period, and minimum n_{min} and maximum n_{max} number of rainfalls per ΔD at assumed discretization levels Δt*

	Time discretization level Δt [min]									
	0	5	10	15	30	60	180	360	720	1440
ΔD days	6.1	6.1	7.0	7.0	7.7	9.6	15.3	19.1	25.5	30.6
$m_{\Delta t}$	25	25	22	22	20	16	10	8	6	5
n_{min}	102	99	99	96	108	98	96	97	114	97
n_{max}	215	200	212	199	185	195	210	189	170	139

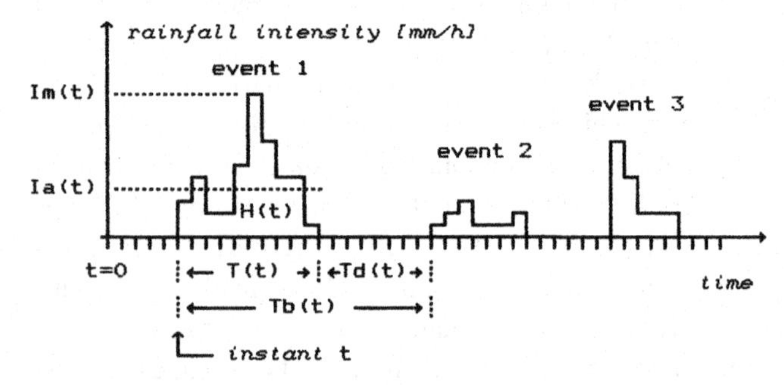

Fig. 1 Illustration of some characteristics of rainfall event. The smallest time interval is equal to Δt.

realizations of the ΔN-process was a sequence of numbers of rainfalls in a given time interval of all successive years.

Two approaches to analysis of a nonstationary stochastic process

An important problem is to choose the method of analysis of a given process. There exist two fundamental approaches (Todorovic & Yevjevich, 1969).

The first approach is a method of separation of deterministic and stochastic part of the investigated process X_t. It is transformed in such a way as to remove all existing deterministic components which are assumed to contain all the nonstationarity of the process. The remaining part X_t is now a pure stochastic process. The most often used transformation is

$$X'_t = \frac{X_t - E(X_t)}{D(X_t)} \quad (3)$$

and it is based on the assumption that the nonstationarity of the process is confined only to $E(X_t)$ and $D^2(X_t)$. Thus the new process X'_t is a stationary one (dependent or not) containing all randomness of the X_t-process.

When taking the second approach, the process X_t is investigated as observed with its deterministic part not separated from the stochastic one. Nonstationarity is being searched in various functions of the process, e.g. in parameters $g_1, \ldots, g_k$ of its PDF $F(x;\ g_1(t), \ldots,\ g_k(t))$. This approach is attractive for intermittent processes and will be applied in the present paper.

Assumptions made

Two assumptions were made. First, that the PDFs of the investigated processes are functions of time through their parameters $g_1, g_2, \ldots, g_k$:

$$P(X_t \leq x) = F(x;\ g_1(t), g_2(t), \ldots, g_k(t)) \quad (4)$$

$t \in (0, D)$, $D = 153$ days, $k \leq 2$. Secondly, that these processes are not highly variable within the year and may be considered stationary within consecutive intervals (named further stationarity intervals) ΔD_i (of the same length ΔD each), $i = 1, 2, \ldots m_{\Delta t}$, where $m_{\Delta t}$ stood for their number at a given TDL. The aim of the analysis was to find the best possible estimation of the function (4) in every ΔD_i.

The important problem for the analysis is the determination of the length ΔD of stationarity intervals at a given TDL. It has to be done in such a way that:

(a) the nonstationarity of the process under investigation is taken into consideration as precisely as possible; and
(b) best possible estimation of the function (4) is obtained.

As a compromise between these conditions the assumption was accepted that the length of a stationarity interval at each given TDL should cover at least about $n_{min} = 100$ rainfalls. This gave different numbers $m_{\Delta t}$ of stationarity intervals at different TDL, from $m_{\Delta t} = 25$ for $\Delta t = 0$ to $m_{\Delta t} = 5$ for $\Delta t = 1440$ minutes (Table 1). Such division ensures, independently of the assumed TDL, approximately constant accuracy (which depend first of all on the sample number) of the estimated parameters of the function (4).

The above-established division was the basis of further investigations which included the following steps:

(1) Proving that the random sample (being a set of data representing the investigated random variable X in an ith stationarity interval ΔD_i, $i = 1, 2, \ldots, m_{\Delta t}$ at a determined Δt) was an independent sample. The proof was given by testing significance of lag-one auto-correlation coefficient within the sample, $r_1(i)$, $i = 1, 2, \ldots, m_{\Delta t}$.

(2) Searching (for each ΔD_i, $i=1,2,\ldots,m_{\Delta t}$, at a determined Δt) for the best probability distribution function $F(x; g_1(t), g_2(t))$ out of a given set of PDFs. This required (i) estimation of the parameters $g_1(i)$, $g_2(i)$, $i=1,2,\ldots,m_{\Delta t}$, of all PDFs, and then (ii) selection of some quantities which will measure the quality of the fit, and calculation of their values.

The first task was accomplished by means of the maximum likelihood method applied to all used PDFs, the second one by the Kolmogorov goodness-of-fit test for continuous variables and the Pearson χ^2 test for discrete ones. These characteristics played an additional role: their values served as the quality indices when choosing the best PDF in the sense that the smaller a test statistic value the better distribution.

The within-the-year variation of g_i and other quantities will not be analysed here, although some information on them will be given herein.

ONE-DIMENSIONAL CHARACTERISTICS OF RAINFALL

In order to illustrate at least qualitatively the influence of time discretization on the chosen processes, they were initially assumed to be stationary and independent within the 153-day period, and (for each process) four statistics were calculated at each TDL, namely the average (av), the standard deviation (sd), the coefficient of variation (cv) and the coefficient of skewness (cs). The results as functions of Δt are presented in Fig. 2. Apart from a few cases, these statistics are highly regular (and mostly monotonic) functions of Δt.

The presented functions, perhaps beside the cs, show expected tendencies. Characteristics whose the values are proportional to time (T, Td, Tb, H) have their averages and standard deviations increasing with Δt, while characteristics with values per time unit (ΔN, Ia, Im) have these statistics decreasing. An interesting case is the Tr-process: its av-function is not monotonic and has unexpectedly high values (30–35%) for lower Δt in relation to the percentage of the total duration of all rainfalls within a year which varies for $\Delta t \leq 60$ min from 5 to 9%. It means that a short rainfall period is typically followed by a short period with no rain. Coefficients of variation are decreasing with Δt and are not high. Coefficients of skewness for the T-processes (T, Tb, Td, Tr) and ΔN are increasing with Δt.

Process $\Delta N(t)$

The realizations of this process were the numbers of rainfalls in a given stationarity interval of all successive years. Usually it is assumed to be Poissonian. This possibility was tested in each stationarity interval at a given TDL by the χ^2 test. As can be seen from Fig. 3 this distribution cannot be accepted at TDLs shorter than 180 minutes. For this reason another distribution was applied, namely two-parameter negative binomial distribution (NBD). The acceptance area of it is in a sense complementary to that of the Poisson PDF. It must be stressed here that the within-the-year variability of the χ^2 statistic is large, especially for the Poisson distribution,

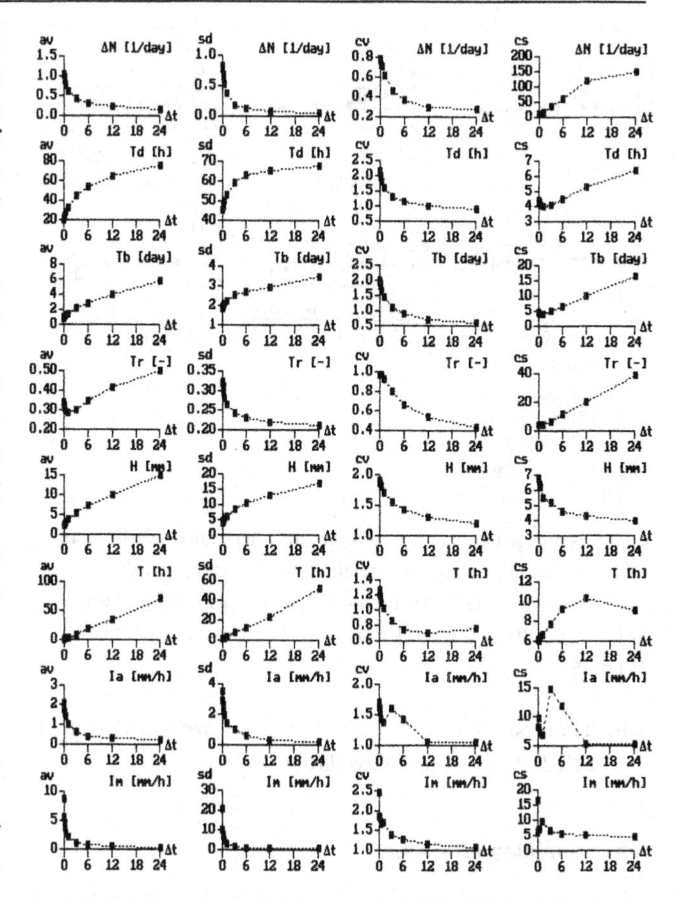

Fig. 2 The influence of the discretization level Δt [hours] on the basic characteristics of the investigated processes assumed to be stationary and independent.

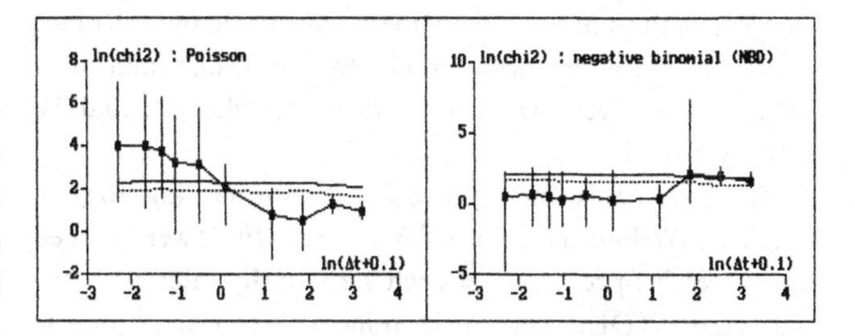

Fig. 3 Average (squares), maximum, and minimum (vertical lines) values of logarithms of the Pearson chi-square goodness-of-fit test statistics for the Poisson and NBD PDFs as functions of the discretization level Δt [hours]. Horizontal lines are 1% (solid) and 5% (dotted) average critical values of the test.

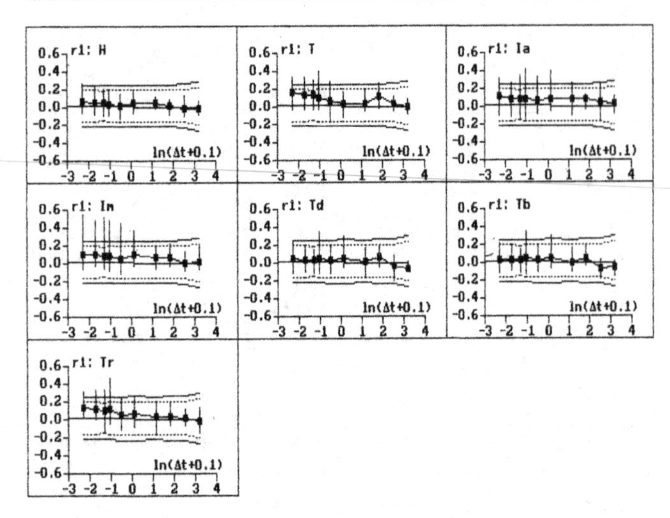

Fig. 4 Average (squares), minimum, and maximum (vertical lines) values of the lag-one autocorrelation coefficient $r1$ as functions of the discretization level Δt [hours] for the investigated processes. Solid and dotted lines limit 99% and 95% average tolerance intervals.

which can be the result of a rather small size (only 25 items per each ΔD_i) of samples available.

The remaining processes

All the remaining characteristics were treated as continuous processes which were stationary in the successive intervals. The independence of these processes was tested by means of the lag-one autocorrelation coefficients $r1(i)$, $i = 1,2,\ldots,m_{\Delta t}$ the average values of which are illustrated in Fig. 4. The extreme values of $r1(i)$ are sometimes outside the tolerance interval – this refers, however, mostly to only one value of all $r1(i)$ for a given Δt. Thus, the independence could be accepted in all cases.

Basing on this finding, five PDFs: normal, log-normal, gamma, Weibull and beta2 (cf. Yevjevich, 1972) were applied to successive processes for each stationarity interval at the successive TDLs. Having in mind the fact that for the Poisson process the distribution of times between events is exponential, this distribution was additionally applied to *Td*- and *Tb*-processes. Parameters of all these distributions were estimated by the maximum likelihood method, and the quality of fit was tested by the λ Kolmogorov-Smirnov test. Average values of λ together with its maximum and minimum at each Δt for the investigated processes are given in Fig. 5. It is clearly seen that beside the *T*- and *Td*-processes, the applied approach makes it possible to use one PDF for all ΔD_i at all TDLs for a given process. The best distribution, from the point of view of its consequency in being significant within the range of Δt-variability, is the log-normal one, although at some TDLs there are distributions with smaller average λ. In some cases (*Ia*, *Im*, *Tr'*) the beta2 PDF is also

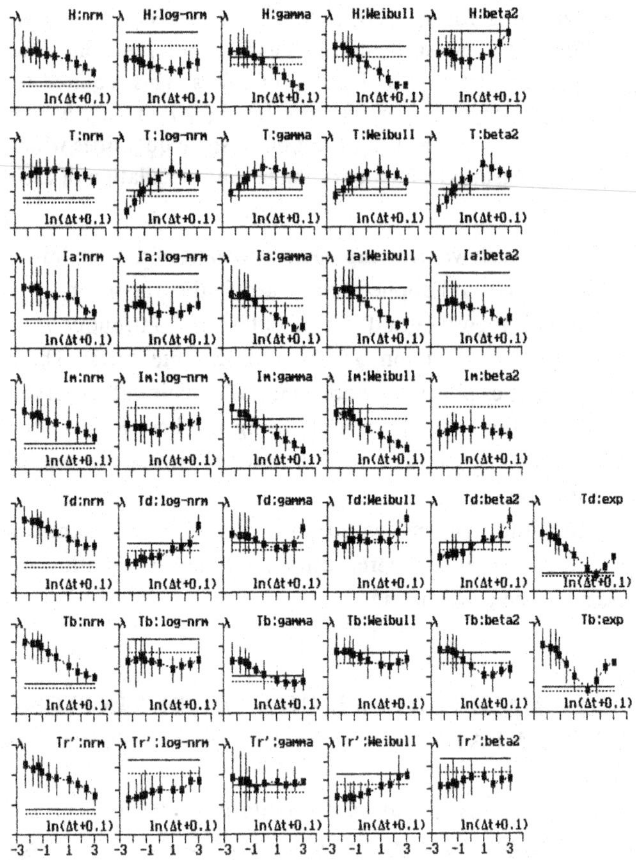

Fig. 5 Average values of the Kolmogorov-Smirnov goodness-of-fit test statistic λ (squares), and its minimum and maximum values (vertical lines) as functions of the discretization level Δt [hours] computed for the PDFs of the investigated processes. Horizontal lines are 1% (solid) and 5% (dotted) critical values of the test.

very good. The remaining distributions (beside the normal one) can be accepted only for some Δts.

The problem is more complicated if one considers the *T*- and *Td*-processes. For small discretization intervals Δt (up to *c*. 10 min for *T* and 60 min for *Td*) the log-normal distribution can be accepted, while for the higher TDLs no PDF can be accepted. The possible reason is the non-adequacy of the continuous description of the variables which are discrete. Accepting that explanation, two discrete PDFs: the Poissonian one and the NBD were applied for those processes at TDLs greater than 10 min for *T*, and greater than 60 min for *Td*. The realizations of the processes were transformed by means of the following formula:

$$X' = [X/\Delta t] - 1 \tag{5}$$

where $X = T$ or $X = Td$ and $[x]$ is the highest integer value not greater than x. The maximum likelihood method was applied to estimate the parameters of the PDFs used. The quality of the fit (measured by the χ^2 statistic values) for both distributions is shown in Fig. 6. The Poisson distribution is not

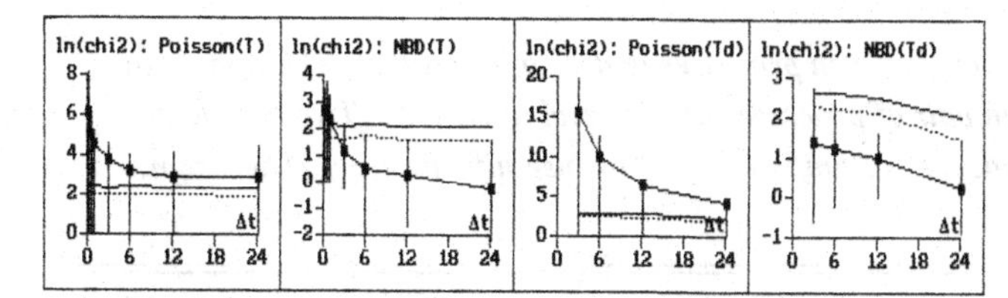

Fig. 6 Average (squares), minimum, and maximum values (vertical lines) of logarithms of the Pearson goodness-of-fit test statistic for Poisson and negative binomial (NBD) PDFs as functions of the discretization level Δt [hours] for the processes T and Td. Horizontal lines are 1% (solid) and 5% (dotted) average critical values of the test.

acceptable in any case, although its fit is getting better with the Δt values. The other distribution, NBD, is very good at all applied TDLs for the Td-process. Unfortunately, it is not the same for the T-process: the NBD is acceptable for $\Delta t \geq 180$ min only. Thus, the Td-process can be described as a nonstationary independent one with its PDF continuous (log-normal) up to $\Delta t = 60$ min and with discrete (negative binomial) PDF for $\Delta t > 60$ min. For the T-process the situation is more complicated because for $\Delta t = 15$, 30 and 60 minutes neither continuous nor discrete distribution is acceptable.

The exponential distribution which is expected for the Td- and T-processes if the ΔN-process is Poissonian is practically not acceptable at any Δt, so the Poisson distribution for $\Delta t \geq 180$ min is a rough approximation.

INTERCORRELATIONS BETWEEN THE DERIVED PROCESSES

As in the previous section, the analysis made here refers to the successive stationarity intervals at each of the ten TDLs. Because of the log-normality of the majority of the one-dimensional PDFs of the investigated processes (i.e. H, T, Ia, Im, Tb, Td, Tr'), the assumption was accepted that all possible pairs of these processes had two-dimensional log-normal PDF within each stationarity interval at each Δt. Then the correlation coefficients between logarithmically transformed processes were computed and their significance tested. The intra-year variability of the coefficients was consistently over or under the critical values so that it was possible to use the average correlation coefficients as indices of interdependence between the processes. There were, however, some cases (i.e. some Δts) for which it was difficult to decide whether the correlation within all the 153-day period was significant or not. This took place when the frequent within-the-year alternation between significance and insignificance occurred. The obtained results are summarized in Table 2, as well as in Fig. 7 where more information is provided.

It can be seen from Table 2, that eight of the correlations were constantly (i.e. at each Δt-level) significant, seven constantly insignificant, and in the remaining six cases the crossing of the tolerance interval limit took place. More details are given in Fig. 7.

The significant interdependence is very strong in most cases with intra-year variability of the correlation coefficient rather low (cf. Fig 7, diagrams A, B, C, F, L, S, U). It is interesting that there is such strong interdependence for the (Ia, H), (Ia, Im), and (Td, Tr') pairs of processes. A rise in Δt affects moderately the average correlation coefficient changing its values at about 0.2 at maximum.

The lack of the dependence (diagrams E, J, M, N, O, Q, R of Fig. 7) means, in turn, that the correlation coefficient is more variable within the year.

In cases where significance/insignificance alternation occurred (i.e. D, G, H, I, P and T in Fig. 7) the influence of Δt is so strong that it changed qualitatively the interdependence, and the within-the-year variability of $r(\Delta t)$-functions was the greatest. The point of Δt equal to one hour seems to separate the interval of almost constant values of $r(\Delta t)$ from the interval with a continuous rise with Δt.

CONCLUSIONS

Based on the 25-year continuous records of rainfall depth at the Kraków gauging station, an analysis of the influence of time discretization on some chosen characteristics of rainfall event was made. The processes were treated as nonstationary stochastic ones that can be investigated within the year in consecutive (stationarity) intervals containing at least ca 100 rainfall events. The following conclusions can be drawn from the analysis:

1. For the investigated processes there exist the significant within-the-year sequential independence which is the better the larger the discretization interval Δt is. It does not concern the ΔN-process whose independence was not investigated.
2. For the majority of the processes investigated the log-normal PDF can be accepted as the best one at each Δt.
3. For the Td- and T-processes the conclusion 2 is valid only for small Δt. When large TDLs (i.e. $\Delta t \geq 180$ min) are considered, these processes have to be treated as the discrete ones with the negative binomial distribution. For $\Delta t = 15$, 30 and 60 min the T-process has no acceptable PDF out of the applied set of PDFs.
4. The ΔN-process can described by the negative binomial distribution for $\Delta t < 180$ min and the Poisson distribution for $\Delta t \geq 180$ min.
5. For almost all cases the interdependence between the

Table 2. *The global results of significancy test of inter-dependence between processes under consideration. The sign + or − without numbers means signifinancy or insignificancy of a given pair of processes at 1% level at each Δt. Two numbers determine an interval, one a Δt value for which the inter-dependence is significant (+) or insignificant (−). The ? sign means lack of decision. Letters A–U refer to Fig. 7*

process	*T*	*Ia*	*Im*	*Tb*	*Td*	*Tr′*
H	+ A	+ B	+ C	− 0–60 + D	− E	+ F
T		− 0 + ? G	− 0 + H	− 0–60 + I	− J	+ K
Ia			+ L	− M	− N	− O
Im				− 0–360? + P	− Q	− R
Tb					+ S	+ − 1440 T
Td						+ U

investigated processes is constantly significant or constantly insignificant within the year at a given Δt, enabling therefore the use of an average correlation coefficient $r(\Delta t)$ as an index of interdependence at a given Δt. When significant, the correlation function $r(\Delta t)$ is very regular.

6. In some cases increasing the Δt-values causes at a certain Δt appearing or disappearing the interdependence.

REFERENCES

Committee on Precipitation, AGU Hydrology Section (1984) A new inter-disciplinary focus on precipitation research. *Eos*, **65**(23), 377–80.

Marsalek, J. & Watt, W. E. (1984) Design storms for urban drainage design. *Can. J. Civ. Eng.*, **11**(3), 574–84.

Todorovic, P. & Yevjevich, V. (1969) *Stochastic process of precipitation, Hydrology Papers*, Colorado State University, Fort Collins, Colorado, U.S.A., Paper No 35.

Valdes, J. B. & Rodriguez-Iturbe, I. (1985) Approximations of temporal rainfall from a multidimensional model. *Water Resour. Res.*, **21**(8), 1259–70.

Woolhiser, P. A. & Osborn H., B. (1985) A stochastic model of dimensionless thunderstorm rainfall. *Water Resour. Res.*, **21**, 511–22.

Yevjevich, V. (1972) *Probability and Statistics in Hydrology*, Water Resources Publications, Fort Collins, Colorado, USA.

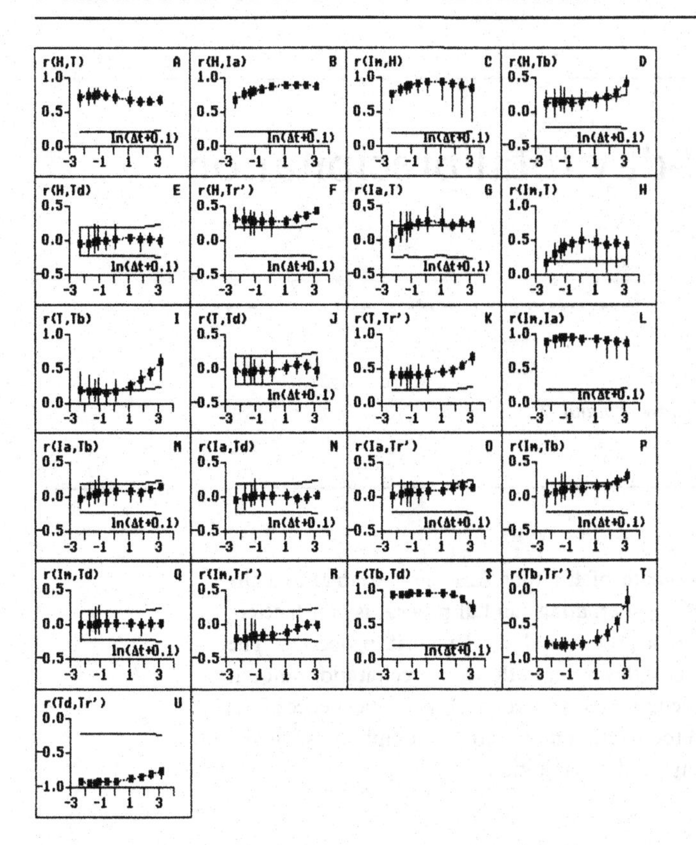

Fig. 7 Average (squares), minimum, and maximum values (vertical lines) of the correlation coefficient between the investigated processes as functions of the discretization level Δt. Horizontal lines are the average limits (upper one, if positive, lower one, if negative, or both) of the 1% tolerance intervals.

4 The distribution of the l-day total precipitation amount

W. JAKUBOWSKI

Chair of Mathematics, Agricultural University of Wrocław, Poland

ABSTRACT On the two-dimensional domain of the time (discrete variable) and the cumulative daily rainfalls (continuous one) an alternating rainfall process is constructed. Methods of renewal theory are used to define the process and to calculate its properties. The fundamental assumption used to derive results of this paper is: the total precipitation amount of the wet days sequence is dependent on the length of this sequence. The independence of the successive dry and wet sequences is assumed too. Based on the above assumptions we finally derive the theoretical distribution of the l-day total precipitation.

INTRODUCTION

Hydrology, from its beginnings, is concerned with rainfall modelling. The importance of precipitation for other hydrological processes caused a stormy development of various models. For a detailed review see for instance the work of Waymire & Gupta (1981). More recent investigations about the properties of daily rainfalls can be divided into three parts:

(a) the seasonal variation of the stochastic model parameters (Woolhiser & Pegram, 1979; Valdes, Rodriguez-Iturbe & Gupta 1985; Yevjevich & Dyer, 1983; Woolhiser & Roldan, 1986);
(b) the description of precipitation occurrence and intensity (Buishand, 1979; Woolhiser & Roldan, 1982);
(c) the construction of models of wet and dry sequences occurrence (Roldan & Woolhiser, 1982; Mimikou, 1983; Jakubowski, 1988).

Unfortunately, though the obtained results are generally more useful than raw observations, they can rarely be applied as an input for simulation models. They do not even give a precise description of the precipitation climatology of a region.

One of the fundamental problems of rainfall modelling is to find a relation between daily precipitation intensity and character of the wet interval. The extensive study on such relation for Dutch and foreign stations has been published by Buishand (1979). Among the conclusions presented there are:

(a) the distribution of daily rainfalls depends on the number of adjacent wet days;
(b) during winter and autumn there is a small, but significant correlation between daily rainfalls on successive wet days.

More generally one can express these conclusions in a following statement: *the daily rainfalls of any wet sequence are dependent and not homogeneous.*

This formulation determines the basic assumption of this paper: the total precipitation of the wet days sequence is dependent on the length of this sequence. The second, generally accepted assumption is that the length of successive dry and wet periods are independent. Finally we restrict our considerations to unchangeable climatology conditions only. These three assumptions let us to employ renewal theory for the theoretical description of the precipitation behavior of some l-day period (because of the third assumption, l should be limited; in practice it does not exceed 30 days). As a final result, the theoretical distribution of the l-day total precipitation amount is presented.

THE RAINFALL PROCESS

The simplest model describing the rainfall phenomenon gives an alternate discrete process. Two states $\{0,1\}$ denote dry and wet days respectively. The properties of such process and its fitting to observed data are often investigated, see for instance Buishand (1979), Jakubowski (1988). We also concentrate on the alternate process, but in other, more general way. The states, the same as for the occurrence process, depend now not only on the time, but also on the sum of daily totals. In this approach the sum is taken as a second variable. The process, called further the rainfall process, is described on the two-dimensional domain of time and amount, where time expressed in days is a discrete variable, while the amount in millimeters is a continuous one.

To define the rainfall process first we concentrate on a random character of the phenomenon. We can determine two sequences of random variables:

for the dry period – B_i – the number of the days;
for the wet period – (A_i, H_i) – A_i is the number of days;

where H_i is the total rainfall of the wet sequence.

Note that i, which indicates the successive dry and wet periods, not only represents the time but also reflects on the variety of cumulative precipitation amounts.

The defined variable (A_i, H_i) is a two-dimensional one. As it is observed in reality (see Buishand, 1979) we have some kind of connection between the length of wet period and it's total rainfall. Therefore in this paper we assume a dependency of A_i on H_i. Hence for the probability we have $\mathrm{P}(A_i, H_i) \neq \mathrm{P}(A_i)\mathrm{P}(H_i)$ and consequently the next defined rainfall process should be two-dimensional.

According to the above assumptions we describe the rainfall phenomenon by a sequence of independent random variables: $\ldots, (A_i, H_i), B_i, (A_{i+1}, H_{i+1}), B_{i+1}, \ldots, i \in \mathbb{N}$. We will investigate it in the equivalent form: $\ldots, (A_i, H_i), (B_i, I_i)$ $(A_{i+1}, H_{i+1}), (B_{i+1}, I_{i+1}) \ldots, i \in \mathbb{N}$, where for any $i \in \mathbb{N}$ $\mathrm{P}(I_i = 0) = 1$. We also assume that $(A_i, H_i) \equiv (A, H)$ and $(B_i, I_i) \equiv (B, I)$ are identically distributed. To make further calculations more clear we restrict our study to the sufficiently smooth distributions only.

Note that for determining the variety in time it is sufficient to study the sequence $\ldots, A_i, B_i, A_{i+1}, B_{i+1}, \ldots, i \in \mathbb{N}$. The sequence $\ldots, H_i, I_i, H_{i+1}, I_{i+1}, \ldots, i \in \mathbb{N}$, which is equivalent to the sequence $\{H_i\}, i \in \mathbb{N}$, defines the successive increments of the wet periods rainfalls. From practical point of view, a renewal process of amount, given by $\{H_i\}$, would not be investigated without its precise location in time.

For $k = 1, 2, \ldots$ let us denote by

$$S_k = A_1 + \ldots + A_k,$$
$$V_k = H_1 + \ldots + H_k,$$
$$T_k = B_1 + \ldots + B_k$$

and take $S_0 = V_0 = T_0 = 0$.

The distributions of (S_k, V_k) and (T_k, I), $k = 0, 1, \ldots$ will be denoted by $F_k(n, h)$ and $G_k(n)$, $n = k, k+1, \ldots h > 0$. The distributions of the sum $(A_1, H_1) + (A_2, H_2) = (S_2, V_2)$ and $(A_1, H_1) + (B_1, I_1) = (A_1 + B_1, H_1)$ are given by the convolutions:

$$\mathrm{P}(S_2 = n, V_2 \leq h) = \int_0^h \sum_{i=1}^{n-1} \mathrm{P}(A_1 = i, H_1 \leq h - s) \mathrm{d}\mathrm{P}(A_2 = n - i, H_2 \leq s)$$

$$\mathrm{P}(A_1 + B_1 = n, H_1 \leq h) = \int_0^h \sum_{i=1}^{n-1} \mathrm{P}(A_1 = i, H_1 \leq h - s) \mathrm{d}\mathrm{P}(B_1 = n - i, I_1 \leq s)$$

$$= \sum_{i=1}^{n-1} \mathrm{P}(A_1 = i, H \leq h) \mathrm{P}(B_1 = n - i)$$

which we will write in the form

$$F_1(n, h) * F_1(n, h) \text{ and } F_1(n, h) * G_1(n)$$

The probability that the kth wet and the kth dry sequences terminated exactly on the nth day with the total precipitation not greater h is given by

$$\Phi_k(n, h) = \mathrm{P}(S_k + T_k = n, V_k \leq h), \quad k = 1, 2, \ldots \quad n = 2k, 2k+1, \ldots \quad h > 0.$$

Taking the sum of Φ_k over k we obtain

$$\Phi(n, h) = \sum_k \Phi_k(n, h) = \sum_{k=1}^{[n/2]} \mathrm{P}(S_k + T_k = n, V_k \leq h), \quad n = 2, 3, \ldots; h > 0$$

as the two-dimensional renewal function.

Note that $\Phi(n, h)$ satisfied the renewal equation

$$\Phi(n, h) = \mathrm{P}(A + B = n, H \leq h) + \int_0^h \sum_{k=2}^{n-1} \mathrm{P}(A + B = n - k, H \leq h - s) \mathrm{d}\Phi(k, s)$$

Having described the random character of the rainfall we can now start to construct the simple rainfall process $Y(t, h)$:

(a) We fix the initial conditions. We assume that process starts from the point (0,0) and the day $t = 1$ is first at the wet days sequence.

(b) To describe the process domain $\mathbb{D}$ we fix an arbitrary point (t_0, h_0), where time $t_0 \in \mathbb{N}$ and cumulative rainfall $h_0 > 0$. We can distinguish:

– t_0 is a wet day and belongs to kth, $k = 1, 2, \ldots$ wet sequence;

consequently we have $V_k < h_0 \leq V_{k+1}$; for other h_0 (i.e. $h_0 \leq V_k$ or $h_0 > V_{k+1}$) points $(t_0, h_0) \notin \mathbb{D}$;

– t_0 is a dry day and belongs to kth, $k = 1, 2, \ldots$ dry sequence;

this forces $h_0 = V_k$, for other h_0 $(h_0 \neq V_k)$ points $(t_0, h_0) \notin \mathbb{D}$.

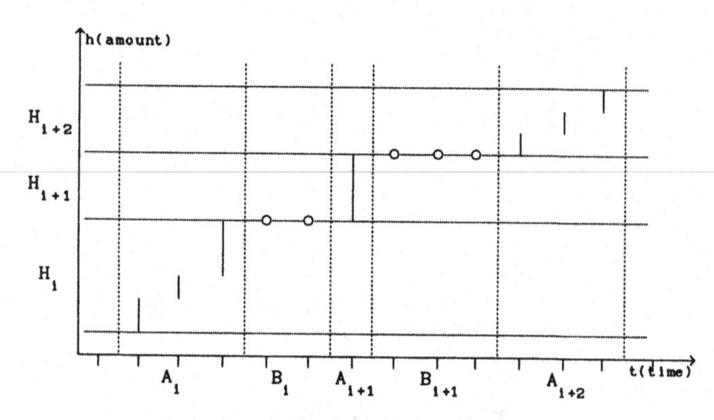

Fig. 1 A realization of the rainfall process domain.

Notice that process domain $\mathbb{D}$ is a random variable. An example of its realization is shown in Fig. 1.

(c) To define a simple rainfall process $Y(t,h)$ for any point $(t,h) \in \mathbb{D}$ we put

$$Y(t,h) = \begin{cases} 1 & \text{where} \quad S_k + T_k < t \leq S_{k+1} + T_k \text{ and } V_k < h \leq V_{k+1} \\ & \text{for some } k = 0,1,\ldots \\ 0 & \text{where} \quad S_{k+1} + T_k < t \leq S_{k+1} + T_{k+1} \text{ and } h = V_{k+1} \end{cases}$$

The process $Y(t,h)$ is used to construct a (stationary) rainfall process $a(t,h)$. For this purpose we have to calculate the limits of the probabilities:

(a) $P(Y(t,h)=1), P(Y(t,h)=0),$ (1)

(b) $P(\gamma(t)=k, \zeta(h)>s \mid Y(t,h)=1), P(\eta(t)=k \mid Y(t,h)=0)$ (2)

when t and h tend to infinity.

The random variable $(\gamma(t), \zeta(h))$ is the two-dimensional residual waiting time and amount of the process $Y(t,h)$ in the state 1, and it is defined as follows.

Let for given t and h, $t \in \mathbb{N}$, $h>0$, $Y(t,h)=1$, then $\gamma(t)$ is equal to the number of successive wet days, which follows tth day, $\gamma(t)=0,1,2,\ldots$ and $\zeta(h)$ is equal to the sum of daily rainfalls which appear in these days (if $\gamma(t)=0$ then $\zeta(h)=0$).

The random variable $\eta(t)$ is the residual waiting time of the process $Y(t,h)$ in the state 0 and it is defined similarly as $\gamma(t)$.

Consider the probabilities (1). It is easily seen that

$$P(Y(t,h)=1) = P(A \geq t, H \geq h) + \int_0^h \sum_{n=2}^{t-1} P(A \geq t-n, H \geq h-s)\,d\Phi(n,s)$$

and

$$P[Y(t,h)=0] = P(A+B \geq t, H \geq h) - P(A \geq t, H \geq h) + \int_0^h \sum_{n=2}^{t-2} [P(A+B \geq t-n, H \geq h-s) - P(A \geq t-n, H \geq h-s)]\,d\Phi(n,s)$$

The functions:

$$M_1(t,h) = P(Y(t,h)=1)$$

and

$$M_0(t,h) = P(Y(t,h)=0)$$

satisfy the renewal equations:

$$M_1(t,h) = P(A \geq t, H \geq h) + M_1(t,h)^* F_1(t,h)^* G_1(t)$$
$$M_0(t,h) = P(A+B \geq t, H \geq h) - P(A \geq t, H \geq h) + M_0(t,h)^* F_1(t,h)^* G_1(t)$$

It follows from the renewal theory (Bickel & Yahav, 1965) that

$$\lim_{t,h\to\infty} M_1(t,h) = \frac{1}{E[(A+B)H]} \int_0^\infty \sum_{t=1}^{\infty} P(A \geq t, H \geq h)\,dh = \frac{E(AH)}{E(AH)+E(BH)}$$

and

$$\lim_{t,h\to\infty} M_0(t,h) = \frac{1}{E[(A+B)H]} \int_0^\infty \sum_{t=1}^{\infty} [P(A+B \geq t, H \geq h) - P(A \geq t, H \geq h)]\,dh = \frac{E[(A+B)H]-E(AH)}{E(AH)+E(BH)} = \frac{E(BH)}{E(AH)+E(BH)},$$

where E(.) denotes the expected values. To simplify notation we denote $E(AH)=EAH$. Taking into account the assumption on the independence of the successive wet and dry periods we have $EBH=EBEH$. (Note that $EAH \neq EAEH$.) Hence the limiting values of the process $Y(t,h)$ are

$$\lim_{t,h\to\infty} P(Y(t,h)=1) = \frac{EAH}{EAH+EBEH}$$

and

$$\lim_{t,h\to\infty} P(Y(t,h)=0) = \frac{EBEH}{EAH+EBEH}$$

For the residual variables from (2) we have:

$$P(\gamma(t)=k, \zeta(h)>s, Y(t,h)=1) = P(A=t+k, H \geq h+s) + \int_0^h \sum_{n=2}^{t-1} P(A=t+k-n, H \geq h+s-q)\,d\Phi(n,q)$$

and

$$P(\eta(t)=k, Y(t,h)=0) = P(A+B=t+k, A<t, H \geq h)$$

$$+\int_0^h \sum_{n=2}^{t-2} \mathrm{P}(A+B=t+k-n, A<t-n, H\geq h-q)\mathrm{d}\Phi(n,q)$$

The functions: $R_0(t,h)=\mathrm{P}(\eta(t)=k,\ Y(t,h)=0)$ $R_1(t,h) =\mathrm{P}(\gamma(t)=k,\ \zeta(h)>s,\ Y(t,h)=1)$, satisfy the renewal equations:

$$R_0(t,h)=\mathrm{P}(A+B=t+k, A<t, H\geq h) + R_0(t,h)^*F_1(t,h)^*G_1(t).$$

$$R_1(t,h)=\mathrm{P}(A=t+k, H\geq h+s)+R_1(t,h)^*F_1(t,h)^*G_1(t).$$

Using once more the renewal theory (Bickel & Yahav, 1965) we have:

$$\lim_{t,h\to\infty} R_0(t,h)=$$

$$\frac{1}{\mathrm{E}[(A+B)H]}\int_0^\infty \sum_{t=2}^{\infty} \mathrm{P}(A+B=t+k, A<t, H\geq h)\mathrm{d}h$$

$$=\frac{1}{\mathrm{E}AH+\mathrm{E}B\mathrm{E}H}\times$$

$$\int_0^\infty \sum_{n=1}^{\infty} \mathrm{P}(A=n, H\geq h) \sum_{t=n+1}^{\infty} \mathrm{P}(B=t+k-n)\mathrm{d}h$$

$$=\frac{1}{\mathrm{E}AH+\mathrm{E}B\mathrm{E}H}\times$$

$$\int_0^\infty \sum_{n=1}^{\infty} \mathrm{P}(A=n, H\geq h)\mathrm{P}(B>k)\mathrm{d}h=\frac{\mathrm{E}H\mathrm{P}(B>k)}{\mathrm{E}AH+\mathrm{E}B\mathrm{E}H}.$$

and

$$\lim_{t,h\to\infty} R_1(t,h)=$$

$$\frac{1}{\mathrm{E}[(A+B)H]}\int_0^\infty \sum_{t=1}^{\infty} \mathrm{P}(A=t+k, H\geq h+s)\mathrm{d}h$$

$$=\frac{1}{\mathrm{E}AH+\mathrm{E}B\mathrm{E}H}\int_s^\infty \mathrm{P}(A>k, H>h)\ \mathrm{d}h$$

These calculations enable us to write:

$$\lim_{t,h\to\infty} \mathrm{P}(\eta(t)=k|\,Y(t,h)=0)$$

$$=\frac{\mathrm{E}AH+\mathrm{E}B\mathrm{E}H}{\mathrm{E}B\mathrm{E}H}\ \frac{\mathrm{E}H\mathrm{P}(B>k)}{\mathrm{E}AH+\mathrm{E}B\mathrm{E}H}=\frac{\mathrm{P}(B>k)}{\mathrm{E}B}$$

and

$$\lim_{t,h\to\infty} \mathrm{P}(\gamma(t)=k,\ \zeta(h)>s|\,Y(t,h)=1)$$

$$=\frac{\mathrm{E}AH+\mathrm{E}B\mathrm{E}H}{\mathrm{E}AH}\ \frac{1}{\mathrm{E}AH+\mathrm{E}B\mathrm{E}H}\int_s^\infty \mathrm{P}(A>k, H>h)\mathrm{d}h$$

$$=\frac{1}{\mathrm{E}AH}\int_s^\infty \mathrm{P}(A>k, H>h)\mathrm{d}h$$

To construct the rainfall process $a(t,h)$ as mentioned earlier we take that for any time t and any cumulated amount h the suitable probabilities of $a(t,h)$ are equal to the limiting values of the process $Y(t,h)$. Therefore for the process $a(t,h)$, $t\in \mathbb{N}$, $h>0$ and residual variables connected with it we put:

$$\mathrm{P}(a(t,h)=1)=\frac{\mathrm{E}AH}{\mathrm{E}AH+\mathrm{E}B\mathrm{E}H}=\beta$$

$$\mathrm{P}(a(t,h)=0)=\frac{\mathrm{E}B\mathrm{E}H}{\mathrm{E}AH+\mathrm{E}B\mathrm{E}H}=\sigma=1-\beta$$

$$\mathrm{P}(\gamma(t)=k, \zeta(h)<s|a(t,h)=1)$$

$$=\frac{1}{\mathrm{E}AH}\int_0^s \mathrm{P}(A>k, H>x)\mathrm{d}x=\mathscr{F}(k,s)$$

$$\mathrm{P}(\eta(t)=k|a(t,h)=0)=\frac{\mathrm{P}(B>k)}{\mathrm{E}B}=\mathscr{G}(k)$$

The random variables A, B, $\gamma(t)$, $\zeta(h)$ and $\eta(t)$ are defined as for simple rainfall process Y, but $\mathscr{F}$ and $\mathscr{G}$ are some distribution functions.

Straightforward computations show that covariance function of $a(t,h)$, $t\in \mathbb{N}$, $h>0$, does not depend on time t, and total precipitation amount h. This proves a second order stationarity of the rainfall process.

Taking the assumption that the length of wet days sequence is independent on its total rainfalls, we obtain a new process $\tilde{a}(t,h)$ – a simplified version of $a(t,h)$.

$$\mathrm{P}(\tilde{a}(t,h)=1)=\frac{\mathrm{E}A}{\mathrm{E}A+\mathrm{E}B};\qquad \mathrm{P}(\tilde{a}(t,h)=0)=\frac{\mathrm{E}B}{\mathrm{E}A+\mathrm{E}B}$$

$$\mathrm{P}(\gamma(t)=k, \zeta(h)>s|\tilde{a}(t,h)=1)$$

$$=\frac{1}{\mathrm{E}A\mathrm{E}H}\int_s^\infty \mathrm{P}(A>k)\mathrm{P}(H>x)\mathrm{d}x$$

$$=\frac{1}{\mathrm{E}A}\mathrm{P}(A>k)\frac{1}{\mathrm{E}H}\int_s^\infty \mathrm{P}(H>x)\mathrm{d}x$$

$$\mathrm{P}(\eta(t)=k|\tilde{a}(t,h)=0)=\frac{\mathrm{P}(B>k)}{\mathrm{E}B}$$

It means, that the process $\tilde{a}(t,h)$ is compound of two independent processes: the alternate occurrence one for time step and the simple renewal – for increments of wet sequence rainfalls.

THE DISTRIBUTION OF THE l-DAY TOTAL PRECIPITATION AMOUNT

To find the distribution of the total precipitation amount W_l in a fixed l-day period let us consider the two-dimensional random variable (N_l, W_l), where $N_l = 0,1,\ldots,l$ denotes the number of wet days in the l-day period.

Because of the rainfall stationarity it is sufficient to accept that after 0th day we have $h=0$ mm and the l-day period starts at the first day on time scale. We consider four events: Q_{ij}, $i,j=0,1$. They describe the states of 0th and lth days. For $i=0$ and $j=1$ the 0th day is dry and the lth one is wet. The other three events are defined similarly. To obtain the distribution of W_l it is enough to calculate:

$$P(W_l \le h) = \sum_{k=0}^{l} \sum_{Q} P(N_l = k, W_l \le h, Q)$$

where $Q \in \{Q_{00}, Q_{01}, Q_{10}, Q_{11}\}$.

To simplify the notation we set $\mathscr{G}_n(k) = \sum_{i=0}^{k-n} \mathscr{G}(i) G_n(k-i)$ for $n=0,1,\ldots; k=n,n+1,\ldots$.

For Q_{00} (by definition) it is clear that

$$P(N_l = l, W_l \le h, Q_{00}) = 0.$$

For k wet days ($k=0,\ldots,l-1$) we have

$$n=0,\ldots,q=\begin{cases} k & \text{when} \quad k<l/2 \\ l-k & \text{when} \quad k \ge l/2 \end{cases}$$

transitions from the dry to the wet state. Thus

$$\begin{aligned} &P(N_l = k, W_l \le h, Q_{00}) \\ &= \begin{cases} \sigma \sum_{n=1}^{q} F_n(k,h) \sum_{j=n-1}^{l-k-1} \mathscr{G}_{n-1}(j) P(B \ge l-k-j) & \text{for } k=1,\ldots,l-1 \\ \sigma \sum_{j=1}^{\infty} \mathscr{G}(j) & \text{for } k=0 \end{cases} \\ &= \begin{cases} \sigma \sum_{n=1}^{q} F_n(k,h) \left[\mathscr{G}_{n-1}(l-k) - \mathscr{G}_n(l-k)\right] & \text{for } k=1,\ldots,l-1 \\ \sigma \sum_{j=1}^{\infty} \mathscr{G}(j) & \text{for } k=0 \end{cases} \end{aligned} \tag{3}$$

For Q_{01} we have $P(N_l = 0, W_l \le h, Q_{01}) = 0$, and for k wet days ($k=1,\ldots,l$) we get

$$n=0,\ldots,r=\begin{cases} k-1 & \text{when} \quad k \le l/2 \\ l-k & \text{when} \quad k > l/2 \end{cases}$$

transitions from the wet to the dry state. Thus

$$\begin{aligned} &P(N_l = k, W_l \le h, Q_{01}) \\ &= \sigma \mathscr{G}_0(l-k) \left[F_1(k,h) + \frac{EAH}{E(H|A>k)} \mathscr{F}(k,h) \right] + \\ &\sigma \sum_{n=1}^{r} \mathscr{G}_n(l-k) \sum_{i=1}^{k-n} \int_0^h \left[F_1(i,h-s) + \frac{EAH}{E(H|A>i)} \mathscr{F}(i,h-s) \right] dF_n(k-i, \le s) \\ &= \sigma \sum_{n=0}^{r} \mathscr{G}_n(l-k) \left[F_1(k,h) + \frac{EAH}{E(H|A>k)} \mathscr{F}(k,h) \right] * F_n(k,h) \end{aligned} \tag{4}$$

To prove the above formula we consider the situation of the lth day.

(a) If the lth day is the last in the $k$day wet sequence, then the distribution of the sequence is obviously $F_1(k,h)$.

(b) If the lth day is not the last, then the situation is much more complex. The formula

$$\frac{\int_0^h P(A>k, H>s)\,ds}{\int_0^\infty P(A>k, H>s)\,ds} = \frac{\int_0^h P(A>k, H>s)\,ds}{E(H|A>k) P(A>k)}$$

gives the distribution of the residual participation amount. Note, that for stationary, one-dimensional, nonperiodic, renewal processes the distributions of residual waiting time and up to now time are equal one to another. Hence the probability of the first k-day part of the wet sequence which is contained in the l-day period is equal to:

$$P(A>k) \frac{\int_0^h P(A>k, H>s)\,ds}{E(H|A>k) P(A>k)} = \frac{EAH}{E(H|A>k)} \mathscr{F}(k,h)$$

For Q_{10} we have $P(N_l = l, W_l \le h, Q_{10}) = 0$, and for k wet days ($k=0,\ldots,l-1$) we get

$$n=0,\ldots,u=\begin{cases} k & \text{when} \quad k<l/2 \\ l-k-1 & \text{when} \quad k \ge l/2 \end{cases}$$

transitions from the dry to the wet state. Thus

$$\begin{aligned} &P(N_l = k, W_l \le h, Q_{10}) = \beta \mathscr{F}(k,h) \left[1 - G_1(l-k)\right] \\ &+ \beta \sum_{n=1}^{u} \sum_{j=n}^{l-k-1} G_n(j)\left[1 - G_1(l-k-j)\right] \times \\ &\sum_{i=n}^{k} \int_0^h F_n(i,h-s)\, d\mathscr{F}(k-i,s) \\ &= \beta \sum_{n=0}^{u} \left[G_n(l-k) - G_{n+1}(l-k)\right] F_n(k,h) * \mathscr{F}(k,h). \end{aligned} \tag{5}$$

For Q_{11} we have $P(N_l = 0, W_l \le h, Q_{11}) = 0$, and for k wet days ($k=1,\ldots,l-1$) we get

$$n=0,\ldots,v=\begin{cases} k & \text{when} \quad k \le l/2 \\ l-k & \text{when} \quad k > l/2 \end{cases}$$

transitions from the wet to the dry state. The considerations concerning the lth day are the same as for the condition Q_{01}. Thus

$$P(N_l=k, W_l \le h, Q_{11}) = \beta G_1(l-k) \times \sum_{i=1}^{k} \int_0^h \left[F_1(i,h-s) + \frac{EAH}{E(H|A>i)} \mathcal{F}(i,h-s) \right] d\mathcal{F}(k-i,s)$$

$$+\beta \sum_{n=2}^{v} G_n(l-k) \sum_{i=n-1}^{k-1} \int_0^h \sum_{j=1}^{k-1} \int_0^{h-z} \left[F_1(j,h-z-s) + \frac{EAH}{E(H|A>j)} \mathcal{F}(j,h-z-s) \right] d\mathcal{F}(k-i-j,s) dF_{n-1}(i,z)$$

$$= \beta \sum_{n=1}^{v} G_n(l-k)\, F_{n-1}(k,h) * \times \left[F_1(k,h) + \frac{EAH}{E(H|A>k)} \mathcal{F}(k,h) \right] * \mathcal{F}(k,h) \quad (6)$$

The case when $k=l$, we will discuss separately. Proceeding similarly as for the condition Q_{01} one can examine two cases of the lth day, and determine the probabilities for both of them. We make a simplification, and take

$$P(N_l=l, W_l \le h, Q_{11}) = \beta \sum_{j=l}^{\infty} \mathcal{F}(j,h).$$

Summing up the probabilities (3) and (4) as well as (5) and (6) over k, $k=0,\dots,l$, (for the details write to the author) we can obtain:

$$P(W_l \le h, Q_{00}+Q_{01}) = \sum_{k=0}^{l} P(N_l=k, W_l \le h, Q_{00}) + P(N_l=k, W_l \le h, Q_{01})$$

$$= \sigma \left\{ 1 + \sum_{n=1}^{r} \sum_{k=n-1}^{l-n} \mathcal{G}_{n-1}(k) \left[\sum_{j=1}^{l-k} F_1(j,h) + \frac{EAH}{E(H|A>l-k)} \mathcal{F}(l-k,h) - 1 \right] * F_{n-1}(l-k,h) \right\} \quad (7)$$

where $r = \begin{cases} [l/2] & \text{for even } l \\ [l/2]+1 & \text{for odd } l \end{cases}$, $l>0$, and

$$P(W_l \le h, Q_{10}+Q_{11}) = \sum_{k=0}^{l} P(N_l=k, W_l \le h, Q_{10}) + P(N_l=k, W_l \le h, Q_{11})$$

$$= \beta \left\{ \sum_{j=0}^{\infty} \mathcal{F}(j,h) + \sum_{n=1}^{q} \sum_{k=n}^{l-n} G_n(k) F_{n-1}(l-k, \le h) * \times \left[\sum_{j=1}^{l-k} F_1(j,h) + \frac{EAH}{E(H|A>l-k)} \mathcal{F}(l-k,h) - 1 \right] * \mathcal{F}(l-k,h) \right\} \quad (8)$$

where $q=[l/2]$ for every $l>0$. The result of the convolution of $F_r(m,h)$ with 1 is

$$\sum_{i=0}^{m-r} \int_0^h 1 dF_r(m-i,s) = \sum_{j=0}^{m} F_r(j,h)$$

Denote by $U(i,h) = \sum_{j=1}^{i} F_1(j,h) + \frac{EAH}{E(H|A>i)} \mathcal{F}(i,h)$.

Notice that for each $i>0$, $U(i,h)$ is a probability distribution. The convolution $U(k,h) * F_r(k,h)$ is taken for $i=0,\dots,k-r$. Notice furthermore that $EBP(\mathcal{G}_{n-1}m) = P(T_{n-1} \le m) - P(T_n \le m)$ (Jakubowski, 1988). Hence the probabilities (7) and (8) can be expressed by:

$$P(W_l \le h, Q_{00}+Q_{01}) = \sigma + \rho \sum_{n=1}^{r} \sum_{k=n-1}^{l-n} [G_{n-1}(k) - G_n(k)][U(l-k,h)-1] * F_{n-1}(l-k,h)$$

and

$$P(W_l \le h, Q_{10}+Q_{11}) = \beta \sum_{j=0}^{\infty} \mathcal{F}(j,h) + \rho \sum_{n=1}^{q} \sum_{k=n}^{l-n} G_n(k) F_{n-1}(l-k,h) * [U(l-k,h)-1] * \frac{EAH}{EH} \mathcal{F}(l-k,h)$$

where $\rho = \frac{EH}{EAH+EBEH}$, and r,q are defined as previously. After the burdensome calculus (for the details write to the author) for the distribution $P(W_l \le h)$ of the $l-$day total precipitation amount we obtain:

$$P(W_l \le h) = \sigma + \beta \sum_{j=0}^{\infty} \mathcal{F}(j,h) - \rho \sum_{k=1}^{l} [1-U(k,h)] + \rho \sum_{n=1}^{[l/2]} \sum_{k=n}^{l-n} G_n(k)\, F_{n-1}(l-k,h) * [1-U(l-k,h)] * [1-\tilde{U}(l-k,h)] \quad (9)$$

where $\tilde{U}(i,h) = \sum_{j=1}^{i} F_1(j,h) + \frac{EAH}{EH} \mathcal{F}(i,h)$.

CONCLUSIONS

1. Theoretical results presented here can become a basic tool for a practical estimation of a week, 10-day or fortnight total rainfall distributions. Though the form of the formula (9) is rather complicated, it is possible to evaluate the required probabilities. It is sufficient, as it is often done in practice, to choose a gamma-type distributions for fitting to all random variables used here. For instance for the wet sequence we can take $F_1(i,h) = P_{q,v}(i)\, \Gamma_{\alpha,v+i-1}(h)$, where $P_{q,v}$ denotes Pascal

distribution describing the length of the wet days sequence, and $\Gamma_{\alpha,\nu+i-1}$ denotes gamma distribution describing the total precipitation amount of the sequence.

2. The distribution of the l-day total precipitation amount, given by (9) is quite general. Taking the precipitation amounts on individual wet days as the independent and identically distributed random variables, independent also on the occurrence process, we can easily obtain the well known formula

$$P(W_l \le h) = P(N_l = 0) + \sum_{k=1}^{l} F^{*k}(h) P(N_l = k)$$

where $F(h)$ is the distribution of daily rainfalls, and $*k$ denotes k-fold convolution. The differences between the results from these two expressions depend mostly on the character of the daily rainfall data. More precisely they are caused by the degree of the dependency and nonhomogeneity of a given daily rainfall record.

3. Taking $h=0$ we obtain the probability of a dry spell in the l-day period.

$$P(W_l = 0) = \sigma + \beta\mathscr{F}(0,0) - \rho l + \rho \sum_{k=1}^{l-1} G_1(k)\, F_0(l-k,h) \\ *[1 - U(l-k,h)] * [1 - \tilde{U}(l-k,h)]\Big|_{h=0}$$

Since $1 - U(0,h) = 0$ for each $h \ge 0$ and $\beta\mathscr{F}(0,0) = \rho$, hence

$$P(W_l = 0) = \sigma + \rho(1-l) + \rho \sum_{k=1}^{l-2} (l-k-1) G_1(k) \\ = \rho\left[EB - \sum_{k=0}^{l-2} P(B > k)\right]$$

It means that the distribution of random variable W_l has a positive value at $h=0$. If we assume the independence of the wet sequence length on its amount, then the probability of a dry spell is identical to the probability of no wet days in the l-day period (cf. Jakubowski (1988)).

REFERENCES

Bickel, P. J. & Yahav, J. A. (1965) Renewal theory in the plane. *Ann. Math. Stat.*, **36**(3), 946–55.

Buishand, T. A. (1977) *Stochastic modeling of daily rainfall sequences*, Mededelingen Landbouwhogeschool Wageningen, Netherlands, pp. 77–3.

Jakubowski, W. (1988) A daily rainfall occurrence process. *Stoch. Hydrol. Hydraul.*, **2**(1), 1–16.

Mimikou, M. (1983) Daily precipitation occurrence modeling with Markov chain of seasonal order. *Hydrol. Sci. J.*, **28**(2), 221–32.

Roldan, J. & Woolhiser, D. A. (1982) Stochastic daily precipitation models; 1. A comparison of occurrence processes. *Water Resour. Res.*, **18**(5), 1451–9.

Valdes, J. B., Rodriguez-Iturbe, I. & Gupta, V. K. (1985) Approximations of temporal rainfall from a multidimensional model. *Water Resour. Res.*, **21**(8), 1259–70.

Woolhiser, D. A. & Pegram, G. G. S. (1979) Maximum likelihood estimation of Fourier coefficient to describe seasonal variations of parameters in stochastic daily precipitation models. *J. Appl. Meteor.*, **18**, 34–42.

Woolhiser, D. A. & Roldan, J. (1982) Stochastic daily precipitation models; 2. A comparison of distribution of amounts. *Water Resour. Res.*, **18**(5), 1461–8.

Woolhiser, D. A. & Roldan, J. (1986) Seasonal and regional variability of parameters for stochastic daily precipitation models, South Dakota, USA. *Water Resour. Res.*, **22**(6), 965–78.

Waymire, E. & Gupta, V. K. (1981) The mathematical structure of rainfall representation; 1. A review of the stochastic rainfall models. *Water Resour. Res.*, **17**(5), 1261–72.

Yevjevich, V. & Dyer, T. G. J. (1983) Basic structure of daily precipitation series. *J. Hydrol.*, **64**, 49–67.

5 Analysis of outliers in Norwegian flood data

L. GOTTSCHALK

Department of Geophysics, University of Oslo, Norway

Z. W. KUNDZEWICZ

Research Centre for Agricultural and Forest Environment Studies, Pol. Acad. Sci., Poznan and Institute of Geophysics, Pol. Acad. Sci., Warsaw, Poland

ABSTRACT Plausibility analysis of annual maximum flows of Norwegian rivers is performed. The data embrace time series of 60 years (1921–80) gathered at 42 observation stations and time series of 30 years (1921–50 and 1951–80) collected at 86 and 83 observation stations, respectively. Six different tests for outliers detection have been used (Shapiro–Wilk, skewness, Student, RST, probability plot coefficient and Anderson–Darling). The tests are based on the assumption of normal distribution, so the normalization (logarithmic or cube root transformation) of the raw data may be a prerequisite. The empirical orthogonal functions approach was used to simulate regional samples with preserved first and second order moments. Outliers analysis of the simulated data was performed and the results were compared with observations.

INTRODUCTION

The existence of outliers in hydrological observation series can possibly explain many of the problems faced in the regional analysis of hydrological data. Figs. 1 and 2 show some examples of hydrographs and probability plots, containing suspicious outliers conceived as observations strongly deviating from the remainder of the data set.

Processing outliers consists of two stages – detection and treatment. Depending on the way the outliers are treated, one can get quite a different representation of the process. In practice outliers are detected and removed in accordance with some intuitive rule. It is so because one finds it difficult to properly choose the theoretical distribution function for an individual observation series and to estimate its parameters. If the parent distribution was known these problems could have been easily solved.

The situation is different if one deals with an ensemble of observation series. The point is, whether or not one is convinced that the outlying observations really reflect the hydrological phenomenon (natural variability) and are not caused by inadequacies in the observation method or in the processing of data (e.g. uncertainty of the rating curve in the high flow zone due to extrapolation). The frequency of outliers in a regional ensemble gives information about the eventual existence of a parent distribution with a heavy right tail. Within modern hydrological literature the interest has mainly been concentrated on the regional behaviour in the third order statistical moment, i.e. the skewness coefficient, describing the symmetry of the distribution of data. A high positive value of this parameter indicates a heavy right tail of the distribution. This parameter, when estimated from observed data is significantly influenced by the eventual existence of outliers in the data set. It is not uncommon that one single outlier in a hydrological observation series can change this parameter by a factor of two. The high regional variability in the skewness can be caused by several outliers in the regional data set.

Hawkins (1980) distinguished two principal mechanisms to explain the occurrence of outliers.

(a) The data come from a heavy tailed distribution. That is an outlying observation need not be erroneous. An 'outlier prone' distribution has a tail slowly tending to zero. A distribution is called absolute outlier prone if (Green, 1976):

$$\forall \varepsilon, \delta > 0 \quad \Pr\{x_{n,n} - x_{n,n-1} > \varepsilon\} \geq \delta \quad \text{for all } n > n_0$$

where $x_{n,i}$ is the ith order statistics based on a sample of size n. A distribution is called relatively outlier prone if:

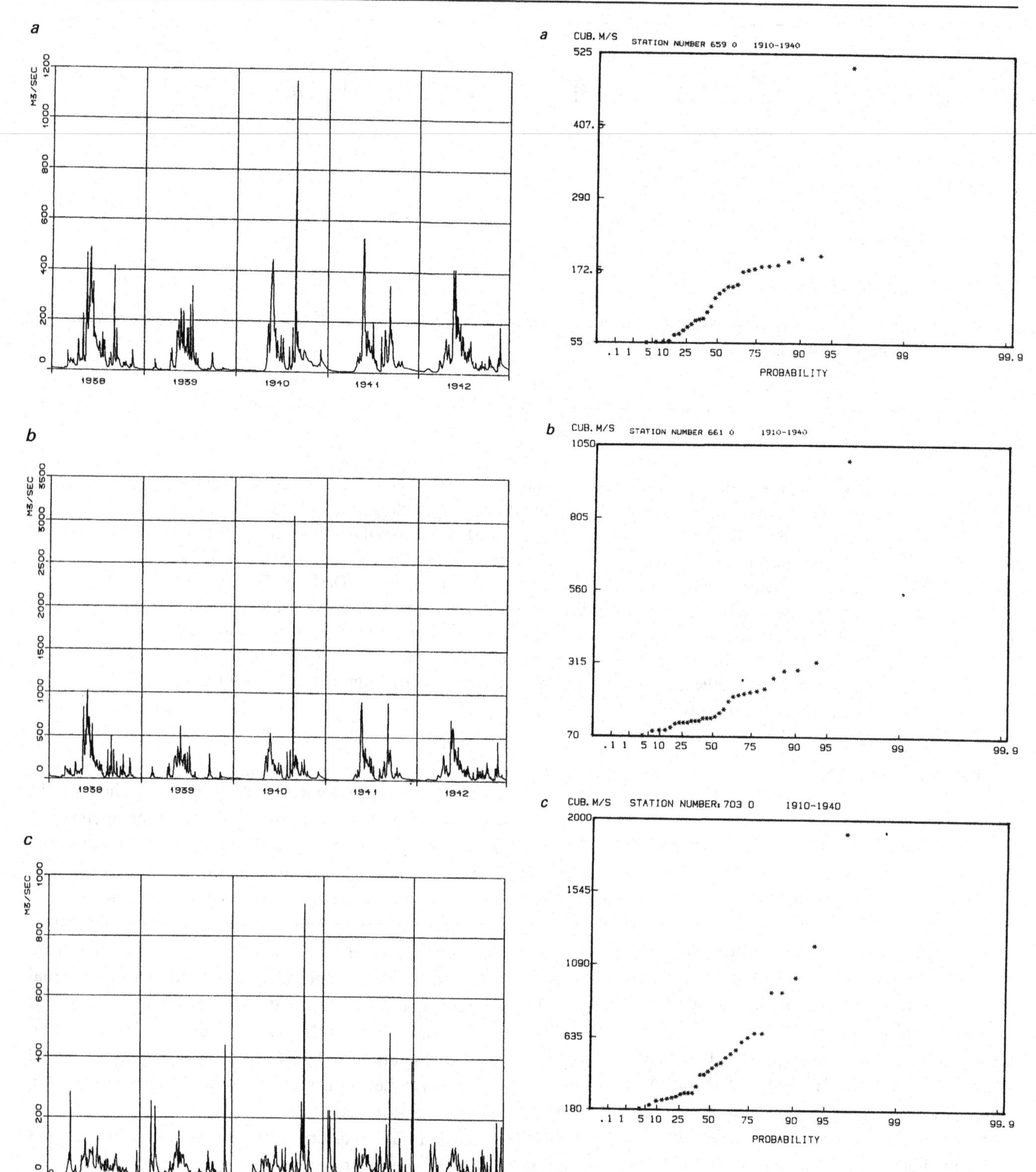

Fig. 1 Examples of hydrographs containing suspicious outliers; (a) Gauge 659 (Bjorset); (b) Gauge 661 (Haga Bru); (c) Gauge 703 (Kapskarmo).

Fig. 2 Examples of probability plots of extreme events containing suspicious outliers; (a) Gauge 659 (Bjorset); (b) Gauge 661 (Haga Bru); (c) Gauge 703 (Kapskarmo).

$$\forall c>1, \delta>0 \quad \Pr\{x_{n,n}/x_{n,n-1}>c\}\geq\delta \quad \text{for all } n>n_0.$$

In other cases a distribution is called absolutely, and relatively outlier resistant, respectively. As examples, it can be mentioned that the normal distribution is classified as absolute outlier resistant, while the gamma distribution is absolutely outlier prone and relatively outlier resistant.

(b) The data arise from two distributions, a basic distribution $f_0(.)$ and a contaminating distribution $f_1(.)$. Two situations can occur:

(i) exactly $n-k$ observations of the total sample size n come from the basic distribution, and k observations arise from the contaminating distribution.

(ii) k is not known. Instead the probability p is given that an observation comes from the contaminating distribution and the probability $(1-p)$ that it comes from the basic one, i.e.:

$$f(.)=(1-p)f_0(.)+pf_1(.)$$

A basic problem then is whether the distribution functions conventionally used in hydrology can explain the frequency with which outliers appear in regional samples or if it is necessary to use the model (b – ii) with some super-population of floods. In both cases it is of interest to identify the existence of outliers. In the first case one needs to take such measures as using estimation methods that are insensitive to outliers (robust) or simply excluding the outliers from the sample. In the second case the aim is to separate the two populations.

TESTS FOR REJECTION OF OUTLIERS —

In the following a review of potentially useful tests for outliers detection will be given. Subsequently all these tests will be applied to flood flows data of Norway. Similar analysis has been applied by Kottegoda (1984) to 51 series of annual maximum flows in North America.

Tests for rejection of outliers are mainly based on the slippage or departure-from-model concepts. In a slippage test (which is a generalisation of the outlier problem) observations $x_{ij}, j=1$ to $m_i, i=1$ to n, which all are independent, are assumed to come from distributions $f_i(.)$. The slippage hypothesis is that while the majority of the $f_i(.)$ are identical to some common distribution $f(.)$, some small number of them have large probabilities in the tail regions of it, and are said to have 'slipped'. Karlin and Truax (1960) have suggested the following test for this situation: Suppose that for $i=1$ to n, x_i has a density $f(.|\theta_i)$, where θ_i is an unknown parameter. The null hypothesis is formulated as:

$$H_0: \theta_i=\theta \quad , i=1 \text{ to } n$$

and the alternative hypothesis is:

$$H_1: \theta_i=\theta+\Delta \quad , \Delta>0$$
$$\theta_j=\theta \quad , j\neq i$$

Three tests that are applied here can be seen as special cases of the slippage problem. The first one is to use the Student ratio:

$$(x_{(n)}-\bar{x})/s \tag{1}$$

where $x_{(n)}$ is the largest value among the total of n observations, and $\bar{x}$ and s are the mean and the standard deviation, respectively, of these observations. A slight modification of this test is to exclude the suspected outlier from the data used for estimation of the mean and the standard deviation $(\bar{x}', s')$. The test statistic has the form (RST-statistic, Rosner (1977)):

$$(x_{(n)}-\bar{x}')/s' \tag{2}$$

The null hypothesis is that the x_i, $i=1,\ldots,n$ belongs to a normal distribution with the same mean. If the test statistic is larger than the critical value, the hypothesis is rejected and $x_{(n)}$ is marked an outlier. There can of course exist more outliers in the sample. The test can then be repeated after having excluded $x_{(n)}$ from the sample.

The sample coefficient of skewness C_s can also be used as a test statistic. The null hypothesis is again that the observations $x_i, i=1,\ldots,n$ come from the same normally distributed population. The hypothesis is rejected if C_s is bigger than some critical value. This test statistics does not actually test whether a specific observation is an outlier. It merely indicates that one or more outliers are present. To turn it into a rule for identifying outliers, it must be supplemented by another rule for rejection. One possibility would be to reject extreme observations until C_s becomes non-significant.

The other alternative is to use 'departure from model' statistics. These tests are not directly designed for the outlier problem, but can give some indication of the existence of outliers in data. Three different tests of this type have been applied in this study.

(a) The Shapiro and Wilk's W-statistic is a goodness of fit test for normality. The test statistic originates essentially from the one by Cramer–Von Mises which is the square of deviations between the theoretical distribution $F_0(x)$ and the empirical one, $F_n(x)$:

$$W=\int_{-\infty}^{\infty}(F_n(x)-F_0(x))^2 dF_0(x_0) \tag{3}$$

For a discrete sample x_i $i=1,\ldots,n$ the statistics has the form

$$W_n=\sum_{i=1}^{m} a_{n,n-i+1}(x_{(n-i+1)}-x_{(i)})^2/\sum_{i=1}^{n}(x_{(i)}-\bar{x}_n)^2 \tag{4}$$

where m is the highest integer less than $n/2, \bar{x}_n$ is the mean and $x_{(i)}$ is the ith value in the ordered set $x_{(1)} < x_{(2)} < \ldots < x_{(n)}$. The weights $a_{n,j}$ are tabulated by, for instance, Shapiro & Wilk (1965). Here the statistics has been calculated from algorithms suitable for computer implementation developed by Royston (1982).

(b) The Anderson–Darling test is also a test for normality and is a special case of the weighted Cramer–Von Mises test (Shapiro, Wilk & Chen, 1968). The test statistics A_n^2 is based on the difference between the sample step function $F_n(x)$ and the underlying theoretical distribution $F_0(x)$:

$$A_n^2 = n \int_{-\infty}^{\infty} \frac{[F_n(x) - F_0(x)]^2}{F_0(x)[1 - F_0(x)]} \, \mathrm{d}F_0(x) \tag{5}$$

For a discrete sample the test statistics can be written as:

$$A_n^2 = n - 1/n \sum_{i=1}^{n} (2i-1)\{\ln u_1 + \ln[1 - u_{(n-i+1)}]\}/n \tag{6}$$

where $u_i = F_0(x_i)$. For the normal distribution u_1 can be approximated by $u_i = (i - .3175)/(n + .365)$ (Blom, 1958).

(c) The probability plot correlation coefficient test (Filliben, 1975) with the test statistic:

$$r = Corr\{x_{(i)}, u_i\} \tag{7}$$

where $Corr$ denotes the correlation.

The null hypothesis is that the x come from a normal distribution. In that case $x_i = g(u_i)$ should form a straight line. The hypothesis is rejected if r is below a critical level. Filliben (1975) has tabulated percentage points of the normal probability plot correlation coefficient.

It should be noted that the interpretation of significance level in these tests is not related to the existence of outliers, and therefore cannot be used in a straightforward manner.

All the enumerated tests are based on the assumption of normally distributed observations. In order to tackle non-normal data one has to use appropriate transformations. Here the simple logarithmic and cube root transformations will be applied. The first one allows testing of lognormally distributed observations. The second one renders two-parameter gamma distributed variates approximately normal (Wilson–Hilferty transformation).

CASE STUDY

The tests were applied to annual flood series data from Norway. Two sets of 30-year records (1921–50, 1951–80) at 86 and 83 observation sites, respectively, and one set of 60 year records (1921–80) at 42 observation sites were utilized. The statistical tests were complemented by visual inspection of all series (cf. Figs. 1 and 2). One of the results is shown in Fig. 3, where years of largest annual extreme values in regional samples for the years 1921–80 are given. The observations classified as outliers are underlined and the suspected ones are marked with brackets. The results of the statistical analysis of the raw data, and after logarithmic and cube root transformations are shown in Table 1.

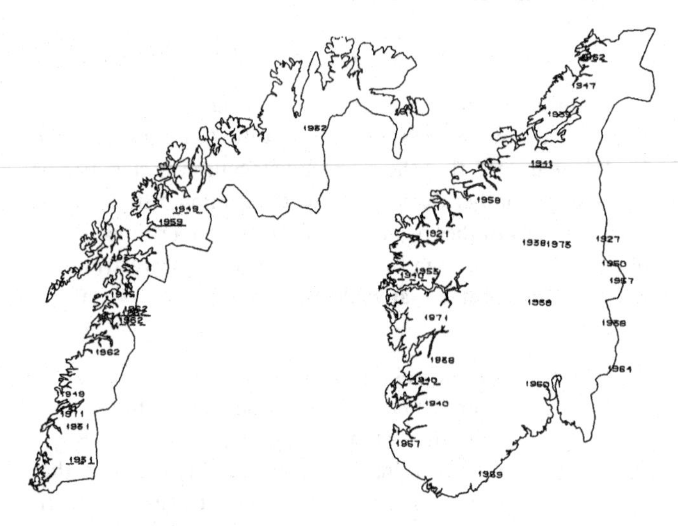

Fig. 3 Years of largest annual extreme values in regional samples of flood data in Norway (1921–1980).

As seen in Table 1, results of different tests do largely differ. This could be expected, however, as the tests are sensitive not only to the existence of outliers but also to extreme observations in both tails of the distribution. Best agreements are found between the Shapiro–Wilk, the Student deviation and the probability plot correlation tests. If the test of the skew coefficient is applied as a one-(right)-sided test, it also gives a good agreement with the three others, as well as with the subjective visual judgment.

A dramatic difference in results between the different transformations has been observed. The large number of rejections when the tests are applied to the raw data, can be interpreted as a rejection of the assumption of normality. The smallest number of rejections are observed in the case of the cube root transformation. Splitting the data set into two populations can be used as a check of stability of the tests and also to see if there is a tendency of the number of outliers to relatively decrease with increasing sample size. In the case of the Shapiro–Wilks, Student, and probability plot correlation tests it is the same number of outlying observations both in the 60 years and in the 30 years series. These tests thus do not give any indication that the number of outliers may decrease with increasing sample size.

The effect of outliers in the analysis of the regional behaviour in statistical parameters is pronounced. Table 2 presents a comparison between the regional parameters calculated from the original regional data sets of 60 years of records at 42 points and from the purified data set, i.e. after the removal of five most significant outliers from the samples.

Table 1. *Number of rejections at 5 and 1% significance levels of six outlier tests*

	60 years series 1921–80			30 year series 1912–50			30 year series 1951–80		
Test	norm	log-norm	$\sqrt[3]{}$ -trans	norm	log-norm	$\sqrt[3]{}$ -trans	norm	log-norm	$\sqrt[3]{}$ -trans
Shapiro–Wilk									
5% sign.	29	14	9	22	5	4	27	8	11
1% sign.	25	5	6	14	0	0	14	6	7
Anderson–Darling									
5% sign.	6	1	0	0	0	0	3	1	0
1% sign.	1	0	0	0	0	0	0	0	0
Skewness									
5% sign.	33	18	15	27	10	7	30	14	14
1% sign.	28	11	7	17	4	1	20	8	9
Student deviation									
5% sign.	22	3	10	17	1	6	20	4	9
1% sign.	16	2	4	11	0	1	11	2	3
RST									
5% sign.	29	7	14	25	5	11	27	9	14
1% sign.	21	3	9	21	2	7	24	5	10
PPC									
5% sign.	31	16	10	18	8	5	23	9	11
1% sign.	22	11	7	11	3	1	13	7	5

Table 2. *Illustration of effect of five most apparent outliers in the regional sample (42 stations observed in 60 years) on regional statistical parameters*

Parameter	Outliers	Mean	Standard deviation
Mean	Included	324	256
	Removed	322	255
C_v	Included	0.42	0.11
	Removed	0.40	0.09
C_s	Included	1.14	0.88
	Removed	0.92	0.46
C_v/C_s	Included	2.58	1.53
	Removed	2.29	1.15

The regional mean is little affected, but already the standard deviation of the coefficient of variation shows a reduction. The standard deviation of the coefficient of skewness is reduced by 30 per cent.

A basic question is whether the number of outliers is in agreement with the standard distributions applied in hydrology. To solve this problem Monte Carlo methods can be applied. Such approaches have their limitation in the difficulty in deriving a proper algorithm for the simulation of regional samples with a certain parent distribution which preserves all the statistical properties of the original samples.

An empirical orthogonal functions (EOF) approach is applied in the present study. The general idea of using EOFs is to produce a set of orthogonal functions via a linear transformation of the original data. The method orders the available information, and excludes redundancies. The type of approach used to derive EOFs has much in common with such methods as principle component analysis and eigenvector analysis. The theory of expansion into empirical orthogonal functions has been treated among others by Holmström (1969) and Obled and Creutin (1986). A concept of the method is presented in the Appendix.

In the present study the idea of EOFs has been used to simulate regional samples of normally and log-normally distributed data. 1000 data sets of 42 series of 10,20,30 and 60 elements, respectively, were generated. The parameters were estimated from the 42 observed series with a record of 60 years. The number of rejections of outliers in the synthetic data sets of 60 years series are given in the Table 3. For comparison the results of the same analysis on the original data set, shown earlier in the Table 1, are repeated.

The simulation model is able to preserve the first and the second order moments (means, variances and covariances). The distribution functions of the regional behaviour of higher order moments and of outliers calculated from Monte Carlo simulations can be used for hypothesis testing. The null hypothesis is that the observed regional sample belongs

Table 3. *Number of rejected outliers in original and synthetic data*

Method		Normal distribution				Lognormal distribution			
		Original	Synthetic			Original	Synthetic		
			mean	95%	99%		mean	95%	99%
Shapiro–Wilk	5%	29	4.2	8	9	14	5.0	9	10
	1%	25	0.9	3	4	5	1.1	3	4
Anderson–Darling	5%	6	0.0	0	0	1	0.0	0	0
	1%	1	0.0	0	0	0	0.0	0	0
Skewness	5%	33	4.6	8	11	18	4.7	9	10
	1%	28	0.9	3	4	11	1.1	3	4
Student	5%	22	2.0	5	6	3	2.1	5	7
	1%	16	0.4	2	3	2	0.5	2	3
RST	5%	29	4.8	9	11	7	4.9	10	12
	1%	21	1.6	4	5	3	1.7	4	6
PCC	5%	31	2.2	5	6	16	2.3	5	7
	1%	22	0.4	2	2	11	0.4	2	3

to a certain parent distribution. If the observed regional mean or variance of the coefficient of skewness C_s and of the ratio C_s/C_v and of the observed number of outliers lie outside some critical level of the respective distributions of these parameters, the hypothesis is rejected. Such a regional test is found to be more reliable than applying goodness-of-fit-test to each individual observed series.

It is difficult to interpret these results. This is due to the fact that the tests are mainly aimed at detecting a departure from model rather than outliers. The tests based on the Student deviation and the RST-statistic indicate most directly the existence of extremely large values. In both these tests the number of rejections in the original data sets is well in agreement with that of the synthetic data with lognormal parent distribution. If the number of outliers is used as a test statistic for the regional sample, the hypothesis that the original data comes from a lognormal distribution cannot be rejected. Based on the results of goodness-of-fit tests, the same hypothesis must be rejected in all cases except for the Anderson–Darling test. The hypothesis of a normal parent distribution is rejected in all cases.

CONCLUSIONS

The study has revealed the problems encountered when trying to detect outliers in a set of maximum annual flow data for a site. As the parent distribution function is unknown, there is indeed no rigorous cookbook for finding outliers in a set of maximum annual flow data. The tests for outlier rejection that perform comparably are: Shapiro–Wilk, Skewness, Student and RST, whereas the performance of the Anderson–Darling test is entirely different. As the original data are very far from normality, some normalizing transformation is necessary, what results in reduction of the number of outliers detected. Removal of one or few of most significant outliers from the population may have a profound effect on the statistical characteristics, and in particular on the skewness coefficient.

ACKNOWLEDGEMENTS

The study was financially supported by the Norwegian National Council for Hydrology. The support is gratefully acknowledged.

REFERENCES

Blom, G. (1958) *Statistical Estimates and Transformed Beta Variables*, Wiley, New York.

Filliben, J. J. (1975) The probability plot correlation coefficient test for normality. *Technometrics*, **17**(1), 111–16.

Green, R. F. (1976) Outlier – prone and outlier – resistant distributions. *Journal of the American Statistical Association*, **71** (354), 502–5.

Hawkins, D. M. (1980) *Identification of Outliers*, Chapman and Hall, London.

Holmström, I. (1969) Extrapolation of meteorological data, SMHI, *Notiser och preliminara rapporter, serie Meteorologi*, nr 22, Stockholm.

Karlin, S. & Truax, D. (1960) Slippage problems. *Annals of Mathematical Statistics*, **31**, 296–324.

Kottegoda, N. T. (1984) Investigation of outliers in annual maximum flow series. *J. Hydrol.*, **2**, 105–37.

Obled, Ch. and Creutin, J. D. (1986) Some developments in the use of empirical orthogonal functions for mapping meteorological fields *Journal of Applied Meteorology*, **15**, 1189–204.

Rosner, B. (1977) Percentage points for the RST many outlier procedure, *Technometrics*, **19**(3), 307–12.
Royston, J. P. (1982) An extension of Shapiro and Wilk's W test for normality of large samples. *Applied Statistics*, **31**(2), 115–24.
Shapiro, S. S. & Wilk, M. B. (1965) An analysis of variance test for normality (complete samples). *Biometrika*, **52**(3), 591–611.
Shapiro, S. S., Wilk, M. B. & Chen, H. J. (1968) A comparative study of various tests for normality. *American Statistical Association Journal*, **63**, 1343–72.

APPENDIX. THE NOTION OF EMPIRICAL ORTHOGONAL FUNCTIONS (EOF)

Consider a set of time series $P_i(t), i=1,\ldots,N$ over a time interval (a,b). $P'_i(t)$ are the corresponding series with respective time averages subtracted. An expansion into EOF has the form:

$$P'_i(t)=\sum_{n=1}^{M} h_{ni} \quad \beta_n(t),\ i=1,\ldots,N \tag{1}$$

where: h_{ni} are weight coefficients (summing to M over $n=1,\ldots,M$) varying between the series but constant in time and β_n are sets of functions common to all series. These functions are called amplitude functions.

Requiring the possibly fast convergence of the series expansion and adding a normalizing condition to the weight coefficients, one gets orthogonal sets of weight coefficients and amplitude functions with the properties:

$$h_{ni}\,h_{mi}=\delta_{nm}M \tag{2}$$

and

$$\frac{1}{T}\int_a^b \beta_n(t)\,\beta_m(t)\,\mathrm{d}t=\delta_{nm}\frac{1}{T}\int_a^b (\beta_n(t))^2\,\mathrm{d}t=\delta_{nm}\lambda_n \tag{3}$$

where: δ_{nm} is the Kronecker delta and λ_n are the eigenvalues of the covariance matrix. The weight coefficients, h_{ni}, are the elements of the eigenvectors of the covariance matrix. The new set of functions created by this expansion is empirical in the sense that they are based on the series themselves and not restricted to any predetermined polynomial form. Normally the sets of EOFs are arranged in descending order according to the proportion of variance explained by each function. An important property is that a small number of functions will be able to reproduce a great part of the total variance. An EOF representation using $M=N$ linearly independent functions is a complete description of the original data.

6 Stochastic modelling of the operation of hydrants in an irrigation network

G. TSAKIRIS and O. MANOLIADIS

Laboratory of Rural Technology, National Technical University of Athens, Greece

ABSTRACT The irrigation system design of pressurized networks is often dimensioned based on the probability of hydrant operation. Simple statistical techniques have been extensively used in the past to model the hydrants operation and to calculate the design capacity of each reach of the network. Such methods as, for example, Clemment's 'on demand' approach, are adopted by designers and agencies in the design of irrigation systems. The objective of this paper is to attempt to model the hydrants operation in the irrigation network using the Alternating Renewal Process in continuous time. Extensive data were gathered from collective irrigation networks in Crete. These data were used to estimate the parameters of the Alternating Renewal Process model. A graphical representation of the results could assist in drawing useful conclusions.

INTRODUCTION

Design of collective pressurized irrigation systems is often based on the probability of hydrants operation. The method known as 'on demand' was introduced by Clemment (1955) for the estimation of peak season discharge requirements. Therefore the probability of operation of hydrants as a design factor has a significant effect on the overall economy of the irrigation project (construction, operation and management). Common practice for the description of the demand pattern when designing an irrigation project is to calculate a probability of hydrant operation based on earlier experience or on assumed pattern of operation.

Numerous studies on modelling irrigation networks can be found in the literature. The difficulty in predicting the pattern of the hydrant operation is, that this pattern is related to the user's behavior which is not stationary in time. According to Svehlik (1977) recording of the fluctuations in demand is the main obstacle for finding the correct design capacity of the irrigation system. To model the variations of irrigation water demand Delclaux (1984) suggested a real time control model used for demand prediction based on discrete time schedule. However, it results from the literature research, that there are no mathematical models characterizing the behaviour of hydrants in continuous time.

The objective of this paper is to use the Alternating Renewal Process theory in continuous time for modelling the operation of hydrants. The renewal theory, used to study probability problems connected with failure and replacement of mechanical components, is used here to model the operation of hydrants of a collective pressurized irrigation network.

It was assumed for the simulation of the model that all hydrants have similar operating characteristics (irrigation area, discharge, pressure etc.). A schematic representation of this process is shown in Fig. 1, where the sequence of 'on' and 'off' intervals is plotted versus time. According to the Renewal Theory, the points in time at which the process is switched from a non-operating ('off') state to an operating ('on') state interval are called the 'renewals'.

ALTERNATING RENEWAL PROCESS FORMULATION

Consider one or more hydrants, each of which is, at any instant, either on or off. Assume that the 'on' and 'off' intervals belong to a continuous stochastic process with two states. These two states are denoted by the two-state stochastic process as follows:

$$Z(t)=\begin{cases}1 & \text{when a hydrant is 'on'}\\ 2 & \text{when a hydrant is 'off'}\end{cases} \quad (1)$$

Two other processes related to the above are now defined:

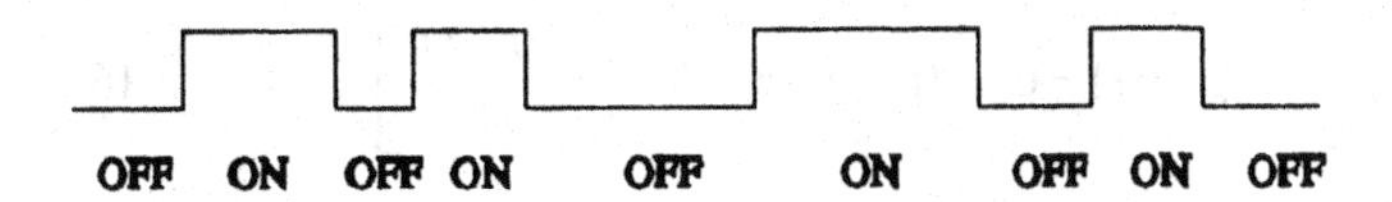

Fig. 1 A schematic representation of a Renewal Process of 'on' intervals.

(a) the process $\{D(t), t>0\}$ as the total duration of the 'off' period during time interval $(0, t)$; and
(b) the process $\{W(t), t>0\}$ as the total duration of the 'on' period during time interval $(0, t)$.

That is $D(t)$ and $W(t)$ represent the total time the process spends in the state $Z=0$ and in the state $Z=1$, respectively. Since the time interval t comprises either 'on' or 'off' periods, then:

$$D(t)+W(t)=t \tag{2}$$

It must be stressed that the three above processes are mathematically completely equivalent. They can be expressed as a function of $Z(t)$ as follows:

$$D(t)=\int_0^t [1-Z(x)]\mathrm{d}x; \qquad t>0 \tag{3}$$

$$W(t)=\int_0^t [Z(x)]\mathrm{d}x; \qquad t>0 \tag{4}$$

The Poisson process

In the renewal theory, the probability density functions are represented by the family of Erlang distribution functions. Erlang functions are the exponential distribution, the Gamma distribution, and the General and the Special Erlangian distribution. In this study the exponential distribution with rate λ is used, that is:

$$f(t)=\lambda \mathrm{e}^{-\lambda t} \tag{5}$$

and the cumulative probability function is:

$$F(t)=1-\mathrm{e}^{-\lambda t} \tag{6}$$

Fig. 2 shows an example of the probability density function and the cumulative density function for an exponential distribution.

THE PROBABILITY OF OPERATION OF HYDRANTS IN THE INTERVAL (0, *t*)

In accordance with this theory the distribution of the total operating period may be found (Takacs, 1957; Barlow & Hunter, 1961; Tsakiris 1988):

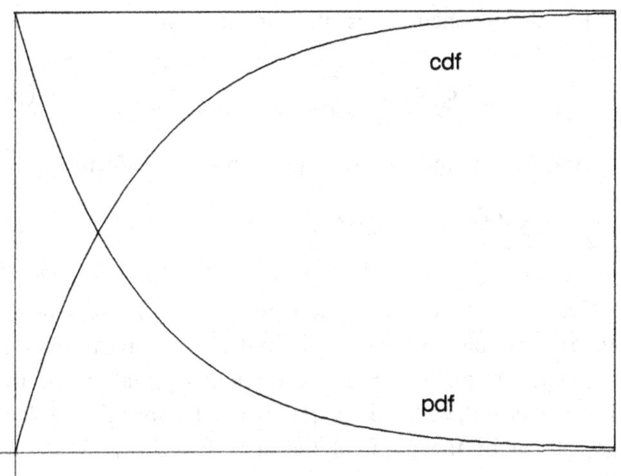

Fig. 2 Probability density function (p.d.f.), and cumulative density function for an exponential distribution.

$$\mathrm{P}\{W(t)\le r\}=\sum_0^{\infty} F_n(r)\left[G_n(t-r)-G_{n+1}(t-r)\right]; \text{ for } t>r \tag{7}$$

in which the term inside the brackets on the right-hand side of equation (7) represents the probability of having n renewals in the time period $(0, t)$ while $F_n(r)$ represents the cumulative density function of the sum of the n separate 'on' intervals.

The probability of having *n* renewals in the time period (0, *t*)

The probability density function represents that at any time t the probability of a renewal occurring before $t+\Delta t$ is $\lambda\Delta t$. Divide the time interval $(0, t)$ into a large number k of small subintervals of length Δt, where $k\Delta t=t$. The probability of a renewal in the subinterval Δt is $\lambda\Delta t$. Furthermore the probability of two or more renewals in the same subinterval is $0\Delta t$. Using the binomial probability law given as equation (7), or

$$\mathrm{Prob}(N_t=m)=\lim \binom{k}{m}(\lambda\Delta t)^m(1-\lambda\Delta t) \tag{8}$$

one can obtain the number of renewals in $(0, t)$ for the exponential distribution.

The sum of the 'on' separate intervals

One of the main mathematical tools used in the renewal theory is the Laplace tranform technique. The Laplace transform $f^*(s)$ is defined by:

$$f^*(s)=\mathrm{E}(\mathrm{e}^{-sx})=\int_0^{\infty} \mathrm{e}^{-sx}f(x)\,\mathrm{d}x \tag{9}$$

For the exponential distribution function

$$f^*(s)=\lambda e^{-\lambda x}e^{-sx}\,dx=\frac{\lambda}{\lambda+s} \quad (10)$$

Another example of use of the Laplace transform is:

$$f^*\left(\frac{(\lambda x)^{n-1}e^{-\lambda x}}{(n-1)!}\right)=\frac{\lambda^{\alpha}}{(\lambda+s)^{\alpha}} \quad (11)$$

The importance of the Laplace transforms in renewal theory is visible for sums of independent random variables. If X_1, X_2, ... X_n are non-negative random intervals when the hydrant is 'on', with p.d.f.s: $f_1(t), f_2(t), f_n(t)$, then the Laplace transform of the p.d.f. of the sum $X_1+X_2+\ldots X_n$ is, by definition

$$E[\exp(-\lambda(X_1+X_2+\ldots X_n))] \\ =E(e^{-\lambda x_1})E(e^{-\lambda x_2})\ldots E(e^{-\lambda x_n}) \quad (12)$$

$$E[\exp(-\lambda(X_1+X_2+\ldots X_n))]=f_1^*(\lambda)f_2^*(\lambda)\ldots f_n^*(\lambda) \\ =f^*(\lambda)^n=\frac{\lambda^n}{(\lambda+s)^n} \quad (13)$$

and using equation (12)

$$E[\exp(-\lambda(X_1+X_2+\ldots X_n))]=\frac{\lambda(\lambda x)^{n-1}e^{-\lambda x}}{(n-1)!}=F_n(t) \quad (14)$$

which is the n-fold convolution and represents the probability of having n renewals in the time period $(0, t)$.

THE PROBABILITY OF OPERATION OF HYDRANTS IN THE INTERVAL (0, t)

The solution of equation (7) depends mainly on the form of $F(x)$ and $G(x)$. The main difficulty encountered is the evaluation of convolutions. In the general case equation (15) may be solved using numerical techniques (McConaloque, 1978). According to Li (1971), the infinitive sum in equation (7) may be approximated by a finite sum up to a number N by adding an error term $\varepsilon(N)$, as follows:

$$P\{W(t)\le r\}=\sum_{0}^{N}F_n(r)\,[G_n(t-r)-G_{n+1}(t-r)] \\ +\varepsilon(N); \quad \text{for } t>r \quad (15)$$

Li (1971) showed that the error $\varepsilon(N)$ is bounded and therefore satisfactory approximations may be obtained.

There are special cases for which the distribution of $D(t)$ can be calculated using equation (15) with the aid of a computer. The computations may be significantly simplified if $F(r)$ and $G(r)$ are the members of the family of Erlang distributions. From equation (8) the propability of having n renewals in the time interval $(0, t)$ may be derived:

$$G_n(t-r)-G_{n+1}(t-r)=e^{-\mu(t-r)}\left[\frac{\mu^n(t-r)^n}{n!}\right] \quad (16)$$

From equations (12) the n-fold convolution may be derived:

$$F_n(y)=\frac{\lambda^n y^{n-1}e^{-\lambda y}}{(n-1)!} \quad (17)$$

where $F_0(y)=1$.

Inserting equations (16) and (17) into equation (15) the latter becomes:

$$P\{W(t)\le r\}=\sum_{0}^{\infty}e^{-\mu(t-r)}\left[\frac{\mu^n(t-r)^n}{n!}\right]\frac{\lambda^n y^{n-1}e^{-\lambda y}}{(n-1)!} \quad (18)$$

Or finally according to Li (1971):

$$P\{W(t)\le r\}= \\ e^{-\mu(t-r)}\left(1+\lambda\mu(t-r)\,e^{-\lambda y}y^{-0.5}I_1(2[\lambda\mu(t-r)y])\right)dy \quad (19)$$

where all the symbols retain the meanings attributed to them previously and I_1 is the modified Bessel function of the first kind. Similar expression may be obtained for the 'off' intervals.

FIELD APPLICATION – DISCUSSION

To apply the methodology presented above, data concerning the operation of hydrants were obtained from an existing trickle irrigation project in Chania – Crete Greece. The irrigation project area is 1.4 km^2 with 350 users and 43 hydrants. The area was considered representative of the region. The collected data are observations of flow meters on a continuous time basis for the peak irrigation season of 1989. The hourly observations were found suitable for representing the continuous process as was proposed previously by Cox (1962).

The data were collected at flow meters and pressure gauges in hourly intervals. An example of on-off intervals is shown in Figs. 3a and 3b.

According to the 16 hour daily shift during the peak irrigation season there were 16×43 daily sampling points on a real time continuous basis. The period of the study started at mid July and ended at the end of August which is considered the peak irrigation season. During the irrigation season the data were found to be homogeneous with respect to the discharge of the hydrant characteristics. The data were then normalized in terms of pressure deviations in the network. The parameter of the exponential distribution is $\lambda=0.527\,h^{-1}$. The exponential distribution is accepted at the 10% significance level using the Kolmogorov–Smirnov test.

Based on equation (15) a computer programme was

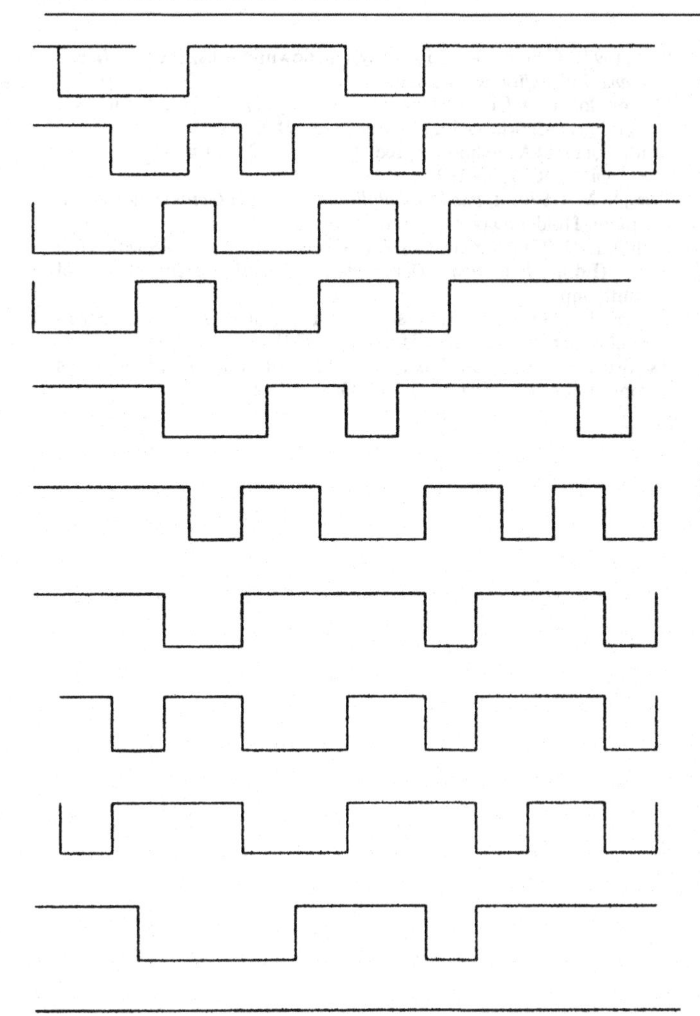

Fig. 3a Input 'on'-'off' states of hydrants operation, Monday 23 July 1989.

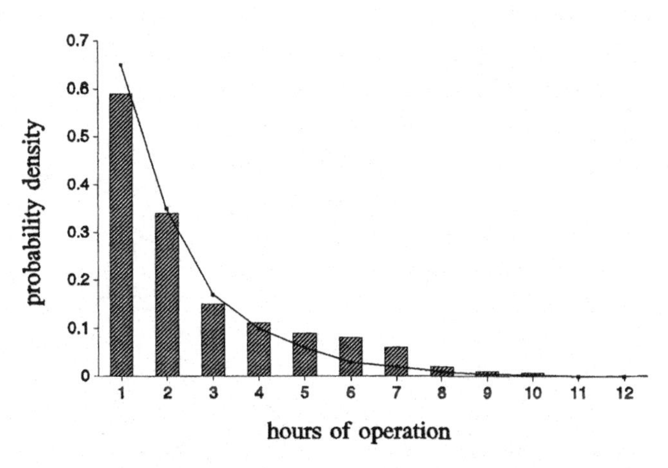

Fig. 3b Exponential probability density function of 'on' intervals.

written for the calculation of the cumulative probability of the operation of hydrants within any interval $(0, t)$. The results are presented in Fig. 4. As can be seen the resulting cumulative probability of operation of hydrants for four hours operation is 92%.

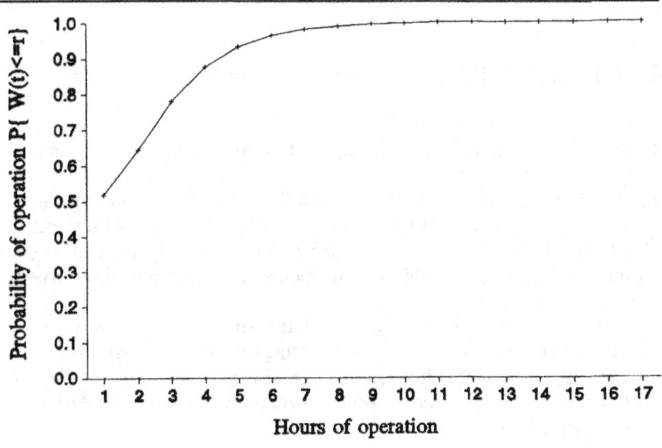

Fig. 4 Cumulative probability of operation $P\{W(t) \leq t\}$ for 18 hours' operation.

Based on the results of Fig. 4 the duration of a hydrant operation may be obtained associated with the corresponding level of risk. The information provided by the above procedure could be useful for the selection of the probability of hydrants operation as a design procedure for the design of collective trickle irrigation networks.

CONCLUDING REMARKS

It may be also concluded that the proposed methodology can be used as the basis for assessing the probability of hydrant operation and the design of pressurized systems of 'on demand' water distribution in areas in which field data as those presented in this study can be collected and analysed. Further extensive field work is needed under a variety of conditions (method of irrigation, farm unit area, farmer's behaviour etc.) in order to make the method operational for a variety of other conditions. As can be seen from the graphical results of the field application analyzed in the previous paragraphh the probability of duration of hydrant operation is a variable to be decided when designing a collective irrigation system. This conclusion is in contrast to the 'on demand' theory which accepts that this probability is calculated from data of the farm unit.

ACKNOWLEDGEMENTS

This study is a by-product of a project carried out at the Laboratory of Rural Technology of NTUA for the Ministry of Agriculture of the Hellenic Republic. Permission to publish this paper is acknowledged.

REFERENCES

Barlow, R. & Hunter, L. (1961) Reliability analysis of a one unit system. *Oper. Res.*, **9**, 200–8.

Buishand, T. A., (1977) *Stochastic modelling of daily rainfall sequences*, Commun., Agricultural University, Wageningen, The Netherlands.

Clemment, R. (1955) *Le calcul des debits dans les canalisations d'irrigation*, Association amicale des Ingenieurs du Genie Rural, Journees d'etudes sur l'irrigation.

Cox, D. R., (1962) *Renewal Theory*, Chapman and Hall, New York.

Delclaux, F. (1984) The role of a forecasting model in the regulation of a water supply system. In: Tsakiris, G. (Ed.), *Proceedings of the 5th International Conference on Water Resources Planning and Management*, Subject 5, pp. 121–35.

Li, T. (1971) On the calculation of system downtime distribution, *IEEE, Trans. Reliability*, R-**20**, 38–9.

McConalogue, D. J. (1978) Convolution integrals involving distribution functions (Algorithm 102). *Computer J.*, **21**, 270–2.

Muth, E. (1968) A method for predicting system downtime. *IEEE Trans. Reliability*, **R-17**, 97–102.

Ross, S. M. (1970) *Applied Probability Models with Optimization Applications*, Holden Day, San Francisco.

Svehlik Z. (1977) Estimation of Water Requirements. In: Rydzewski, J. R. (Ed.), *Irrigation Development Planning*, University of Southampton.

Takacs, L. (1957) On certain sojourn time problems in the theory of stochastic processes. *Acta Math.*, **8**, 169–91.

Tsakiris, G. (1988) Stochastic modelling of rainfall occurences in continuous time. *Hydrol. Sci. J.*, **33**(5), 437–47.

7 Order and disorder in hydroclimatological processes

K. KOWALSKI

Research Centre of Agricultural and Forest Environment Studies, Polish Academy of Sciences, Poznań, Poland

ABSTRACT The order of hydroclimatological processes has been evaluated with the help of the correlation time. Correlation time is a synthetic measure of temporal consistency of a signal. This measure has several interesting mathematical features. It allows the signal-to-noise-ratio and the spectral bandwidth to be determined. These features were used in the analysis of long time series of flows of a few European rivers, temperature and precipitation in Poznań.

INTRODUCTION

Autocorrelation function is one of important characteristics used in hydroclimatic series analysis. A correlation time lends itself as a convenient synthetic measure of order in the autocorrelation function. It allows the analyst to check if the order of a signal has changed in time and to compare orders of different processes.

The correlation time has interesting mathematical features both in temporal and in frequency domains. It allows the signal-to-noise and the spectral bandwidth to be determined.

Fig. 1 presents the autocorrelation functions of such hydroclimatic processes as river flows (Warta, Poznań and Göta, Sweden), precipitation in Poznań and solar activity, expressed by sunspot numbers.

THE CORRELATION TIME: DEFINITION AND MATHEMATICAL FEATURES

The correlation time will be understood in the present work as:

$$\tau_c = \int_0^{\infty} [\rho(\tau)]^2 d\tau \qquad (1)$$

where $\rho(\tau)$ is the autocorrelation function.

This follows one of alternative definitions of the correlation time (cf. Cempel, 1979, Kowalski, 1982). Alternatively, the use of the absolute value of the autocorrelation function rather than squared $\rho(\tau)$ in the integral (1), is advocated. Bendat (1981) called this measure *the correlation duration*, while Gottschalk (1977) used the term *time scale*.

The advantage of the formulation given as equation (1) is the existence of a simple relationship between the representations in the temporal and the spectral domains.

In the frequency domain the correlation time can be expressed with the help of the Parseval theorem as:

$$\tau_c = [1/(2^*\sigma^4)] * \int_{-\infty}^{\infty} G^2(f)\, df \qquad (2)$$

Assuming that the process considered is normal and with constant spectral density function with limited bandwith, ΔF, one can determine the spectral bandwidth as:

$$\Delta F = 1/[2^*(\tau_c)] \qquad (3)$$

In the case of sine function (complete order, $\Delta F = 0$), the correlation time is infinite, $\tau_c = \infty$, whereas in the other extreme case, of the white noise, the correlation time is $\tau_c = 0$.

The ratio of signal to noise can be easily expressed by the correlation time according to the formula:

$$r = S/N = -\tau_c/(\tau_c - c) \qquad (4)$$

where: c is the correlation time for ordered function and r is the signal-to-noise ratio.

The correlation time is a convenient aggregated measure of autocorrelation properties, that lend themselves well to comparison of deterministic parts of processes. very useful in detecting deterministic processes in noise. This deterministic part is considered as the order of a process. Correlation time is a of measure of a 'memory' of a signal. In the sequel some properties will be discussed of the correlation time given as equation (1) and of its sensitivity (5) in comparison to such measures as normalized moments of higher order:

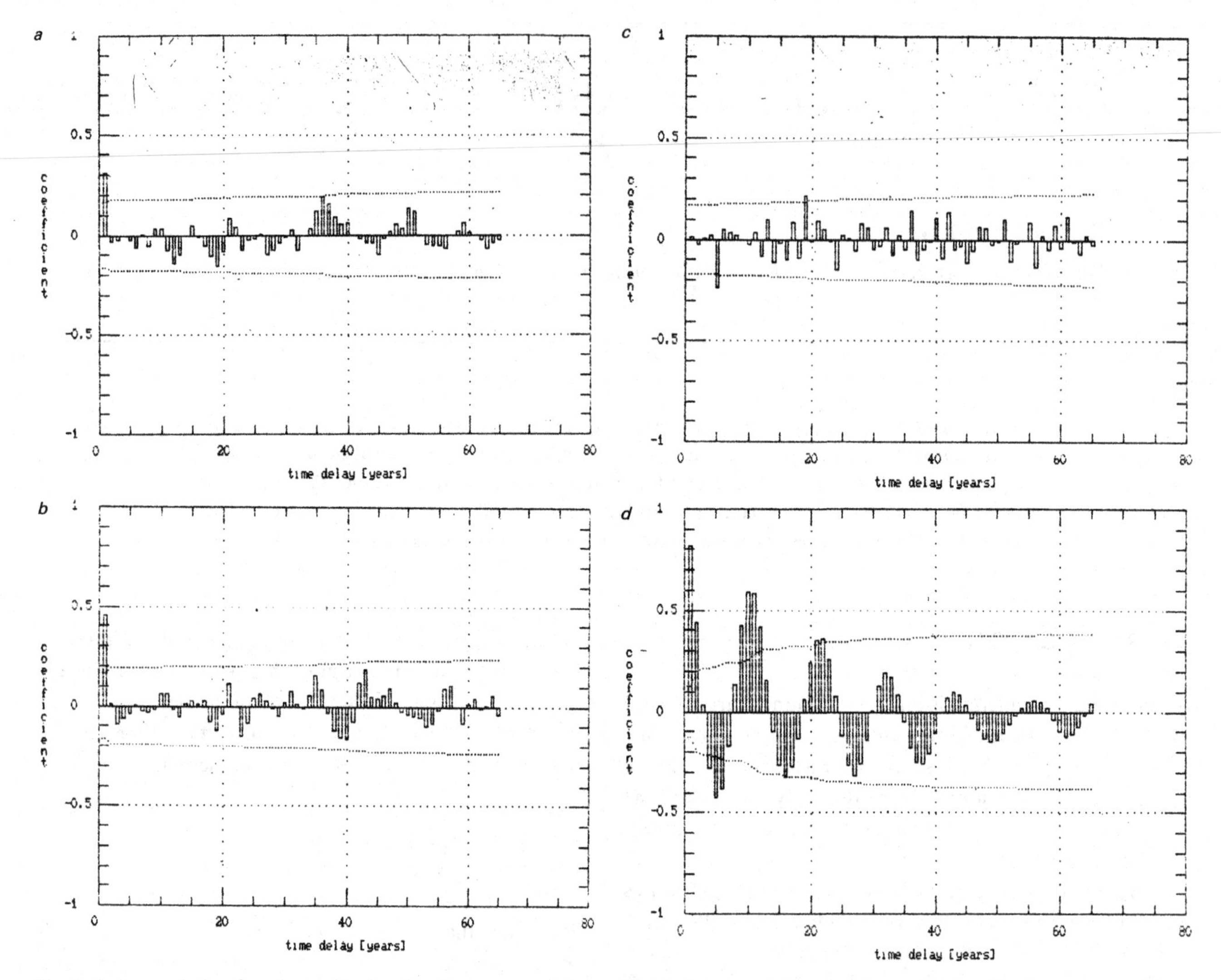

Fig. 1 Autocorrelation functions of hydroclimatic processes: (a) flows of the river Warta in Poznań, Poland; (b) flows of the river Göta in Sjotorp-Vannersburg, Sweden.

Fig. 1 Autocorrelation functions of hydroclimatic processes: (c) precipitation in Poznań; (d) solar activity expressed by sunspot number.

$$S_r^{\tau_c} = \frac{\partial \ln \tau_c}{\partial \ln r} \quad (5)$$

The sensitivity of correlation time function has the following properties:

(a) $\lim_{r \Rightarrow 0} S_r^{\tau_c} = 1;$ (6)

(b) $\lim_{r \Rightarrow \infty} S_r^{\tau_c} = 0;$ (7)

(c) sensitivity function $S_r^{\tau_c}$ is a monotonic function of the variable $r = S/N$;

(d) it does not depend on c.

Sensitivity functions of other statistical measures like moments of higher order $\gamma^{(n)}$, (e.g. skewness or kurtosis) do not have these features but rather (Kowalski, 1986):

(a) $\lim_{r \Rightarrow 0} S_r^{\gamma^{(n)}} = 0;$ (8)

(b) dependence of $S_r^{\gamma^{(n)}}$ is not a monotonic function of the variable $r = S/N$;

(c) $\lim_{r \Rightarrow \infty} S_r^{\gamma^{(n)}} = 0$ (9)

These differences between time correlation τ_c and other statistical measures $\gamma^{(n)}$ are clearly illustrated in Fig. 2. It can be seen that the correlation time performs well in detecting deterministic signal in noise, particularly for a low signal-to-noise ratio. Other statistical measures like normalized moments higher of order $\gamma^{(n)}$ do not have these features.

The correlation time depends on spectral moments like derivatives of different order of Rice frequency (Hjorth, 1972):

$$\tau_c = f(f_x, \ldots, f_x(n), l) \quad (10)$$

where

$$f_x(n) = 1/(2^*\pi)*[\sigma_x(n+1)]/[\sigma_x(n)] \quad (11)$$

is the Rice frequency of the process being nth order derivative of the process $x(t)$ and l is the integration interval.

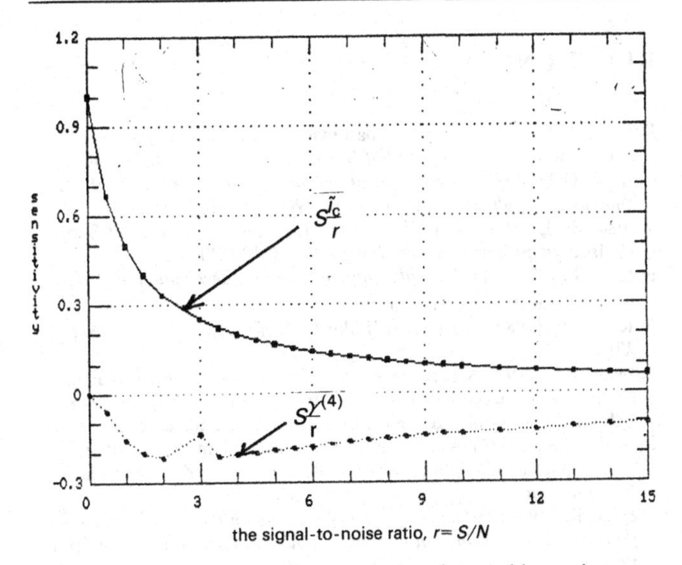

Fig. 2 Sensitivity function for correlation time and kurtosis as a function of the signal-to-noise ratio for a harmonic signal.

The important properties of the correlation time, related to the Rice frequencies and the integration interval read:

$$\lim_{f_x(n)\Rightarrow 0} S^{\tau_c}_{f_x(n)} = 0 \tag{12}$$

$$\lim_{l\Rightarrow 0} S^{\tau_c}_{l} = \infty \tag{13}$$

$$\lim_{l\Rightarrow \infty} S^{\tau_c}_{l} = 0 \tag{14}$$

ORDER AND DISORDER IN HYDROCLIMATIC PROCESSES

Using the correlation time as measure of order to a few long time series of geophysical processes such as: river flows, temperature, precipitation, and of solar activity, the results given in Table 1 were obtained. The correlation time was calculated with the integration interval l equal to 65 years. The data on river flows (except Warta) stem from Yevjevich (1963), the data on flows of the river Warta are taken from Olejnik (1985), and the data on solar activity originate from Vitinskij (1973).

Considering the geophysical processes mentioned in Table 1 one can note that the best ordered processes are the flows of the river Göta, whereas the flows of the Lithuanian river Nemunas are least ordered. Climatic processes in Poznań have similar correlation time to hydrological processes. An example of non-geophysical time series with very high correlation time is the solar activity, expressed as sunspot number.

Spectral bandwidth is largest for the flows of the river Rhine and smallest for the flows of the river Göta and for sunspot number.

It can be seen in Table 1 that the ratio S/N is about 1–2%.

Table 1. *Correlation time, the signal-to-noise ratio and spectral bandwidth for different hydroclimatic processes and solar activity*

Hydroclimatic process	Correlation time, τ_c [years]	The signal-to-noise ratio	Spectral bandwidth
river Warta in Poznań, Poland (1822–1988)	1.477	0.009	1.05
river Göta in Sjotorp Vannersburg, Sweden (1807–1957)	1.6	0.011	0.83
river Rhine in Basle, Switzerland (1807–1957)	1.3547	0.007	1.35
river Nemunas in Smalinakai, Lithuania (1811–1941)	1.32	0.007	1.28
precipitation in Poznań, Poland (1848–1988)	1.407	0.007	1.26
temperature in Poznań, Poland (1848–1988)	1.39	0.007	1.28
solar activity expressed as sunspot number (1848–1988)	6.5	0.11	0.09

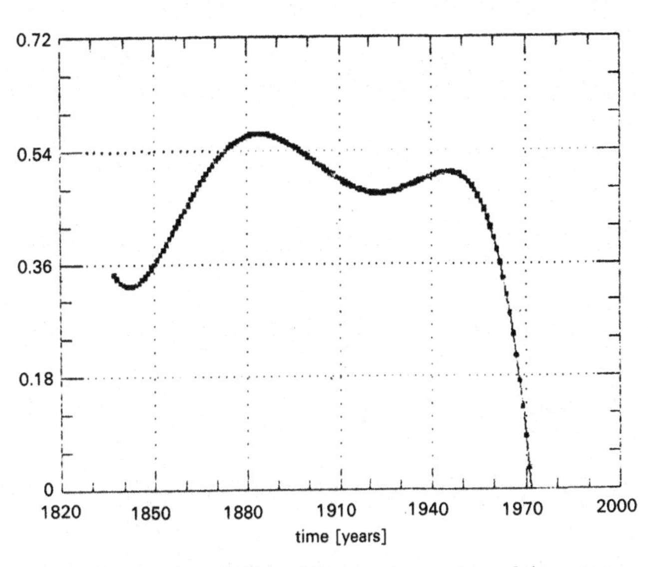

Fig. 3 Temporal variability of the correlation time of the process of flow of the river Warta.

This order of magnitude agrees with the results of analysis of levels of the Lake Victoria, where periodic component is not greater than 3%. (Kite, 1982).

That is, the 'ordered' processes are small parts of signals, and therefore a toolbox of subtle methods is required to study regularities and irregularities.

Fig. 3 shows the temporal variability of the correlation time of flow in the river Warta. The variability became stronger after the year 1950. It is not clear to what extent it has been caused by the climatic variability accompanying changes in chemical composition of the atmosphere (CO_2, SO_2 and other trace gases).

CONCLUSIONS

Correlation time seems to be a convenient aggregate measure of order of hydroclimatic processes. It lends itself well as a synthetic measure for comparison.

Hydroclimatic processes do not have a 'long memory' and their ratio of signal to noise is small (less than 2% of the power of the signal).

The order of hydroclimatic processes changes in time. The observed changes of the correlation time of the river Warta were particularly strong after about 1950, what may have something in common with the chemical changes in the atmosphere.

REFERENCES

Bendat, J. S. (1981) Spectral bandwidth, correlation duration and uncertainty relation. *Journal of Sound and Vibration*, **76**(1), 146–9.

Cempel, C. (1979) *Self-estimates of Vibroaccoustic Processes and Their Diagnostic Applications*, Ossolineum, Wrocław (in Polish).

Gottschalk L. (1977) Correlation Structure and Time Scale of Simple Hydrologic Systems. *Nordic Hydrology*, **8**, 129–40.

Hjorth, B. (1972) *The physical significance of time descriptions in E. E. G. analysis*, Electroencephalograph.

Kite, G. W. (1982) Analysis of Lake Victoria Levels. *Hydrol. Sci. J.*, **27**(2).

Kowalski, K. (1982) Correlation time and spectral bandwidth as measures of a vibroaccoustical signal, Abstract – *Proc. X Symposium on Vibrations in Physical Systems*, Błażejewko, Poland (in Polish).

Kowalski, K. (1986) Application of normalized moments of higher orders in accoustical diagnostics, Abstract – *Proc. XII Symposium on Vibrations in Physical Systems* (in Polish).

Olejnik, K. (1985) *Influence of cyclic processes on the nonstationarity of flows of the river Warta in Poznań*, Habilitation thesis (manuscript in Polish).

Yevjevich, V.M. (1963) *Fluctuation of wet and dry years*, Part I Papers, Colorado State University, Fort Collins.

Vitinskij, Ju. I. (1973) *Cyclicity and forecasts of solar activity*, Nauka, Leningrad, USA (in Russian).

8 Towards the physical structure of river flow stochastic process

J. J. NAPIÓRKOWSKI

Institute of Geophysics, Polish Academy of Sciences, Warsaw, Poland

W. G. STRUPCZEWSKI

Department of Civil Engineering, University of Calgary, Alberta, Canada T2N 1N4

ABSTRACT The transformation of white noise and Markov processes through the simplified St. Venant flood routing model is examined. This model has been derived from the linearized St. Venant equation for the case of a wide uniform open channel flow with arbitrary cross-section shape and friction law. The only simplification results in filtering out the downstream boundary condition. The cross-correlation and normalized autocorrelation functions are determined in analytical way.

INTRODUCTION

The development of water resources research has created the need for an extension of mathematical analysis of hydrological data. An awareness of the stochastic structure of hydrologic processes is necessary for modelling water resources systems. The aim of the paper is to investigate the physical structure of the process of outflow from a river reach.

The widely accepted assumption about a structure of an inflow process is that it can be considered as a sum of deterministic and random components. It is assumed that the input signal is weekly stationary (stationarity of the first two moments).

It is assumed that the system behaves linearly. This is the crude simplification granting the compromise between simplicity and accuracy. The structure of the random component transformed by some conceptual linear flood routing models (linear reservoir, Nash, Muskingum) was examined by Strupczewski *et al.* (1975a, b). Some of their results are easily available (e.g. Singh, 1988, p. 240). In the present paper the structure of the random component transformed by the flood routing model based on the St. Venant equations will be analyzed.

The rigorous hydrodynamic description of open channel flow (St. Venant model), requires two boundary conditions and in the case of a tranquil flow one of these is at the downstream end of the channel. In practical flood routing the influence of downstream controls is typically neglected and the routing takes part only in a downstream direction.

The hydrodynamic model used in this paper, called the rapid flow model (RFM), was developed by filtering out the downstream boundary condition to approximate the diffusion term in the St. Venant equations.

DERIVATION OF THE RAPID FLOW MODEL (RFM) FROM THE LINEARIZED ST. VENANT EQUATION

The findings presented in this paragraph borrow heavily from Strupczewski & Napiórkowski (1990). The linearized St. Venant equation for one-dimensional unsteady flow in uniform channel with arbitrary cross-section shape and either of the common friction laws may be written as (Dooge *et al.*, 1987a):

$$(1-F^2)g\bar{y}_0\frac{\partial^2 Q}{\partial x^2}-2v_0\frac{\partial^2 Q}{\partial x\partial t}-\frac{\partial^2 Q}{\partial t^2} = gA_0\left(\frac{\partial S_f}{\partial Q}\frac{\partial Q}{\partial t}-\frac{\partial S_f}{\partial A}\frac{\partial Q}{\partial x}\right) \quad (1)$$

where Q is the perturbation of flow about an initial condition of steady uniform flow Q_0, A_0 is the cross-sectional area corresponding to this flow, F is the Froude number, S_f is the friction slope, $\bar{y}_0$ is the hydraulic mean depth, v_0 is the mean velocity, S_0 is the bottom slope, x is the distance from the upstream boundary, t is the elapsed time and derivatives of the friction slope S_f are evaluated at the reference conditions.

The variation of the friction slope with discharge at the reference condition for either of the common frictional formulas for rough turbulent flow could be expressed as:

$$\frac{\partial S_f}{\partial Q}=2S_0/Q_0 \tag{2}$$

Define an auxilliary parameter m as the ratio of the kinematic wave speed to the average velocity of flow

$$m=c_k/(Q_0/A_0) \tag{3}$$

where the kinematic wave speed c_k is as given by Lighthill & Whitham (1955)

$$c_k=\frac{dQ}{dA}=-\frac{\partial S_f/\partial A}{\partial S_f/\partial Q} \tag{4}$$

The parameter m is a function of the shape of channel and of a friction law parameter.

Substituting equations (2)–(4) into equation (1) one obtains

$$(1-F^2)\frac{\bar{y}_0}{2mS_0}\frac{\partial^2 Q}{\partial x^2}-F^2\frac{\bar{y}_0}{c_k S_0}\frac{\partial^2 Q}{\partial x \partial t}-F^2\frac{my_0}{2c_k^2 S_0}\frac{\partial^2 Q}{\partial t^2} =\frac{\partial Q}{\partial x}+\frac{1}{c_k}\frac{\partial Q}{\partial t} \tag{5}$$

The linear equation (5) is a hyperbolic one, i.e. it has two real characteristics. The direction of these characteristics gives the celerity of both the primary and secondary waves. For Froude numbers less than 1, the celerity of secondary wave is in an upstream direction. In order to filter out the downstream boundary condition the small convective term (the first term in equation 5) can be neglected entirely. It provides the exact solution for Froude number equal to one. However, in order to increase the accuracy for the value of the Froude number close to one, one can represent the convective term in equation (5) on the basis of lower order approximation to the solution of the equation. This low order approximation is given by neglecting all terms on the left-hand side of equation (5) to obtain kinematic wave equation

$$\frac{\partial Q}{\partial x}=-\frac{1}{c_k}\frac{\partial Q}{Qt} \tag{6}$$

This lower order solution can be used to approximate the first term on the left-hand side of equation (7) in terms of: the second term:

$$\frac{\partial^2 Q}{\partial x^2}=-\frac{1}{c_k}\frac{\partial^2 Q}{\partial x \partial t} \tag{7}$$

the third term:

$$\frac{\partial^2 Q}{\partial x^2}=\frac{1}{c_k^2}\frac{\partial^2 Q}{\partial t^2} \tag{8}$$

or by the linear combination of the second and third terms:

$$\frac{\partial^2 Q}{\partial x^2}=-C_1\frac{1}{c_k}+C_2\frac{1}{c_k^2}\frac{\partial^2 Q}{\partial t^2} \tag{9}$$

where C_1 and C_2 are coefficient to be determined.

Note, that:

(i) equation (7) is a special case of equation (9) for $C_1=1$ and $C_2=0$;

(ii) equation (8) is a special case of equation (9) for $C_1=0$ and $C_2=1$;

(iii) the approximation based on entirely neglecting the diffusion term is for $C_1=0$ and $C_2=0$.

Substitution of approximations (9) into equation (5) gives the Rapid Flow Model (RFM) in the form

$$-\alpha\frac{\partial^2 Q}{\partial x \partial t}-\beta\frac{\partial^2 Q}{\partial t^2}=\frac{\partial Q}{\partial x}+\frac{1}{c_k}\frac{\partial Q}{\partial t} \tag{10}$$

On general grounds one could expect that the models based on the approximation of the diffusion term through the kinematic wave approximation would be preferable to the one in which this term is neglected. These general considerations are reinforced by comparing some properties of equation (10) with other known results in open channel hydraulics (Strupczewski & Napiórkowski, 1990). All forms of the RFM discussed will exactly predict the first moment or lag of the Linear Channel Response (LCR), i.e. the solution of the equation (1) for semi-infinite channel and for $F<1$. To get equivalence of second moments of the RFM and the LCR the coefficients C_1 and C_2 should fulfill the relation $C_1+C_2=1$, while for the additional equivalence of third moments $C_1=2$. It is suggested that any discussion of the applicability of the RFM should be confined to this form that preserves all three moments of the complete linear equation. Therefore the final values of the parameters α and β in equation (10) are:

$$\alpha=\frac{1}{mc_k}[1+(m-1)F^2]\frac{y_0}{S_0} \tag{11}$$

$$\beta=\frac{1}{2mc_k^2}[1+(m^2-1)F^2]\frac{y_0}{S_0} \tag{12}$$

Since the downstream boundary condition was filtered out from the St. Venant equation only upstream boundary condition $Q_u(t)=Q(0,t)$ is required to solve equation (10). Hence, all transfer properties of the hydrodynamic model described by equations (10)–(12) can be described by the impulse response given in the Fourier transform domain as:

$$H^{RFM}(x,j\omega)=\exp\left(-\Delta j\omega-\lambda+\frac{\lambda}{1+\alpha j\omega}\right) \tag{13}$$

where

$$\lambda=\frac{m}{2}\frac{1-(m-1)^2F^2}{[1+(m-1)F^2]^2}\frac{S_0}{y_0}x \tag{14}$$

$$\Delta = 0.5 \frac{1+(m^2-1)F^2}{1+(m-1)F^2} \frac{x}{c_k} \qquad (15)$$

The RFM impulse response in the time domain has a clear conceptual interpretation being the total of the products of the Poisson distribution

$$P_k(\lambda) = \frac{\lambda^k}{k!} \exp(-\lambda) \qquad (16)$$

and the impulse response of cascade of k-linear reservoirs (CLR) with a time constant α (equation 11)

$$h_{k,\alpha}^{\mathrm{CLR}}(t) = \frac{1}{\alpha(k-1)!} (t/\alpha)^{k-1} \exp(-t/\alpha) \qquad (17)$$

shifted in time by a time delay Δ given in equation (15). This impulse response is thus given by (cf. Strupczewski & Napiórkowski, 1990):

$$h^{\mathrm{RFM}}(x,t) = P_0(\lambda)\delta(t-\Delta) + \sum_{k=1}^{\infty} P_k(\lambda)\, h_{k,\alpha}^{\mathrm{CLR}}(t-\Delta)\, 1(T-\Delta) \qquad (18)$$

The upstream boundary condition is delayed by a linear channel with time lag Δ, divided according to the Poisson distribution with the mean λ, and then transformed by parallel cascades of equal linear reservoirs (with time constant α) of varying lengths.

The Rapid Flow Model can be considered as both, a conceptual and a physical one. On one hand it is a conceptual model with physically derived parameters. On the other it is a rigorous simplification of the linearized St. Venant equations. This simplification results in reducing the number of model parameters and filtering out the downstream boundary condition. The RFM can be applied to any length of channel reach. However, the quality of the Linear Channel Response approximation by the RFM depends on the type of motion, as discussed in Strupczewski & Napiórkowski (1990).

TRANSFORMATION OF STOCHASTIC PROCESSES IN THE RFM

In this section the transformation of stationary random processes in the RFM will be analyzed. This class of processes is important because stationarity provides the possibility of learning the statistical properties under various ergodicity hypotheses. Also the amount of information required to statistically describe stationary processes is greatly reduced. Finally, frequency-domain methods can be used in the analysis of the RFM with stationary input processes. White noise and Markovian noise are the processes assumed as the input stochastic processes in the analysis. They are commonly used in stochastic hydrology due to their simplicity and existing relationship to real processes.

If the stationary random process $X(t)$ is fed to a linear shift-invariant system with the impulse response $h(t)$, then the output random process can be expressed as the convolution integral

$$Y(t) = \int_{-\infty}^{+\infty} h(\tau)X(t-\tau)\mathrm{d}\tau \qquad (19)$$

Computing the cross-correlation function between the input process and the output process one finds:

$$R_{YX}(\tau) = \int_{-\infty}^{+\infty} h(\alpha)R_X(\tau-\alpha)\mathrm{d}\alpha = h(\tau)*R_X(\tau) \qquad (20)$$

Taking the Fourier transforms of both sides of equation (20) one obtains the frequency domain representation of the cross-correlation function

$$S_{YX}(\omega) = H(\omega)S_X(\omega) \qquad (21)$$

Finally, the autocorrelation function of the output is expressed as

$$R_Y(\tau) = \int_{-\infty}^{+\infty} R_{YX}(\tau+\alpha)h(\alpha)\mathrm{d}\alpha = R_{YX}(\tau)*h(-\tau) \qquad (22)$$

while in the spectral domain via Fourier transformation it becomes

$$S_Y(\omega) = S_{YX}(\omega)H^*(\omega) \qquad (23)$$

Combining the preceding results, one obtains a fundamental equation relating the autocorrelation function of the output to the autocorrelation function of the input

$$R_Y(\tau) = h(\tau)*R_X(\tau)*h(-\tau) \qquad (24)$$

which in the spectral domain takes the form:

$$S_Y(\omega) = |H(\omega)|^2 S_X(\omega) \qquad (25)$$

TRANSFORMATION OF WHITE NOISE IN THE RFM

Consider the RFM with a white noise input (X_1). The correlation function of the white noise process contains the Dirac delta impulse, i.e.:

$$R_{X_I}(\tau) = \sigma^2\delta(\tau) \qquad (26)$$

so that its power spectral density defined as its Fourier transform is a constant

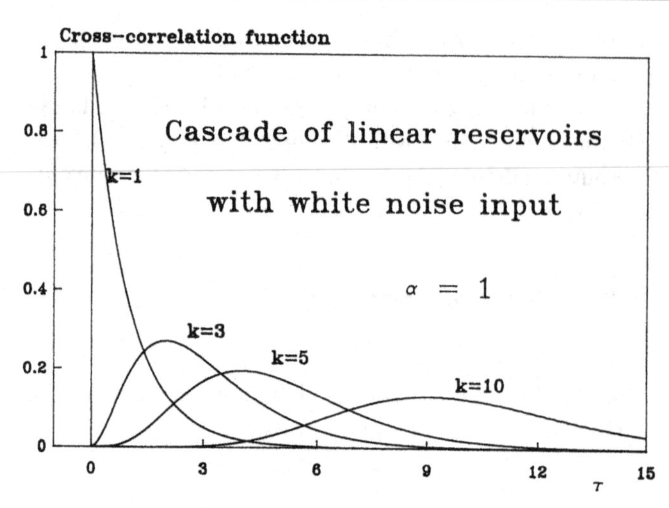

Fig. 1 Cross-correlation function of the cascade of linear reservoirs with white noise input.

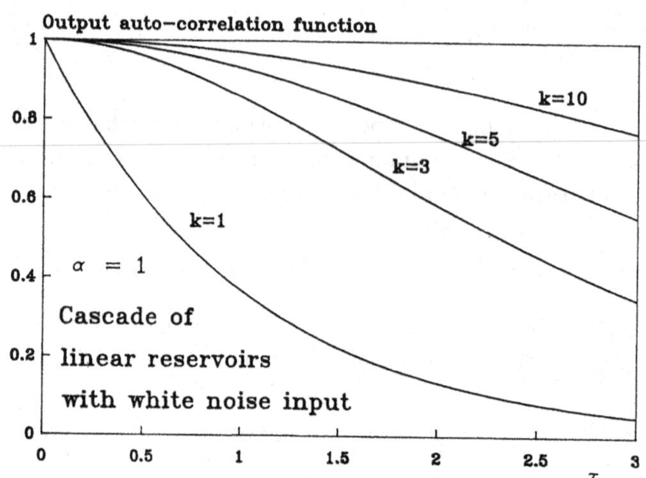

Fig. 2 Normalized auto-correlaton function of the cascade of linear reservoirs with white noise input.

$$S_{X_1}(\omega)=\sigma^2, \qquad -\infty<\omega<+\infty \tag{27}$$

For the RFM transfer function (transform of the impulse response in the Fourier domain) given by equation (13) one gets the following cross- and output-power spectral densities

$$\begin{aligned} S_{YX_1}^{\mathrm{RFM}}(x,\omega) &= H^{\mathrm{RFM}}(x,\omega)S_{X_1}(\omega) \\ &= \sigma^2 \exp\left(-\Delta j\omega-\lambda+\frac{\lambda}{1+\alpha j\omega}\right) \end{aligned} \tag{28}$$

$$S_{Y_1}^{\mathrm{RFM}}(x,\omega)=|H^{\mathrm{RFM}}(x,\omega)|^2 S_{X_1}(\omega)=\sigma^2 e^{-2\lambda}\exp\left(\frac{2\Lambda}{1+\alpha^2\omega^2}\right) \tag{29}$$

It is shown in Appendix A that:

(i) the inverse Fourier transform of equation (28) yields the cross-correlation function

$$R_{YX_1}^{\mathrm{RFM}}(x,\tau)=\begin{cases} P_0(\lambda)\delta(t-\Delta)+\sum_{k=1}^{\infty} P_k(\lambda)h_{k,\alpha}^{\mathrm{CLR}}(t-\Delta) & \tau\geq\Delta \\ 0 & \tau<\Delta \end{cases} \tag{30}$$

where $h_{k,\alpha}^{\mathrm{CLR}}(t)$, given by equation (17) and plotted in Fig. 1, is the impulse response of the CLR, that is cross-correlation function of the CLR with white noise input;

(ii) the inverse Fourier transform of equation (29) yields the output correlation function

$$R_{Y_1}^{\mathrm{RFM}}(x,\tau)=\sigma^2 P_0(2\lambda)\delta(\tau)+\sum_{k=1}^{\infty} P_k(2\lambda)\, R_{Y_1}^{\mathrm{CLR}}(k,\alpha,\tau) \tag{31}$$

where $P_k(2\lambda)$ is a Poisson distribution (equation 16) and

$$R_{Y_1}^{\mathrm{CLR}}(k,\alpha,\tau)=\sigma^2\frac{\exp(-|\tau|/\alpha)}{\alpha(k-1)!2^k}\sum_{i=0}^{k-1}\frac{(k+i-1)!}{(k-i-1)!2^i i!}(|\tau|/\alpha)^{k-i-1} \tag{32}$$

is the autocorrelation function of the output from the CLR with the white noise input (see Fig. 2).

It is interesting that the RFM cross-correlation function given by equation (30) is 0 for $\tau<\Delta$. This means that the output Y in time instant t is orthogonal to values of the input X_1 for $t\in(-\Delta,+\infty)$, which is a white noise. This occurs because of three reasons: The part of the model responsible for the modulatory performance is causal, another part of the model can be interpreted as a time shift (pure delay) and the input is a white noise. The system causalty requires that the output does not depend directly on future inputs but only depend directly on present and past inputs. The whiteness of the input X_1 guarantees that the past and present inputs will be uncorrelated with future inputs. Combining three conditions we see that there will be no cross-correlation between the present output and the inputs in time interval $(-\Delta,+\infty)$. If we assume additionally that the input is Gaussian, then the input process is an independent process and the output becomes independent of all future inputs and those in time interval $(-\Delta,0)$. So, the causality of the system prevents the direct dependence of the present output on future inputs, and the independent process input prevents any indirect dependence. These ideas are important to the theory of Markov process in next section.

It is convenient to illustrate the results in terms of dimensionless independent variables defined with the help of bottom slope S_0, the depth y_0, and the velocity v_0 for the steady uniform reference conditions about which perturbation are taken. Thus we can write:

$$x'=x\frac{S_0}{y_0} \tag{33}$$

$$t'=t\frac{v_0 S_0}{y_0} \tag{34}$$

Hence, the dimensionless parameters of transfer function are given respectively by:

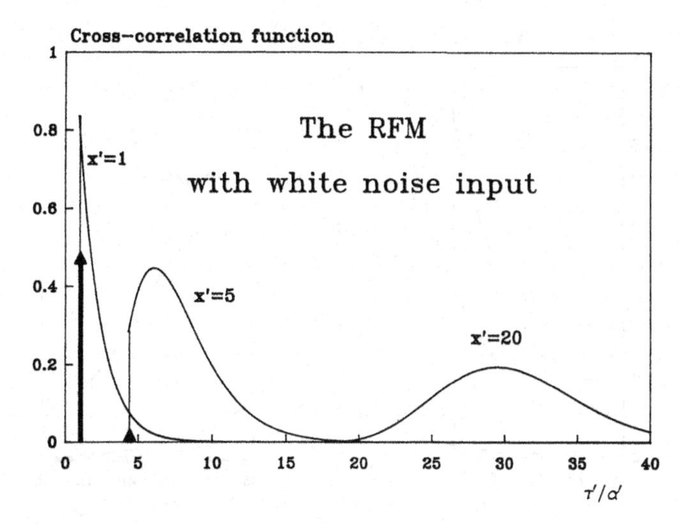

Fig. 3 Cross-correlation function of the Rapid Flow Model with white noise input.

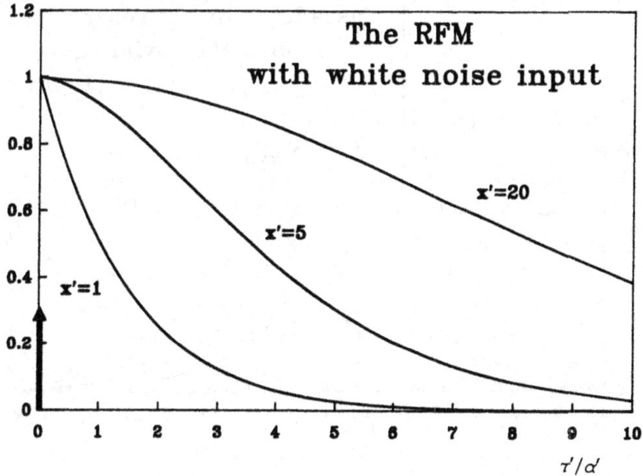

Fig. 4 Normalized auto-correlaton function of the Rapid Flow Model with white noise input.

$$\alpha' = \frac{1}{m^2}[1+(m-1)F^2 \tag{35}$$

$$\lambda' = \frac{m[1-(m-1)^2F^2]}{2[1+(m-1)F^2]^2}x' \tag{36}$$

$$\Delta' = \frac{1+(m^2-1)F^2}{2m[1+(m-1)F^2]}x' \tag{37}$$

For illustration the flow in a broad rectangular channel with Manning friction ($m = 5/3$) and the Froude number $F = 0.3$ ($\alpha' = 0.381$) will be considered. Figs. 3 and 4 show the cross- and output-correlation functions of the Rapid Flow Model described by eqs.30 and 32 for wide rectangular channel of dimensionless lengths: $x' = 1$ (short channel), $x' = 5$ (intermediate channel), and $x' = 20$ (long channel). Both figures are drawn in function of dimensionless time τ'/α'.

TRANSFORMATION OF MARKOVIAN NOISE IN THE RFM

In this section the RFM with Markovian noise input is considered. The correlation function of normal Markovian noise (X_2) is given by:

$$R_{X_2}(\tau) = D^2\exp(-c|\tau|), \qquad -\infty < \tau < +\infty \tag{38}$$

where D^2 is the variance of the input process, so its power spectral density is described by

$$S_{X_2}(\omega) = \frac{2D^2c}{c^2+\omega^2}, \qquad -\infty < \omega < +\infty \tag{39}$$

Accordingly, we have the following cross- and output-power spectral densities

$$S_{YX_2}^{\text{RFM}}(x,\omega) = H^{\text{RFM}}(x,\omega)S_{X_2}(\omega) = \exp\left(-\Delta j\omega - \lambda + \frac{\lambda}{1+\alpha j\omega}\right)\frac{2D^2c}{c^2+\omega^2} \tag{40}$$

$$S_{Y_2}^{\text{RFM}}(x,\omega) = |H^{\text{RFM}}(x,\omega)|^2 S_{X_2}(\omega) = e^{-2\lambda}\exp\left(\frac{2\lambda}{1+\alpha^2\omega^2}\right)\frac{2D^2c}{c^2+\omega^2} \tag{41}$$

It is shown in Appendix B that:

(a) the inverse Fourier transform of equation 40 yields the cross-correlation function

$$R_{YX_2}(x,\tau) = P_0(\lambda)D^2e^{-c|\tau-\Delta|} + \sum_{k=1}^{\infty} P_k(\lambda)R_{YX_2}^{\text{CLR}}(k,\alpha,\tau-\Delta) \tag{42}$$

where

$$R_{YX_2}^{\text{CLR}}(k,\alpha,\tau) = \begin{cases} \dfrac{D^2e^{-c\tau}}{(1-\alpha c)^k} + D^2e^{-\tau/\alpha}\displaystyle\sum_{i=0}^{k-1}\frac{(\tau/\alpha)^i}{i!}\left[\frac{1}{(1+\alpha c)^{k-i}} - \frac{1}{(1-\alpha c)^{k-i}}\right] & \tau \geq 0 \\ \dfrac{D^2e^{c\tau}}{(1+\alpha c)^k}; \quad \tau < 0 \end{cases} \tag{43}$$

is the cross-correlation function for the cascade of linear reservoirs with Markovian noise input and is plotted in Fig. 5;

(b) the inverse Fourier transform of equation (41) yields the output correlation function

$$R_{Y_2}^{\text{RFM}}(x,\tau) = P_0(2\lambda)D^2e^{-c|\tau|} + \sum_{k=1}^{\infty} P_k(2\lambda)\, R_{Y_2}^{\text{CLR}}(k,\alpha,\tau) \tag{44}$$

where $P_k(2\lambda)$ is a Poisson distribution and

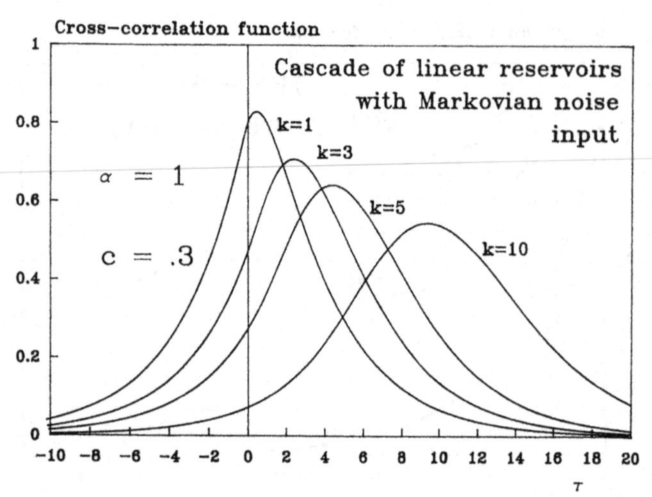

Fig. 5 Cross-correlation function of the cascade of linear reservoirs with Markovian noise input.

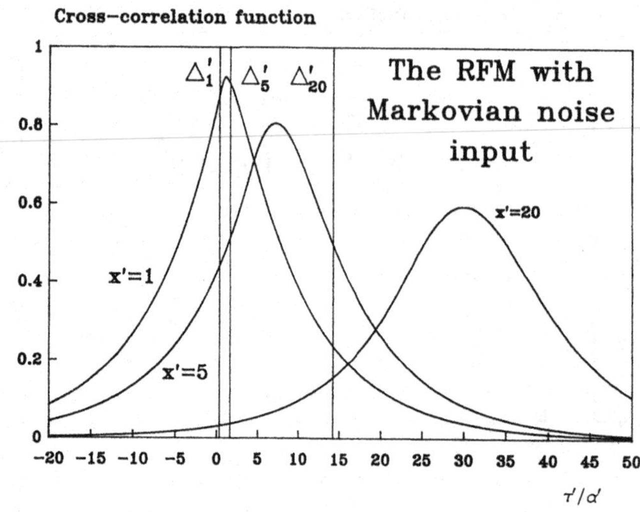

Fig. 7 Cross-correlation function of the Rapid Flow Model with Markovian noise input.

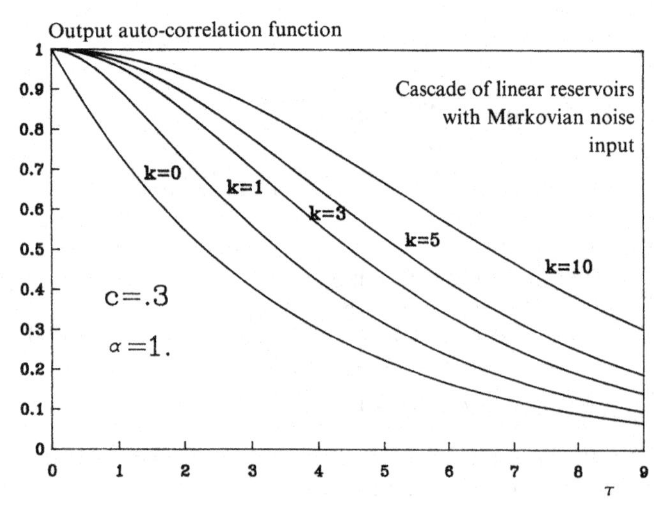

Fig. 6 Normalized auto-correlation function of the cascade of linear reservoirs with Markovian noise input.

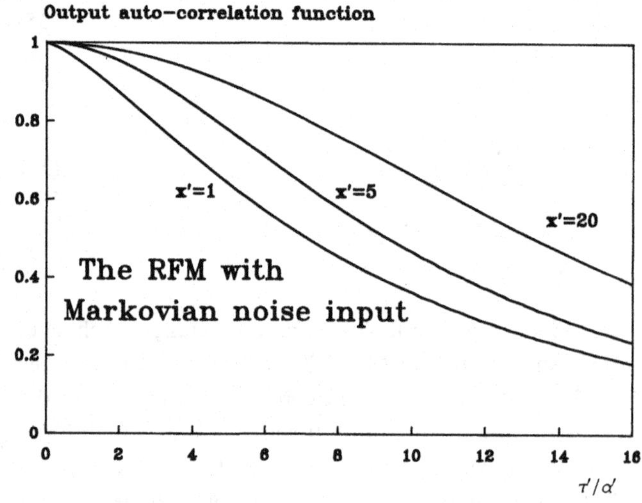

Fig. 8 Normalized auto-correlaton function of the Rapid Flow Model with Markovian noise input.

$$R_{Y_2}^{\mathrm{CLR}}(k,\alpha,\tau)=\frac{D^2 e^{-c|\tau|}}{(1-\alpha c)^k(1+\alpha c)^k}+\frac{D^2 e^{-|\tau|/\alpha}}{(k-1)!}\sum_{i=0}^{k-1}\left[\frac{1}{(1+\alpha c)^{k-i}}-\frac{1}{(1-\alpha c)^{k-i}}\right]\sum_{j=0}^{i}\frac{(k+j-1)!(|\tau|/\alpha)^{i-j}}{(i-j)!j!\,2^{k+j}} \tag{45}$$

is the output-correlation function of the cascade of linear reservoirs with Markovian noise input (see Fig. 6).

Figs. 7 and 8 show the cross- and output-correlation functions of the Rapid Flow Model described by equations (42) and (44) for wide rectangular channel of dimensionless lengths: $x'=1$ (short channel), $x'=5$ (intermediate channel), and $x'=20$ (long channel) with the Manning friction $m=5/3$, Froude number $F=0.3$, and dimensionless parameter of Markovian noise $c'=0.3$. Both figures are drawn in function of dimensionless time τ'/α'.

CONCLUSIONS

The cross- and auto-correlation functions are derived in the analytical way for the simplified linearized St. Venant model with upstream control only, i.e. for the Rapid Flow Model, with white noise and Markovian inputs. Obtained functions are much more complicated than those of the cascade of linear reservoirs. It is also possible to obtain time averaged results in an analytical way.

In the case of the white noise input one can see that the output autocorrelation function is considerably weaker than the one observed in nature (Fig. 4). Consider a wide rectangular channel with depth $y_0 = 1.4$ m, the Manning friction coefficient $n = 0.03$ and the Froude number $F = 0.3$, i.e. the dimensionless α' parameter equal to 0.381. Then the mean velocity and the bottom slope are respectively equal to $v_0 = 1.1$ m/s, and $S_0 = 0.0007$. Accordingly, independent variables x and t are related to the dimensionless variables as follows

$$x = 2000\, x'[\mathrm{m}] \tag{33a}$$

$$t = 12\, t'[\mathrm{min}] \tag{34b}$$

Hence the values of the dimensionless lengths: $x' = 1$, $x' = 5$ and $x' = 20$ in Fig.4 correspond to 2 km, 10 km and 40 km respectively, and the unit of dimensionless time $\tau'/\alpha' = \tau/\alpha$ corresponds to 12 min.

For example, for hourly observations and 40 km length of a reach (above values correspond approximately to $x' = 20$ and $\tau'/\alpha' = 5$) the value of the output auto-correlation function is less than 0.7, while in a real system it is generally greater than 0.99. For the analyzed Markovian noise input and the same flow conditions the value of normalized autocorrelation function is 0.9. It proves, that channel impact on autocorrelation function of river process is strongly limited to a short time. Therefore the long term auto-correlation observed in the river processes must be caused by other contributing processes like watershed alimentation, surface and subsurface runoff.

Streamflow data used in the time series analysis are usually pulse data. In order to make theoretical and empirical results comparable Strupczewski *et al.* (1975b) working on conceptual flood routing models covered also such case. At the cost of additional algebra it is possible to account it also for the RFM with white noise and Markovian inputs and to obtain the cross- and auto-correlation functions for any length of the discretization interval.

More realistically, one can consider a stream network rather than a single river reach with multidimensional input, correlated in space or both in space and time, or alternatively the RFM by complete linear St.Venant equation solved for semiinfinite channel or one can try to tackle with a complex system having rainfall as input. It is obvious that any attempt to an extension of generalization of the presented model would lead to further complication of analytical solution.

Responding to the question, whether the aspiration to establish the physical structure of the stochastic process of river flow is justified, is left to the reader.

REFERENCES

Dooge, J. C. I., Napiórkowski, J. J. & Strupczewski, W. G. (1987) The linear downstream response of a generalized uniform channel, *Acta Geophysica Polonica*, **35**(3), 279–93.

Lighthill, M. H. & Whitham, G. B. (1955) On kinematic waves, 1. Flood movements in long rivers. *Proc. R. Soc. Lond.*, **A229**, 281–316.

Napiórkowski, J. J. & Dooge, J. C. I (1988) Analytical solution of channel flow model with downstream control. *Hydrol. Sci. J.*, **33**, 3.

Singh V. P. (1988) *Hydrologic Systems, Vol. I, Rainfall-runoff Modeling*, Prentice-Hall, Englewood Cliffs, New Jersey, USA.

Stark, H. & Woods, J. W. (1986) *Probability, Random Processes, and Estimation Theory for Engineers*. Prentice-Hall, Englewood Cliffs, New Jersey 07632, USA.

Strupczewski, W. G., Kiczko, R. J., Kundzewicz, Z. W., Napiórkowski, J. J. & Mitosek, H. T. (1975) Stochastic properties of the processes transformed in linear hydrological systems, *Proceedings of the Bratislava Symposium on Application of Mathematical Models in Hydrology and Water Resources Systems*, IAHS Publ. No. 115, pp. 231–7.

Strupczewski, W. G., Kiczko, R. J., Kundzewicz, Z. W., Napiórkowski, J. J. & Mitosek, H. T. (1975) Transformation of the processes in the linear hydrological systems, *Proceedings of the Second World Congress, International Water Resources Association*, New Delhi, India, **5**, 33–43.

Strupczewski, W. G. & Napiórkowski, J. J., 1990. Linear flood routing model for rapid flow. *Hydrol. Sci. J.*, **35**(1), 2.

APPENDIX A. DERIVATION OF CROSS- AND OUTPUT-CORRELATION FUNCTION FOR THE RFM WITH WHITE NOISE INPUT

Cross-correlation function

For the case of the RFM with white noise input X_1 the following equation for the cross-power spectral density (equation 28) holds:

$$S_{YX_1}^{\mathrm{RFM}}(x,\omega) = \sigma^2 \exp\left(-\Delta j\omega - \lambda + \frac{\lambda}{1+\alpha j\omega}\right) \tag{A1}$$

Expanding equation (A1) into a convergent series and operating on it term by term one obtains

$$S_{YX_1}^{\mathrm{RFM}}(x,\omega) = = \sigma^2 e^{-\lambda} e^{-\Delta j\omega} \sum_{k=0}^{\infty} \frac{\lambda^k}{k!} \frac{1}{(1+\alpha j\omega)^k} \tag{A2}$$

Recalling the definition of the Poisson distribution (equation 16) and the system function of the cascade of k-linear reservoirs, namely

$$H_{k,\alpha}^{\mathrm{CLR}}(\omega) = \frac{1}{(1+j\omega\alpha)^k} \tag{A3}$$

equation (A2) may be rewritten in the form

$$S_{YX_1}^{\mathrm{RFM}}(x,\omega) = \sigma^2 P_0(\lambda) e^{-\Delta j\omega} \sum_{k=1}^{\infty} \sigma^2 P_k(\lambda)\, e^{-\Delta j\omega} H_{k,\alpha}^{\mathrm{CLR}}(\omega) \tag{A4}$$

Applying the translation theorem one gets the cross-correlation function

$$R_{YX_1}^{\mathrm{RFM}}(x,\tau) = \begin{cases} \sigma^2 P_0(\lambda)\delta(\tau-\Delta) + \sum_{k=1}^{\infty} \sigma^2 P_k(\lambda) h_{k,\alpha}^{\mathrm{CLR}}(\tau-\Delta) & \tau \geq \Delta \\ 0 & \tau < \Delta \end{cases} \tag{A5}$$

where the impulse response of the cascade of k-linear reservoirs $h_{k,\alpha}^{\mathrm{CLR}}(t)$ is given by equation (17).

Output-correlation function

For the case of the RFM with white noise input the following equation for the power spectral density of the output signal (equation 29) holds:

$$S_{Y_1}^{RFM}(x,\omega)=\sigma^2 e^{-2\lambda}\exp\left(\frac{2\lambda}{1+\alpha^2\omega^2}\right) \quad (A6)$$

Expanding equation (A6) into a series one gets

$$S_{Y_1}^{RFM}(x,\omega)=\sigma^2 e^{-2\lambda}\sum_{k=0}^{\infty}\frac{(2\lambda)^k}{k!}\frac{1}{(1+\alpha^2\omega^2)^k} \quad (A7)$$

Since the spectral density of the output signal for the cascade of linear reservoirs with white noise input is given by

$$S_{Y_1}^{CLR}(k,\alpha,\omega)=\sigma^2|H_{k,\alpha}^{CLR}(\omega)|^2=\frac{\sigma^2}{(1+\alpha^2\omega^2)^k} \quad (A8)$$

equation (A7) may be rewritten in a way similar to cross-correlation function as

$$S_{Y_1}^{RFM}(x,\omega)=\sigma^2 P_0(2\lambda)+\sum_{k=1}^{\infty}P_k(2\lambda)\, S_{Y_1}^{CLR}(k,\alpha,\omega) \quad (A9)$$

and the RFM with white noise input has the autocorrelation function of the output signal given by

$$R_{Y_1}^{RFM}(x,\tau)=\sigma^2 P_0(2\lambda)\delta(\tau)+\sum_{k=1}^{\infty}P_k(2\lambda)\, R_{Y_1}^{CLR}(k,\alpha,\tau) \quad (A10)$$

Hence, it remains to invert equation (A8) from the Fourier-transform domain to original domain.

Applying the residue method to evaluate the inverse Fourier transformation (Stark & Woods, 1986), rewrite the right hand side of equation (A8) in terms of complex variable $j\omega$ to obtain

$$S_{Y_1}^{CLR}(k,\alpha,j\omega)=\frac{\sigma^2}{[1-\alpha^2(j\omega)^2]^k} \quad (A11)$$

Replacing $j\omega$ by s one extends the function $S(j\omega)$ to the entire complex plane (two-sided Laplace transform of the correlation function)

$$S_{Y_1}^{CLR}(k,\alpha,s)=\frac{\sigma^2}{(1+\alpha s)^k(1-\alpha s)^k} \quad (A12)$$

Evaluating the residues for positive τ (kth order pole, counterclockwise traversal of the contour), one gets

$$\begin{aligned}R_{Y_1}^{CLR}(k,\alpha,\tau)&=\mathrm{Res}[S_{Y_1}^{CLR}(k,\alpha,s)e^{s\tau};\ s=-1/\alpha]\\&=\frac{1}{(k-1)!}\frac{d^{k-1}}{ds^{k-1}}\left[S_{Y_1}^{CLR}(k,\alpha,s)e^{s\tau}(s+1/\alpha)\right]_{s=-1/\alpha}\\&=\sigma^2\frac{\exp(-\tau/\alpha)}{\alpha(k-1)!2^k}\sum_{i=0}^{k-1}\frac{(k+i-1)!}{(k-i-1)!\,2^i i!}(\tau/\alpha)^{k-i-1}\end{aligned} \quad (A13)$$

while for negative τ (clockwise traversal of the contour) one gets

$$\begin{aligned}R_{Y_1}^{CLR}(k,\alpha,\tau)&=-\mathrm{Res}[S_{Y_1}^{CLR}(k,\alpha,s)e^{s\tau};\ s=1/\alpha]\\&=-\frac{1}{(k-1)!}\frac{d^{k-1}}{ds^{k-1}}\left[S_{Y_1}^{CLR}(k,\alpha,s)e^{s\tau}(s-1/\alpha)^k\right]_{s=1/\alpha}\\&=\sigma^2\frac{\exp(\tau/\alpha)}{\alpha(k-1)!2^k}\sum_{i=0}^{k-1}\frac{(k+i-1)!}{(k-i-1)!\,2^i i!}(-\tau/\alpha)^{k-i-1}\end{aligned} \quad (A14)$$

combining the results into a single formula, one gets

$$R_{Y_1}^{CLR}(k,\alpha,\tau)=\sigma^2\frac{\exp(-|\tau|/\alpha)}{\alpha(k-1)!2^k}\sum_{i=0}^{k-1}\frac{(k+i-1)!}{(k-i-1)!\,2^i i!}(|\tau|/\alpha)^{k-i-1} \quad (A15)$$

APPENDIX B. DERIVATION OF CROSS- AND OUTPUT-CORRELATION FUNCTION FOR THE RFM WITH MARKOVIAN NOISE INPUT

Cross-correlation function

For the RFM with Markovian noise input X_2 one gets the following cross-power spectral density (equation 40)

$$S_{YX_2}^{RFM}(x,\omega)=\exp\left(-\Delta j\omega-\lambda+\frac{\lambda}{1+\alpha j\omega}\right)\frac{2D^2c}{c^2+\omega^2} \quad (B1)$$

Expanding equation (B1) into a series one gets

$$S_{YX_2}^{RFM}(x,\omega)=e^{-\lambda}e^{-\Delta j\omega}\sum_{k=0}^{\infty}\frac{\lambda^k}{k!}\frac{1}{(1+\alpha j\omega)^k}\frac{2D^2c}{(c+j\omega)(c-j\omega)} \quad (B2)$$

Since the cross-power spectral density for the cascade of linear reservoirs with Markovian noise input is given by

$$S_{YX_2}^{CLR}(k,\alpha,\omega)=\frac{1}{(1+\alpha j\omega)^k}\frac{2D^2c}{(c+j\omega)(c-j\omega)} \quad (B3)$$

equation (B3) may be rewritten as

$$S_{YX_2}^{RFM}(x,\omega)=P_0(\lambda)e^{-\Delta j\omega}S_{X_2}(\omega)+\sum_{k=1}^{\infty}P_k(\lambda)\,e^{-\Delta j\omega}S_{YX_2}^{CLR}(k,\alpha,\omega) \quad (B4)$$

Applying the translation theorem one gets the cross-correlation function for the RFM with Markovian noise input

$$R_{YX_2}^{RFM}(x,\tau)=P_0(\lambda)D^2e^{-c|\tau-\Delta|}+\sum_{k=1}^{\infty}P_k(\lambda)\,R_{XY_2}^{CLR}(k,\alpha,\tau-\Delta) \quad (B5)$$

It remains to invert equation (B3) from the Fourier-transform domain to original domain. Applying the residue method to evaluate the inverse Fourier transformation, we replace $j\omega$ by s to extend the function $S(j\omega)$ to entire complex plane

$$S_{YX_2}^{CLR}(k,\alpha,s)=\frac{1}{(1+\alpha s)^k}\frac{2D^2c}{(c+s)(c-s)} \quad (B6)$$

Evaluating the residues for positive τ (one first order pole, one kth order pole, counterclockwise traversal of the contour), one gets

$$R_{YX}^{CLR}(k,\alpha,\tau)=\mathrm{Res}[S_{YX}^{CLR}(k,\alpha,s)e^{s\tau};\ s=-1/\alpha]+\mathrm{Res}[S_{YX_2}^{CLR}(k,\alpha,s)e^{s\tau};\ s=-c] \quad \text{(B7)}$$

Evaluating the residue for $s=-c$, one gets

$$\mathrm{Res}[S_{YX_2}^{CLR}(k,\alpha,s)e^{s\tau};\ s=-c]=\frac{D^2}{(1-\alpha c)^k}e^{-c\tau} \quad \text{(B8)}$$

while for $s=-1/\alpha$ one has

$$\begin{aligned}\mathrm{Res}[S_{YX_2}^{CLR}(k,\alpha,s)e^{s\tau};\ s=-1/\alpha]&=\frac{1}{(k-1)!}\frac{d^{k-1}}{ds^{k-1}}\{S_{YX_2}^{CLR}(k,\alpha,s)\,e^{s\tau}(s+1/\alpha)^k\}_{s=-1/\alpha}\\&=\frac{2D^2c}{(k-1)!\alpha^k}\frac{d^{k-1}}{ds^{k-1}}\{(c+s)^{-1}(c-s)^{-1}e^{s\tau}\}_{s=-1/\alpha}\\&=D^2e^{-\tau/\alpha}\sum_{i=0}^{k-1}\frac{(\tau/\alpha)^i}{i!}\left[\frac{1}{(1+\alpha c)^{k-i}}-\frac{1}{(1-\alpha c)^{k-i}}\right]\end{aligned} \quad \text{(B9)}$$

Evaluating the residue for negative τ (one first order pole, clockwise traversal of the contour), one gets

$$R_{YX_2}^{CLR}(k,\alpha,\tau)=-\mathrm{Res}[S_{YX_2}^{CLR}(k,\alpha,s)e^{s\tau};\ s=c]=\frac{D^2e^{c\tau}}{(1+\alpha c)^k} \quad \text{(B10)}$$

Combining equations (B7)–(B9) one gets the cross-correlation function of the cascade of linear reservoirs with Markovian noise input as

$$R_{YX_2}^{CLR}(k,\alpha,\tau)=\begin{cases}\dfrac{D^2e^{-c\tau}}{(1-\alpha c)^k}+D^2e^{-\tau/\alpha}\displaystyle\sum_{i=0}^{k-1}\frac{(\tau/\alpha)^i}{i!}\left[\frac{1}{(1+\alpha c)^{k-i}}-\frac{1}{(1-\alpha c)^{k-i}}\right] & \tau\geq 0\\[2ex] \dfrac{D^2e^{c\tau}}{(1+\alpha c)^k}; & \tau<0\end{cases} \quad \text{(B11)}$$

Output-correlation function

For the case of the RFM with Markovian noise input the following output spectral density is obtained (equation 41):

$$S_{Y_2}^{RFM}(x,\omega)=e^{-2\lambda}\exp\left(\frac{2\lambda}{1+\alpha^2\omega^2}\right)\frac{2D^2c}{c^2+\omega^2} \quad \text{(B12)}$$

Expanding equation (B12) into a series one gets

$$S_{Y_2}^{RFM}(x,\omega)=\frac{2D^2c}{c^2+\omega^2}e^{-2\lambda}\sum_{k=0}^{\infty}\frac{(2\lambda)^k}{k!}\frac{1}{(1+\alpha^2\omega^2)^k} \quad \text{(B13)}$$

Since the output-power spectral density for the cascade of linear reservoirs with Markovian noise input is given by

$$S_{Y_2}^{CLR}(k,\alpha,\omega)=|H_{k,\alpha}^{CLR}(\omega)|^2S_{X_2}(\omega)=\frac{1}{(1+\alpha^2\omega^2)^k}\frac{2D^2c}{c^2+\omega^2} \quad \text{(B14)}$$

equation (B13) may be rewritten as

$$S_{Y_2}^{RFM}(x,\omega)=P_0(2\lambda)S_{X_2}(\omega)+\sum_{k=1}^{\infty}P_k(2\lambda)\,S_{Y_2}^{CLR}(k,\alpha,\omega) \quad \text{(B15)}$$

and the RFM with Markovian noise input has the correlation function of output given by

$$R_{Y_2}^{RFM}(x,\tau)=P_0(2\lambda)D^2e^{-c|\tau|}+\sum_{k=1}^{\infty}P_k(2\lambda)\,R_{Y_2}^{CLR}(k,\alpha,\tau) \quad \text{(B16)}$$

It remains to invert equation (B14) from the Fourier-transform domain to original domain. Applying the residue method to evaluate the inverse Fourier transformation, one can replace $j\omega$ by s to extend the function $S(j\omega)$ to entire complex plane

$$S_{Y_2}^{CLR}(k,\alpha,s)=\frac{2D^2c}{(c+s)(c-s)}\frac{1}{(1+\alpha s)^k(1-\alpha s)^k} \quad \text{(B17)}$$

Evaluating the residues for positive τ (one first order pole, one kth order pole, counterclockwise traversal of the contour), one gets

$$R_{Y_2}^{CLR}(k,\alpha,\tau)=\mathrm{Res}[S_{Y_2}^{CLR}(k,\alpha,s)e^{s\tau};\ s=-1/\alpha]+\mathrm{Res}[S_{Y_2}^{CLR}(k,\alpha,s)e^{s\tau};\ s=-c] \quad \text{(B18)}$$

Evaluating the residue for $s=-c$, one gets

$$\mathrm{Res}[S_{YX_2}^{CLR}(k,\alpha,s)e^{s\tau};\ s=-c]=\frac{D^2e^{-c\tau}}{(1-\alpha c)^k(1+\alpha c)^k} \quad \text{(B19)}$$

while for $s=-1/\alpha$

$$\begin{aligned}\mathrm{Res}[S_{Y_2}^{CLR}(k,\alpha,s)e^{s\tau};\ s=-1/\alpha]&=\frac{1}{(k-1)!}\frac{d^{k-1}}{ds^{k-1}}\left\{S_{Y_2}^{CLR}(k,\alpha,s)e^{s\tau}\left(s+\frac{1}{\alpha}\right)^k\right\}_{s=-\frac{1}{\alpha}}\\&=\frac{2D^2c}{(k-1)!\alpha^k}\frac{d^{k-1}}{ds^{k-1}}\{(c+s)^{-1}(c-s)^{-1}(1-\alpha s)^{-k}e^{s\tau}\}_{s=-1/\alpha}\\&=\frac{D^2e^{-\tau/\alpha}}{(k-1)!}\sum_{i=0}^{k-1}\left[\frac{1}{(1+\alpha c)^{k-i}}-\frac{1}{(1-\alpha c)^{k-i}}\right]\sum_{j=0}^{i}\frac{(k+j-1)!(\tau/\alpha)^{i-j}}{(i-j)!j!\,2^{k+j}}\end{aligned} \quad \text{(B20)}$$

Evaluating the residues for negative τ (one first order pole, one kth order pole, clockwise traversal of the contour), and combining equations (B19) and (B20) one can find the output-correlation function of the cascade of linear reservoirs with Markovian noise input as

$$R_{Y_2}^{CLR}(k,\alpha,\tau)=\frac{D^2e^{-c|\tau|}}{(1-\alpha c)^k(1+\alpha c)^k}+\frac{D^2e^{-|\tau|/\alpha}}{(k-1)!}\sum_{i=0}^{k-1}\left[\frac{1}{(1+\alpha c)^{k-i}}-\frac{1}{(1-\alpha c)^{k-i}}\right]\sum_{j=0}^{i}\frac{(k+j-1)!(|\tau|/\alpha)^{i-j}}{(i-j)!j!\,2^{k+j}} \quad \text{(B21)}$$

VI

Risk, reliability and related criteria

1 Stochastic approach to non-point pollution of surface waters

E. J. PLATE

Institute for Hydrology and Water Resources Planning, University of Karlsruhe, Germany

ABSTRACT An outline of conceptual stochastic models for describing the concentration of pollutants from non-point sources in a river is presented. Pollutants are assumed to have originated from agricultural fields and to have reached a river attached to suspensa eroded from a watershed. The model consists of three parts: a module describing runoff of pollutants from the land into the river, a module for transport in the river, and a decision module which quantifies the consequences of the river pollution. The model serves as a guide for structuring an experimental programme being conducted at the University of Karlsruhe.

INTRODUCTION

Surface runoff from fertilized fields is an important source of pollution of surface waters. In order to remedy a potentially critical pollution, it is necessary to quantify the amount of pollutants carried by the waters. A quantification of the pollutant load must include random aspects, as crops, agricultural chemicals used and hydrological variables vary in space and time. A stochastic approach requires a long term simulation, which is feasible only if the physical situation is suitably simplified. That is, only a class of pollutants is typically considered. One can distinguish three basic classes, that is substances that adhere to the soil particles (e.g. phosphates), well soluble substances that act like simple tracers (e.g. NaCl), and those that interact chemically with the soil and with other substances (e.g. nitrates).

Further simplifications depend on the time and space scales of the model. A model for annual pollutant load can be based on less detailed components, than a model used for determination of the exceedence of a critical concentration in a river. An event based stochastic model is outlined in the present contribution.

The structure of a river pollution stochastic model

Consider the situation depicted in Fig. 1. The river basin studied is partly covered with agricultural fields. It is the farmers who decide what crops to grow and what kind and what amount of agricultural chemistry products (fertilizers, insecticides, pesticides and herbicides) to use. These substances may pollute surface waters alimented by the runoff from the field in question. The paths of pollutants to the point of water extraction – denoted by point Q in Fig. 1 – and the concentration c at Q depend not only on natural topographic and geological conditions of the area, but also on the soil, on the climate and on the anthropogenic impacts (agricultural practices). The combination of these factors results in a time variable concentration of pollutants in surface waters.

A stochastic non-point pollution model for describing this situation must consist of three parts. The first module is the input model, providing the pollutant loading into the river. The information needed by the input model consists of rainfall fields, providing the runoff for pollutant transport, and of fertilizer inputs into each field, identified as M_i for the ith field. The input model also includes process models for the fields, which convert rainfall into runoff, and fertilizer mass into fertilizer concentration c_i in the runoff from field i.

The second module is a transport model for the river, by means of which transport and mixing in the river are described. This process model incorporating the transport by convection and by diffusion yields a time series of the concentration $c(t)$ and of discharge $Q(t)$ in the river at the water extraction point.

The third module is called the decision model. In its most elementary form, the model consists of mass balance relationships with statistical parameters, and decisions at any level have thus to be made on a statistical basis. The problem associated with decision making on the basis of pollutant concentration in a river may then be taken as a design

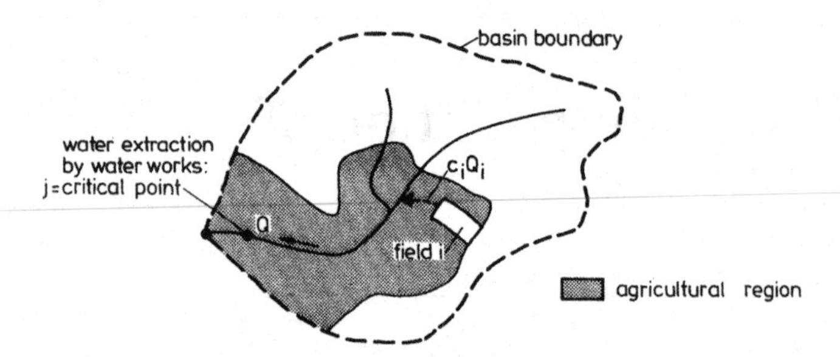

Fig. 1 Schematic presentation of pollutant transport in a river.

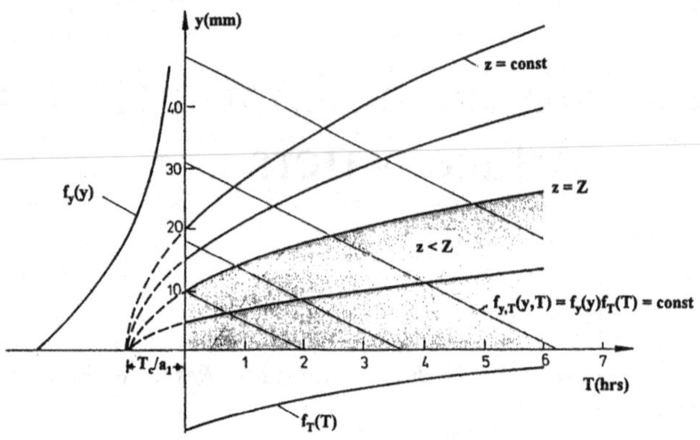

Fig. 2 Schematic presentation of dependence of $F(z)$ on the bivariate density distribution $f(y, T)$.

problem based on failure probability, to which standard techniques of reliability analysis (Duckstein *et al.*, 1986) or of stochastic design (Ang & Tang, 1984; Plate & Duckstein, 1988) can be applied. Consider the three modules as they are integrated into a stochastic water pollution model.

THE INPUT MODEL

The purpose of the input model is to provide the time distribution of pollutant at any point along the river, for a given rainfall and pollutant input.

Consider an event based model supported by experimental results. It seems that the Universal Soil Loss Equation (USLE) appears to be most suitable for modelling erosion-driven pollution. This useful equation in its original form is valid for calculating annual sediment yield from small agricultural plots, and its originator Wischmeyer expressedly warned against application of the USLE for shorter times. However, for the lack of other similarly well documented erosion models, the USLE has been used in a modified form given by Williams (1975) for shorter time periods as well. The modified USLE, or MUSLE, has been used in several stochastic sediment studies (Smith *et al.*, 1977; Bogardi *et al.*, 1985). Smith *et al.* (1977), for example, have combined the MUSLE with the SCS method of calculating rainfall abstractions. They obtained the following expression for the sediment yield $z(r)$ per rainfall event r:

$$z(r) = w\left[\frac{a_0 y_r^4}{(y_r+f)^2(a_1 T_r + T_c)}\right]^{0.56} \tag{1}$$

where w and a_0 are conversion factors incorporating size of field, land use etc, y_r is the effective rainfall depth per event r (in mm), for $r = 1,2,.. N$, where N is the number of events per season, f is the constant describing total infiltration (in mm), a_1 is a dimensionless constant, T_r is the rainfall event duration (in hours), and T_c is the concentration time, i.e. the time which elapses between the beginning of the rainfall and the occurrence of the peak runoff. For details and for numerical values, the original papers by Smith *et al.* (1977) or Bogardi *et al.* (1985) should be consulted.

The structure of equation (1) makes it clear that the input for this model must be an event based rainfall. The actual rainfall event is approximated by a block rainfall with the basis T_r and the height y_r. In this way, the rainfall event is described by the two variables T and y, and thus, the random distribution of rainfall events can be approximated by the joint probability density $f(y, T)$. It has been solved numerically by Smith *et al.* (1977) with rainfall event inputs which have a bivariate exponential distribution for rainfall intensity and rainfall duration. The number N of events per season is Poisson distributed. It has also been solved by simulation in Bogardi *et al.* (1985).

If the function $f(y, T)$ is known either theoretically, or experimentally, then the method for finding the erosion rate is as schematically illustrated in Fig. 2. The random variables y and T are the coordinates of a plane in which the isolines of the joint pdf, $f(y, T)$ are indicated. In this plane, the condition $z' = $ constant produces a family of curves for different values of the constant (Fig. 2). The cumulative distribution $F_z(z)$ is calculated for each $z' = Z$ by integration of $f(P_r, t_r)$ over the area in which $z < Z$, as is indicated by the hatched region of Fig. 2. Finally the probability density function $f_z(z')$ is found from $F_z(z')$ by differentiation.

Equation (1) can also be used with experimentally determined rainfall functions $f(y, T)$ which have been extensively studied in urban hydrology (cf. Xanthoupolos, 1990). An example of a bivariate relationship for German conditions (station Essen) is shown in Fig. 3.

Water quality parameters depending directly on the rate of surface erosion are obtained by combining the soil loss equations with an elementary water quality model. As an example the phosphorus load model of Bogardi & Duckstein (1978) may be mentioned. However, uncertainty is introduced by the fact that such a model is based on the assumption that the pollutant is distributed uniformly throughout the top soil layer.

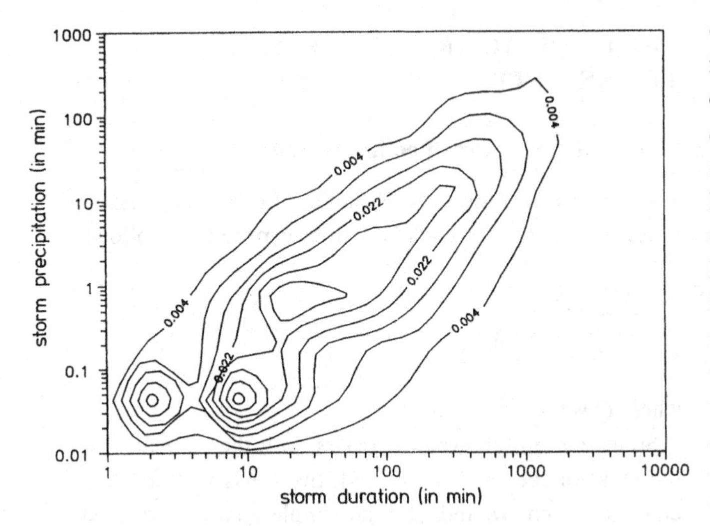

Fig. 3 Typical empirical joint probability distribution of y and T for the station Essen, Germany (courtesy of A. Bardossy).

As an alternative to MUSLE a more physically-based model or a unit pollutograph approach may be used. An example of the former category is the model devised by Julien (Julien & Frenette, 1986, Julien & Dawod, 1987), who used a more elaborate input hydrograph and an overland erosion model based on hillslope hydraulics.

A 'pollutograph' is a function constructed in the same manner as a unit hydrograph. The pollutant applied to the land is considered as the input to be convoluted with the impulse response function, $hs(t)$ called pollutograph (Jury *et al.*, 1986). As far as the properties of a pollutograph are concerned, it is often noticed that the peak of the pollutant wave arrives faster than that of the flood wave. That is, a unit pollutograph may have a shorter rise time than the unit hydrograph.

In general, unit hydrographs and 'pollutographs' are used deterministically, i.e. both the input function and the impulse response (unit hydrograph) are deterministic. However, it is not difficult to use a deterministic $h(t)$ with random inputs, by employing simulation techniques. A special case is a random input consisting of a time series with normally distributed input magnitudes. In this case the output is also normally distributed, with a variance which can be determined via the spectrum function of the input process and the Fourier transform of the impulse response. Some use of these relations has been reported for groundwater models, see for example Geldner (1980).

In order to proceed with the pollutant model for a river one needs the information on the delivery ratio in addition to the soil erosion from individual fields. That is, one needs an estimate of the losses of eroded sediment in the course of its transport from the field to the river. Attempts have been made to develop such models, although to the knowledge of this author, there is no truly physically-based model capable of operating without calibration. Erosion models have been included into catchment models consisting of a network of channels and unit hydrographs of the type described by Plate *et al.* (1988) in order to calculate the rate of erosion from soil surfaces and deposition in reservoirs. Studies of this kind have been reported by Bogardi *et al.* (1985) and Hrissanthou (1986). Both authors used simulation methods to obtain the historical sequence of event-based erosion yields, which were then summed over the year. They subdivided the basins into many small, approximately field size subbasins, to each of which the MUSLE has been applied. An annual precipitation index K was used. The event dependence was introduced by the use of a daily value of effective rainfall from which the sediment inflow into the reservoir was calculated. The differences between the two studies were in the way the MUSLE was used for the subbasins, and how they were linked for the total yield. For each partial area the assumptions made by Hrissanthou (1986) were about the same as for the model of Smith *et al.* (1977), which has been described above. However, instead of using the SCS method for determination of the effective rainfall, he used a modification of the SCS method developed by Lutz (see Plate *et al.*, 1988). Furthermore, the MUSLE included a channel sediment routing subroutine (Williams, 1975). In both cases the results were compared with measurements of annual reservoir deposition and good agreement was found for long term average deposition but considerable scatter for the individual values.

Input model for pollutant input along the river

One has to determine the distribution of the pollutant input along the river as a function of distance and time. This is not directly obtained from the erosion model for a field. The point load M of sediment which leaves the field during the rainfall event, has to be transformed into the pollutant load for the river. One needs to know how much of the sediment which is eroded from the field will reach the river, where it will reach the river, and how it will be distributed over time. One needs to have component models combining many fields into inputs. That is, in addition to the soil erosion z_i from individual fields the delivery ratio D_i should be known. One needs an estimate of the losses of eroded sediment in the course of the transport from field to river.

Pollutants adhering to the eroded material enter the river from many fields as indicated schematically in Fig.1. The river originates in an unpolluted area, so that zero concentration in the river upstream from the agricultural areas can be assumed. The rural area consists of many different fields, each of which is fertilized. However, application of fertilizer, land use, and fertilizer uptake by crops are not constant. The rates administered may vary from field to field in a random manner.

A complete simulation procedure which should be based on such input models requires not only that the temporal variability of all input variables are appropriately considered, but also that physical processes contributing to the conversion of the input into the pollutant loaded runoff be well known and quantifiable. In general, it is impossible to exactly quantify all the relevant processes, and the result may show a high degree of uncertainty. One therefore needs a certain amount of stochasticity in all water quality models. Stochastic water quality models have been reviewed in Plate & Duckstein (1990).

A feasible concept employing such a model is to describe the fertilizer runoff model for daily mean quantities, and to relegate all other dynamic effects to an uncertainty band around the mean. It is likely that such a strictly statistical model can provide results of similar quality as physically-based deterministic models, provided it can be calibrated with reliable data.

It is furthermore assumed that only little runoff of fertilizer takes place during dry conditions: critical conditions for concentration c to exceed concentration c_p occur only during or after rainfall events. Let c_i be the concentration in the runoff Q from field i during a rainfall event, so that the total amount of pollutant from field i which reaches the river at point j is $c_{ij}Q_{ij}$. The total amount M of pollutant substance reaching point j on the river is also a random variable, which, with given discharge Q and average concentration c, and with the subscript j dropped, can be expressed as:

$$M=\sum_{i=1}^{I} c_iQ_i=\sum_{i=1}^{I} D_ip_iz_i \tag{2}$$

where D_i is the delivery ratio; p_i is the pollutant concentration in sediment from the field; and the summation has to be extended over all I fields.

Since the number of fields involved in producing M is assumed to be large, one may invoke the central limit theorem of statistics and assume that M is a normally distributed random variable, with probability density function:

$$f_M(M)=\frac{1}{\sigma_M\sqrt{2\pi}}\exp\left[\frac{(M-\mu_M)^2}{2\sigma_M^2}\right] \tag{3}$$

Experimental evidence, for example by Marani & Bendoricchio (1986), indicates that event based runoff from individual fields can be adequately represented by a gamma distribution, or a generalized gamma distribution. Clarke (1990) used the log-normal distribution for describing both Q_i and c_i, in which case c_iQ_i is also log-normally distributed. It results from the central limit theorem that a sum of many variables of the same distribution (e.g. one of the distributions considered above) attains the normal distribution.

PROCESS MODELS FOR RIVER TRANSPORT

Continuous processes of pollutant transport

The average concentration of pollutant in a river caused by runoff of sediment from fields can be expressed as follows:

$$c=\frac{\sum_{i=1}^{I} D_ip_iz_i}{Q}=M/Q \tag{4}$$

where Q is the river flow.

Since the rainfall event increases both the river flow and the mass of the eroded material, there exists some dependence between M and Q. The simplest type of a model describing such a dependence is a functional relationship between an input variable x and an output variable y:

$$y=g(x) \tag{5}$$

Such a relation is found for many water quality processes. It usually connects the variable y describing the transport of a pollutant with the variable x describing the flow of a transporting agent. The variable x can pertain directly to water flow, or to sediment runoff, thus indirectly to water flow. The established relations for process models of pollutants are of the form:

$$y=ax^n \tag{6}$$

where a and n are typically empirical parameters.

For the case of dissolved pollutants, x is the flow rate or the rainfall, $n=1$, and y is the output concentration (Haith & Tubbs, 1981). For the case of particulate pollutants, x is the eroded soil, and y is the outflow of particulate pollutants, and $n=1$. In the case of pollutant adhering to the soil and transported as suspended load by a river one has to use a relationship between the river discharge ($x=Q$) and the suspended sediment load ($y=QS$). This relationship is often found to agree with equation (6), with n ranging from 1.5 to 3, (see for example Walling, 1977). Thus one can describe the pollutant outflow as:

$$M=aQ^s \tag{7}$$

where M is the mass of pollutant per time unit.

The average concentration, c, during the time unit, expressed as M/Q, reads:

$$c=aQ^{s-1} \tag{8}$$

In order to improve the fit of experimental data to this curve, it may be useful to consider a to be a random variable as well. In this case the result c is a product of two random variables.

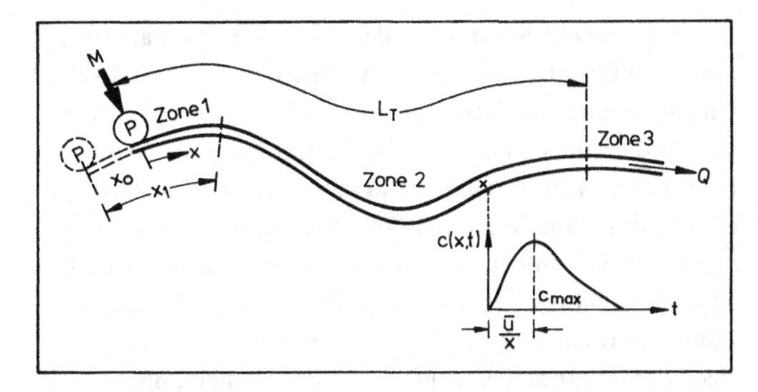

Fig. 4 Zones of dispersion in a river downstream of a point source of pollutant input.

Short duration input processes

Assume the input of pollutants to be of the point-source type and of short duration (Fig. 4). Then the pollutant input M is conveyed through the river at the discharge Q. Due to convection and dispersion the input moves downstream, spreading in the longitudinal and transverse (lateral) direction.

The concentration $c(x, t)$ downstream from the point P in Fig. 4 can be found by means of a model of the process of convection and dispersion for a given discharge Q and a given mass M. The output of this model is the concentration $c(x, t)$ and its input is the two-dimensional input vector (MQ).

Consider the traditional diffusion model with a point source. Such a model assumes that diffusion is a linear process, which is not likely to hold in natural rivers. Nevertheless, in order to illustrate the dispersion effect, it is useful to consider the dispersion from a point source, and to make use of the convenient property of superposition. After this has been done, one is better prepared to determine whether dispersion should be considered or not.

Spreading of pollutants from a point source takes place in three zones depicted in Fig. 4. In each of these zones, diffusion is governed by different processes of fluid mechanics. In the first zone near the source diffusion is governed by the configuration of the source, and by the rate of lateral spreading. In particular when immission takes place from the shore, it takes a considerable distance before the pollutants are uniformly distributed in a cross-section of the river (Naudascher & Fink, 1983). This process is modelled by assuming a virtual origin to exist for the source, which is located at a virtual distance x_0 upstream from the actual point of immission.

Both the second and the third zones have the property that the concentration is constant across the river, so that only longitudinal dispersion needs to be considered. It results from two different effects, that is from diffusion caused by small scale turbulence generated on banks and river bottom, and from dispersion caused by velocity gradients in the vertical and horizontal directions. The effect of dispersion has first been formulated by Taylor (1954) for pipe flow. Fischer (1973) applied Taylor's model to rivers, including the transfer processes induced by horizontal velocity gradients. However, these models yield very crude approximations to observed dispersion in actual rivers, which might deviate by more than three orders of magnitude from the theoretical predictions. This is due to the occurrence of large eddies generated by large scale bottom features (Grimm-Strele, 1983), and exchange processes in dead water zones, caused by river training structures, by natural bays (Valentine & Wood, 1977), or by overbank flow over wooded flood plains (Pasche & Rouvé, 1986). Furthermore, the curvature of the rivers also has an effect (Friedrich & Plate, 1973).

Determination of failure conditions

The decision model allows one to determine if the water quality standards expressed in terms of the permitted concentration c_p are exceeded so frequently that toxicity problems come about. In order to define the effect of this exceedance, one has to know the corresponding probability function. A situation in which the standards are not met, i.e. if the river or creek is in the state $c > c_p$ is called a 'failure'. A reliability based decision model starts from the assumption that it is not necessary to completely avoid the 'failure' condition, but that intolerable damage, i.e. health hazards or damage to the ecosystem, occurs if the failure probability $PF = \mathrm{P}\{c > c_p\}$ is higher than some critical value. The failure condition $c > c_p$ is a stochastic event, which in terms of the naturally random variables Q and the permissible mass $M = c_pQ$ can be rewritten as:

$$M > c_pQ \tag{9}$$

The left hand side of equation (9) is a random variable due to changes in the runoff from fields and in the concentration of substances. The right hand side is a random variable due to variability of discharge Q. The model expressed by equation 9 is equivalent to the classical case of calculating the probability of failure for random resistances and loads, $r = c_p\,Q$ and $s = M$, as illustrated in Fig. 5. A two-dimensional p.d.f. $f_{rs}(r, s)$ for resistance and load is shown in Fig. 5 for the case where the existence of statistical dependence between both variables is assumed. Marginal distributions $f_r(r)$ and $f_s(s)$ are also plotted on the axes. It is to be noted that $f_r(r)$ is identical to the probability density of Q except for the scaling factor c_p, whereas $f_s(s)$ is the density of the pollutant mass loading into the river.

The condition $r = s$, corresponding to the failure surface, is a straight line bisecting the $s - r$ quadrant. The probability of failure is that of finding a combination of r and s in the

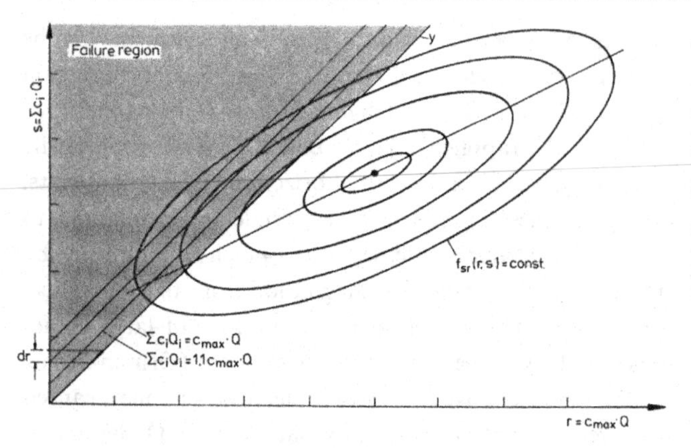

Fig. 5 Probability diagram for resistance r and load s, with failure surface indicated by $z=0$.

shaded region of the $r-s$ quadrant. It is calculated from the probability distribution of the pollutant excess z defined as:

$$z=s-r \tag{10}$$

with the mean value:

$$u_z=u_s-u_r \tag{11}$$

The equation for $f_z(z)$ can be obtained by integration of $f_{rs}(r,s)$ over the region $-\infty<z<0$ as shown in Fig. 5. It reads (Papoulis, 1965):

$$f_z(z)=\int_{-\infty}^{\infty} f_{rs}(r,s^*)\left|\frac{ds^*}{dz}\right| ds \tag{12}$$

where s^* is the transformation of $s(z)$ for the condition $z=$constant. Here s^* is simply:

$$s^*=z+r \tag{13}$$

Finally, one obtains:

$$f_z(z)=\int_{-\infty}^{\infty} f_{r,s}[(z+r)|r]\,f_r(r)\,dr \tag{14}$$

where $f_{r,s}(s|r)$ is the conditional probability density for s given r. Equation 14 cannot be analytically solved for the general case, and the solution usually has to be found by numerical integration.

As an approximation, the pollution level for continuous inputs can be obtained if one assumes that the central limit theorem holds for the mass M contributed by all the fields along the river, so that M is distributed normally. For simplicity, one may also assume Q to be normally distributed. Since discharge and mass are caused by the same rainfall conditions, a certain degree of correlation exists between the two variables, and variables r and s have a correlation coefficient q. The calculation of the probability of failure for this case can be done in a closed form (Plate, 1991). One obtains a formulation which is entirely equivalent to the standard expression used in the Second Moment Analysis, i.e. the approach to reliability analysis based on the normal distribution (Ang & Tang, 1984). Therefore, methods developed in the context of structural safety analysis can be applied, including uncertainty analysis. Also, if the assumption of normality of distributions or that of perfect (error free) data cannot be maintained, then other approaches towards refined second moment analysis can also be used.

DECISION MODELS FOR RIVER POLLUTION

Because the general problem illustrated in Fig.1 is of extreme complexity, simplifications are needed if the design or management decisions must be made. Decision models have components whose outputs can be modified by means of decision variables, being influenced directly or indirectly by human actions. The type of model to be used in water quality decision processes depends on the kind of decisions in question. For example, a model used for prediction of the peak concentrations of pesticides in a river at a particular point can have a different structure than a model used for prediction of the amount of pesticides penetrating into the groundwater in a subarea of the basin.

The generalized risk, as defined for example in Duckstein *et al.* (1987), which quantifies such decisions, is therefore an important figure of merit for pollutant transport models.

Risk as figure of merit

In the context of stochasticity of decision models one can accept the following definition of risk (Berger, 1985). Let $y=(y_1, y_2, \ldots y_I)$ be those variables from the output vector of a WQM which can be manipulated by decisions, so that their values are conditional on the decisions $d=(d_1, d_2, \ldots d_J)$, where d is the vector of decision variables. Let these variables occur in combinations determined by the joint probability density function (joint pdf) given by:

$$f(y,d)=f(y_1, y_2, \ldots y_I, \ldots d_1, d_2, \ldots d_J) \tag{15}$$

Furthermore, let $K(y,D)$ be the function describing the consequences of the occurence of the combination $y_1, y_2 \ldots$ for a given decision vector $d=D$. Then the risk is defined in general as the expected value of K over the (conditional) pdf $f(y|D)$, or:

$$RI(D)=\int_{-\infty}^{\infty} K(y,D)\,f(y|D)\,dy \tag{16}$$

where the integration has to be performed over all the elements of the vector y.

The risk is a measure of the gamble which one takes, making the decision D. It is a single valued number, called a Figure of Merit, FM (Duckstein *et al.*, 1987), which permits to judge the value of the decision. Note that there might be other figures of merit associated with any decision process. FMs may be based on different criteria using output variables. Also, different types of risk may occur, depending on the definition of the consequence function. For details reference is made to Duckstein *et al.* (1987).

Risk is a FM that evaluates the consequences of nonsatisfactory system behaviour. It serves therefore two goals. First, it helps evaluate the alternatives and the necessary actions to optimize decisions (e.g. which region should be held free of fertilizers, or in which areas the amount of fertilizer application should be reduced, and by how much). Second, a decision analysis can be used to select a model for a given purpose, and to determine how far one should go in improving simple models.

The probability of failure is a particular FM, based on a consequence function $K(.,.)=1$. It may not be a useful criterion for assessing the merit for pollution control, (e.g. by changing the prevalent practice of fertilizer application), because it does not include the consequences of exceeding c_p.

Uncertainty in decision models

Because uncertainty of different kinds may limit the quality of the information which can be obtained from a model, it is useful to consider the tradeoff of model uncertainty against natural variability. Obviously, if the variance of the pdf for the model uncertainty is small compared to that of the sample or the measurement, then it is not worthwhile to further improve the model.

More generally, the value of a model can best be assessed if a preference value is assigned to the consequences of a decision which is based on the model. If a clearly advantageous decision can be made independently of the quality of the model, then there is no operational sense in further improving the model. If, on the other hand, costly investments would be necessary for preparation against possible but uncertain consequences, then model improvements, or improvements in the data basis may be in order.

In general, stochastic models are subject to many uncertainties. Uncertainty has the effect that the risk $RI(d)$ becomes a random variable. It is a conditional function $f(RI\{d|PAR\})=g(d|PAR)$, also called goal function, where the vector PAR consists of all the uncertain parameters on which RI may depend. From this pdf one must estimate the best value of RI to be used for decision making. It could be for example the expected value.

CONCLUSIONS AND OUTLOOK

A concept was presented by means of which the failure probability for a concentration of substances in a river can be determined from a stochastic model. The complete model consists of an input part, a river part, and a decision part. For some cases, the problem can be formulated as a problem of determining the failure probability, or the exceedence probability. The use of normal distributions simplifies this problem, but it is not a necessary condition for the applicability of the concept. An important aspect of this formulation, and also of deterministic formulations, is that the transfer functions which describe the process of rainfall conversion to mass runoff M need to be quantified. Natural variability of processes is a very important factor in uncertainty considerations. This is caused by the stochasticity of the river discharge Q, and by the randomness of the input of pollutants. Statistical properties of Q have been extensively studied, whereas the input from pollutants requires investigation of the pollutant sources. The parameters determining the mass of fertilizers entering the river from a field i depend on the amount of pollutant administered or generated in soil and groundwater, and on the processes determining the chemical and biological changes of the pollutants during their path from the source to the river.

It is, however, not very likely that a simple single input model can be used to describe a real situation. A typical rural region may show the diversity depicted in Fig. 6, representing a typical small agricultural and silvicultural area in rural Germany, which may cover an area of a few sqare kilometers. Pollutant content of runoff from such an area reflects the diversity of land use, and different fertilizing patterns for different crops. It also reflects the variety of soils that are found in the region, and the geological structure of the underlying subsoil, determining the amount of fertilizer-carrying capacity of water percolating into the groundwater. It is also influenced by the topography of the region, with runoff characteristics of flat lands quite different from those of steeper areas.

It is obvious that the rigorous description of runoff characteristics of water and pollutants from such an area would require much more elaborate process models than the simple approaches described in this paper. Such in-depth studies, however, would last long and be prohibitively costly. Therefore it is unlikely that one can study each situation of pollution potential in detail, and even the construction of models from building blocks of the kind used in the flood studies (Plate *et al.*, 1988) would be too costly for general applications. This is in contrast to the great need that exists for an operational model on which agricultural or other land use decisions can be based. Such a model should be capable

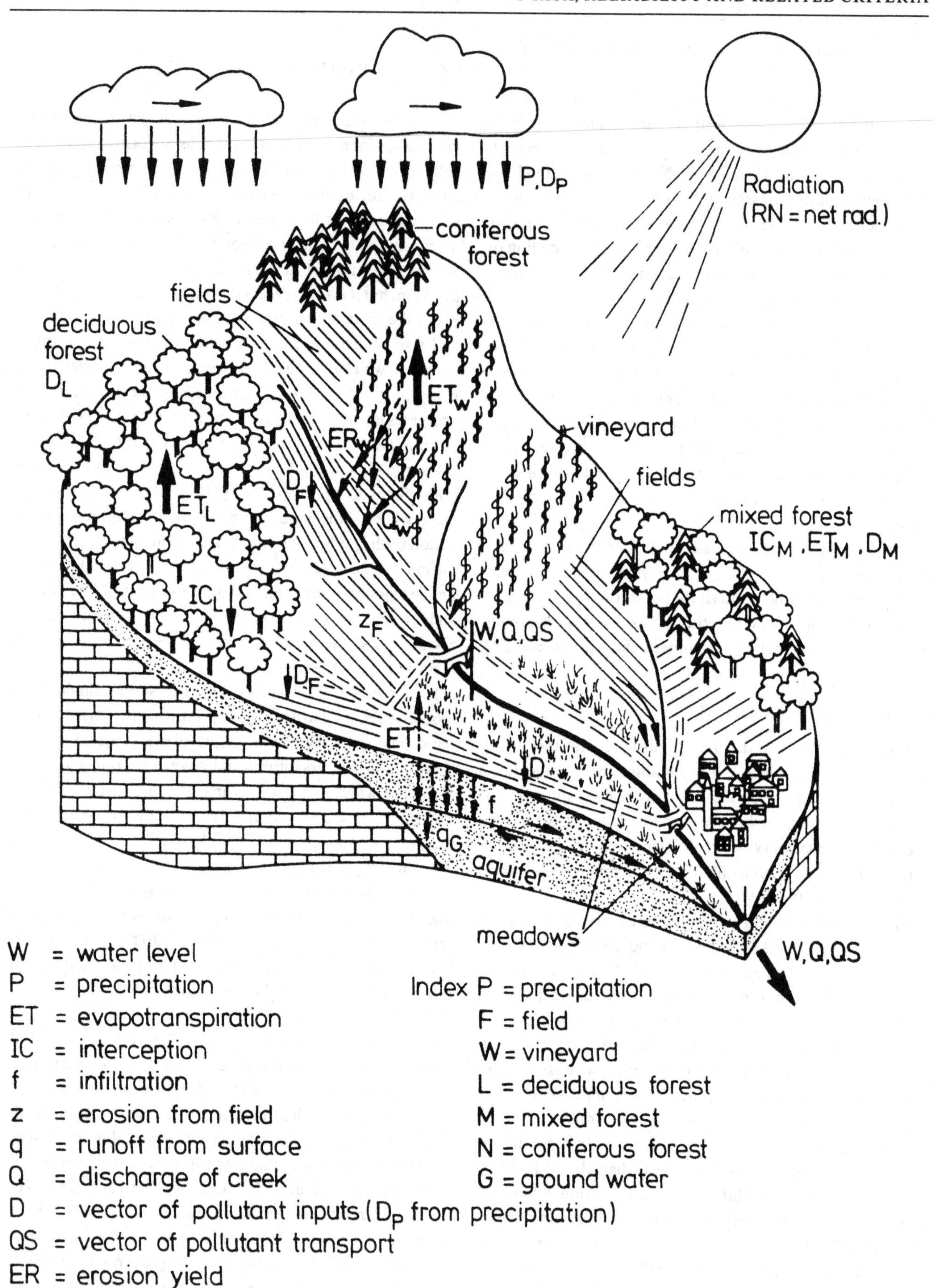

Fig. 6 Sketch of a typical rural area, with different land use and soil characteristics.

of evaluation of effects of feedlots or accidental spills as point sources, or of acid rain, fertilizer application and other external spatially distributed sources of pollutants. Therefore it is necessary to develop simple models, which retain the essential features of the prototype, but which are simple enough to be formulated and calibrated in practical applications, including emergency situations.

It seems that numerous field studies are required to determine the variability of the processes involved in the problem. Such a study whose ultimate purpose is to develop a simple model for describing the transport of agricultural pollutants from fields to a river is being carried out at the University of Karlsruhe. Fifteen institutes cooperate in this study, financed by the German Ministry of Science and Technology. The procedure accepted starts from a possibly complete description of the process of non-point water pollution, and then simplifies the model to the practical form, where only the most essential aspects for the decision situation are left. In an exemplary manner, this model is being developed in a combination of theoretical and field studies, under the collective name of the Weiherbach Project. The project is carried out in the catchment of Weiherbach creek (6 square kilometers area) situated in the Kraichgau area, North East of the city of Karlsruhe. Within the framework of the project the transport processes from rainfall and fertilizer application on every field to runoff and concentrations of substances ranging from fertilizers to pesticides and heavy metals in Weiherbach creek, are measured and mathematically modelled.

In the initial phase the Weiherbach situation should be described by a model, which is as deterministic as possible. However, the long term objective is to use this model as a test case for statistical models calibrated and tested against the 'ground truth', or against the results from an accurately calibrated and accurately formulated deterministic model. It will give the opportunity to find out which are the essential processes contributing to the mass transfer from land to water, and which processes can be considered secondary and thus relegated to statistical variability.

The structure of a model of the Weiherbach type is hierarchical, as schematically shown in Fig. 7. A number of different levels exist, for which the information has to be stored and prepared in suitable form.

(a) *The input level.* The input level in Fig. 7 consists of parameters that describe both the area and its transport features. Basic variables needed for models at any time scale are those describing the natural properties of the area, such as topography, geology, soil and land use characteristics. These sort of data are usually available from topographic and geological maps, and from information collected by the Agricultural Service. It is likely that in future much of this information can be obtained from remote sensings, from satellites or from aerial photographs. The data are stored in computer based information systems, such as a Geographical Information System (GIS). These basic data are sufficient to specify the properties of water and matter transport for long term averages.

Superimposed on these basic parameters are parameters pertaining to a seasonal time scale, which are needed to determine average seasonal balances. Both seasonal and permanent data are likely to be stored in a GIS and can be represented on maps. They also provide background information for the event based models, whose time scale is of the order of hours to days.

The actual dynamics of the transport must be generated from event based information, such as historical rainfall events, or time series of rainfall events to be generated by Monte Carlo techniques. Such process models are needed to provide space time variable inputs into event based hydrological decision models. The event input includes also information on the input of matter into the hydrological system, such as fertilizer inputs, or inputs through accidental releases of toxic or harmful substances.

The input parameters describing the boundary conditions and the local parameters of all hydrological models are constant for all time steps, and should be available in the GIS of the area. The time invariant parameters of the permanent inputs will be the same for all types of models. However, not all will be needed for the different types of problems that are to be solved, and also, additional parameters pertaining to the time variability of the processes to be considered may be required if shorter time scales are selected. We distinguish different types of models associated with different levels of resolution.

(b) *The model level.* The model level of Fig. 7 includes all possible models for describing the water and pollutant transport processes. It is likely that different tasks will require models based on different time scales. Long term averages, required for example for annual balances, can be aggregated from short term models, but it is likely that simpler models are useful for long term or large area averages, as for example in the use of hydrological models for climate research (Becker & Nemec, 1987). Consequently, three types of models can be identified.

The first is the *basic hydrological site model*, in which topography, soil properties, land use and other factors are analysed and combined. It incorporates all permanent and time invariant information, from which an output consisting of maps and tables can be obtained. This model requires access to tables of parameters (in terms of an expert system, or simply by drawing on information supplied from local field studies), which represent local properties needed for transport models.

The second is the *seasonal or climate model*, which combines the site model with seasonally varying parameters, such as meteorological information on seasonally varying mean temperatures, precipitation, average cloudiness and the like, and seasonal variations of the land use,

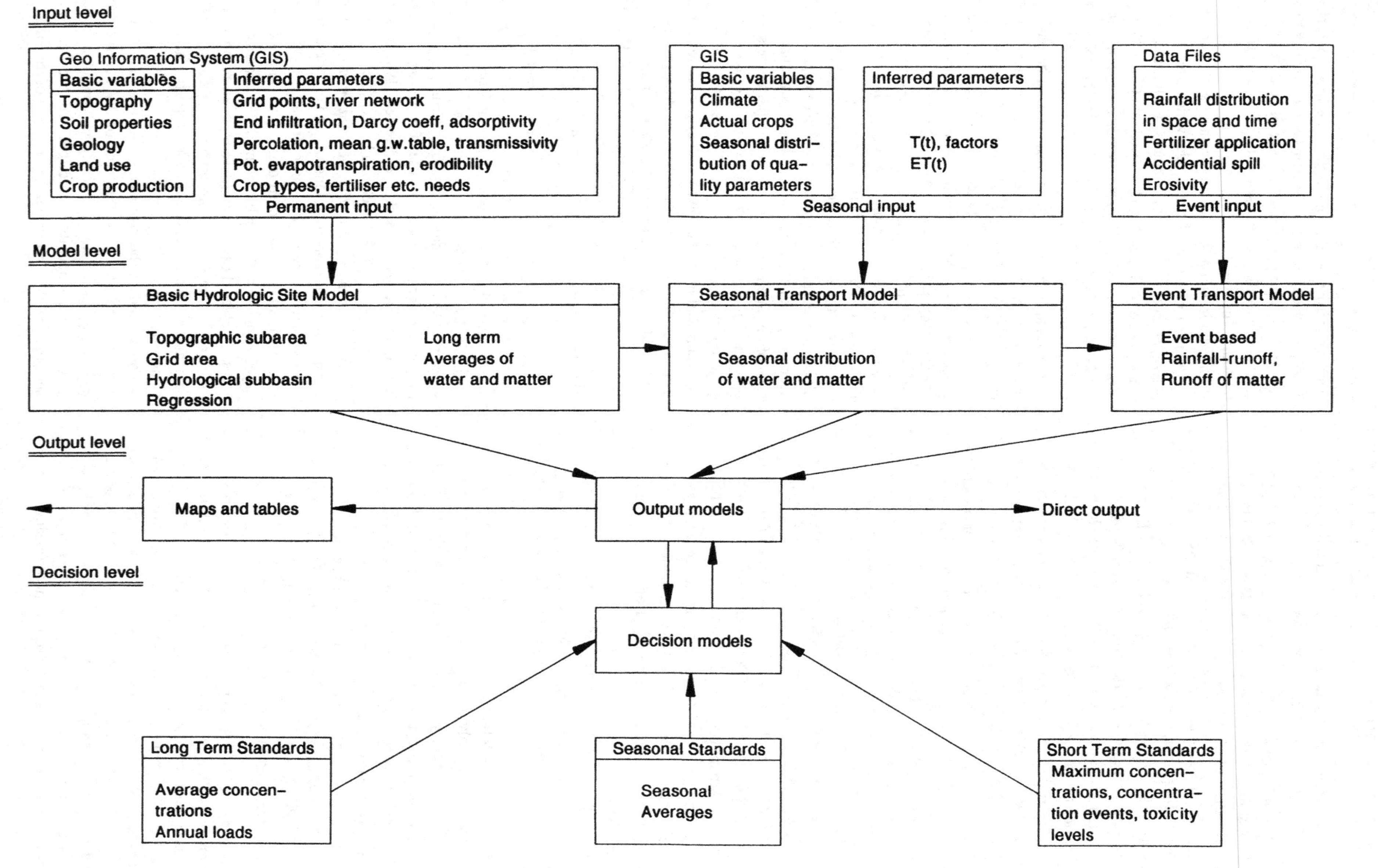

Fig. 7 Levels of model building for general river pollution models.

such as potential evapotranspiration, crop production factors, average use of fertilizers.

The third class is the *event based model*, capable of yielding the most detailed output information, but requiring the most extensive input information. If this type of model is used for long term application, one would have to deterministically calculate the flow of water and matter at one or more points of the site to obtain a continuous record of output functions. It is obvious that a model for long term simulation of occurrences of polluting episodes is expensive to use and time consuming to construct. Therefore it can be applied only in exceptional circumstances, e.g. during pilot studies (Weiherbach project).

(c) *The output level.* The output level consists of the results. The basic hydrological information is presented in maps and graphs, and may be subjected to statistical analysis. Naturally, the output level may consist of outputs from all kinds of models, and can be used for all special purposes to which the model is to be applied. Therefore, the output also includes information obtained during the last stage of the four stage process described in Fig. 7: the decision level.

(d) *The decision level.* Actual operational models are likely to be simplified, whereas the degree of simplification depends on the purpose of the model. Output performance of the model can be measured in terms of the performance criteria. Monitoring models, which serve the purpose of verifying the observed data, require different performance criteria than models which are used for making decisions on improving water quality. We find it important, therefore, to incorporate a decision level into the system model, as has been discussed.

Only some of the ideas that are used in stochastic modelling can be outlined in the present contribution. In spite of great efforts made in many parts of the world, one is still far away from arriving at a universal model of general usefulness. Therefore field studies are of paramount importance and the Weiherbach project is just one of many needed.

ACKNOWLEDGEMENT

The research reported in the present contribution has been supported by the German Ministry of Science and Technology.

REFERENCES

Ang, A. H. S. & Tang, W. H. (1984) *Probability Concepts in Engineering Planning and Design: Vol. 2*, J.Wiley, New York.

Becker, A. & Nemec, J. (1987) Macroscale hydrologic models in support of climate research. In: *The Influence of Climate Change and Climatic Variability on the Hydrologic Regime and Water Resources*, (Proceedings of the Vancouver Symposium, August 1987) IAHS Publication No. 168, pp. 431–45.

Berger, J. B. (1985) *Statistical Decision Theory and Bayesian Analysis*, Springer, New York.

Bogardi, I. & Duckstein, L. (1978) Input for a stochastic control model of P-loading. *Ecological Modelling*, **4**, 173–95.

Bogardi, I., Bardossy, A. & Duckstein, L. (1985) Effect of parameter uncertainty on calculated sediment yield. *Adv. in Water Resour.*, **8**, 96–101.

Clarke, R. T. (1990) Statistical characteristics of some estimators of sediment and nutrient loadings. *Water Resour. Res.*, **26**, 2229–34.

Duckstein, L., Plate, E. J. & Benedini, M. (1987) Water engineering reliability and risk: a system framework. In: Duckstein, L. & Plate, E. J. (Eds), *Engineering Reliability and Risk in Water Resources*, NATO Advanced Study Series, Nijhoff, pp. 1–20.

Geldner, P. (1981) *Deterministische und stochastische Methoden zur Bestimmung der Selbstdichtung von Gewässern*, Mitteilungen No. 49, Institute for Hydraulic Engineering, Stuttgart University, FRG.

Grimm-Strehle, J. (1983) Diffusion und Dispersion in Flüssen. *Die Wasserwirtschaft*, **73**, 380–5.

Fischer, H. B. (1973) Longitudinal dispersion and turbulent mixing in open channel flow, *Annual Reviews of Fluid Mechanics*, **5**, 59–77.

Friedrich, R. & Plate, E. J. (1973) The effect of bends on longitudinal dispersion of floating particles, *Proceedings, XVth Congress International Association for Hydraulic Research*, Istanbul, Vol. **2**, pp. 61–9.

Haith, D. A. & Tubbs, L. J. (1981) Watershed loading functions for non-point sources. *J. Environmental Engineering Division Amer. Soc. of Civil Eng.*, **107**, 127–37.

Hrissanthou, V. (1986) Computation of sediment yield from a large watershed in middle Europe, *Proceedings, Third Internat. Symposium on River Sedimentation*, Jackson, Miss., USA.

Julien, P. Y. & Dawod, A. M. (1987) On predicting upland erosion losses from rainfall depth. *Stochastic Hydrology and Hydraulics*, **1**, 127–34.

Julien, P. Y. & Frenette, M. (1985) Modeling of rainfall erosion. *J. Hydraul. Div. Amer. Soc. of Civil Eng.*, **111**, 1344–59.

Jury, W. A., Sposito, G. & White, R. E. (1986) A transfer model of solute transport through soil: 1. Fundamental concepts. *Water Resour. Res.*, **22**, 243–7.

Marani, A. & Bendoricchio, G. (1986) Models of statistical distributions for NPSP (non-point source pollution) concentrations. In: Giorgini, A. & Zingales, F. (Eds.), *Agricultural Non-point Source Pollution: Model Selection and Application*. Elsevier, Amsterdam.

Naudascher, E. & Fink, L. (1983) Ausbreitung bei Einleitung in Strömungen. *Die Wasserwirtschaft*, **73**, 369–79.

Pasche, E. & Rouvé, G. (1985) Overbank flow with vegetatively roughened flood plains. *J. Hydraul. Div. Amer. Soc. of Civil Eng.*, **111**, 1262–78.

Papoulis, A. (1965) *Probability, Random Variables and Stochastic Processes*, McGraw Hill Book Co., NY.

Plate, E. J. (1991) *Probabilistic modelling of water quality in rivers*. In: Ganoulis, J. (Ed.), *Water Resources Engineering Risk Assessment*, NATO ASI Series, Vol. G29, Springer, Berlin, pp. 137–66.

Plate, E. J. & Duckstein, L. (1988) Reliability based design concepts in hydraulic engineering. *Water Resour. Bull., AWRA*, **24**, 235–45.

Plate, E. J. & Duckstein, L. (1990) Stochastic aspects of water quality modeling for non-point sources. In: DeCoursey, D. (Ed.), *Proceedings of International Symposium on Water Quality Modelling of Agricultural Non-point Sources*, Part 2, US Agriculture Research Service, ARS, pp. 631–54.

Plate, E. J., Ihringer, J. & Lutz, W. (1988) Operational models for flood calculations. *J. Hydrol.*, **100**, 489–506.

Smith, J. H., Davis, D. R. & Fogel, M. (1977) Determination of sediment yield by transferring rainfall data. *Water Resour. Bull.*, **13**, 529–41.

Taylor, G. I. (1954) The dispersion of matter in turbulent flow through a pipe. *Proc. R. Soc. Lond.*, **A233**, 446–68.

Valentine, E. M. & Wood, I. R. (1977) Longitudinal dispersion with dead zones. *J. Hydraulics Division Amer. Soc. of Civil Eng.*, **103**, 975–80.

Walling, D. E. (1977) Assessing the accuracy of suspended sediment rating curves for a small basin. *Water Resour. Res.*, **13**, 531–8.

Williams, J. R. (1975) *Sediment yield predictions with universal equation using runoff energy factor*, Agricultural Research Service ARS-S-40, USDA, Wash. USA, pp. 244–52.

Xanthoupoulos, C. (1990) *Methode zur Entwicklung von Modellregensspektren für die Schmutzfrachtberechnung*, Schriftenreihe des Instituts für Siedlungswasserwirtschaft, University of Karlsruhe, 57.

2 Statistically safe sets methodology for optimal management of reservoirs in risk situations

A. KARBOWSKI

Institute of Automatic Control, Warsaw University of Technology, Warsaw, Poland

ABSTRACT The paper concerns the problems of optimal control of reservoirs when subsequent inflows are represented as independent random variables of known distributions or the Markov chains, and apart from performance index other control goals are present. They may have the form of chance constraints or constraints on the expected value of certain functions of stage variables (i.e., controls or states in given periods). A method of conversion of the global constraints to the stage ones is also presented. The statistically safe sets methodology is explained on the background of other methods of determining reservoir control strategies such as linear decision rules, reliability programming, penalty function method or the method of the Lagrange multipliers.

INTRODUCTION

For many years the specialists in designing control algorithms for water reservoirs have agreed that the problems of risk connected with uncertainty of inflows are of primary importance. It is almost impossible to enumerate all techniques proposed in the literature. The models that account risk through imposing lower constraints on the probability of desirable events and upper constraints on the probability of undesirable events (so-called 'reliability' or 'chance' constraints) are most popular. These events may be considered in any given period, a specific control horizon or during the entire life of the system.

The 'reliability' or 'chance' constraints define an admissible area in which the optimal release curve, maximizing or minimizing an objective function, is sought. Sometimes they are replaced with constraints on expected value of some variables or functions.

The typical formulation is as follows:

$$\max_{r_k} \mathrm{E}\left\{\sum_{k=1}^{N} B_k(s_k, r_k, i_k)\right\} \qquad (1)$$

$k=1,..,N$

subject to:

$$s_{k+1}=s_k+i_k-r_k \qquad k=1,..,N-1 \qquad (2)$$

$$s_1=\bar{s} \qquad (3)$$

$$0 \leq s_k \leq s_{max} \qquad (4)$$

$$r_k \in R_k(s_k) \qquad (5)$$

$$\mathrm{P}(s_k \in S_k) \geq \alpha_k \qquad k=2,..,N \qquad (6)$$

$$\mathrm{P}(g_k(s_k, r_k, i_k) \in G_k) \geq \beta_k \qquad k=1,..,N \qquad (7)$$

$$\mathrm{P}(h_k(s_k, r_k, i_k) \in H_k, \qquad k=1,..,N) \geq \gamma \qquad (8)$$

where:

E is the expectation operator on the probability distributions of inflows;
$[1, N]$ is the control horizon;
$B_k(.,.,.)$ are the benefits in the period k;
s_k is the storage level at the start of the period k;
r_k is the release in the period k;
i_k is the inflow in the period k, a random variable with a known probability distribution (it is assumed that subsequent inflows are independent random variables);
$\bar{s}$ is the initial storage of the reservoir;
s_{max} is the maximal storage capacity;
S_k is the set of desirable storage levels in the period k;
α_k is the minimal admissible reliability of belonging the storage s_k in the period k to the set S_k;
g_k, h_k are the real-valued functions of storage, release and inflow in the period k;
G_k is the set of desirable values of the function g_k;
β_k is the minimal admissible reliability of belonging the value of function g_k to the set G_k;
H_k is the set of desirable values of function h_k;
γ is the minimal admissible reliability (probability) that the values of the functions h_k belong to the sets H_k, calculated over the whole control horizon;

$R_k(s_k)$ is the set of admissible releases in the period k dependent on the current storage s_k.

Only upper constraints on probabilities are considered. Lower constraints on probabilities can be transformed into the upper constraints in an obvious way.

Constraints (4) and (6) are used alternatively. The former one is usually incorporated into the state equation. That leads to the following transition rule:

$$s_{k+1} = \max(0, \min(s_{max}, s_k + i_k - r_k)) \tag{9}$$

Of course, this trick guarantees the fulfillment of the state constraint (4), but it can be shown (Karbowski, 1991), that such simplification causes, even in deterministic case, the departure of the solution from the optimal to a suboptimal one.

Sets S_k are usually time-invariant and have the form:

(a) $[s_{min}, \infty]$ the corresponding constraint $P(s_k \geq s_{min}) \geq \alpha_k$
(b) $[-\infty, s_{max}]$ ——————”—————— $P(s_k \leq s_{max}) \geq \alpha_k$
(c) $[s_{min}, s_{max}]$ ——————”—————— $P(s_{min} \leq s_k \leq s_{max}) \geq \alpha_k$

Similarly, sets G_k are in most cases constant intervals. A typical example of the constraint (8) is the lower constraint on reliability to supply the target release r_k during the whole control horizon. In this case the functions h_k and sets H_k are stationary and have the form:

$$h_k(s_k, r_k, i_k) = h(s_k, r_k, i_k) = s_k + i_k - r_k \tag{10}$$

$$H_k = H = [0, \infty) \tag{11}$$

As it was mentioned earlier, all reliability constraints (6)–(8) may be replaced with constraints on expected value of the same expressions, i.e.,

$$\mathrm{E}\, s_k \in S_k \tag{12}$$

$$\mathrm{E}[g_k(s_k, r_k, i_k)] \in G_k \tag{13}$$

$$\mathrm{E}\left[\sum_{k=1}^{N} \rho_{H_k}(h_k(s_k, r_k, i_k))\right] \in HS \tag{14}$$

where ρ is the characteristic set function of the form:

$$\rho_A(x) = \begin{cases} 1 & x \in A \\ 0 & x \notin A \end{cases} \tag{15}$$

and HS is the set of the desired number of successes.

SHORT REVIEW OF METHODS USED FOR DETERMINATION OF CONTROL STRATEGIES IN THE PRESENCE OF THE CHANCE CONSTRAINTS

At the beginning the optimal control problems with chance constraints were solved in a simplified way under the assumption, that the release in every period is a linear (or more precisely affine) function of storage (or storage and past inflows) and inflows in subsequent time periods are independent random variables with known distributions (cf. ReVelle *et al.*, 1969; Loucks, 1970). This assumption allowed expressing the stochastic control problem in terms of the deterministic linear or nonlinear programming, for linear or nonlinear objective function, respectively. Unfortunately the solutions, although elegant from a mathematical point of view, belonged to a very narrow class of functions, were often far from the optimum (Stedinger, 1984).

In the turn of seventies and eighties a new method of a similar kind was proposed (Colorni & Fronza, 1979; Simonovic & Marino, 1980; Marino & Mohammadi, 1983). In the so-called 'reliability programming' (RP) the demand of linearity of the control rule was rejected and the probability levels restricting risk were optimized as additional decision variables, rather than being fixed *a priori*. Because of that the objective function was augmented with components representing losses related to these probabilities.

The first impression was that this approach is more general than the previous one, based on linear decision rules. However, as noted by Strycharczyk & Stedinger (1987), in contrast to the linear decision rule (LDR) models, there were no 'allowance for variations in response to actual inflow and demand levels' in the RP. One can add that there were no variations in response to the reservoir storage level, either. In terms of the control theory, the RP models proposed open-loop control scheme. Their results were therefore worse than these of the LDR models, because the class of the control rules that are constant functions is a subset of the class of these control rules that are functions of measurements, even as simple functions as linear (affine).

From the stochastic control theory point of view both approaches are far from the optimal one. The main reasons are the following:

1. They use very small pieces of the information concerning the distribution of inflows, only one statistics of the distribution, namely a quantile of a specified order (α_k, β_k, γ or $1-\alpha_k$, $1-\beta_k$, $1-\gamma$). The rest is completely ignored. That is, the proposed releases may be identical for various distribution curves, provided their quantiles of specified orders are the same.
2. They do not take into account the information how the real-time control system works. The most important features of such systems are as follows:
   (a) at time k, while calculating a control for the nearest period there is uncertainty about future inflows, but the past inflows and the current state are known
   (b) the current intervention is not the last one (unless k is the ending time of the control horizon), and at times $k+1, k+2, k+3$ one gets new measurements and may calculate the suitable control. This is the main difference between the open-loop feedback control (OLFC)

and the closed-loop feedback control (CLFC) schemes (see Bertsekas, 1976).

The mechanisms that ignore this wealth of information cannot be effective. That is why their performances in the tests conducted by Strycharczyk & Stedinger (1987) and Stedinger (1984) were so bad.

The only method of solving this problem that makes use of its sequential nature, as well as of the sequential nature of the real-time control process, is the stochastic dynamic programing (SDP).

The first algorithm was proposed by Askew (1974a, b). He analyzed the problem of deriving the optimal operating policies that maximize the expected net benefits after a 50 year design period for a water supply system. Its statement included the performance index (1), the state equation (9) with initial condition (3) for $\bar{s} = s_{max}$, the release constraint (5) and the global risk constraint of the type (8) or (14), where functions h_k and the sets H_k were defined by (10) and (11). That means, the risk of the deficit in the next period did matter the most. Askew proposed a combination of dynamic programing and simulation. The main idea consisted in augmenting the right hand side of the recursive equation of the dynamic programing (DP) with a penalty function taking only two values: $W < 0$ if the deficit of water is expected at the next stage (that is if $h_k(s_k, r_k, i_k) \notin H_k$) and 0 otherwise. By heuristic, interactive searching over a range of values for W and performing the whole DP algorithm for each value, a policy was found that offered the maximum expected net benefit and complied with the reliability constraint. Unfortunately, such iterative process might converge very slowly and, as Sniedovich & Davis remarked (1975), not necessarily to the optimal solution of the original problem.

The latter difficulty was overcome by Rossman (1977) who proposed more effective and formally justifiable algorithm, yet still based on the penalty function approach. The algorithm for problems with constraints of type (14) was developed within the context of the Lagrangian duality theory of nonlinear programing. In this connection the objective function (1) was augmented with a linear penalty term representing the cost of violating the reliability constraint (14) and all stage problems were augmented with appropriate stage penalty terms. Optimal value of the Lagrange multiplier λ was sought, due to duality theorem, via minimization of the total cost-to-go (that is the cost after performing one – for a specific λ – DP backward iteration for $k = N$, $N-1$, $N-2, .., 1$) by any method of scalar minimization.

After two years Sniedovich (1979) generalized the Rossman approach for problems with probability constraints of type (8). The method proposed by him has the same drawbacks as those of Askew and Rossman. It requires multiple solving of the control problem by DP method for different Lagrange multipliers to achieve the saddle point of the Lagrange function. Some convexity conditions must be satisfied to achieve this, otherwise the duality gap may appear (see Rossman, 1977).

The question arises if there is a method of solving this problem directly by passing the DP recursive scheme only once. To the present author's knowledge, supported by the opinion of Yeh (1985): '*The reliability-constrained or chance-constrained formulation* [of DP] *has been solved with the penalty function approach or the Lagrangian duality theory of nonlinear programming*', until now there has been no other investigations on this problem.

Although Yakowitz (1982) stated that: '*It does not seem possible to fit such chance-constraint reservoir problem exactly into the optimal control problems format*', there are some possibilities to imbed this problem into the classical DP formulation. They were perceived by Sniedovich & Davis (1975), but without presentation of a concrete algorithm and results, and as it will be explained below, with some shortcomings.

At last some remarks should be made on very popular in recent years algorithms applying differential DP to reservoir management problems with stochastic inflows (Georgakakos & Marks, 1987, Trezos & Yeh, 1987). Although they could deal with problems of very high dimension, from control theory point of view they are suboptimal, because they realize open-loop feedback control scheme. We will not analyze them here, ending with the remark that they take the risk constraints into account through barrier function (Georgakakos & Marks, 1987) or through guessing a feasible initial trajectory and considering these constraints in some suboptimal way, independently while every DP iteration (Trezos & Yeh, 1987).

STATISTICALLY SAFE SETS METHODOLOGY

The main idea underlying the statistically safe sets methodology stems from the observation, that difficulties which appear while solving the problems (1)–(8) by combination of DP and penalty function technique are related to twofold character of each stage optimization. At every time period we look for the control which assures the best trade-off between the benefits expressed by the objective function and losses expressed by the penalty term. The result is dependent on the penalty coefficient (W in the Askew's method or λ in Rossman's method). During many iterations the solutions, i.e., the optimal strategies $r_k^l = \hat{r}_k(s_k, W^l)$ or $r_k^l = \hat{r}_k(s_k, \lambda^l)$, where l is the number of the penalty coefficient iteration, are inadmissible from the constraints' (8) or (14) point of view. On the other hand, many iterations may produce strategies giving very low value of the objective function, yet being admissible.

The question arises if these two requirements can be separated. The answer is positive, we can achieve it by using *statistically safe sets methodology*. This methodology consists in imposing additional deterministic restrictions on the sets R_k of admissible releases (5), in order to assure that all reliability constraints (6)–(8) are satisfied. These constraints are calculated recursively in the same manner (i.e., passing backward) as the Bellman function.

The proposed approach may be treated as the relaxed version of the Yakowitz's (1969) 'sets of strategies associated with an adaptive control problem'. He analyzed the optimal control problems with statistic law of motion, the finite horizon (he called them 'truncated Markovian adaptive control processes') and deterministic constraints on state variable having the form

$$s_k \in S_k \qquad \forall k \in \overline{1, N}$$

To ensure that these conditions are satisfied for each k, at every stage Yakovitz (1969) restricted the set of admissible controls $R_k(s_k)$ to

$$R_k^*(s_k) = \{r \in R_k(s_k)\text{: } P(s_{k+1} \in S_{k+1} | s_k, r, k) = 1\} \tag{16}$$

In statistically safe sets methodology formula (16) is modified on account of probabilistic, not deterministic as above, nature of constraints (6) and the necessity to ensure that all controls from the set R_k^* (constraint (5)) lead the system (of course in a probabilistic sense) to such states s_{k+1} in time period $k+1$ for which any admissible control strategy exist.

More precisely, at each stage k we consider two sets, namely the set of statistically safe storages and the set of statistically safe releases. The first one comprises these storages s_k for which there exists at least one sequence of releases r_k, $r_{k+1}, \ldots, r_N$ belonging respectively to sets R_k, $R_{k+1}, \ldots, R_N$, which, together with the resulting state trajectory, satisfies the reliability constraints (6) in periods k, $k+1, \ldots, N$.

The set of statistically safe releases in period k includes these releases satisfying constraint (5), that lead from the current to the next (in period $k+1$) set of statistically safe storages. It should be emphasized, that both definitions correspond to the behaviour of the release and storage trajectories, which means, that the attention is paid not only to current values of these variables in every period k, $k+1, \ldots, N$, but also to transitions between subsequent periods. The latter thing is quite often neglected while constructing reservoirs' control rules.

The method used to evaluate the safe sets is based, as optimization in DP, on recursion.

Let us denote as X_k the set of statistically safe storages in period k and $\bar{R}_k(s_k)$ – the set of statistically safe releases in period k, when the storage at the start of this period is equal to s_k. The recursive formula accounting both the deterministic constraint on release (5) and the reliability constraints (6), (7) is as follows

$$\bar{R}_k(s_k) = \{r \in R'_k(s_k)\text{: } \mathrm{P}(s_k + i_k - r \in S_{k+1} \cap X_{k+1}) \geq \alpha_k\} \tag{17}$$

$$R'_k(s_k) = R_k(s_k) \cap \tilde{R}_k(s_k) \tag{18}$$

$$\tilde{R}_k(s_k) = \{r\text{: } \mathrm{P}(g_k(s_k, r, i_k) \in G_k) \geq \beta_k\} \tag{19}$$

$$X_k = \{s_k\text{: } \bar{R}_k(s_k) \neq \varnothing\} \tag{20}$$

Expressions (18) and (19) mean, that as far as we assume that inflows in subsequent time periods are independent random variables, the mixed reliability constraint (7) on current storage and release can be treated as an additional deterministic constraint on release defining set $\tilde{R}_k(s_k)$. On the other hand, sets X_{k+1} in expression (17) additionally restrict sets S_{k+1} to ensure not only the fulfillment of the risk constraints (6) but also transitions between safe sets in subsequent time periods. In this way the presented approach may be treated as stochastic version of the 'adaptive corridor width selection' (Yakowitz, 1969; Murray & Yakowitz, 1979).

Expressions (17)–(19) can be written commonly as:

$$\begin{aligned}\bar{R}_k(s_k) = \{r \in R_k(s_k)\text{: } &\mathrm{P}(g_k(s_k, r, i_k) \in G_k) \geq \beta_k, \\ &\mathrm{P}(s_k + i_k - r \in S_{k+1} \cap X_{k+1}) \geq \alpha_k\}\end{aligned} \tag{21}$$

The compact form of the above definitions was obtained due to the property that during control we use feedback and while determining the release for period k we will know the value of the current storage. Because of that and the independence assumption, the probabilities in (6), (7) and (17), (19) or (21) are calculated with regard only to current inflow i_k.

The global constraint (8) has a different character since the probability does not concern the inflow i_k in one period k, but is calculated over the sequence of inflows i_1, $i_2, \ldots, i_N$. The only method of incorporating it into DP procedure is based on introducing an additional state variable accounting the probability of not failure up to the beginning of the period under consideration. Denote this variable as y_k. Its state equation is:

$$\begin{aligned} y_{k+1} = f_k(y_k, s_k, r_k, i_k) = y_k \cdot \mathrm{P}(h_k(s_k, r_k, i_k) \in H_k), \\ k = 2, \ldots, N \qquad y_1 = 1 \end{aligned} \tag{22}$$

Now the constraint (8) can be written as

$$y_{N+1} \geq \gamma \tag{23}$$

To apply the safe sets methodology, we should add the suitable terms to definitions (17) or (21). For example the expression (17) will take the form:

$$\begin{aligned}\bar{R}_k(s_k, y_k) = \{r \in R'_k(s_k)\text{: } &\mathrm{P}(s_k + i_k - r \in S_{k+1} \cap XS_{k+1}(v)) \geq \alpha_k, \\ &v = y_k \cdot \mathrm{P}(h_k(s_k, r, i_k) \in H_k) \in XY_{k+1}\}\end{aligned} \tag{24}$$

$$X_k = \{(s_k, y_k): \bar{R}_k(s_k, y_k) \neq \varnothing\}, \qquad k = 1, \ldots, N \tag{25}$$

$$X_{N+1} = S \times [\gamma, 1] \tag{26}$$

$$XS_{k+1}(v) = \left\{ s: \begin{bmatrix} s \\ v \end{bmatrix} \in X_{k+1} \right\} \tag{27}$$

$$XY_{k+1} = \Pi_Y(X_{k+1}) = \left\{ y: \exists s \begin{bmatrix} s \\ y \end{bmatrix} \in X_{k+1} \right\} \tag{28}$$

The sets X_k are now defined on the cartesian product of spaces of storages S and probabilities Y. The boundary condition (26) for the iterative process of determining these sets results from the global probability constraint of type (8) and assures that this constraint (expressed also by the new state variable y_k and the inequality (23)) will be satisfied. The set XY_{k+1} is a projection (Π is the operator of projection) of the set X_{k+1} on the space Y and $XS_{k+1}(v)$ is a set of safe storages corresponding to the probability of success equal v. The sense of all these sets is explained in Fig. 1.

In a similar way we can define the safe sets for the expected value type of constraints. The equivalent of (17)–(19) is

$$\bar{R}_k(s_k) = \{r \in R'_k(s_k): \mathrm{E}\,[s_k + i_k - r] \in S_{k+1} \cap X_{k+1}\} \tag{29}$$

$$\tilde{R}_k(s_k) = \{r: \mathrm{E}\,[g_k(s_k, r, i_k)] \in G_k\} \tag{30}$$

and the equivalent of (24)

$$\bar{R}_k(s_k, y_k) = \{r \in R'_k(s_k): \mathrm{E}\,[s_k + i_k - r] \in S_{k+1} \cap XS_{k+1}(v)), \quad v = y_k + \mathrm{E}\,[\rho_{H_k}(h_k(s_k, r, i_k))] \in XY_{k+1}\} \tag{31}$$

In the latter expression y_k is an additional state variable satisfying the state equation:

$$y_{k+1} = f_k(y_k, s_k, r_k, i_k) = y_k + \mathrm{E}\,[\rho_{H_k}(h_k(s_k, r_k, i_k))], \quad k = 2, .., N \quad y_1 = 0 \tag{32}$$

and the final condition

$$y_{N+1} \in HS$$

In this case the variable y_k accounts the number of successes (of belonging the value of the functions $h_k(s_k, r_k, i_k)$ to sets H_k for subsequent k) up to the beginning of the period under consideration.

The definitions of sets X_k, $XS_k(v)$, XY_k remain the same as above (see formulas: (25), (27), (28)) as well as the set $R'_k(s_k)$ (see (18)). Only the boundary condition on the sets of admissible states (both storages and the expected number of successes) will be different, namely:

$$X_{N+1} = S \times HS \tag{33}$$

Let us notice, that all the above presented sets, both the sets of the safe states (physical, i.e., storages and artificial probabilities or the expected number of successes) can be determined before the optimization, that is before the essential iteration of the DP algorithm. Of course, the Bellman

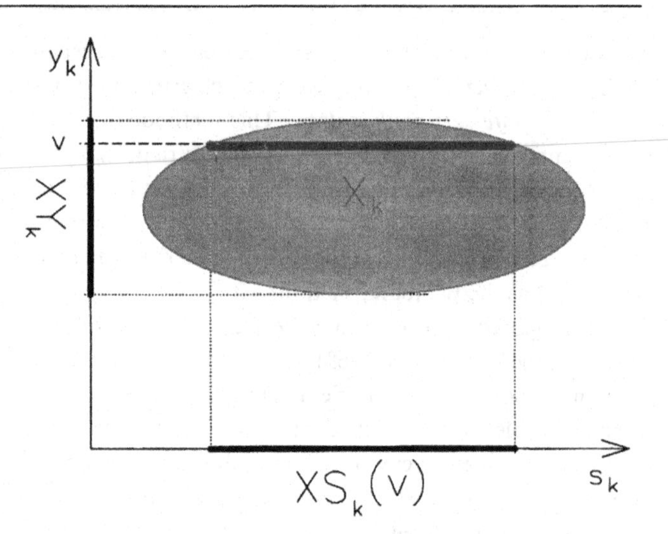

Fig. 1 Sets of safe states in problems where the global risk constraints are present.

functions (functions of the 'optimal cost-to-go') and the control rules $\hat{r}_k(s_k, y_k)$ will be calculated only for points (s_k, y_k) belonging to sets X_k for all k.

Hence, the backward DP algorithm will be as usual:

$$F_k(s_k, y_k) = \max_{r_k \in \bar{R}_k(s_k, y_k)} \mathrm{E}\,\{B_k(s_k, r_k, i_k) + F_{k+1}(s_k + i_k - r_k, f_k(y_k, s_k, r_k, i_k)\}, \qquad k = N-1, .., 1 \tag{34}$$

with

$$F_N(s_N, y_N) = \max_{r_N \in \bar{R}_N(s_N, y_N)} \mathrm{E}\,\{B_N(s_N, r_N, i_N)\} \tag{35}$$

where functions f_k are defined by (22) or (32), and the expectation is calculated with respect to the stage inflow i_k.

As it is seen from the recurrence scheme (34), (35), all constraints in the general problem (1)–(8) or (1)–(5), (12)–(14) can be transformed, owing to statistically safe sets methodology, to deterministic constraints on release. Moreover, this process can be performed before the optimization (i.e., before DP backward recurrence scheme (35),(34)). Because of that, while we restrict our attention only to sets X_k, for which the Bellman functions F_k as well as the optimal control strategies $\hat{r}_k(s_k, y_k)$ are calculated, there is no necessity to check if 'chance' or 'expected value type' constraints are satisfied. It is simply guaranteed by choosing releases r_k from sets $\bar{R}_k(s_k, y_k)$.

To explain better the application and possibilities of stochastically safe sets methodology, in the next section its application to two case studies will be presented. The first example is taken from the classical literature (Askew, 1974a, b; Rossman, 1977; Sniedovich, 1979, see the review section), while the second one concerns real-time winter flood control problem.

CASE STUDIES

Example 1

Askew (1974a, b), Rossman (1977) and Sniedovich (1979) studied a single reservoir water supply system. The basic goal was to determine the release rule which maximized the expected value of discounted net benefits subject to constraints on the probability of failure to supply the target release or on the expected number of failures over the lifetime of the system. More precisely, the problem was as follows:

$$\max_{r_k} E\left\{\sum_{k=1}^{N}\left[B_k(s_k, r_k, i_k)=\frac{1}{(1+p)^{k-1}}\cdot(B(a_k)-c\cdot r_k)\right]\right\}$$

$$k=1,..,N \quad (36)$$

$$s_{k+1}=\max(0, \min(s_{\max}, s_k+i_k-r_k)) \quad (37)$$

$$s_1=s_{\max} \quad (38)$$

$$r_k\in[0, s_k+i_{\max}] \quad (39)$$

$$\mathrm{P}(s_k+i_k-r_k<0, \quad k=1,..,N)\leq\delta \quad (40a)$$

or

$$\mathrm{E}\left[\sum_{k=1}^{N}\rho_{[-\infty,0)}(s_k+i_k-r_k)\right]\leq NF \quad (40b)$$

where:

N is the number of years under consideration; lifetime of the reservoir system (it was assumed in calculations that $N=50$);

i_k is the inflow to the reservoir, a discrete random variable with identical and independent distributions given by the following table

i	P(i)
0	0.0062
1	0.0606
2	0.2417
3	0.3830
4	0.2417
5	0.0606
6	0.0062

$i_{\max}$ is the maximal annual inflow, $i_{\max}=6$;
B_k is the annual net dollar benefits function;
$a_k=\min(s_k+i_k, r_k)$ is the actual annual release;
$s_{\max}$ is the maximal admissible storage (the assumed value: $s_{\max}=3$);
s_k is the storage at the beginning of the kth year, a discrete variable, $s_k\in S=\{0, 1, 2, 3\}$;
r_k is the release in kth year, a discrete variable;
p is the discount rate;
c is the cost parameter (the assumed value: $c=4$);
$B(a_k)$ is the undiscounted revenue function, defined by the following table:

a	$B(a)$ – case A	$B(a)$ – case B
0	0	0
1	10.5	5.0
2	20.5	10.5
3	30.0	16.5
4	39.0	23.0
5	47.5	30.0
6	55.5	37.5
7	63.0	45.5
8	70.0	54.0
9	76.5	63.0

δ is the maximal admissible probability of failure;
NF is the maximal admissible number of failures;
ρ is the characteristic set function (see def. (15)).

To express this problem in terms of our general formulation (1)–(8) (or (1)–(5), (12)–(14)) we should take:

$$R_k(s_k)=R'(s_k)=\{0,1,..,s_k+6\} \quad (41)$$

$$h_k(s_k, r_k, i_k)=s_k+i_k-r_k \quad (42)$$

$$H_k=[0,\infty) \quad (43)$$

$$\gamma=1-\delta \quad (44)$$

$$HS=\{N-NF, N-NF+1,\ldots, N\} \quad (45)$$

and add a new artificial state variable y_k, satisfying state equations, respectively, (22) in the case of the chance constraint (40a) or (32) in the case of the 'expected value' type constraint (40b).

Moreover, let us notice, that due to the definition of success (the situation when there is no *necessity* of the deficit at the next stage), and the form of the release constraint (41), for every storage $s_k\in\{0, 1, 2, 3\}$, and every y_k admissible, there exists at least one release, that guarantees that y_{k+1} will be also admissible (that is the global constraints (40a) or (40b) will not be violated by an action performed at the next stage). This special release is equal 0. But that means, that the safe sets will be time-invariant, and will have the form of a box

$$X_k=X=X_{N+1}=S\times[\gamma,1] \quad (46)$$

for the global constraint (40a), or

$$X_k=X=X_{N+1}=S\times HS \quad (47)$$

for the constraint (40b).

Because of that the definitions (24) and (31) will take the form:

$$\bar{R}_k(s_k, y_k)=\{r\in R(s_k): y_k\cdot\mathrm{P}(h_k(s_k, r, i_k)\in H_k)\in[\gamma,1]\} \quad (48)$$

$$\bar{R}_k(s_k, y_k)=\{r\in R(s_k): y_k+\mathrm{E}[\rho_{H_k}(h_k(s_k, r, i_k))]\in HS\} \quad (49)$$

which is equivalent to introducing additional stage constraints on variable y_k of a form:

$$y_k\geq\gamma \qquad k=2,3,..,N \quad (50)$$

or

$$y_k \geq N - NF \qquad k=2,3,..,N \tag{51}$$

It should be emphasized, however, that such transformation can be applied only in the case when the sets of safe states have the form (46), or (47), which in our problem is simply a consequence of possibility of implementing the zero release from the reservoir (i.e., $\forall s_k \in S\ 0 \in R(s_k)$). In general, the formulation (36)–(39), (22), (50) or (36)–(39), (32), (51) is not equivalent to (36)–(39), (40a) or (36)–(39), (40b).

Because of that, Sniedovich & Davis (1975) in general were not right. Their state variable y_k together with constraints of (50) or (51) type does not lead, except the specific situation that was mentioned above, to optimal solution of the reservoir control problem with risk constraints. Thus the only method transforming general reservoir's management problems with risk constraints (1)–(8) or (1)–(5), (12)–(14) to the form where the standard dynamic programming algorithm can be applied is based on statistically safe sets methodology.

There are also some misunderstandings with the interpretation of the state variable y_k. Though artificial, it is deterministic and proper state variable, because it depends on the history only through its last value y_{k-1} and the last storage s_{k-1}. The probability of success (not failure) can take values from the interval $[\gamma,1]$. So that is not true what Yakowitz (1982) wrote, that '[while process] the only possible value for the probability are 0 and 1'. During the control process the probability of success y_k (the same with the expected number of successes) should be calculated all the time due to formula (state equation) (22) (or (32)). Of course, its value is changing after each period. And there is no relation between this value and the real number of successes (or failures) that occur during the control process. The feedback with the process is realized only through the measured value of storage s_k. It influences y_k by the state equation (22) or (32).

Two numerical experiments were performed. The first one with the chance constraints (40a), for the probability of failure $\delta=0.1$, and the benefit function $B(a)$ as in case A, with the discount rate $p=0.1$, and the second one with the expected value constraints of type (40b), for the number of failures $NF=1$, and the benefit function $B(a)$ as in case B, without discounting (i.e., $p=0$). In both cases the artificial variable y_k was discretized into 50 intervals. The results were as follows: in the first case, 191.2275, and in the second case, 242.7426. When we compare these results with the results reported by Sniedovich (1979), and Rossman (1977) it turns out that in the first case our result is a little worse (Sniedovich obtained 191.8, but author could not achieve it for the optimal strategy described by Sniedovich), but in the second case our result was better (the best result reported by Rossman for randomized policy was equal to 242.1).

The computations took about 2 minutes on IBM PC/AT-compatible computer with 8 MHz clock.

Example 2

This example concerns the reservoir management during thaw flood. It is presented here to show that the statistically safe sets methodology is quite a flexible tool and can be applied not only to problems expressed in terms of (1)–(8) or (1)–(5), (12)–(14), but also to other water reservoir management problems.

The main difference between the model considered in this section and the previous one consists in other than summation performance index (it is of the minimax type now, which means that we assume, that flood damages are proportional to the culminant flow at a cross-section just below the reservoir) and in the first order Markovian model of inflow (for higher order processes the argumentation is analogous). Of course, the latter thing is a big simplification, but we present this problem rather to illustrate the possibilities of statistically safe sets methodology than to solve it for a concrete system.

It is obvious, that when a Markovian model of inflow is used, the state vector contains one or several (in the case of a higher order of the Markov process) additional coordinates (corresponding to past inflows) and their number is equal to the order of the Markov process. Thus, in our case the state vector will have the form:

$$x_k = [s_k, i_{k-1}] \tag{52}$$

We want to work out a unified control strategy, that is a sequence of the reservoir's control rules for each time stage during the whole flood period. Obviously, these rules should depend on the reservoir storage and on the latest inflow (or several latest inflows according to the order of the Markov chain).

The problem can be formulated as follows

$$\min_{r_k} E\{\max_{k\in[1,N-1]} r_k\} \tag{53}$$

$$k=1,\ldots,N-1$$

subject to:

$$s_{k+1} = s_k + i_k - r_k, \qquad k=1,..,N \tag{54}$$

$$r_k \in R(s_k) = [r_{\min}, r_{\max}(s_k)] \qquad \text{''} \tag{55}$$

$$\mathrm{P}(s_k \in S_k = [s_{\min}, s_{\max}]) \geq \alpha_k, \qquad k=2,..,N \tag{56}$$

The quantity α_k is a given *a priori* level of reliability that our control will not evoke the violation of the reservoir storage constraints (56).

Because our structure will be closed-loop, the control as a function of the state variables (or the state vector) will be also a random variable and there is a point in minimizing its expected value (more precisely the expected value of a functional of it).

According to the first order Markovian assumption, dependencies between inflows in subsequent time instants can be represented by transition probability matrices $[p_k^{mn}]$, where p_k^{mn} is the probability that the inflow, or more precisely the level of inflow, i^n in period $k+1$ follows the inflow i^m in period k.

Before the synthesis of the optimal control we should determine for every stage k of the control horizon the set of safe controls $\bar{R}_k(s_k, i_{k-1}^m)$ as well as the set of safe storages and inflows X_k from the relationship:

$$\bar{R}_k(s_k, i_{k-1}^m) = \{r \in R(s_k): \\ P(s_{k+1} = s_k + i_k - r \in XS_{k+1}(i_k) \cap S_{k+1} | i_{k-1} = i^m) \geq \alpha_k \tag{57}$$

$$X_k = \{(s_k, i_{k-1}): \bar{R}_k(s_k, i_{k-1}) \neq \varnothing\} \tag{58}$$

$$XS_{k+1}(v) = \left\{s: \begin{bmatrix} s \\ v \end{bmatrix} \in X_{k+1}\right\} \tag{59}$$

Taking into account that the inflow is a Markov chain of the first order, we can calculate the probability in definition (57) due to the formula

$$P(s_{k+1} = s_k + i_k - r \in XS_{k+1}(i_k) \cap S_{k+1} | i_{k-1} = i^m) \\ = \sum_n p_{k-1}^{mn} \cdot \rho_{A_{n,k+1}}(s_k + i_k^n - r) \tag{60}$$

where ρ denotes the characteristic set function defined by (15) and

$$A_{n,k+1} = XS_{k+1}(i_k^n) \cap S_{k+1}$$

As the final condition we take:

$$X_N = \{(s, i): s \in S_N, i \in I\} \tag{61}$$

where I denotes the range of possible inflows.

If at a stage k the set X_k or $\bar{R}_k(.,.)$ proves to be empty, it will signify that there exists no control assuring the maintenance of the reservoir in the admissible area with the reliability greater than α_k.

After the determination of the sets of safe storages and releases, we will calculate the optimal control strategy, that is such a sequence of admissible control rules

$$\hat{r}_k(s_k, i_{k-1}), k = 1, 2, \ldots, N-1 \tag{62}$$

where

$$\hat{r}_k(s_k, i_{k-1}) \in \bar{R}_k(s_k, i_{k-1}) \tag{63}$$

which asserts the lowest value of the performance index (53). The expected value in the performance index (53) is calculated with respect to densities of random variables $i_1, i_2, \ldots, i_{N-1}$ forming the Markov chain with the given vector of initial probabilities

$$p_0 = [0, 0, \ldots, 0, 1, 0, \ldots, 0]$$

where the figure one is put at the mth position according to the measured level of the initial inflow $i_0 = i^m$.

For this problem the Bellman recursion scheme will be as follows:

$$F_{N-1}(s_{N-1}, i_{N-2}^m) = \min_{r_{N-1} \in \bar{R}(s_{N-1}, i_{N-2}^m)} r_{N-1} \tag{64}$$

$$F_k(s_k, i_{k-1}^m) \\ = \min_{r_k \in \bar{R}_k(s_k, i_{k-1}^m)} \mathop{E}_{i_k}\left[\max(r_k, F_{k+1}(s_k + i_k - r_k, i_k)) | i_{k-1} = i^m\right] \\ = \min_{r_k} \sum_n p_{k-1}^{mn} \cdot \max\left(r_k, F_{k+1}(s_k + i_k^n - r_k, i_k^n)\right) \\ \text{for } k = N-2, N-3, \ldots, 3, 2, 1 \tag{65}$$

The Bellman function for the last period of the control horizon does not contain the expectation operator, because, due to our convention the control r_{N-1} at the stage $N-1$ is independent of the inflow at this stage.

Performing at each step k the minimization (65) for all pairs $\begin{bmatrix} s_k \\ i_{k-1}^m \end{bmatrix} \in X_k$ we obtain a sequence of control rules (62), that is the optimal control strategy. Applying them during control horizon at each time instant k we assure the minimization of the performance index (53) and at the same time not violation of the given (instantaneous) reliability level α_k of maintaining the reservoir within constraints. It will not protect us, however, completely from going out of the sets of admissible states X_k or S_k. It is a consequence of our admission that the reliability α_k is less than one. Moreover, the inflow model as each model of the physical phenomenon, is not ideal. The most sensible thing as may be done in such, failure situation is to realize a control out of the set $\bar{R}_k(s_k, i_{k-1})$ but from the set $R(s_k)$, leading the storage to set $X_{k+1}(i_k)$ at the next time instant $k+1$. Then we will be able to apply the rules (62) again.

CONCLUSIONS

The presented methodology is a general tool for solving reservoir management problems in risk conditions, when inflows are random variables and some 'chance' or 'expected value' constraints (imposed on stage values or on trajectories of storages or storages together with releases and inflows) have to be satisfied. If the optimal control strategy exists, it could be calculated with the help of this methodology. Its main advantage results from the fact that it separates two inherent components of each reservoir control problem: optimization of a performance index and assurance that all reliability constraints are satisfied. The second task, consisting in the determination of the sets of safe storages (and sometimes other state variables) and releases, can be solved before the essential DP recursion. This recursion is the standard DP backward stagewise optimization performed for particular values of state variables with properly modified sets of admissible releases depending on states.

While using statistically safe sets methodology we can determine the optimal control strategy after one iteration of the DP algorithm, there is no necessity, as in other approaches, to perform many DP iterations to find the optimal value of a penalty coefficient. There is no danger of getting into the 'duality gap' (compare Rossman, 1977), either.

Statistically safe sets methodology can be applied both in discrete DP and DP with approximations of the stagewise cost-to-go functions and the optimal control rules (see Foufoula-Georgiou & Kitanidis, 1988), as well as in problems with independent and Markovian inflow models.

It can be also useful in some suboptimal approaches stemming from differential DP idea (see Georgakakos & Marks, 1987; Trezos & Yeh, 1987). In that case subsequent iterations would be performed only to minimize the performance index, they would not be lost on looking for an admissible (i.e., satisfying risk constraints) solution.

REFERENCES

Askew, A. J. (1974a) Optimum reservoir operating policies and the imposition of a reliability constraint. *Water Resour. Res.*, **10**(1), 51–6.

Askew, A. J. (1974b) Chance-constrained dynamic programing and the optimization of water resource systems. *Water Resour. Res.*, **10**(6), 1099–106.

Bertsekas, D. P. (1976) *Dynamic Programming and Stochastic Control*, Academic Press, New York.

Colorni, A. & Fronza, G. (1976) Reservoir management via reliability programming. *Water Resour. Res.*, **12**(1), 85–8.

Foufoula-Georgiou, E. & Kitanidis, P. K. (1988) Gradient dynamic programming for stochastic optimal control of multidimensional water resources systems. *Water Resour. Res.*, **24**(8), 1345–59.

Georgakakos, A. P. & Marks, D. H. (1987) A new method for the real-time operation of reservoir systems. *Water Resour. Res.*, **23**(7), 1376–90.

Karbowski, A. (1991) Optimal control of single retention reservoir during flood; Analytical solution of deterministic, continuous-time problems. *Journal of Optimization Theory and Applications*, **69**(1), 55–81.

Loucks, D. P. (1970) Some comments on linear decision rules and chance constraints. *Water Resour. Res.*, **6**(2), 668–71.

Mariño, M. A. & Mohammadi, B. (1983) Reservoir management: A reliability programming approach. *Water Resour. Res.*, **19**(3), 613–20.

Murray, D. M. & Yakowitz, S. J. (1979) Constrained differential dynamic programming and its application in multireservoir control. *Water Resour. Res.*, **15**(5), 1017–27.

ReVelle, C., Joeres, E. & Kirby, W. (1969) The linear decision rule in reservoir management and design, 1. Development of the stochastic model. *Water Resour. Res.*, **5**(4), 767–77.

Rossman, L. A. (1977) Reliability-constrained dynamic programing and randomized release rules in reservoir management. *Water Resour. Res.*, **13**(2), 247–55.

Simonovic, S. P. & Mariño, M. A. (1980) Reliability programming in reservoir management, 1. Single multipurpose reservoir. *Water Resour. Res.*, **16**(5), 844–8.

Sniedovich, M. (1979) Reliability-constrained reservoir control problems, 1. Methodological issues. *Water Resour. Res.*, **15**(6), 1574–82.

Sniedovich, M. & Davis, D. R. (1975) Comment on 'Chance-constrained dynamic programing and optimization of water resource systems' by Arthur J. Askew. *Water Resour. Res.*, **11**(6), 1037–8.

Stedinger, J. R. (1984) The performance of LDR models for preliminary design and reservoir operation. *Water Resour. Res.*, **20**(2), 215–24.

Strycharczyk, J. B. & Stedinger, J. R. (1987) Evaluation of a 'Reliability Programming' reservoir model. *Water Resour. Res.*, **23**(2), 225–9.

Trezos, T. & Yeh, W. W.-G. (1987) Use of stochastic dynamic programming for reservoir management. *Water Resour. Res.*, **23**(6), 983–96.

Yakowitz, S. (1969) *Mathematics of Adaptive Control Processes*, Elsevier, New York.

Yakowitz, S. (1982) Dynamic programming applications in water resources. *Water Resour. Res.*, **18**(4), 673–96.

Yakowitz, S. (1988) Theoretical and computational advances in differential dynamic programming. *Control and Cybernetics*, **17**(2–3), 173–89.

Yeh, W. W.-G. (1985) Reservoir management and operations models: A state-of-the-art review. *Water Resour. Res.*, **21**(12), 1797–818.

3 Risk assessment in control of reservoir systems

A. KOZŁOWSKI and A. ŁODZIŃSKI
Institute of Geophysics, Polish Academy of Sciences, Warsaw, Poland

ABSTRACT Decision making in the process of control of water storage reservoirs is always combined with risk, whose evaluation is of utmost practical importance. Define the risk as the probability of failure within an operational control process, in the sense of water deficit or water surplus. The control is the sequence of interventions (releases) on future intervals. The risk is estimated on the basis of probability distributions of the total inflows within the horizon of an intervention. The stochastic process of total inflow is conceptualized as a non-stationary Markov chain of first or second order, under discrete time. The methodology is illustrated at the example of the water supply system of a cascade of reservoirs on the river Soła.

INTRODUCTION

In a real decision making problem a decision is made under the conditions of uncertainty, i.e. the decision maker does not have complete information about all elements that influence his decision.

In the water resources management the estimation of the quality of control is carried out *a posteriori* as a result of analysis of water management system performance within a long period of time, usually several years. A probabilistic estimation of failure in the control process is conducted through the determination of the periods of assurance of various desirable outflows. Recently some specific performance indices of water resources system, such as reliability, vulnerability, resilience and robustness have been considered (cf. Cohen, 1982; Haimes *et al.*, 1984; Hashimoto *et al.*, 1982a, b).

The present paper deals with the probabilistic estimation of failure, *C* understood as the impossibility to meet the planned outflows), within an operational control process, being a sequence of interventions planned over some control horizons, which may have several weeks. This estimation is made at the beginning of each individual intervention. Thus it is an *a priori* estimation, concerning the future situations. A main task of the proposed method is evaluating of the risk of failure to maintain the planned outflow in the course of the time horizon considered. This value of risk is useful for the decision maker.

The present paper illustrates the proposed method of risk assessment by its application to keeping up the planned outflow from the cascade of the Tresna, Porąbka, Czaniec storage reservoirs on the river Soła in Southern Poland.

SOURCES OF RISK

Even though risk has not been precisely defined, it is assumed that it is a function of two variables:

(a) the probability of failure;
(b) the magnitude of loss.

Human activity is liable to risk, because within our environment we always deal with uncertainty and we may expect some negative consequences.

The present paper deals only with the first risk index – that of failure probability. The other index – the magnitude of loss – depends on the utility function. It is difficult to determine the appropriate formula for this function in water management; in particular, as one does not have all necessary data. Failure probability can be estimated on the basis of sequences of many years of observed inflows.

METHOD OF RISK ESTIMATION USED IN THE DECISION MAKING PROCESS

While determining a decision making model, which is to serve as a decision making instrument, the random factors can be taken into account in various ways.

According to the way of considering uncertainty in the

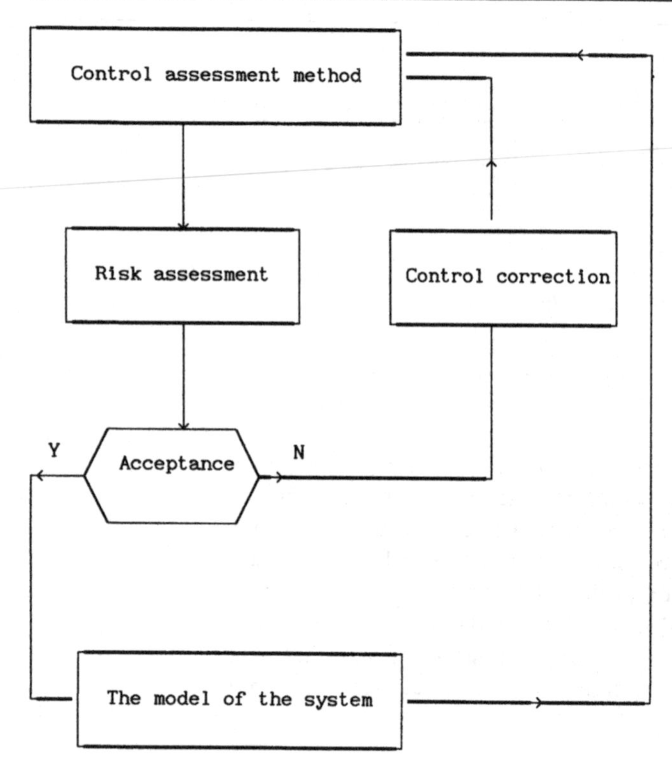

Fig. 1 Scheme of decision making.

decision making process, various approaches are possible. They range from the optimistic ones, where favourable conditions are assumed, to the pessimistic ones, where the worst possible conditions are assumed.

Such different methods reflect different risk perceptions of the decision maker (risk prone or risk averse attitude).

So, it would be worthwhile to support the decision maker operation by introducing an estimation of the chances of success to the decision making scheme. The method of decision making is presented in Fig. 1.

The process of control consists of a sequence of repeated control and risk assessment. The control horizon is divided into K intervals. At the kth moment – the beginning of intervention, $k=1,\ldots,K$, the control for the interval of the length H (horizon of intervention), i.e. for the times $t\in[k,k+H]$ is assigned. For this control the value of risk is estimated. If the decision maker accepts the calculated value of risk, he accepts also the assigned control. The non-acceptance involves the necessity of control correction, which is made by modifications of some parameters in the method of control assessment, by application of another method, etc. The decision chosen is applied to the system over a part of the intervention horizon only, $[k,k+1]$, rather than over the whole horizon $[k,k+H]$. Such procedure is continued until the moment $K-1$, when the control for the last interval is assessed and applied over $[K-1,K]$.

Within this method of control the decision maker is constantly being informed about the consequences of its choices.

While introducing risk into the decision making process, one has to introduce two characteristic quantities:

(a) acceptable risk level R_d;
(b) tolerated departure from the planned control, that is, tolerated relative deficit or surplus.

As long as the risk of given control is not greater than R_d, no correction of the control value is made. In the opposite case the second threshold is taken into account: the tolerated departure from the planned control, i.e. tolerated failure value. While doing so we want to decrease the risk of impossibility of realization of the accepted decision.

RISK IN A STORAGE RESERVOIR SYSTEM

A characteristic feature of a reservoir system is the stochastic character and inpredictability of the inflow process in a deterministic sense. The reservoir inflow is a random variable. While making decisions concerning the maintenance of the outflow on a desirable (non-random) level one is faced with a danger of violating conventional or physical limits of the reservoir storage capacity. In order to avoid this, one sometimes has to give up the desirable outflow levels. This results in undesirable disturbances in water alimentation of the user – in deficits or in undesirable surplus. Assessment of the probability of the fact that the planned outflow will be not realised (probability of failure) is of utmost importance.

A formula for the risk estimation in a discrete time case will be derived in the sequel.

The following notation is assumed:

$i=0,1,2,\ldots,h$ – index of the time interval within the horizon of intervention;
$h=0,1,2,\ldots,H$ – number of time intervals within the horizon of control under consideration;
S_0 – storage value at the beginning of an intervention (at the beginning of the $k+1$ time interval, $k=1,\ldots,K$);
S_h – storage at the end of the horizon of an intervention of the length h;
q_i – average inflow within the ith time interval;
d_i – average planned outflow within the ith time interval;
u_i – average actual outflow within the ith time interval;
δ_i – relative water deficit within the ith time interval;

$$\delta_i=\frac{d_i-u_i}{d_i}$$

η_i – relative water surplus within the ith time interval;

$$\eta_i=\frac{u_i-d_i}{d_i}$$

e_i – evaporation and filtration within the ith time interval;
V_h^l – lower physical or economic limit of the capacity of the reservoir at the end of the horizon of the length h;
V_h^u – upper physical or economic limit of the capacity of the reservoir at the end of the horizon of the length h;
$Q_h = \sum_{i=1}^{h} q_i$ – total inflow over the horizon of the length h;
$D_h = \sum_{i=1}^{h} d_i$ – total planned outflow over the horizon of the length h;
$U_h = \sum_{i=1}^{h} u_i$ – total realized outflow over the horizon of the length h;
$E_h = \sum_{i=1}^{h} e_i$ – total evaporation and filtration over the horizon of the length h.

On the basis of the reservoir equation of mass conservation, the volume at the end of the horizon of the length h is expressed by the formula:

$$S_h = S_0 + Q_h - D_h - E_h \tag{1}$$

The occurrence of water deficit before the end of the horizon of the length h time is expressed by the inequality:

$$S_h < V_h^l \tag{2}$$

The occurrence of water surplus before the end of the horizon of the length h time is expressed by the inequality:

$$S_h > V_h^u \tag{3}$$

Regarding (1) and neglecting filtration and evaporation for simplicity, these inequalities can be put as follows:

$$S_0 + Q_h - D_h < V_h^l \tag{4}$$

and

$$S_0 + Q_h - D_h > V_h^u \tag{5}$$

or

$$Q_h < V_h^l - S_0 + D_h \tag{6}$$

and

$$Q_h > V_h^u - S_0 + D_h \tag{7}$$

Inequalities (6)–(7) determine the division of the total inflow area Ω into three parts: deficit inflow, desirable inflow and surplus inflow, which are marked in Fig. 2, respectively by F_d, S, F_s. In the case of $V_i^u = \text{const}$, $V_i^l = \text{const}$ (denoted as V^d, V^l) and $d_i = \text{const}$, $i = 1, \ldots, h$ the lines marking the division are straight. In the general case they are broken.

When the total inflow Q_h at the end of the intervention horizon of the length h takes the values above the upper line, the case can be called water surplus (the storage volume greater than V_h^u value), whereas for Q_h below the lower line one gets the water deficit (the storage volume less than V_h^l). In both situations the outflow u_i instead of d_i is realized.

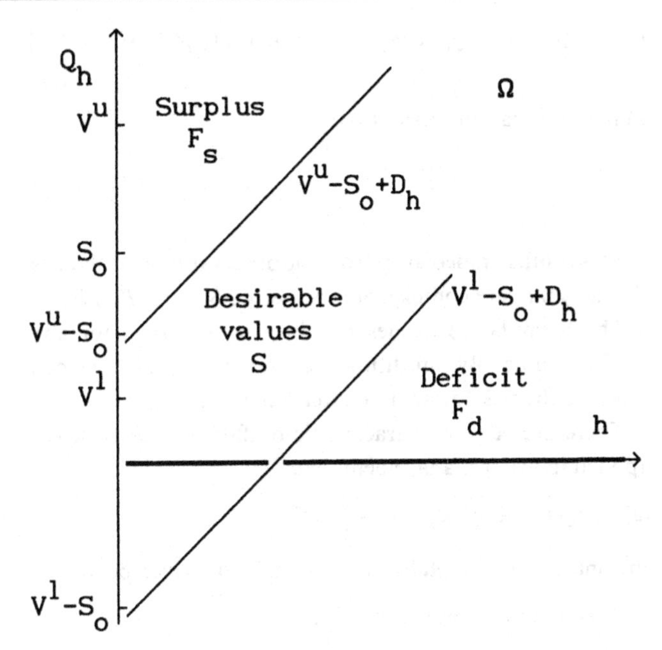

Fig. 2 Division of the total inflow area into deficit inflow, desirable inflow, and surplus inflow.

The risk considered, i.e. the probability that the planned outflow d_i, $i = 1, \ldots, h$ over the horizon of the length h will not be realized, may be calculated as the probability that the total inflow takes a value from the area F_d or F_s in any time during this horizon of the length h:

$$\begin{aligned} r_h = \mathrm{P}\, \{&[Q_0 = q_0 \wedge (Q_1 < V^l - S_0 + D_1 \vee Q_1 > V^u - S_0 + D_1)] \\ &\vee \ldots \vee [Q_0 = q_0 \wedge V^l - S_0 + D_1 < Q_1 < V^u - S_0 + D_1 \wedge \ldots \\ &\wedge V^l - S_0 + D_{h-1} < Q_{h-1} < V^u - S_0 + D_{h-1} \wedge \\ &(Q_h < V^l - S_0 + D_h \vee Q_h > V^u - S_0 + D_h)]\} \end{aligned} \tag{8}$$

This is the probability of the event that either the total inflow Q_i is too small to maintain the planned outflow within the horizon of the length h of intervention on the d_i level without violating V^l, or the total inflow Q_i is too great to maintain the planned outflow on the d_i level without violating V^u, $i = 1, \ldots, h$. Thus it is the forecast of the total control failure with the horizon of the length h.

Using the symbols S, F_s and F_d of the respective desirable, surplus and deficit areas one can reformulate (8) as follows:

$$\begin{aligned} r_h = \mathrm{P}\, \{&[Q_0 = q_0 \wedge (Q_1 \in F_d \vee Q_1 \in F_s)] \vee \\ &[Q_0 = q_0 \wedge Q_1 \in S \vee (Q_2 \in F_d \vee Q_2 \in F_s)] \vee \ldots \\ &\vee [Q_0 = q_0 \wedge Q_1 \in S \wedge Q_2 \in S \wedge \ldots \wedge Q_{h-1} \in S \wedge \\ &(Q_h \in F_d \vee Q_h \in F_s)]\} \end{aligned} \tag{9}$$

In order to calculate the risk r_h one transforms the formula (9) into the form of sum of the following components, $i = 1, \ldots, h$:

$$r_i = \mathrm{P}\left[Q_0 = q_0 \wedge Q_1 \in S \wedge \ldots \wedge Q_{i-1} \in S \wedge (Q_i \in F_d \vee Q_i \in F_s)\right] \quad (10)$$

In fact, it is easy to verify that:

$$r_h = \sum_{i=1}^{h} r_i \quad (11)$$

This results immediately from the observation that the sets S and $F_d \cup F_s$ are complementary, i.e. $x \in S \Leftrightarrow x \notin F_d \cup F_s$.

The formula (8) defines risk in general terms, for any outflow value. By substituting concrete releases, one can estimate the risk for an individual situation.

There are certain characteristic outflow values typically applied in water management, like

(a) target draft d_i^{td} and $D_h^{td} = \sum_{i=1}^{h} d_i^{td}$;

(b) maximum acceptable flow value in the river below the reservoir d_i^{max} and $D_h^{max} = \sum_{i=1}^{h} d_i^{max}$;

(c) minimum acceptable flow in the river below the reservoir (for environmentally essential needs only) d_i^{min} and $D_h^{min} = \sum_{i=1}^{h} d_i^{min}$.

where $i = 1, \ldots, h$, $h = 1, \ldots, H$.

By substituting the above values into the general formula, one arrives at formulas for $h = 1, \ldots, H$ describing the following situations:

(a) the risk of not sustaining the target draft value:

$$\begin{aligned} r_h^{td} = \; & \mathrm{P}\{[Q_0 = q_0 \wedge (Q_1 < V^l - S_0 + D_1^{td} \vee Q_1 > V^u - S_0 + D_1^{td})] \\ & \vee \ldots \vee [Q_0 = q_0 \wedge V^l - S_0 + D_1^{td} < Q_1 < V^u - S_0 + D_1^{td} \wedge \\ & \ldots \wedge V^l - S_0 + D_{h-1}^{td} < Q_{h-1} < V^u - S_0 + D_{h-1}^{td} \wedge \\ & (Q_h < V^l - S_0 + D_h^{td} \vee Q_h > V^u - S_0 + D_h^{td})]\} \end{aligned} \quad (12)$$

(b) the risk of not sustaining the planned release within the limits of the economically acceptable outflow:

$$\begin{aligned} r_h^{el} = \; & \mathrm{P}\{[Q_0 = q_0 \wedge (Q_1 < V^l - S_0 + D_1^{td} \vee Q_1 > V^u - S_0 + D_1^{max})] \\ & \vee \ldots \vee [Q_0 = q_0 \wedge V^l - S_0 + D_1^{td} < Q_1 < V^u - S_0 + D_1^{max} \wedge \\ & \ldots \wedge V^l - S_0 + D_{h-1}^{td} < Q_{h-1} < V^u - S_0 + D_{h-1}^{max} \wedge \\ & (Q_h < V^l - S_0 + D_h^{td} \vee Q_h > V^u - S_0 + D_h^{max})]\} \end{aligned} \quad (13)$$

(c) the risk of transgressing the minimum acceptable flow:

$$\begin{aligned} r_h^{min} = \; & \mathrm{P}\left[(Q_0 = q_0 \wedge Q_1 < V^l - S_0 + D_1^{min}) \vee \ldots \vee (Q_0 \right. \\ & = q_0 \wedge Q_1 \geq V^l - S_0 + D_1^{min} \wedge \ldots \\ & \left. \wedge Q_{h-1} \geq V^l - S_0 + D_{h-1}^{min} \wedge Q_h < V^l - S_0 + D_h^{min})\right] \end{aligned} \quad (14)$$

(d) the risk of transgressing the maximum flow in the river below the reservoir:

$$\begin{aligned} r_h^{max} = \; & \mathrm{P}\left[(Q_0 = q_0 \wedge Q_1 > V^u - S_0 + D_1^{max}) \vee \ldots \vee (Q_0 \right. \\ & = q_0 \wedge Q_1 \leq V^u - S_0 + D_1^{max} \wedge \ldots \\ & \left. \wedge Q_{h-1} \leq V^u - S_0 + D_{h-1}^{max} \wedge Q_h > V^u - S_0 + D_h^{max})\right] \end{aligned} \quad (15)$$

If the risk calculated for the planned outflow is too high then the planned outflow must be changed – regarding, however, a given, acceptable level of alteration. This acceptable level, assumed by the system users, is defined by some concrete values: relative deficit δ_i and relative surplus η_i. Thus, instead of the planned value d_i, the following values u_i of the outflow are realized:

(a) in a deficit case:

$$u_i = d_i(1 - \delta_i), \qquad i = 1, \ldots, h, \qquad U_h^d = \sum_{i=1}^{h} u_i$$

('deficit outflow');

(b) in a surplus case:

$$u_i = d_i(1 + \eta_i), \qquad i = 1, \ldots, h, \qquad U_h^s = \sum_{i=1}^{h} u_i$$

('surplus outflow').

For these new values of the outflow one can calculate the new value of risk. The formulas for estimation of the risk of transgressing the tolerated failure values are set up analogically as in (14) and (15):

(a) the risk of transgressing the tolerated deficit outflow:

$$\begin{aligned} r_h^{trd} = \; & \mathrm{P}\left[(Q_0 = q_0 \wedge Q_1 < V^l - S_0 + U_1^d) \vee \ldots \vee (Q_0 = q_0 \wedge \right. \\ & Q_1 \geq V^l - S_0 + U_1^d \wedge \ldots \\ & \left. \wedge Q_{h-1} \geq V^l - S_0 + U_{h-1}^d \wedge Q_h < V^l - S_0 + U_h^d)\right] \end{aligned} \quad (16)$$

(b) the risk of transgressing the tolerated surplus outflow:

$$\begin{aligned} r_h^{trs} = \; & \mathrm{P}\left[(Q_0 = q_0 \wedge Q_1 > V^u - S_0 + U_1^s) \vee \ldots \vee (Q_0 = q_0 \wedge \right. \\ & Q^1 \leq V^u - S_0 + U_1^s \wedge \ldots \\ & \left. \wedge Q_{h-1} \leq V^u - S_0 + U_{h-1}^s \wedge Q_h > V^u - S_0 + U_h^s)\right] \end{aligned} \quad (17)$$

(c) the risk that the planned outflow does not fall within the limits of the tolerated outflow:

$$r_h^t = r_h^{trd} + r_h^{trs} \quad (18)$$

Consider now the method of effective calculation of the risk r_h with the help of equations (10) and (11). These formulas deal with the probability distributions of the inflows Q_i, $i = 1, \ldots, h$; namely with the probability of its occurrence in desirable S or undesirable F_s or F_d areas. The probability distributions of Q_i, $i = 1, \ldots, h$ are not stationary. These distributions depend on the initial moment (beginning of intervention), the initial inflow value q_0 and the previous inflows Q_j, $j = 1, \ldots, i-1$. As time goes by, this dependence disappears.

The above suggests that r_i can be represented with the aid of respective conditional probabilities. So, r_i is noted as follows:

$$r_i = \mathrm{P}(Q_0 = q_0 \wedge Q_1 \in S \wedge \ldots \wedge Q_{i-1} \in S \wedge Q_i \in F_d \cup F_s)$$
$$= \mathrm{P}(Q_1 \in S | Q_0 = q_0) \cdot \ldots \cdot \mathrm{P}(Q_{i-1} \in S | Q_0 = q_0 \wedge Q_1 \in S \wedge \ldots \wedge Q_{i-2} \in S) \cdot \mathrm{P}(Q_i \in F_d \cup F_s | Q_0 = q_0 \wedge Q_1 \in S \wedge \ldots \wedge Q_{i-1} \in S) \quad (19)$$

where, for instance, the last conditional probability can be defined by the formula:

$$\mathrm{P}(Q_i \in F_d \cup F_s | Q_0 = q_0 \wedge Q_1 \in S \wedge \ldots \wedge Q_{i-1} \in S)$$
$$= \frac{\mathrm{P}(Q_0 = q_0 \wedge Q_1 \in S \wedge \ldots \wedge Q_{i-1} \in S \wedge Q_i \in F_d \cup F_s)}{\mathrm{P}(Q_0 = q_0 \wedge Q_1 \in S \wedge \ldots \wedge Q_{i-1} \in S)} \quad (20)$$

This probability should be evaluated on the basis of large number of observed hydrological records. However, the number of these records is typically not adequate. For this reason one often takes the assumption that the stochastic process Q_h is a Markov process. This allows one to use for example the following relationship, $i = 1, \ldots, h$:

$$\mathrm{P}(Q_i \in F_d \cup F_s | Q_0 = q_0 \wedge Q_1 \in S \wedge \ldots \wedge Q_{i-1} \in S)$$
$$= \mathrm{P}(Q_i \in F_d \cup F_s | Q_{i-1} \in S) = \frac{\mathrm{P}(Q_{i-1} \in S \wedge Q_i \in F_d \cup F_s)}{\mathrm{P}(Q_{i-1} \in S)} \quad (21)$$

So one may transform the relationship (19) into the following form:

$$\mathrm{P}(Q_0 = q_0 \in S \wedge \ldots \wedge Q_{i-1} \in S \wedge Q_i \in F_d \cup F_s)$$
$$= \mathrm{P}(Q_0 = q_0) \cdot \mathrm{P}(Q_1 \in S | Q_0 = q_0) \cdot \ldots \cdot \mathrm{P}(Q_{i-1} \in S | Q_{i-2} \in S) \cdot [1 - \mathrm{P}(Q_i \in S | Q_{i-1} \in S)] \quad (22)$$

Finally, the stochastic process Q_h, $h = 1, \ldots, H$ can be approximated by a non-stationary Markov chain of first or second order and considered for the discrete time. At the same time the total inflow area Ω is divided into F_d, S and F_s subareas. Thus one arrives at a stochastic, triple Markov's chain. In practice one increases a number of chain states in order to make the appearance of initial inflow within the various states. In connection with it subareas F_d, S and F_s can comprise more than one chain status.

EXAMPLE OF CALCULATIONS

The results of the application of the suggested method to the risk estimation are given for the system of the storage reservoirs on the river Soła (Tresna, Porabka and Czaniec storage reservoirs), a tributary of the Upper Vistula. Capacities of these reservoirs are considered jointly.

The upper constraint of the conservation capacity value of these reservoirs (without spillway capacity and flood control capacity) is 101.7 million m³, the lower constraint is 18.2 million m³ (dead capacity). The calculation is conducted for different initial storage values S_0 – (20.0, 40.0, 80.0 million m³), different initial inflow values Q_0 – (3.0, 15.0, 43.0 m³) and for the fixed target release – 11.8 m³/s.

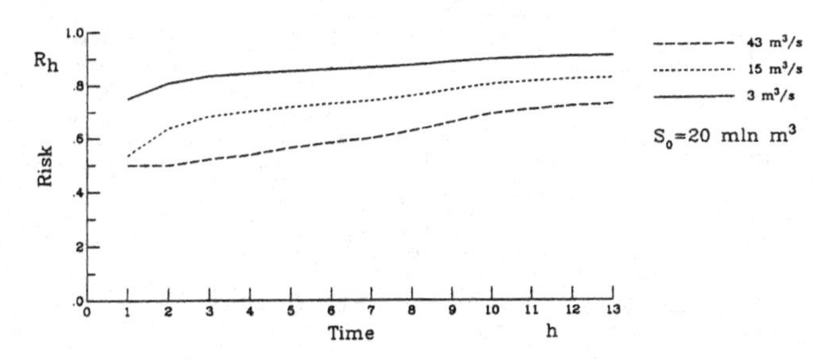

Fig. 3 Risk as a function of time for initial storage value $S_0 = 20$ mln m³.

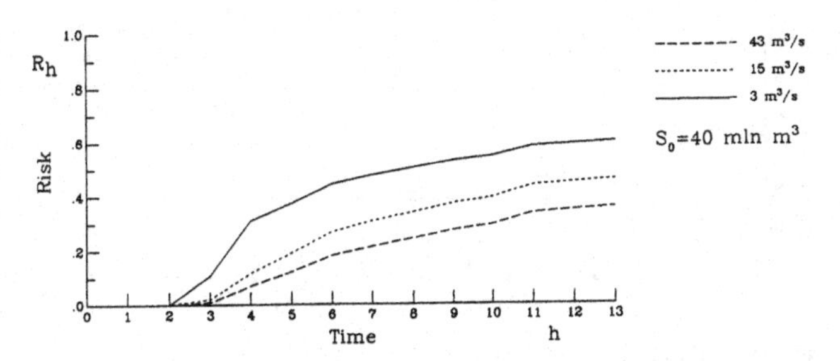

Fig. 4 Risk as a function of time for initial storage value $S_0 = 40$ mln m³.

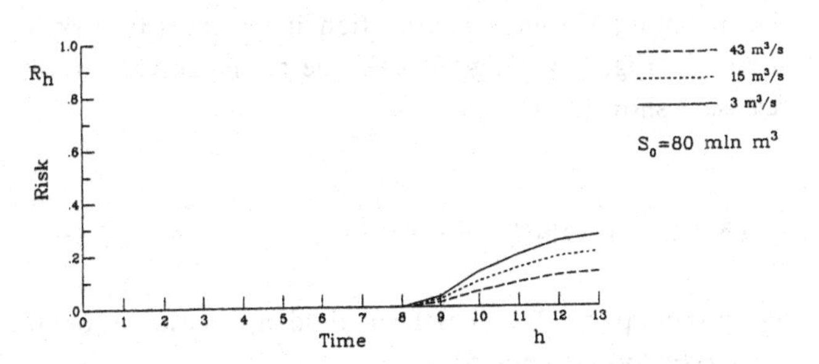

Fig. 5 Risk as a function of time for initial storage value $S_0 = 80$ mln m³.

Figs. 3, 4, 5 represent the alteration of the risk that the planned outflow is not sustained for an intervention started at the end of the third interval ($h = 3$). They present the influence of the storage value S_0 and of the inflow Q_0 on the value of r_h, i.e. on the risk that the planned release trajectory cannot be sustained in the intervention horizon of the length h. The influence of the storage, S_0 on the risk value is remarkably greater than that of the inflow, Q_0. The latter is significant for smaller initial storage values, S_0 (Figs. 3, 4).

Fig. 6 represents the correction of the planned release, necessary for the risk not to exceed the acceptable value during of sequence of 13 interventions. The correction refers to the run of inflow of the fall of 1987 and initial storage 43.12 million m³. Fig. 7 illustrates corresponding changes of the storage trajectory.

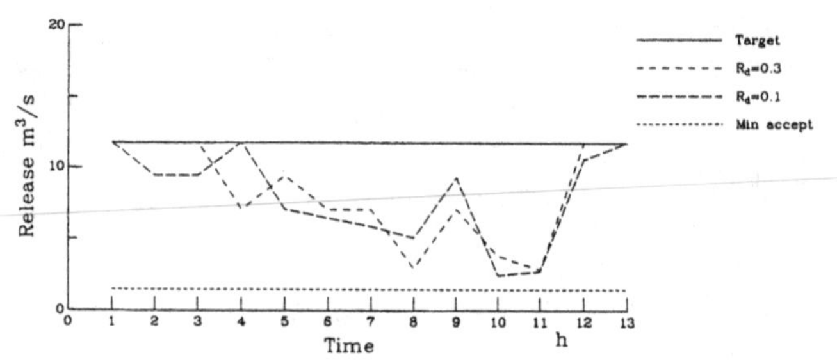

Fig. 6 Correction of the planned release.

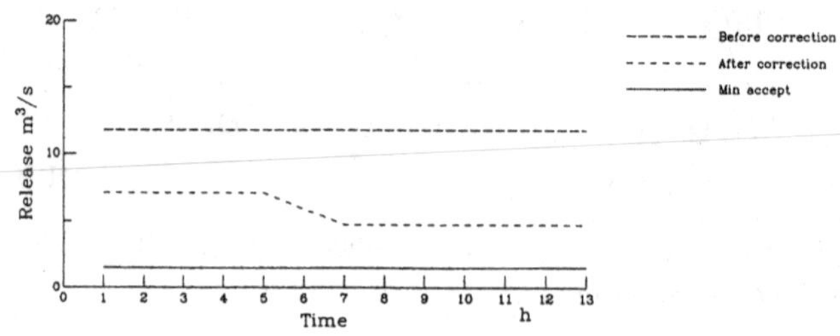

Fig. 8 Change of target release due to excessive risk.

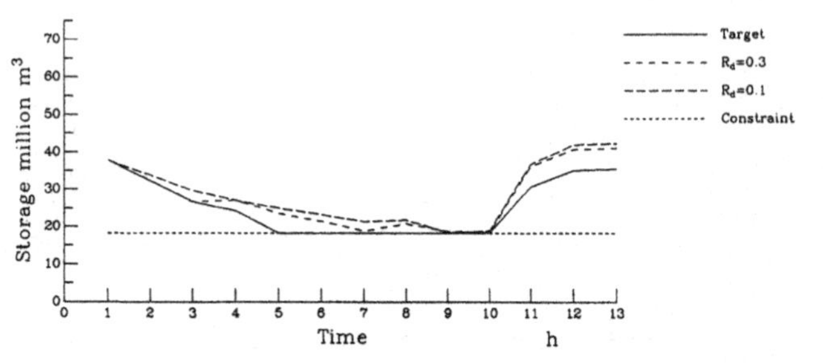

Fig. 7 Change of storage trajectory.

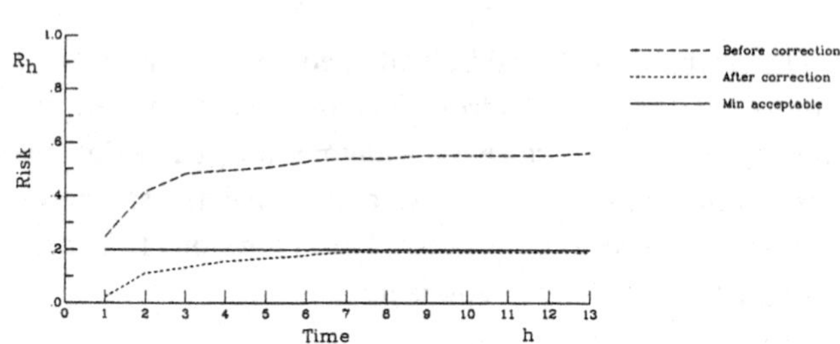

Fig. 9 Change of risk due to correction of release.

An example of the change of the target release planned at the beginning of a single intervention due to excessive risk is shown in Fig. 8. Change of risk due to the correction of release is shown in Fig. 9.

CONCLUSIONS

The present paper describes the method of risk estimation in a reservoir system control.

Failure probability is the information concerning possible future dangers, of direct significance for water system users. However, it says nothing about the scale of a possible failure, hence it makes no distinction between the failures with the same probability of occurrence and different magnitude of loss.

Were both risks indexes, i.e. failure probability and loss magnitude, considered in the decision making process, then the information concerning possible future dangers would be significantly more complete. The suggested method is particularly useful in low-water seasons.

REFERENCES

Asbeck, E. & Haimes, Y. (1984) The partitioned multiobjective risk method. *Large Scale Systems*, **6**(1).

Bharucha-Reid, A.T. (1960) *Elements of the Theory of Markov Processes and Their Application*, McGraw-Hill, New York.

Cohen, A. (1982) The nature of decisions in risk management. In: Griffits, R. (Ed.), *Dealing with Risk*, Wiley & Sons, New York.

Cox, D. R. & Miller, H.D. (1965) *The Theory of Stochastic Processes*, John Wiley & Sons, New York.

Feller, W. (1970) *Wstęp do rachunku prawdopodobieństwa*, PWN, Warsaw, Poland.

Fiering, M. B. & Holling, C. S. (1974) Management and standards for perturbed ecosystems. *Agroecosystems*, **1**(4).

Haimes, Y. (1984) Integrated risk and uncertainty assessment in water resources within a multiobjective framework. *J. Hydrol.*, **68**.

Haimes, Y. & Leach, M. (1984) Risk assessment and management in multiobjective framework. In: *Proceedings of the Conference on Decision Making with Multiobjectives*, Cleveland, Ohio.

Hashimoto, T., Loucks, D. P. & Stedinger, J. R. (1982) Robustness of water resource systems. *Water Resour. Res.*, **18**.

Hashimoto, T., Stedinger, J. R. & Loucks, D. P. (1982) Reliability, resiliency and vulnerability criteria for water resource systems. *Water Resour. Res.*, **18**.

Kaczmarek, Z. (1970) *Metody statystyczne w hydrologii i meteorologii*. WKiŁ, Warsaw, Poland.

Kaczmarek, Z. 1984 Kryteria sterowania systemami wodnogospodarczymi. *Przegląd Geofizyczny*, **4**.

Kaplan, S. & Garrick, B. (1981) On the quantitative definition of risk. *Risk Analysis*, **1**(1).

Loucks, D. P. & Stedinger, J. & Haith, D. A. (1981) *Water Resource Systems Planning and Analysis*. Prentice-Hall, Englewood Cliffs, Inc.

Lowrance, W. (1976) *Of Acceptable Risk*. William Kaufmann, Inc., Los Altos, California.

Okrent, D. (1980) Comment on societal risk. *Science*, **208**.

Rowe, W. (1977) *An Anatomy of Risk*. Wiley & Sons, New York.

Shamir, U. (1987) Reliability of water supply systems. In: Duckstein, L. & Plate, E. J. (Eds.), *Engineering Reliability and Risk in Water Resources*, NATO ASI Series, M. Nijhoff, Dordrecht, The Netherlands.

Sage, A. & White, E. (1980) Methodologies for risk and hazard assessment: a survey and status report. *IEEE Transactions on Systems, Man and Cybernetics*, SMC-10, **8**.

Starr, C. & Whipple, C. (1980) Risks of risk decisions. *Science*, **208**.

Wai-See Moy, Cohon, J. L. & ReVelle, C. S. (1986) A programming model for analysis of the reliability, resilience and vulnerability of a water supply reservoir. *Water Resour. Res.*, **22**, 4.

Yen, B. C. (1986) *Stochastic and Risk Analysis in Hydraulic Engineering*. Littleton, Colorado: Water Resources Publications.

4 Reliability-related criteria in water supply system studies

Z. W. KUNDZEWICZ

Institute of Geophysics, Polish Academy of Sciences, Warsaw, and Research Centre of Agricultural and Forest Environment Studies, Poznań, Poland

A. ŁASKI

HYDROPROJEKT Consulting Engineers, Warsaw, Poland

ABSTRACT In the study of water supply system expansion in two areas of Poland, the design of a number of water storage reservoirs is considered. The reliability studies reported are based on a system simulation performed for a historical time series of observations of river flows, with and without the hypothetical reservoirs, for future water demands given in two variants. Periods of non-satisfactory and of satisfactory system performance and values of maximum and of cumulative deficits are analyzed. Frequency, duration and severity of non-satisfactory system performance are assessed as reliability, resilience, and vulnerability. Assumption of exponential distribution of periods of non-satisfactory and satisfactory performance allows straightforward links between several criteria to be established.

INTRODUCTION

In order to evaluate various aspects of non-satisfactory performance of a technical, or natural system, it is necessary to use multiple performance criteria. The use of a set of criteria to describe the possible variants, scenarios and policies offers a perspective that cannot be achieved with a single objective.

The criteria of concern pertain to time-related characteristics (duration and frequency) of the non-satisfactory system performance and also to the amplitude-related characteristics (maximum instantaneous entry into the state of non-satisfactory performance and the cumulated entry). The ultimate criteria, combining both above types of characteristics measure various aspects of severity and consequences of the non-satisfactory performance. Characteristics of this kind have been recently given a more systematic form via a number of criteria (risk, safety, reliability, resilience, vulnerability and robustness) discussed in the water resources literature in the last decade (cf. Hashimoto *et al.*, 1982; Fiering, 1982; Kaczmarek, 1984; Duckstein & Plate, 1985; Moy *et al.*, 1986; Kindler & Tyszewski, 1989).

One possible formulation of these criteria is presented in the present contribution. They measure such properties of the time series of variables representing the system behaviour as duration of periods of satisfactory and non-satisfactory performance, and maximum and cumulated deficits. The methodology is applied to a problem of water supply expansion in two drainage basins in Poland.

MATHEMATICAL FORMULATION: TIME-RELATED CHARACTERISTICS

Assume, that there is a single variable deciding whether the system performance is satisfactory or not. This may be approximately correct in a number of technical applications, including water supply expansion problems considered here.

Define the satisfactory behaviour of a system in the load-resistance framework, originating from the structural engineering, that is according to the inequality:

$$\text{load} < \text{resistance} \tag{1a}$$

The interpretation in the water supply case is

$$\text{water demand} \leq \text{water supply} \tag{1b}$$

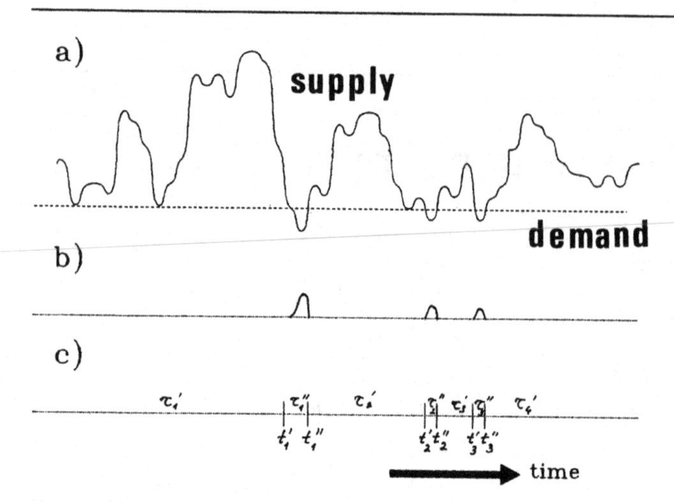

Fig. 1 Load-resistance analogy in the time series representation; (a) time series of constant load (water demand) and variable resistance (water supply, i.e. river flow rate); (b) time series of deficits; (c) renewal stream.

If, in a particular time instant, the system fulfills the above inequality, it is performing satisfactorily.

Fig. 1a shows an example of time series of constant water demand (load) and non-stationary water supply, i.e. flow rate in a river (resistance). Fig. 1b illustrates the time series of deficits and Fig. 1c – the so called renewal stream. The term renewal means a comeback of the system from a state of non-satisfactory performance to a state of satisfactory performance. It has been found convenient (Kundzewicz, 1989) to use the renewal theory framework (Solovjev, 1983) embracing different measures of reliability and resilience criteria. The notation used in Fig. 1 reads:

- τ_i' – periods of system sojourn in the state of satisfactory system performance;
- τ_i'' – periods of system sojourn in the state of non-satisfactory system performance;
- t_i' – time instant of system passage from the state of satisfactory system performance to the state of non-satisfactory system performance;
- t_i'' – time instant of system passage from the state of non-satisfactory system performance to the state of satisfactory system performance.

Assume that:

(a) All variables τ_i' and τ_j'' are independent.
(b) All variables τ_i' are identically distributed:

$$P(\tau_i' < t) = F(t) \tag{3}$$

$$f(t) = dF(t)/dt \tag{4}$$

$$E(\tau_i') = T_S \tag{5}$$

$$D_i^2(\tau') = \sigma_S^2 \tag{6}$$

(c) All variables τ_j'' are identically distributed:

$$P(\tau_i'' < t) = G(t) \tag{7}$$

$$g(t) = dG(t)/dt \tag{8}$$

$$E(\tau_i'') = T_{NS} \tag{9}$$

$$D_i^2(\tau_{NS}'') = \sigma^2 \tag{10}$$

A special case of the renewal process dwells on the concept of a two-state homogeneous Markov chain for which the past history of the process is completely summarized in the current state. Therefore the distributions of the system sojourns in satisfactory and non-satisfactory states must be memory-less, that is exponential:

$$f(t) = \lambda \exp(-\lambda t) \tag{11}$$

$$g(t) = \mu \exp(-\mu t) \tag{12}$$

It is possible to obtain characteristics of various aspects of reliability via simple analytic manipulations with two parameters (one for each exponential distribution), λ and μ (Kundzewicz, 1989). The inverse of T_{NS} is tantamount to Hashimoto's definition of resiliency (Hashimoto *et al.*, 1982). It is easy to determine the renewal density $m(t)$ that allows calculation of the recurrence, or expected number of renewals $M(t)$, i.e. of cycles of satisfactory and non-satisfactory performances in the time interval $(0, t)$:

$$M(t) = \int_0^t m(t)\, dt \tag{13}$$

where, under the assumptions (11)–(12)

$$m(t) = \lambda\mu/(\lambda+\mu) - \lambda\mu/(\lambda+\mu) \exp[-(\lambda+\mu)t] \tag{14}$$

The instantaneous availability $A(t)$, i.e. the satisfactory performance of the system in the time instant t reads:

$$A(t) = \mu/(\lambda+\mu) + \lambda/(\lambda+\mu) \exp[-(\lambda+\mu)t] \tag{15}$$

The stationary (limiting) availability is consistent with a common notion of reliability (also called temporal certainty in Klemes, 1969).

$$\lim_{t \Rightarrow \infty} A(t) = \mu/(\lambda+\mu) = T_S/(T_{NS}+T_S) \tag{16}$$

It may be problematic, whether the parameters λ and μ can be indeed considered constant in time, what is necessary in order to justify the assumption of exponential distributions of the sojourns in particular states. However, it is possible (Kundzewicz, 1989) to accommodate the nonstationarity in the above Markovian scheme, e.g. by allowing the explicit changes in the parameters of the exponential distributions, λ and μ.

CASE STUDIES

The problem analyzed in the case studies is the one of expansion of the existing water supply capacities in two river basins. The case studies embrace the municipal systems of Radom in Central Poland, and of the Lublin conurbation in the South-Eastern Poland, located in the basins of the River Radomka, and of the River Wieprz, respectively. The existing water supply systems seem insufficient for the foreseeable water needs in the near future. The possible expansion of the systems consists primarily of creating water storage reservoirs, Domaniów and Oleśniki, in the Radom and the Lublin systems, respectively.

The description of network schemes, hydrological data, water demand characteristics and hierarchy of fulfilling the water needs of particular users of the system are described in a series of progress reports by HYDROPROJEKT Consulting Engineers (1988–90).

The simulation was performed with the help of a network model software package developed by HYDROPROJEKT. The hierarchy of fulfilling water demands gives greater priority to the municipal needs than to the ones of local industries. The idea of the operating policy is to provide the target release for as long as possible (under no account of forecast, and without hedging).

Hydrological records (ten-day streamflow values) and projections of water demands were used to generate resistance and load variables (cf. equation (1)). The analysis was carried out for several conditions on supply (without reservoir, with reservoir) and on demand (several scenarios of demand growth, typically intensive and moderate development, resulting from different forecasts of population, regional development and water consumption standards). The drought management policies have been also considered, where the household consumption was reduced to 200 (in emergency, even 150) liters per capita per day. One of the reasons of growth of water demand (in addition to the increase of population and to the regional development – agriculture, industry) is the mandatory limitation to the groundwater use in the future. Excessive exploitation of groundwater in the Radom area caused the water level to drop considerably. Reconstruction of groundwater reserves seems urgently necessary and the temporal schedule of reduction of groundwater use has been set up.

There were water deficits occuring in most analyzed variants. The data resulting from the simulations were used to construct the deficit series and renewal streams allowing the reliability-related criteria to be evaluated.

Examples of deficit series and renewal streams for the water supply expansion study in the Radom area are shown in Table 1 for the horizon of the year 2020, under assumptions of intensive development, with reservoir. The notation used is: time – time instant, TS – duration of the system sojourn in the state of satisfactory performance, TNS – duration of the system sojourn in the state of non-satisfactory performance, whereas DEFMAX and DEFCUM denote the amplitude of the deficit and the cumulative deficit, respectively. The values of time, TS and TNS are given in decades, whereas DEFMAX in m^3/d and DEFCUM in tens of m^3.

Table 1. *Example of renewal stream with deficit series*

time [decades]	TS [decades]	TF [decades]	DEFMAX [m^3/d]	DEFCUM [$\times 10m^3$]
139	139			
142		3	124	340
145	3			
153		8	108	516
222	69			
226		4	83	235
229	3			
233		4	108	211
307	74			
310		3	48	82
643	333			
646		3	114	286
649	3			
653		4	74	262
654	1			
657		3	43	111
658	1			
663		5	70	201
664	1			
667		3	109	138
668	1			
682		14	115	911
685	3			
693		8	117	458
694	1			
696		2	92	175
704	8			
718		14	79	521
720	2			

The statistical processing of the simulation data in the sense of fitting a distribution was not always possible, due to a small amount of events of non-satisfactory system performance, observed in some cases. However, good fit of a simple one-parametric exponential distribution, plausible for use with scarce data was generally obtained. This is true not only for TNS, what is intuitively expected, but also for TS. A couple of illustrations of the fit are given in Figs. 2a–4a for TNS and in Figs. 2b–4b for TS, respectively. The particular Figs. pertain to the Radom system:

(a) Case 1 (Figs. 2a and 2b) – variant presented in Table 1 (horizon 2020, intensive development, with reservoir);
(b) Case 2 (Figs. 3a and 3b) – horizon of the year 2010, intensive development, no reservoir;

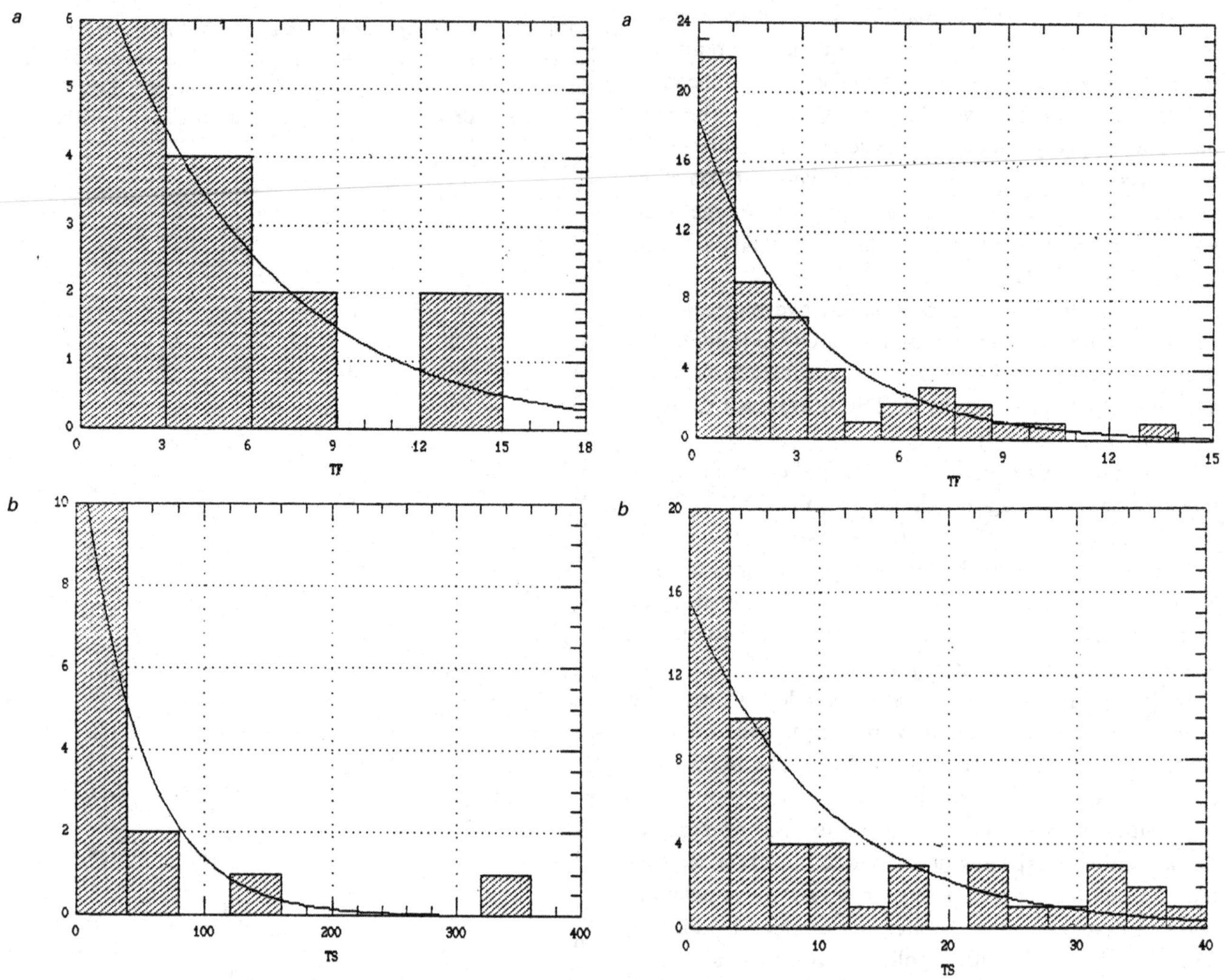

Fig. 2 Frequency histogram of durations of system sojourn in the states of: (a) non-satisfactory performance; (b) satisfactory performance (Case 1, horizon 2020, intensive development, with reservoir).

Fig. 3 Frequency histogram of durations of system sojourn in the states of: (a) non-satisfactory performance; (b) satisfactory performance. (Case 2, Radom, horizon 2010, intensive development, no reservoir).

(c) Case 3 (Figs. 4a and 43b) – horizon of the year 2010, moderate development, no reservoir.

These illustrations correspond to the controlled regime (discharge from the reservoir, cf. Figs. 2a–2b) and to the natural regime (no reservoir, cf. Figs 4a–4b). In the controlled situation (with reservoir) the value of TS and of temporal reliability is much higher than for the case without reservoir.

As can be expected, the sojourn time in the state of non-satisfactory performance and the cumulated deficit volume are strongly correlated (cf. Fig. 5). The values of correlation coefficients between the variables TNS, DEFMAX AND DEFCUM for the three illustrative cases are given in Table 2.

A set of reliability-related criteria analyzed in the present work was:

(a) reliability (temporal reliability, or stationary, i.e limiting availability), i.e. the ratio of the number of decades in which the system did not leave the state of satisfactory performance to the total number of decades considered;

(b) resilience (reciprocal of the mean system sojourn in the non-satisfactory state);

(c) volumetric reliability (cf. Klemes, 1969), understood as the portion of required volume of water supplied in the time period considered, with the rate less than or equal to the demanded volume;

(d) mean maximum deficit, as a simple measure of vulnerability. For the sake of consistency with the former three measures the reciprocal of the mean maximum deficit was used.

Ranking of cases, according to the four above criteria can be done by analyzing the four axes diagram given in Fig. 6. The further the point is from the zero point of the ordinate system, the more welcome the situation. Presentation of Fig.

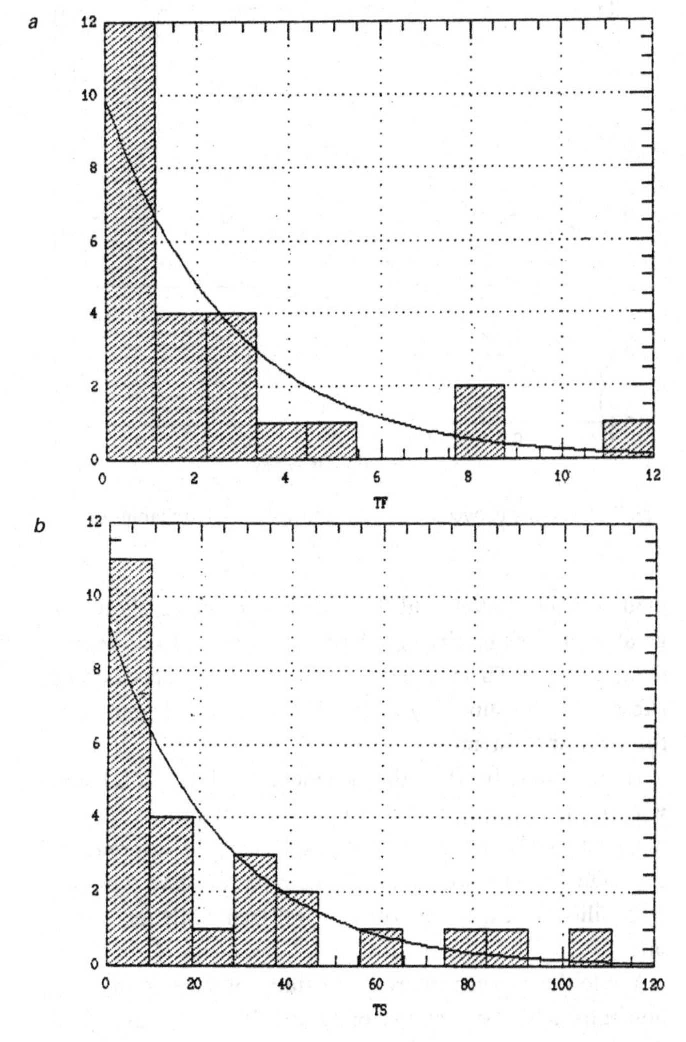

Fig. 4 Frequency histogram of durations of system sojourn in the states of: (a) non-satisfactory performance; (b) satisfactory performance. (Case 3, Radom, horizon 2010, moderate development, no reservoir).

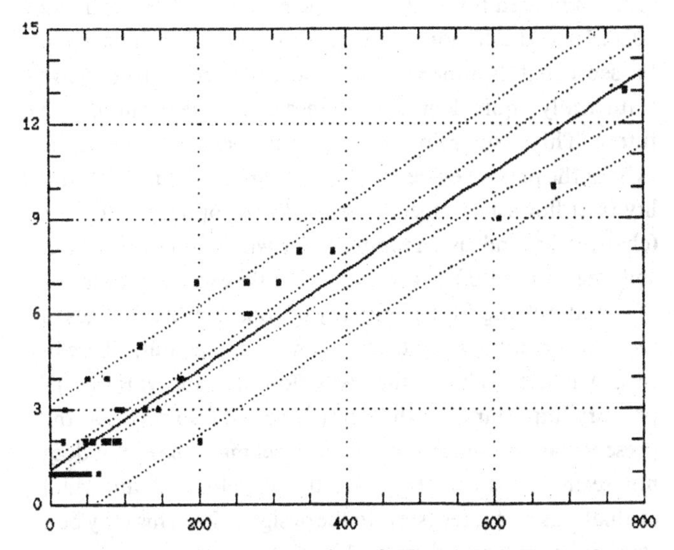

Fig. 5 Example of linear regression of TF and DEFCUM. (Case 2, Radom, horizon 2010, intensive development, no reservoir).

Table 2. *Sample correlations between the variables TF, DEFMAX and DEFCUM for the cases 1–3 corresponding to Figs. 2–4*

	TF	DEFMAX
Case 1	TF	DEFMAX
DEFMAX	.2952	
DEFCUM	.8787	.5734
Case 2	TF	DEFMAX
DEFMAX	.6458	
DEFCUM	.9359	.7572
Case 3	TF	DEFMAX
DEFMAX	.5919	
DEFCUM	.9072	.8095

6 offers versatility of description of the non-satisfactory system performance and helps evaluating the trade-offs combined with alternatives.

It is instructive to trace particular lines in Fig. 6. The line A-F pertains to the time horizon 2020 and links the variant of intensive development with reservoir (A) and without reservoir (F). The line A-E links the cases of the time horizon 2020, with reservoir, under the assumption of intensive (A) and moderate (E) developments. The line F-G pertains to two time horizons: 2010 (G) and 2020 (F), under the assumption of intensive development, and no reservoir. Finally, the broken line A-B-C-D links four variants of the horizon 2020 and intensive development. Variant (A) is the standard operating rule, (B) – drought management with emergency reduction of household consumption (200 liters per capita per day), (C) – as (B) with full cut-off of water supply to industry if the contracted volume of municipal supply cannot be released and (D) – drought management with emergency reduction of household consumption down to 150 liters per capita per day.

CONCLUSIONS

It seems that the set of four criteria (temporal reliability, resilience, volumetric reliability and vulnerability) characterizing the non-satisfactory system performance is a versatile means of description of various aspects of the system performance in the cases analyzed. However, in general case of arbitrary system the above set of four criteria may be insufficient. This is particularly true for extreme situations ('tight' system). Then the set of criteria accepted would depend on the concrete situation. It may happen that the behaviour of some criteria is counter-intuitive – large changes in the system may only slightly alter the criterion values. The sensitivity of particular criteria may be examined

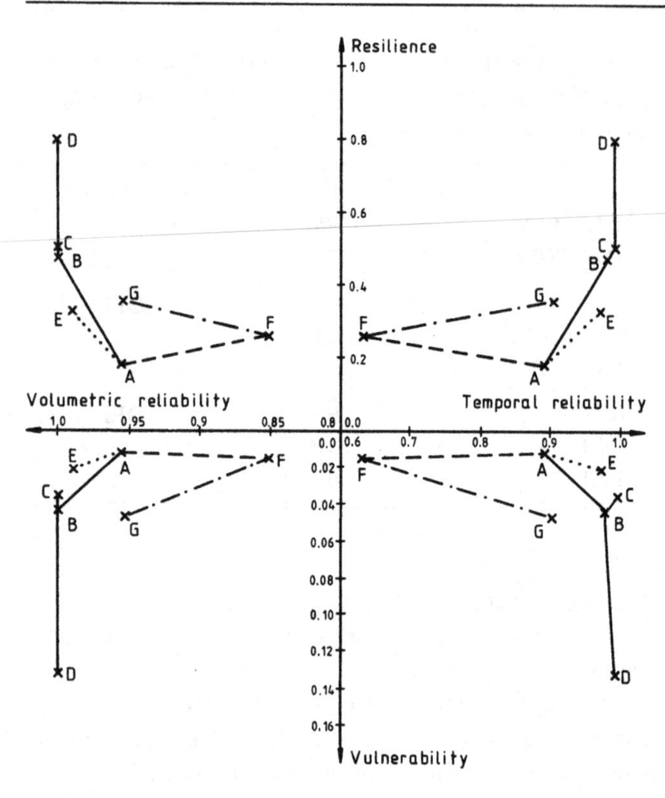

Fig. 6 Four-axes diagram of evaluations of multiple criteria for particular cases.

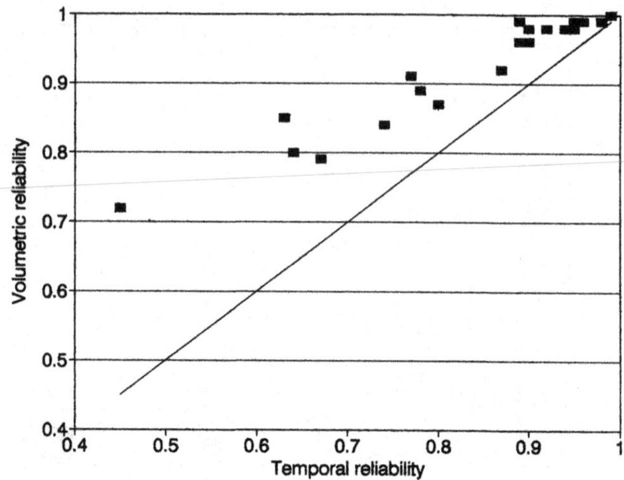

Fig. 7 Relations between temporal and volumetric reliabilities.

in Fig. 6. When considering two variants of strongly differing severities (e.g. the cases with reservoir and without reservoir, line A-F) there is a large difference in the values of both temporal and volumetric reliabilities, but counter-intuitive differences of resilience and vulnerability. It is clear that the sum of times of system sojourn in the non-satisfactory state for the case with existing reservoir is significantly shorter than for the case without reservoir. The same pertains also to the sum of deficits, and, consequently to temporal and volumetric reliabilities. Such a statement cannot be issued, however, for a mean time of system sojourn in the non-satisfactory state nor for a mean amplitude of deficits. This is so, as the number of system sojourns in the non-satisfactory state in the case with existing reservoir is lower than in the case without reservoir. Long sojourns in the non-satisfactory state get shorter and short ones disappear, but no general corollary on the mean time of a single system sojourn in the non-satisfactory state can be issued.

There is a number of candidate reliability-related criteria that may be useful in particular cases, when the four above criteria are not sufficient (e.g. the ones reviewed by Kundzewicz, 1989). A measure of frequency of occurrence of maximum deficit (e.g. deficits of particular exceedence probability) is an example of possibly useful criterion. It violates the simplicity of the approach though.

The comparison of temporal and volumetric reliabilities made for the Radom and the Lublin case studies showed a great deal of similarity in behaviour, that is also of dependence (Fig. 7). This property, however, may be specific to a subset of water supply systems rather than to hold generally for arbitrary systems.

It may well be that the dependence of temporal and volumetric reliability is stronger in extreme ranges. This is intuitively obvious that the temporal reliability approaches the volumetric reliability if the system performs very poorly (reliability tending to zero) or very well (reliability tending to one).

While the sojourn times in the states of satisfactory and non-satisfactory system performances fit well the exponential distributions, it is possible to analytically extend the set of reliability-related criteria by manipulation with two parameters, using the results of the renewal theory.

Available time series of flows observed within some twenty years were used for the future time horizons 2000, 2010 and 2020. This is tantamount to the assumption of stationarity in the sense that the time series of flows observed in the past are statistically equivalent to the ones that will occur in the future. This assumption may be considered dubious in the light of the present evidence. Apart from the natural variability river flows are subject to several anthropogenic impacts (changes in land use, manipulation with the land surface, soil, vegetation etc.). Moreover, under the expected scenarios of climatic change, the possibility of considerable changes in temperature and precipitation cannot be ruled out. These in turn will heavily alter the river flow, i.e. the variable of primary importance in the study reported. However, at the present stage of understanding of the climate change, it does not seem realistic to translate the sketchy and uncertain evaluations and forecasts into the design rules. This may be a strong growth area of research in the years to come.

It is not the scope of the present contribution to optimize

the reservoir operating policies. This is a separate, challenging and difficult task. In the present contribution only a small set of standard operating policies were considered, differing in the rules for management of droughts.

The study clearly shows, that it is justified to pose the problem of reduction of controlled discharges in the times of troubles that allows attaining more acceptable values of reliability criteria (attempt to meet reduced targets). The practical question of this kind reads – how big reduction of the delivered amount of water is necessary for the actual standards of long-term reliability-related criteria to be fulfilled.

ACKNOWLEDGEMENTS

The work reported has been carried out within the framework of the Central Programme of Fundamental Research, CPBP 03.09.06.01. The collaboration of Ms. Anna Malinowska and Mr. Tomasz Daab in performing the simulation of system operation with a network model is gratefully appreciated.

REFERENCES

Duckstein, L. & Plate, E. J. (1985) A system framework for hydrologic reliability with application to the design of structures. In: *Scientific Procedures Applied to the Planning, Design and Management of Water Resources Systems* (Proc. of the Hamburg Symposium, Aug. 1983), IAHS Publ. No. 147.

Fiering, M. B. (1982) Alternative indices of resilience. *Water Resour. Res.*, **18**, 33–9.

Hashimoto, T., Stedinger, J. R. & Loucks, D. P. (1982) Reliability, resiliency and vulnerability criteria for water resource system performance evaluation. *Water Resour. Res.*, **18**, 14–20.

HYDROPROJEKT Consulting Engineers (1988–1990) *Criteria of evaluation of risk and reliability of water resources system; definition, measures, examples of applications*, series of unpublished CPBP reports prepared under supervision of A. Laski (in Polish).

Kaczmarek, Z. (1984) *Criteria of control of water resources system*, Przeglad Geofizyczny, XXIX (in Polish).

Kindler, J. Tyszewski, S. (1989) Multicriteria evaluation of decision rules in the design of a storage reservoir. In: *Closing the Gap Between the Theory and Practice*, Proc. IAHS Third Scientific Assembly, 10–19 May 1989, Baltimore, Maryland, USA, IAHS Publ.

Klemes, V. (1969) Reliability estimates for a storage reservoir with seasonal input. *J. Hydrol.*, **7**, 198–216.

Kundzewicz, Z. W. (1989) Renewal theory criteria of evaluation of water-resource systems: reliability and resilience. *Nordic Hydrology*, **20**, 215–30.

Moy Wai-See, Cohon, J. L. & ReVelle, C. (1986) A programming model for analysis of the reliability, resilience and vulnerability of a water supply reservoir. *Water Resour. Res.*, **22**, 489–98.

Solovjev, A. D. (1983) *Analytical Methods in Reliability Theory*, WNT, Warsaw (in Polish).

5 Reliability analysis of reservoir operation

J. J. BOGARDI and A. VERHOEF

Wageningen Agricultural University, Department of Water Resources, Wageningen, The Netherlands

ABSTRACT Optimization (mainly dynamic programming) based operation of reservoir systems has proven its superiority over traditional techniques based on the concept of rule curve, at least in terms of the selected objective function. Stochasticity of the system can be considered in the optimization by using stochastic dynamic programming to derive long-term, expectation oriented optimal policies. Since the optimality of the operation can only be realized during an infinite operational period, the performance of a stochastic system should not be characterized for practically available time series alone by the expected (annual) value of the objective function. In addition to the traditional output figures a number of performance indices (PI) can be derived to describe the operational behaviour (especially reliability) of the system upon the application of a certain operational policy. These performance indices can be estimated by simulation of the system operation according to the operational policy to be tested.

The inclusion of these PIs in the overall judgment creates a multicriterion framework for decision making. Furthermore, these performance indices are believed to be more sensitive in reflecting the impact of certain constraints than the value of the objective function alone.

The validity of this hypothesis will be tested on a multiunit multipurpose reservoir system, the Victoria, Randenigala, Rantembe reservoir cascade situated on the river Mahaweli in Sri Lanka.

Based on the results of simulation, the PIs will be analyzed for their viability in practical applications.

INTRODUCTION

The consideration of reliability aspects is rapidly becoming one of the major issues in engineering. Changing attitude of the public towards scrutiny of new projects and the dwindling tolerance to accept failures of infrastructure and technical systems have led to efforts to analyze reliability aspects in a systematic context. Besides structural reliability, the operational reliability of existing systems, here in particular that of storage reservoir systems is of importance. The use of refined operation policies (mainly optimization based) is one way to avoid, or at least to postpone the physical extension of the systems. However, optimal operation may 'push' the system to the limit of its performance capability, thus reducing the inherent redundancy.

The subject of this study is the investigation of the performance indices characterizing the behaviour of reservoirs following an optimal operational policy. While reliability in general is covered in great detail in many technical publications, like Henley & Kumamoto (1981), in water resources management the papers of Hashimoto *et al.* (1982) and Fiering (1982) were among the first publications advocating the use of performance (reliability) indices. An in-depth review of the literature by Budhakooncharoen (1990) revealed certain inconsistencies in definitions and the terminology used. It appeared that it would be more appropriate to talk about performance indices (only few of them are in fact directly related to reliability). The term 'reliability analysis', being narrower in its scope, has historical background.

In this paper a number of PIs will be derived, based on the work of Duckstein & Plate (1987).

The hypothesis that performance indices may reflect better the influence of constraints than the optimal value of the objective function, applied to derive the optimal operational policy, will be tested on the case study of the three reservoirs (Victoria, Randenigala and Rantembe: V + R + R) subsys-

Table 1. *Salient features of the Victoria, Randenigala, Rantembe subsystem of the Mahaweli Ganga Development Scheme*

Reservoir	Active storage MCM	Live stor. upper lower limits MCM	Installed capacity MW	Average net head m	Turbine discharge m^3/s	Francis turbines	Monthly firm energy GWh
Victoria	686.0	720/ 34	210	190	140	2	37.0
Randenigala	580.0	875/295	126	78	180	2	25.0
Rantembe	17.0	21/ 4	49	32.7	180	2	14.5

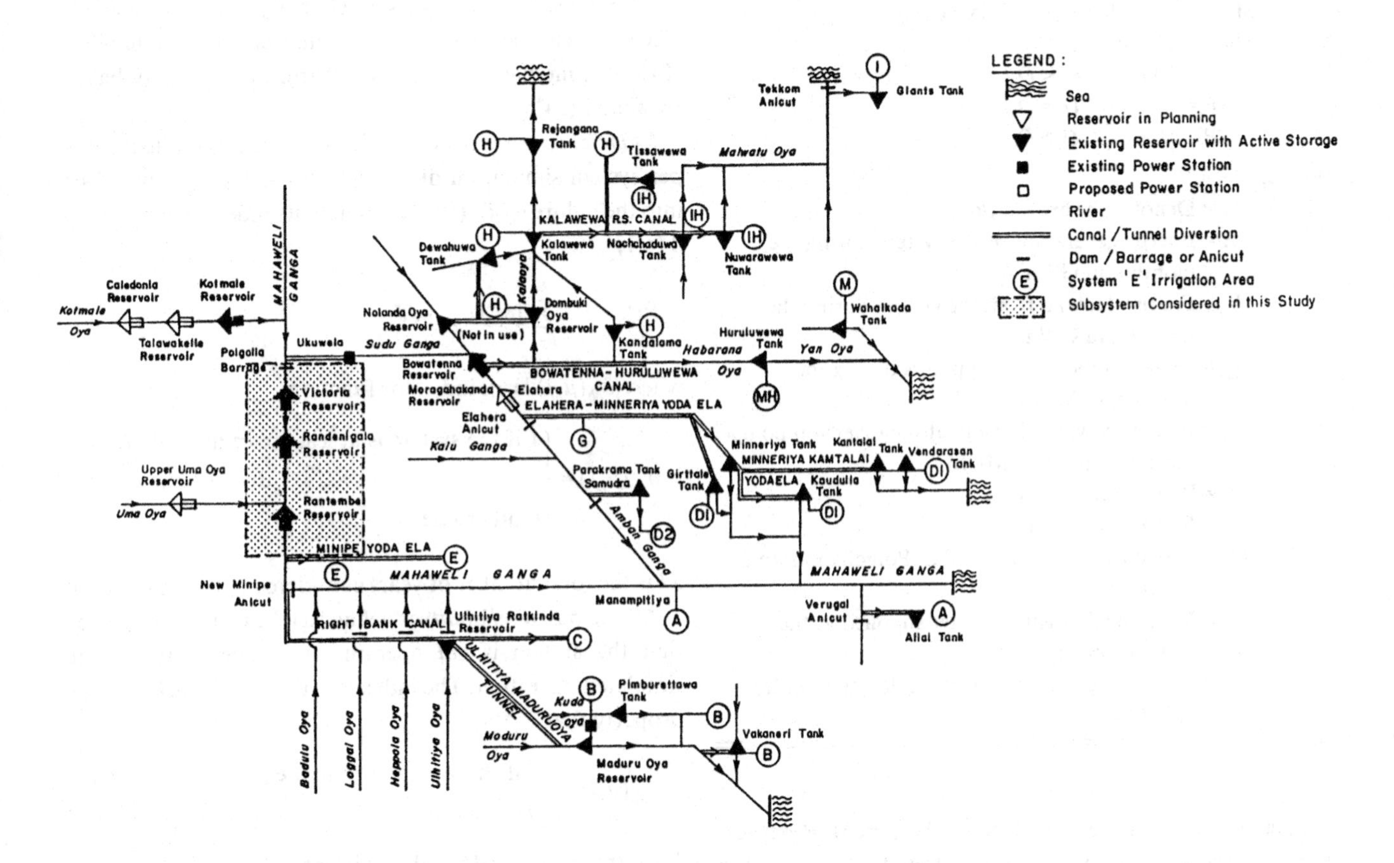

Fig. 1 Schematic layout of Mahawelli Ganga Development Scheme.

tem of the Mahaweli River Development Scheme in Sri Lanka (see Fig. 1 and Table 1). Operational policies will be derived by means of stochastic dynamic programming (SDP) following the version of this technique as recommended by Loucks *et al.* (1981). It is described by the Bellman recursive equation as follows:

$$f_t^n(k,i) = \max\,[B_{k,i,l,t} + \sum_{j=1}^{JM} p_{ij}^t f_{t+1}^{n-1}{}^*(l,j)] \quad \text{for feasible } l \quad (1)$$

where

k, l – are reservoir storage classes at the beginning of subsequent time steps, t and $t+1$;

i, j – inflow classes characterizing the inflows during time steps t and $t+1$;

p_{ij}^t – transitional probability specifying the occurrence of inflow class j during time step $t+1$, given that during time step t class i occurred;

$l = l(k,i,t)$ – operational policy specifying the targetted storage class l as function of k, i and t;

$B_{k,i,l,t}$ – system output (increment of the value of the objective function during stage t, given k, i and l);

JM – number of inflow classes to be considered during $t+1$.

The increasing superscript, n and the decreasing subscript, t indicate the backward DP approach.

Program coding and the simplification of the subsystem, i.e. considering the Rantembe reservoir as a simple run-of-the-river hydropower plant by neglecting its storage capacity

Table 2. *Objective criteria used for SDP optimizations*

No.	Objective function
1	Min. $\varepsilon(\Sigma(E_i - E_{di})^2)$
2	Min. $\varepsilon(\Sigma((E_i - E_{di})^2 + (0.72)^2 (Q_i - Q_{di})^2_{pol}))$
3	Min. $\varepsilon(\Sigma((E_i - E_{di})^2 + (1.00)^2 (Q_i - Q_{di})^2_{pol}))$
4	Min. $\varepsilon(\Sigma((E_i - E_{di})^2 + (1.50)^2 (Q_i - Q_{di})^2_{pol}))$
5	Min. ε $(\Sigma(Q_i - Q_{di})^2_{pol+min})$
6	Max. $\varepsilon(\Sigma E_i)$; Demand constraints
7	Max. $\varepsilon(\Sigma E_i + 0.5 \times Q_{pi} + 0.25 \times Q_{mi}))$
8	Max. $\varepsilon(\Sigma SP_i)$;
	$SP_i = 10000 \quad Q_i < Q_{di}$
	$SP_i = \quad 0 \quad Q_i = Q_{di}$
	$SP_i = Q_i - Q_{di} \quad Q_i > Q_{di}$

Where,

ε = Denotes the expectation,
E_i = Energy generation of the system during the month i, in GWh
E_{di} = Energy demand from the system during the month i, in GWh,
Q_i = Inflow to the diversion points during the month i, in MCM,
Q_{di} = Irrigation water demand at the diversion points during month i, in MCM,
pol = Polgolla diversion,
min = Minipe diversion,
Q_{pi} = Volume of water diverted at Polgolla month i, in MCM,
Q_{mi} = Volume of irrigation water supplied at the Minipe anicut,
SP_i = The spillages at the diversion during month i, in MCM

are according to the results of Nandalal (1986). Reservoir operation is derived by using monthly time steps. The hypothesis of the relative insensitivity of the system output (like total energy output) to the use of different objective criteria and constraints stems from the results of Kularathna & Bogardi (1989). As Tables 2 and 3 reveal the average annual energy output is virtually independent of the type of the selected objective function, whereas the mean of the annual shortages also does not seem to provide an adequate measure of the system performance. The large standard deviation figures (see Table 3) refer to few extreme shortage events within generally satisfactory fulfilment of the downstream irrigation water demand.

PERFORMANCE INDICES

In order to avoid confusion in terminology, the nine performance indices (PI) applied in this study will follow the notations used by Duckstein & Plate (1987).

These PI may be developed for both types of system outputs, being subject of optimization or being enforced through a (stochastic) constraint.

Number of failures. Indicates the number of times the system fails to deliver the target output. One failure may last several months.

Number of failure months. Number of months the system is operating in failure mode (i. e. substandard, compared to the target output level).

PI^4 reliability. Three types of reliability can be considered. They are referred to as: annual occurrence-based reliability (PI^4ob), time-based reliability (PI^4tb), and quantity-based reliability (PI^4qb).

* *Annual occurrence-based reliability* is the fraction of years per system simulation during which the system never flips into the failure mode (μ). This reliability index will vary from 0 to 1,

$$PI^4_{ob} = 1 - \sum_{j=1}^{JJ} [\delta_{ob}(\mu, j)]/JJ \tag{2}$$

where $\delta_{ob}(\mu, j)$ is an indicator function

$$\delta_{ob}(\mu, j) = \begin{cases} 1 \text{ if the system is at least once in mode } \mu \text{ during year } j \\ 0 \text{ otherwise} \end{cases} \tag{3}$$

JJ is the total number of years considered in the simulation.

* *Time-based reliability* is defined as the relative frequency that the system is *not* operating in mode μ during the simulated operation. The indicator function $\delta_{tb}(\mu, k)$ has the property that

$$\delta_{ob}(\mu, j) = \begin{cases} 1 \text{ if the system is in mode } \mu \text{ during month } k \\ 0 \text{ otherwise} \end{cases} \tag{4}$$

Then the time-based reliability is obtained by

$$PI^4_{tb} = 1 - \sum_{k=1}^{JK} [\delta_{tb}(\mu, k)]/JK \tag{5}$$

where JK is the number of time steps (months) in the simulated operational time horizon ($JK = 384$ in this particular study). Due to its similarity to the annual occurrence-based reliability concept, the time-based reliability (in case of monthly time steps) may be called monthly occurrence-based reliability.

* *Quantity-based reliability.* This third reliability indicator differs from the above described types in that it can take values lower and higher than 1.0. Its theoretical minimum value is 0.0, but this will never occur in practice since this refers to a permanently empty reservoir system (except for some dead storage). For the supply oriented V+R+R reservoir system Verhoef (1990) found the maximum values of $PI^4_{qb} = 2.3$ (for irrigation) and $PI^4_{qb} = 1.7$ (for energy). High values can be regarded as indicators of oversized systems.

Table 3. *Simulation results according to the different objective functions*

Objective function	Kotmale annual energy (GWh)	Polgolla annual shortage (MCM)	Victoria Randen and Rantembe annual energy (GWh)	Minipe annual shortage (MCM)	Std. Dev. of total annual energy	Std. Dev. of total annual shortage
1	370.5	19.9	1531.7	7.3	351.9	47.3
2	372.1	9.1	1529.9	2.7	357.8	23.1
3	369.5	9.0	1532.1	2.7	364.4	23.1
4	368.6	10.4	1535.5	2.7	365.4	23.8
5	370.0	15.2	1539.7	4.2	375.7	29.1
6	396.2	2.8	1593.7	0.0	407.7	9.9
7	399.0	2.3	1596.7	0.0	422.4	9.4
8	383.2	9.4	1539.3	0.0	392.5	23.3

$II^4_{qb} > 1.0$ implies that the accumulated system output exceeds the accumulated demand (target output).

However, P^4_{qb} values over 1.0 may not exclude shortages. If this reliability < 1.0, then the reservoir system is having a substandard performance. Negotiations between the users will be necessary to accomplish new agreements about lower demand.

$$PI^4_{qb} = \sum_{k=1}^{JK} Q_k / \sum_{k=1}^{JK} T_k \qquad (6)$$

Q_k indicates simulated system output for month k, whereas T_k is the corresponding target output.

In the present study the aggregate target may be defined as the sum of monthly firm power targets stipulated by the power company.

Period of incident, PI^5. The period of incident (a better name would be period between incidents) is the mean interevent time (in days) between the subsequent failures of a simulation experiment. It is also called average recurrence time.

If $d_5(\mu, n), n \geq 1$, denotes the duration of the nth interevent time then:

$$PI^5 = \frac{1}{N-1} \sum_{n=1}^{N-1} d_5(\mu, n) \qquad \text{[days]} \qquad (7)$$

where N is the number of incidents during the time horizon J. Such an incident lasts one month or more.

Because the time horizon is finite (in this example 32 years or 384 months), the question arises whether the periods before occurrence of the first failure and after occurrence of the last failure should not be considered in this equation. With increasing number of failures the influence of neglecting these head and tail periods will be small. However, for experiments with only a few events the value of the mean interarrival period is a non-informative and misleading parameter. Tables 6 and 8 display both values.

Repairability, PI^8. Average length of time (in days) that the reservoir system stays in mode μ during a system simulation.

$$PI^8 = \frac{1}{N} \sum_{n=1}^{N} d_8(\mu, n) \qquad \text{[days]} \qquad (8)$$

where $d_8(\mu, n)$ is the duration of the nth incident ($n = 1, 2, \ldots, N$). The inverse of this index measures how quickly the system bounces back to a satisfactory state. It is termed the resiliency.

Vulnerability, PI^9. Besides assessing the frequency and duration indicators of failures, two indicators can be applied to estimate the severity of the mode μ.

Mean vulnerability can be defined as the average of the accumulated shortages per failure incident.

$$PI^9_{mean} = \frac{1}{N} \sum_{n=1}^{N} \sum_{t=1}^{D(\mu,n)} DEF_t \qquad (9)$$

where $D(\mu, n)$ is the duration of the nth operational period in failure mode (in months) and DEF_t is the achievement deficit to meet the target in month t. The maximum vulnerability index PI^9_{max} is equal to the largest monthly deficit observed during the simulation period

$$PI^9_{max} = \max\{DEF_{n\max}\} \qquad (10)$$

where $DEF_{n\max}$ is the largest monthly deficit of failure n.

Using for example constant target monthly firm power output, the maximum vulnerability cannot exceed this value. Following the above definitions the numerical value of PI^9_{max} may exceed the instantaneous PI^9_{mean}.

OBJECTIVE FUNCTIONS, SIMULATION –

The V + R + R subsystem of the Mahaweli scheme has two major objectives: the first one is to maximize hydropower output, whereas the predetermined firm energy (constraint 1)

Table 4. *Summary of optimizations and simulations*

SDP policy	*without* firm energy constraint but *with* irrigation constraints	*with* firm energy and *with* irrigation constraints	*with* firm energy but *without* irrigation constraints	*without* firm energy and *without* irrigation constraints
based on 32 years data	8 experiments IF = 1.0–1.5;2.0;2.5	8 experiments IF = 1.0–1.5;2.0;2.5	1 exper. IF = 1.0	1 exper. IF = 1.0
based on 16 years data	3 experiments IF = 1.0;1.2;1.4	3 experiments IF = 1.0;1.2;1.4		

should be supplied as given in Table 1. Simultaneously the aggregate irrigation water demand at Minipe diversion (constraint 2) should be met. Given the expectation oriented characteristics of SDP, the optimization can be carried out by maximizing the expected annual total energy generation,

$$\max E \sum_{i=1}^{12} TEP_i$$

subject to the firm energy and irrigation constraints. Alternatively the SDP optimization can be carried out without one or both constraints.

There are 32 years of historical inflow records and estimated irrigation demands for the present size of irrigated area available. The SDP-based policies will be derived by using the whole set of data to derive the transitional probability matrices. Alternatively the policy will be derived by using only the data of the first 16 years of record. This is done to test the robustness of the SPD-based policy. The potential extension of irrigation is conceived by multiplying the estimated present demand figures by an irrigation factor (IF) varying from 1.0 to 1.5 (0.1 increments) as well as using factors 2.0 and 2.5. Subsequent simulations will always be carried out, based on the same 32-year-long data sets.

Table 4 summarizes the different optimization and simulation runs conducted to assess the PIs.

RESULTS

Tables 5–14 summarize the PIs obtained by simulation, according to the set of experiments as given in Table 4.

DISCUSSION

Long term based (LTB) experiments

The results presented in a condensed form in Tables 5–14 seem to confirm the original hypothesis. The pairwise comparisons of the Tables 5 and 6, and respectively, 7 and 8 reveal the following.

Table 5. *Firm, total and dump water supplies for the 'without firm energy' (policy (a); and the 'with firm energy policy' (b). All values are based on a period of 32 years (b is expressed as percentage of a)*

irrigation factor IF	total demand [MCM] a	b%	total supply [MCM] a	b%	dump supply [MCM] a	b%
1.0	49742.6	100	113580.1	100.95	63837.5	101.69
1.1	54716.9	100	113616.8	100.98	58899.9	101.88
1.2	59691.1	100	114214.1	100.50	54523.0	101.04
1.3	64665.4	100	114248.9	100.54	49583.5	101.24
1.4	69639.6	100	114568.6	100.38	44929.0	100.97
1.5	74613.9	100	114629.2	100.42	40015.3	101.21
2.0	99485.2	100	114910.0	100.00	15424.8	100.01
2.5	124356.5	100	112654.0	100.30	(−)11702.5	97.11

Note:
(−) total supply is less than demand: (negative dump supply component)

Even a 100% increase of the annual irrigation demand (a constraint) had virtually no impact upon the total irrigation supply (approximately 1% increase). By reducing the dump component of the total release the increased demand could be met. A further increase of the irrigation factor (IF = 2.5) implies the underachievement of this high demand. In addition the total release is decreasing in absolute terms. Quantity based reliability is less than 1.0. The system is 'over-challenged'. The SDP policy encounters a large number of infeasible releases, thus the performance is sharply deteriorated. The incorporation of an additional constraint (firm power) seems to increase (within 1%) the total irrigation supply compared to the no firm power constraint case. Since the corresponding increases in dump water releases are somewhat higher (Table 5) than that of the total supply, a slight reduction of the necessary releases, intended to meet the irrigation water demand can be expected for the operation considering firm power requirements. This can be seen in the pronounced differences between the corresponding PIs (except for the time- and quantity-based reliabilities) in Table 6a and 6b. The additional firm power constraint has a

Table 6. *Performance indices for irrigation water supply obtained by a policy derived without firm energy constraints (a) and (b): performance indices for irrigation water supply obtained by a policy with firm energy constraints*

a.	Without firm energy: Irrigation							
Irrigation factor IF	1.0	1.1	1.2	1.3	1.4	1.5	2.0	2.5
Annual reliability	1.00	1.00	1.00	0.94	0.81	0.59	0.03	0.00
Time-based reliability	1.00	1.00	1.00	0.99	0.98	0.95	0.69	0.53
Quantity-based reliability	2.28	2.08	1.91	1.77	1.65	1.54	1.16	0.91
Period of incident [days]	—	—	—	11460.0	2256.0	782.1	150.0	106.8
without head/tail periods				690.0	2022.0	775.7	149.4	103.7
Repairability [days]	—	—	—	30.0	40.0	38.0	66.1	93.6
Mean vulnerability [MCM]	—	—	—	64.7	66.8	79.8	371.5	750.8
Max. vulnerability [MCM]	—	—	—	93.4	110.8	283.2	529.8	680.2
Number of failures	0	0	0	2	6	15	54	58
Months of failures	0	0	0	2	8	19	119	181

b.	With firm energy: Irrigation							
Irrigation factor IF	1.0	1.1	1.2	1.3	1.4	1.5	2.0	2.5
Annual reliability	1.00	0.91	0.78	0.66	0.53	0.34	0.06	0.00
Time-based reliability	1.00	0.98	0.96	0.94	0.90	0.86	0.64	0.51
Quantity-based reliability	2.30	2.10	1.92	1.78	1.65	1.54	1.16	0.91
Period of incident [days]	—	3780.0	1585.7	900.0	517.5	331.0	133.6	104.5
without head/tail periods	—	324.0	747.7	875.0	504.0	328.0	133.1	101.3
Repairability [days]	—	45.0	52.5	55.4	55.7	51.3	74.5	99.5
Mean vulnerability [MCM]	—	107.6	121.3	131.8	175.9	173.0	425.2	787.8
Max. vulnerability [MCM]	—	194.8	234.9	259.0	283.2	307.3	529.8	680.2
Number of failures	0	4	8	13	21	31	56	57
Months of failures	0	6	14	24	39	53	139	189

Table 7. *Firm, total and dump energy generation for the 'without firm energy' policy (a); and the 'with firm energy policy' (b). All values are based on a period of 32 years (b is expressed as percentage of a)*

irrigation factor IF	firm energy [GWh] a	b%	total energy [GWh] a	b%	dump energy [GWh] a	b%
1.0	29376	100	50117	94.85	20741	87.57
1.1	29376	100	49921	94.15	20545	85.79
1.2	29376	100	49728	93.62	20352	84.42
1.3	29376	100	49203	92.79	19827	82.10
1.4	29376	100	48372	91.75	18996	78.99
1.5	29376	100	47806	89.74	18430	73.39
2.0	29376	100	39678	96.05	10302	84.77
2.5	29376	100	35605	99.12	6229	94.88

dramatic deteriorating effect on most of the performance indices. Furthermore the dependence of most of the PIs on increasing IF values is clearly seen, whereas the consequence of these demand changes remains almost undetected in Table 5 (except for IF = 2.5, see negative dump supply). However, a different picture emerges in the case of energy generation. The total energy output decreases with increasing IF values, indicating that additional releases may be needed to meet irrigation demand even during full capacity utilization of the turbines. The impact of the additional firm power constraint deteriorates the total energy output even more. While the total generated energy decreases by up to 10%, compared to the corresponding energy output figures according to an operational policy without firm energy constraint, the drop in dump (Table 7) energy output (within 5–26%) is even more pronounced. In fact this difference implies an increase of the firm energy output. This is duly reflected by the generally improving PI values for the option relying on firm power constraint. Except the quantity-based reliability, the repairability and mean vulnerability (Table 8 a and 8 b), all PIs show consistently better performance. Similar to the

Table 8. *Performance indices for energy generations obtained by a policy derived without firm energy constraints (a) and (b): performance indices for energy generation obtained by a policy with firm energy constraints*

a.	Without firm energy: Energy generation							
Irrigation factor IF	1.0	1.1	1.2	1.3	1.4	1.5	2.0	2.5
Annual reliability	0.06	0.06	0.06	0.09	0.03	0.03	0.03	0.03
Time-based reliability	0.72	0.71	0.73	0.72	0.73	0.73	0.66	0.58
Quantity-based reliability	1.71	1.70	1.69	1.67	1.65	1.63	1.35	1.21
Period of incident [days]	150.0	151.7	155.0	145.3	147.4	161.0	136.1	115.9
without head/tail periods	147.3	150.6	153.9	144.2	146.3	158.1	132.9	114.8
Repairability [days]	58.4	60.5	57.3	55.9	53.8	59.4	68.4	81.4
Mean vulnerability [MCM]	64.2	71.0	59.8	58.8	57.9	63.2	79.8	95.2
Max. vulnerability [MCM]	72.3	72.3	72.3	73.6	73.6	72.3	73.4	73.4
Number of failures	56	55	55	58	58	53	57	59
Months of failures	109	111	105	108	104	105	130	160

b.	Without firm energy: Energy generation							
Irrigation factor IF	1.0	1.1	1.2	1.3	1.4	1.5	2.0	2.5
Annual reliability	0.31	0.25	0.34	0.31	0.22	0.25	0.06	0.06
Time-based reliability	0.85	0.84	0.82	0.80	0.78	0.77	0.62	0.58
Quantity-based reliability	1.62	1.60	1.58	1.55	1.51	1.46	1.30	1.20
Period of incident [days]	349.3	310.6	324.8	270.9	248.3	260.3	130.4	121.6
without head/tail periods	342.9	304.8	319.7	265.6	243.3	255.9	127.6	121.1
Repairability [days]	60.0	59.1	70.0	66.0	69.7	76.3	77.7	86.3
Mean vulnerability [MCM]	59.8	64.8	74.7	66.0	74.4	86.7	91.7	103.8
Max. vulnerability [MCM]	68.1	69.3	69.4	69.3	69.3	69.3	73.4	73.4
Number of failures	29	32	30	35	37	35	56	56
Months of failures	58	63	70	77	86	89	145	161

Table 9. *Comparison of irrigation and energy performance indices for simulations based on policies with and without considering irrigation demand as a constraint in SDP. With and without firm energy constraints for irrigation factor IF = 1.0*

Irrigation	without firm energy constraint		with firm energy constraint	
	without irrigation constraint	with irrigation constraint	without irrigation constraint	with irrigation constraint
Annual reliability	0.09	1.00	0.13	1.00
Time-based reliability	0.79	1.00	0.86	1.00
Quantity-based reliability	2.28	2.28	2.30	2.30
Period of incident*	200.7	—	251.5	—
Repairability	51.5	—	40.5	—
Mean vulnerability	159.4	—	81.9	—
Max. vulnerability	262.9	—	201.6	—
Number of failures	46	0	40	0
Months of failures	79	0	54	0

Table 9. *(cont.)*

Energy	without energy constraint	with energy constraint	without energy constraint	with energy constraint
Annual reliability	0.06	0.06	0.38	0.31
Time-based reliability	0.72	0.72	0.86	0.85
Quantity-based reliability	1.71	1.71	1.67	1.62
Period of incident*	120.0	147.3	349.3	342.9
Repairability	46.5	58.4	53.8	60.0
Mean vulnerability	57.8	64.2	56.5	59.8
Max. vulnerability	74.0	72.3	68.4	68.1
Number of failures	69	56	29	29
Months of failures	107	109	52	58

Note:
*without head and tail periods

Table 10. *Comparison of total irrigation and energy outputs for simulations [in %]. Irrigation factor IF = 1.0*

	Total energy generation		Total irrigation supply	
	without firm energy constraint in SDP	with firm energy constraint in SDP	without firm energy constraint in SDP	with firm energy constraint in SDP
with irrigation demand as constraint in SDP	100%	94.85%	100%	100.95%
with irrigation demand as constraint in SDP	100.34%	97.7%	99.7%	100.74%

Table 11. *Irrigation-related performance indices for a 'without' (a) and 'with' (b) firm energy policy, respectively. Operating policy derived from years 1–16. Simulation executed for years 1–32*

	IF = 1.0		IF = 1.2		IF = 1.4	
Irrigation factor	a	b	a	b	a	b
Annual reliability	1.00	0.94	0.97	0.75	0.78	0.47
Time-based reliability	1.00	0.99	0.99	0.95	0.97	0.89
Quantity-based reliability	2.31	2.31	1.92	1.92	1.65	1.65
Period of incident [days]	—	11430.0	—	1098.0	1395.0	443.5
without head/tail periods	—	270.0	—	918.0	1248.8	430.4
Repairability [days]	—	45.0	30.0	49.1	40.0	55.0
Mean vulnerability [MCM]	—	66.8	11.6	137.0	84.0	200.9
Max. vulnerability [MCM]	—	113.3	11.6	234.9	196.7	306.5
Number of failures	0	2	1	11	9	24
Months of failures	0	3	1	18	12	44
Cumulative irrigation water releases [MCM] (32 years)	113830	114668	114398	114903	114607	115095

Table 12. *Energy-related performance indices for a 'without' (a) and 'with' (b) firm energy policy, respectively. Operating policy derived from years 1–16. Simulation executed for years 1–32*

	IF = 1.0		IF = 1.2		IF = 1.4	
Irrigation factor	a	b	a	b	a	b
Annual reliability	0.03	0.31	0.09	0.31	0.06	0.25
Time-based reliability	0.70	0.82	0.72	0.80	0.73	0.77
Quantity-based reliability	1.69	1.60	1.67	1.54	1.64	1.47
Period of incident [days]	127.6	277.9	163.5	290.0	182.6	259.4
without head/tail periods	125.2	271.8	162.4	291.3	181.3	254.1
Repairability [days]	54.4	59.1	61.2	70.3	66.4	77.1
Mean vulnerability [MCM]	54.4	58.0	59.3	75.2	67.0	89.2
Max. vulnerability [MCM]	70.1	71.1	68.7	71.2	73.6	73.4
Number of failures	64	35	52	32	47	35
Months of failures	116	69	106	75	104	90
Cumulative energy generation [GWh] (32 years)	49549	46967	49019	45153	48167	43317

Table 13. *Comparison of short-term-based (16 years) STB and long-term-based (32 years) LTB policies. Total irrigation releases for 'without' (a) and 'with' (b) firm energy constraint. IF = 100; LTB a = 100%*

Periods of	IF = 1.0				IF = 1.2				IF = 1.4			
comparison (years)	LTBa	STBa	LTBb	STBb	LTBa	STBa	LTBb	STBb	LTBa	STBa	LTBb	STBb
1–16	100	99.79	100.17	100.37	100.04	99.82	100.43	101.44	100.36	100.35	101.46	102.24
17–32	100	100.71	101.85	101.63	101.16	101.76	101.79	100.84	101.46	101.55	101.01	100.29
1–32	100	100.22	100.95	100.96	100.56	100.72	101.06	101.16	100.87	100.90	101.25	101.33

Table 14. *Comparison of the short-term-based (16 years) STB and the long-term-based (32 years) LTB policies. Total energy generation for 'without' (a) and 'with' (b) firm energy constraint. IF = 1.0; LTB a = 100%*

Periods of	IF = 1.0				IF = 1.2				IF = 1.4			
comparison (years)	LTBa	STBa	LTBb	STBb	LTBa	STBa	LTBb	STBb	LTBa	STBa	LTBb	STBb
1–16	100	99.82	96.41	95.18	99.32	99.19	94.13	93.35	97.81	97.82	91.32	89.35
17–32	100	97.77	93.05	92.02	99.11	96.21	91.47	86.33	95.02	94.05	85.35	83.05
1–32	100	98.87	94.85	93.71	99.22	97.81	92.90	90.10	96.52	96.11	88.55	86.43

irrigation PIs the set of performance indices allow a more detailed assessment of the operation than the simulated O.F. and constraint values alone.

Table 9 reveals some interesting characteristics as far as the impact of an irrigation demand constraint is concerned.

By not incorporating this constraint in the SDP optimization of the operational policy, most of the PIs react with disastrous figures as far as the irrigation water supply is concerned. In the meantime most of the PIs show the opposite effect for the energy output. However the rates of change are quite moderate.

Table 10 is not only a confirmation of the previous comparison but also indicates that the 'with' or 'without' irrigation constraint options do not differ significantly as far as the two essential outputs – total energy generation and irrigation water supply are concerned.

Short term based (STB) experiments

The system operational experiments are repeated for IF = 1.0; 1.2; and 1.4; respectively. The operational policies, however, have been derived by relying only on the first 16 years of record to assess the transitional probabilities for the SDP computations. In spite of the limited data base, the operational policies have proven to be compatible with the ones, derived by using the full 32-year-long inflow time series. Table 13 indicates that irrigation releases vary between 99.79% and 102.24% of the value obtained for the long-term bases (LTB) SDP-policy without considering firm energy constraints.

While the shorter inflow time series do not seem to affect the total irrigation release (Table 13), there is a marked deterioration in terms of PIs. The comparisons of the relevant columns of Tables 6 and 11 reveal that the performance indices for the short-term based (STB) policies are worse. In addition the incorporation of the firm energy constraint appears to aggravate the situation even more as shown in Table 11. An observation which can not be made for the LTB experiment in Table 6 for IF = 1.0, and only to a lesser degree for IF = 1.2 and IF = 1.4.

The energy related performance indices derived for the same experiments (Table 12) show, in comparison with the respective figures of Table 8 less pronounced deterioration than in the case of irrigation water supply. As expected the firm energy constraint influences the energy related PIs less than it was the case for the respective irrigation performance indices.

These results are astonishing, especially in the light of the results displayed in Table 14. Compared to the LTBa for IF = 1.0 (100%) reference simulation results all other combinations yielded less total energy. With increasing IF this drop went up to 17% compared to the reference simulation run. Even within the IF = 1.4 class the deviations between LTBa and STBb can be as high as 12%. One would have expected a corresponding 'landslide' as far as PIs had been concerned.

CONCLUSIONS

As the discussion of the above results reveals, some of the conclusions might be only tentative, reflecting specific characteristics of the V + R + R system.

Some conclusions, however, can be drawn with considerable certainty.

- The hypothesis that the simulated value of the objective function and the output figure related to the constraints are not sufficient for an adequate characterization of the operational performance seems to be proven.
- It is still disputable how many and which performance indices are indispensable to add more information for the assessment of the system performance. Quantity-based reliability, being closely related to the 'traditional' output indicators seems to be a redundant PI.
- The choice of other indices, or the deletion of some may depend on psychological factors (risk averse or gambler behaviour of the decision maker), or on the particular aspects of the purpose. It is conceivable that municipal water supply can be better characterized for example by maximum vulnerability and frequency of failures, than by duration or mean vulnerability.
- SDP seems to be quite insensitive as far as the available streamflow data are concerned. This result seems to confirm similar findings by Bogardi & He (1991). Rather, the policy seems to be sensitive versus exaggerated constraints (IF = 2.5).
- Due to the relatively limited impact as far as energy generation output is concerned, it seems to be appropriate to include the irrigation demand constraint in the SDP computations. The question, whether firm power requirements could be omitted as a constraint is rather a practical than a scientific one.

REFERENCES

Bogardi, J. J. & He, Q. (1991) Impact of streamflow data accuracy on the operational performance of large water resources systems. Poster presented at the *IAHS Symposium S1: Hydrology for Water Management of Large River Basins*, XX General Assembly of the IUGG, Vienna, 11–24 August 1991.

Budhakooncharoen, S. (1990) *Interactive multi-objective decision making in reservoir operation.* Dissertation, Division of Water Resources Engineering, Asian Institute of Technology, Bangkok, Thailand.

Duckstein, L. & Plate, E. J. (1987) Water engineering reliability and risk: a system framework. In: Duckstein, L. & Plate, E. J. (Eds.), *Engineering Reliability and Risk in Water Resources*, NATO ASI Series E, No. 124, Nijhoff.

Fiering, M. B. (1982) Estimates of resiliency indices by simulation. *Water Resour. Res.*, **18**(1), 41–50.

Hashimoto, T., Stedinger, J. R. & Loucks, D. P. (1982) Reliability, resiliency and vulnerability criteria for water resources system performance evaluation. *Water Resour. Res.*, **18**(1), 14–20.

Henley, E. J. & Kumamoto, H. (1981) *Reliability and Risk Assessment*, Prentice Hall Inc., Englewood Cliffs, New Jersey, USA.

Kularathna, M. D. U. P. & Bogardi, J. J. (1989) A stochastic dynamic programming based multiobjective analysis of a water resources system using a simplified configuration. *Proc. International Seminar-Workshop on Conflict Analysis in Reservoir Management*, Asian Institute of Technology, Bangkok, Thailand, pp. 445–53.

Loucks, D. P., Stedinger, J. R. & Haith, D. A. (1981) *Water Resource Systems Planning and Analysis.* Prentice Hall Inc., Englewood Cliffs, New Jersey, USA.

Nandalal, K. D. W. (1986) *Operation policies for two multipurpose reservoirs of the Mahaweli Development Scheme in Sri Lanka*, M. Eng. Thesis No. WA-86-9, Asian Institute of Technology, Bangkok, Thailand.

Verhoef, A. (1990) *Evaluation of performance indices: a tool in the decision making process of a multipurpose reservoir control problem.* Ir.-Thesis, Department of Hydrology, Soil Physics and Hydraulics, Wageningen Agricultural University, The Netherlands.

6 Composite risk model of Ogee type spillway

M. SOWINSKI

Department of Water Resources and Environmental Engineering, Ahmadu Bello University, Zaria, Nigeria

M. I. YUSUF

Civil Engineering Programme, Abukabar Tafawa Balewa University, Bauchi, Nigeria

ABSTRACT In the existing design methods consideration is given to stochastic processes of flood flows imposed on the structure in order for the hydrological uncertainty to be accounted. However, there are also various uncertainties associated with flood conveyance structures which have to be considered in the design of hydraulic structures. A static model integrating hydrological and hydraulical uncertainties in the design of the Ogee spillway is devised in the present contribution. Results show that the conventional method of the evaluation of risk, where hydrological risk only is accounted, produces underestimation of the risk of failure. This becomes particularly significant if the return period and the safety factor are large.

INTRODUCTION

There are many parameters and variables subject to uncertainty in the process of design of a flood conveyance structure. These uncertainties have been classified as hydrological, hydraulic, structural and economical ones (cf. Tung & Mays, 1980). In the conventional method of design of a spillway structure it is considered that the annual flood flow input is a stochastic process and that the capacity of the structure is deterministic. This approach underestimates the risk of failure and consequent economic losses. The present work deals with hydrological and hydraulic uncertainties in the development of a composite risk model, as applied to an Ogee type spillway. The relationships among various risk levels, safety factor, expected service life of the structure, and design return period are derived using the design data of Dadin-Kowa dam and the flow data of the river Gongola in Bauchi State of Nigeria.

LITERATURE REVIEW

A few efforts have been made until recently to integrate the stochastic process of flood flows and other uncertainties. Yen & Ang (1971) incorporated the inherent randomness of hydrological events among the hydraulic uncertainty in the performance of a hydraulic structure. Later, Tang *et al.* (1975) incorporated this risk procedure into a dynamic programming approach for the optimal risk-based design of storm sewers. Other researchers (e.g. Mays, 1979; Tung & Mays, 1980, 1981a,b) have developed composite risk models for flood conveyance structures such as culverts and flood levees. In those papers, however, the hydrological parameter uncertainty arising from insufficient data was neglected. The importance of the uncertainty was treated recently by Tung (1987) in the risk models by constructing confidence intervals for flood frequency analysis under normal and log-normal condition, using the sampling distributions developed by Stedinger (1983). This is a step forward in considering the total hydrological risk advocated by Benjamin & Cornell (1970) in the design of a hydraulic structure.

In the present contribution, the inherent randomness of flood flows is integrated with the hydraulic uncertainty to develop the static model for an Ogee type spillway.

HYDROLOGICAL UNCERTAINTY AND FLOOD FLOWS PROBABILITY DISTRIBUTION

In order to determine flood flows probability distribution, the frequency analysis was performed using the data records

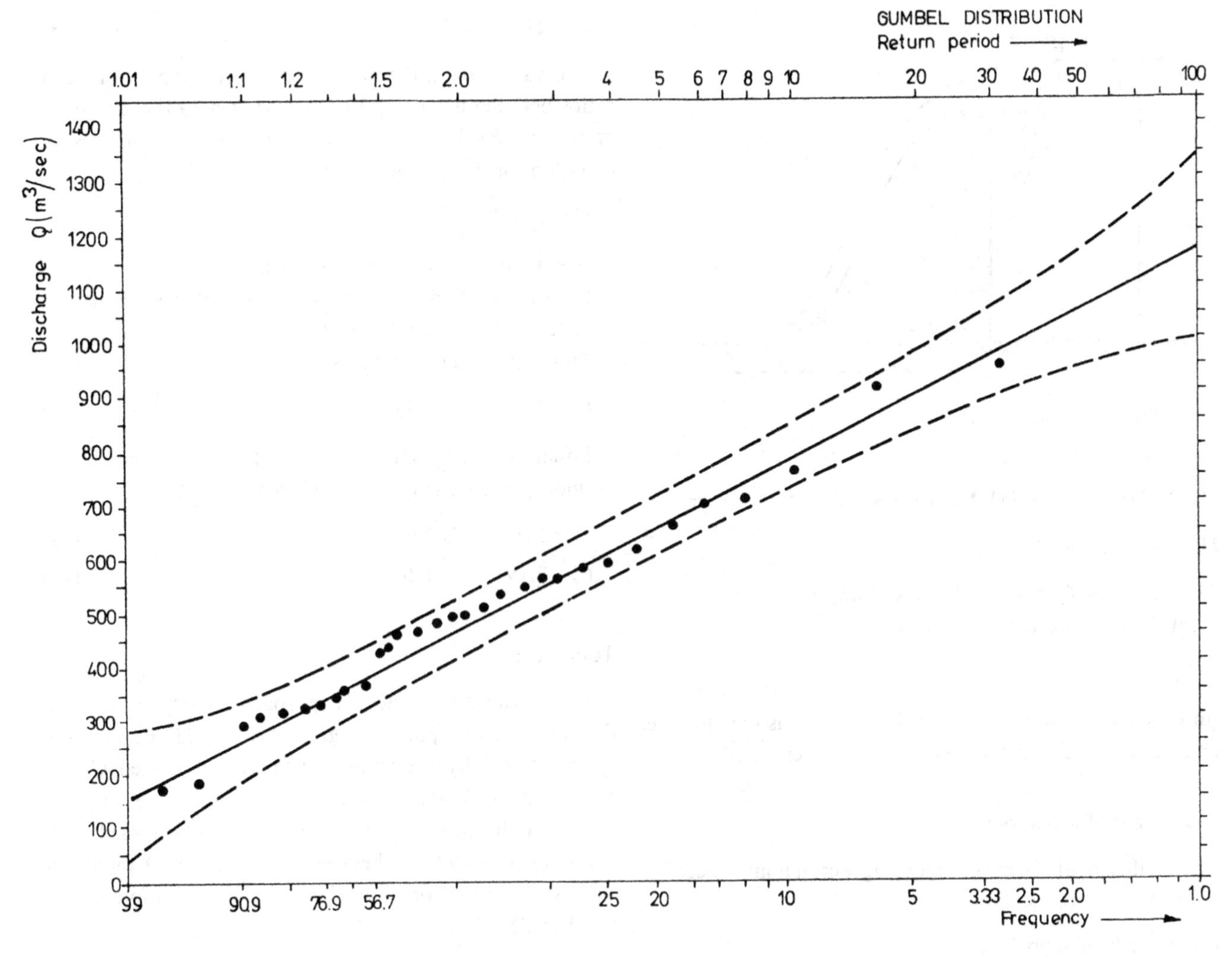

Fig. 1 Flood frequency curve for annual flood values of the river Gongola at Dadin-Kowa, Bauchi State. Dashed lines show the zone where the true value for that occurrence interval lies (with 90% probability).

of maximum annual flood for the river Gongola at Dadin-Kowa for the period of 1956–87. Statistical model verification test (Kolmogorov–Smirnov) was performed on the three selected distributions: log-Pearson III, lognormal, and Gumbel (extreme value I) distribution. The latter distribution performed satisfactorily both with regard to the model verification test and to the confidence band curves placed on the frequency curves (Fig. 1). The approximate control curves placed on the frequency curves were those proposed by Beard (1962).

The two-parametric Extreme Value I probability density function of floods is given by

$$f(Q_1)=\frac{1}{\alpha_2}\exp\left[-\frac{Q_d-\alpha_1}{\alpha_2}-\exp\left(-\frac{Q_d-\alpha_1}{\alpha_2}\right)\right] \tag{1}$$

where α_1 and α_2 are parameters

$$\alpha_1=\bar{Q}_d-\gamma\alpha_2 \tag{2}$$

$\gamma=0.577$ (Euler constant)

$$\alpha_2=\frac{\sqrt{6}}{\pi}\sigma_{Q_d} \tag{3}$$

$\bar{Q}_d$ is the mean annual flood flow, and σ_{Q_d} is the standard deviation of Q_d.

From the hydrological data, $\bar{Q}_d=499$ m^3/s, and $\sigma_{Q_d}=187.14$. Substituting these values in equations (2) and (3), the following parameter values were obtained:

$\alpha_1=145.9$, $\alpha_2=415.5$.

The cumulative probability function of the Gumbel distribution is expressed as

$$F(Q_d)=\exp\left[-\exp\left(\frac{Q_d-\alpha_1}{\alpha_2}\right)\right] \tag{4}$$

and the flood magnitude for any return period T_r may be computed from the equation:

$$Q_d(T_r)=-\alpha_2\ln\left[-\ln\left(1-\frac{1}{T_r}\right)\right]+\alpha_1 \tag{5}$$

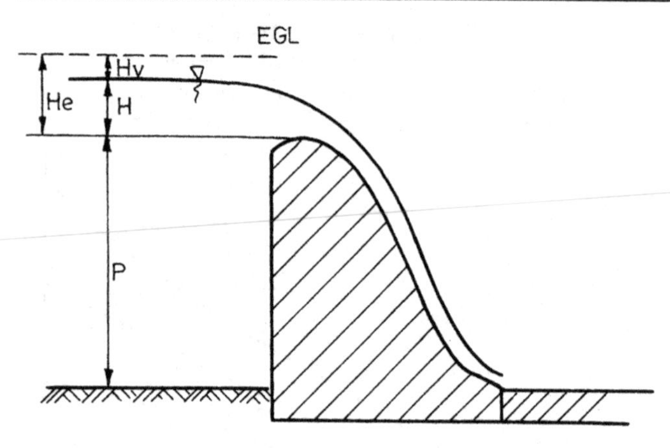

Fig. 2 Profile of Ogee spillway.

HYDRAULIC UNCERTAINTY

Hydraulic model

The discharge Q_c over the Ogee spillway may be computed from the well known weir formula

$$Q_C = C L H_e e^{-3/2} \tag{6}$$

(in which C is the coefficient of discharge, L is the effective crest length and H_e is the total head on the crest.

Coefficient of discharge

The coefficient of discharge for the Ogee crest is influenced by a number of factors, like:

(a) velocity of approach;
(b) upstream face slope;
(c) downstream submergence; and
(d) head differing from the design head (Fig. 2).

Assuming that the depth of approach (equal to the height of the spillway) is greater than 2.5 H_d (where H_d is the design head of the crest), the effect of this velocity on the coefficient of discharge may be neglected (Chow, 1959). Considering only a crest of spillway with the vertical upstream face the effect of upstream face slope is not taken into account. Finally, the effect of submergence is not considered.

The effect of the head on the crest differing from the design head H_D is considered below. The relationship between the coefficient of discharge and the head is usually presented in the dimensionless form as the ratio of coefficients of discharge C/C_D against the ratio of heads, $H_r = H/H_D$

$$C/C_D = f(H_r) \tag{7}$$

where the subscript D refers to the design head.

This relationship was plotted by Chow (1959), US Bureau of Reclamation (1974), and Creager (1964), but in the latter case the left hand side of equation (7) is not dimensionless (it is not divided by C_D), so equation (7) becomes:

$$C = f(H) \tag{8}$$

The values from the above mentioned graphs for head ratios between 0.3 and 1.3 were used as data for determination of the coefficient of discharge. The proposed regression equation for this coefficient is:

$$C = A_C + B_C \ln H_r \tag{9}$$

where A_c and B_c are regression constants.

For the expected value of C the values of these constants were found to be $A_c = 2.193$ and $B_c = 0.25$.

Thus equation (9) becomes

$$C_P = 2.193 + 0.250 \ln H_r \tag{10}$$

Equations describing the lower C_l and the upper C_u boundary of changes of the coefficient C read:

$$C_l = 2.167 + 0.283 \ln H_r \tag{11}$$

$$C_u = 2.193 + 0.250 \ln H_r \tag{12}$$

Total head

The total head H_e includes the approach velocity head. When the height of the spillway is greater than 1.33 H_D, then the approach velocity is so small, that it may be neglected (Chow, 1959). This case was considered in the following analysis. Therefore the symbol H_e in equation (6) was replaced by H denoting the difference between the elevation of the water surface upstream from the spillway and the elevation of crest (see Fig. 2).

Crest length

The effective crest length may be computed by reducing the net crest length as follows

$$L = L_t - 2(N \cdot KP - KA)\, H_e \tag{13}$$

where L_t is the net (total) length of the crest between abutments, N is the number of piers, KP is the pier contraction coefficient and KA is the abutment contraction coefficient.

The pier contraction coefficient KP, varies with nose shape of the pier and head ratio. This relationship was plotted on a graph by Chow (1959). For the purpose of this analysis, the pier nose shape between round-nose (type 2) and oval-pointed-nose (type 3) was chosen. The values of KP for the type 2 were designated as lower boundary values and for the type 3, as the upper boundary values for determining the interval of changes of value of KP. The expected values were computed at the middle of the above specified intervals assuming symmetric, triangular probability distribution for a random variable, KP. Regression analysis was performed to fit the equation for the lower, the upper and the expected

values of the pier contraction coefficient. The following results were obtained:
for the lower boundary

$$KP_L = 0.0025 - 0.045 \ln H_r \quad (14)$$

for the upper boundary

$$KP_u = 0.0098 - 0.054 \ln H_r \quad (15)$$

and for the expected value

$$KP_P = 0.0047 - 0.054 \ln H_r \quad (16)$$

The general equation for ith pier coefficient may then be written as:

$$KP = A_P - 0.054 \ln H_r \quad (17)$$

where A_p is the coefficient, which assumes the values indicated in equations (14), (15) and (16).

Model error

Because hydraulic conditions for flow over any spillway are non-uniform and unsteady, equation (6) yields an approximate discharge value. To take into account the possibility of an error, a model error term, λ_m, should be incorporated into equation (6). Introducing λ_m and substituting equations (9), (13), and (17) in equation (6), and neglecting the abutment contraction coefficient, yields:

$$Q_C = \lambda_m (A_C + B_C \ln H_r) \left[L_t - 2NH(A_p - 0.054 \ln H_r) \right] H^{3/2} \quad (18)$$

The expected value for model error, λ_m, may be assumed 1.0 since there is no tendency of over- or underestimation of the discharge over the spillway by equation (6).

First-order analysis of uncertainty

The uncertainty of an independent random variable may be fully described by its distribution. Often this distribution is not known due to insufficient information and an approximate method based on statistical moments (mean and variance) is used. This approach allows the determination of the uncertainty of the dependent variable, W, which is a function of independent variables:

$$W = f(w_1, w_2, \ldots, w_n) \quad (19)$$

in terms of these moments. First-order analysis can be applied for this purpose and it is based on Taylor expansion of the functional relationship (19) around the expected value, as:

$$W = f(\mu) + \sum_{i=1}^{n} \left. \frac{\partial f(W)}{\partial w_i} \right|_{W=\mu} (w_i - \mu_i) + R \quad (20)$$

in which μ is the vector of expected values of $w_1, w_2, \ldots, w_n$

$$\mu = f(\mu_1, \mu_2, \ldots, \mu_n)$$

and R is the remainder term.

Assuming that all variables w_i are statistically independent, the expected value of W is equal to the first term of the right hand side of this equation, since

$$E(w_i - \mu_i) = 0$$

thus:

$$E(W) = f(\mu) \quad (21)$$

Since

$$\mathrm{var}[f(\mu)] = 0,$$

the variance of W is obtained by transforming the second term of equation (20), remembering that

$$\mathrm{var}[cw_i] = c^2 \, \mathrm{var}[w_i]$$

where c is a constant;

$$\mathrm{var}(W) = \sum_{i=1}^{n} \left(\left. \frac{\partial f(W)}{\partial w_i} \right|_{W=\mu} \right)^2 \mathrm{var}[w_i] \quad (22)$$

Introducing the coefficient of variation defined as

$$\Omega_w^2 = \frac{\mathrm{var}(W)}{E(W)} \quad (23)$$

one can rewrite equation (22) as

$$\Omega_w^2 = \frac{1}{[E(W)]^2} \sum_{i=1}^{n} \left(\left. \frac{\partial f(W)}{\partial w_i} \right|_{W=\mu} \right)^2 \mu_i^2 \, \Omega_{w_i}^2 \quad (24)$$

Inserting equation (21) to (18) and remembering that the independent variables are λ_m, A_c, B_c, L_t, and A_p, (where the head H, and the head ratio, H_r, are treated as parameters), one gets the mean discharge over the spillway as:

$$\bar{Q}_C = \bar{\lambda}_m (\bar{A}_C + \bar{B}_C \ln H_r) \left[L_t - 2NH(\bar{A}_p - 0.054 \ln H_r) \right] H^{3/2} \quad (25)$$

or

$$Q_C = \lambda_m \left[L_t (A_C + B_C \ln H_r) - 2NH(A_C + B_C + \ln H_r) A_p + 2N(0.054) H (A_C + B_C \ln H_r) \ln H_r \right] H^{3/2} \quad (26)$$

Taking the partial derivatives of the independent random variables of equation (26), substituting the results in (24), and simplifying, one gets:

$$\Omega_{Q_C}^2 = \Omega_{\lambda_m}^2 + \frac{\bar{A}_C^2}{\beta^2} \Omega_{A_C}^2 - \frac{\bar{B}_c^2 (\ln H_r)^2}{\beta^2} \Omega_{B_C}^2 - \frac{\bar{L}_t^2}{\gamma^2} \Omega_{L_t}^2 + \frac{4N^2 H^2 \bar{A}_P^2}{\gamma^2} \Omega_{A_P}^2 \quad (27)$$

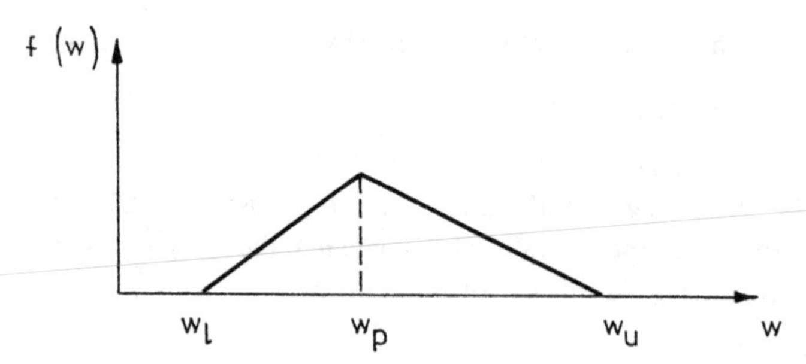

Fig. 3 Triangular probability distribution for variable w.

where

$$\beta = A_C + B_C \ln H_r \quad (28)$$

and

$$\gamma = L_t - 2NH(A_P - 0.054 \ln H_r) \quad (29)$$

Coefficient of variation of independent variables and spillway capacity

In order to perform further computations it is necessary to specify the geometric characteristics of the spillway, i.e. the net length of the crest, L_t, the design head, H_D, and the maximum head, H_{max}. The data taken for the case study of Dadin Kowa dam on the river Benue (Northern Nigeria) are as follows:

$$L_t = 18.0 \text{ m}, H_D = 10.0 \text{ m}, H_{max} = 12.0 \text{ m}.$$

A tolerance limit of linear measurement of length of the crest was assumed as $+0.05$ m. Thus the lower and upper boundary values were determined as 17.95 m, and 18.05 m, respectively. It seemed reasonable to assume, after Tung & Mays (1980), triangular distributions for all independent variables, using symbols from Fig. 3, where: w_l is the lower boundary, w_u is the upper boundary, and w_p is the expected value of the variable.

The general formulae for the mean, $\bar{w}$, and the coefficient of variation (c.o.v.) Ω_w of the random variable, w, are:

$$\bar{w} = \frac{1}{3}(w_l + w_p + w_u) \quad (30)$$

$$\Omega_w = \left(\frac{1}{2} - \frac{w_l w_u + w_u w_p + w_l w_p}{\sigma \bar{w}^2}\right)^{\frac{1}{2}} \quad (31)$$

In the case of equal distances of the lower and the upper boundaries from the expected value, equation (30) is simplified and the mean becomes

$$\bar{w} = w_p \quad (30a)$$

The values of mean and coefficient of variation were computed using the last two equations for boundary limits assumed earlier. The results are listed in Table 1.

Table 1. *Boundary limits and coefficient of variation of independent variables*

Variable	Lower boundary w_l	Probable value w_p	Upper boundary w_u	c.o.v. Ω_w
A_c	2.167	2.193	2.215	0.013
B_c	0.230	0.250	0.263	0.046
A_p	0.025	0.0047	0.0098	0.630
L_t [m]	17.95	18.00	18.05	0.001

Table 2. *Mean capacity and coefficient of variation of the spillway as a function of head ratio*

Head ratio $H_r = H/H_D$	Head H [m]	Mean capacity $\bar{Q}_c$ [m³/s]	C.o.v. Ω_{Q_c}
0.3	3.0	174	0.02320
0.4	4.0	277	0.02200
0.5	5.0	406	0.02138
0.6	6.0	531	0.02104
0.7	7.0	688	0.02074
0.8	8.0	856	0.02064
0.9	9.0	1043	0.02062
1.0	10.0	1244	0.02060
1.1	11.0	1486	0.02059
1.2	12.0	1727	0.02055

Due to lack of information, the value of the coefficient of variation of model error, λ_m, was assumed as Ω_λ. Substituting the obtained values of means and coefficients of variation into equations (26) and (27), one gets the mean and the coefficient of variation of the spillway capacity. The results obtained for different heads on the spillway are given in Table 2.

Computation of probability distribution parameters

Log-normal distribution was assumed for the spillway capacity, so its density function is given by the formula

$$f(Q_C) = \frac{1}{\sqrt{2\pi}\, Q_C \sigma_{\ln Q_c}} \exp\left[-\frac{1}{2}\left(\frac{\ln Q_C - \mu_{\ln Q_c}}{\sigma_{\ln Q_c}}\right)^2\right] \quad (32)$$

where

$\mu_{\ln Q_c}$ – the mean of transformed variable $\ln Q_C$

$\sigma_{\ln Q}$ – the standard deviation of this variable

The relation between the standard deviation $\sigma_{\ln Q_c}$ and the coefficient of variation, Ω_{Q_c}, is given (Benjamin & Cornell, 1970), as:

$$\sigma_{\ln Q_c} = [\ln(\Omega_{Qc}^2 + 1)]^{1/2} \quad (33)$$

Table 3. *Risk-safety factors for the statical composite model*

H_r	0.3	0.4	0.5	0.6	0.7	0.8	0.9	1.0	1.1	1.2
$\mu \ln Q_c$	5.12	5.59	5.75	6.25	6.51	6.73	6.92	1.10	7.29	7.42
$\sigma \ln Q_c$	.0232	.0220	.0214	.0210	.0207	.0206	.0206	.0206	.0206	.0205
risk	.996	.94	.71	.39	.16	.054	.016	.004	.0007	.0002
Safety factors										
$T_r=2$	0.37	0.59	0.84	1.14	1.47	1.84	2.25	2.69	3.17	3.70
$T_r=5$	0.26	0.42	0.60	0.81	1.04	1.30	1.59	1.90	2.23	2.62
$T_r=10$	0.22	0.35	0.50	0.68	0.86	1.09	1.33	1.60	1.88	2.19
$T_r=20$	0.19	0.30	0.43	0.58	0.75	0.94	1.15	1.38	1.62	1.89
$T_r=30$	0.17	0.28	0.40	0.54	0.70	0.87	1.06	1.28	1.50	1.75
$T_r=50$	0.16	0.26	0.34	0.50	0.65	0.77	0.98	1.18	1.39	1.62
$T_r=100$	0.14	0.23	0.33	0.45	0.58	0.72	0.88	1.05	1.24	1.45

and between the mean $\mu_{\ln Q_c}$ and the mean $\bar{Q}_c$, as

$$\mu_{\ln Q_c} = \ln Q_C - \tfrac{1}{2}\sigma^2_{\ln Q_c} \tag{34}$$

The values of the mean, $\mu_{\ln Q_c}$, and the standard deviation, $\sigma_{\ln Q_c}$, were computed using equations (33) and (34) for different head ratios. The results are given in Table 3.

Static reliability model

In the development of the static reliability model, when the annual flood is described by Gumbel distribution and the capacity of the structure by lognormal distribution, the reliability can be computed from the formula (Tung & Mays, 1980):

$$R = \int_0^\infty \frac{1}{\sqrt{2\pi}\, Q_C \sigma_{\ln Q_c}} \exp\left[-\tfrac{1}{2}\left(\frac{\ln Q_C - \mu_{\ln Q_c}}{\sigma_{\ln Q_c}}\right)^2\right] \cdot \exp\left\{-\exp\left[-\left(\frac{\bar{Q}_C - \alpha_1}{\alpha_2}\right)\right]\right\} dQ_C \tag{35}$$

This equation was solved by numerical integration for different head ratios, $\mu_{\ln Q_c}$ and $\sigma_{\ln Q_c}$, and Gumbel parameters $\alpha_1 = 415.52$ and $\alpha_2 = 143.9$. For each head and the corresponding spillway capacity, the safety factor was computed for different return periods, T_r, from the relation:

$$SF = \frac{Q_C}{Q_d(T_r)} \tag{36}$$

in which the flood discharge $Q_d(T_r)$ was computed from equation (5). The risk values were computed from the relation:

$$\text{risk} = 1 - R \tag{37}$$

The results of the computations were listed in Table 2 and illustrated in Fig. 4.

Table 4. *The hydraulic risk values with various head ratios*

Head ratio, H_r	Discharge, Q_c [m³/s]	Hydrological risk, $\bar{R}_H$	δR[%] from equation (39)
0.3	174	0.995	0.1
0.4	277	0.92	2.1
0.5	405	0.66	7.0
0.6	531	0.36	11.7
0.7	688	0.14	12.7
0.8	856	0.047	13.0
0.9	1043	0.0138	13.75
1.0	1244	0.0034	15.0
1.1	1486	0.00057	18.6
1.2	1727	0.00012	37.5

When only hydrological uncertainty is considered, the expression for hydrological risk is given by equation:

$$R_H = 1 - \exp\left[-\exp\left(-\frac{Q_C - \alpha_1}{\alpha_1}\right)\right] \tag{38}$$

in which Q_c is expressed as in equation (6), and α_1 and α_2 are as defined earlier. The result for the risk values are shown in Table 4. The percentage contribution of hydraulic risk to the composite risk computed from the formula

$$\delta R = \frac{R - R_H}{R} \cdot 100\ (\%) \tag{39}$$

are listed in the last column of Table 4.

CONCLUDING REMARKS

There are various uncertainties involved in the design of flood conveyance structures. The methodology presented in this paper considers a composite risk resulting from the hydrological uncertainty in the development of the static risk

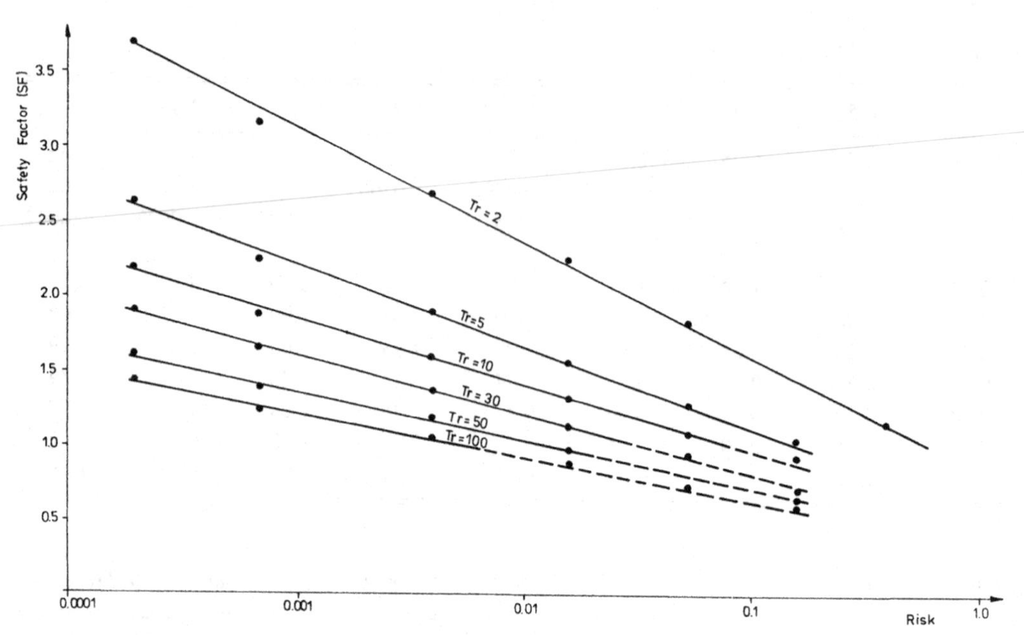

Fig. 4 Risk-safety factor relationship for a static model.

model for an Ogee type spillway. Hydrological model verification was performed using the Kolmogorov–Smirnov test and confidence band was constructed around the flood frequencey curve to select the probability distribution that best fits the flood flows of the river Gongola. Among the selected competing hydrological models, Gumbel distribution was found to satisfy both the Kolmogorov–Smirnov test and the confidence reliability band.

The relationship among the risk values, the safety factor and the return period has been established, which can facilitate the risk-based optimal design of spillway structures.

The example illustrated that considering only hydrological uncertainty underestimates the risk of flood exceeding the spillway capacity, as compared to a composite risk which takes into account the hydrological and hydraulic uncertainties. The percentage contribution of hydraulic risk to the composite risk was found to range between 0.1% and 37.5% for the range of the safety factor between 1.04 and 3.70, and the return period between 2 and 100 years.

The static model developed does not take into account the repeatability of flood events. Repeated loadings on a hydraulic structure are characterized by the time each flood value is applied and the behaviour of time intervals between the application of flood values. The consideration of this time-dependent composite risk will form a subject of a future work.

ACKNOWLEDGEMENTS

The work reported was sponsored by the Ababakar Tafawa Balew University, Bauchi. The writers would like to thank Professor Salihu Mustaph of Ababakar Tafawa Balewa University, Bauchi, for his constructive advice. Gratitude is also extended to the Upper Benue River Basin Development Authority, Bauchi, for the provision of the necessary data for this work.

REFERENCES

Beard, L. R. (1962) *Statistical Methods in Hydrology*, US Army Corps of Engineers, Civil Engineering Works Investigation, Sacramento District.

Benjamin, J. R. & Cornell, C. A. (1970) *Probability, Statistics, and Decision for Civil Engineers*, Mc Graw-Hill, New York.

Chow, V. T. (1959) *Open Channel Hydraulics*, McGraw-Hill, New York.

Creager, W. P., Hinds, J. & Justin, J. D. (1964) *Engineering for AMS*, Vol. II, Wiley, New York.

Mays, L. W. (1979) Optimal design of culverts under uncertainties, *J. Hydraul. Div. Amer. Soc. of Civil Eng.*, **105**(5), 443–60.

Rao, N. S. L. (1975) Theory of weirs. *Advances in Hydroscience*, **10**, 309–406.

Stedinger, J. R. (1983) Confidence intervals for design events. *J. Hydraul. Div. Amer. Soc. of Civil Eng.*, **109**(1), 13–27.

Tang, W. H., Mays, L. W. & Yen, B. C. (1975) Optimal risk-based design of storm sewer networks. *J. Envir. Engr. Div. Amer. Soc. of Civil Eng.*, **103**(3), 381–98.

Tung, Y. K. (1987) Effect of uncertainties on optimal risk-based design of hydraulic structures. *J. Water Resour. Plan. and Management*, **113**(5), 709–22.

Tung, Y. K. & Mays, L. W. (1980) Risk analysis for hydraulic design, *J. Hydraul. Eng. Amer. Soc. of Civil Eng.*, **106**(5), 893–913.

Tung, Y. K. & Mays, L. W. (1981a) Risk models for flood levee design, *Water Resour. Res.*, **17**(4), 833–41.

Tung, Y. K. & Mays, L. W. (1981b) Optimal risk-based design of flood levee systems, *Water Resour. Res.*, **17**(4), 843–52.

U.S. Bureau of Reclamation (1974) *Design of Small Dams*. U.S. Govt. Printing Office, Washington, D.C.

Yen, B. C. & Ang, H. S. (1971) Risk analysis in design of hydraulic projects, *Proc. 1st Intern. Symp. on Stochastic Hydro.*, pp. 694–709.

Yen, B. C. & Tang, W. H. (1973) Risk-safety factor relation for storm-sewer design, *J. Envir. Eng. Amer. Soc. of Civil Eng.*, **104**(4), 509–16.

THE FRONTIERS COLLECTION

The books in this collection are devoted to challenging and open problems at the forefront of modern science, including related philosophical debates. In contrast to typical research monographs, however, they strive to present their topics in a manner accessible also to scientifically literate non-specialists wishing to gain insight into the deeper implications and fascinating questions involved. Taken as a whole, the series reflects the need for a fundamental and interdisciplinary approach to modern science. Furthermore, it is intended to encourage active scientists in all areas to ponder over important and perhaps controversial issues beyond their own speciality. Extending from quantum physics and relativity to entropy, consciousness and complex systems – the Frontiers Collection will inspire readers to push back the frontiers of their own knowledge.

Information and Its Role in Nature
By J. G. Roederer

Relativity and the Nature of Spacetime
By V. Petkov

Quo Vadis Quantum Mechanics?
Edited by A. C. Elitzur, S. Dolev, N. Kolenda

Life – As a Matter of Fat
The Emerging Science of Lipidomics
By O. G. Mouritsen

Quantum–Classical Analogies
By D. Dragoman and M. Dragoman

Knowledge and the World
Challenges Beyond the Science Wars
Edited by M. Carrier, J. Roggenhofer, G. Küppers, P. Blanchard

Quantum–Classical Correspondence
By A. O. Bolivar

Mind, Matter and Quantum Mechanics
By H. Stapp

Quantum Mechanics and Gravity
By M. Sachs

Extreme Events in Nature and Society
Edited by S. Albeverio, V. Jentsch, H. Kantz

The Thermodynamic Machinery of Life
By M. Kurzynski

The Emerging Physics of Consciousness
Edited by J. A. Tuszynski

Weak Links
Stabilizers of Complex Systems from Proteins to Social Networks
By P. Csermely

Mind, Matter and the Implicate Order
By P.T.I. Pylkkänen

Quantum Mechanics at the Crossroads
New Perspectives from History, Philosophy and Physics
By J.Evans, A.S.Thomdike

Particle Metaphysics
A Critical Account of Subatomic Reality
By B.Falkenburg